COMPREHENSIVE CHEMICAL KINETICS

COMPREHENSIVE

Section 1.	THE PRACTICE AND THEORY OF KINETICS (3 volumes)
Section 2.	HOMOGENEOUS DECOMPOSITION AND ISOMERISATION REACTIONS (2 volumes)
Section 3.	INORGANIC REACTIONS (2 volumes)
Section 4.	ORGANIC REACTIONS (5 volumes)
Section 5.	POLYMERISATION REACTIONS (3 volumes)
Section 6.	OXIDATION AND COMBUSTION REACTIONS (2 volumes)
Section 7.	SELECTED ELEMENTARY REACTIONS (1 volume)
Section 8.	HETEROGENEOUS REACTIONS (4 volumes)
Section 9.	KINETICS AND CHEMICAL TECHNOLOGY (1 volume)
Section 10.	MODERN METHODS, THEORY AND DATA

CHEMICAL KINETICS

EDITED BY

R.G. COMPTON
M.A., D. Phil. (Oxon.)
Oxford University
The Physical and Theoretical Chemistry Laboratory
Oxford, England

G. HANCOCK
Oxford University
The Physical and Theoretical Chemistry Laboratory
Oxford, England

VOLUME 37

APPLICATIONS OF KINETIC MODELLING

1999
ELSEVIER
AMSTERDAM–LAUSANNE–NEW YORK–OXFORD–SHANNON–SINGAPORE–TOKYO

ELSEVIER SCIENCE B.V.
Sara Burgerhartstraat 25
P.O. Box 211, 1000 AE Amsterdam, The Netherlands

First edition 1999

Library of Congress Cataloging in Publication Data

Applications of kinetic modelling / edited by R.G. Compton, G. Hancock.
p. cm. -- (Comprehensive chemical kinetics ; v. 37)
Includes bibliographical references and index.
ISBN 0-444-50164-9
1. Chemical kinetics--Mathematical models. I. Compton, R. G. II. Hancock, G. (Gus) III. Series.
QD501.B242 vol. 37
[QD502]
541.3'94 s--dc21
[541.3'94'015118] 99-17977
CIP

ISBN: 0 444 41631 5 (Series)
ISBN: 0 444 50164 9 (Vol. 37)

The paper used in this publication meets the requirements of ANSI/NISO Z39.48-1992 (Permanence of Paper).
Printed in The Netherlands.

COMPREHENSIVE CHEMICAL KINETICS

Volumes in the Series

Section 1. THE PRACTICE AND THEORY OF KINETICS (3 volumes)

Volume 1	The Practice of Kinetics
Volume 2	The Theory of Kinetics
Volume 3	The Formation and Decay of Excited Species

Section 2. HOMOGENEOUS DECOMPOSITION AND ISOMERISATION REACTIONS (2 volumes)

Volume 4	Decomposition of Inorganic and Organometallic Compounds
Volume 5	Decomposition and Isomerisation of Organic Compounds

Section 3. INORGANIC REACTIONS (2 volumes)

Volume 6	Reactions of Non-metallic Inorganic Compounds
Volume 7	Reactions of Metallic Salts and Complexes, and Organometallic Compounds

Section 4. ORGANIC REACTIONS (5 Volumes)

Volume 8	Proton Transfer
Volume 9	Addition and Elimination Reactions of Aliphatic Compounds
Volume 10	Ester Formation and Hydrolysis and Related Reactions
Volume 12	Electrophilic Substitution at a Saturated Carbon Atom
Volume 13	Reactions of Aromatic Compounds

Section 5. POLYMERISATION REACTIONS (3 Volumes)

Volume 14	Degradation of Polymers
Volume 14A	Free-radical Polymerisation
Volume 15	Non-radical Polymerisation

Section 6. OXIDATION AND COMBUSTION REACTIONS (2 volumes)

Volume 16	Liquid-phase Oxidation
Volume 17	Gas-phase Combustion

Section 7. SELECTED ELEMENTARY REACTIONS (1 volume)

Volume 18	Selected Elementary Reactions

Contributors to Volume 37

J.E.T. ANDERSEN	Department of Chemistry, Technical University of Denmark, DK-2800 Lyngby, Denmark
N.H. ANDERSEN	Department of Chemistry, Technical University of Denmark, DK-2800 Lyngby, Denmark
H.L. BANDEY	Department of Chemistry, University of Leicester, Leicester LE1 7RH, United Kingdom
P.N. BARTLETT	Department of Chemistry, Southampton University, Highfield, Southampton SO17 1BJ, United Kingdom
W.R. BOWEN	Centre for Complex Fluids Processing, Department of Chemical and Biological Process Engineering, University of Wales Swansea, Swansea, SA2 8PP, United Kingdom
C. BOXALL	Centre for Photochemistry, Department of Chemistry, University of Central Lancashire, Preston, Lancashire PR1 2HE, United Kingdom
C.M.A. BRETT	Departamento de Química, Universidade de Coimbra, 3049 Coimbra, Portugal
S. BRUCKENSTEIN	Department of Chemistry, University at Buffalo, Buffalo, NY 14260, USA
S. BUCAK	School of Chemical Sciences, University of East Anglia, Norwich, Norfolk NR4 7TJ, United Kingdom
E.J. CALVO	INQUIMAE, Departamento de Química Inorgánica, Analítica y Química Física, Facultad de Ciencias Exactas y Naturales, Universidad de Buenos Aires, AR-1428 Buenos Aires, Argentina
R.A. ETCHENIQUE	INQUIMAE, Departamento de Química Inorgánica, Analítica y Química Física, Facultad de Ciencias Exactas y Naturales, Universidad de Buenos Aires, AR-1428 Buenos Aires, Argentina
P.R. FISK	Peter Fisk Associates, 9 St Swithins Road, Whitstable, Kent, CT5 2HT, United Kingdom
E.P. FRIIS	Department of Chemistry, Technical University of Denmark, DK-2800 Lyngby, Denmark

M.G. FORD	Biological Sciences, University of Portsmouth, Hampshire, United Kingdom
J. HADGRAFT	The Welsh School of Pharmacy, University of Wales, Cardiff, Cardiff, CF1 3XF, United Kingdom
B.G.D. HAGGETT	Research Centre, University of Luton, The Spires, 2 Adelaide Street, Luton, Bedfordshire, LU1 5DU, United Kingdom
A. HAMNETT	Department of Chemistry, Bedson Building, Newcastle University, Newcastle upon Tyne, NE1 7RU, United Kingdom
A.R. HILLMAN	Department of Chemistry, University of Leicester, Leicester LE1 7RH, United Kingdom
N. HILAL	Centre for Complex Fluids Processing, Department of Chemical and Biological Process Engineering, University of Wales Swansea, Swansea, SA2 8PP, United Kingdom
M.L. HITCHMAN	Department of Pure and Applied Chemistry, University of Strathclyde, 295 Cathedral Street, Glasgow G1 1XL, United Kingdom
M. JAIN	Centre for Complex Fluids Processing, Department of Chemical and Biological Process Engineering, University of Wales Swansea, Swansea, SA2 8PP, United Kingdom
P. JONATHAN	Shell Research and Technology Centre, Thornton, Cheshire, CH1 3SH, United Kingdom
R.W. LOVITT	Centre for Complex Fluids Processing, Department of Chemical and Biological Process Engineering, University of Wales Swansea, Swansea, SA2 8PP, United Kingdom
J.V. MACPHERSON	Department of Chemistry, University of Warwick, Coventry, CV4 7AL, United Kingdom
L.L. MADSEN	Department of Manufacturing Engineering, Technical University of Denmark, DK-2800 Lyngby, Denmark
R.A. MARCUS	Department of Chemistry, California Institute of Technology, Pasadena, CA 91125, USA
M. MISRAN	School of Chemical Sciences, University of East Anglia, Norwich, Norfolk NR4 7TJ, United Kingdom
A.W. MOHAMMAD	Centre for Complex Fluids Processing, Department of Chemical and Biological Process Engineering, University of Wales Swansea, Swansea, SA2 8PP, United Kingdom
P. MØLLER	Department of Manufacturing Engineering, Technical University of Denmark, DK-2800 Lyngby, Denmark

A.R. MOUNT	Department of Chemistry, University of Edinburgh, The Joseph Black Building, West Mains Road, Edinburgh EH9 3JJ, United Kingdom
A.M. OLIVEIRA BRETT	Departamento de Química, Universidade de Coimbra, P-3049 Coimbra, Portugal
L.M. PETER	Department of Chemistry, University of Bath, Bath BA2 7AY, United Kingdom
J.A.P. PIEDADE	Departamento de Química, Universidade de Coimbra, P-3049 Coimbra, Portugal
M.T. ROBERTSON	Department of Chemistry, University of Edinburgh, The Joseph Black Building, West Mains Road, Edinburgh EH9 3JJ, United Kingdom
B.H. ROBINSON	School of Chemical Sciences, University of East Anglia, Norwich, Norfolk NR4 7TJ, United Kingdom
S.H.P. SERRANO	Departamento de Química, Universidade de Coimbra, P-3049 Coimbra, Portugal
A.O. SHARIF	Centre for Complex Fluids Processing, Department of Chemical and Biological Process Engineering, University of Wales Swansea, Swansea, SA2 8PP, United Kingdom
M.H. THUESEN	Department of Chemistry, Technical University of Denmark, DK-2800 Lyngby, Denmark
J. ULSTRUP	Department of Chemistry, Technical University of Denmark, DK-2800 Lyngby, Denmark
E.N.K. WALLACE	Department of Chemistry, Southampton University, Highfield, Southampton SO17 1BJ, United Kingdom
P. WATSON	Biological Sciences, University of Portsmouth, Hampshire, United Kingdom
D.E. WILLIAMS	Department of Chemistry, University College London, 20 Gordon St, London WC1H 0AJ, United Kingdom
P.M. WILLIAMS	Centre for Complex Fluids Processing, Department of Chemical and Biological Process Engineering, University of Wales Swansea, Swansea, SA2 8PP, United Kingdom
C.J. WRIGHT	Centre for Complex Fluids Processing, Department of Chemical and Biological Process Engineering, University of Wales Swansea, Swansea, SA2 8PP, United Kingdom

Preface

Volume 37 is concerned with the use and role of modelling in chemical kinetics and seeks to show the interplay of theory or simulation with experiment in a diversity of physico-chemical areas in which kinetics measurements provide significant physical insight. Areas of application covered within the volume include electro- and interfacial chemistry, physiology, biochemistry, solid state chemistry and chemical engineering.

A leading contributor to this general area has been Professor W. John Albery, FRS to whom the contributors and editors dedicate this book. An appreciation of his work, written by Professor M.L. Hitchman, appears in the preceding page.

Oxford
July 1998

R.G. COMPTON
G. HANCOCK

Professor W. John Albery, F.R.S.

John Albery

One enduring memory of John Albery acolytes is the unfailing and unerring way he arrived at approximate analytical solutions for complex physical chemical problems, often with the aid of the biblical-like tome of Abramowitz and Stegun. Another enduring memory many will harbour is of working on a delicate piece of instrumentation and having a laboratory visitation from the Master, which also unfailingly led to a definitive result, although not always as accurate as that achieved theoretically. This bi-functional characteristic of John Albery of being able to use theoretical models of elegant simplicity and to drive forward a research programme which also had a strong practical emphasis was celebrated in September of 1996 when a catholic group of scientists gathered at University College to mark the year of his 60 birthday. The meeting within the space of just two days was a remarkable event, for many aspects of John Albery's career were covered. His theatrical talents were marked with a review after the conference dinner involving, among others, his musical collaborator John Gould and a former scientific collaborator, Mary Archer displaying her considerable range of vocal and thespian talents. Those who have worked with John Albery over the years have also experienced his tremendous joie de vivre and have come to recognise him not only as an outstanding scientist, but also a bon viveur. And these characteristics were celebrated in gourmet and bacchanalian style. However, first and foremost in the two days, there were the scientific contributions from colleagues, students, and postdocs of the last 30 years or so. A good number of these have contributed to this volume.

John Albery's research career started off in R.P. Bell's group in the Physical Chemistry Laboratory in Oxford. This was in the early 60s and at that time major advances were being made in the study of fast reactions in solution. Ronnie Bell's interest in fast proton transfers led to John Albery working on an electrochemical technique for measuring these reactions using a rotating disk electrode. The first chapter in this volume by Rudy Marcus who, through his Honorary Fellowship at University College, also has had a long standing relationship with John Albery links into John's kinetic interests and particularly highlights the interaction between theory and experiment which, as indicated above, has characterised so much of his career. John's doctoral work with rotating disks developed into his lifetime passion for electrochemistry and many of the other chapters in this volume reflect the impact he has had on electrochemical theory and practice in the second half of the twentieth century. John Albery has more than 150 publications on electrochemical related topics. Many of these papers have been seminal. For example, the early papers on ring-disk electrodes in the 1960s laid the basis for the extensive development of a powerful tool for electrochemical diagnostics. Papers in the 1970s on electrochemistry in ESR opened up new possibilities of monitoring electrochemical processes. And in the 1980s some of the studies of modified electrodes have thrown new light on these important electrode systems. Furthermore, John Albery's contributions have not only been at the fundamental level, but also in the application of principles to practical problems, especially in the area of sensors. All of these topics are dealt with to a greater or lesser extent in this volume.

John Albery has not just restricted the use of his talents to electrochemistry, for his interests extend into other areas of physical chemistry and beyond, for example with his very significant excursions into the chemical and biological spheres, with investigations of diffusion and transport process in two phase systems and of enzyme kinetics.

John Albery's contributions to electrochemistry, in particular, and to broader areas of physical chemistry, in general, have been recognised both nationally and internationally. He has held a number of distinguished visiting positions overseas, has been a Tilden Lecturer of the Royal Society of Chemistry, and was elected to Fellowship of the Royal Society in 1985. This volume is a small additional recognition of his achievements. It is, though, much more personal for it is dedicated to John Albery from the large number of his former students, co-workers and colleagues in the UK and worldwide for the many, many ways he has helped and furthered their professional lives and careers.

M.L. Hitchman

Contents

Chapter 1

Interaction of Theory and Experiment in Reaction Kinetics

RUDOLPH A. MARCUS

1.1 INTRODUCTION

The interaction of experiment and theory clearly delights John Albery—a colleague and friend to whom this chapter is dedicated. That interaction is the principal theme of this contribution. The examples given below are taken from the experiences of my research group in a number of fields in, or related to, reaction kinetics:

(1) high-resolution spectroscopic studies and intramolecular vibrational energy migration;
(2) long-range electron transfer (ET) in proteins;
(3) ET in a broad range of systems;
(4) ion, atom and group transfers;
(5) photosynthetic reaction center ET;
(6) solvent dynamics and the chemists' delayed use of Kramers' theory;
(7) unimolecular reaction rates and products' distribution of rotational-vibrational quantum states;
(8) unimolecular reaction rates in clusters;
(9) vibrational adiabaticity and reaction coordinates;
(10) graph theory and additivity in delocalized (aromatic) molecules.

New experiments have been very rich in stimulating theoretical ideas and each of the above topics concerns some particular aspect of the interaction between experiment and theory. Comments on the latter are made at the end of each topic treated in this chapter.

Some of the examples involve the derivation of equations, the predictions from which can be tested experimentally. The choice of which particular examples to cite was not easy. Others that could have been used include: ETs across liquid-liquid and other interfaces; chemiluminescent ETs; various CH overtone spectra; semiclassical dynamics; reaction path Hamiltonians; and translational energy distribution of the products of unimolecular reactions. Several of the topics described below are also discussed in more detail in Reference [1], a recent survey of transition state (TS) theory and its applications and in Reference [2].

1.2 EXAMPLES OF INTERACTION OF THEORY AND EXPERIMENT

1.2.1 High-resolution overtone spectroscopy of many-atom systems

Some years ago, Scoles and coworkers [3] measured the acetylenic CH overtone spectra ($\upsilon = 0 \rightarrow 1$, $\upsilon = 0 \rightarrow 2$) in $(CH_3)_3$ YC≡CH and in their deuterated methyl counterparts:

$$(CH_3)_3YC{\equiv}CH(\upsilon = 0) + h\nu \rightarrow (CH_3)_3YC{\equiv}CH(\upsilon = 1, 2)(Y = C, Si)\,. \tag{1.1}$$

The spectral lines for $\upsilon = 0 \rightarrow 1$ and for $\upsilon = 0 \rightarrow 2$, were much narrower for the Si than for the C compound. For the C compound, the line was broad enough to determine its shape which was found to be Lorentzian. These results are illustrated in Fig. 1.1.

In one view, the result that the C system has a broader spectrum may seem surprising. The width of a homogeneously broadened line is related, in the real-time domain, to the lifetime of the excited zeroth-order state (here the $\upsilon = 1$, or $\upsilon = 2$ acetylenic CH state): the broader the spectral line the shorter the lifetime, according to the uncertainty principle. According to one view, the CH vibrational quantum transfers its energy more rapidly to the vibrations of the rest of the molecule when there is a greater number of those vibrational acceptor quantum states per unit of energy. The Si compound, because of the larger mass of Si compared with C, has a thirty-fold higher density of vibrational states. However, the paradox is that its acetylenic CH spectrum is much narrower rather than broader.

Another possibility is that the heavier mass of the Si may tend to block the kinetic energy transfer between the two parts of the molecule. (There

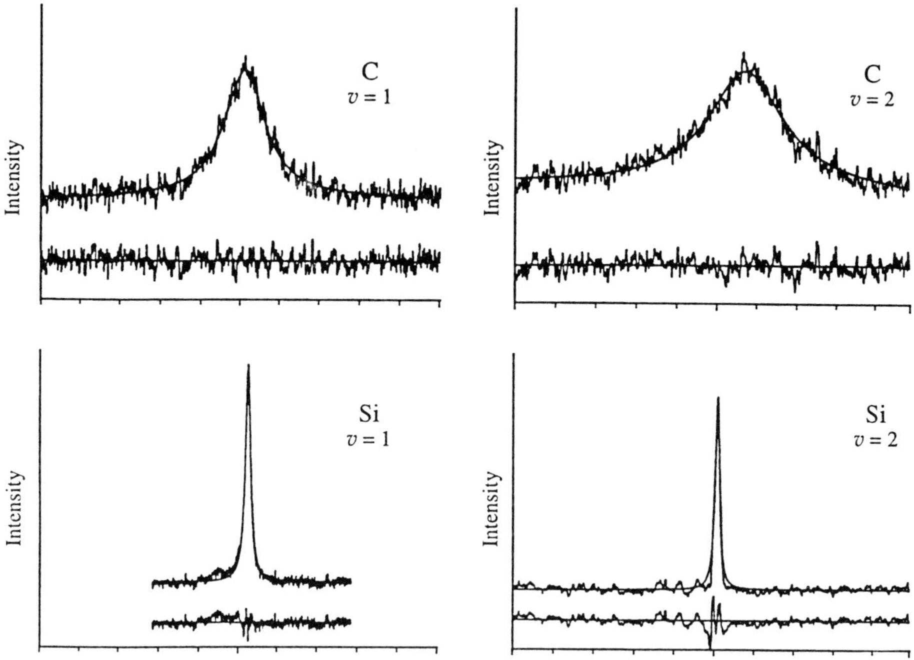

Fig. 1.1. Acetylenic CH overtone spectra of $(CH_3)_3$ C C≡C—H and $(CH_3)_3$ SiC≡C—H. The solid line in the former is the Lorentzian function [3].

had been some earlier controversy on heavy mass blocking, e.g., in the case of a fluorinated tetraallyl tin radical [4].) Some classical and semiclassical calculations were available, but only on a small hypothetical model system. However, a classical trajectory study did not seem desirable for the present system because of the low value of v of the acetylenic CH and the very large number of coordinates. Unphysical results are known to arise for such systems [5]. The system also seemed much too large to undertake semiclassical (semiclassical quantization) studies, our large previous investment in several-dimensional (two- and three-dimensional) semiclassical theory notwithstanding. Accordingly, a quantum mechanical approach seemed best, but the question was how. There are 42 vibrational modes in $(CH_3)_3$ YC≡CH, with potentially millions of zeroth-order acceptor vibrational quantum states.

We consider first the Lorentzian aspect and a sufficient but not necessary condition for a line to be Lorentzian: We recall here the well-known "discrete state in a continuum" problem [6], the discrete state in the present case being the zeroth-order acetylenic CH $\upsilon = 1$ or $\upsilon = 2$ state, and the continuum being the vibrational quantum states of the rest of the molecule. In the idealized problem of a discrete state coupled to a uniform continuum of states, whose density of states is ρ and with a coupling constant V, the spectral absorption intensity at the frequency ν is $I(\nu)$.

$$\frac{I(\nu)}{I(\nu_0)} = \frac{(\Gamma/2)^2}{(\Gamma/2)^2 + (\nu - \nu_0)^2}, \tag{1.2a}$$

where

$$\Gamma = \frac{4\pi}{\hbar}\rho V^2. \tag{1.2b}$$

The ν_0 is the frequency associated with the $\upsilon = 0 \to 1$ or $\upsilon = 0 \to 2$ transition and both ρ and V may be slowly energy-dependent and so vary with ν. When Γ is treated as a constant, the spectral line given by equation (1.2) is a Lorentzian. A well-known consequence is that if the zeroth-order CH state is prepared optically (by a laser pulse broader than the width Γ), its temporal decay to the states of the rest of the molecule would be a single exponential with a decay constant Γ.

On the other hand, when the density of the "continuum" of vibrational states of the system is small, it would have to be treated as being finite rather than as a continuum. Then $I(\nu)$ would be "spiky" rather than Lorentzian. Two questions that arise are how can such a simple Lorentzian expression characterize the real and very complicated $(CH_3)_3$ CC≡CH system and why does the Si compound have the narrower spectral line?

In the actual molecule the "bath" of zeroth-order vibrational states does not really form an exactly uniform continuum, each with a constant or at most energy-dependent coupling element. The exact vibrational quantum states of the system of all states but the cited acetylenic state υ are irregularly but extremely closely spaced and can be expected to have quite variable couplings to that state. However, if we divide the relevant energy range of the spectrum into very small cells, each cell containing numerous vibrational states of the system, there will be a mean value of the coupling element V to the cited υ state and a mean density of states in each cell and each mean may vary slowly from cell to cell. With such a coarse-graining we can see how the simple expression in equation (1.2), with an approximately constant Γ, might apply even for this very complicated system. Similar to the case of the statistical mechanics of bulk systems, the sheer number of contributing states can yield a certain simplicity.

To treat the actual problem, an approximate description for the bath was used involving zeroth-order vibrational quantum states of the system of all vibrational coordinates (apart from the state υ of the acetylenic CH stretch) [7, 8]. Approximate generic anharmonicities were introduced coupling those modes with each other and with the state υ. A tier structure connecting the initial acetylenic CH state υ to the first tier of vibrational quantum states was introduced. In turn, this tier was coupled to the next and so on for more than 30 tiers. With an artificial intelligence search method, the most important states in this tier structure could be selected. Most of the coupling occurred via off-resonant vibrational states and we termed the overall coupling process vibrational superexchange, by analogy with the usual electronic superexchange for coupling of two electronic states via off-resonant states.

Ultimately, a Lorentzian lineshape emerged in the calculation for the C compound [7, 8]. It was also found that while the density of states for the Si molecule was indeed some thirty-fold higher than that of the C molecule, the density of the coupled states in the first five or so tiers, which were particularly important in determining the overall spectral width, was substantially greater for the C compound (Fig. 1.2). A variety of other effects were uncovered, such as the need for using enough couplings (e.g., quartics in some cases) to avoid "dead end states", which give rise to spurious peaks. Ultimately, good agreement was obtained without adjustable parameters. Some questions still remain to be investigated, including the detailed vibrational state paths for the transfer. Perhaps, they can be addressed, as Dr. Stuchebrukhov suggested, by using his new method [9] for following *tunneling paths*.

Comment: In this example an experimental paradox prompted the theoretical treatment. The simplicity of the experimental lineshape also prompted some thought as to the reason for such simplicity. With an approximate force field, and assumed generic anharmonicities, the method itself can be used to make predictions for other molecules.

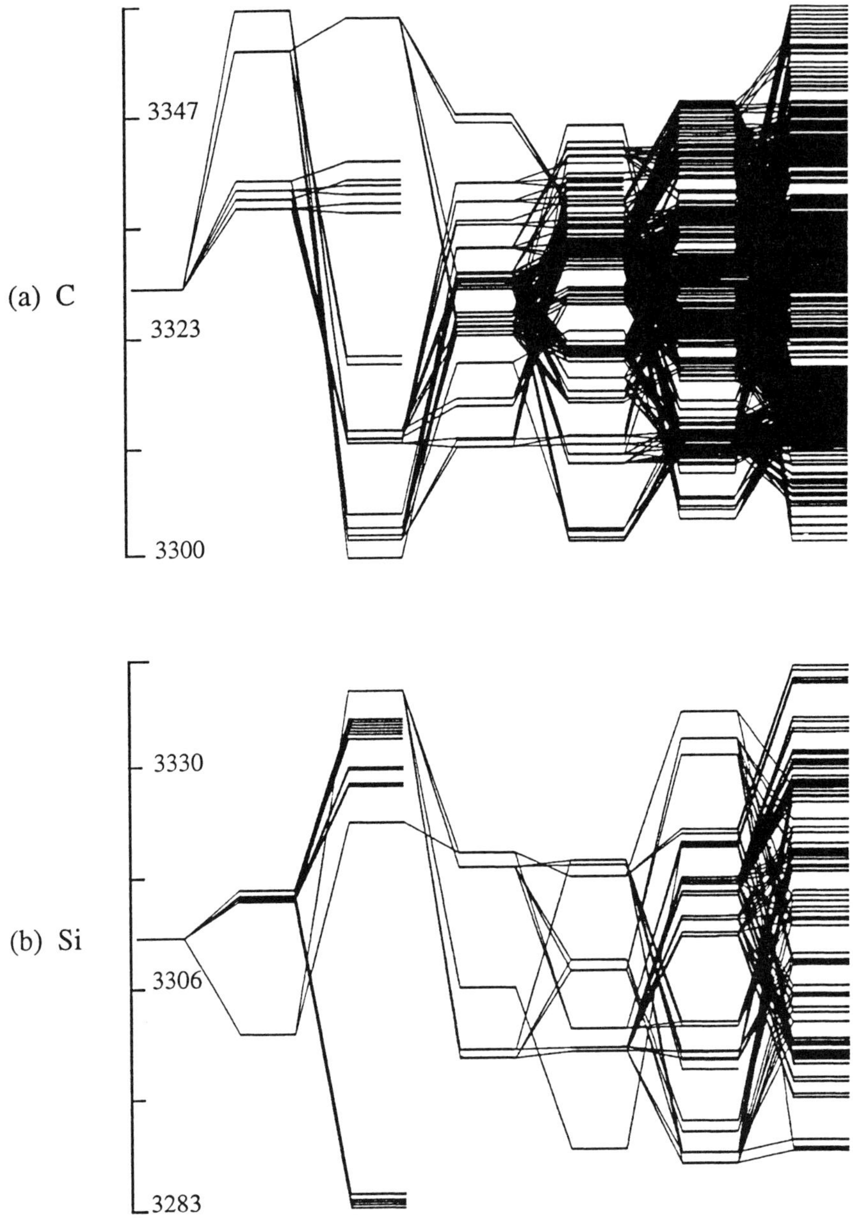

Fig. 1.2. Tier structure connecting zeroth-order vibrational quantum states for the molecules in Fig. 1.1 [7].

1.2.2 Long-range ET in proteins

Gray and coworkers described an interesting set of studies on the long-range ET in proteins, such as cytochrome c or azurin [10]. The position of an electron acceptor A relative to that of a donor D, both in the protein, was varied, using site-directed mutagenesis and ET rates were determined. Two factors affect the reaction rate as shown in equations (1.3)–(1.4). One is an activation free energy ΔG^*, and the other the matrix element H_{DA} coupling the donor and acceptor electronic orbitals for long-range electron coupling [11]:

$$k_{rate} = \frac{2\pi}{\hbar} |H_{DA}|^2 \frac{\exp(-\Delta G^*/k_B T)}{(4\pi\lambda k_B T)^{1/2}}. \tag{1.3}$$

Here, λ is the "reorganization energy" and

$$\Delta G^* = (\lambda/4)(1 + \Delta G^o/\lambda)^2, \tag{1.4}$$

where ΔG^o is the standard free energy of reaction, $D + A \rightarrow D^+ + A^-$. (The actual ionic charges in this one-ET typically differ from those indicated.)

There is now a well-defined theory-based method for separating these two factors: By using suitable substituents, ΔG^o can be varied. Either by extrapolation or otherwise, a condition can be attained where $\Delta G^o \cong -\lambda$, and then the $\Delta G^* \cong 0$. The resulting rate constant in equation (1.3) then provides directly the value of $|H_{DA}|^2$, after an approximate estimate of λ is made from the data on the ΔG^o dependence.

In this way, Gray and coworkers [10] obtained results for $|H_{DA}|^2$'s for the protein system (containing about 100 amino acids for cytochrome c). Similarly, Dutton and coworkers [12] obtained $|H_{DA}|^2$'s for quinone reactions in a photosynthetic reaction center. The next theoretical problem was how to calculate the coupling for comparison with Gray's data. The difficulty was that the protein is quite non-uniform in its composition and, being so large, contains numerous electronic orbitals coupling the electron donor and the acceptor. Using the known molecular structure of the protein, we applied a particular molecular orbital method [13] that we had tested previously against experiment for long-range ET in simple systems

[14]. Employing an artificial intelligence search technique to locate the most important amino acid paths, we found the individual amino acids orbitals contributing most to an electronic superexchange (i.e., off-resonant) mechanism [15, 16]. This smaller subsystem was then used to calculate H_{DA} and so compare with the experimental data [15, 16]. Later, a more all-embracing approach was developed by Dr. Stuchebrukhov in our group, in which the entire protein was treated and "pruned" to locate the most important amino acid paths [17]. He has since extended the theoretical analysis further to obtain the distribution of tunneling paths [9]. Theoretical treatments of other ET groups are cited in refs. [15–17].

There is a strong parallel between the vibrational energy coupling in Section 1.2.1 and the electronic state coupling here, as we have already noted. In the former, the zeroth-order states are excited normal mode vibrational quantum states of the molecule, while in the latter they are zeroth-order electronic orbitals of the amino acids. In one case, the mechanism is a "vibrational superexchange", namely, a coupling via off-resonant vibrational states, while in the other case, it is an "electronic superexchange", i.e., a coupling of D and A via off-resonant electronic orbitals of the amino acids. Ultimately, in either case, if the density of the intermediate coupled states were to become very large, the original localized donor and acceptor orbitals would become delocalized and any transfer of vibrational energy or of the electron would become "conductive" rather than being of the off-resonant type [18]. Such a high density of states did not occur in the examples mentioned.

Comment: In this electronic example, the data did prompt the theory but did not pose a paradox in themselves, since off-resonant ET was or could be anticipated. Instead, the problem was how to treat this extremely large electronic system. Ultimately, two ways were found to do this, and comparison of $|H_{DA}|$'s was then made with the measured ET rates, after an analysis based on equations (1.3) and (1.4) to obtain the $|H_{DA}|$'s.

1.2.3 ET, an early puzzle

In the late 1940s, the 1950s and later, radioactive isotopic tracers were used extensively to study the mechanistic paths of chemical reactions. One such use was in isotopic exchange reactions [19], such as the reaction, equation (1.5), in an aqueous medium,

$$Fe^{+2} + Fe^{+3*} \rightarrow Fe^{+3} + Fe^{+2*}, \tag{1.5}$$

The ET rate was measured by following the appearance of the isotope (the starred atom) in the other valence state, using selective precipitation or other techniques to separate the two redox forms at various reaction times.

In such studies, it appeared that the larger the reactants the faster the rate [19]. This enhancement was not merely a factor of two or so expected from a larger collision diameter, but much larger. To explain this effect, Libby introduced the Franck–Condon principle [20]: An electron jumps so rapidly from one reactant to the other that the nuclei, being much heavier, have no time to move during that instant. Libby argued that each newly formed ion would, therefore, find itself in the wrong molecular solvent environment—one appropriate to the previous charge of the ion rather than to its new charge. He noted that the resultant high energy of the system would then yield an energy barrier to the ET rate. Since the electrostatic ion-solvent dipole interactions increase considerably with decreasing ionic size (e.g., Born model for the calculation of the ion solvation free energy), such a barrier would be much greater for small rather than for large ions and so the ET rate would be slower, in agreement with the experiment.

In 1955, I happened to read Libby's (1952) paper, which was part of a symposium held at the University of Notre Dame and included isotopic exchange reactions [19]. His use of the Franck–Condon principle caught my eye: I had not seen its application to reaction rates before, although its use in electronic spectroscopy, for which it was originally designed in the mid-1920s, was well known. However, the energy barrier itself in Libby's application seemed mysterious, since the law of conservation of energy would be violated by a sudden apparent increase in potential energy with no change in kinetic energy. (According to the full Franck–Condon principle the nuclear momenta are also unchanged.)

I realized that the answer to this problem was that suitable fluctuations of orientations of the solvent dipoles in the neighborhood of the reactants had to precede the ET. In the TS, the ensemble of configurations of solvent dipoles would then be neither an equilibrium one for the reactants, nor an equilibrium one for the products. Instead, for appropriate configurations of the ensemble of solvent molecules the electron could transfer and satisfy both the Franck–Condon principle and the energy conservation.

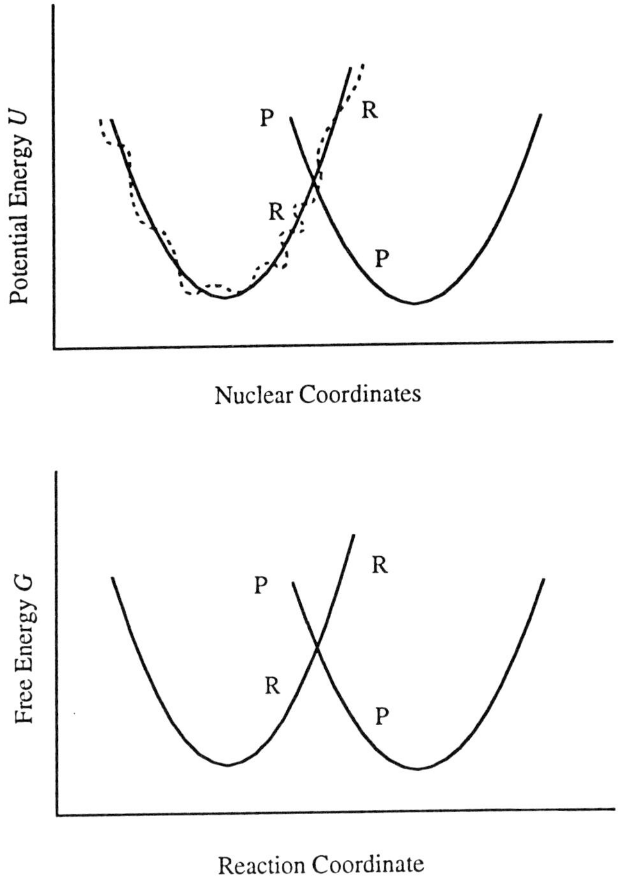

Fig. 1.3. Upper: Schematic view (dotted line) of cross-section of many-dimensional highly anharmonic potential energy surfaces for reactants plus solution (R) and (dotted line omitted) for products plus solution (P). TS occurs at the intersection. Lower: Plot of free energy G for the above R and P systems vs. the reaction coordinate ΔU.

A cross-section of the many dimensional potential energy surface for reactants in solution (R) and that for products in solution (P), is depicted schematically in Fig. 1.3 (upper).

The main task was how to characterize and determine this highly non-equilibrium state of the solvent in the TS, calculate its free energy and, using it, the rate of the ET. In the 1956 papers, these problems were solved by treating the solvent as a dielectric continuum [21]. The corresponding statistical mechanical solution, including changes in reactants' bond lengths [22], was given in the 1960 and 1965 papers [23, 24]. Equation (1.1.3) was obtained in this way, apart from using [21] an "adiabatic" pre-exponential factor instead of the non-adiabatic one appearing in equation (1.3). The non-adiabatic one is appropriate for weak electronic coupling, e.g., long-range ET. Expressions were given for the λ in equation (1.3) in terms of the properties of the reactants and the solvent [21, 23, 24]. Quantum effects for high-frequency vibrational modes were introduced by several authors (e.g., see ref. [25]).

With new developments in technology, increasingly more rapid reactions can be investigated and many chemical, electrochemical and biological systems were studied. Consequently, the ET field developed in many directions as depicted in Fig. 1.4 [2]. Isotopic exchange reactions or, as they are now more generally termed, self-exchange reactions (since non-

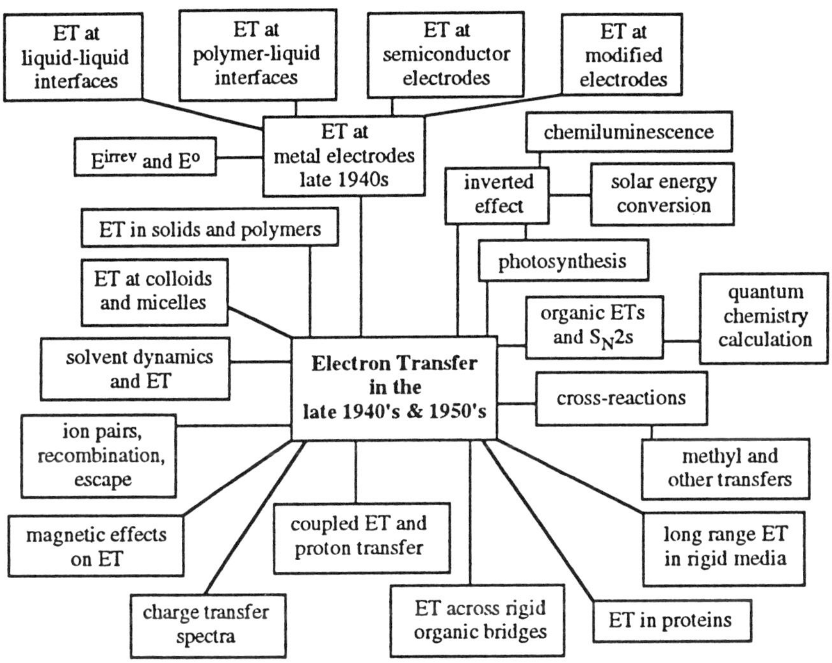

Fig. 1.4. Developments in the electron transfer field [1].

isotopic methods are also used for their study), proved to be particularly important for understanding the factors influencing the rates of ET reactions: Simple ET reactions form the simplest class of reactions in all chemistry, since no chemical bonds are broken or formed. Furthermore, self-exchange ET reactions are the simplest members of this class. In chemical reactions, the products usually differ from the products in their stability, i.e., the ΔG^{o} is usually non-zero and, for any series of reactants, the free energy barrier to the reaction rate usually depends strongly on this ΔG^{o}. For example, there have been numerous scientific conferences on "linear free energy relations" between rate constants and equilibrium, or related constants in a reaction series. Bronsted plots for acid or base-catalyzed reactions form one example. However, for self-exchange reactions this major factor, ΔG^{o}, which can obscure the presence of other factors, is absent. Thus, the effect of other factors, such as the size of the reactants on the ET rate could be observed, with the consequences for theory and interaction with the experiment mentioned above.

There were many predictions arising from the theory and its extension to electrochemical and other systems [11, 23, 24]. One such prediction, the "cross-relation", was based on the relation between the λ for reactions between two different redox systems, λ_{12}, to the λ's of the self-exchange reactions, λ_{11} and λ_{12}, for each of the two systems ($\lambda_{12} \cong 1/2(\lambda_{11} + \lambda_{22})$). The result for k_{12}, the rate constant for the cross-reaction, is

$$k_{12} \cong (k_{11} k_{22} K_{12} f_{12})^{1/2}, \tag{1.6}$$

where k_{11} and k_{22} are the rate constants for the two self-exchange reactions; K_{12} is the equilibrium constant; and f_{12} is a known function of them, usually close to unity.

Most of the predictions were confirmed experimentally (apart from some anomalous cases), but one prediction, the "inverted effect", (the decrease of rate constant with increasingly favorable (increasingly negative) ΔG^0, when $|\Delta G^0|$ is very large) was counterintuitive, but clearly evident in equation (1.4) and took 25 years before it was confirmed [26]. This story has been told many times and so I will not describe it here. However, the inverted effect is believed to have implications for efficiency of solar energy conversion in photosynthetic systems, as discussed elsewhere. Also, the ET theory had spin-offs for ion, atom and group transfer reactions considered in the next section.

One interesting aspect concerns the reason why equation (1.4) is so simple, even though the actual solvent medium for the ET reaction is very complex. In the actual system, the potential energy surface for the pair of reactants and their environment is a function of thousands of molecular solvent coordinates (positions, orientations, vibrations). Furthermore, the interactions of those solvent molecules with each other and with the charged reactants, are highly anharmonic (outside the inner coordination shell). Nevertheless, when the bulk system is in the so-called linear response regime, e.g., when its local dielectric polarization responds linearly to a change in local electric field, the free energy G of the reactants in the complex system becomes a simple quadratic function of a fluctuation coordinate, here the reaction coordinate. A similar remark applies to the products. This simple dependence is depicted in Fig. 1.3 (lower) and led to the simple quadratic expression given by equation (1.4).

The reaction coordinate used as the abcissa in Fig. 1.3 is rather unusual. It is, for any value $\mathbf{q}$ of the entire set of thousands of coordinates of the system, the difference of the products' and reactants' potential energy function, $\Delta U(\mathbf{q})$, both including the solvent's and reactants' coordinates [23, 27]. For each fixed value of ΔU, an equilibrium ensemble of systems of this *constrained* system can be used to calculate the two free energies G that appear in Fig. 1.3 (lower) for the reactants in solution (R) and for the products in solution (P). In this way, the important behavior of the system on each highly anharmonic potential energy surfaces with thousands of coordinates can be re-expressed in terms of a quadratic function, the free energy G, of only one coordinate ΔU, the reaction coordinate. One of the interesting extensions of equation (1.4) was by Savéant to ETs accompanied by the rupture of a chemical bond [28].

Comment: In formulating the theory in 1956, I had no idea as to its wider implications or that it might later be involved in many applications such as those shown in Fig. 1.4. It was merely an interesting problem stimulated by Libby's [20] paper, which in turn was stimulated by experiment.

1.2.4 Ion, atom and group transfers

In the simple ETs just discussed, there are no bond ruptures and usually only a relatively weak electronic coupling of the two reactants, perhaps of the order of k_BT, or even much less when the reactants are far apart.

Since there is no rupture or formation of a chemical bond, the extensive computation normally needed for quantum chemistry calculations for such bond breaking-bond-forming reactions is now absent. For reactions that do involve a concerted bond rupture and formation, the intersecting parabolas of Fig. 1.2 are normally inappropriate.

Only if there were a transfer of, say, an H^+, H^-, or $H^\bullet$ in a reaction, $AH + B \rightarrow A + HB$ (charges not indicated), at a fairly large AB separation distance, would the situation be rather analogous to that of ETs. The H transfer would occur at an approximately fixed position of A and B, fixed because of the substantially larger masses of A and B compared with that of H. That is, an approximate version of the Franck–Condon principle would apply. Under such conditions of an H transfer, the description of the reaction via two intersecting approximate parabolas would be a reasonable first approximation.

The reaction coordinate could again be chosen to be the ΔU of the products' and reactants' many-dimensional potential energy surfaces, each with its own equilibrium distribution of the position of the H. Similar to the ET case, the electron has an equilibrium distribution within a reactant for each value of the coordinates $\mathbf{q}$. The wave functions of H for the reactants, and that for the products at the given $\mathbf{q}$, could each be obtained by solving the respective Schrodinger equation for the H motion at that $\mathbf{q}$. The "splitting" of the two many-dimensional potential energy surfaces in their intersection region could be obtained from the off-diagonal elements of that solution, as could the splitting, $2|H_{DA}|$, for the electronic system in the ET case. The H in such a system might either tunnel (as the electron does typically in ETs), or surmount the barrier along the H coordinate, depending on the energy.

Next, we consider the more typical case where A and B are close enough so that the interaction is sufficiently strong that the intersecting parabolas description is no longer appropriate. Here, at any $\mathbf{q}$ (and so at any given AB distance significant for the TS), the splitting at the intersection is so large that the barrier height is substantially less than it would have been if the many-dimensional potential energy surfaces intersected. Nevertheless, an approximate the Franck–Condon principle still describes the transfer of the light H particle in this strong interaction case. An example is seen from a classical trajectory in Fig. 1.5 for an H transfer in a simple model collinear AHB system [29], where A and B are substantially heavier than H.

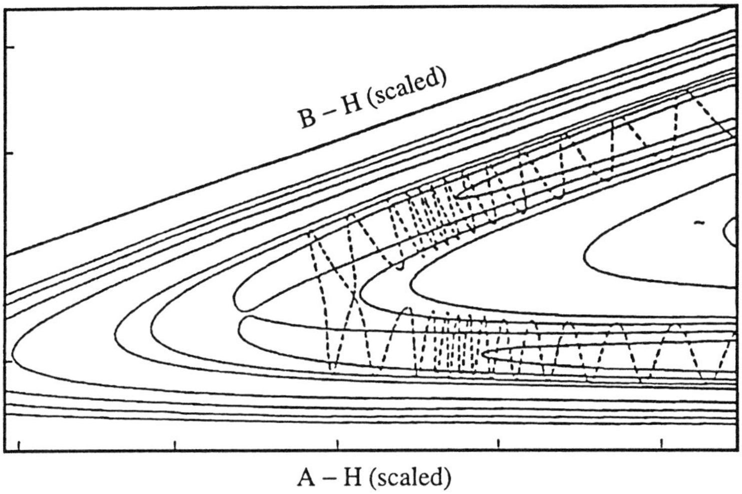

Fig. 1.5. Classical trajectory for an H atom transfer, AH + B → A + HB.

The diagram in Fig. 1.5 is the usual one, with its axes skewed so that the kinetic energy of the three centers has no cross-terms and has a single reduced mass to describe their internal motion [30]. (This single mass can be used even though there are two internal coordinates in this three-center collinear system [30].) One skewed-axis is a scaled AH–B distance and the other is a scaled A–HB distance. When polar coordinates are introduced, with the origin of the skewed axes as center, the polar angle coordinate serves as the H-transfer coordinate, while the radial coordinate is the scaled AB distance. A transfer occurring at a fixed AB distance and so obeying the Franck–Condon principle, occurs at a fixed radius in the symmetric system, B = A. The classical trajectory in Fig. 1.5, which passes over the top of the barrier rather than tunneling through it, is seen to satisfy this approximate Franck–Condon restriction. Although the model system in Fig. 1.5 is symmetric, a Franck–Condon-like behavior can also be expected in asymmetric systems.

The cross-relation (equation (1.6)), has also been applied successfully to transfers of CH_3 [31] and to transfers of H^- [32–33], while equation (1.4) has been used to treat proton transfers [34] and proton bound dimers [34d]. As already noted, the intersecting parabolas of Fig. 1.3 would not be applicable and so some other treatment was needed to understand the cross-relation or the use of equation (1.4) in this case—a treatment which takes into account the strong interaction of AH and B. In a 1968 article [35], the author derived an equation which was shown to reduce to equation (1.3) when $|\Delta G^0/\lambda|$ is not too close to unity. The additivity relation for λ given earlier was also obtained, after a "symmetrization" approximate for the potential energy surface. Equation (1.6) then followed under these conditions. The final result deviated from equation (1.3) and, hence, from equation (1.6) when $|\Delta G^0/\lambda|$ approached unity.

Comment: This example illustrates how equations derived for one set of reactions, which permitted a very detailed theoretical analysis, were intuitively applied by experimentalists to other reactions for which they were not originally designed, but how a different-based derivation yielded equations of a similar form, within certain limits.

1.2.5 Photosynthetic reaction center ET

Nature has constructed an efficient mechanism for the conversion of solar energy into useful chemical energy. At the heart of this process is the photosynthetic reaction center [36]. It is known from picosecond laser experiments that upon excitation of the special pair $(BChl)_2$ of bacteriochlorophylls in the bacterial reaction center, for example, the electron transfers to a pheophytin BPh occurs within 3 ps, indicating a strong electronic coupling between $(BChl)_2^*$ and BPh. (In the next 200 ps the electron transfers from this BPh^- to a quinone Q near the other side of the membrane.) However, in a different kind of measurement, the effect of small magnetic fields on a competitive reaction for the BPh^-, $BPh^- + (BChl)_2^+ \rightarrow BPh + (BChl)_2$ (triplet) in the absence of Q, indicated that the electronic coupling between $(BChl)_2^+$ and BPh^- was very weak. To reconcile these two conflicting results on the coupling between the (BChl) and BPh centers, it was postulated [37] that there was an intermediate site for the electron, which we denote here by X, and that the transfer from $(BChl)_2^*$ to BPh was a two-step process, first forming an intermediate X^-, which then underwent an ET with BPh to form BPh^-.

The X^- has since been identified as an accessory BChl, situated more or less between the $(BChl)_2$ and the BPh, but off to one side. The determination of the crystal structure of the photosynthetic reaction center

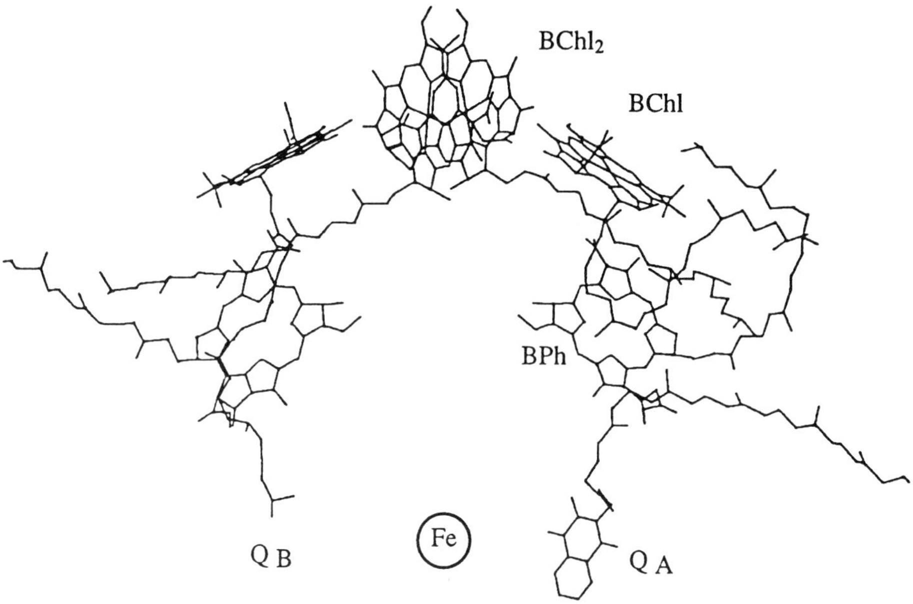

Fig. 1.6. Photosynthetic bacterial reaction center for *Rsp. viridis*. The chromophores are indicated but not the protein part of the structure, helices, etc., holding the whole unit together within the membrane.

(Fig. 1.6) defined the physical positions of each of these components more precisely [38]. For the past decade or more, there has been some dispute as to whether the accessory BChl serves as an actual intermediate, or as an off-resonant site, for the ET from $(BChl)_2^*$ to BPh. At present, the data favor the physical intermediate, $BChl^-$, over the superexchange mechanism [39].

Comment: In this system, it was again experiments that stimulated theoretical treatments of the reaction center and, they in turn, stimulated further experiments. It is also an example where the simple ET theory found application in biological systems.

1.2.6 Solvent dynamics and the delayed recognition of Kramer's theory

With the advent of picosecond and subsequently femosecond laser techniques, it became possible to study increasingly fast chemical reactions, as well as related rapid solvent relaxation processes. In 1940, the famous Dutch physicist, Kramers [40], published an article on frictional effects on chemical reaction rates. Although the article was occasionally cited in chemical kinetic texts, it was largely ignored by chemists until about 1980. This neglect was perhaps due mostly to the absence or sparsity of experimental data to test the theory. Even computer simulation "experiments" for testing the theory were absent for most of the intervening period.

The situation changed dramatically with the application of picosecond and, later, faster techniques. One stimulating study was that of Kosower and Huppert [41]. They found that the reaction time for a particular intramolecular charge transfer in a series of alcoholic solvents was equal to the respective slowest longitudinal dielectric relaxation time of the solvent. It was later pointed out that this equality of the reaction and dielectric relaxation times would apply for barrierless reactions ($\Delta G^* \cong 0$) or, more precisely, for the reactions where the relevant solvent dielectric relaxation, or its fluctuation, are the slow step, i.e., slower than the reaction would be in the absence of any slow solvent relaxational process.

Beginning in 1980 or earlier, numerous theoretical and experimental investigations were undertaken. Kramers' theory was also extended in several ways, such as introducing a frequency-dependent friction to replace a constant one [42] and introducing multidimensional effects [43–46]. Both of these effects caused deviations from Kramers' theory, but his work remained the seminal paper.

Electron transfer reactions are accompanied by significant solvation changes contributing to the reorganization energy term λ, and frequently also by changes in the reactants' vibrational bond lengths and, sometimes, bond angles [11]. Vibrational relaxation is expected to be a fast motion, while the reorientation of solvent molecules is typically slower [47]. This separation of timescales was made in the two-dimensional extension of Kramers' theory for ligand dissociation [43], and for ET [44, 45]. The extension for ETs has been extensively applied to experiments and extended further by including a quantum treatment for the high-frequency vibrational coordinates (see e.g. [48–50]). The latter extension is needed

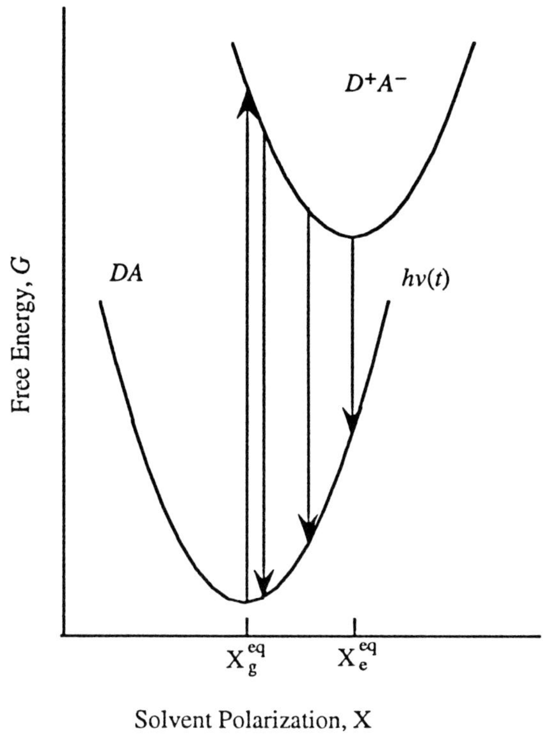

Fig. 1.7. Solvent relaxation and time-dependent fluorescence.

whenever high vibrational frequency modes are important contributors to crossing the TS region, particularly in the inverted regime [51].

A more direct way of exploring the dynamics of polar solvent in the presence of a solute is by the optical excitation of a solute to an intramolecular charge transfer state and then observing the time-dependent fluorescence, as shown in Fig. 1.7 [47, 52]. To follow the time-dependent fluorescence at short times, faster than the usual fluorescent lifetime of nanoseconds, lasers plus an up-conversion technique were used. A quantity frequently measured is the dynamic Stokes shift S(*t*),

$$S(t) = [\nu(t) - \nu(\infty)]/[\nu(0) - \nu(\infty)] , \tag{1.7}$$

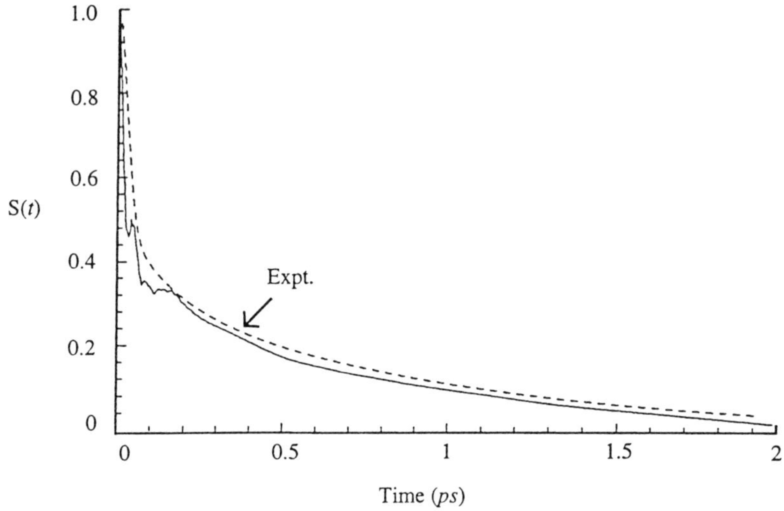

Fig. 1.8. Dynamic Stokes shift S(t) vs. time t for coumarin 343 in water. The theoretically calculated quantity is indicated by a solid line.

where $\nu(t)$ is the frequency at the maximum of the fluorescence spectrum at time t. Typically, the behavior of S(t) is not a single exponential (cf. Fig. 1.6). In one expression for S(t), a dielectric continuum treatment was used for the solvent and S(t) was expressed in terms of its dielectric constant function $\epsilon(\omega)$ [52–54]. Experimental results for coumarin 343 in water are compared in Fig. 1.8 with a theoretical calculation that contains no adjustable parameters [53]. There are other types of experiments, such as photon echoes and the Raman-induced Kerr effect, in which further information is obtained on the dynamics [55]. (Extensive references to the literature are given in refs. [53, 54].) Computer simulations have also played a significant role in the interaction of theory and experiment. However, these are limited mainly by the quality of the model Hamiltonian used for the solvent. Several limitations of some of the models have been noted [56].

Comment: In the case of solvent dynamics in chemical reactions, the theory of Kramers was already available, but there were no experiments for its application at that time and perhaps as a result the theory remained largely unused by chemists for many decades. Later, however, experiments did lead to extensions of the theory. In the related field of solvent relaxation in a photoexcited system, some early experimental work was available in the nanosecond (fluorescence lifetime) [57] and tens of picosecond [58] regimes. Ultrafast (subpicosecond, femtosecond) lasers have now permitted its extension to very short times where interesting multiexponential relaxation properties of the solvent are observed and accompanied by further theoretical developments.

1.2.7 Unimolecular reaction rates and products' quantum states distribution

The field of unimolecular reaction rates had an interesting history beginning around 1920, when chemists attempted to understand how a unimolecular decomposition N_2O_5 could occur thermally and still be first-order, $A \rightarrow$ products, even though the collisions which cause the reaction are second-order ($A + A \rightarrow$ products). The explanation, one may recall, was given by Lindemann [59], i.e., that collisions can produce a vibrationally excited molecule A^*, which has a finite lifetime and can form either products ($A^* \rightarrow$ products), or be deactivated by a collision ($A^* + A \rightarrow A + A$). At sufficiently high pressures of A, such a scheme involving a finite lifetime produces a thermal equilibrium population of this A^*. The reaction rate is proportional to A^*, which would then be proportional to A and so the reaction would be first-order. At low pressures, the collisions of A to form A^* are inadequate to maintain an equilibrium population of A^*, because of the losses due to reaction. Ultimately, the reaction rate at low pressures was predicted to become the bimolecular collisional rate for formation of A^* and, hence, second-order.

The development of a theory of unimolecular reactions proceeded rapidly in the mid-1920s, initiated by Hinshelwood with an A^* whose collision-free lifetime for reaction was approximated by an energy-independent one. The analysis was much elaborated by Rice and Ramsperger [60] and Kassel [61], known later as the RRK theory, where now the lifetime was, as it is in modern times, energy-dependent [62]. These theoretical works of the 1920s stimulated many measurements of the unimolecular rates of dissociation of organic compounds as a function of the gas pressure. Within a few years, however, this entire field collapsed or, more precisely, evolved into a new field: It was shown experimentally that the "unimolecular reactions", assumed originally to consist of only one chemical step, involved the formation of free radical intermediates instead, which in turn reacted with the parent compound, or decomposed, or recombined. In this way, the new field of gas-phase free radical chemistry evolved. Even the N_2O_5 dissociation ultimately shared a similar fate [63]. Photochemical techniques were subsequently extensively used to prepare free radicals and study the individual elementary steps of their reactions.

As a post-doctoral researcher in E.W.R. Steacie's laboratory in the National Research Council of Canada in the late 1940s, I was involved in experiments on several such free radical reaction steps [64, 65] and the data prompted me to wonder about their theoretical interpretation. A second post-doctoral under the tutelage of Oscar Rice led to the formulation in 1951–1952 of what later became known as the RRKM theory [62, 66–68]. Here, I blended the statistical ideas of the RRK theory of the 1920s with the concepts of the TS theory of the mid 1930s.

In the RRKM theory, the microcanonical rate constant $k(E, J)$ at a given E and total angular momentum quantum number J is given by [62, 68],

$$k(E, J) = N^{\dagger}(E, J)/h\rho(E, J) \,, \tag{1.8}$$

where $\rho(E, J)$ is the density of vibrational states of the dissociating molecule at the given E and J. When there is a distribution $p_0(J)$ of J's in the given experiment, the microcanonical rate constant $k(E)$ is given by

$$k(E) = \sum_J k(E, J) p_0(J) \,. \tag{1.9}$$

Although the theory was initially developed in 1952 and had been partly prompted by my prior experimental work, there were very few experimental data to which it could be applied. Around 1959 and subsequent years, B.S. Rabinovitch and coworkers used this theory to interpret their data on chemical activation [62, 69]. It may be recalled that chemical activation produced a narrower energy distribution of dissociating molecules than that in thermal unimolecular reactions and, hence, is better for testing the theory.

Since that time, laser and other techniques were introduced for producing molecules with high vibrational energy [70, 71] and this unimolecular

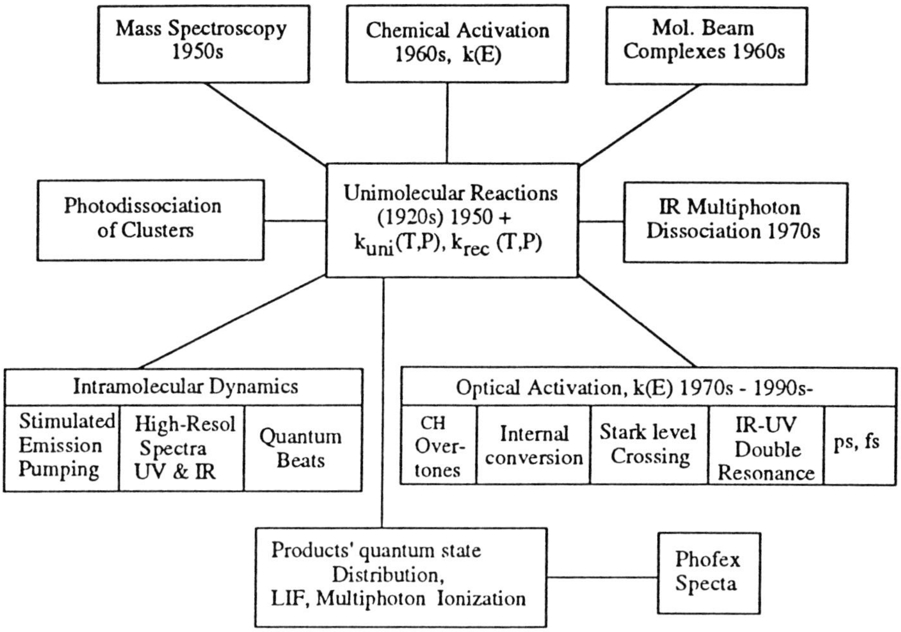

Fig. 1.9. Developments in unimolecular reaction rates and related areas [1].

field expanded in many different directions, as shown in Fig. 1.9. This figure also includes related areas such as that described earlier in Section 1.2.1. One theoretical development was a variational form of the RRKM theory [68, 72]. This theory was prompted by the need to treat experimental data for reactions that had no potential energy maximum to define the position of the TS. The TS occurs where the reactive flux (and thereby the number of states) is a minimum for any given energy E and total angular momentum quantum number J [68, 72–74]. Furthermore, by a convolution of a Monte Carlo evaluation of an expression for the density of classical states of the low frequency modes of the TS with a summation over high-frequency modes, the N^+ in equation (1.8) could be determined, even with highly anharmonic hindered rotations occurring in the TS [68].

The introduction of laser technology also permitted the determination of related data, such as the distribution of rotational-vibrational quantum states of the products of the unimolecular dissociation [75, 76]. It was found that in the threshold energy region of a unimolecular dissociation, a region where no vibrationally excited products could form, phase space theory (PST) [77] provided an excellent description of the rotational quantum state distribution of the products [76, 78]. In PST, the rotations of two fragments of the dissociating molecule are assumed to be free in the TS. The puzzle was that although PST gave good agreement for the rotational quantum state distribution of the products, the calculated unimolecular rate constant at typical energies was much larger by a substantial factor [79]. The reconciliation of these two sets of observations became the stimulus for a development in the theory, namely, an extension of the RRKM theory, so as to calculate the products' quantum state distribution in addition to calculating the rate constants $k(E, J)$ or $k(E)$ [80, 81].

Extending the theory to interpret or predict the rovibrational state distribution of the products of the unimolecular dissociation, requires some postulate about the nature of the motion after the unimolecularly dissociating system leaves the TS on its way to form products. For systems with no potential energy maximum in the exit channel, the higher frequency vibrations will tend to remain in the same vibrational quantum state after leaving the TS. That is, the reaction is expected to be "vibrationally adiabatic" for those coordinates in the exit channel (we return to vibrational adiabaticity in Section 1.2.9). The hindered rotations and the translation along the reaction coordinate were assumed to be in statistical equilibrium in the exit channel after leaving the TS until an outer TS, the "PST TS", is reached. With these assumptions, the products' quantum state distribution was calculated. (After the system leaves the PST TS, there can be no further dynamical interactions, by definition.)

The results served to explain the paradox described above and led to the prediction that a PST-calculated distribution of states would show discrepancies with experiment when the energy was high enough that vibrationally excited products could form. This latter result also agreed with subsequent data and predictions were made and tested on the distribution. The ideas were extended to treat photofragmentation excitation spectra of Moore and coworkers [75]. A two-TS description (the inner TS and the PST TS) in the threshold region [82] played a significant role in this description [81]. While the PST TS was dominant in the immediate threshold region, the inner TS dominated at higher energies.

For a different system $HO_2 \rightarrow H + O_2$, a very interesting model was introduced [83], in which the distribution in the products was assumed to be the projection of the distribution in the (inner) TS on the states of the products. It would be interesting to explore this model for other mass arrangements also—this one being light + (heavy – heavy), while the previous system was heavy + (heavy – heavy).

Comment: In this section we have seen that unimolecular reaction rate theory was prompted by experiment and earlier theoretical work, which in turn, was prompted by experiment. In the case of the products' quantum state distribution, an extension of the theory was stimulated by a seeming paradox involving existing data and both the agreement and disagreement with a very useful theory, PST, appropriate for a "loose TS". Although the actual system is not loose at typical energies, the TS does move outward in the exit channel when the energy E is reduced and the TS then becomes loose. This explains the applicability of PST in the immediate threshold region.

1.2.8 Unimolecular reaction rates in clusters

In experiments on the isomerization of *cis*-stilbene in clusters of n-hexane molecules (n varying from 1 to 4), the microcanonical rate constant at any given energy outside of the threshold region was observed to decrease with increasing number n [84]. Two effects may enter. First, there is extra frictional effect exerted by the hexane molecules on the *cis-trans*-torsional motion (the reaction coordinate). A second effect is the sharing of the excess vibrational energy by the extra coordinates due to the solvent molecules—at least due to the solvent molecules' librational coordinates. This sharing reduces the possibility of a critical energy becoming localized in the twisted configuration, the TS, of the double bond. In terms of equation (1.8), the ρ increases, as usual, much more than the $N\dagger$ when extra coordinates are introduced. A treatment was given which combined the RRKM theory with an adaptation of solvent frictional models to this microcanonical system [85, 86]. Two limits, net transfer of energy to the solvent molecules and no such transfer, were included. Since the reaction time at a typical energy ($\sim$3000 cm^{-1}) is of the order of 500–1000 ps for $n = 2$ or 3 and the expected time for transfer of energy to the librations of the solvent molecules is expected to be of the order of several ps, this energy sharing has to be an important source of the smaller rate constant at a given E.

Comment: Experiments point to a gap in the existing theory, namely, the need for a microcanonical theory which takes some account of the solvent frictional effects. What is desirable next is a detailed comparison of theory and experiment.

1.2.9 Vibrational adiabaticity and reaction coordinate

The term "vibrational adiabaticity" was introduced [87] to describe some results of the "computer experiments" of Wall *et al.* [88]. They were pioneers in the use of classical mechanical trajectories of the atoms to study chemical reactions theoretically, using electronic computers. Using classical trajectories for the collinear transfer of an H atom, $H + H_2 \rightarrow H_2 + H$, they found that when the vibrational energy of H_2 was equal to $(v + 1/2)h\nu$, and the energy barrier to reaction decreased by an amount $(v + 1/2)(h\nu - h\nu^{+})$, where v is 0 or unity; ν is the H_2 vibration frequency; and ν^{+} is the symmetric stretch H—H—H frequency in the TS. The question was how to explain this quantum-like result in a purely classical trajectory calculation.

Had the calculations been quantum mechanical rather than classical, the above result would have implied that somehow a vibrational quantum number v remained constant during the motion along the reaction coordinate, even though the nature of that vibrational motion changed drastically, from an H_2 vibration to a symmetric stretch in the H_3 TS and then to an H_2 vibration in the product H_2.

The classical counterpart of v is the action variable I, which equals $\oint p\,dq$, where the integral is over one vibrational cycle of the vibration. In old quantum theory (or, later, the WKB theory), I is related to v by $I = (v + 1/2)h$. Thus, it occurred to me that the above lowering of the energy barrier for the motion along the reaction coordinates could be rewritten as $I(\nu - \nu^{\ddagger})$ and so the results of Wall imply that the classical vibrational action I was constant along the reaction coordinate in this system.

It seemed appropriate to term this behavior "vibrational adiabaticity". Indeed there were also earlier quantum mechanical results for this reaction [89], for which I found a similar behavior [87]. Furthermore, many years before, Hirschfelder and Wigner [90] suggested the equivalent of a vibrational adiabaticity for reactions. The main question, however, was how to treat this approximate dynamical behavior in a physical way. For this purpose, I introduced a coordinate system which passed smoothly from

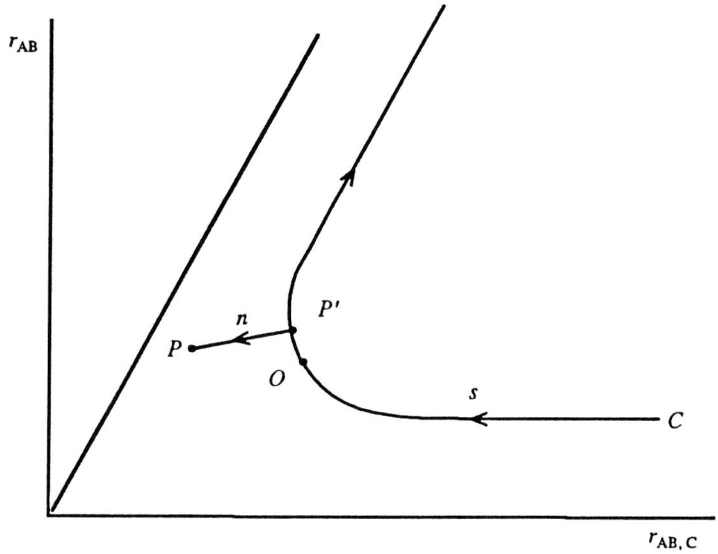

Fig. 1.10. Description of "natural collision coordinates" for a reaction AB + C → A + BC, s and n, for the collinear case. (Those for the three-dimensional problem are described in ref. [53].) The s is the reaction coordinate, measured from any fixed point O on C to the foot P' of the perpendicular from the point P. The n is the vibrational coordinate, i.e., the distance from P to P'.

reactants through the TS and on to products [91–94]. Such a curvilinear coordinate system is shown in Fig. 1.10 and the corresponding Hamiltonian was also introduced. The zeroth-order solution, when the curvature of the curve C in Fig. 1.10 is not large, was, indeed, the desired vibrationally adiabatic solution. Non-adiabatic corrections could then be calculated. There have been various extensions, e.g., the reaction path Hamiltonian [95] and the applications of this approach.

These coordinates in Fig. 1.10, or their extension [53] to three dimensions, are appropriate when the curvature of C is small. Non-adiabatic vibrational transitions then arise from both the curvature and from any rapid change of vibrational frequency as the system moves along the reaction coordinate s [91, 92]. However, in reactions such as the H transfer in AH + B → A + HB in Fig. 1.5, the curvature is so large that there is a tendency for the H to cross from one valley (reactants) to the other (products) at a constant AB distance (the Franck–Condon principle) and even bypass the saddle point [29]. For a symmetric reaction (B = A), the v of the product HB would equal that of AH. However, if the reaction is largely exothermic and the depth of the well in the products' valley is much more than that in the reactants' at the typical AB distance in the TS, the v for BH would be much larger than that for AH, as a result of this application of the Franck–Condon principle. Such a reaction is highly vibrationally non-adiabatic.

Comment: These systems provide two markedly different examples of chemical dynamics and coordinate systems suited to their dynamics can be selected. For one of these systems it was computer experiments, rather than real experiments, that served as the stimulus for some of these theoretical ideas.

1.2.10 Graph theory and additivity in delocalized (aromatic) molecules

In some early applications [96] to ETs involving compounds with quinone-like compounds Q, $Q \xrightarrow{e} Q^- \xrightarrow{e} Q^=$, it was necessary to examine some data on the formation constants of the semiquinone $QH^{\bullet}$, $Q + QH_2 \rightarrow 2QH^{\bullet}$. (The H is typically attached to an O or an N.) My impression, after looking at available data, was that the equilibrium constant was approximately the same, provided all three species had the same charge. Looking at the structures, one could see that every atom in a molecule on the left in this reaction corresponded to one on the right that had the same nearest and next nearest neighbor. I then looked at many examples of other pairs of compounds, which I termed conformal pairs and found that the total of heat of combustion of a pair was approximately the same as its conformal pair [97].

To gain some insight into this type of additivity I examined the results of LCAO calculations for these aromatic compounds [97]. Although individual orbital energies in one conformal pair showed no similarity to those in the other pair, the total energies of the occupied orbitals of the two pairs were approximately equal. Using a contour integral formula for the three total energies and relating the secular determinant appearing in it to the numbers of self-returning random walks, it was possible to gain some insight into this approximate additivity of the conformal pairs, their high delocalized (aromatic) nature notwithstanding. A recent survey of developments in graph theory, with a discussion of the above results, is given in ref. [98].

Comment: In chemistry it is nice to see generalizations from experiment and to find them in unexpected places. The approximate additivity of delocalized systems, described above, provides one such example. With new developments in electronic structure treatments, such as density functional methods, one might hope for a yet deeper insight.

ACKNOWLEDGMENTS

It is a real pleasure to dedicate this article to John Albery and to recognize his influential and joyous contributions to this interaction of experimentalists and theorists. It is a pleasure too to acknowledge the many contributions of my coworkers whose work is described in this chapter. This research was supported by the National Science Foundation and by the Office of Naval Research and I am very pleased to acknowledge that support.

References

The following references are given as examples, rather than being comprehensive.

[1] R.A. Marcus, In *Femtochemistry & Femtobiology*: *Ultrafast Reaction Dynamics at the Atomic-Scale Resolutions*. *Nobel Symposium* 101, V. Sundström, ed. (London: Imperial College Press, 1997), p. 54.
[2] R.A. Marcus, J. Electroanal. Chem. 438 (1997) 251.
[3] (a) E.R.Th. Kerstel, K.K. Lehmann, T.F. Mentel, B.H. Pate and G. Scoles, J. Phys. Chem. 95 (1991) 8282; (b) J.E. Gambogi, R.P. L'Esperance, K.K. Lehmann, B.H. Pate and G. Scoles, J. Chem. Phys. 98 (1993) 1116.
[4] (a) P.J. Rogers, J.I. Selco and F.S. Rowland, Chem. Phys. Lett. 89 (1982) 9; (b) S.P. Wrigley and B.S. Rabinovitch, Chem. Phys. Lett. 98 (1984) 386.
[5] (a) D.H. Lu and W.L. Hase, Chem. Phys. Lett. 142 (1987) 187; (b) D.H. Lu and W.L. Hase, J. Phys. Chem. 92 (1988) 3217.
[6] C. Cohen-Tannoudji, B. Diu and F. Laloë, Quantum Mechanics, Vol. 2 (John Wiley, New York, 1977).
[7] A.A. Stuchebrukhov and R.A. Marcus, J. Chem. Phys. 98 (1993) 6044.
[8] A. A. Stuchebrukhov, A. Mehta and R.A. Marcus, J. Phys. Chem. 97 (1993) 12491.

[9] (a) A.A. Stuchebrukhov, J. Chem. Phys. 104 (1996) 8424; (b) A.A. Stuchebrukhov, J. Chem. Phys. 105 (1996) 10818.
[10] (a) R. Langen, J. Chem, J.P. Gertnanas, J.P. Richards, J.H. Winkler and H. B. Gray, Science 269 (1995) 1733; (b) J.R. Winkler and H.B. Gray, Chem. Rev. 92 (1992) 369.
[11] R.A. Marcus and N. Sutin, Biochim. Biophys. Acta 811 (1985) 265.
[12] C.C. Moser, J.M. Keske, K. Warncke, R.S. Farid and P.L. Dutton, Nature 355 (1992) 796.
[13] (a) S. Larsson, J. Am. Chem. Soc. 103 (1981) 4034; (b) S. Larsson, J. Chem. Soc. Faraday Trans. 2, 79 (1983) 1375.
[14] P. Siddarth and R.A. Marcus, J. Phys. Chem. 94 (1990) 2985.
[15] P. Siddarth and R.A. Marcus, J. Phys. Chem. 97 (1993) 6111.
[16] P. Siddarth and R.A. Marcus, J. Phys. Chem. 97 (1993) 13078.
[17] (a) A.A. Stuchebrukhov and R.A. Marcus, J. Phys. Chem. 99 (1995) 7581; (b) J.N. Gehlen, I. Daizadeh, A.A. Stuchebrukhov and R.A. Marcus, Inorg. Chim. Acta 243 (1996) 271.
[18] (a) D.E. Logan and P.G. Wolynes, J. Chem. Phys. 93 (1990) 4994; (b) D.M. Leitner and P.G. Wolynes, Phys. Rev. Lett. 76 (1996) 216.
[19] See, for examples, articles in J. Phys. Chem. 56 (1952) 801–920.
[20] W.F. Libby, J. Phys. Chem. 56 (1952) 863.
[21] (a) R.A. Marcus, J. Chem. Phys. 24 (1956) 966; (b) R.A. Marcus, J. Chem. Phys. 24 (1956) 979.
[22] R.A. Marcus, Trans. N.Y. Acad. Sci. 19 (1957) 423.
[23] R.A. Marcus, Discussions Chem. Soc. 29 (1960) 21.
[24] R.A. Marcus, J. Chem. Phys. 43 (1965) 679.
[25] (a) N.R. Kestner, J. Logan and J. Jortner, J. Phys. Chem. 78 (1974) 2148; (b) J. Ulstrup and J. Jortner, J. Chem. Phys. 63 (1975) 4358; (c) V.G. Levich and R.R. Dogonadze, Proc. Acad. Sci. USSR, Phys. Chem. Sect. 124 (1959) 9; (d) V.G. Levich and R.R. Dogonadze, Proc. Acad. Sci. USSR, Phys. Chem. Sect. 133 (1960) 591; (e) P. Siders and R.A. Marcus, J. Am. Chem. Soc. 103 (1981) 748.
[26] J. R. Miller, L. T. Calcaterra and G.L. Closs, J. Am. Chem. Soc. 106 (1984) 3047.
[27] A. Warshel, J. Phys. Chem. 86 (1982) 2218.
[28] J.-M. Savèant, J. Am. Chem. Soc. 109 (1987), 6788.
[29] V.K. Babamov and R.A. Marcus, J. Chem. Phys. 74 (1981) 1790.
[30] S. Glasstone, K.J. Laidler and H. Eyring, The Theory of Rate Processes (McGraw-Hill, New York, 1941).
[31] (a) W.J. Albery and M.M. Kreevoy, Adv. Phys. Org. Chem. 16 (1978) 87; (b) E.S. Lewis, J. Phys. Org. Chem. 3 (1990) 1; (c) E.S. Lewis, T.A. Douglas and M.L. McLaughlin, Adv. Chem. Ser. 215 (1987); (d) M.J. Pellerite and J.I. Brauman, J. Am. Chem. Soc. 105 (1983) 35; (e) B.D. Wladkowski and J.I. Brauman, J. Phys. Chem. 97 (1993) 13158.
[32] M.M. Kreevoy, D. Ostovic, I.S.H. Lee, D.A. Binder and G.W. King, J. Am. Chem. Soc. 110 (1988) 534.
[33] I.S.H. Lee, E.H. Jeoung and M.M. Kreevoy, J. Am. Chem. Soc. 119 (1997) 2722.
[34] (a) A.J. Kresge, Chem. Soc. Rev. 2 (1973) 475; (b) M.M. Kreevoy and S.-W Oh, J.

Am. Chem. Soc. 95 (1973) 4805; (c) W.J. Albery, Ann. Rev. Phys. Chem. 31 (1980) 227; (d) D.E. Magnoli and J.R. Murdoch, J. Am. Chem. Soc. 103 (1981) 7465; (e) D.L. Leussing, J. Org. Chem. 55 (1990) 666; (f) J.P. Guthrie, J. Am. Chem. Soc. 118 (1996) 12878.

[35] R.A. Marcus, J. Phys. Chem. 72 (1968) 891.

[36] (a) S. Franzen and J.L. Martin, Ann. Rev. Phys. Chem. 46 (1995) 453; (b) A.J. Hoff and J. Deisenhofer, Phys. Reports 287 (1997) 1.

[37] R. Haberkorn, M.E. Michel-Beyerle and R.A. Marcus, Proc. Natl. Acad. Sci. 76 (1979) 4185.

[38] J. Deisenhofer, O. Epp, K. Miki, R. Huber, H. Michel, J. Mol. Biol. 180 (1984) 385.

[39] P. Hamm and W. Zinth, J. Phys. Chem. 99 (1995) 13537.

[40] (a) H.A. Kramers, Physica, 7 (1940) 284; (b) G.R. Fleming and P. Hänggi, eds. Activated Barrier Crossing: Applications in Chemistry, Physics and Biology (World Scientific, Singapore, 1993).

[41] (a) D. Huppert, H. Kanety and E.M. Kosower, Faraday Disc. Chem. Soc. 74 (1982) 161; (b) E.M. Kosower and D. Huppert, Chem. Phys. Lett. 96 (1983) 433.

[42] J. T. Hynes, J. Phys. Chem. 90 (1986) 3701.

[43] (a) N. Agmon and J. Hopfield, J. Chem. Phys. 78 (1983) 6947; (b) N. Agmon and J. Hopfield, J. Chem. Phys. 79 (1983) 2042; 80 (1984) (E) 592.

[44] H. Sumi and R.A. Marcus, J. Chem. Phys. 84 (1986) 4894.

[45] W. Nadler and R.A. Marcus, J. Chem. Phys. 86 (1987) 3096.

[46] (a) A.M. Berezhkovskii and V. Yu. Zitserman, Physica A. 166 (1990) 585; (b) A.M. Berezhkovskii, S.A. Dudko and V.U. Zitserman, Physica A, 187 (1994) 275.

[47] M.L. Horng, J.A. Gardecki, A. Papazyan and M. Maroncelli, J. Phys. Chem. 99 (1995) 17311.

[48] (a) Y. Nagasawa, A.P. Yartsev, K. Tominaga, P.B. Bisht, A.E. Johnson and K. Yoshihara, J. Phys. Chem. 99 (1995) 653; (b) Y. Nagasawa, A.P. Yartsev, K. Tominaga, A.E. Johnson and K. Yoshihara, J. Chem. Phys. 101 (1994) 5717.

[49] (a) G.C. Walker, E. Akesson, A.E. Johnson, N.E. Levinger and P.F. Barbara, J. Phys. Chem. 96 (1992) 3728; (b) G.C. Walker, P.F. Barbara, S K. Doorn, Y. Dong and J.T. Hupp, J. Phys. Chem. 95 (1991) 5712; (c) E. Akesson, G.C. Walker and P.F. Barbara, J. Chem. Phys. 95 (1991) 4188.

[50] D.K. Phelps and M.J. Weaver, J. Phys. Chem. 96 (1992) 7187.

[51] N. Lang, J.R. Miller and G.L. Gloss, J. Am. Chem. Soc. 112 (1990) 5353.

[52] (a) R. Jiminez, G.R. Fleming, P.V. Kumar and M. Maroncelli, Nature 369 (1994) 471; (b) S.J. Rosenthal, X. Xie, M. Du and G.R. Fleming, J. Chem. Phys. 95 (1991) 4715; (c) M. Maroncelli and G.R. Fleming, J. Chem. Phys. 86 (1987) 6221.

[53] C.-P. Hsu, X. Song and R.A. Marcus, J. Phys. Chem. B. 101 (1997) 2546.

[54] (a) Y. Georgievskii, C.-P. Hsu and R.A. Marcus, J. Chem. Phys. 108 (1998) 5978; (b) C.-P. Hsu, Y. Georgievskii and R.A. Marcus, J. Phys. Chem. 102 (1998) 2658.

[55] S. Mukamel, Principles of Nonlinear Optical Spectroscopy (Oxford University Press, New York, 1995).

[56] X. Song and R.A. Marcus J. Chem. Phys. 99 (1993) 7768.

[57] (a) B.G. Bakhshiev, Opt. Spectrosc. (USSR) 16 (1964) 446; (b) Y. T. Mazurenko,

Opt. Spectrosc. (USSR) 48 (1980) 388; (c) Y.T. Mazurenko and B.G. Bakshiev, Opt. Spectrosc. (USSR) 28 (1970) 490.; (d) W.R. Ware, S.K. Lee, G.J. Brant and P.P. Chow, J. Chem. Phys. 54 (1971) 4729.
[58] T. Okamura, M. Sumitani, K. Yoshihara, Chem. Phys. Lett. 94 (1983) 339.
[59] F. A. Lindemann, Trans. Faraday Soc. 17 (1922) 598.
[60] O. K. Rice and H.C. Ramsperger, J. Am. Chem. Soc. 49 (1927) 1617.
[61] L.S. Kassel, J. Phys. Chem. 32 (1928) 225, 1065.
[62] (a) R.G. Gilbert and S.C. Smith, Theory of Unimolecular and Recombination Reactions (Blackwell, Oxford, 1990); (b) T. Baer and W.L. Hase, Unimolecular Reaction Dynamics: Theory and Experiments (Oxford, New York, 1996); (c) K.A. Holbrook, M.J. Pilling and S.M. Robertson, Unimolecular Reactions, 2nd edn. (Wiley-Interscience, New York, 1996); (d) I.W.M. Smith, Kinetics and Dynamics of Elementary Gas Phase Reactions (Butterworths, London, 1980); (e) W. Forst, Theory of Unimolecular Reactions (Academic Press, New York, 1973).
[63] (a) R.P. Ogg, Jr. J. Chem. Phys. 15 (1947) 337, 613; (b) R.L. Mills and H.S. Johnston, J. Am. Chem. Soc. 73 (1951) 938.
[64] R.A. Marcus, B. de B. Darwent and E.W.R. Steacie, J. Chem. Phys. 16 (1948) 987.
[65] R.A. Marcus and E.W.R. Steacie, Z. Naturforsch. 4a (1949) 332.
[66] R.A. Marcus and O.K. Rice, J. Phys. Colloid Chem. 55 (1951) 894.
[67] R.A. Marcus, J. Chem. Phys. 20 (1952) 359.
[68] D.M. Wardlaw and R.A. Marcus, Adv. Chem. Phys. 70, Part 2 (1988) 231.
[69] I. Oref and B.S. Rabinovitch, Acc. Chem. Res. 12 (1979) 166.
[70] (a) L.R. Khundkar, J.L. Knee and A.H. Zewail, J. Chem. Phys. 87 (1987) 77; (b) E.D. Potter, M. Gruebele, L.R. Khundkar and A.H. Zewail, Chem. Phys. Lett. 164 (1989) 463.
[71] (a) D.J. Nesbitt, H. Petek, M.F. Foltz, S.V. Filseth, D.J. Bamford and C.B. Moore, J. Chem. Phys. 83 (1985) 223; (b) H. Bitto, D.P. Gwyer, W.F. Polik and C.B. Moore, Faraday Disc. Chem. Soc. 81 (1986) 149.
[72] R.A. Marcus, J. Chem. Phys. 45 (1966) 2138.
[73] (a) M. Quack and J. Troe, Ber. Bunsenges, Physik. Chem. 81 (1977) 329; (b) J. Troe, Adv. Chem. Phys. 82 (1992) 485.
[74] (a) D.G. Truhlar, B.C. Garrett and S.J. Klippenstein, J. Phys. Chem. 100 (1996) 12771; (b) D.G. Truhlar and B.C. Garrett, Ann. Rev. Phys. Chem. 35 (1984) 159.
[75] W.H. Green, C.B. Moore and W.F. Polik, Ann. Rev. Phys. Chem. 43 (1992) 591.
[76] (a) C.X. Qian, M. Noble, I. Nadler, H. Reisler and C.J. Wittig, Chem. Phys. 83 (1985) 5573; (b) C. Wittig, I. Nadler, H. Reisler, M. Noble, J. Catanzarite and G. Radhakrishnan, J. Chem. Phys. 83 (1985) 5581; 85 (1986) 1710; (c) C.X.W. Quon, A. Ogai, H. Reisler and C. Wittig, J. Chem. Phys. 90 (1989) 209.
[77] P. Pechukas and J.C. Light, J. Chem. Phys. 42 (1965) 3281.
[78] (a) C.X. Qian, M. Noble, I. Nadler, H. Reisler and C. Wittig, J. Chem. Phys. 83 (1985) 5573; (b) C.X.W. Quon, A. Ogai, H. Reisler and C. Wittig, J. Chem. Phys. 90 (1989) 209.
[79] (a) S.J. Klippenstein and R.A. Marcus, J. Chem. Phys. 91 (1989) 2280; (b) S.J. Klippenstein, L.R. Khundkar, A.H. Zewail and R.A. Marcus, J. Chem. Phys. 89 (1988) 4761.

[80] R.A. Marcus, Chem. Phys. Lett. 144 (1988) 208.
[81] S.J. Klippenstein and R.A. Marcus, J. Chem. Phys. 93 (1990) 2418.
[82] W.H. Miller, J. Chem. Phys. 65 (1976) 2216.
[83] A.J. Dobbyn, M. Stumpf, H.-M. Keller and R. Schinke, J. Chem. Phys. 104 (1996) 8357.
[84] (a) A.A. Heikal, S.H. Chong, J.S. Baskin and A.H. Zewail, Chem. Phys. Lett. 242 (1995) 380; (b) C. Lienau, A.A. Heikal and A.H. Zewail, Chem. Phys. 175 (1993) 171.
[85] R.A. Marcus, Chem. Phys. Lett. 244 (1995) 10.
[86] R.A. Marcus, J. Chem. Phys. 105 (1996) 5446.
[87] R.A. Marcus, J. Chem. Phys. 43 (1965) 1598.
[88] F.T. Wall, L.A. Hiller, Jr. and J. Mazur, J. Chem. Phys. 29 (1958) 255.
[89] E.M. Mortimer and K.S. Pitzer, Chem. Soc. (London) 16 (1962) 57 (special issue).
[90] (a) J.O. Hirschfelder and E. Wigner, J. Chem. Phys. 7 (1939) 616; (b) H.M. Hulburt and J.O. Hirschfelder, J. Chem. Phys. 11 (1943) 276; (c) M.A. Eliason and J.O. Hirschfelder, J. Chem. Phys. 30 (1959) 1426.
[91] R.A. Marcus, J. Chem. Phys. 45 (1966) 4493.
[92] R.A. Marcus, J. Chem. Phys. 45 (1966) 4500.
[93] R.A. Marcus, J. Chem. Phys. 49 (1968) 2617.
[94] (a) G.A. Natanson, Mol. Phys. 46 (1982) 481; (b) G.A. Natanson, Chem. Phys. Lett. 190 (1992) 209, 215.
[95] W.H. Miller, N.C. Handy and J.E. Adams, J. Chem. Phys. 72 (1980) 99.
[96] R.A. Marcus, J. Chem. Phys. 26 (1957) 872.
[97] R.A. Marcus, J. Chem. Phys. 43 (1965) 2643.
[98] M. Randic and N. Trinajstic, Croat. Chem. Acta. 67 (1994) 1.

Chapter 2

The Application of Approximate Analytical Models in the Development of Modified Electrodes for NADH Oxidation

PHILIP N. BARTLETT and EMMA N. K. WALLACE

2.1 WHY BOTHER WITH APPROXIMATE ANALYTICAL MODELS?

The modelling of kinetics at modified electrodes has received much attention over the last 10 years [1–11], mainly due to the interest in the potential uses of chemically modified electrodes in analytical applications. The first treatment published by Andrieux *et al.* [5] was closely followed by a complimentary treatment by Albery and Hillman [1, 2]. Both deal with the simplest basic case, that is, the coupled effects of diffusion and reaction for a second-order reaction between a species freely diffusing in the bulk solution and a redox mediator species trapped within the film at the modified electrode surface. The results obtained by the two treatments are essentially identical, although the two approaches are slightly different.

In their work, Andrieux and Savéant analyse the behaviour in terms of the characteristic current densities for each process. This has the advantage that, in general, the expressions for these characteristic current densities are fairly straightforward and obvious and, therefore, easily derived. In the case where the different processes involved are strictly first-order, the reciprocal of total current density can be obtained by combining the reciprocals of the current densities for the individual steps. For example, the limiting current density, I_L, for a reaction at the surface of a rotating disc coated by a permeable membrane is given by

$$1/I_{\mathrm{L}} = 1/I_{\mathrm{S}} + 1/I_{\mathrm{A}}, \tag{2.1}$$

where I_{S} is the limiting current density for diffusion through the film

$$I_S = FD_{film}K_S c_\infty / L, \tag{2.2}$$

and I_A is the limiting current density for mass transport in the bulk solution to the membrane surface

$$I_A = FD_{soln} c_\infty / X_D. \tag{2.3}$$

Here D_{film} and D_{soln} are the diffusion coefficients in the film and solution, respectively; K_S is the partition coefficient for the species into the film; c_∞ is the bulk concentration; L is the film thickness; and X_D is the thickness of the diffusion layer at the electrode surface. For more complex systems, such as the modified electrode case where the processes are non-linear, the combination of the different limiting current densities is less straightforward (see [8] for examples), and the resulting equations are considerably more complex. Nevertheless, it is possible to identify different simple limiting behaviours which apply under particular conditions, for example, when mass transport in solution is rate-limiting, or when reaction between the substrate from the solution and the mediator within the film is rate-limiting, and so on. Andrieux and Savéant represent the inter-relationship between these different possible limiting behaviours in terms of zone diagrams where the axes are the logarithm of the ratios of the limiting current densities (see [8] for examples).

In contrast, Albery and Hillman adopt a different formalism in which the problem is treated in terms of the characteristic lengths associated with the different processes. Since the limiting currents and characteristic lengths are directly related (see equations (2.1) and (2.3) for example), the two approaches are mathematically equivalent. The advantage of using the lengths is that it allows the concentration profiles for the reactants and products to be directly visualised, and it is more directly linked to the experimental variables such as film thickness, rotation speed (hence diffusion layer thickness) and so on. This makes it easier to link the mathematical results to the corresponding physical situations and experimental variables. Again, for simple first-order processes it is easy to derive expressions by combining the expressions for each individual step, whereas, when non-linear processes are involved the expressions are more complex (see [2] for example). The inter-relationship between the different limiting behaviours can be expressed in terms of a case diagram in which the different axes are the logarithms of the ratios of different characteristic lengths. Thus, the zone diagrams of Andrieux and Savéant and the case diagrams of Albery and Hillman are equivalent. In this chapter, we follow the characteristic length formalism used extensively by Albery.

It may be argued that approximate analytical treatments of the type developed independently by Andrieux and Savéant, and by Albery and Hillman for reactions at modified electrodes, have been rendered redundant by the availability of low-cost powerful computers combined with the developments in rapid digital simulation techniques. However this is not true. Although numerical simulations can help to verify semianalytical treatments and are important in deriving results in those cases where the problem is made more complex by the geometry of the electrode, or because no one process is clearly rate-limiting (see [12] for example), the simulation itself does not provide insight into the nature of the rate-limiting processes, their interplay nor the ways in which the behaviour of the system can be manipulated by changes in the experimental variables. By careful study of the experimental data, in combination with the appropriate approximate analytical treatment, it is possible to gain a preliminary understanding of the reaction mechanism. This preliminary understanding can then be used to tailor subsequent experiments towards further exploration of the system leading to a complete kinetic analysis. Once the model has been shown to successfully fit all the experimental data, the experimenter can determine exactly how the system will behave under the full range of possible experimental conditions. Thus, the approximate analytical treatments provide an efficient method for determining the optimum operating conditions for any particular application and is far superior to the "hit and miss" approach which must otherwise be used. The modelling of kinetics also provides fundamental insights into the operation of the catalytic system. These insights can often be the starting point for the design and development of new, and better, catalytic systems. In the field of biosensors, the advantages of this systematic approach towards design and improvement are often overlooked in favour of a more heuristic approach. Nevertheless, the necessary models are available for many situations [13] and they are continually being improved and extended and should be more widely used.

In this chapter, we present an overview of the detailed kinetic analysis of the oxidation of NADH at a poly(aniline) modified electrode, both to demonstrate the basic approach which can be used, and to illustate the type of information which can be extracted. In many ways the design of modified electrodes for NADH oxidation is a paradigm for the design of modified electrodes to catalyse particular processes because much is known about the mechanism of NADH oxidation and the requirements for efficient modified electrodes to catalyse this reaction. In addition, the oxidation of NADH to enzymatically active NAD^+ is a problem which has attracted, and continues to attract, much attention because of the large number of dehydrogenase enzymes utilising this, or the closely related NADPH, coenzyme [14]. Consequently, an efficient modified electrode for the oxidation of NAD(P)H could form the basis for the development of a wide range of amperometric biosensors based on NAD(P)H-dependent dehydrogenase enzymes.

2.2 NADH OXIDATION

2.2.1 Nicotinamide adenine dinucleotide (NAD^+) [15, 16]

Nicotinamide adenine dinucleotide is a coenzyme which is only loosely bound to the active site of the enzymes with which it interacts and is free therefore, to dissociate from the enzyme during the catalytic cycle. The role of the dehydrogenase enzyme is to bring together the substrate and the NAD^+ in the correct orientation for the two to react. These NAD^+-dependent enzymes are known as dehydrogenases. They work in conjunction with NAD^+ to oxidise substrates by the transfer of $1H^+$ and $2e^-$ from the substrate to the 4-position of the nicotinamide ring of the NAD^+ (see Fig. 2.1). The overall reaction is the equivalent of a hydride transfer and is commonly referred to as such. NAD^+-dependent enzymes are primarily involved in respiration (NAD^+ occurs in significant amounts in mitochondria), whereas, $NADP^+$-dependent coenzymes are primarily involved in the transfer of electrons from intermediates in catabolism.

Nearly all NAD^+-dependent dehydrogenases studied follow an ordered bisubstrate mechanism. In this mechanism, the oxidation of a substrate proceeds in a sequential manner: first, NAD^+ binds in the active site of the dehydrogenase; then the substrate binds; next a "hydride equivalent" is transferred in a chemical step from the bound substrate to the bound NAD^+, hence, oxidising the substrate and reducing the NAD^+ to NADH; the oxidised substrate is then released from the active site and is finally followed by the NADH.

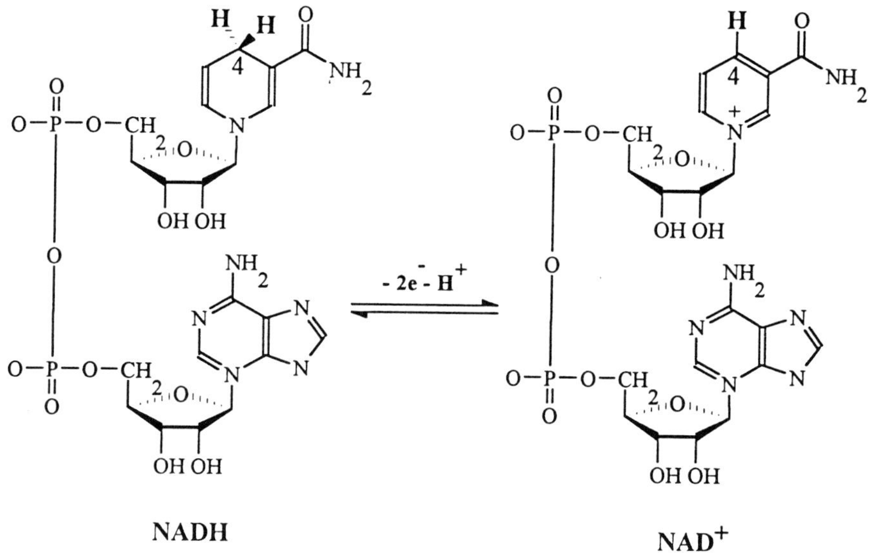

Fig. 2.1. The structures of NAD^+ and NADH and the interconversion between the two.

Currently, a major problem in the construction of successful NADH/dehydrogenase biosensors is the inability to produce a system that will oxidise NADH in a stable and reproducible manner at potentials where there are not significant problems from interferent species. In practice, this means below around +0.3 V vs. SCE at pH 7. To date, many of the most promising systems suffer from short working lifetimes due to leaching of the mediator or poisoning of the electrode surface.

2.2.2 Direct electrochemistry

The formal potential of the NAD^+/NADH redox couple is −0.56 V vs. SCE at pH 7 [15, 17]. However, at platinum and glassy carbon electrodes NADH, oxidation occurs at 0.7 V and 0.6 V vs. SCE, respectively [18]. From these oxidation potentials, it is clear that the direct electrochemical oxidation of NADH requires a substantial overpotential. In nature, NADH oxidation is thought to occur by a one-step hydride transfer. However, on bare electrodes the reaction has been shown to occur via a different and higher energy pathway which produces $NAD^{\bullet}$ radicals as intermediates.

The large amount of energy required for formation of NAD$^•$ radicals is the origin of the large overpotential required at bare electrodes. As a result, the direct electrochemical oxidation of NADH is complicated by the variety of products produced, electrode fouling and poor reproducibility. The direct electrochemical oxidation has been shown to produce a mixture of products including enzymatically active NAD$^+$, $(NAD)_2$ dimers and products from the side reactions of the NAD$^•$ radicals. The fouling of the electrode surface can occur by two mechanisms. First, the NAD$^•$ radicals interact with oxide functionalities on the electrode surface [17] and second, the reaction product, NAD$^+$, adsorbs on the electrode surface [19]. Both these processes are irreversible and result in the gradual blocking of the electrode surface to continued oxidation. It is this electrode fouling that results in the irreproducibility of the electrochemical responses at bare electrodes. If this method was used to analyse biological samples for NADH content, the high overpotentials required would result in the response being a combination of signals from the oxidation of both common interferents and NADH.

In conclusion, the direct electrochemical oxidation of NADH is of little practical use in analytical applications due to electrode fouling, and also in biosensor applications because of the low efficiency for the generation of enzymatically active NAD$^+$ due to the different possible side reactions.

2.2.3 Common approaches

The problems encountered during the direct oxidation of NADH can be overcome by using mediators. A mediator is a small catalytic molecule that can transfer charge from the substrate to the electrode, i.e., the direct electrochemistry of the substrate is bypassed. The use of a mediator has three potential advantages when compared to the direct oxidation of a substrate. First, if the substrate has irreproducible electrochemistry due to electrode fouling, or competing chemical reactions, this can be avoided. Second, the rate of electron transfer between the substrate and the electrode can be enhanced resulting in an increased response. Finally, a mediator can be selected with a redox potential that will limit the effects of common interferents on the signal response.

In the context of NADH oxidation, the mediator interacts with NADH in a chemical step to oxidise it and is itself reduced (Fig. 2.2, reaction a). The mediator is then reoxidised at the electrode surface (Fig. 2.2, reaction

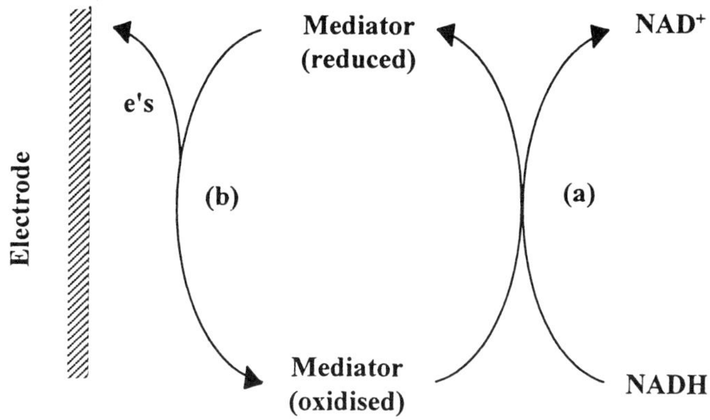

Fig. 2.2. Scheme for of the mediated electrochemical oxidation of NADH.

b) and, thus, produces an electrocatalytic current. This chemical oxidation of NADH bypasses the problems associated with the electrochemical oxidation. The advantages of using this technique are, first, virtually 100% of the oxidation product is enzymatically active NAD^+, and second, the electrode surface is not fouled by reaction products, although in some cases NAD^+ still adsorbs at the electrode surface. The choice of a mediator depends principally on two factors: how fast are the kinetics of chemical oxidation of NADH, and the nature and redox chemistry of the potential interferents. The rate at which NADH is chemically oxidised by the mediator will depend on the potential difference between the formal potential of the $NADH/NAD^+$ couple and the redox potential of the mediator. The greater this potential difference, the greater the driving force for the reaction and, hence, the greater the rate of the chemical oxidation in general. As discussed above, the interest in the production of a catalytic NADH oxidation system stems from its potential as the basis, when combined with suitable dehydrogenase enzymes, for the development of amperometric biosensors. For these applications (e.g., in the analysis of bodily fluids), it is important to limit interference from other, electrochemically active, molecules present in the sample. Many biological samples contain L-ascorbic acid (vitamin C), uric acid and acetoamidophenol (paracetamol). These have the respective redox potentials of 0.4, 0.34 and 0.35 V vs. SCE on glassy carbon electrodes [20]. Thus, to remove interferences from these substrates, mediators should be chosen that have redox potentials of 0.25 V vs. SCE or lower. Hence, when choosing a mediator there is a compromise to be struck between increasing the rate of the chemial oxidation by increasing the driving force on the one hand, and avoiding interferences by reactions with other electroactive species present in the solution on the other.

A reasonably detailed picture for the mechanism of oxidation of NADH in homogeneous solution emerges from the work of Miller and colleagues [21–26], who studied the oxidation of NADH using a range of oxidants at different solution pH and made kinetic isotope studies. This work is important for our present purposes because it forms the basis on which we can build in the design of modified electrodes for NADH oxidation. Figure 2.3 is a compilation of the kinetic results from Miller and coworkers, along with more recent data from other groups [27, 28]. The results are presented as a linear free energy relationship by plotting the logarithm of the second-order rate constant for the oxidation of NADH against the electrode potential of the oxidant at pH 7 [27].

From Fig. 2.3, we can see that the oxidants fall into two distinct groups: group (A) (top left) contains only two-electron oxidants while group (B) (bottom right) contains only one-electron oxidants. The redox potentials of the one-electron systems are greater than most of the potentials for the two-electron systems. Hence, the thermodynamic driving force for the oxidation of NADH will be greater for the one-electron oxidants. It is interesting, therefore, to find that the one-electron oxidants react with NADH at similar or slower rates than the two-electron oxidants. This implies that the mechanism of oxidation is different in the two cases, and there is a kinetic barrier to the oxidation of NADH by one-electron oxidants. From their studies, and based on the absence of any significant primary kinetic isotope effect or pH dependence, Carlson, Miller and coworkers [22, 23] showed that for the one-electron oxidants, electron transfer to give $NADH^{+\bullet}$ was the rate-limiting step and the redox potential for the $NADH/NADH^{+\bullet}$ couple was very positive. These results are consistent with the relatively slower rate of oxidation by the one-electron oxidants and the high overpotential for NADH oxidation at bare-metal electrodes. In contrast, for the two electron oxidants they found that the oxidation occurred either by H-atom transfer followed by electron or hydride transfer, that is, either

$$NADH + Med \rightarrow NAD^{\bullet} + MedH^{\bullet} \rightarrow NAD^{+} + MedH^{-},$$

or

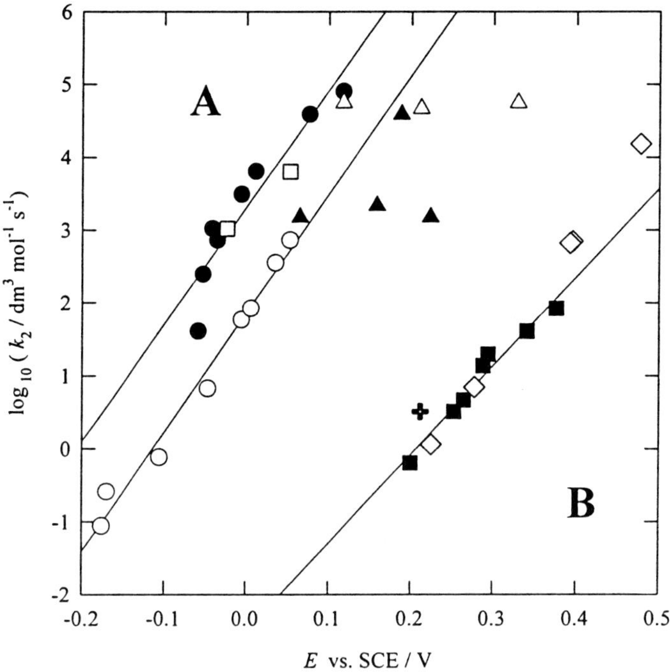

Fig. 2.3. Plot of the second order rate constants for homogeneous oxidation of NADH by a variety of different mediators as a function of their electrode potential. The graph uses results taken from work by Miller and coworkers [21–26] and from [28]. The different mediator groups are: (□) aminopyrimidines; (▲) 1,4-diaminobenzenes; (△) 1,2-diaminobenzenes; (●) *o*-quinones; (○) *p*-quinones; (■) ferrocenes; (+) ferrocyanide; and (◇) heteropolyanions.

$$\mathrm{NADH} + \mathrm{Med} \rightarrow \mathrm{NAD}^{+} + \mathrm{MedH}^{-}.$$

Whichever the mechanism, the data indicate that these two-electron acceptors are potentially excellent NADH oxidants, although to eliminate possible problems from interferents, it is best to choose systems with working potentials of 0.25 V vs. SCE or below. Based on this kinetic analysis and the structures of known mediators, it is proposed that for a molecule to be a good mediator for NADH oxidation, it must have a number of attributes [27]. First, it should have a redox potential more positive than −0.56 V vs. SCE, so that the oxidation is thermodynamically favourable. Second, it must have the ability to accept two-electrons in

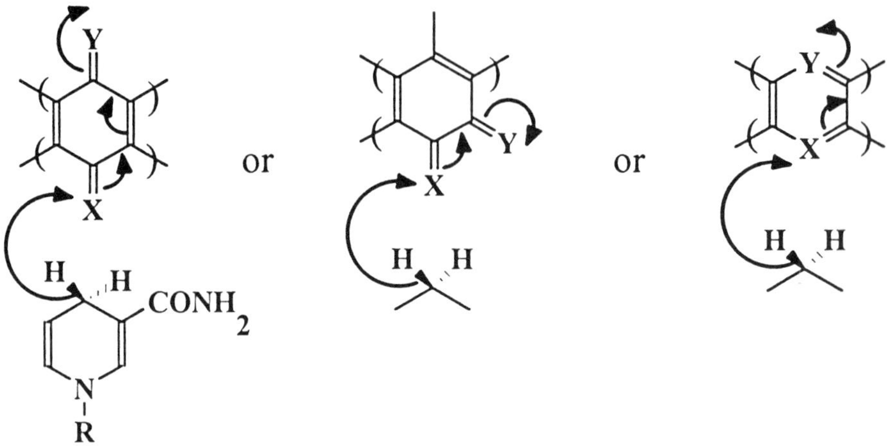

Fig. 2.4. Proposed mechanistic scheme for NADH oxidation by reaction with a redox mediator, where X is a group capable of accepting a hydride and Y is an electron withdrawing group. Adapted from [25].

order to avoid the formation of reactive NADH radical intermediates. Third, it should provide a site for hydride transfer. Fourth, it should be able to delocalise the negative charge within the molecule. The proposed structural features for good mediators and the possible mechanism for their reduction by NADH are shown in Fig. 2.4.

One group of NADH oxidants, which does not fit the proposed reaction scheme in Fig. 2.4 are the metal complexes. Examples of this type include nickel hexacyanoferrate deposited on porous nickel electrodes [29], gold electrodes modified with cobalt hexacyanoferrate films [30] and adsorbed 1,10-phenanthroline-5,6-dione complexes of ruthenium and osmium [31]. It is unclear how these systems work and no mechanism has been proposed to date. It may be worth noting that dihydronicotinamide groups have been shown to reduce aldehydes in a non-enzymatic reaction when the reaction is catalysed by zinc, a metal ion [15]. In a reaction between 1,10-phenanthroline-2-carboxaldehyde and *N*-propyl-1,4-dihydronicotinamide, no reaction was seen in the absence of zinc but when added to the system, the aldehyde was reduced and the nicotinamide was oxidised. This implies that either coordination to, or close proximity of, the metal ion activates this reaction. The proposed mechanism for this reaction is shown in Fig. 2.5.

2.2.4 Chemically modified electrodes for NADH oxidation [32]

Currently, the most popular method for producing an electrochemical NADH oxidation system is to chemically modify an electrode surface with a known mediating species. This immobilisation of the mediator has the advantage that the system is self-contained and, thus, the electrode can be used for analyses without the addition of a soluble mediator to the sample. For these chemically modified electrodes to be analytically useful, they should ideally have the following three characteristics: (1) stability; the chemically modified electrode should be stable typically for months without loss of mediator by leaching or chemical degradation and there should be no irreversible reaction between the modifier and NADH or NAD^+; (2) selectivity; the electrode should only respond to NADH and not to other solution species; and (3) fast kinetics; the chemical reaction between NADH and the mediator must be fast, the transfer of charge within the mediating layer must be fast, and electron transfer between the mediator and the electrode surface must be fast. The main classes of immobilised system are listed in Table 2.1 (see [27] for a more complete survey).

Gorton and coworkers have been particularly active in this field and produced an excellent review of the methods and approaches used for the successful chemical modification of electrodes for NADH oxidation [33]. They concentrated mainly on the adsorption onto electrode surfaces of mediators which are known to oxidise NADH in solution. The resulting systems were based on phenazines [34], phenoxazines [35, 36] and phenothiazines [32]. To date, this approach has produced some of the most successful electrodes for NADH oxidation. However, attempts to use similar mediators attached to poly(siloxane) films at electrode surfaces have proved less successful. Kinetic analysis of the results indicates that this is because of the slow charge transfer between the redox centres within the film so that the catalytic oxidation of NADH is restricted to a thin layer nearest the electrode surface [37, 38]. This illustrates the importance of a charge transfer between mediator groups in polymer modified electrodes.

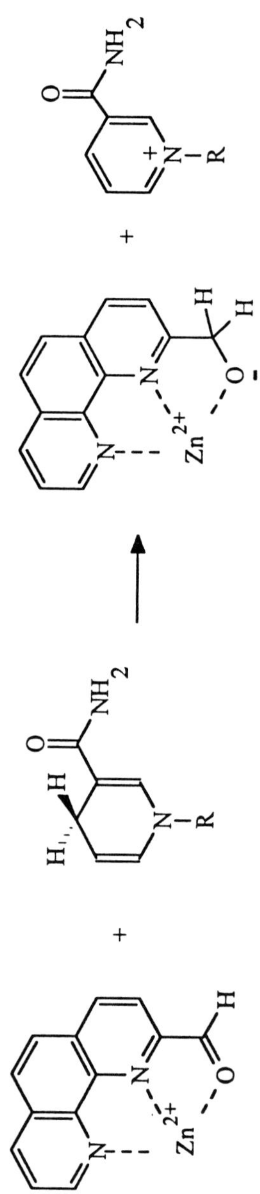

Fig. 2.5. Proposed mechanism for NADH oxidation in the presence of 1,10-phenanthroline-2-carboxaldehyde catalysed Zn^{2+} [15].

TABLE 2.1
Types of modified electrode for NADH oxidation

Methods of mediator incorporation	Comments	Ref.
Adsorption onto electrode surface	Mediator desorbs from electrode surface with time	[35, 36]
Activate glassy carbon by potential cycling	The carbonyl moieties on the electrode surface do not carry out selective oxidation. Intermediates from the oxidation reaction irreversibly inhibit the reaction	[19]
Electrode surface modified by the addition of a polymer containing mediating groups	The polymers used are non-conducting, hence, devices suffer from poor charge transport	[56]
Conducting polymer electropolymerised on the electrode surface	Two approaches in the literature. Precise mechanism for NADH oxidation unclear	[57, 58]
Hexacyanoferrate films	These systems show good stability but suffer from high overpotentials (~0.5 V vs. SCE) which in biological systems will result in signals form interferents such as ascorbic acid	[30, 59]

2.2.5 The common approach to modelling NADH oxidation

In the literature, NADH oxidation at chemically modified electrodes is most commonly suggested to occur via a two-step reaction mechanism. In the first step, NADH forms a complex with an oxidised mediator site, and in the second step, this complex breaks down producing product and a reduced mediator site, Fig. 2.6.

This type of mechanism has been proposed for a variety of different NADH oxidation systems, including phenoxazine dyes [35, 36], phenothiazine dyes [39] and a conducting organic salt [40]. Experimental evidence for the formation of a complex during the chemical oxidation of NADH has been provided by Fukuzumi *et al.* [41]. These authors showed that a mixture of an NADH model compound and a quinone derivative formed a charge-transfer complex in solution as determined by UV/Vis spectros-

$$\text{NADH} + \text{Mediator}_{(\text{oxidised})} \underset{k_{-1}}{\overset{k_{+1}}{\rightleftharpoons}} [\text{NADH} - \text{Mediator}] \xrightarrow{k_{+2}} \text{NAD}^{+} + \text{Mediator}_{(\text{reduced})}$$

Charge Transfer Complex

Fig. 2.6. The two-step reaction mechanism commonly proposed for NADH oxidation at chemically modified electrodes.

copy. They detected identical transient absorption bands during hydride transfer from the NADH model compound to the quinone derivatives and, hence, hypothesised that charge-transfer complexes are intermediates in the "hydride transfer" during NADH oxidation.

Due to the formation of an intermediate complex, this type of reaction mechanism was described as being analogous to Michaelis-Menten kinetics [39]. A common error made when examining the behaviour of systems of this type is to use the Koutecky-Levich equation to analyse the rotation speed-dependence of the current. This is incorrect because the Koutecky-Levich analysis is only applicable to surface reactions obeying strictly first-order kinetics. Applying the Koutecky-Levich analysis to situations where the surface kinetics are non-linear, as in this case, leads to erroneous values for the rate constants. Below, we present the correct treatment for this problem based on an extension of a model originally developed by Albery *et al.* [42]

2.2.6 Kinetic modelling at electrodes modified by films

In this section, we present a kinetic model for the oxidation of NADH within a polymer film at a rotating disc electrode. In this model, we assume that the rate of electron transfer between the mediating species in the film and the electrode is sufficiently fast so that it is not rate limiting [43]. We also assume that the substrate diffuses into the film where it is adsorbed at the reaction sites within the film and where it undergoes oxidation to the product. The possibility that the product is able to compete for these reaction sites with the film in a reversible manner is also considered.

Thus the steps in the reaction scheme are:

Mass transport to the electrode surface

$$\mathrm{NADH}_\infty \underset{k'_{\mathrm{D,S}}}{\overset{k'_{\mathrm{D,S}}}{\rightleftharpoons}} \mathrm{NADH}_0,$$

where the NADH_∞ and NADH_0 represent species in the bulk of the solution, and at the electrode surface, respectively; $k'_{\mathrm{D,S}}$ is the mass transport rate constant for substrate at the rotating disc electrode ($k'_{\mathrm{D,S}} = 1.554 D_S^{2/3} \nu^{-1/6} W^{1/2}$, where ν is the kinematic viscosity and W the rotation speed in Hz);

partition and diffusion within the film

$$\mathrm{NADH}_0 \overset{K_\mathrm{S}}{\rightleftharpoons} \mathrm{NADH}_{\mathrm{film}},$$

where $\mathrm{NADH}_{\mathrm{film}}$ represents substrate within the bulk of the film, and K_S is the partition coefficient for substrate into the film;

binding to reaction site

$$\mathrm{NADH}_{\mathrm{film}} + \{\mathrm{site}\} \overset{K_\mathrm{M}}{\rightleftharpoons} \{\mathrm{NADH}\},$$

where {site} represents a vacant adsorption site within the polymer, and K_M is the equilibrium constant for binding the substrate;

oxidation of bound NADH

$$\{\mathrm{NADH}\} \xrightarrow{k_{\mathrm{cat}}} \{\mathrm{NAD}^+\} + \mathrm{H}^+ + 2e,$$

where k_{cat} is the rate constant for the oxidation of the adsorbed substrate;

desorption of product from site

$$\{\mathrm{NAD}^+\} \underset{}{\overset{K_i}{\rightleftharpoons}} \mathrm{NAD}^+{}_{\mathrm{film}}{}^+ + \{\mathrm{site}\},$$

where K_i is the equilibrium constant for binding the product;

diffusion of product and partition into the film

$$\mathrm{NAD}^+_{\mathrm{film}} \overset{K_P}{\rightleftharpoons} \mathrm{NAD}^+_0,$$

where K_P is the partition coefficient for the product in the film;

diffusion of product in solution

$$\mathrm{NAD}^+_0 \underset{k'_{D,P}}{\overset{k'_{D,P}}{\rightleftharpoons}} \mathrm{NAD}^+_{\mathrm{bulk}},$$

where $k'_{D,P}$ is the mass transport rate constant for the product in solution.

This type of diffusion/reaction mechanism has been treated semi-analytically by Albery *et al.* [42, 44, 45], under steady-state conditions and its applications to amperometric chemical sensors has been described by Lyons *et al.* [46]. In both models, only diffusion and reaction within a boundary layer is considered, while the effect of concentration polarisation in the solution is neglected. Thus, to apply the model to an experimental system it is necessary to be able to accurately determine the concentration of substrate at the polymer/solution interface. Assuming that the system is in the steady state, the use of the rotating disc electrode allows simple determination of the substrate concentration at the interface from the bulk concentration and the experimentally determined flux using [47]

$$[\mathrm{NADH}_0] = [\mathrm{NADH}_\infty] - i/nFAk'_{D,S}. \tag{2.4}$$

The approximate analytical treatment described by Albery can now be applied. In the absence of any inhibition, and assuming rapid desorption of product, an expression for the flux can be determined by combining equations (6) and (23) from Albery *et al.* and substituting current, i, for flux (j) to give

$$i = \frac{nFAK_M D_{S,film} y}{L}, \tag{2.5}$$

and

$$y = \{2\epsilon[\alpha - \ln(1+\alpha)]\}^{1/2} \tanh\left[\frac{\epsilon^{1/2}\alpha}{(1+\alpha)\{2[\alpha - \ln(1+\alpha)]\}^{1/2}}\right], \tag{2.6}$$

where L is the thickness of the film, $D_{S,film}$ is the diffusion coefficient of substrate in the film and in our notation

$$\epsilon = \frac{L^2[\{site\}]k_{cat}}{D_{S,film} K_M}, \tag{2.7}$$

and

$$\alpha = K_S[NADH_0]/K_M. \tag{2.8}$$

Equations (2.5)–(2.8) describe four different limiting cases for the behaviour of the system. Before proceeding, it is necessary to define the "reaction layer thickness" (X_k):

$$X_k = \left(\frac{D_{S,film}}{k_{cat}}\right)^{1/2}. \tag{2.9}$$

This is the distance that the substrate diffuses into the layer before it reacts with the mediator within the film as illustrated in Fig. 2.7. The four limiting forms will now be discussed in turn.

Case I: No concentration polarisation within the polymer layer

In case I, the diffusion within the layer is fast, hence, as soon as the product is produced it diffuses out of the film. Thus, there is virtually no product in the layer and the concentration of substrate is uniform throughout the film. This is only valid when the film is thin ($\epsilon < 1$) and

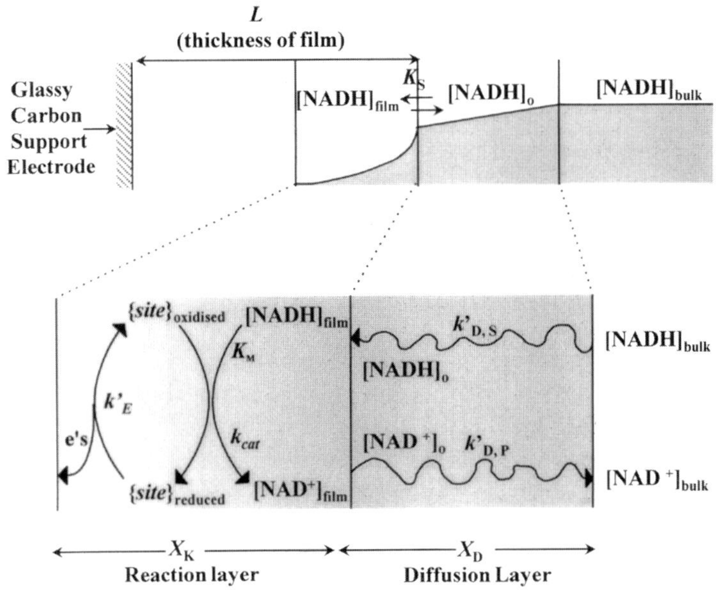

Fig. 2.7. Schematic representation of proposed reaction mechanism for oxidation of NADH by poly(aniline)/poly(vinylsulfonate) composite films (the symbols are defined in the text). *Note*: There is no differentiation between the substrate/{site} complex formation and the chemical reaction, k_{cat}, that occur at the site.

the concentration of substrate is insufficient to saturate the reaction sites ($\alpha < 1$). In case I, the current is given by

$$i_{\text{I}} = nFAk_{\text{cat}}\,[\{\text{site}\}]LK_{\text{S}}[\text{NADH}_0]/K_{\text{M}}. \tag{2.10}$$

Equation (2.10) is valid for all substrate concentrations. Overall, the reaction is first order in substrate and occurs throughout the film.

Case II: Low substrate concentration, unsaturated kinetics

In case II, the concentration of substrate is again insufficient to saturate the sites ($\alpha < 1$), but now the film is thick ($\epsilon > 1$) and, therefore, all the substrate is consumed in a first-order reaction layer (X_k) at the outside of the film and the current is given by,

$$i_{II} = nFAK_S[\text{NADH}_0](k_{cat}[\{\text{site}\}]D_S/K_M)^{1/2}. \quad (2.11)$$

In this case, the flux will be independent of film thickness.

Case III: High substrate concentration, saturated kinetics

Now the concentration of substrate is sufficient to saturate the sites ($\alpha > 1$) and the film is sufficiently thin ($\epsilon < 2\alpha$) so that the reaction of the substrate occurs with zero-order kinetics throughout the whole of the film. The expression for the current in this case is,

$$i_{III} = nFAk_{cat}[\{\text{site}\}]L. \quad (2.12)$$

The current is now dependent on the film thickness and the concentration of sites within the polymer film, but it is independent of substrate concentration.

Case IV: Intermediate substrate concentration

In case IV ($\alpha > 1$; $\epsilon > 2\alpha$), the substrate concentration is sufficient to saturate the sites at the outside of the film but then falls as it is consumed within the film, so that the kinetics become unsaturated further into the film. For this mixed case the current is given by,

$$i_{IV} = nFA(2k_{cat}[\{\text{site}\}]D_SK_S[\text{NADH}_0])^{1/2}. \quad (2.13)$$

The current is independent of film thickness and is half-order with respect to the concentration of sites within the film and the concentration of substrate.

The different current expressions and reaction orders for each case are summarised in Table 2.2. The relationships between these four cases and sketches of the corresponding concentration profiles are shown in Fig. 2.8.

2.2.6.1 Product inhibition

So far, the effect of product inhibition has not been considered. Inhibited enzyme electrodes have been discussed in a collection of publications by Albery *et al.* [42, 44, 45]. Solving our equations by applying a steady-state analysis, and including the effects of product inhibition within the film, results in the following expression for the homogeneous reaction flux, *f*, within the polymer layer [48],

TABLE 2.2
Summary of current expressions for each case and the reaction orders with respect to the different reaction variables

Case	Current expression	Reaction order			Eqn
		$[NADH_\infty]$	$[\{site\}]$	L	
I	$i_I = nFAk_{cat}[\{site\}]LK_S[NADH_0]/K_M$	1	1	1	(2.10)
II	$i_{II} = nFAK_S[NADH_0](k_{cat}[\{site\}]D_S/K_M)^{1/2}$	1	$\frac{1}{2}$	0	(2.11)
III	$i_{III} = nFAk_{cat}[\{site\}]L$	0	1	1	(2.12)
IV	$i_{IV} = nFA(2k_{cat}[\{site\}]D_SK_S[NADH_0])^{1/2}$	$\frac{1}{2}$	$\frac{1}{2}$	0	(2.13)

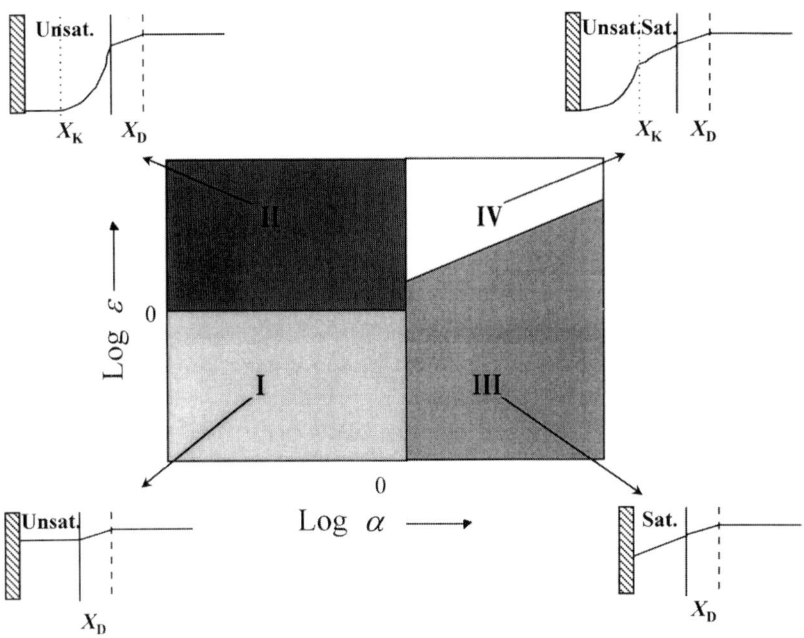

Fig. 2.8. Case diagram summarising the four possible cases and the corresponding concentration profiles that arise during electrocatalysis at the chemically modified electrode.

$$\frac{[\{site\}]}{f} = \frac{K_M}{k_{cat}[NADH_{film}]} + \frac{1}{k_{cat}} + \frac{K_M[NAD^+_{film}]}{K_i k_{cat}[NADH_{film}]}, \tag{2.14}$$

where $[NADH_{film}]$ and $[NAD^+_{film}]$ are the concentrations of the species at any point in the film. From the previous discussion, the concentration of product at the outside of the film for the rotating disc electrode is given by,

$$[NAD^+_0] = [NAD^+_{bulk}] + i/nFAk'_{D,P}. \tag{2.15}$$

Then, replacing $[NAD^+_{film}]$ by $K_P[NAD^+_0] + K_S[NADH_0] - [NADH_{film}]$ and rearranging gives,

$$f = \frac{k'_{cat}[\{site\}][NADH_{film}]}{K'_M + [NADH_{film}]}, \tag{2.16}$$

where

$$k'_{cat} = k_{cat}/(1 - K_M/K_i), \tag{2.17}$$

and

$$K'_M = K_M\left\{\frac{1 + (K_P[NAD^+_0] + K_S[NADH_0])/K_i}{(1 - K_M/K_i)}\right\}. \tag{2.18}$$

Equation (2.16) clearly shows that the inhibited kinetics are still of the Michaelis-Menten type. Thus, extending the argument of Albery *et al.* [42] we can use our original equations (2.5)–(2.12) to analyse the data simply by substituting in the new definitions for k'_{cat} and K'_M. From equations (2.17) and (2.18), it is clear that when K_i is large, corresponding to little product inhibition, $k'_{cat} = k_{cat}$ and $K'_M = K_M$ and the inhibited solution reverts to the solution found previously for the situation where there was no product inhibition as required.

2.3 POLY(ANILINE)

2.3.1 *Introduction*

As discussed previously, the literature indicates that the best NADH oxidation systems are based on two-electron/one-proton acceptors with redox potentials less that 0.25 V vs. SCE. On this premise, it was proposed

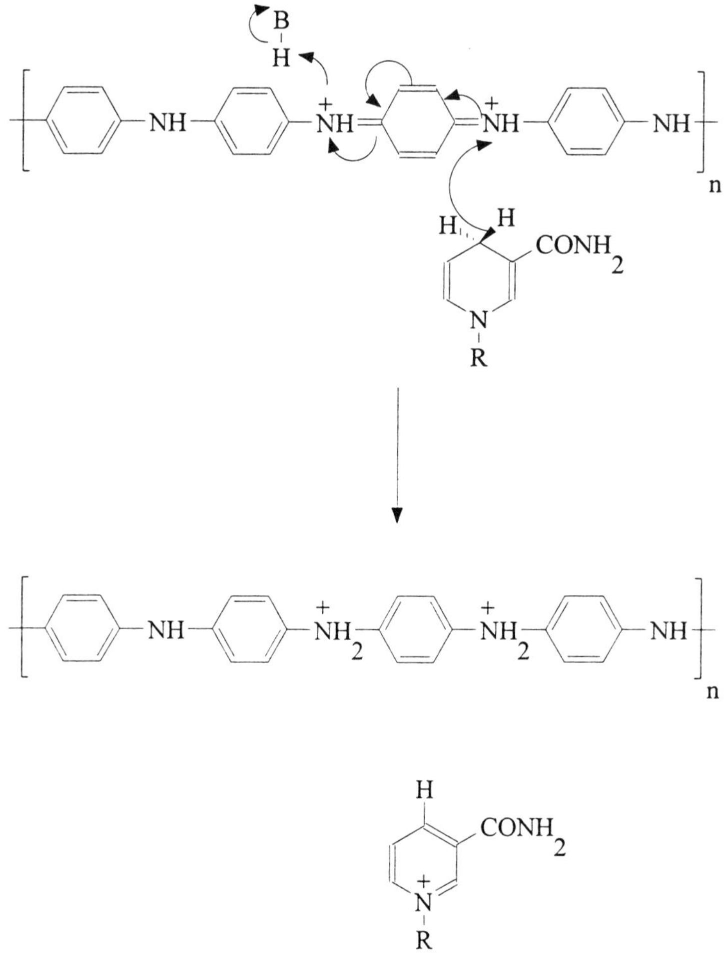

Fig. 2.9. A postulate mechanism for the oxidation of NADH by poly(aniline).

that poly(aniline), which consists of diimine units well documented to be excellent NADH mediators, may be an electrocatalytic surface for NADH oxidation. Figure 2.9 shows a possible mechanism for the proposed reaction.

Since at pH above 5 poly(aniline) films deprotonate and become insulating in all oxidation states, Fig. 2.10, preliminary experiments were carried out at pH 5. Figure 2.11 shows the effect of the addition of

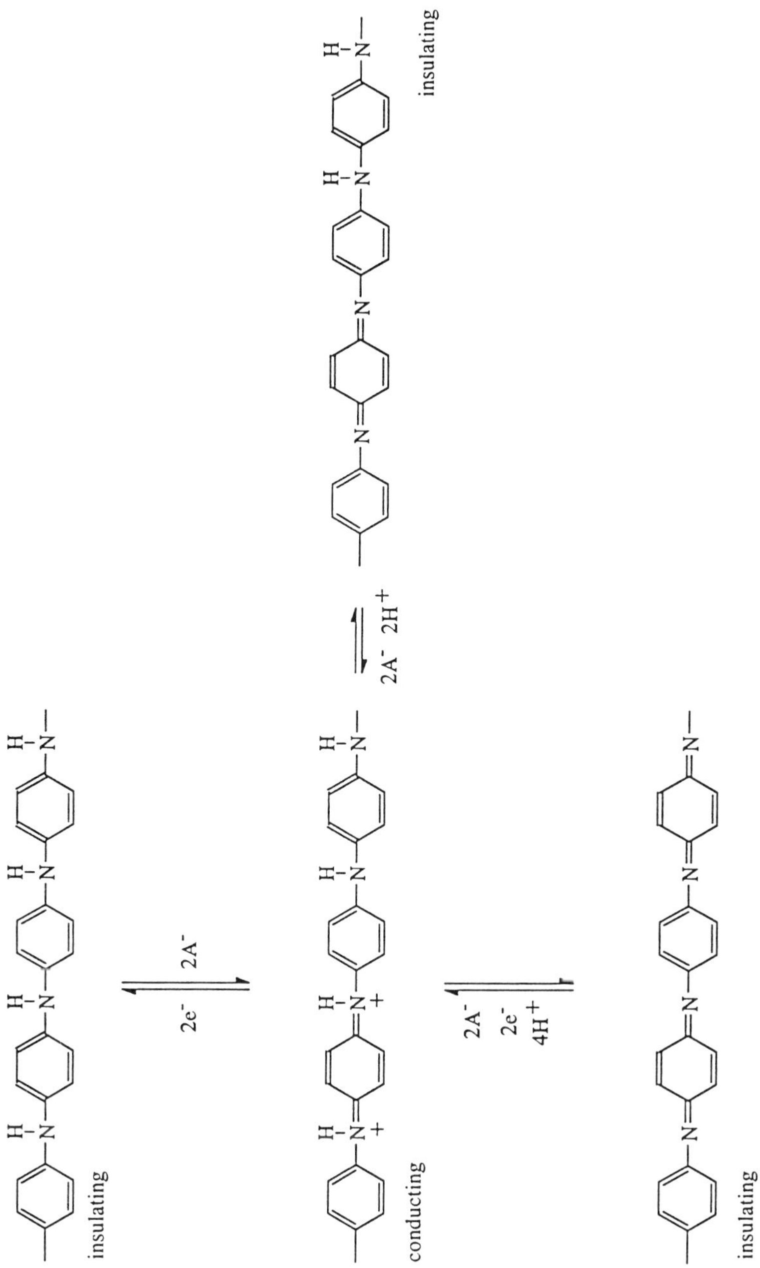

Fig. 2.10. The different protonation states and redox states for poly(aniline).

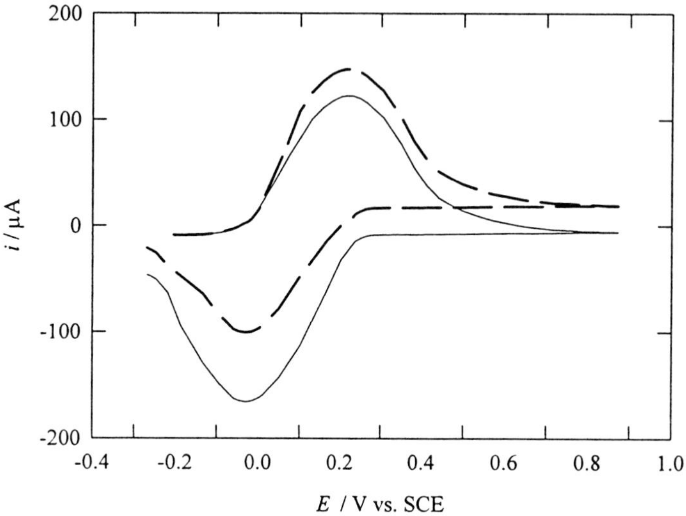

Fig. 2.11. Cyclic voltammograms of a poly(aniline)-coated glassy carbon electrode (deposition charge 150 mC, geometric area 0.38 cm^2), recorded at 5 mV s^{-1} in oxygen-free 0.1 mol dm^{-3} citrate/phosphate buffer at pH 5 in the absence (—), and in the presence (- - -), of 1 mmol dm^{-3} NADH. Before each scan the electrode was held at −0.3 V for 3 min to ensure complete reduction of the film.

1.0 mmol dm^{-3} NADH on the voltammetry of poly(aniline) at pH 5. Figure 2.11 clearly shows that NADH causes an increase in the anodic current concurrent with a decrease in the cathodic current indicating that poly(aniline) mediates the electrochemical oxidation of NADH.

Having shown that poly(aniline) films can mediate NADH oxidation, studies of the effect of altering the applied potential and the rotation rate of the electrode were undertaken. Preliminary results from these studies showed that the maximum current response was obtained when the applied potential was 0.2 V vs. SCE and that the currents were two orders of magnitude higher for poly(aniline) modified electrodes when compared to a bare electrode indicating that poly(aniline) is a good catalytic surface for the oxidation of NADH. However, studies of the effect of rotation speed carried out at pH 5 show a decline in current with time (see Fig. 2.12).

Any decrease in current response with time greatly complicates the kinetic analysis and makes the system unsuitable for use in a biosensor. It is important, therefore, to determine the reason for the poor stability and

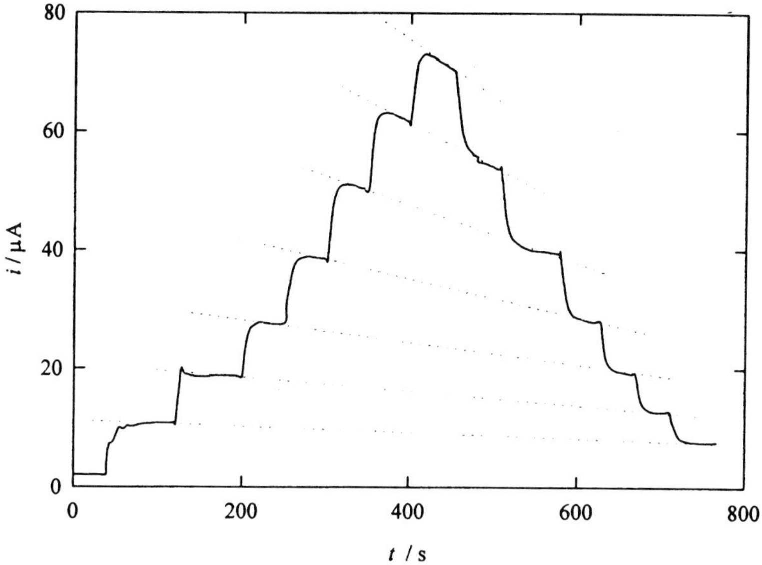

Fig. 2.12. Plot of the current as a function of time for the oxidation of 4 mmol dm^{-3} NADH at 0.2 V at a poly(aniline)-coated rotating disc electrode (area 0.38 cm^2, deposition charge 150 mC) in 0.1 mol dm^{-3} citrate/phosphate buffer, pH 5. The rotation speed of the electrode was increased in the sequence 1, 4, 9, 16, 25, 36 and 49 Hz and reduced in sequence back to 1 Hz. The broken line connects segments of the curve corresponding to the different rotation speeds. *Note*: The current decays more rapidly at the higher rotation speeds and responds rapidly to changes in rotation speed.

seek to overcome the problem before proceeding. Although there are many possible factors which could be responsible, such as dissolution of the polymer or inactivation of the polymer, further studies showed that the decrease in current was caused by the acid catalysed decomposition of the NADH and not by any change in the film itself.

2.3.2 Acid catalysed decomposition of NADH

Braun *et al.* [49] showed that NADH decomposes much more rapidly under acidic than neutral conditions. They found that the decomposition was first-order in both cases with the respective rate constants being: $5.9 \times 10^{-4}\ s^{-1}$ at pH 4.1 and $0.11 \times 10^{-4}\ s^{-1}$ at pH 7.1 for a 0.62 mmol dm^{-3} NADH solution in the Tris buffer. The authors propose that the decomposition reaction occurs via hydration of the 5,6-double bonds in the pyridine ring of NADH, Fig. 2.13.

To determine the effect of acid catalysed decomposition of NADH on the electrochemical response in our experiments, the decrease in oxidation current for NADH was recorded as a function of time. The results of this experiment were compared with the decrease in NADH concentration as spectrophotometrically determined. The rates of decrease of the current and the concentration of NADH are both first-order and occur on similar timescales (Fig. 2.14). Analysis of the data for the two experiments provide first-order rate constants of 1.68 and $1.16 \times 10^{-4}\ s^{-1}$ for the electrochemical and spectrophotometric measurements, respectively. The small difference between these two constants can be explained by the additional consumption of NADH by reaction at the electrode during the electrochemical measurement. This electrochemical process is also a first-order rate process, and the extent of the effect can be determined by using the treatment of Hitchman and Albery [50] for electrolysis using a rotating disc electrode. The results are consistent with the observed difference in the two rate constants.

To determine whether, or not, it would be possible to remove the effect of acid catalysed decomposition of NADH by working at neutral pH, the decomposition rate was measured at pH 7. The decomposition was again found to be first-order, but with a rate constant of $8.6 \times 0^{-6}\ s^{-1}$. Comparing this to the value of $1.16 \times 10^{-4}\ s^{-1}$, determined for the decomposition in pH 5 buffer, shows that the decomposition rate has decreased by an order of magnitude.

2.3.3 *Poly(aniline) at pH 7*

Although poly(aniline) is a catalytic surface for the oxidation of NADH at pH 5, without modification this approach is of little practical use because NADH undergoes rapid decomposition in the solution at this pH. NADH decomposition is significantly slower at pH 7 but poly(aniline) is insulating in all oxidation states at this pH. A solution to this conundrum is suggested by the work of Asturias *et al.* [51], who showed that composites of poly(aniline) and polyanions remained conducting at high pH. We have exploited this effect by growing poly(aniline) films in the presence of poly(vinylsulfonate) to produce films which are electroactive at pH 7 and remain catalytic for NADH oxidation [48]. We now discuss the characterisation

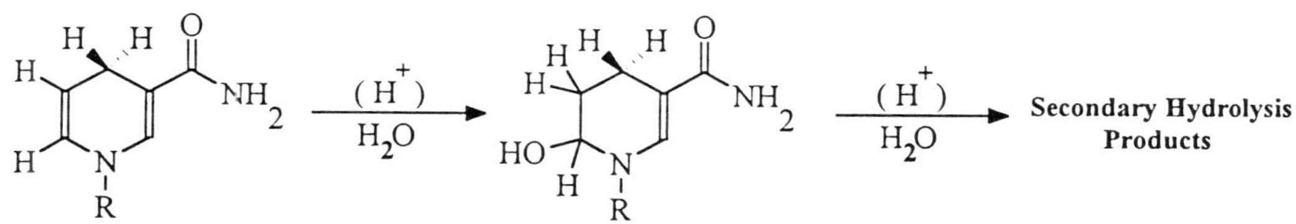

Fig. 2.13. Mechanism of the acid catalysed decomposition reaction of NADH [15].

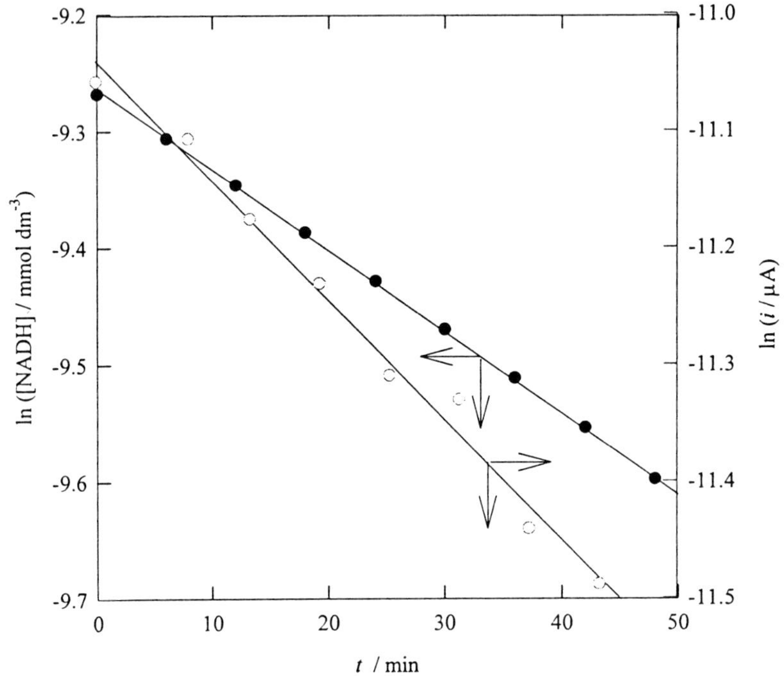

Fig. 2.14. Plots of the first-order decomposition of NADH in pH 5 buffer measured electrochemically at a poly(aniline) coated electrode (○), and spectrophotometrically (●).

of NADH electrocatalysis at these poly(aniline)/poly(vinylsulfonate) composite films.

2.3.4 *Experimental results*

2.3.4.1 *Catalysis of NADH oxidation*

Figure 2.15 shows the first cycle response of two different poly(aniline)/ poly(vinylsulfonate) composite films in 0.1 mol dm^{-3} pH 7 citrate/phosphate buffer with and without added NADH.

From the observed increase in the oxidation current, and decrease in the reduction current, it is clear that these modified films are catalytic surfaces for NADH oxidation at pH 7. With time, the current decay seen at pH 5 is no longer evident at neutral pH, as predicted if the degradation is mainly due to acid catalysed NADH decomposition. In Fig. 2.16, the

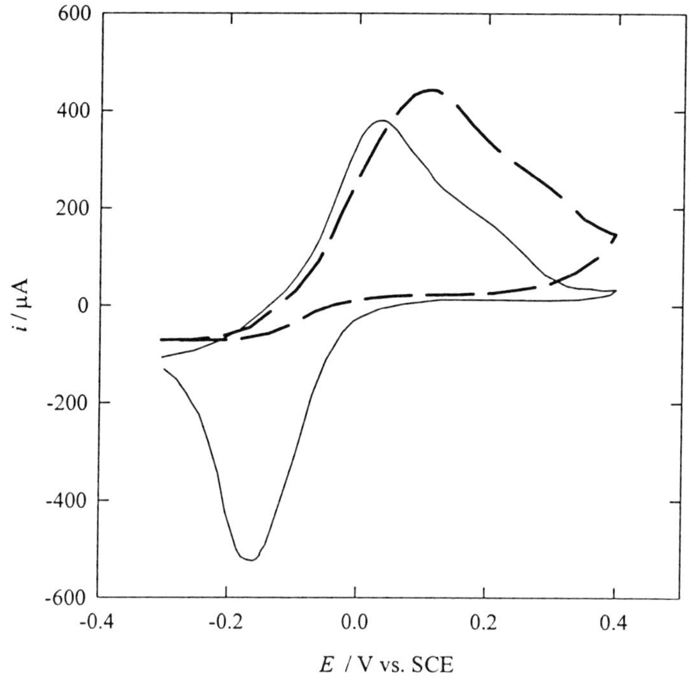

Fig. 2.15. The first cycles at two identical poly(aniline)/poly(vinylsulfonate)-coated glassy carbon electrodes (deposition charge 150 mC, geometric area 0.38 cm^2) recorded at 2 mV s^{-1} in oxygen-free 0.1 mol dm^{-3} citrate/phosphate buffer at pH 7, in the absence of NADH (—), and in the presence of 4.4 mmol dm^{-3} NADH (- - -). Before each scan, the electrode was held at −0.3 V for 3 min to ensure complete reduction of the film.

amperometric responses to NADH of three different films grown from different batches of aniline and poly(vinylsulfonate) over a six-month period, are presented. It is clear from the figure that the amperometric responses of these modified poly(aniline) electrodes to NADH are reproducible between electrodes fabricated at different times, using different batches of reagents. Figure 2.16 also confirms the electrocatalytic nature of the polymer for NADH oxidation. The inset shows the response to NADH under the same conditions for a bare glassy carbon electrode. Comparing this with the response obtained using poly(aniline) modified electrodes, the currents are three orders of magnitude lower for the bare electrode thus demonstrating that the poly(aniline) film mediates NADH

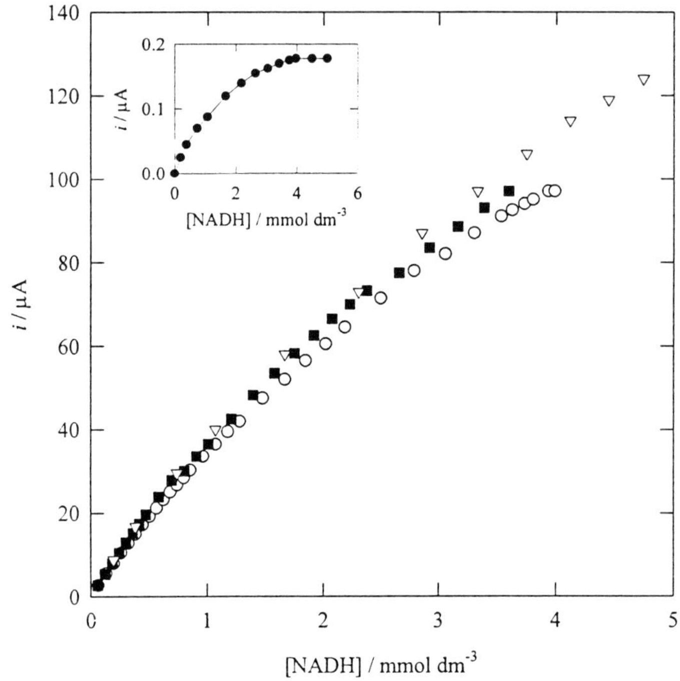

Fig. 2.16. Plot of the current recorded at +0.1 V at a poly(aniline)/poly(vinylsulfonate)-coated glassy carbon electrode (deposition charge 150 mC, geometric area 0.38 cm^2) rotated at 9 Hz in 0.1 mol dm^{-3} citrate/phosphate buffer at pH 7 as a function of the NADH concentration. The currents are corrected for the background current (<0.2 μA) in each case. Data for three different films prepared under identical conditions are shown. The inset shows the current for oxidation of NADH at +0.1 V at an uncoated glassy carbon electrode treated identically to the coated electrode except that no aniline monomer was added to the growth solution.

oxidation. Figure 2.17 shows that the responses are stable and linear over the range 0–1.0 mmol dm^{-3} NADH.

The time required for the electrode to attain a steady-state response after addition of an aliquot of NADH is of the order of seconds (~10 s for a film held at 0.1 V). Comparing this response time with those obtained in the mass transport-limited regime for a standard couple at a bare electrode, it is clear that the response time of the modified poly(aniline) electrodes is of the order of the mixing time for the system.

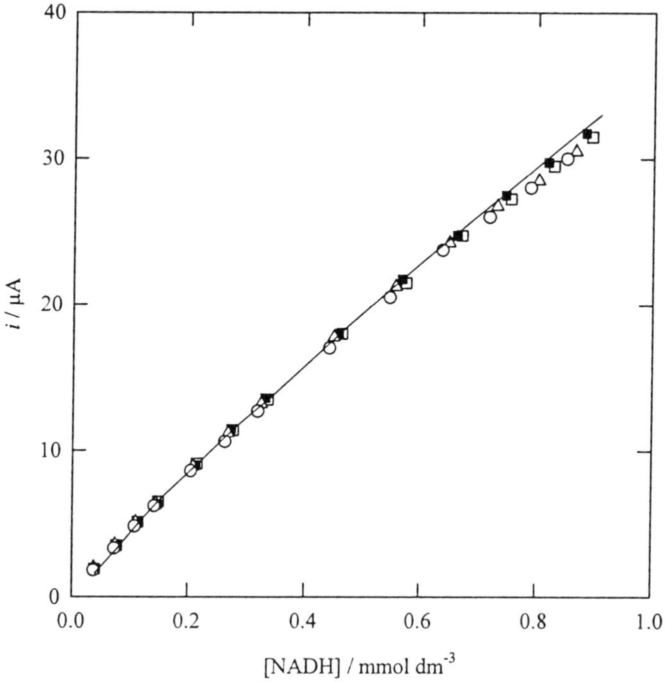

Fig. 2.17. Plots of the current at +0.1 V for a poly(aniline)/poly(vinylsulfonate)-coated glassy carbon electrode (deposition charge 150 mC, geometric area 0.38 cm^2) rotated at 9 Hz in 0.1 mol dm^{-3} citrate/phosphate buffer at pH 7 as a function of the NADH concentration showing the stability of the electrode response. Four replicate calibration curves recorded in succession over 4 h using the same electrode are shown: (□) run 1; (■) run 2; (△) run 3; and (○) run 4. The solid line is drawn as a guide for the eye.

The excellent stability and reproducibility of these modified poly(aniline) electrodes for NADH oxidation allows detailed kinetic studies of the system to be completed. To determine the mechanism of NADH oxidation the effects of film thickness, electrode potential, rotation rate and NAD^+ concentration have been investigated.

First, the effect of film thickness on the amperometric response was studied. The effect of film thickness has been assessed for eight different film thicknesses at four different NADH concentrations. For thin films, Fig. 2.18 shows that the response increases with film thickness reaching a

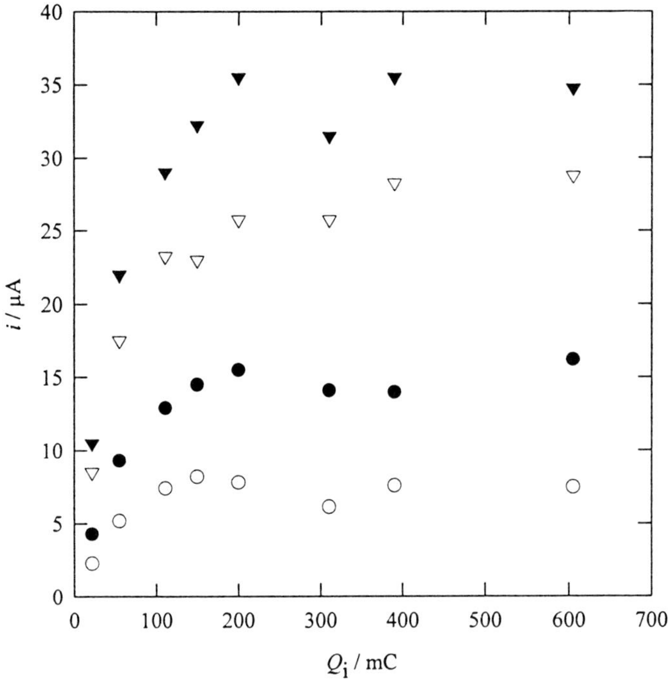

Fig. 2.18. Plots of the current at +0.1 V for poly(aniline)/poly(vinylsulfonate)-coated glassy carbon electrodes (geometric area 0.38 cm^2) rotated at 9 Hz in 0.1 mol dm^{-3} citrate/phosphate buffer at pH 7 as a function of charge required to grow the film, Q_i. The results are shown for four different NADH concentrations: (○) 0.15 mmol dm^{-3}; (●) 0.3 mmol dm^{-3}; (▽) 0.6 mmol dm^{-3}; and (▼) 0.8 mmol dm^{-3}.

plateau above which film thickness has no effect. This behaviour implies that the reaction occurs throughout the whole of the film.

Having ascertained where the reaction occurs, the effect of altering the potential applied to the poly(aniline) film on the amperometric response to NADH was studied. The resistance characteristics of the poly(aniline)/poly(vinylsulfonate) at this pH show that within the range −0.10 to +0.15 V vs. SCE, the polymer is sufficiently conducting to ensure that the system has reasonable response times. Figure 2.19(A) shows the response of a poly(aniline)/poly(vinylsulfonate) film to NADH at seven different potentials, within this chosen range. It was found that increasing the applied potential increases the current for NADH oxidation—the linear

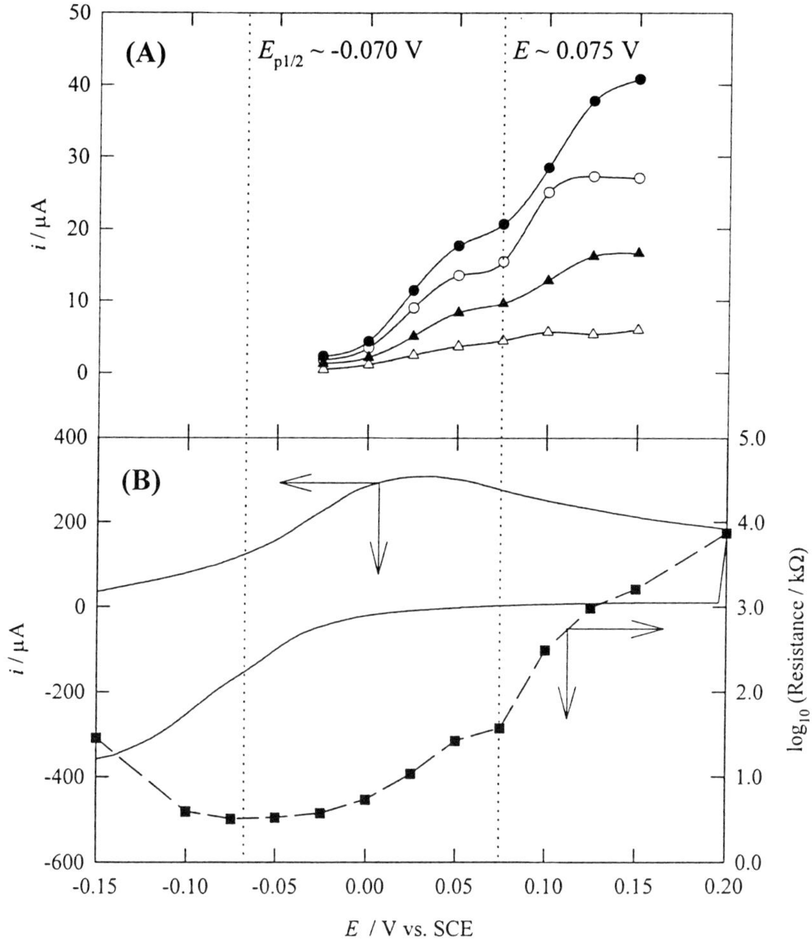

Fig. 2.19. (A) The effect of potential on the observed electrocatalytic response obtained at poly(aniline)/poly(vinylsulfonate) electrode (geometric area 0.38 cm^2, Q_i ~ 150 mC) rotated at 9 Hz in 0.1 mol dm^{-3} citrate/phosphate pH 7 buffer solution. Responses have been recorded at four different concentrations of NADH: (▲) 0.12 mmol dm^{-3}; (□) 0.3 mmol dm^{-3}; (●) 0.6 mmol dm^{-3}; and (○) 0.8 mmol dm^{-3}. (B) Cyclic voltammetry of an identical poly(aniline)/poly(vinylsulfonate) electrode in 0.1 mol dm^{-3} citrate/phosphate pH 7 buffer scanned at 5 mV s^{-1}, superimposed on this trace is the measured resistance for the same film (--■--).

response to NADH concentrations being extended to higher concentrations at higher potentials. For potentials above 0.1 V, the response time of the electrode becomes very slow (typically of the order of minutes rather than seconds as observed for all potentials below 0.1 V). For potentials below −0.025 V, the response of the electrode was negligible in comparison to the background currents which increased with decreasing potential. The increased background current is due to the polymer becoming insulating, resulting in it requiring longer times to respond to the applied potential.

Returning to Fig. 2.19(A), we can compare the observed potential dependence for the oxidation of NADH with the electrochemistry and changes in conductivity of the poly(aniline) film over the same potential range plotted in Fig. 2.19(B). Comparing Figs. 2.19(A) and (B), it is clear that there is no response to NADH when the film is at its maximum conductivity—between −0.075 and −0.05 V vs. SCE. This supports the idea that the oxidation of NADH is a chemical reaction mediated by particular structural units within the poly(aniline) film, rather than the film acting as a metallic electrode. In the discussion below, we will return to the potential dependence once the mechanism has been established.

The effect of altering the rate of mass transport to the electrode surface was also studied (see Fig. 2.20). At low rotation rates, the reaction is mass transport-controlled but as the rotation speed is increased, the current tends to a rotation speed-independent value indicating that the current becomes limited by some other process.

Finally, the effect of added NAD^+ on the amperometric response to NADH was studied. The currents for NADH oxidation were measured in the absence of NAD^+ and in the presence of NAD^+ in the bulk solution. To establish that the changes were reversible, the electrode was retested in NAD^+-free solution at the end of the experiment. It was found that with the addition of NAD^+ the bulk solution inhibits the oxidation of NADH (Fig. 2.21). When the NAD^+ is removed from the system, the currents for NADH oxidation return to their original values, indicating that the inhibition is reversible.

2.3.4.2 Interferents

The effects of three common interferent species, ascorbic acid, uric acid and acetoamidophenol, were investigated. For both uric acid and acetoamidophenol the oxidation currents at +0.1 V at the poly(aniline)

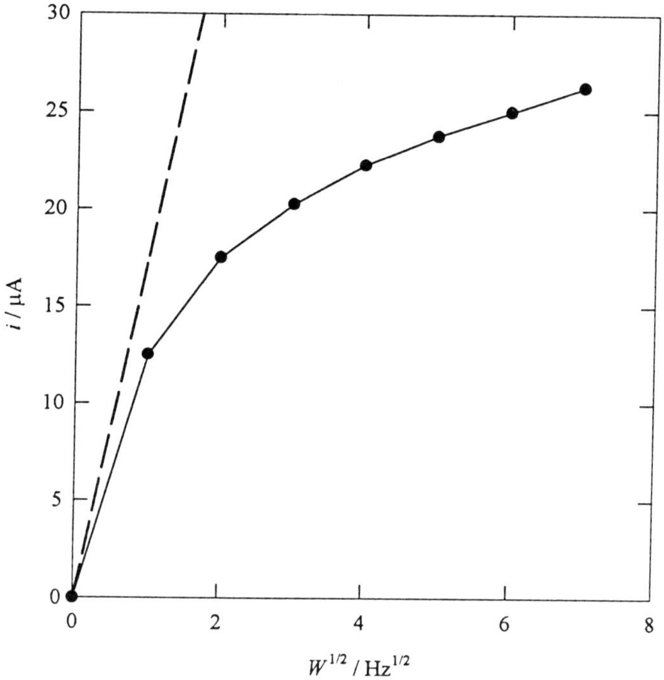

Fig. 2.20. Plot of the current for poly(aniline)/poly(vinylsulfonate)-coated glassy carbon electrode (deposition charge 150 mC, geometric area 0.38 cm^2) held at +0.1 V in 0.1 mol dm^{-3} citrate/phosphate buffer at pH 7 as a function of the square root of the rotation rate. The results are shown for 0.4 mmol dm^{-3} NADH in solution; (---) Levich line for mass transport limited current.

modified electrode were negligible (<1 μA) for concentrations up to 10 times their maximum physiological concentrations (0.48 and 0.17 mmol dm^{-3} [52, 53]). For ascorbic acid, in contrast, significant oxidation currents were obtained at +0.1 V, indicating that the poly(aniline)/poly(vinylsulfonate) film is also a good catalyst for ascorbate oxidation and that interference from ascorbate could be a significant problem for this electrode.

2.3.4.3 NADPH Oxidation

NADPH and NADH are closely related structurally, differing only by the addition of a phosphate group to the ribose sugar of NADH. One would expect, therefore, that a film which was a good electrocatalyst for

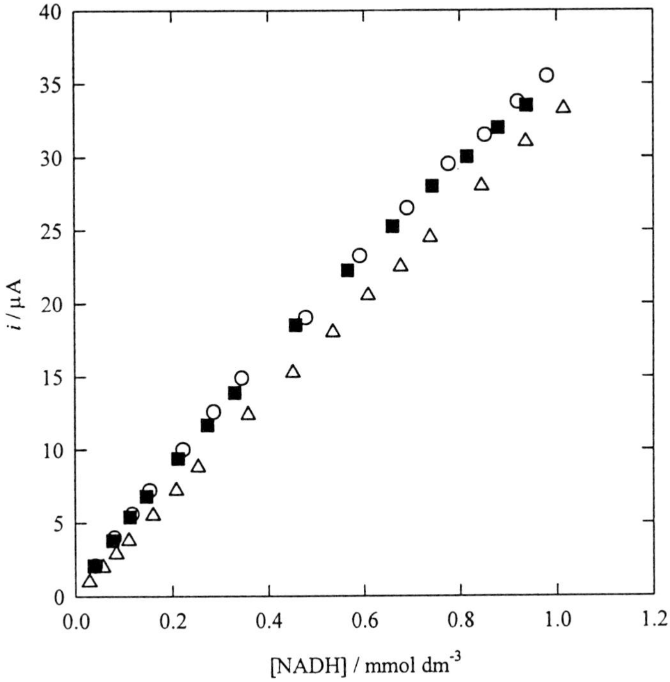

Fig. 2.21. Plots of the current for a poly(aniline)/poly(vinylsulfonate)-coated glassy carbon electrode (deposition charge 150 mC, geometric area 0.38 cm^2) held at +0.1 V and rotated at 9 Hz in 0.1 mol dm^{-3} citrate/phosphate buffer at pH 7 as a function of NADH concentration with, and without, added NAD^+: (○) run 1, no added NAD^+; (△) run 2, with NAD^+ added to the bulk solution; and (■) run 3, no NAD^+.

NADH oxidation to function equally well for the oxidation of NADPH. Figure 2.22 shows a direct comparison of the responses of a poly(aniline)-coated electrode to NADH and NADPH. At low concentrations, the currents are identical within experimental error but at higher concentration (in this case above 0.6 mmol dm^{-3}), the currents for NADPH fall below those for NADH. These results clearly show some saturation of the current at high NADPH concentrations.

2.3.5 *Kinetic model*

The results presented above show that the poly(aniline)/poly(vinylsulfonate) composite is an electrocatalytic surface for NADH oxidation at

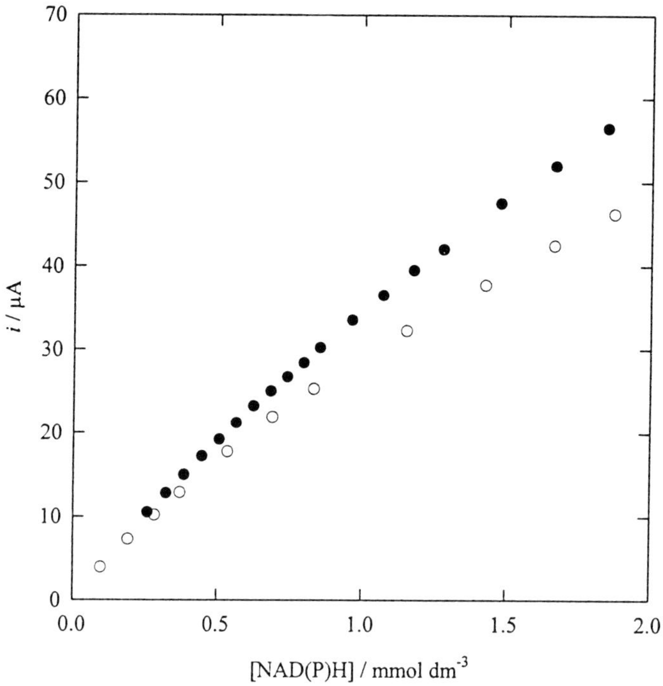

Fig. 2.22. Plots of the current for poly(aniline)/poly(vinylsulfonate)-coated glassy carbon electrodes (deposition charge 150 mC, geometric area 0.38 cm^2) held at +0.1 V and rotated at 9 Hz in 0.1 mol dm^{-3} citrate/phosphate buffer at pH 7 as a function of both: (○) NAD(P)H; and (●) NADH concentration.

pH 7. It would be useful to know more about the system and determine how it would respond to different concentrations of NADH, at what concentrations the response saturates, how thick the film has to be for all the NADH to be used up in a thin layer near the polymer/solution boundary, under what conditions the current is proportional to the concentration of NADH and so on. To answer these questions, the experimental data has been fitted to a detailed kinetic model. This analysis is now presented.

Considering the results presented so far for the oxidation of NADH at poly(aniline)/poly(vinylsulfonate) films, it is clear that the kinetic model must include the reaction occurring throughout the whole of the film, the potential dependence and inhibition by the product. This model was presented in Section 2.2.6. To apply the model to our data, we need to know the film thickness and the mass transport-rate constants for the reactant and product in solution.

It has been shown that electropolymerisation of aniline from 1.0 mol dm^{-3} HCl solution produces uniform films [54]. Hence, we make the assumption that the film thickness is linearly related to the charge passed to grow the film, Q. Hence, L (film thickness) can be replaced by

$$L = \sigma Q, \tag{2.19}$$

where σ is a constant. The diffusion coefficient for NADH in our electrolyte was found to be $2.4 \pm 0.1 \times 10^{-6}$ cm^2 s^{-1} from rotating disc measurements at an uncoated carbon electrode. The value for NAD^+ was assumed to be the same.

To test the validity of our model, it is necessary to determine expressions for the current from equations (2.5)–(2.8) which are applicable across the case boundaries. These expressions are given in Table 2.3. We can then analyse the experimental data using the appropriate expression. Since these equations do not include concentration polarisation in the solution, equations (2.4) and (2.15) are used to determine the concentrations of NADH and NAD^+ at the solution/film interface. The case, or cases, that the experimental data span are determined by inspection and using Table 2.2. Non-linear least squares fitting was used to fit the experimental data to the appropriate equations. The resulting best-fit parameters are critically

TABLE 2.3
Approximate analytical expressions for the currents across the different case boundaries

Case boundary	Approximate analytical expressions for the current	Eqn
I/II	$nFAK_SD_S\sqrt{\frac{k_{cat}[\{site\}]}{K_MD_D}}[NADH_0]\tanh\left(\sqrt{\frac{k_{cat}[\{site\}]}{K_MD_S}}L^2\right.$	(2.20)
II/IV	$nFA\sqrt{2k_{cat}[\{site\}]K_MD_S\left\{\left(\frac{K_S[NADH_0]}{K_M}\right)-\ln\left(1+\frac{K_S[NADH_0]}{K_M}\right)\right\}}$	(2.21)
IV/III	$nFA\sqrt{2k_{cat}[\{site\}]D_SK_S[NADH_0]}\tanh\sqrt{\frac{k_{cat}[\{site\}]L^2}{2K_SD_S[NADH_0]}}$	(2.22)
I/III	$nFAk_{cat}[\{site\}]L\left(\frac{K_S[NADH_0]}{K_S[NADH_0]+K_M}\right)$	(2.23)

evaluated to assess the quality of the agreement between theory and experiment.

2.4 ANALYSIS OF EXPERIMENTAL DATA

2.4.1 The effect of film thickness

The results of experiments carried out at a fixed rotation speed for nine different film thicknesses at a range of NADH concentrations up to 5 mmol dm^{-3}, have been analysed. The data, which are clearly in the unsaturated case (concentration of NADH less than 0.8 mmol dm^{-3}), were fitted to the analytical expression for the current across the Case I/II boundary (equation (2.20) of Table 2.3 and equations (2.4) and (2.19)) using a commercial non-linear least mean squares program to determine the best-fit values of the two independent parameters. In this analysis, we have ignored the effects of inhibition by NAD^+ because under these conditions they are negligible (see below). Figure 2.23 shows the good quality of the agreement between the experimental data shown by the points, and the best fit to the theory shown by the lines. The values for the two best-fit parameters, $(k_{cat}[\{site\}]\sigma^2/K_M D_S)^{1/2}$ and $(K_S D_S[\{site\}]k_{cat}/K_M)^{1/2}$, are given in Table 2.4. It is clear that the parameters from the four separate fits, carried out at different concentrations, agree quite closely indicating the validity of this simple kinetic model.

2.4.2 The effect of electrode potential

The data in Fig. 2.24 refer to a single film thickness corresponding to a thick film ($\epsilon > 1$). Therefore, these data have been analysed by fitting the results to the analytical expression for the current across the Case II/IV boundary (equation (2.21) in Table 2.3) using equation (2.4) to determine the concentration of NADH at the polymer/solution interface. The lines in Fig. 2.24 represent the non-linear least squares best fits of the NADH responses to the equation. Again the agreement between experiment and theory is highly satisfactory. The values for the two parameters $k_{cat}[site]D_S K_S$ and K_M/K_S derived from these fits are given in Table 2.5. It can be seen that the potential has a significant effect on the value of $k_{cat}[\{site\}]D_S K_S$, whereas, the effect of increasing potential on K_M/K_S is

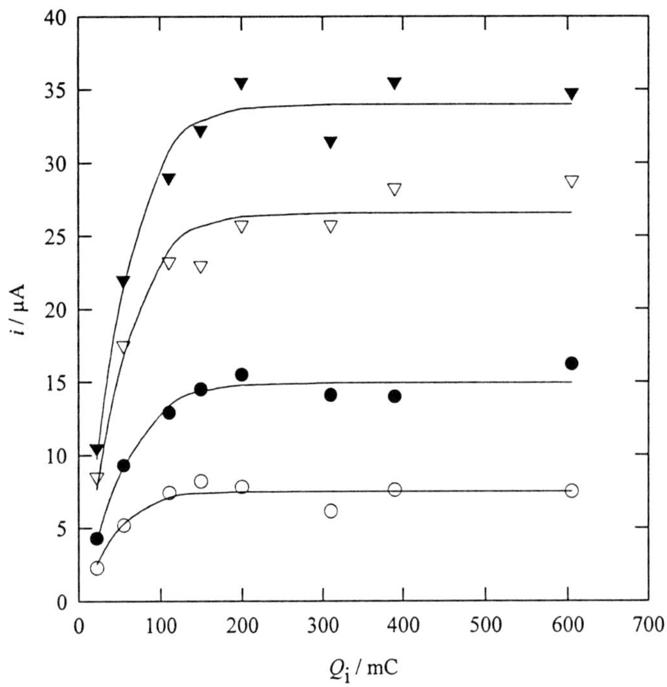

Fig. 2.23. Plots of the current at +0.1 V vs. SCE for a poly(aniline)/poly(vinylsulfonate)-coated glassy carbon electrodes (geometric area 0.38 cm^2) rotated at 9 Hz in 0.1 mol dm^{-3} citrate/phosphate buffer at pH 7 as a function of the total charge passed during polymer growth (Q_i). Results are shown for four different concentrations of NADH: (○) 0.15 mmol dm^{-3}; (●) 0.3 mmol dm^{-3}; (▽) 0.6 mmol dm^{-3}; and (▼) 0.8 mmol dm^{-3}. The solid lines were calculated from equations (2.4), (2.19) and (2.20) using the best-fit parameters derived from the analysis of each dataset (given in Table 2.4).

much less marked. In Fig. 2.25, we compare the potential dependence of $(k_{cat}/K_M)([\{site\}]D_S K_S^2)$ with that found for the homogeneous oxidation of NADH for 2e redox mediators from Fig. 2.3. Comparing the two, it is interesting that the shapes of the two plots are very similar. If we assume that we can equate the second-order rate constants k_2 and k_{cat}/K_M for the two cases then we find that $([\{site\}]D_S K_S^2)$ is ca. 10^{-12} mol cm^{-1} s^{-1}. This is not an unreasonable value, given the typical values for the diffusion coefficient within polymer films (10^{-9} cm^2 s^{-1}), it corresponds to a site concentration of the order of 1 mol dm^{-3}.

TABLE 2.4
Best-fit parameters from the analysis of the data in Fig. 2.23 for the currents for NADH oxidation in the absence of added NAD^+ at poly(aniline)/poly(vinylsulfonate)-modified electrodes. In each case, the data were fitted to equation (2.20) for the I/II boundary (n is the number of data points used)

Potential/V vs. SCE	$[NADH_\infty]$/ mmol dm^{-3}	$(k_{cat}[\{site\}]\sigma^2/K_M D_S)^{1/2}/C^{-1}$	$(K_S^2 D_S[\{site\}]k_{cat}/K_M)^{1/2}$/cm s^{-1}	n
0.100	0.15	16.9 2.9	$(6.93 \pm 1.82) \times 10^{-4}$	9
0.100	0.30	13.2 1.4	$(6.60 \pm 1.06) \times 10^{-4}$	9
0.100	0.60	13.5 1.7	$(5.98 \pm 1.14) \times 10^{-4}$	9
0.100	0.80	13.6 1.3	$(5.86 \pm 8.40) \times 10^{-4}$	9

2.4.3 The effect of rotation rate

The effect of variation of the rotation speed on the current response is shown in Fig. 2.26. The results are for a film of the same thickness as those in Fig. 2.24. We therefore fit the data to the equation for the Case II/IV boundary (equation (2.21) in Table 2.3). Again, the effect of rotation rate on the NADH concentration at the film/solution interface was determined and accounted for by using equation (2.4). The best fits of the theory to experiment are shown by the lines in Fig. 2.26, and the corresponding values of k_{cat}[site]$D_S K_S$ and K_M/K_S are given in Table 2.6 along with results from a separate replicate experiment.

Comparing the values for k_{cat}[site]$D_S K_S$ and K_M/K_S obtained from all the different experiments (Tables 2.5 and 2.6), we find that the agreement between the different experiments is good and confirms the validity of the proposed model and demonstrates the excellent reproducibility and stability of these films for NADH oxidation.

2.4.4 The effect of the addition of NAD^+ to the system

Figure 2.27 shows a set of responses to NADH recorded for different concentrations of NAD^+ in the bulk solution. In this case, we analyse the data taking account of the effects of product inhibition. The solid lines in Fig. 2.27 are the best fits of the data to the expression for the current across the Case II/IV boundary taking account of product inhibition (that

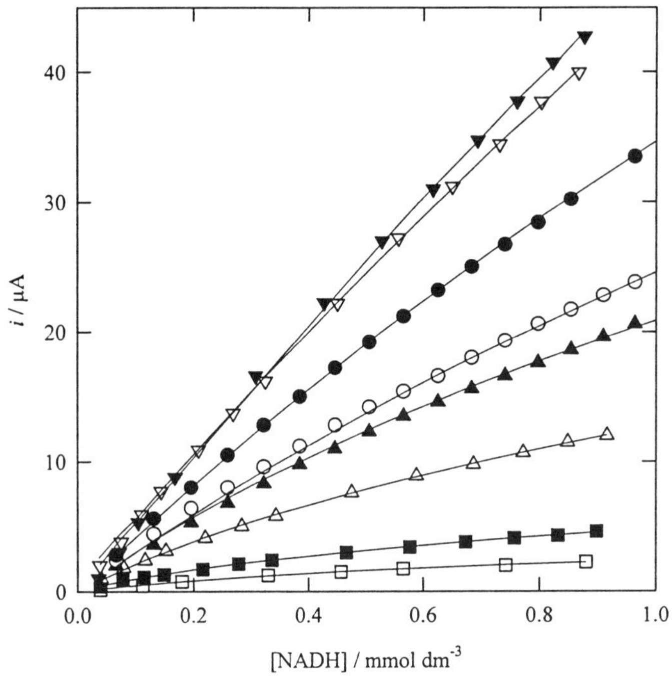

Fig. 2.24. Plots of the current for poly(aniline)/poly(vinylsulfonate)-coated glassy carbon electrodes (deposition charge 150 mC, geometric area 0.38 cm^2) rotated at 9 Hz in 0.1 mol dm^{-3} citrate/phosphate buffer at pH 7 as a function of the NADH concentration recorded at eight different potentials: (□) −0.025 V; (■) 0.000 V; (△) +0.025 V; (▲) +0.050 V; (○) +0.075 V; (●) +0.100 V; (▽) +0.125 V; and (▼) +0.150 V. The solid lines in each case are calculated from the best fits of the data to equations (2.4) and (2.21) and the resulting fitting parameters are given in Table 2.5.

is using equation (2.21) in Table 2.3 with equations (2.4), (2.15), (2.17) and (2.18)).

Because of the inclusion of product inhibition this is now a three parameter fit and the three fitting parameters, $k_{cat}[\{site\}]D_SK_S$, K_M/K_S, and K_i/K_P, are given in Table 2.7. Again, the agreement between these parameters, and those previously obtained, is further confirmation of the validity of the model. It is notable that the value for K_i/K_P is relatively large confirming that the effects of inhibition are only significant at high NAD^+ concentrations and justifies our earlier neglect of the effects of product

TABLE 2.5
Best-fit parameters from analysis of the data in Fig. 2.24 for the currents for NADH oxidation in the absence of added NAD^+ at poly(aniline)/poly(vinylsulfonate)-modified electrodes. In each case, the data were fitted to equation (2.21) for the II/IV boundary (n is the number of data points used)

Potential/V vs. SCE	K_M/K_S/mmol dm^{-3}	$k_{cat}[\{site\}]D_SK_S$/mol cm^{-1} s^{-2}	n
−0.025	0.12 ± 0.03	(7.85 ± 0.73) × 10^{-16}	8
0.000	0.09 ± 0.01	(2.93 ± 0.09) × 10^{-15}	13
0.025	0.17 ± 0.01	(2.56 ± 0.07) × 10^{-14}	13
0.050	0.26 ± 0.02	(8.77 ± 0.14) × 10^{-14}	46
0.075	0.50 ± 0.02	(1.70 ± 0.02) × 10^{-13}	37
0.100	0.55 ± 0.16	(4.13 ± 0.04) × 10^{-13}	33
0.125	0.56 ± 0.04	(8.68 ± 0.48) × 10^{-13}	13
0.150	0.93 ± 0.35	(1.46 ± 0.43) × 10^{-12}	12

inhibition in the absence of added NAD^+. We note that there is no significant difference in the values for $k_{cat}[\{site\}]D_SK_S$ and K_M/K_S obtained from analysis of the experiments carried out without added NAD^+ when the effects of inhibition are included.

2.4.5 Global fit

Finally, the data for all the experiments carried out at fixed rotation rate for nine film thicknesses, with and without added NAD^+ at a range of NADH concentrations up to 5 mmol dm^{-3} (in all a total of 343 separate data points), were fitted to equations (2.4)–(2.8) for the uninhibited fit, with the addition of equations (2.15), (2.17), (2.18) and (2.19) for the inhibited fit using three adjustable parameters. Figure 2.28 shows a three-dimensional representation of the data, shown by the points, and the best fit to the theory, shown by the surface. The resulting best-fit parameters are given in Table 2.8. The position of the case boundary and the different cases were calculated from the best-fit parameters and are shown in Fig. 2.28.

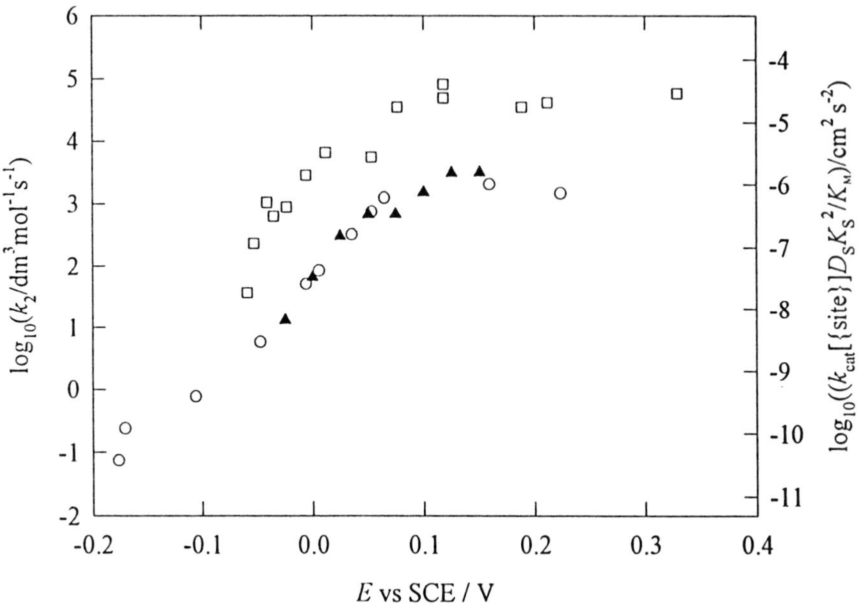

Fig. 2.25. A comparison of the potential dependence for the electrochemical oxidation of NADH at a poly(aniline)/poly(vinylsulfonate)-coated electrode, (k_{cat}[site]$D_S K_S^2/K_M$), and the second-order rate constant for the homogeneous oxidation of NADH by a range of two-electron mediators, k_2. The open symbols correspond to homogeneous oxidation by ortho, □, and para, ○, substituted quinones and diaminobenzenes, the filled triangles, ▲, are the electrochemical data.

2.4.6 *Uninhibited and inhibited models*

The similarity between the parameters for the global inhibited and uninhibited fits is a strong indication that, as previously stated, inhibition in the absence of added NAD^+ is small. To demonstrate this proposition, the difference between the two fits to experimental points (without additional NAD^+) are shown in Fig. 2.29. In the figure, the lines have been calculated using the best-fit parameters given in Table 2.8. As we can see, the differences between the two calculated lines are small and are most significant at high NADH concentrations. There are two reasons for this. First, K_i/K_P is larger than K_M/K_S. From our global analysis we find that K_M/K_S = 1.22 ± 0.08 mmol dm^{-3} and K_i/K_P = 2.49 ± 0.28 mmol dm^{-3} so that if K_S

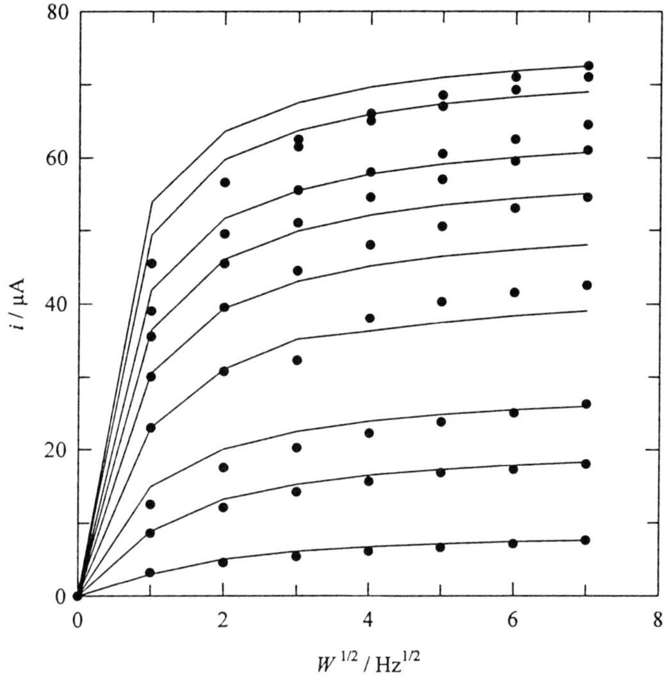

Fig. 2.26. Plots of the current for poly(aniline)/poly(vinylsulfonate)-coated glassy carbon electrodes (deposition charge 150 mC, geometric area 0.38 cm^2) held at +0.1 V in 0.1 mol dm^{-3} citrate/phosphate buffer at pH 7 as a function of the square root of the rotation rate. The results are shown for nine different NADH concentrations. The solid lines through the points are calculated from the best fits of the data to equations (2.4) and (2.21), and the resulting fitting parameters are given in Table 2.6.

TABLE 2.6
Best-fit parameters from analysis of the data in Fig. 2.26 for the currents for NADH oxidation in the absence of added NAD^+ at poly(aniline)/poly(vinylsulfonate)-modified electrodes. In each case, the data were fitted to equation (2.21) for the II/IV boundary (*n* is the number of data points used)

Potential/V vs. SCE	K_M/K_S/mmol dm^{-3}	$k_{cat}[\{site\}]D_SK_S$/mol cm^{-1} s^{-2}	n
0.100	0.225 ± 0.054	(2.51 ± 0.22) × 10^{-13}	54
0.100	0.099 ± 0.024	(2.89 ± 0.13) × 10^{-13}	72

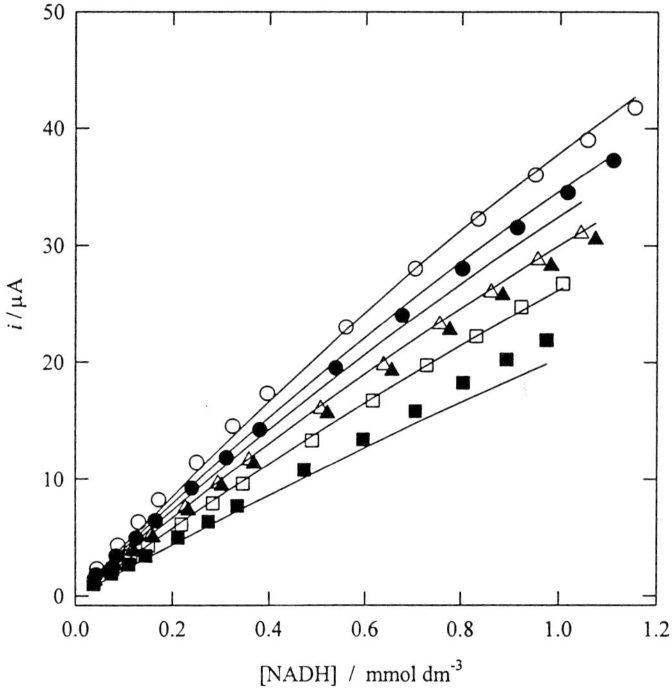

Fig. 2.27. Plots of the current for poly(aniline)/poly(vinylsulfonate)-coated glassy carbon electrodes (deposition charge 150 mC, geometric area 0.38 cm^2) held at +0.1 V and rotated at 9 Hz in 0.1 mol dm^{-3} citrate/phosphate buffer at pH 7 as a function of NADH concentration with differing NAD^+ concentrations: (○) 0 mmol dm^{-3} NAD^+; (●) 0.91 mmol dm^{-3}; (△) 1.82 mmol dm^{-3}; (▲) 2.73 mmol dm^{-3}; (□) 4.55 mmol dm^{-3}; and (■) 9.09 mmol dm^{-3}. The solid lines are calculated from equations (2.4), (2.15), (2.17), (2.18) and (2.21) using the best-fit parameters in Table 2.7.

TABLE 2.7

Best-fit parameters from analysis of the data in Fig. 2.27 for the currents for NADH oxidation in the presence of added NAD^+ at poly(aniline)/poly(vinylsulfonate)-modified electrodes. In this case, the data was fitted to equations (2..4), (2.15), (2.17), (2.18) and (2.21) for the II/IV case boundary (n is the number of data points used)

Potential/V vs. SCE	K_M/K_S/ mmol dm^{-3}	$k_{cat}[\{site\}]D_SK_S$/ mol cm^{-1} s^{-2}	K_i/K_P/ mmol dm^{-3}	n
0.100	0.773 ± 0.1060	(6.21 ± 0.56) × 10^{-13}	1.70 ± 0.10	78

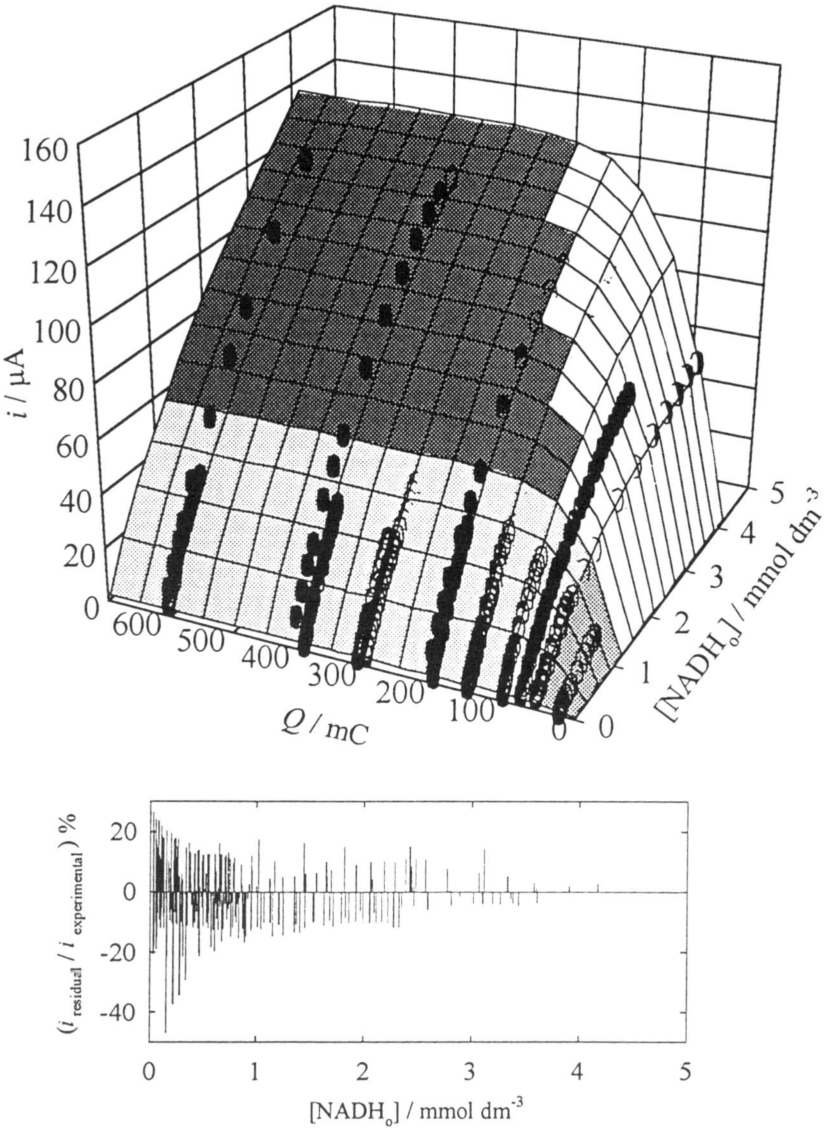

Fig. 2.28. Case diagram for NADH oxidation at a glassy carbon electrode coated with a poly(aniline)/poly(vinylsulfonate) film. The points are the experimental data. The surface and case boundaries have been determined from the inhibited fit parameters given in Table 2.8. The residuals are shown as a function of the concentration of NADH below the plot.

TABLE 2.8
Best-fit parameters from the analysis of the currents for all NADH oxidation data fitted at once using both the uninhibited and inhibited fits at poly(aniline)/poly(vinylsulfonate)-modified electrodes. The values were obtained by non-linear least squares fits of the experimental data to equations (2.4)–(2.8) for the uninhibited fit, with the addition of equations (2.15), (2.17) and (2.18) for the inhibited fit. In each case n is the total number of data points used in the fit

Potential/V vs. SCE	K_M/K_S/ mmol dm^{-3}	$k_{cat}[\{site\}]D_S K_S$/ mol cm^{-1} s^{-2}	K_i/K_P/ mmol dm^{-3}	n
Uninhibited fit				
0.100	0.87 ± 0.06	(6.05 ± 0.43) × 10^{-13}	N/A	278
Inhibited fit				
0.100	1.22 ± 0.08	(8.93 ± 1.16) × 10^{-13}	2.49 ± 0.28	343

and K_P, the partition coefficients for NADH and NAD^+ are comparable, then NADH binds twice as strongly as NAD^+. Second, since the electrocatalytic responses to NADH are much lower than the predicted mass transport-limiting currents, the concentration of NAD^+ produced within the film will be significantly lower than the bulk concentration of NADH. In our experiments, we have calculated that the NAD^+ concentration reaches a maximum value of 0.7 mmol dm^{-3}, the average being 0.2 mmol dm^{-3}. These values are significantly lower than K_i, implying that few of the sites will be inhibited by bound NAD^+.

2.4.7 NADPH oxidation

It was shown in Section 2.3.4.3 that poly(aniline)/poly(vinylsulfonate) films are also electrocatalytic surfaces for the oxidation NADPH. Assuming that the kinetics of this system are similar to those for the NADH system, the data have been analysed using the uninhibited kinetic model. From comparison to the NADH data, this film is thick ($\epsilon > 1$) and the data span both Cases II and IV. Therefore, all the data were fitted using the expression for the Case II/IV boundary (equation (2.21) Table 2.4) using equation (2.4) to calculate the concentration of NAD(P)H at the film/solution interface. The fit is shown in Fig. 2.30 and the parameters generated by this fit are given in Table 2.9.

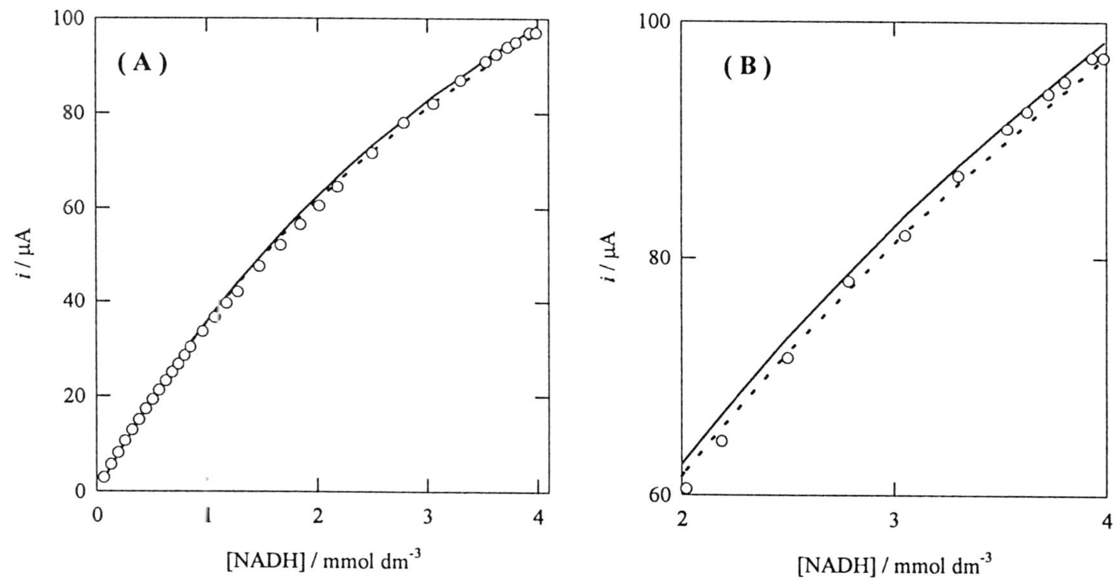

Fig. 2.29. Experimental data from poly(aniline)/poly(vinylsulfonate) film deposited on a glassy carbon electrode ($0.38\,cm^2$, $Q_i \sim 150\,mC$). The response to increasing NADH concentration as shown by the points. The lines represent the uninhibited (——) and the inhibited (. . .) curve fits using the two sets of parameters given in Table 2.9. (B) is an expansion of (A) to show the actual difference between the inhibited and uninhibited fits.

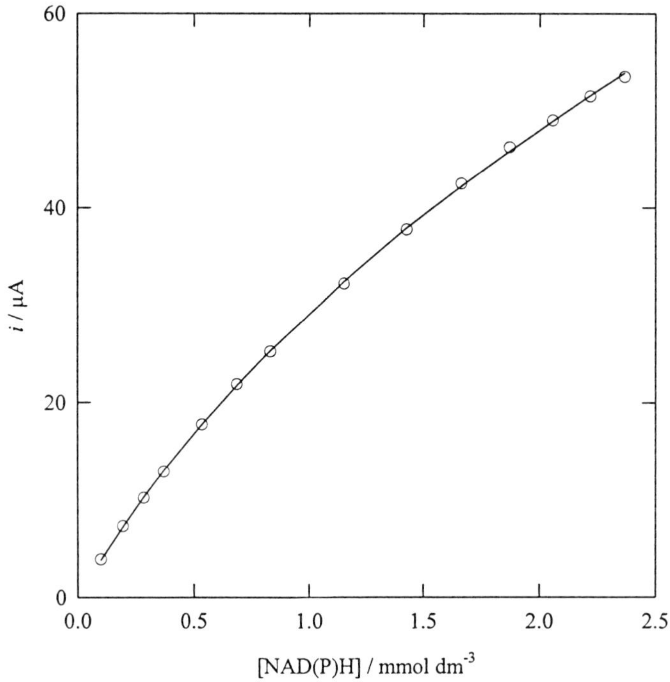

Fig. 2.30. Plots of the current for poly(aniline)/poly(vinylsulfonate)-coated glassy carbon electrodes (deposition charge 150 mC, geometric area 0.38 cm^2) held at +0.1 V and rotated at 9 Hz in 0.1 mol dm^{-3} citrate/phosphate buffer at pH 7 as a function of (○) NADPH concentration. The lines through the points were calculated from equations (2.4) and (2.21) for the Case II\IV boundary (see Table 2.3) using the best-fit parameters given in Table 2.9.

Comparison with the corresponding values for NADH oxidation shows that the parameters are quite similar as expected. Since the experiment was carried out using the same polymer film, the fact that the parameters obtained are not identical indicates slight differences in the kinetics for the two substrates.

2.5 CONCLUSIONS

Starting from a knowledge of the homogeneous mechanism and kinetics, we have shown in this chapter how we can identify possible approaches to

TABLE 2.9
Best-fit parameters from the analysis of the currents for NAD(P)H oxidation using the uninhibited 2 parameter fit at poly(aniline)/poly(vinylsulfonate)-modified electrodes. The values were obtained by non-linear least squares fits of the experimental data in Fig. 2.30 to equations (2.4) and (2.21). Also shown are the best-fit parameters from the uninhibited 2 parameter fit analysis of the currents for NADH oxidation at the same poly(aniline)/poly(vinylsulfonate)-modified electrode (n is the number of data points used)

Potential/V vs. SCE	K_M/K_S/mmol dm^{-3}	$k_{cat}[\{site\}]D_SK_S$/mol cm^{-1} s^{-2}	n
NADPH			
0.100	0.17 ± 0.01	$(2.04 \pm 0.03) \times 10^{-13}$	14
NADH			
0.100	0.33 ± 0.02	$(3.27 \pm 0.11) \times 10^{-13}$	33

the design of new modified electrodes for a target analyte. Then, by combining experimental studies and detailed analysis of the experimental data, and relying heavily on the approximate analytical solutions of the diffusion/reaction problem to guide our experiments, we have been able to establish the mechanism and kinetics for the reaction at the modified electrode. Kinetic parameters characterising the reaction were obtained from the experimental data by non-linear least squares fitting to the appropriate approximate analytical expressions.

The power of this approach is that it gives us considerable insight into the operation of the electrocatalyst and allows us explore the effects of changing experimental conditions and predict the behaviour of the system. We can then proceed to use the model to guide the design poly(aniline)/poly(vinylsulfonate) composite electrodes for use in NAD(P)H-dependent dehydrogenase-based biosensors. The model can also be used as a starting point in understanding the transient behaviour for the modified electrode, based on our knowledge of the case diagram and rate-limiting processes for the system (for an example of the transient treatment, see Lyons *et al.* [55]).

This type of approach is essential in establishing a sound understanding of the interplay of diffusion and reaction in this type of modified electrode and more generally in amperometric biosensors.

ACKNOWLEDGEMENTS

ENKW thanks the EPSRC for a postgraduate studentship. The work was supported in part by EPSRC Grant No. GR/K39431.

APPENDIX: NOTATION

A	geometric electrode area
c_∞	concentration in bulk solution
D_{film}	substrate diffusion coefficient in the film
D_{sol}	substrate diffusion coefficient in the solution
F	Faraday constant; 96480 C mol^{-1}
i	current
I_A	limiting current density for mass transport in the bulk solution to the membrane surface
I_L	total limiting current density for a reaction occurring at the surface of a rotating disc coated by a permeable membrane
I_S	limiting current density for diffusion through the film
k_2	the second order rate constant for the oxidation of NADH
k_{cat}	the rate constant for the oxidation of the adsorbed substrate
k'_{cat}	the apparent rate constant for the oxidation of the adsorbed substrate
$k'_{\text{D,P}}$	the mass transport rate constant for the product at the rotating disc electrode
$k'_{\text{D,S}}$	the mass transport rate constant for the substrate at the rotating disc electrode
K_i	the equilibrium constant for binding the product
K_M	the equilibrium constant for binding the substrate
K'_M	the apparent equilibrium constant for binding the substrate
K_P	the partition coefficient for the product into the film
K_S	partition coefficient of substrate into the film
L	the thickness of the film deposited on the electrode
n	number of electrons involved in oxidation of substrate
$[\text{NAD}^+_\infty]$	concentration of NAD^+ in bulk solution
$[\text{NAD}^+_0]$	concentration of NAD^+ at the electrode surface
$[\text{NADH}_\infty]$	concentration of NADH in bulk solution
$[\text{NADH}_0]$	concentration of NADH at the electrode surface

$\{NAD^+\}$	NAD^+ adsorbed at a site within the polymer
{NADH}	NADH adsorbed at a site within the polymer
Q	the charge passed to deposit the film
{site}	a vacant adsorption site within the polymer
[{site}]	concentration of vacant adsorption sites within the polymer
W	the rotation speed in Hz
X_D	the thickness of the diffusion layer at the electrode surface
X_k	the reaction layer thickness
ϵ	a dimensionless constant
α	a dimensionless constant
σ	conversion factor between growth charge and film thickness
ν	the kinematic viscosity
ω	the rotation speed in rad s^{-1}

References

[1] W.J. Albery and A.R. Hillman, Annu. Rep. Prog. Chem. Section C (1981) 377.
[2] W.J. Albery and A.R. Hillman, J. Electroanal. Chem. (1984).
[3] C.P. Andrieux and J.-M. Savéant. J. Electroanal. Chem. 142 (1982) 1.
[4] C.P. Andrieux and J.-M. Savéant, J. Electroanal. Chem. 134 (1982) 163.
[5] C.P. Andrieux, J.-M. Dumas-Bouchiat and J.-M. Savéant, J. Electroanal. Chem. 131 (1982) 1.
[6] C.P. Andrieux and J.-M. Savéant, J. Electroanal. Chem. 171 (1984) 65.
[7] C.P. Andrieux, J.-M. Dumas-Bouchiat and J.-M. Savéant, J. Electroanal. Chem. 169 (1984) 9.
[8] C.P. Andrieux and J.-M. Savéant, in: Molecular Design of Electrode Surfaces, Vol. XXII, p. 207, ed. R. Murray (John Wiley, New York, 1992).
[9] M.E.G. Lyons, in: Electroactive Polymer Electrochemistry, Part 1, Fundamentals, ed. M.E.G. Lyons (Plenum Press, London, 1994).
[10] M.E.G. Lyons, D.E. McCormack and P.N. Bartlett. J., Electroanal. Chem. 261 (1989) 51.
[11] M.E.G. Lyons, D.E. McCormack, O. Smith and P.N. Bartlett, Faraday Discuss. Chem. Soc. 88 (1989) 139.
[12] P.N. Bartlett and K.F.E. Pratt, J. Electroanal. Chem. 397 (1995) 61.
[13] P.N. Bartlett and K.F.E. Pratt, Biosensors & Bioelectronics 8 (1993) 451.
[14] M. Dixon and E.C. Webb, The Enzymes (Longman, 1979).
[15] P.J. Elving, C.O. Schmakel and K.S.V. Santhanam, Crit. Rev. Anal. Chem. 6 (1976) 1.
[16] J. Darnell, H. Lodish and D. Baltimore, Molecular Cell Biology (Scientific American Books, New York, 1990)

[17] B. Persson, PhD Thesis (Lund, 1990).
[18] Z. Samec and P.J. Elving, J. Electroanal. Chem. 144 (1983) 217.
[19] N. Cenas, J. Rozgaite, A. Pocius and J. Kulys, J. Electroanal. Chem. 154 (1983) 121.
[20] P.N. Bartlett and D.J. Caruana, Analyst 117 (1992) 1287.
[21] L.L. Miller and J.R. Valentine, J. Am. Chem. Soc. 110 (1988) 3982.
[22] B.W. Carlson and L.L. Miller, J. Am. Chem. Soc. 105 (1983) 7453.
[23] B.W. Carlson, L.L. Miller, P. Neta and J. Grodkowski, J. Am. Chem. Soc. 106 (1984) 7233.
[24] B.W. Carlson and L.L. Miller, J. Am. Chem. Soc. 107 (1985) 479.
[25] A. Kitani and L.L. Miller, J. Am. Chem. Soc. 103 (1981) 3595.
[26] A. Kitani, Y.-H. So and L.L. Miller, J. Am. Chem. Soc. 103 (1981) 7636.
[27] P.N. Bartlett, P. Tebbutt and R.G. Whitaker, Prog. React. Kinet. 16 (1991) 55.
[28] B. Keita, K. Essaadi, L. Nadjo, R. Contant and Y. Justum, J. Electroanal. Chem. 404 (1996) 271.
[29] B.F.Y.Y. Hin and C.R. Lowe, Anal. Chem. 59 (1987) 2111.
[30] C.-X. Cai, H.-X. Ju and H.-Y. Chen, J. Electroanal. Chem. 397 (1995) 185.
[31] C.A. Goss and H.D. Abruna, Inorg. Chem. 24 (1985).
[32] B. Persson, J. Electroanal. Chem. 287 (1990) 61.
[33] L. Gorton, J. Chem. Soc. Faraday Trans. I, 82 (1986) 1245.
[34] A. Torstensson and L. Gorton, J. Electroanal. Chem. 130 (1981) 199.
[35] L. Gorton, A. Torstensson, H. Jaegfeldt and G. Johnasson, J. Electroanal. Chem. 161 (1984) 103.
[36] L. Gorton, G. Johansson and A. Torstensson, J. Electroanal. Chem. 196 (1985) 81.
[37] B. Persson, H.S. Lee, L. Gorton, T. Skotheimand P. Bartlett, Electroanalysis, 7 (1995) 935.
[38] Z.W. Huan, B. Persson, L. Gorton, S. Sahni, T. Skotheim and P. Bartlett, Electroanalysis 8 (1996) 575.
[39] B. Persson, J. Electroanal. Chem. 287 (1990) 61.
[40] W.J. Albery and P.N. Bartlett, J. Chem. Soc. Chem. Commun. (1984) 234.
[41] S. Fukuzumi, N. Nishizawa and T. Tanaka, J. Org. Chem. 49 (1984) 3571.
[42] W.J. Albery, A.E.G. Cass and Z.X. Shu, Biosensors and Bioelectronics 5 (1990) 367.
[43] F.C. Anson, J. Phys. Chem. 84 (1980) 3336.
[44] W.J. Albery, A.E.G. Cass and Z.X. Shu, Biosensors and Bioelectronics 5 (1990) 379.
[45] W.J. Albery, A.E.G. Cass, B.P. Mangold and Z.X. Shu, Biosensors and Bioelectronics 5 (1990) 397.
[46] M.E.G. Lyons, C.H. Lyons and A. Michas, Analyst 117 (1992) 1271.
[47] W.J. Albery, Electrode Kinetics (Clarendon Press, Oxford, 1975).
[48] P.N. Bartlett, P.R. Birkin and E.N.K. Wallace, J. Chem. Soc. Faraday Trans. I, 93 (1997) 1951.
[49] R. Braun, K. Santhanam and P. Elving, J. Am. Chem. Soc. 97 (1975) 2591.
[50] M.L. Hitchman and W.J. Albery, Electrochim. Acta 17 (1972) 787.
[51] G.E. Asturias, G.-W. Jang, A.G. MacDiarmid, K. Doblhofer and C. Zhong, Ber. Bunsenges. Phys. Chem. 95 (1991) 1381.
[52] R.J. Geise, J.M. Adams, N.J. Barone and A.M. Yacynych, Biosensors and Bioelectronics 6 (1991) 151.

[53] G. Pallesch, M. Rahni, G.J. Lubrano, J.N. Ngwainbi and G.G. Guilbault, Anal. BioChem. 159 (1986) 114.

[54] J. DeSilvestro and W. Scheifele, J. Mater. Chem. 3 (1993) 263.

[55] M.E.G. Lyons, C.H. Lyons, C. Fitzgerald and P.N. Bartlett, J. Electroanal. Chem. 365 (1994) 29.

[56] C. Degrand and L.L. Miller, J. Am. Chem. Soc. 102 (1980) 5728.

[57] N.F. Atta, A. Galal, A.E. Karagozler, H. Zimmer, J.F. Rubinson and J.H.B. Mark, J. Chem. Soc. Chem. Commun. (1990) 1347.

[58] M. Somasundrum and J.V. Bannister, J. Chem. Soc. Chem. Commun. (1993) 1629.

[59] C.-X. Cai, H.-X. Ju and H.-Y. Chen, Anal. Chim. Acta 310 (1995) 145.

Chapter 3

Electrochemistry of DNA

ANA MARIA OLIVEIRA BRETT, SÍLVIA H.P. SERRANO and
ANTÓNIO JOSÉ P. PIEDADE

3.1 INTRODUCTION

DNA (deoxyribonucleic acid) was discovered in 1869 by Friedrich Miescher but only much later was recognized its genetic role. Avery and coworkers' [1] discoveries in 1944 about nucleic acid desoxyribose, and the studies of Hershey and Chase in 1952, were great contributions to the identification of DNA as the macromolecule responsible for the storage and expression of genetic information. DNA molecules are the largest of the macromolecules found in a cell. DNA has to be packaged very efficiently to fit into the cell nucleus. However, only the discovery by Watson and Crick [2] in 1953 of the double helical structure of DNA could clarify the conservation and expression mechanisms of genetic information. The genetic information is stored in the form of nucleotide sequences in DNA and this information is copied into RNA before it can be used. From that time onwards, a great interest in the nucleic acids enabled the finding of other structural forms for DNA. But although a triple-stranded model for nucleic acid, was proposed by Pauling and Corey [3] in 1953 on the basis of X-ray data, the existence of triplex nucleic acid structures was first shown by Felsenfeld *et al.* [4] in 1958. At present, it is assumed that the complex organization of the chromossomic DNA, its mechanisms of replication, transcription and recombination, are the consequence of a dynamic interaction of its different structural forms. DNA is a structurally polymorphic macromolecule which, depending on nucleotide sequence and environmental conditions, can adopt a variety of conformations.

Many methods have been used to study nucleic acids but here the goal is to review the electrochemical behaviour of DNA. For more than 30 years electrochemical methods have been used to study the electron-transfer reactions of nucleic acids using mercury electrodes. Recently, however, the use of other electrode materials, such as glassy carbon or carbon paste, enabled the development of new voltammetric and potentiometric methods with enhanced sensitivity, as well as DNA-modified electrodes and DNA biosensors. These DNA-modified electrodes allow the study of the mechanism of interaction with DNA of metals, drugs and pollutants, and consequently, a wide range of analytical and biotechnological applications can be foreseen.

3.2 THE DOUBLE HELIX DNA

The double helical structure of DNA consists of two strands, each of them on the outside of the double helix and formed by alternating phosphate and pentose groups in which phosphodiester bridges provide the covalent continuity. The two chains of the double helix are held together by hydrogen bonds between purine and pyrimidine bases (Fig. 3.1), and it is possible to identify a major and a minor groove, the latter with a higher negative density charge. The purine: adenine (A) and guanine (G), and the pyrimidine: cytosine (C) and thymine (T), bases of the nucleotide units carry genetic information, whereas, the sugar and phosphate groups perform a structural role. The bases are always paired: adenine with thymine and guanine with cytosine.

This specificity of the base-pairing is the most important aspect of the DNA double helix, since one strand of the double helix is the complement of the other. The covalent structure of the DNA polynucleotide chains is constant throughout the molecule but the base sequence is variable. The precise sequence of bases carries the genetic information and in 1950, Chargaff [5], found that in DNA the quantity of adenine is always equal to the quantity of thymine and that the quantity of guanine is always equal to the quantity of cytosine; this was very important to the development of the double-helix model. The base-pairing (Fig. 3.1), first proposed by Watson and Crick [2] explained the available data on DNA structure in terms of a double-helical structure which is now known as the B-form. This structure is characterized to be right-handed with 10 nucleotides per turn and is stabilized by bound water that fits perfectly into the minor groove.

However, in 1972 Arnott and Hukins [6] discovered that when DNA is

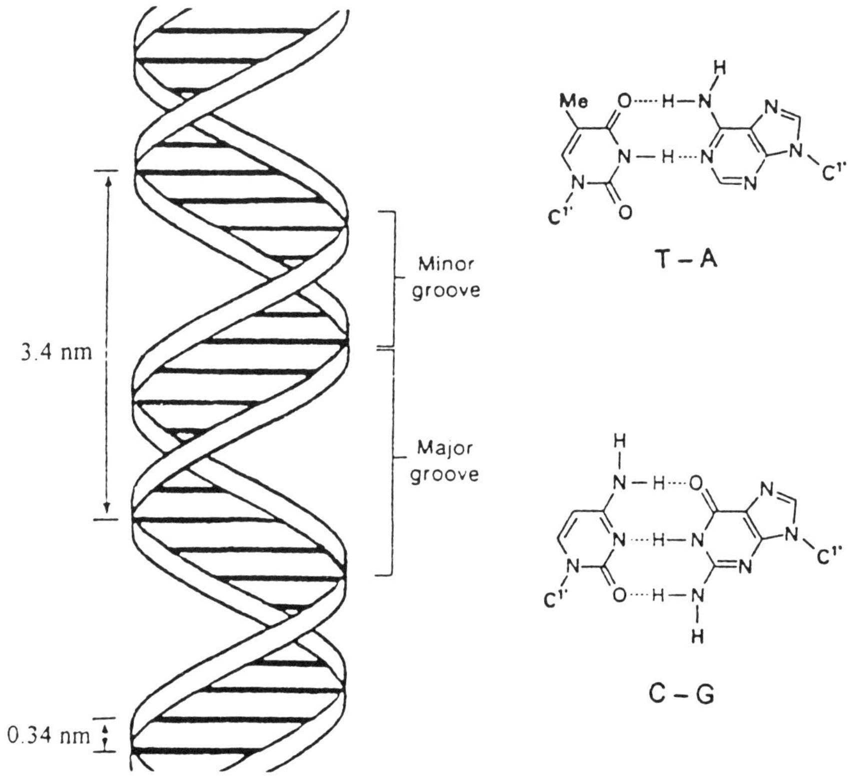

Fig. 3.1. Schematic representation of double-stranded DNA.

strongly dehydrated (relative humidity <75%), a structural alteration occurs due to a greater electrostatic response between the phosphate groups and this form was designated by A-DNA; it is right-handed, has 11 nucleotides per turn, and differs from B-DNA because the pentoses of the nucleotides are in a conformation that causes an increase in the double-helix diameter and a decrease in its length.

In 1979, following studies of X-ray diffraction of double chain crystals with a base sequence d(CpG)n, Rich and collaborators [7] were surprised to discover a left-handed double helix with 12 nucleotides per turn, only one groove and with the phosphate groups lying on a zig-zag line which gave rise to the name Z-DNA.

The different structural forms of the double helix lead to different dynamic interactions, and the geometry of the grooves is important in allowing or preventing access to the bases. Electrostatic interactions play a crucial role in the transition between right-handed B-DNA and left-handed Z-DNA [8], which is one of the best characterized conformational changes in double-stranded DNA.

The physical properties of double helical DNA in solution can change when put under conditions of extremes of pH or heat, referred to as denaturation. However, during denaturation of DNA no covalent bonds are broken: only the unwinding and separation of the double helical DNA structure occurs. This is because either or both of the two sets of forces responsible for maintaining the two strands of a DNA helix: hydrogen bonding between base-pairs and stacking interactions between successive bases, are broken. So the native, double-helix three-dimensional structure of DNA is disrupted into a random form called single-stranded DNA. Denaturation is physiological, so structures ought not to be too stable and, indeed, double-stranded DNA must come apart to replicate and for RNA synthesis. Heat denaturation of DNA is called melting because the transition from native to the denatured state occurs over a narrow temperature range. The reassociation of complementary strands of DNA is a spontaneous process: the renaturation process, also called annealing. This reversibility of the denaturation process is of great importance for achieving the biological functions of DNA.

Double-stranded right-handed B-DNA has a highly repetitive, negatively charged polyphosphate surface, single-stranded nucleic acids that do not fold up to compact structures are much more flexible than the double-stranded nucleic acids and the predominantly hydrophobic bases are much more exposed. A common feature of all the A-, B- and Z-DNA conformations is that the two chains are antiparallel.

3.3 THE TRIPLE HELIX DNA

Some research over a number of years has been devoted to the existence of triple helix DNA, which results from the interaction of native and single stranded DNA [9–11]. The triple helix complex formed between two strands of poly(uridylic acid) and one strand of poly(adenylic acid) was first described in 1957 by Felsenfeld [4]. Later in 1959, Hoogsteen [12] explained the hydrogen bonds in the triple base-pairing by considering the third strand positioned in the major groove of a Watson-Crick base-pairing. The ability to form triplex structures is greatly influenced by modification at the phosphate centre [13] and a major challenge concerns the characterization of the triple pairing alignments and structures in genetic recombination [14]. Regions of DNA with triplex characteristics are found in control regions for genes and triplex formation prevents expression of the gene.

Triple-stranded DNA can be generated intermolecularly or intramolecularly and can only be formed if one strand of the original B-helix is all purines (A and G) and the corresponding region of the other strand is all pyrimidines. It is more stable at low pH and has a second polypyrimidine strand wound in the major grove of B-DNA where it forms Hoogsteen base-triples with a complementary polypurine strand, Fig. 3.2. Triple-

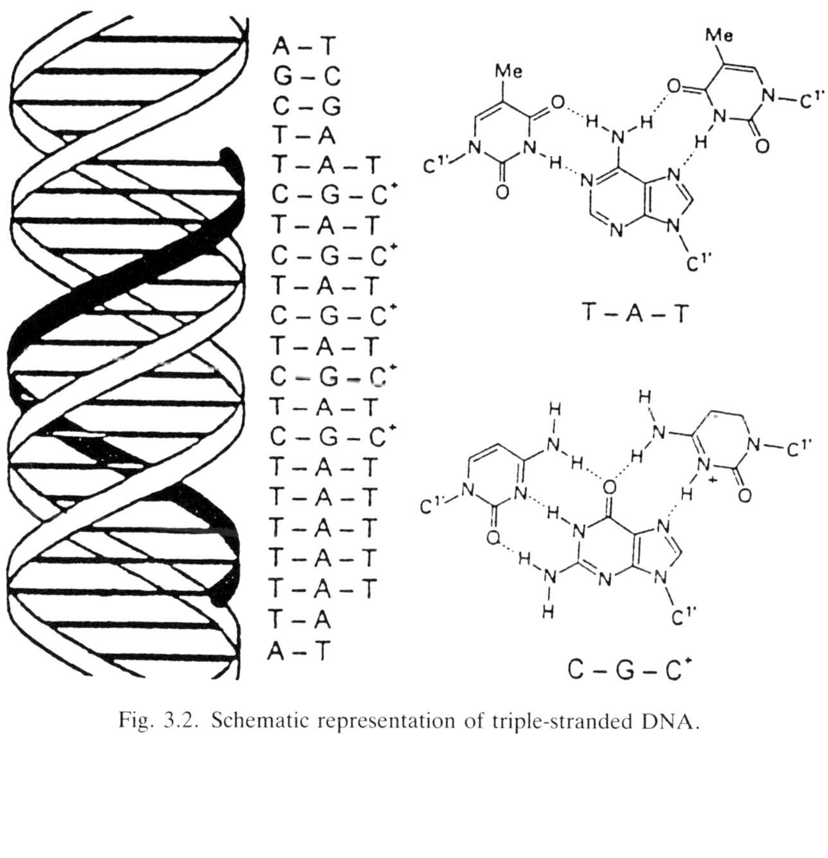

Fig. 3.2. Schematic representation of triple-stranded DNA.

stranded nucleic acids use Hoogsteen hydrogen bond base-pairing to add a polypyrimidine third strand to a polypurine.polypyrimidine duplex.

In 1985, Lyamichev *et al.* [15] called H-DNA to the intramolecular triplex of sequence PyPuPy in order to indicate the high H^+ concentration of the media where this triplex exists. This requires protonation of one cytosine in each base-triplet and the two pyrimidine strands must run antiparallel. These form has undergone much research recently [16, 17], as it is very important to understand what is the role of triplex DNA in the nucleic acid characteristics and biotechnological applications may be of great relevance.

On the other hand, the formation of intramolecular triplexes of PyPuPu sequence is favoured by the presence in neutral media of multivalent cations that stabilize the triple helix [18, 19].

The identification of *in vivo* intramolecular triple helixes have been made by immunological techniques [20], the use of chemical probes [21] and by detection of uncoupled DNA chains sensitive to the action of nucleases [22].

3.4 DNA ELECTROCHEMISTRY

The electrochemical behaviour of DNA and adsorption at different types of electrodes has been investigated for a number of years [23–25], first using a dropping mercury electrode and more recently solid electrodes.

Electrochemical reduction of natural and biosynthetic nucleic acids at a dropping mercury electrode [26, 27] showed that adenine and cytosine residues [28, 29] as well as guanine residues [30, 31] in a polynucleotide chain are reducible and that the DNA structure significantly influences their reducibility. This was further investigated using differential pulse and linear sweep voltammetry of samples of double-stranded DNA (dsDNA) containing single-stranded (ssDNA) ends of varying length [32] and of damages caused to DNA by different treatments: heat, ionization radiation and chemical [33].

Electrochemical oxidation of natural and synthetic nucleic acids at carbon electrodes: pyrolytic graphite [34], paraffin-wax-impregnated spectroscopic graphite [35] and glassy carbon [36, 37] has been studied. It was shown that at pH 4.5 the electrochemical activity of nucleic acids is conditioned by the presence of purine residues in polynucleotide chains.

The proposed oxidation mechanisms for guanine and adenine are in Fig. 3.3.

Moreover, natural nucleic acids give rise to two well-separated oxidation peaks in differential pulse voltammograms, which can be used to probe individual adenine-thymine (AT) and guanine-cytosine (GC) pairs in double helical DNA during its conformational changes [38]. Differences in signals obtained at carbon electrodes were observed according to whether, or not, the DNA was denatured [39].

A very different detection approach was used [40] based on the electrocatalytic oxidation of sugars and amines at a copper electrode surface. The ssDNA and dsDNA were detected in the picomolar concentration range. The electrochemical signal due to dsDNA was higher than due to ssDNA owing to the larger number of easily accessible sugars on the outer perimeter of dsDNA double helix compared to those on a ssDNA of the same size, in contrast to most electrochemical studies based on the electroactivity of the bases.

Voltammograms [41] obtained with native and denatured DNA illustrate the greater difficulty for the transition of electrons from the inside of the double-stranded rigid form of DNA to the electrode surface than from the flexible single-stranded form where the guanine and adenine residues are in close proximity to the electrode surface. The roughness of a solid electrode surface means that double-helix DNA has some difficulty in following the surface contours, whereas, denatured unwound DNA molecules fit more easily into the grooves on the surface, because of their greater flexibility.

However, flow injection direct amperometric detection at carbon paste electrodes [42] was used for measurements of nucleic acids down to the 0.5 ng level without time-consuming derivatization or hydrolysis steps.

3.4.1 Adsorption at electrode surfaces

The study of the adsorption of DNA at electrode surfaces is of fundamental interest, since the interaction of DNA with charged surfaces can be expected in biological systems. In fact, a number of studies of the adsorption of DNA were conducted at mercury electrodes [43–51] and carbon electrodes [52] and in general weaker adsorption was observed with native DNA than with denatured DNA.

Recently, the electrochemical reduction of DNA has been reexamined

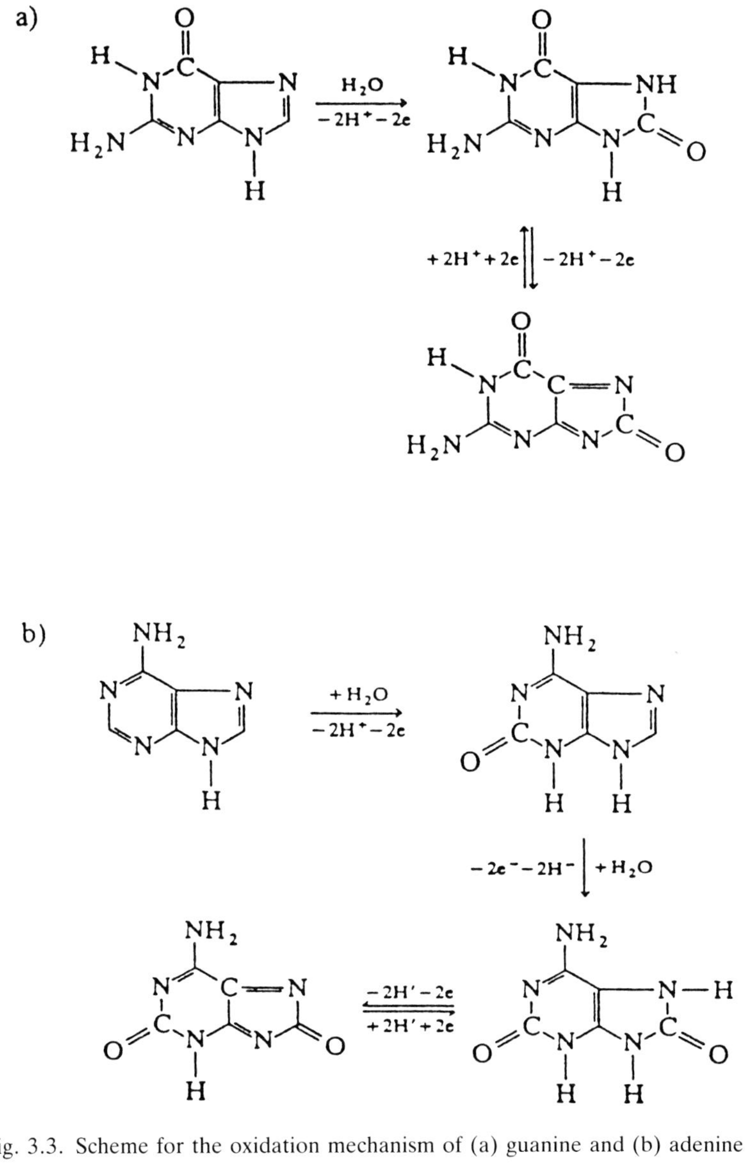

Fig. 3.3. Scheme for the oxidation mechanism of (a) guanine and (b) adenine.

with a view to more fundamental studies of the adsorption phenomenon [53, 54], and the possibility of using adsorption to preconcentrate DNA on a mercury electrode surface, thus increasing sensitivity, enabling lower detection limits [55]. This method has been referred to as adsorptive transfer stripping voltammetry (AdTSV) [56], by analogy with the common adsorptive stripping voltammetry, and it is based on the adsorptive preconcentration of a layer of DNA on an electrode followed by the transfer of the mercury DNA-modified electrode into a new solution containing background electrolyte only, where voltammetric analysis takes place.

In adsorptive transfer stripping voltammetry (AdTSV) the behaviour of ssDNA and dsDNA differs in the same way as was observed previously by conventional voltammetry; thus, AdTSV can also be used for structural analysis purposes [57, 58]. Since adsorption can be performed from a drop of DNA solution it was possible to reduce the volume of the sample to be analyzed by two orders of magnitude [59]. Using the hanging mercury drop electrode the limit of detection of single-stranded DNA was below 0.1 $\mu g\ cm^{-3}$ and, as the adsorption was from a 10-μl drop of DNA solution, subnanogram quantities of single-stranded DNA were sufficient for voltammetric analysis [60]. Chronopotentiometric stripping analysis (CPSA) of bases, DNA and RNA have been performed at mercury [61–63] and carbon paste [64, 65] electrodes.

Osmium tetroxide is an electroactive marker of the polynucleotide chain and is a good probe of the DNA structure since dsDNA is modified by osmium to a much lesser extent than single-stranded polynucleotides. The limit of detection of osmium-labelled DNA was below 5 $ng\ cm^{-3}$ after 2 min accumulation time [66]. Adsorptive stripping linear sweep voltammetry of osmium tetroxide labelled DNA at a mercury electrode [66, 67] was shown to be a good sensor for hybridization of DNA.

Some similar features were observed concerning the adsorption and electrochemical oxidation of DNA on glassy carbon and tin oxide electrodes [68]. Differential pulse voltammograms were recorded in buffer solution without DNA after adsorption of DNA onto the electrode surface during a predetermined time at a fixed potential suggesting the possibility of using adsorption to preconcentrate DNA on solid electrode surfaces and use this DNA-modified electrode for analytical purposes.

A recently developed small-volume sono-voltammetric cell [69] in which the glassy carbon electrode is exposed to high intensity ultrasound was used in order to permit reliable simultaneous voltammetric determinations

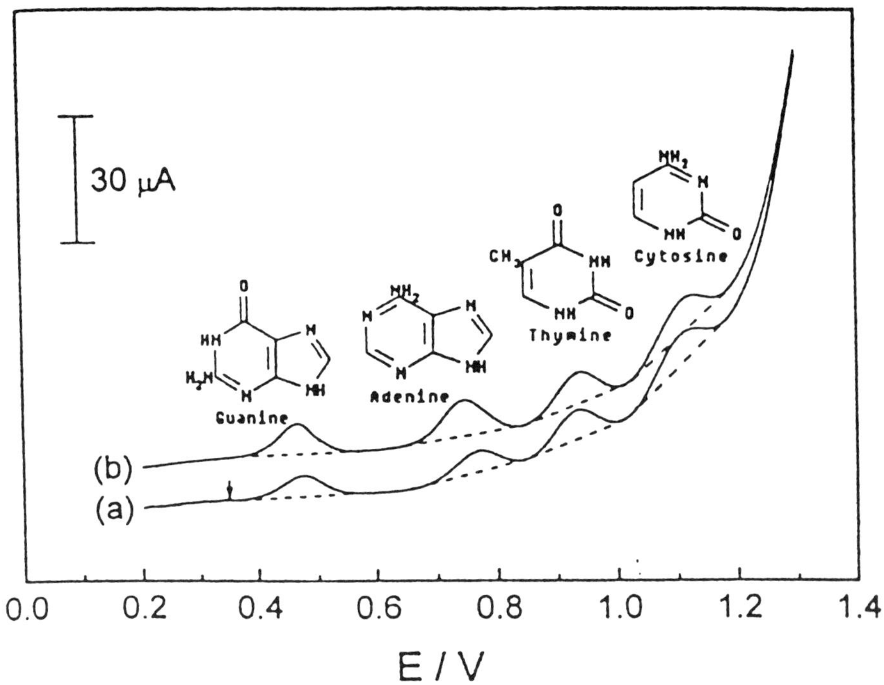

Fig. 3.4. Differential pulse voltammetric determination of purine and pyrimidine bases guanine (2×10^{-5} M), adenine (3×10^{-5} M), thymine (3×10^{-4} M) and cytosine (3×10^{-4} M) in borate buffer (pH = 10.02) (a) with ultrasonic pretreatment (power intensity 72 W/cm^2, horn tip-electrode separation 5 mm, (b) successive scan without ultrasonic pretreatment. DPV conditions: scan rate 5 mV/s, amplitude 50 mV. (Reprinted from ref. [68] with permission.)

of the four bases: guanine, adenine, thymine and cytosine [70], Fig. 3.4. This figure shows for the first time that the pyrimidine bases thymine and cytosine undergo oxidation at glassy carbon electrodes albeit for a high pH value of 10.0 and very positive potentials. The main effects of ultrasound are the large enhancement of mass transport [71–73] and electrode surface modification or cleaning which prevents progressive electrode fouling [74, 75].

Spectroscopic techniques such as surface enhanced Raman spectroscopy (SERS) have also been used to investigate the adsorption of DNA onto the electrode surfaces [76, 77].

Images by scanning tunneling microscopy (STM) and atomic force microscopy (AFM) of DNA molecules adsorbed onto the gold electrode surfaces showed that at positive charges DNA can be reversibly adsorbed and undergo conformation transitions [78–81]. It was shown that electrochemical deposition and *in situ* potential control during imaging can be used to obtain reproducible STM images of the internal structure of adsorbed ssDNA and dsDNA.

3.4.2 Glassy carbon-modified electrodes

Evidence of the electrical conductivity of DNA and of its important mechanisms has been discussed for a long time and has led to a theory of electron conduction in biopolymers [25, 82]. From this it appeared that the major carrier of conductivity is either electronic or ionic, depending on the temperature of the sample, the water content, and the fact that the conductivity of native samples is higher than that of denatured samples. Following electrochemical oxidation of dsDNA and ssDNA in electrolyte solutions over a wide range of pH, interesting electrochemical properties of a glassy carbon electrode with dsDNA or ssDNA adsorbed on the electrode surface were observed [68].

The electrochemical oxidation of DNA at a glassy carbon electrode at pH 4.5 occurs in the guanine group at a potential of +0.84 V and in the adenine group at a potential of +1.16 V. However, the peak currents are very small for dsDNA compared with those obtained for ssDNA. This has been justified by the stereochemistry of the DNA double helix in comparison with the ribbon-like feature of the ssDNA that can follow the electrode contours more easily. A modified electrode containing dsDNA immobilized by adsorption on the surface of a glassy carbon electrode was developed and its behaviour in solutions containing ssDNA characterized. The glassy carbon electrode covered by adsorbed dsDNA had to be conditioned in a solution containing ssDNA, Fig. 3.5.

The ssDNA was prepared by treating a sample of calf thymus DNA of approximately 4 mg of dsDNA with 0.5 mL of 60% pure perchloric acid; after dissolution, 0.5 mL of 9 M NaOH were immediately added to neutralize the solution followed by 9 mL of pH 4.5 acetate buffer. The DNA-modified glassy carbon electrode was prepared by covering a glassy carbon electrode with 3 mg of dsDNA dissolved in 80 μL of pH 4.5 acetate buffer and leaving the electrode to dry. The DNA layer was ~0.5 mm thick and

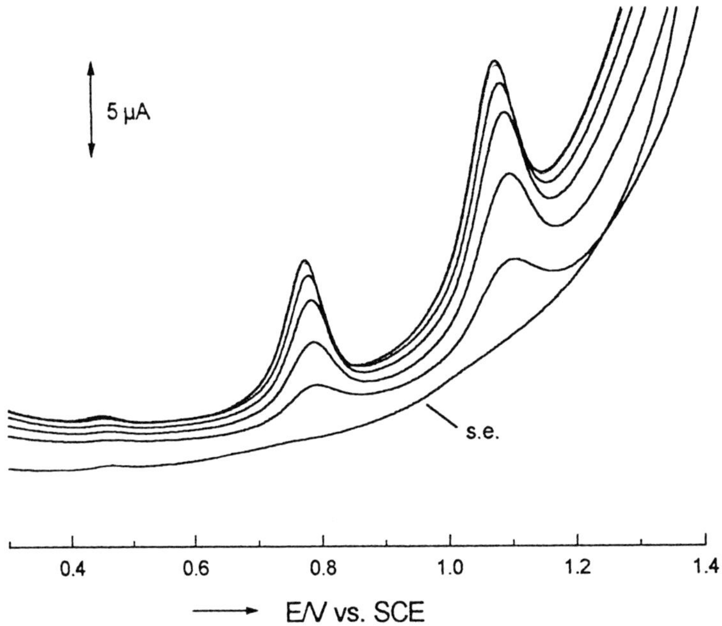

Fig. 3.5. Successive differential pulse voltammograms of the glassy carbon electrode with dsDNA adsorbed, immersed in the ssDNA solution during modification; s.e. = supporting electrolyte: pH 4.5 acetate buffer 0.1 M. Pulse amplitude 50 mV, pulse width 70 ms, scan rate 5 mV s^{-1}.

covered all the glassy carbon electrode. After drying the electrode was conditioned in acetate buffer solution at +1.4 V and was transferred to a solution containing ssDNA where voltammograms in the potential range 0 to +1.4 V were recorded. Between each scan, the electrode was maintained in the ssDNA solution at 0 V during 2 min under constant stirring. The electrode was then cleaned in buffer solution by application of 0 V during 5 min.

This conditioning process was repeated until a stable surface modification was obtained, i.e., until when plots of I_{ox} vs. adsorption time, corresponding to the currents of guanine and adenine peaks did not change, Fig. 3.6. After this treatment the electrode reached maximum activity and it was not necessary to repeat this procedure again. The currents of guanine and adenine peaks obtained in differential pulse voltammetry during conditioning increased reaching saturation.

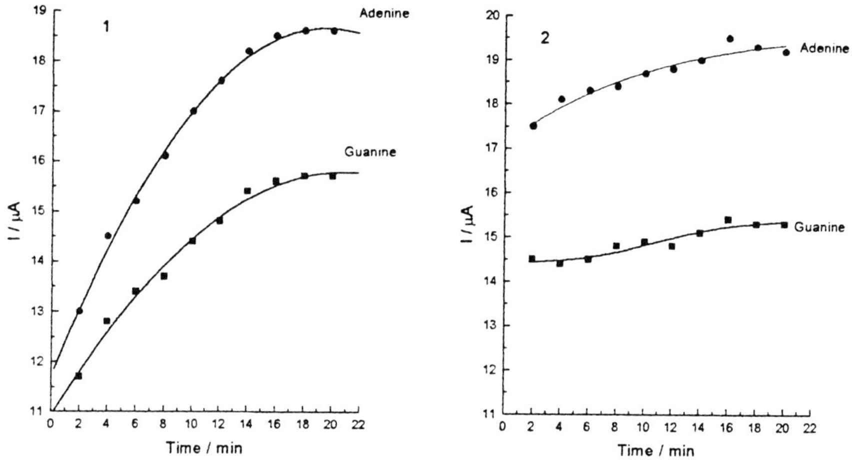

Fig. 3.6. Curves obtained for I vs. adsorption time corresponding to the guanine (■), and adenine (●) peaks during surface modification: (1) first conditioning; (2) last conditioning.

The electrochemical pretreatment of the DNA-modified electrode was essential to obtain reproducible results. During the conditioning procedure the dsDNA film interacted with the ssDNA in bulk solution. The attainment of saturation, i.e., maximum values for the peak currents in the DP voltammograms, showed that there is a limit to the possibilities for the formation of the interchain bonds. Probably the groups participating in the networks yield stronger hydrogen bonds in the film after its interaction with ssDNA than they would in either ssDNA or dsDNA alone. Clearly this favours electron transfer within the DNA itself via electron hopping, or self-exchange, a process called the redox conduction mechanism [83]. Normally the electron is considered to undergo hops in three dimensions, i.e., in any of six directions, only one of which is towards the measuring electrode [84].

These interesting observations could be explained through the formation of triple helix DNA on the electrode surface. Side- and main-chain atoms may interact with a particular DNA base-pair of another chain or with more than one base-pair supporting each other in interwoven hydrogen-bonding networks that stabilize the contacts between the bases, forming a triple helix on the DNA adsorbed to the electrode surface [9–11]. The triple helix DNA is formed by interaction of the ssDNA in the bulk solution with the dsDNA on the electrode surface. The fact that saturation occurred suggests that only a certain number of interactions is possible and the contribution of the oxidized bases to the total conduction process should also be considered.

If the conditioning procedure is carried out on a bare glassy carbon electrode immersed in the ssDNA solution, without adsorbed dsDNA on the electrode surface, the currents in the differential pulse voltammograms during modification increase continuously and never reach saturation, Fig. 3.7, but the adsorbed species do not block the electrode. Blockage of the surface of the bare glassy carbon was only observed in adenosine 5′-monophosphate and adenine solutions but never in guanosine 5′-monophosphate and guanine solutions.

This modified electrode was used to evaluate the electrochemical oxidation of guanosine and adenosine 5′-monophosphates. A similar conclusion

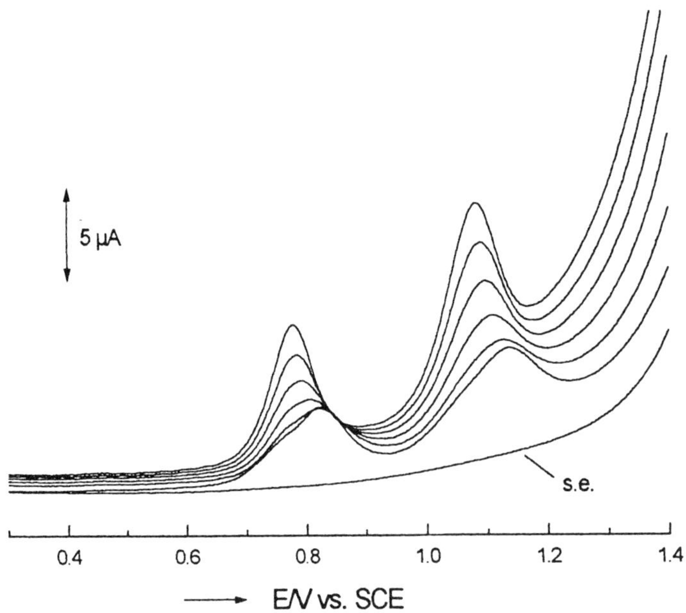

Fig. 3.7. Successive differential pulse voltammograms of the clean glassy carbon electrode, immersed in the ssDNA solution during modification; s.e.-supporting electrolyte: 0.1 M acetate buffer pH 4.5. Pulse amplitude 50 mV, pulse width 70 ms, scan rate 5 mV s^{-1}.

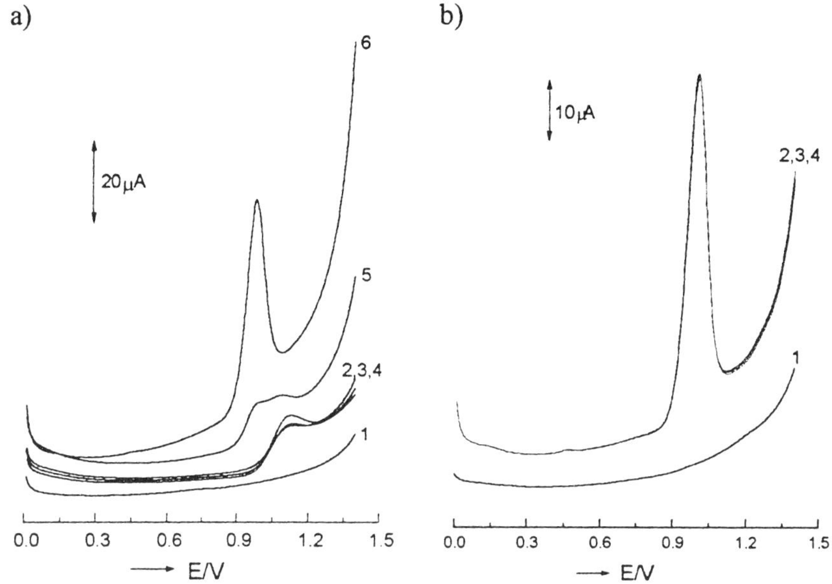

Fig. 3.8. Differential pulse voltammograms of 1 mM guanosine 5′-monophosphate in 0.2 M acetate buffer: (a) Bare glassy carbon electrode: 1 – s.e.; 2, 3, 4 – first, second and third scans in the guanosine solution; 5–10th scan; 6–20th scan; (b) DNA-modified glassy carbon electrode: 1 – s.e.; 2, 3, 4 – first, second and third scans in the guanosine solution. Pulse amplitude 50 mV, pulse width 70 ms, scan rate 5 mV s^{-1}. s.e.=supporting electrolyte.

concerning the DNA triple helix mediating the electron-transfer reaction was reached.

If we consider Figs. 3.8 and 3.9, where the results for the oxidation of guanosine and adenosine 5′-monophosphates at a bare glassy carbon electrode and at a DNA-modified electrode are presented, several facts are worth mentioning when comparing voltammograms at both electrodes.

Whereas, for the oxidation of guanosine 5′-monophosphate at the glassy carbon electrode the peak currents were very small at the beginning and a succession of scans (at least 20) was necessary to obtain a reasonably high peak but never reaching a constant value (Fig. 3.8a) by the twentieth scan, for the DNA-modified glassy carbon electrode the peaks were of equal height from the first scan and the current higher than in the former case and at the same potential, (Fig. 3.8b) first, second and third scans. The oxidation of guanosine 5′-monophosphate at the bare glassy carbon

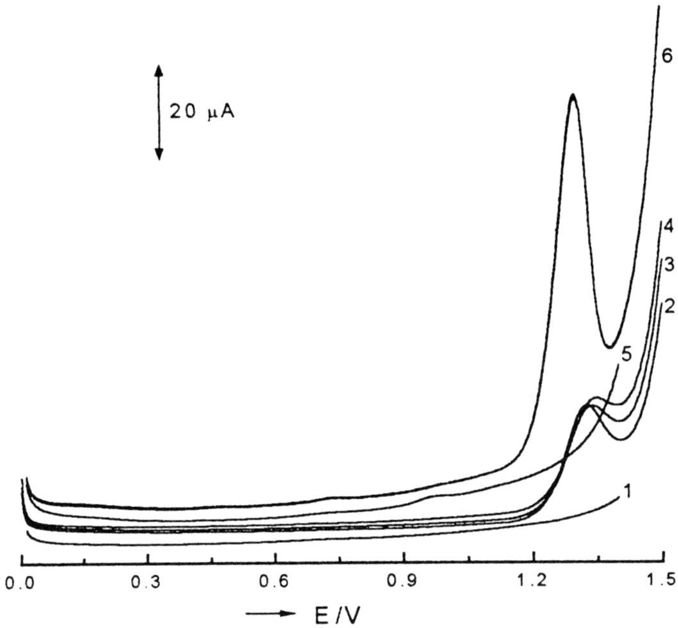

Fig. 3.9. Differential pulse voltammograms of 1 mM adenosine 5′-monophosphate in 0.2 M acetate buffer: Bare glassy carbon electrode: 1 – s.e.; 2, 3, 4 – first, second and third scans in the adenosine solution; DNA-modified glassy carbon electrode: 5 – s.e.; 6 – first, second and third scans in the adenosine solution. Pulse amplitude 50 mV, pulse width 70 ms, scan rate 5 mV s^{-1}. s.e. = supporting electrolyte.

electrode was probably accompanied by reorientation of the molecule and reorganization of the solvent at the electrode surface, as the peak potential is changing and shifts from more positive to less positive values until it reaches the value obtained already in the first scan when using the DNA-modified glassy carbon electrode.

In the case of adenosine 5′-monophosphate, oxidation at a bare glassy carbon electrode leads to the rapid formation of products that adsorb irreversibly on the electrode surface and consequently the peaks diminish on successive scans (see Fig. 3.9, curves 2, 3, 4). On the contrary, the current obtained when using the DNA-modified glassy carbon electrode (Fig. 3.9, curve 6), is five times higher and there are no signs of electrode blocking by adsorption.

The results obtained enable us to conclude that in the DNA-modified glassy carbon electrode, although the groups that undergo electrochemical oxidation are guanine and adenine, the mechanism of oxidation is greatly dependent on other groups in the DNA molecule that in our view act as conducting bridges enabling fast electron flow along strands to reach the electrode.

In fact, the film formed on the GC electrode surface by the ssDNA acts purely as a conducting path such as in a conducting polymer. Alternatively, if we only use the dsDNA adsorbed on the glassy carbon electrode without conditioning by the ssDNA to investigate an electron-transfer process it is not possible to obtain reproducible results.

The possibility of reversible electron transfer within the modified DNA film was tested by carrying out an electrochemical study [85] of the redox couple Fe(III)/Fe(II) which has reasonably fast electrode kinetics, and which are dependent on electrode material. The oxidation of $Fe(CN)_6^{4-}$ in 0.4 M K_2SO_4 aqueous solution contacting the DNA-modified glassy carbon electrode showed virtually the same reaction rate as when using the bare glassy carbon electrode, Fig. 3.10, and the results were comparable to

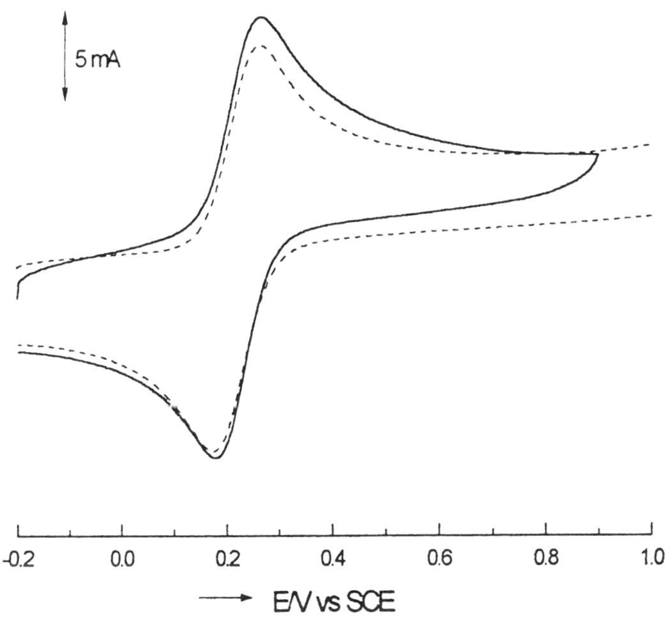

Fig. 3.10. Cyclic voltammograms of: ------ Bare glassy carbon electrode in a solution of 2 mM $Fe(CN)_6^{4-}$ in 0.4 M K_2SO_4; —— DNA-modified glassy carbon electrode in a solution of 2 mM $Fe(CN)_6^{4-}$ in 0.4 M K_2SO_4. Scan rate 100 mV s^{-1}.

those obtained at bare platinum electrodes. It can be deduced that the modified DNA film is able to easily conduct the electrons from the solution to the glassy carbon substrate, meaning that the electron-transfer reaction was mediated by the dsDNA modified by ssDNA to form the triple helix DNA. From calculations the rate of the electron transfer for the oxidation of ferrocyanide using the DNA-modified glassy carbon electrode, (Fig. 3.10, curve 2), is 1.6×10^{-3} cm s^{-1}.

The conditioning procedure in ssDNA solution can lead to the formation of a triple helix that works like a wire mediating the electron-transfer reactions: the electronic levels arising from the ssDNA allow specific binding to the major groove of a DNA duplex and its reorganization on the electrode surface that can facilitate electron transfer between the glassy carbon electrode surface and the solution. Since the modification of the dsDNA and interchain crosslinking with ssDNA takes place on the electrode surface at 0 V this is a mechanism that can occur *in vivo*, the biological consequences and implications being vast.

3.5 DNA BIOSENSORS

The objective in developing the DNA-modified glassy carbon electrode was to use it to evaluate and to predict DNA interactions and damage by health hazardous compounds based on their binding to nucleic acids. Thus, the fabrication of a DNA biosensor can be achieved consisting in fact of producing a glassy carbon electrode covered by dsDNA containing portions of triple helix DNA on the surface. Strong evidence for this was the fact that the electrode performed efficiently only after conditioning.

DNA biosensors are also of great interest for the future development of microelectronic sensors for the detection of biological compounds and antigens. DNA would act as a promoter between the electrode and the biological molecule under study. Recently, some electrochemical research has been done in this direction with the aim of studying the interaction of DNA immobilized on the electrode surface with substances in solution.

A membrane for a DNA-probe of a microelectronic biosensor based on Langmuir-Blodgett (LB) diacetylene film with covalently bonded DNA has been developed [86]. Diacetylene films formed on the surface of oxidized silicon by the LB method were used for covalent immobilization of DNA molecules, using the standard *N,N'*-dicyclohexylcarbodiimide

(DCC) method, by formation of linker structures between the amino group of the DNA base and the surface COOH group of the LB film.

The possibilities of biosensor applications of Langmuir-Blodgett and self-assembly (SA) films were monitored on binding of specific reagents for DNA by DNA-containing films or mononucleotides by a complementary single-stranded polynucleotide immobilized on a positively charged solid support [87].

Due to their simplicity of construction, ease of modification, electrical methods of detection, fast response time and the fact that they are the principal structural component of all biomembranes, conventional bilayer lipid membrane (BLM) arises as an ideal system for biosensor technology [88] and they have been studied regarding the possibility of developing DNA biosensors consisting of a glassy carbon electrode-modified by a BLM with incorporated ssDNA [89].

The development of a biosensor capable of selectively detecting a point mutation in human DNA would enable an early diagnosis of inherited human diseases. With this aim, a prototype of a sequence-selective DNA biosensor based on hybridization indicators was evaluated [90]. For the covalent immobilization of single-stranded DNA onto the carboxylic groups present on the surfaces of oxidized glassy carbon, water-soluble carbodiimide and *N*-hydroxysuccinimide coupling reagents were used.

Evidence that double-stranded DNA molecules are adsorbed in such a way that the helical axis becomes parallel to the electrode surface, the base-pairs being vertically oriented against the electrode surface [45] leads to the conclusion that the thickness of a monolayer of adsorbed DNA at the electrode surface is less than 2 nm. This fact has been applied to use DNA adsorbed at a glassy carbon electrode as an effective electron promoter enabling electron transfer via hopping conduction through electrode-/base-pair/cytochrome *c* [91].

Another sensor based on a fiber-optic-based spectroelectrochemical probe uses a DNA/ethidium bromide system to take advantage of the biological recognition processes [92]. The concept of immobilizing electrochemical reagents on the end of an optical fibre is a useful addition to the field of bioanalytical sensors. Before this development, optical and electrochemical detection of DNA were performed separately. Optical and electrochemical detection of DNA are suitable for a DNA detection system [93, 94] and these techniques will enable a production of a cheap DNA biosensor with a rapid and quantitative response.

The utilization of a DNA matrix to aid enzyme immobilization is a promising alternative for the development of new biosensors and its application in flow systems. A DNA-tyrosinase carbon paste electrode [95] showed excellent performance for detection of cathecol in a FIA system and suggests that other enzymatic biosensors can benefit from the presence of DNA.

3.6 ELECTROCHEMISTRY FOR PROBING DNA INTERACTIONS

There are hundreds of compounds that bind and interact with DNA. These reactions cause changes in the structure of DNA and the base sequence leading to perturbation on DNA replication. It is very important to explain the factors that determine affinity and selectivity in binding molecules to DNA, because a quantitative understanding of the reasons that determine selection of DNA reaction sites is useful in designing sequence-specific DNA binding molecules for application in chemotherapy and in explaining the mechanism of action of neoplastic drugs [96]. Research on metal ion nucleic acid complexes was advanced when antitumour activities of platinum(II) compounds were discovered.

Chemical modification of DNA bases is called mutagenesis. Mutations are caused by substitutions, deletions and insertions in the base sequence of DNA [97]. Exposure to toxic chemicals is the cause of many human cancers; these chemical carcinogens act by chemically damaging the DNA. Thus it is important to identify these compounds and ascertain their potency so that human exposure to them can be minimized.

A metal ion have four different potential coordination sites for binding with DNA: the negatively-charged phosphate oxygen atoms, the ribose hydroxyls, the base ring nitrogens and the exocyclic base keto groups. Metal ions such as Na^+, K^+, Mg^{2+} and Ca^{2+}, exist in the body in high concentrations and the nucleic acids and nucleotides occur as complexes coordinated with these ions.

Electrochemical methods [85] have shown great potential for studying the adsorption and reactions of biological molecules at electrified interfaces. They have been used by a number of researchers studying metal ion-DNA interactions [98].

Studies of metallointercalation agents [99–101] in the presence of dsDNA have been done with solutions containing the redox complexes

$Co(phen)_3^{3+/2+}$, $Fe(phen)_3^{2+/3+}$, $Co(bpy)_3^{3+/2+}$, $Fe(bpy)_3^{2+/3+}$ and $Os(bpy)_3^{2+/3+}$ (where phen = 1,10-phenanthroline and bpy = 2,2′-bipyridyl). The binding was interpreted in terms of the interplay of electrostatic interactions of the metal coordination complexes with the charged sugar-phosphate backbone and the intercalative, hydrophobic, interactions within the DNA helix, i.e., the stacked base-pairs. Chronopotentiometric detection of the $Co(bpy)_3^{3+}$ indicator was used [102] to monitor the hybridization onto a screen-printed carbon electrode of short DNA sequences from *E. coli* pathogen in environmental water samples. The couple $Co(bpy)_3^{3+/2+}$ was also used to evaluate the adsorption of DNA on gold electrodes [103–104].

The binding of ligand and ligand substituents in complexes of ruthenium(II) [105] has been investigated in a systematic fashion and the binding parameters for the series compared in order to determine the different ligand functionalities and sizes in the binding with DNA, i.e., intercalation and surface binding. It was found that if one compares the various factors that contribute to stabilizing the metal complexes of ruthenium on the DNA helix it appears that the most significant factor is that of molecular shape. The complexes that fit most closely to the DNA helical structure, those in which the van der Waals interactions between complex and DNA are maximized, display highest binding affinity.

A doubly metallated 15 base-pair double helix containing ruthenium and rhodium at each end of the strands [106] showed the efficiency of DNA for coupling electron donors and acceptors over a very long range, greater than 40 Å. The DNA double helix was found to behave like a piece of molecular wire with fast electron-transfer rates ($\geqslant 10^{10}$ s^{-1}) for the photoinduced electron transfer between the metallointercalators [107–109] and semiempirical Hartree-Fock calculations of H_{AB} for DNA mediated electron transfer [110] were described.

Oxidative damage to DNA was demonstrated to depend [111] on oxidation potential and to be promoted from a remote site as a result of an electron hole migration through the DNA π-stack, the hole migrates down the double helix to damage guanine, a site sensitive to oxidative nucleic acid damage within the cell.

During the transfer of genetic information the interactions between DNA and the divalent ions Mg^{2+}, Mn^{2+}, Co^{2+}, Cu^{2+}, Zn^{2+}, Ni^{2+}, Cr^{2+}, Cd^{2+} and Pb^{2+}, play an essential role in promoting and maintaining the nucleic acid functionalities. Some are recognized for their carcinogenicity as they damage DNA molecules and alter the fidelity of DNA synthesis [112, 113]. Nickel, as well as chromium and cadmium, seems to be recognized as the most effective carcinogen. Many inorganic nickel compounds have been tested and their effect on a cell or tissue has been established as an interaction with the base donor systems, especially within unwound parts of nucleic acids [114–116]. Damage of DNA structure has also been described due to the effects of Zn [117], Cd [117] and Pb [118]. The interaction modes of metallopophyrins with DNA [119] present different electrochemical behaviour depending on the metal ions: Cu, Ni, Zn or Cd, and the porphyrin ligands.

The adsorptive and voltammetric characteristics of Cu(II) complexes with guanine, guanosine and adenosine were exploited [120] in order to detect these bases after separation by capillary zone electrophoresis, and the enzyme-mimic catalytic activity of a DNA-Cu^{2+} complex [121] was used to develop an amperometric quinacrine sensor using an oxygen electrode covered by the complex entrapped in polyacrylamide gel.

The cytostatic activity of various platinum drugs has shown that platinum coordination complexes cause irreversible inhibition of DNA synthesis due to covalent binding with DNA [122]. This causes the treatment to be often accompanied by adverse reactions. Differential pulse voltammetry with the static mercury drop electrode has been used to investigate the interactions of a group of eight anticancer-active Pt(II) and Pt(IV) complexes in solution with DNA, and the conformational alterations induced by the binding of the drugs [123–128]. It was found that the binding of platinum drugs causes local distortions in the DNA molecule with the formation of interstrand cross-links.

The antitumour action of the natural antibiotic bleomycin is thought to involve the aerobic degradation of DNA by the Fe^{2+}-bleomycin complex. In order to probe the mechanism of antitumour action of bleomycin, the 4-ethylamido[5,(2′-thienyl)-2-thiophene] imidazole iron(II) complex was synthesized [129]. It was studied in non-aqueous solution using cyclic voltammetry and showed antitumour activity *in vitro*, its action causing cleavage of the double helical DNA.

The DNA interacting drugs prevent cell growth, not only cancer cell growth and the cytotoxic effect blocks, but also the growth of normal cells. The lack of selectivity of cancer drugs is one of the main problems in cancer chemotherapy [130, 131] and the chemical and biological mechanism of drugs active against cancer cells must be thoroughly investigated.

Spectroscopic methods have been complemented by electrochemical methods for investigating the binding strength and specificity of antineoplasic drugs with DNA at the molecular level, as well as explaining their toxic action. Some tumour cells, but not normal cells, have nucleic acids associated with the cell surface that can suppress many cell immunological reactions.

The anticancer activity and host toxicity of compounds with a quinone group was investigated. There is considerable death of tumour cells, together with the induction of breaks in DNA single and double strands, although a low binding value for the semiubiquinone with DNA was found [132].

The binding affinities of a group of cancerostatic anthracycline antibiotics to DNA were evaluated by differential pulse and other voltammetric techniques using mercury electrodes [133] and the formation of complexes with DNA in relation to the sugar residue was measured. It was found that binding was higher for derivatives with two basic sugars and lower for derivatives with one neutral sugar. Nevertheless, the cytostatic properties of these anthracycline antibiotics also depend on the influence of rates of dissociation, splitting and transport to the DNA of the tumour.

The potent antineoplasic, antiviral and antibacterial agent, the basic oligopeptide netropsin, is known to increase the stiffening of double helical DNA segments to which it is bound. In agreement with this, oxidation voltammetric peaks at graphite electrodes of a DNA solution containing netropsin were smaller for the peak corresponding to the oxidation of the adenine residues than in solutions without netropsin [134]. These results show the effect of the A.T-specific drug netropsin; electrochemistry has the advantage of a direct observation of changes in the A.T and G.C peaks in double-stranded DNA solutions with and without netropsin. This is a particularly good example which demonstrates the relevance of electrochemistry for research into specific binding of biologically active ligands to DNA.

The detection of chemicals that cause irreversible damage to DNA is very important, as this may lead to hereditary or carcinogenic diseases. A short-time screening test for carcinogens based on ac voltammetric measurements was developed to study *in vitro* damage to DNA caused by the action of alkylating mutagens [135]. Nitroimidazoles are among the most important nitroheterocyclic drugs of interest in cancer chemiotherapy. It was observed that adenine and guanine interact with intermediates generated during nitroimidazole reduction [136], causing irreversible damage to DNA and suggesting mutagenic properties of these compounds.

The mechanism of reduction of a group of nitroimidazoles: metronidazole, secnidazole and tinidazole was investigated using the DNA biosensor [137–139] in a new way by preconcentrating the analyte on the electrode surface containing DNA and studying either their reduction or oxidation of their reduction products retained on the electrode surface. It was possible to follow their reduction, the reversible oxidation of the hydroxylamine derivative (RNHOH) to the corresponding nitroso derivative (RNO), the condensation reaction between the hydroxylamine and nitroso derivatives to form the azoxycompound (RNO:NR) and the interaction with H-DNA. The DNA biosensor contained a small amount of H-DNA segments that were detected by the hydroxylamine—a well-known chemical probe for single strands in H-DNA [140].

The DNA-modified electrodes have been used for trace measurements of toxic amine compounds [141], and for trace measurements of phenothiazine compounds with neuroleptic and antidepressive action [142], as well as detection of radiation-induced DNA damage [143].

Conventional cancer chemotherapy is as toxic as it is beneficial. Antisense and triplex-forming oligonucleotides are currently being developed as therapeutic agents for cancers and viral infections [144]. They are highly effective and safe. The recognition of the potential biological roles of H-DNA and the interest in the triple-helical nucleic acid research also includes genetic applications.

3.7 CONCLUSIONS

Electrochemical research on DNA is of great relevance to explain many biological mechanisms. The DNA-modified electrode is a very good model for simulating the nucleic acid interaction with cell membranes, potential environmental carcinogenic compounds and to clarify the mechanisms of action of drugs used as chemotherapeutic agents.

Development of an electrochemical DNA biosensor opens wide perspectives using a particularly sensitive and selective method for the detection of specific interactions. The possibility of foreseeing the damage that these compounds cause to DNA integrity arises from the possibility of preconcentration of either the starting materials, or the redox reaction products on the DNA biosensor surface, thus permitting the electrochemical probing of the presence of short-lived intermediates and of their damage to DNA.

References

[1] O.T. Avery, C.M. Macleod and M. McCarty, J. Exp. Med. 79 (1944) 137.
[2] J.D. Watson and F.H.C. Crick, Nature 171 (1953) 737.
[3] L. Pauling and R.B. Corey, Proc. Natl. Acad. Sci. (USA) 39 (1953) 84.
[4] G. Felsenfeld, D.R. Davies and A. Rich, J. Am. Chem. Soc. 79 (1957) 2023.
[5] E. Chargaff, Biochem. Biophys. Acta 9 (1952) 402.
[6] S. Arnott and D.W.L. Hukins, Biochem. Biophys. Res. Comm. 47 (1972) 1504.
[7] A.H.-J. Wang, G.J. Quigley, F.J. Kolpak, J.L. Crawford, J.H. van Boom, G. van der Marel and A. Rich, Nature 282 (1979) 680.
[8] V.K. Misra and B. Honig, Biochemistry 35 (1996) 1115.
[9] V.N. Potaman and R.R. Sinden, Biochemistry 34 (1995) 14885.
[10] C. Shin and H.-S. Koo, Biochemistry 35 (1996) 968.
[11] V.N. Soyfer and V.N. Potaman, Triple-Helical Nucleic Acids (Springer, New York, 1995).
[12] K. Hoogsteen, Acta Crystallogr. 12 (1959) 822.
[13] E. Uhlman and A.Peyman, Chem. Rev. 90 (1990) 544.
[14] I. Radhakrishnan and D.J. Patel, Biochemistry 33 (1994) 11405.
[15] V.I. Lyamichev, S.M. Mirkin and M.D. Frank-Kamenetskii, J. Biomol. Struct. Dyn. 3 (1985) 327.
[16] N.T. Thuong and C. Hélène, Angew. Chem. Int. Ed. Engl. 32 (1993) 666.
[17] S.O. Doronina and J.-P. Behr, Chem. Soc. Rev. (1997) 63.
[18] J. Bernues, R. Beltran, J.M. Casasnovas and F. Azorin, EMBO J. 8 (1989) 2087.
[19] J. Bernues, R. Beltran, J.M. Casasnovas and F. Azorin, Nucl. Acids Res. 18 (1990) 4067.
[20] Y.M. Agazie, J.S. Lee and G.B. Burkholder, J. Biol. Chem. 269 (1994) 7019.
[21] R.R. Sinden and D.W. Ussery, Methods Enzymol. 212 (1992) 319.
[22] G. Yagil, Crit.Rev. Biochem. Mol. Biol. 26 (1991) 475.
[23] G. Dryhurst and P.J. Elving, Talanta 16 (1969) 855.
[24] E. Palecek, in Topics in Electrochemistry and Bioenergetics, Vol. 5, ed G. Milazzo (John Wiley, London, 1983) p. 65.
[25] H. Berg, in Comprehensive Treatise of Electrochemistry, Vol. 10, Bioelectrochemistry, eds S. Srinivasan, Y.A. Chizmadzev, J.O'M. Bockris, B.E. Conway and E. Yeager (Plenum, New York, 1985, Chap. 3) p. 189.
[26] E. Palecek, Bioelectrochem. Bioenerg. 8 (1981) 469.
[27] P. Valenta and H.W. Nürnberg, J. Electroanal. Chem. 49 (1974) 55.
[28] E. Palecek, Nature 188 (1960) 656.
[29] E. Palecek, Biochim. Biophys. Acta 51 (1961) 1.
[30] E. Palecek, F.Jelen and L. Trnková, Gen. Physiol. Biophys. 5 (1986) 315.

[31] F. Jelen and E. Palecek, Biophys. Chem. 24 (1986) 285.
[32] E. Lukásová, M. Vojtisková and E. Palecek, Bioelectrochem. Bioenerg. 7 (1980) 671.
[33] F. Jelen, M. Fojta and E.Palecek, J. Electroanal. Chem. 427 (1997) 49.
[34] V. Brabec and G. Dryhurst, J. Electroanal. Chem. 89 (1978) 161.
[35] V. Brabec, Bioelectrochem. Bioenerg. 7 (1980) 69.
[36] V. Brabec and J. Koudelka, Bioelectrochem. Bioenerg. 7 (1980) 793.
[37] T. Tao, T. Wasa and S. Mursha, Bull. Chem. Soc. (Japan) 51 (1978) 1235.
[38] V. Brabec, Bioelectrochem. Bioenerg. 8 (1981) 437.
[39] V. Brabec, Biophys. Chem. 9 (1979) 289.
[40] P. Singhal and W.G. Kuhr, Anal. Chem. 69 (1997) 4828.
[41] E. Palecek, Bioelectrochem. Bioenerg. 15 (1986) 275.
[42] J. Wang, L. Chen and M. Chicharro, Anal. Chim. Acta 319 (1996) 347.
[43] I. R. Miller, J. Mol. Biol. 3 (1961) 229.
[44] B. Malfoy and J.A. Reynaud, J. Electroanal. Chem. 67 (1976) 359.
[45] M.W. Humphreys and R. Parsons, J. Electroanal. Chem. 75 (1977) 427.
[46] B. Malfoy, J. M. Sequaris, P. Valenta and H. W. Nürnberg, J. Electroanal. Chem. 75 (1977) 455.
[47] H. Kinoshita, S.D. Christian and G. Dryhurst, J. Electroanal. Chem. 83 (1977) 151.
[48] V. Brabec and E. Palecek, J. Electroanal. Chem. 88 (1978) 373.
[49] V. Brabec, M.H. Kim, S.D. Christian and G. Dryhurst, J. Electroanal. Chem. 100 (1979) 111.
[50] J.A. Reynaud, Bioelectrochem. Bioenerg. 7 (1980) 267.
[51] G.C. Barker, J. Electroanal. Chem. 214 (1986) 373.
[52] V. Brabec, Bioelectrochem. Bioenerg. 11 (1983) 245.
[53] E. Palecek, V. Kolár, F. Jelen and U. Heinemann, Bioelectrochem. Bioenerg. 23 (1990) 285.
[54] E. Palecek, Bioelectrochem. Bioenerg. 28 (1992) 71.
[55] E. Palecek, F. Jelen, C. Teijeiro, V. Fucik and T.M. Jovin, Anal. Chim. Acta 273 (1993) 175.
[56] E. Palecek and I. Postbieglova, J. Electroanal. Chem. 214 (1986) 359.
[57] E. Palecek, Bioelectrochem. Bioenerg. 20 (1988) 179.
[58] F. Jelen, M. Fojta and E. Palecek, J. Electroanal. Chem. 423 (1997) 141.
[59] E. Palecek, Anal. Biochem. 170 (1988) 421.
[60] E.Palecek, Electroanalysis 8 (1996) 1.
[61] W. Jin, X. Zhao and W. Gao, Anal. Chim. Acta 347 (1997) 263.
[62] E. Palecek, M. Tomschik, V. Stankova and L. Havran, Electroanalysis 9 (1997) 990.
[63] J. Wang, D.H. Grant, M. Ozsoz, X. Cai, B. Tin and J.R. Fernandes, Anal. Chim. Acta 349 (1997) 77.
[64] J. Wang, X. Cai, G. Rivas and H. Shiraishi, Anal. Chim. Acta 326 (1996) 141.
[65] J. Wang, X. Cai, J.R. Fernandes, D.H. Grant and M. Ozsoz, Anal. Chem. 69 (1997) 4056.
[66] E. Palecek and M.A. Hung, Anal. Biochem. 132 (1983) 236.
[67] J.M. Hall, J. Moore-Smith, J.V. Bannister and I.J. Higgins, Biochem. Mol. Biol. Inter. 32 (1994) 21.

[68] C.M.A. Brett, A.M. Oliveira Brett and S.H.P. Serrano, J. Electroanal. Chem. 366 (1994) 225.
[69] A.M. Oliveira Brett and F.-M. Matysik, Electrochim. Acta 42 (1997) 945.
[70] A.M. Oliveira Brett and F.-M. Matysik, J. Electroanal. Chem. 429 (1997) 95.
[71] R.G. Compton, J.C. Eklund, S.D. Page, G.H.W. Sanders and J. Booth, J. Phys. Chem. 98 (1994) 12410.
[72] C.R.S. Hagan and L.A. Coury, Anal. Chem. 66 (1994) 399.
[73] D.J. Walton, S.S. Phull, A. Chyla, J.P. Lorimer, T.J. Mason, L.D. Burke, M. Murphy, R.G. Compton, J.C. Eklund and S.D. Page, J. Appl. Electrochem. 25 (1995) 1083.
[74] H. Zhang and L.A. Coury, Anal. Chem. 65 (1993) 1552.
[75] A.M. Oliveira Brett and F.-M. Matysik, Bioelectrochem. Bioenerg. 42 (1997) 111.
[76] J.M. Sequaris, Electrochemistry, Sensors and Analysis, eds M.R. Smyth and J.G. Vos (Elsevier, Amsterdam, 1986) pp 191–203.
[77] C. Otto, F.P. Hoeben and J. Greve, J. Raman Spectrosc. 22 (1991) 791.
[78] S.M. Lindsay, T. Thundat, L. Nagahara, U. Knipping and R.L. Rill, Science 244 (1989) 1063.
[79] S.M. Lindsay, N.J. Tao, J.A. DeRose, P.I. Oden, Yu. L. Lyubchenko, R.E. Harrington and L. Shyakhtenko, Biophys. J. 61 (1992) 1570.
[80] N.J. Tao, J.A. DeRose and S.M. Lindsay, J. Phys. Chem. 97 (1993) 910.
[81] S.M. Lindsay, N.J. Tao, STM and SFM in Biology, eds M. Amrein and O. Marti (Academic Press, London, 1993, Chap. 5) pp. 229–0257.
[82] H. Berg, Electrochemistry of Biopolymers (Plenum Press, New York, 1985).
[83] P.C. Pickup and R.W. Murray, J. Am. Chem. Soc. 105 (1983) 4510.
[84] R.P. Buck, J. Phys. Chem. 92 (1988) 4196.
[85] C.M.A. Brett and A.M. Oliveira Brett, Electrochemistry. Principles, Methods and Applications (Oxford University Press, Oxford, 1993).
[86] M.A. Karymov, A.A. Kruchinin, Yu. A. Tarantov, I.A. Balova, L.A. Remisova, N.G. Sukhodolov, A.I. Yanklovich and A.M. Yorkin, Sensors and Actuators B, 6 (1992) 208.
[87] G.B. Skhorukov, M.M. Montrel, A.I. Petrov, L.I. Shabarchina and B.I. Sukhorukov, Biosen. Bioelectron. 11 (1996) 913.
[88] H.T. Tien and A.L. Ottova, Electrochim. Acta 43 (1998) 3587.
[89] C.G. Siontorou, A.M. Oliveira Brett and D.P. Nikolelis, Talanta 43 (1996) 1137.
[90] K.M. Millan and S.R. Mikkelsen, Anal. Chem. 65 (1993) 2317.
[91] O. Ikeda, Y. Shirota and T. Sakurai, J. Electroanal. Chem. 287 (1990) 179.
[92] D.A. Van Dyke and H.-Y. Cheng, Anal. Chem. 61 (1989) 633.
[93] M.E.A. Downs, P.J. Warner, A.P.T. Turner and J.C. Fothergill, Biomaterials 66 (1988) 9.
[94] M.E.A. Downs, Biochem. Soc. Trans. 19 (1991) 39.
[95] P. Dantoni, S.H.P.Serrano, A.M. Oliveira Brett and I.G.R. Gutz, Anal. Chim. Acta 366 (1998) 137.
[96] I.K. Larsen, A Textbook of Drug Design and Development, eds P. Krosgaard-Larsen and H. Bundgaard (Harwood Academic Publishers, Germany, 1991) p. 192.

[97] W. Saenger, Principles of Nucleic Acid Structure, Springer Advanced Texts in Chemistry, ed. C.R. Cantor (Springer-Verlag, New York, 1984).
[98] J. Swiatek, J. Coord. Chem. 33 (1994) 191.
[99] M.T. Carter and A.J. Bard, J. Am. Chem. Soc. 107, (1987) 7528.
[100] M.T. Carter, M. Rodriguez and A.J. Bard, J. Am. Chem. Soc. 11 (1989) 8901.
[101] M. Rodriguez and A.J. Bard, Anal. Chem. 62 (1990) 2658.
[102] J. Wang, G. Rivas and X. Cai, Electroanalysis 9 (1997) 395.
[103] D.-W. Pang, M. Zhang, Z.-L. Wang, Y.-P. Qi, J.-K. Cheng and Z.-Y. Liu, J. Electroanal. Chem. 403 (1996) 183.
[104] Y.-D. Zhao, D.-W. Pang, Z.-L. Wang, J.-K. Cheng and Y.-P. Qi, J. Electroanal. Chem. 431 (1997) 203.
[105] A.M. Pyle, J.P. Rehmann, R. Meshoyrer, C.V. Kumar, N.J. Turro and J.K. Barton, J. Am. Chem. Soc. 111 (1989) 3051.
[106] C.J. Murphy, M.R. Arkin, Y. Jenkins, N.D. Ghatlia, S.H. Bossmann, N.J. Turro and J.K. Barton, Science 262 (1993) 1025.
[107] D. Clery, Science 267 (1995) 1270.
[108] D. Paterson, Scientific American, (May, 1995) 22.
[109] T.J. Meade and J.F. Kayyem, Angew. Chem. Int. Ed. Engl. 34 (1995) 352.
[110] T.L. Netzel, J. Chem. Educ. 74 (1997) 646.
[111] D.B. Hall, R.E. Holmlin and J.K. Barton, Nature 382 (1996) 731.
[112] L.A. Loeb and R.A. Zakour, Nucleic Acid-Metal Interations, ed T.G. Spiro (John Wiley, New York, 1980) p. 115.
[113] G.L. Eichhorn, Metal Ions in Genetic Information Transfer, eds G.L. Eichhorn and L.G. Marzilli (Elsevier, New York, 1981) p. 1.
[114] T. Pawlowski, J. Swiatek, K. Gasiorowski and H. Kozlowski, Inorg. Chim. Acta 136 (1987) 185.
[115] H. Elzanowska and J.H. van de Sande, Bioelectrochem. Bioenerg. 19 (1988) 425.
[116] H. Elzanowska and J.H. van de Sande, Bioelectrochem. Bioenerg. 19 (1988) 441.
[117] J. Swiatek, T. Pawlowski, K. Gasiorowski and H. Kozlowski, Inorg. Chim. Acta 138 (1987) 79.
[118] J.-M. Séquaris and J. Swiatek, Bioelectrochem. Bioenerg. 26 (1991) 15.
[119] F. Qu and N.-Q. Li, Electroanalysis 9 (1997) 1348.
[120] W. Jin H. Wei and X. Zhao, Anal. Chim. Acta 347 (1997) 269.
[121] Y. Hasebe, K. Hamada, T. Hirano and S. Uchiyama, Anal. Commun. 34 (1997) 153.
[122] L.A. Zwelling, Cancer Chemotherapy, Vol. 8, eds H.M. Pinedo and B.A. Chabner (Elsevier, Amsterdam, 1986) p. 97.
[123] O. Vrana and V. Brabec, Anal. Biochem. 142 (1984) 16.
[124] D. Kasparova, O.Vrana, V. Kleinwachter and V. Brabec, Biophys. Chem. 28 (1987) 191.
[125] O. Vrana and V. Brabec, Bioelectrochem. Bioenerg. 19 (1988) 145.
[126] V. Brabec, V. Vetterl, V. Kleinwachter and J. Reedijk, Bioelectrochem. Bioenerg. 21 (1989) 199.
[127] S.D. Kim, O. Vrana, V. Kleinwachter, K. Niki and V. Brabec, Anal. Lett. 23(8) (1990) 1505.

[128] A.M. Oliveira Brett, S.H.P. Serrano, T.A. Macedo, D. Raimundo, M.H. Marques and M.A. La-Scalea, Electroanalysis 8 (1996) 992.
[129] C.W. Ong and H.-C. Lee, Inorg. Chim. Acta 125 (1986) 203.
[130] The Chemotherapy Source Book, eds. M.C. Perry (Williams and Wilkins, 1991, Baltimore).
[131] Nucleic acids in Chemistry and Biology, 2nd edn, eds C. M. Blackburn and J. Gait Oxford University Press, Oxford, 1996).
[132] R.S. Schrebler, A. Arratia, S. Sánchez, M. Haun and N. Durán, Bioelectrochem. Bioenerg. 23 (1990) 81.
[133] H. Berg, G. Horn, U. Luthardt and W. Ihn, Bioelectrochem. Bioenerg. 8 (1981) 537.
[134] V. Brabec and Z. Balcarova, Bioelectrochem. Bioenerg. 8 (1982) 245.
[135] J.-M. Séquaris and P. Valenta, J. Electroanal. Chem. 227 (1987) 11.
[136] P.J. Declerck and C.J. De Ranter, J. Chem. Soc., Faraday Trans. 1, 83 (1987) 257.
[137] A.M. Oliveira Brett, S.H.P. Serrano, I. Gutz and M.A. La-Scalea, Bioelectrochem. Bioenerg. 42 (1997) 175.
[138] A.M. Oliveira Brett, S.H.P. Serrano, I. Gutz and M.A. La-Scalea, Electroanalysis 9 (1997) 110.
[139] A.M. Oliveira Brett, S.H.P. Serrano, I. Gutz, M.A. La-Scalea and M.L. Cruz, Electroanalysis 9 (1997) 1132.
[140] B.H. Johnston, Methods in Enzymology, eds. D.M.L. Lilley and J.E. Dahlberg (Academic Press, New York, 1992) p. 180.
[141] J. Wang, G. Rivas, D. Luo, X. Cai, F.S. Valera and N. Dontha, Anal. Chem. 68 (1996) 4365.
[142] J. Wang, G. Rivas, X. Cai, H. Shiraishi, P.A.M. Farias, N. Dontha and D. Luo, Anal. Chim. Acta 332 (1996) 139.
[143] J. Wang, G. Rivas, M. Ozsoz, D.H. Grant, X. Cai and C. Parrado, Anal. Chem. 69 (1997) 1457.
[144] S. Abdulla, Chem. Brit. (July, 1997) 30.

Chapter 4

Kinetic Modelling and the Skin

JONATHAN HADGRAFT

4.1 INTRODUCTION

The objective of this chapter is to show how the concepts of chemical kinetics can be applied to the passage of molecules across the skin. The barrier function of the skin is still not totally understood but application of kinetic principles has allowed us to gain a better understanding of the barrier and how penetration enhancers may modulate it. The application of kinetics has allowed a mechanistic interpretation and therefore the development of predictive models. These can assist in the identification and design of novel transdermal drugs, and in the optimal design of topical formulations.

The skin is the largest organ of the body and the one that is most accessible. Despite this, it is perhaps surprising that it is not better understood. It has evolved over the millennia to be an extremely good barrier to both the ingress of xenobiotics and the egress of body fluids such as water.

4.2 THE ROUTE OF PENETRATION

The major barrier of the skin is the outermost "dead" layer, the stratum corneum. A number of routes of penetration of a drug, across this region can be identified. First, the appendages, the pilosebaceous and eccrine glands, form a potential shunt route across the stratum corneum. The relative surface area of these is very small ($<0.1\%$) and there has been little conclusive evidence to suggest that this is a major route. Second, the penetrant could diffuse across the entire stratum corneum through the dead cells, the corneocytes. A large surface area is available but the

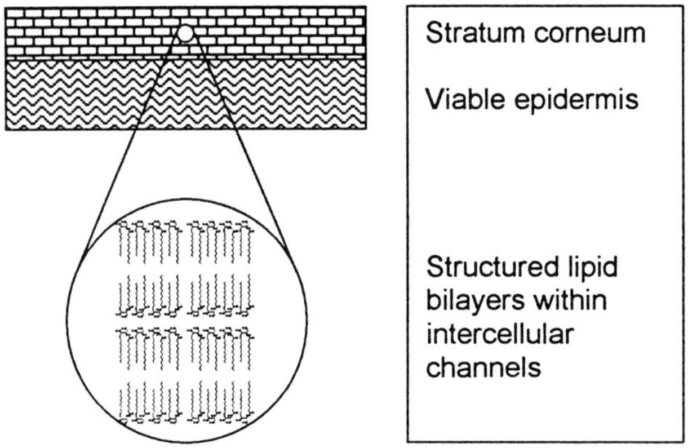

Fig. 4.1. A schematic diagram of the skin.

cells contain dense keratinous material through which diffusion would be anticipated to be very inefficient. The diffusional pathlength would be comparatively short (~15 μm), the thickness of the stratum corneum. Third, the permeant can diffuse through the intercellular spaces where the medium is less dense but the route is far more tortuous. Using a "bricks and mortar" model an estimated diffusional pathlength of 350 μm can be calculated [1]. A diagrammatic representation of the skin is shown in Fig. 4.1. The outer layer, the stratum corneum is represented as a "bricks and mortar" structure. The underlying tissue, the viable epidermis (150 μm thick) resembles and aqueous protein gel in diffusional resistance. A section of the "mortar" is expanded to reveal a complex array of structured lipid bilayers.

Despite the fact that the skin is a heterogeneous membrane, Fick's laws of diffusion have been successfully used to analyze skin permeation data. Solutions to the second law have been used in mechanistic interpretations (see later) and in considering concentration profiles within the skin. Fick's first law has been used to analyze steady-state diffusion rates and in the development of predictive models for skin permeability.

4.3 FICK'S FIRST LAW

The normal form of the first law for skin permeation studies is written,

$$J = \frac{KDA\Delta C}{l}, \tag{4.1}$$

where J is the flux over an area A, K is the partition coefficient between the skin and the applied solution, D is the diffusion coefficient within the skin and l is the diffusional pathlength. ΔC is the difference in concentration between the donor (C_d) and receptor solutions (C_r). Under normal experimental conditions, and *in vivo*, $C_r \ll C_d$ and often Fick's first law is expressed as an approximated form:

$$J = Ak_p c_d, \tag{4.2}$$

where $k_p(=KD/l)$ is a permeability coefficient that has the units of a heterogeneous rate constant (e.g., cm h^{-1}).

Potts and Guy [2] examined a large dataset of permeability coefficients (~90) and have determined a simple relationship between k_p and two fundamental physicochemical constants—the octanol water partition coefficient (K_{oct}) and the molecular weight of the permeant (MW).

$$\log k_p = -2.7 + 0.71 \log K_{oct} - 0.0061\,\text{MW}. \tag{4.3}$$

In simplistic terms, the inclusion of the K_{oct} term reflects the significance of the partitioning process in skin absorption. The pre-multiplier coefficient for MW is an indication that the size of the molecule has a limited contribution (over the molecular weight range of the dataset) to diffusion within the skin. K_{oct} and MW are not independent variables and it is therefore difficult to give an exact interpretation of their significance. However, equation (4.3) does provide a useful means of predicting skin permeability from simple and easily accessible physicochemical parameters. If the partition coefficient is not known experimentally it can be predicted from standard equations with some confidence. The maximum flux of the permeant will be obtained from a saturated solution (where its chemical potential is at its highest). Aqueous solubilities (S_w), at a temperature (T), can be predicted from basic physicochemical parameters, e.g., Valvani *et al.* [3] provide an expression relating the solubility (S_w) to melting point (mp), the entropy of fusion (ΔS_f) and the octanol water partition coefficient:

$$\log(S_w) = -\log(K_{oct}) - 1.11\,\frac{\Delta S_f(\text{mp} - T)}{4.58(273 + T)} + 0.54\,. \tag{4.4}$$

It can be seen that the maximum flux, J_{max}, across the skin will increase with K_{oct} as a result of an increase in k_p. However, in competition with this, J_{max} will decrease as a result of decreasing S_w with increasing K_{oct}. As is often found with biological systems, which involve membrane transfer, there is a parabolic dependency between flux (or amount absorbed over a given time period) and log K_{oct}. For the skin, the optimum value appears to be value appears to be around a log K_{oct}=2. This has been demonstrated for a range of non-steroidal anti-inflammatory agents [4] where

$$\log(\%\,\text{dose absorbed}) = 0.42 + 1.14\log(K_{oct}) - 0.23(\log K_{oct})^2\,. \tag{4.5}$$

Drugs that are absorbed well percutaneously have relatively balanced partition coefficients and good solubilities.

Equation (4.4) shows that the melting point is an important determinant in solubility calculations, the lower the melting point, the lower the intermolecular forces, and therefore the easier it is for the drug to dissolve in the intercellular lipids of the skin. It is interesting to note that two of the successful transdermally delivered drugs, nitroglycerin and nicotine, are low melting point materials.

Pugh and Hadgraft [5] have further analyzed the permeability coefficient dataset and have identified the contribution that various functional groups have on permeability coefficients. Using a fragmental approach, correlation analysis has demonstrated that it is possible to assign values to individual functional groups and produce relationships with as good a statistical significance ($r^2 = 0.69$) as equation (4.3). The presence of carbon atoms increases permeability coefficient whereas oxygen atoms reduce it. A more detailed analysis has separated k_p into K and D/l [6, 7]. From these analyses, it is clear that the diffusional parameter is significantly related to the number and types of hydrogen bonding groups in the permeant. In general terms, the diffusion coefficient is reduced by an order of

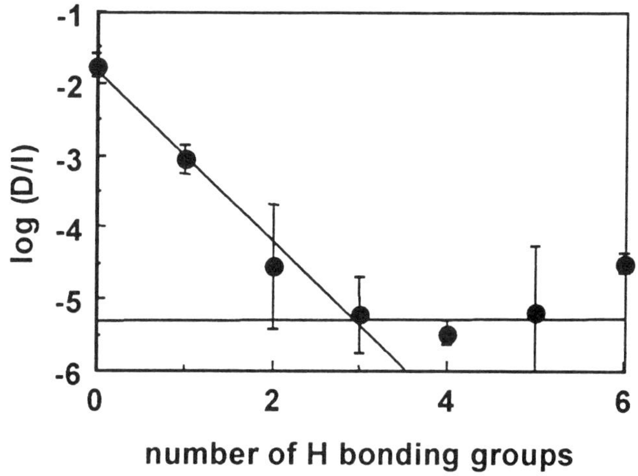

Fig. 4.2. The relationship between the estimated diffusion parameter (D/I) in the stratum corneum and the number of H-bonding groups attached to the permeant. (Data adapted from ref. [6].)

magnitude for each added H-bonding group up to three, thereafter, there is little effect (Fig. 4.2).

If a series of monofunctional permeants, taken from the permeability coefficient dataset, is examined, it is possible to gain a more specific idea of the effects of the functional group on K and D/I [6, 8]. Solvatochromic constants α (the H-bonding donor ability) and β (the H-bonding acceptor ability) can be considered in this type of analysis. It is seen that both an increase in α or β decreases permeability with β having the greater effect (the coefficients being -1.37 and -4.53, respectively). In terms of D/I, the histogram in Fig. 4.3 shows the relative contributions of some of the more common functional groups. The presence of a carboxylic acid group has the largest effect on retarding diffusion. It has not been possible to rationalize the magnitudes of more than one group per molecules due to the paucity of data and the complexity in looking at proximity effects.

4.4 FICK'S SECOND LAW

Albery and Hadgraft [9, 10] obtained solutions to Fick's second law of diffusion to examine the mechanism by which simple nicotinate esters

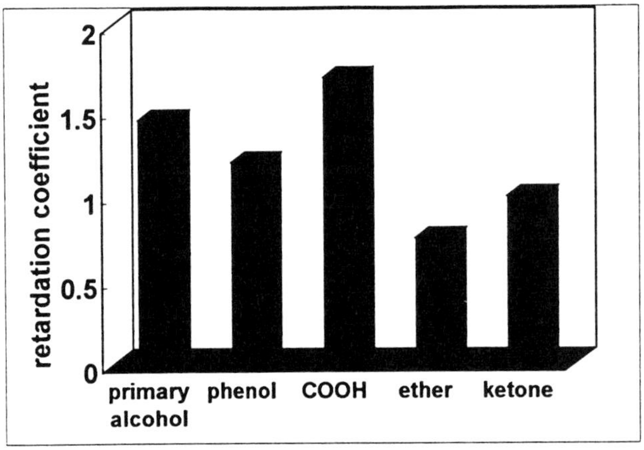

Fig. 4.3. The relative magnitudes of the retardation coefficients for monofunctional compounds. (Data from ref. [7].)

penetrate the skin. This group of compounds was chosen since they rapidly cross the skin and a known triggering concentration is required to elicit a vasodilatation of the underlying blood vessels. This manifests itself as a visible red response, an erythema. The time of onset of the erythema (t_e) is a characteristic of the nicotinate, its concentration and the formulation in which it is applied. A detailed analysis showed that the experimental data were best fit by equation (4.6)

$$\log\left(\frac{n_e}{I}\right) = \log(ct_e^{3/2}) + \log\left(\frac{K}{1+K\sqrt{p}}\right) + \log\left[\frac{8}{\sqrt{\pi}}\left(\frac{D}{I^2}\right)^{3/2}\right] - \frac{I^2}{9.2Dt_e}, \tag{4.6}$$

where n_e is the number of moles of nicotinate required to trigger erythema, I is the diffusional pathlength, K is the skin-applied phase partition coefficient, and p is the ratio of the diffusion coefficients in the skin (D) and the applied phase. The fit of the data to this equation indicated that the route of permeation was via the intercellular channels and that a tortuous route was involved. This has been confirmed subsequently for other compounds. Visualization studies on butanol [11] and mercury chloride [12] have shown localization of these substrates in the intercellular channels. A thermodynamic evaluation of the transport of water across skin has also indicated a very tortuous route for this small permeant [13]. A recent mathematical model of skin permeation has also shown the significance of the tortuosity in transdermal drug penetration [14].

The advances in analytical techniques (sensitivity and separation) have allowed identification of the contents of the intercellular channels. It has revealed that the intercellular domains contain a complex mixture of lipids that is atypical of normal biological membranes. There are no phospholipids present but there is a mixture of ceramides, cholesterol (and its sulphate), and free fatty acids (and their esters). Importantly, these arrange themselves into stacked bilayer arrays [15] that have a rigid structure (see Fig. 4.1). (The microviscosity has been examined by ESR [16] and fluorescence spectroscopy [17]; it has been demonstrated that the methylene groups close to the head groups of the ceramides form a comparatively solid environment. It is thought that the ceramide head groups associate strongly through hydrogen bonding. This will hold together the alkyl chains on adjacent lipids).

The reasons, therefore, for the excellent barrier properties of the skin are the tortuous route and that a penetrant has to cross sequentially a large number of rigid bilayers. The role of the lipids in topical delivery, how they control water transfer through the skin, and how biophysical techniques have been used to examine them have been reviewed [18–20].

4.5 PENETRATION ENHANCEMENT

To improve topical therapy, it is advantageous to use formulation additives (penetration enhancers) that will reversibly and safely modulate the barrier properties of the skin. Fick's first law of diffusion shows that two potential mechanisms are possible. The two constants that could be altered significantly are the diffusion coefficient in the stratum corneum and the concentration in the outer regions of the stratum corneum. Thus, one of mechanisms of action of an enhancer is for it to insert itself into the bilayer structures and disrupt the packing of the adjacent lipids, thereby, reducing the microviscosity. The diffusion coefficient of the permeant will increase This effect has been observed using ESR and fluorescence spectroscopy [16, 17].

Inspection of equation (4.6) shows how the effect of an enhancer on the diffusional barrier may be quantified. If an experiment is conducted such that the boundary conditions for the equation are met, a graph can be constructed of log ($ct^{3/2}$) against ($1/t$). The gradient of the line will indicate changes in the diffusional properties of the membrane. This technique has been used to study the influence of urea on skin permeation [21].

Further solutions of Fick's second law can be used to probe the way in which the concentration of the permeant varies as a function of position in the tissue. Relatively straightforward solutions can be obtained for a homogeneous membrane but there is evidence that the barrier function may vary with depth into the stratum corneum. ATR-FTIR techniques have been used to examine the –CH symmetric and asymmetric stretching frequencies of the intercellular lipids. These should provide a measure of the lipid viscosity. Bommannan *et al.* [22] stripped the skin repeatedly using adhesive tape. After three or four tape strips, the wavenumber of the –CH stretch frequency decreased by $2\,cm^{-1}$—thereafter, there was little change. This implies an ordering of the lipids with increasing depth into the stratum corneum. Watkinson *et al.* [23] used this information to show how this may affect concentration profiles in the skin. Similarities were shown between the predicted profile and that gained experimentally for a novel psoriatic agent MZPES. In general, however, it has proved sufficient to model the stratum corneum as if it were a homogeneous membrane. The wide biological variability in skin barrier properties often precludes exact mechanistic evaluations.

A slightly different kinetic analysis has been conducted to differentiate the mechanism of action of penetration enhancers. Briefly the following ATR-FTIR experiment was conducted. The skin sample was placed on a ZnSe crystal mounted in a FTIR spectrometer. A cell is placed on top of this that contains a saturated solution of the permeant in a suitable solvent. If, for example, a cyano-compound is presented in propylene glycol, both will permeate the membrane and their appearance will be detected by the increasing signals associated with –CN and –OH with time. Since a saturated solution is used, the absorbance time profiles are sigmoidal in shape with the long time plateau value being proportional to the solubility of the permeant in the membrane. Solutions to Fick's second law of diffusion can be fit to the data using appropriate software and give both diffusional and partitioning parameters. It is clear that two types of solvents exist, those that influence diffusional properties and those that influence partitioning. The use of this technique has allowed, for the first time, the differentiation of the mechanism of action of two well-known penetration enhancers, Azone® and Transcutol®. The former acts by inserting itself into the skin lipids and renders them more fluid; the latter changes the solubility parameter of the skin lipids and acts by improving drug solubility in the stratum corneum [24].

A kinetic consideration of Fick's first law also shows that it is possible to obtain synergy between the two types of enhancement strategies. If the skin is treated with an enhancer that increases D by a factor of five, and one that will increase the concentration gradient across the skin by a factor of five, there should be an overall 25-fold increase in the drug flux. This type of synergy has been shown; for example, metronidazole is synergistically enhanced by the presence of Azone and propylene glycol [25] and prostaglandins by mixtures of Azone and propylene glycol [26].

Care has to be taken when considering simple concentrations of the permeant since the driving force for diffusion is really the chemical potential gradient. As stated above the maximum flux should occur for a saturated solution of the permeant. However, if supersaturated solutions are applied to the skin, it is possible to obtain enhanced fluxes [27]. This can only be true if the outer skin lipids are capable of sustaining a supersaturated state of the diffusant. Figure 4.4 shows the linear increase in skin permeation with degree of supersaturation, and Fig. 4.5 demonstrates

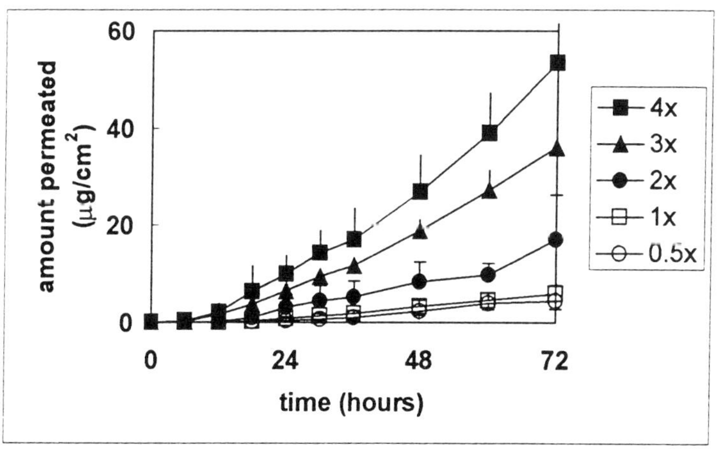

Fig. 4.4. Cumulative amounts of piroxicam diffused across full-thickness skin from different degrees of supersaturation (mean ± se). (Data adapted from ref. [27].)

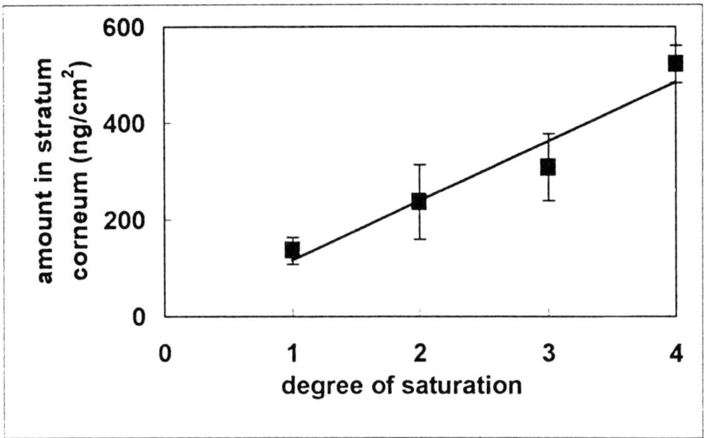

Fig. 4.5. The amount of piroxicam that accumulates in the stratum corneum after application of supersaturated solutions for 24 h (mean ± sd). (Data adapted from ref. [28].)

the uptake of the permeant into the skin [28]. It is possible to combine supersaturation with enhancer strategies that increase D and also obtain synergistic effects [29].

4.6 *IN VITRO–IN VIVO* MODELLING

It would be very useful to be able to estimate skin permeation *ab initio*, examining the physicochemical properties of the permeant and the formulation into which it is incorporated, and estimating the amounts that would get into the body. Although simple approaches can be used, such as those described above to predict the permeability coefficient, it is not possible to calculate, with confidence, permeation from formulations. These types of calculations are required in order to assess topical and transdermal drug delivery and to obtain risk assessments for cosmetic materials and agrochemicals etc. *In vitro* techniques have been developed and, coupled with an appropriate kinetic analysis, these can provide reasonable *in vivo* predictions. It is important to conduct a well designed *in vitro* experiment in order to obtain a measure of the steady-state flux of the drug across the skin [30]. Also it is important to recognize any limitations in the model. For example, it is possible that the xenobiotic will be subject, *in vivo*, to metabolism as it crosses the skin. *In vitro* this effect may be reduced or more likely absent. Once a measure of the steady-state flux has been obtained, this is equivalent to the input rate into the body. This can be equated to known clearance kinetics of the xenobiotic, and in the same way that is used for sequential first order chemical kinetics, the steady-state concentration in the middle compartment (the plasma) is calculated. This approach has been successfully used in a number of instances for predicting the steady-state plasma levels that are achievable following transdermal delivery. One such example is for nitroglycerin [31].

4.7 CONCLUSIONS

Transfer rates of molecules across the skin can be modelled using basic kinetic equations and appropriate solutions to Fick's Laws of diffusion. They have been applied to elucidate the mechanism by which molecules cross the skin and how the barrier function may be modulated. It is possible to absorb formulation components into the outer layers of the skin such that they enhance or retard penetration [32]. Even though considerable effort has been given to understanding these mechanisms of action, the precise route has still not been unequivocally identified. Part of the problem is the inherent variability of the skin. Despite this, predictive models have been obtained that have considerable utility in risk assessment and in the development of topical and transdermal medicines and their formulations.

REFERENCES

[1] A.S. Michaels, C. Chandrasekaran and J.E. Shaw, AIChE J. 21 (1975) 985–996.
[2] R.O. Potts and R.H. Guy, Pharm. Res. 9 (1992) 663–669.
[3] S.C. Valvani, S.H. Yalkowsky and T.J. Roseman, Pharm. Sci. 70 (1981) 502–507.
[4] T. Yano, A. Nakagawa, M. Tsuji and K. Noda, Life Science 39 (1986) 1043–1050.
[5] W.J. Pugh and J. Hadgraft, Int. J. Pharm. 103 (1994) 163–178.
[6] M.S. Roberts, W.J. Pugh and J. Hadgraft, Int. J. Pharm. 132 (1996) 23–32.
[7] W.J. Pugh, M.S. Roberts and J. Hadgraft, Int. J. Pharm. 138 (1996) 149–165.
[8] M.S. Roberts, W.J. Pugh, J. Hadgraft and A.C. Watkinson, Int J. Pharm. (1995) 219–233.

[9] W.J. Albery and J. Hadgraft, J. Pharm. Pharmacol. 31 (1979) 140–147.
[10] W.J. Albery and J. Hadgraft, J. Pharm. Pharmacol. 31 (1979) 129–139.
[11] M.K. Nemanic and P.M. Elias, J. Histochem. Cytochem. 28 (1980) 573 –578.
[12] H.E. Bodde, I. Vandenbrink, H.K. Koerten and F.H.N. Dehaan, J. Control. Release 15 (1991) 227–236.
[13] R.O. Potts an M.L. Francoeur, Proc. Natl. Acad. Sci. 87 (1990) 3871–3873.
[14] M. Heisig, R. Lieckfeldt, G. Wittum, G. Mazurkevich and G. Lee, Pharm. Res. 13 (1996) 421–426.
[15] K.C. Madison, D.C. Swartzendruber, P.W. Wertz and D.T. Downing, J. Invest. Derm. 88 (1987) 714–718.
[16] C.L. Gay, T.M. Murphy, J. Hadgraft, I.W. Kellaway, J.C. Evans and C.C. Rowlands, Int. J. Pharm. 49 (1989) 39–45.
[17] R.O. Potts, N.A. Azimi, T.S. Spencer, F.E. Lyttle and D.A. Chen, Acta Pharm. Nordica 4 (1992) 121–121.
[18] J. Hadgraft, K.A. Walters and R.H. Guy, Semin. Dermatol. 11 (1992) 139–144.
[19] P.M. Elias and K.R. Feingold, Semin. Dermatol. 11 (1992) 176–182.
[20] R.O. Potts and M.L. Francoeur, Semin. Dermatol. 11 (1992) 129–138.
[21] J. Beastall, R.H. Guy, J. Hadgraft and I. Wilding, Pharm. Res. 3 (1986) 294–297.
[22] D. Bommannan, R.O. Potts and R.H. Guy, J. Invest. Derm. 95 (1990) 403–408.
[23] A.C. Watkinson, A.L. Bunge and J. Hadgraft, Naik, A. Int. J. Pharm. 87 (1992) 175–182.
[24] J.E. Harrison, A.C. Watkinson, D.M. Green, J. Hadgraft and K. Brain, Pharm. Res. 1996, 13, 542–546.
[25] P.K. Wotton, B. Mollgaard, J. Hadgraft and A. Hoelgaard, Int. J. Pharm. 24 (1985) 19–26.
[26] A.C. Watkinson, J. Hadgraft and A. Bye, Int. J. Pharm. 74 (1991) 229–236.
[27] M.A. Pellett, A.F. Davis and J. Hadgraft, Int. J. Pharm. 111 (1994) 1–6.
[28] M.A. Pellett, M.S. Roberts and J. Hadgraft, Int. J. Pharm. 151 (1997) 91–98.
[29] M.A. Pellett, A.C. Watkinson, K.R. Brain and J. Hadgraft, in Perspectives in Percutaneous Penetration, Vol. 5a, eds K.R. Brain, V.J. James and K.A. Walters (STS Publishing, Cardiff. 1997).
[30] J.P. Skelly, V.P. Shah, H.I. Maibach, R.H. Guy, R.C. Wester, G. Flynn and A. Yacobi, Pharm. Res. 4 (1987) 265–267.
[31] J. Hadgraft, D. Beutner and H.M. Wolff, Int J. Pharm. 89 (1993) R1–R4.
[32] J. Hadgraft, J. Peck, D.G. Williams, W.J. Pugh and G. Allan, Int. J. Pharm. 141 (1996) 17–25.

Chapter 5

Electron Transport and Two-Dimensional Organization of Metalloprotein Adsorbates Investigated by Cyclic Voltammetry and *In Situ* Scanning Tunnelling and Atomic Force Microscopy

ESBEN P. FRIIS, JENS E.T. ANDERSEN, LARS L. MADSEN, PER MØLLER, MARIANNE H. THUESEN, NIELS H. ANDERSEN and JENS ULSTRUP

5.1 INTRODUCTION

Functional electron transfer (ET) patterns of metalloproteins in homogeneous solution have been characterized in great detail [1–10]. Target systems have been small single-metal redox proteins, i.e., the cytochromes, blue copper proteins or iron-sulphur proteins [1–7], and multicentre proteins and protein complexes such as cytochromes c_4 [8] and c_3 [9], copper oxidases [10–12], cytochrome *c* oxidase [13–15], nitrogenases [16] and the bacterial photosynthetic reaction centres [17, 18]. This has also warranted new theoretical work where features such as driving force and environmental effects [5, 19–21], directional long-range electron tunnelling through "matter" [5, 6, 22–26], fluctuational tunnel effects [6, 27], vibrational coherence [28, 29], and ET coupled to critical behaviour of the environment [30] have come to stand forward. Protein function is also broadly associated with the adsorbed state at solid–liquid interfaces [31]. Most central in adsorbed metalloprotein ET function has been interfacial electrochemical ET [32–34]. The adsorbed redox metalloprotein state is characterized here in much less detail than the solute state. Investigations are fraught with principal and preparative difficulties such as inhomogeneous adsorption patterns, protein instability, electric fields at the interface etc. In addition, theoretical approaches particularly to the electronic tunnel factor of molecular ET processes in bulk solution and at electrochemical interfaces, are rooted in the different areas of molecular quantum mechanics and surface density functional theory, respectively [6, 35–38]. Prospects for molecular resolution of composite electrochemical interfacial ET patterns in multicentre redox proteins, such as metalloenzymes are, however, also beginning to emerge here [39, 40].

Scanning tunnelling microscopy (STM) and atomic force microscopy (AFM) have opened new perspectives for characterization of large adsorbate molecules including biological macromolecules in real time and space, to molecular or even submolecular resolution [41–45]. So far, nearly all reports on biological macromolecules, such as proteins and DNA, refer to the *ex situ* mode, i.e., vacuum or ambient gas environment. Aqueous solution is, however, the natural medium for protein function, and water an integrated element of all soluble proteins. This has raised recognized fundamental and technological issues. The former relate to the nature of the tunnel phenomenon in the structurally fluctuating medium [46], the role of the metal atoms in resonance, coherent and activated tunnel channels [47–50], and the translational and internal mobility of the "soft" target adsorbate molecules [51, 52]. Technical issues are related to the independent, bipotentiostatic control of substrate and tip potentials, and to the coating of the tip. Both have the aim of suppressing parasitic Faradaic currents that would interfere with the very small (pA to nA) tunnel currents [53].

In return, *in situ* STM/AFM holds exciting prospects for imaging and controlling chemical and biological processes at the molecular and submolecular level. This would itself be a new dimension in chemistry. In situ STM images of DNA bases [54], metalloporphyrins [55–57], and other large aromatic molecules, to submolecular or almost atomic resolution, have indeed been achieved. Cruder, but still molecular, resolution has been achieved for small redox metalloproteins and will be discussed in the present chapter. *In situ* AFM of several proteins to molecular resolution has also been reported and even extended to observation of conformational changes induced by enzyme-substrate complex formation in enzyme catalysis [44, 45]. Other important, technology-related perspectives of *in situ* STM/AFM of biological macromolecules would be associated, e.g., with (a) immunological antibody-antigene interactions; (b) mapping of immobilized redox metalloproteins *in parallel* with electrochemical function in biosensor and enzyme electrode environments; (c) mapping of functionally important macromolecular adsorption induced by biological liquids in contact with solid surfaces. Microbial corrosion is a large-scale example here; and (d) spatial and functional mapping of biomolecular interfacial interactions in the context of *molecular nanoscale science* is, finally, an area where *in situ* STM/AFM would be indispensable and add substantially to the disclosure of molecular function.

In this chapter, we overview first some recent examples of interfacial electrochemical ET of composite metalloproteins where molecular mechanistic detail has in some way been achieved. We discuss next some theoretical issues regarding *in situ* STM of large molecules, where resonance or environmentally activated tunnel channels are opened by the redox metal centre. This is followed by an overview of some recent achievements in the area of *in situ* STM/AFM of the single-metal proteins cytochrome *c* and azurin on polycrystalline and single-crystal platinum and gold surfaces. Such an integrated approach offers new perspectives for experimental and theoretical characterization of metalloproteins at solid surfaces in contact with the natural aqueous medium for metalloprotein function.

5.2 SOME OBSERVATIONS ON THE BEHAVIOUR OF PROTEINS AT SOLID–LIQUID INTERFACES

Protein function at solid–liquid interfaces holds a structural and a dynamic perspective [31]. The structural perspective addresses macroscopic adsorption, molecular interactions between the protein and the surface, collective interactions between the individual adsorbed protein molecules, and changes in the conformational and hydration states of the protein molecules induced by these physical interactions. Interactions caused by protein adsorption are mostly non-covalent but strong enough to cause drastic functional transformations. All these features are, moreover, affected by the double layer and the electrode potential at electrochemical interfaces. Factors that determine protein adsorption patterns have been discussed in detail recently, both in the broad context of solute proteins at solid surfaces [31], and in specific contexts of interfacial metalloprotein electrochemistry [34]. Some important elements that can also be *modelled* in suitable detail would be:

(1) The surface charge distribution of the protein. This is determined by

the primary structure but also by the environment (pH, ionic strength etc.).

(2) The hydrophobic or hydrophilic nature of the protein surface.

(3) The secondary and tertiary protein folding patterns. This could, for example control the conformational flexibility and the resistance of proteins to unfolding. Conformationally "soft" proteins would, for example, be expected to generally exhibit stronger surface affinity than "rigid" proteins.

(4) The protein solvation, and the internal and surface hydrogen bond structure.

Such controlling factors are matched by complementary properties of the solid surface, i.e., the hydrophobic or hydrophilic surface character, porosity and topology, charge, hydration, and the presence and composition of surface groups. Particularly for electrochemical surfaces intrinsic catalytic reactivity of the surface groups formed spontaneously or by potentiostatic control also follow. Both in the context of electrochemical protein reactivity, and in the broader areas of proteins at surfaces, surface control and modification to structural and functional compatibility with the proteins are key issues. The use of electrochemical promoters, illustrated below, is one such example [32–34].

Protein *dynamics* at surfaces exhibit a variety of patterns ranging from almost full activity to complete inactivation. Close to full activity, for example, is retained in covalently immobilized metalloenzymes such as alcohol dehydrogenases [58], in adsorbed monolayers of the laccases and [59–61] other composite multicentre metalloenzymes [32–34, 61], and in some covalently immobilized single-ET proteins such as cyt *c* [52]. In the context of protein interfacial electrochemistry, reversible or kinetically controlled interfacial ET in non-catalytic and electrocatalytic modes, is first a conspicuous reflection that protein activity is indeed retained on adsorption. This applies to a range of enzymes such as laccases, peroxidases, and flavoenzymes among which glucose oxidase is the most extensively studied [32, 33, 61–64]. Resolution of the molecular mechanisms of interfacial ET of composite enzymes can also be addressed. The identification of competing intramolecular ET channels and the coupling of a sequence of protonation equilibria to the catalytic and non-catalytic electrochemical ET modes of fumarate reductase and succinate dehydrogenase are important examples [39]. Other examples that qualify mechanistically in some way for the notion "molecular" are ET patterns of glucose oxidase modified by ET relays of ferrocenes to ensure facile electronic contact between the electrode and the flavin centres in the enzyme [63].

In the following, we proceed to several new aspects regarding molecular resolution of adsorbed metalloproteins. In Section 5.3, we first show that close to molecular resolution of the interfacial ET patterns of some *two-centre* proteins is within reach. The number of intermetallic interactions and resulting microscopic reduction potentials and rate constants in metalloproteins with more than two centres is, however, prohibitively large for such resolution. In Section 5.4 we address molecular and supramolecular adsorption. We show here that *in situ* STM and AFM, indeed do hold exciting new perspectives for this important and central structural aspect of proteins at surfaces.

5.3 APPROACHES TO ELECTROCHEMICAL ET MECHANISMS OF MULTICENTRE METALLOPROTEINS

We illustrate these aspects of metalloprotein dynamics at surfaces by two specific proteins. One is the *two-centre* bacterial di-heme protein cyt c_4 from *Pseudomonas stutzeri*, and the other is the fungal four-centre redox enzyme laccase from *Polyporus versicolor*.

5.3.1 Properties and cyclic voltammetry of *P. stutzeri* cytochrome c_4

Cytochrome c_4's are believed to constitute parts of bacterial respiratory chains. In the present context, the following properties specific to *P. stutzeri* cyt c_4 are important [8, 40]:

(a) The MW ≈ 19 kDa.
(b) The recently reported 2.2 Å X-ray crystal structure discloses two domains connected by a single peptide string (Fig. 5.1) [65]. The heme groups are, interestingly, in H-bond contact via two propionates, and 30° tilted. This motive is encountered in another di-heme protein, *chromatium vinosum* flavocytochrome *c* sulphide dehydrogenase [67]. His and Met are axial ligands in both hemes.
(c) The protein is strongly dipolar (pH ≈ 7) with four excess negative

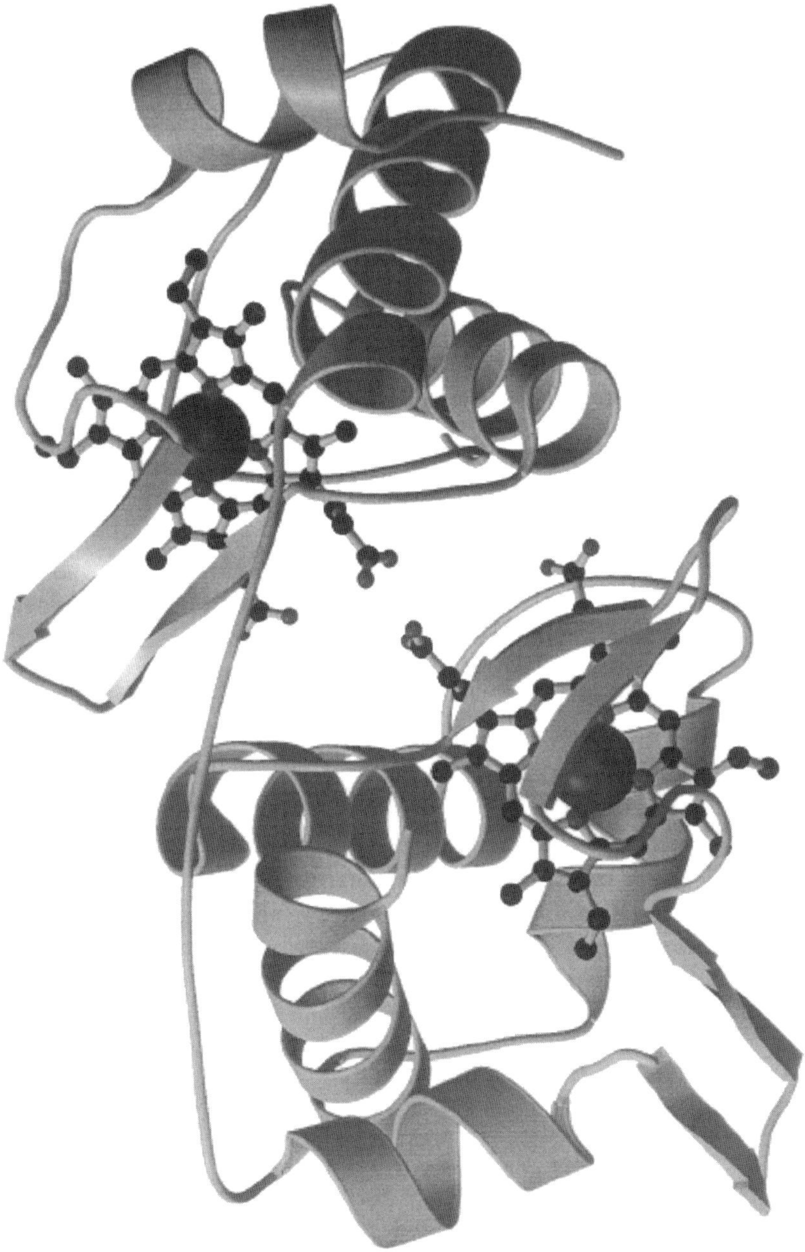

Fig. 5.1. 2.2 Å X-ray crystal structure of *P. stutzeri* cyt c_4 [65]. Molscript graphic representation [66].

charges in the N-terminal and one excess positive charge in the C-terminal domain. This is important for the protein orientation at solid surfaces and compatibility with electrochemical promoters (cf. below).

(d) The protein has been characterized by UV/VIS, NMR, and resonance Raman spectroscopy [8, 68, 69]. The oxidized form is, importantly, a low-spin/high-spin equilibrium mixture while the reduced form is pure low-spin. Therefore, spin conversion might add nuclear reorganization in the ET processes. This expectation has received some support from Gibbs free energy relations for ET between cyt c_4 and several transition metal complexes and flavine radicals spanning a free energy range of almost 1 eV [21]. In spite of the close heme group proximity intramolecular, ET is not, therefore, necessarily fast.

(e) The kinetics of ET between cyt c_4 and external reaction partners represented both by transition metal complexes [8] and organic radicals [21] (cf. above) has been characterized. Intermolecular ET is multiexponential and in the μs-ms time range. The macroscopic reduction potentials are ≈240 and ≈320 mV (SHE), with the lower reduction potential in the negatively charged N-terminal domain. Microscopic interaction potentials between the heme groups are negative and in the 30–40 mV range.

(f) The kinetics points to ET at both heme groups. An open question is the importance of intramolecular ET between the heme groups. The kinetic and NMR spectral data indicate that the rate constants k_{int} (s^{-1}) accord either with $k_{int} \ll 10^2\ s^{-1}$, or with $10^2\ s^{-1} \ll k_{int} \ll 10^5\ s^{-1}$. In the following, we show that cyclic voltammetry of cyt c_4 holds a clue here.

Metalloprotein voltammetry often requires organic or inorganic chemical promoters [32–34]. A common promoter structure is X–Y where X adsorbs to the electrode and Y links to the protein by H-bond or other contacts. By this surface modification, the protein is prevented from denaturation at the surface. Broad classes or promoters are, moreover, specific *either* to positively *or* negatively charged proteins. Figure 5.2 shows three promoters used in a recent study of cyt c_4 voltammetry [40]. Cystamine is specific to negatively and *bis*-4,4′-pyridyl-bisulphide (4,4′bpySS) to positively charged proteins at gold surfaces. These molecules are adsorbed via the sulphur atoms. The 3-pyridinylmethylene-hydrazinecarbothiamide (3-PMHC) is a usable promoter for either group of proteins.

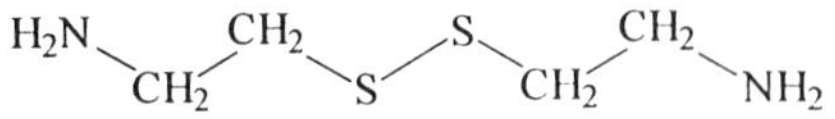

Cystamine
(Negatively charged proteins)

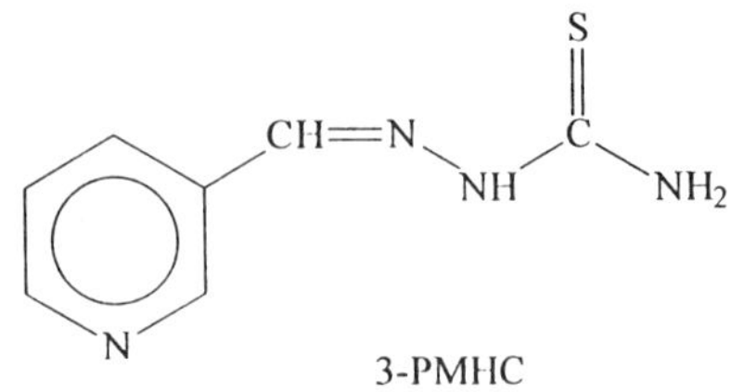

3-PMHC
(Either protein group)

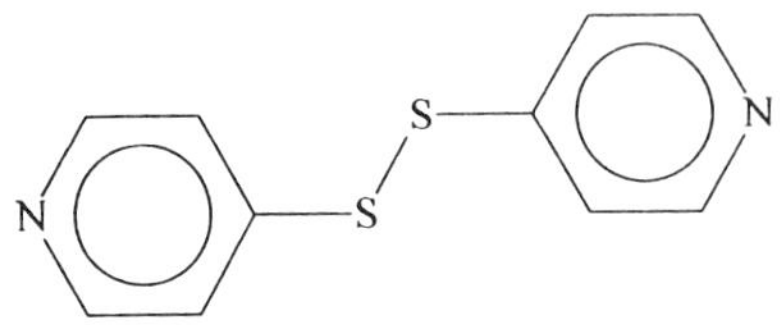

4,4'bpySS
(Positively charged proteins)

Fig. 5.2. Organic promoters for cyclic voltammetry of metalloproteins.

There is no cyt c_4 voltammetry at gold electrodes in promoter-free solution, but stable voltammograms are observed (Fig. 5.3) when the electrode has been pretreated with any of the three promoters shown in Fig. 5.2. Two closely spaced, nearly reversible, peaks are observed at high ionic strength (0.1 M). This is indicative that the electrostatic component of the promoter effect is screened, while other effects still ensure protein surface stabilization. A rather different character emerges at low ionic strength (0.01 M). The voltammograms are now partly kinetically controlled and also show interesting asymmetry in the presence of the protein charge specific promoters cystamine and 4,4′bpySS. The former promoter gives *two* peaks in the *anodic* scan but a *single*, higher peak in the *cathodic* scan. The latter promoter gives a similar pattern but with the opposite peak distribution.

The observations best reflect the view that cyt c_4 is adsorbed in a promoter-specific orientation at low ionic strength. The negatively charged, low-potential N-terminal domain then points towards the electrode when cystamine is present. As the potential is scanned anodically, the adjacent heme group is oxidized first, followed by oxidation of the remote group at a higher potential, by ET via the adjacent group. As the scan is reversed, reduction of the remote, high-potential heme is hampered, however, by the adjacent group until the reduction potential of the latter is reached. *Two* merging single-ET steps then occur within a narrow potential range. A similar, suitably reversed pattern emerges when 4,4′bpySS is the pro-

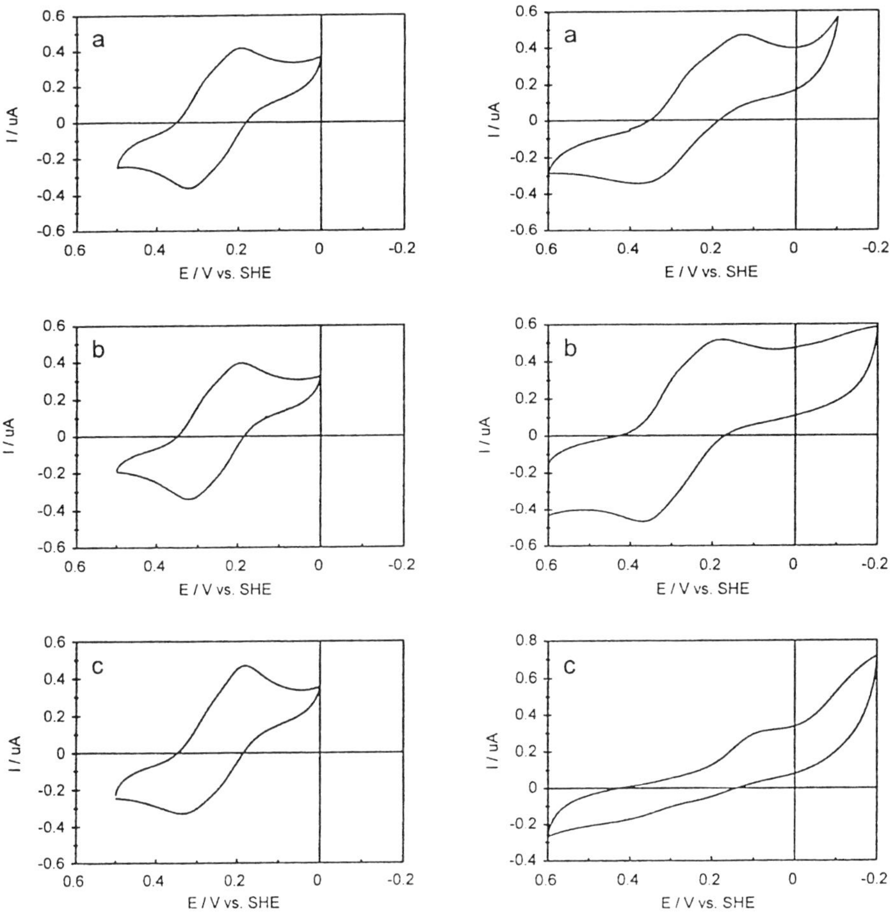

Fig. 5.3. Cyclic voltammograms of cyt c_4 on a gold electrode [40]. Top: 3-PMHC; middle: 4,4′bpySS; bottom: cystamine (cf. Fig. 5.2). Left column: 0.1 M phosphate. Right column: 0.01 M phosphate. pH 7.5. Scan rate: $10\,mV\,s^{-1}$.

moter. The voltammetric *shapes* are, furthermore, determined by the intramolecular ET rate when the latter is in the same time range as the potential scans.

Data such as those in Fig. 5.3 have been addressed by numerical methods from which the reduction potentials, interfacial ET rate constants *and* the intramolecular ET rate constants can be obtained [40]. We note particularly that:

(1) Such analysis is feasible because as a *two-centre* protein cyt c_4 is "simple" enough for fair mechanistic resolution but still holds the fundamental features of intramolecular ET and other interaction between the heme groups.
(2) The intramolecular ET rate constants are 1600 s^{-1} (low- to high-potential heme) and 40 s^{-1} (high- to low-potential heme). As expected from the accompanying spin change (cf. above) and the small driving force, these values are low compared with many data for strongly exothermic photoinduced intramolecular ET processes in metalloproteins [5, 19, 20].

The rate constants also hold a useful clue to concentration and time ranges for new investigations of intermolecular ET processes of cyt c_4 in solution, from which the intramolecular rate constants may be substantiated. Voltammetry, therefore, stands forward as a powerful tool towards otherwise elusive ET dynamics in composite metalloproteins.

5.3.2 Direct and catalytic voltammetry of P. versicolor laccase

The laccases are tree and fungal copper oxidases that are catalysts for the conversion of molecular dioxygen to water,

$$O_2 + 4H^+ + 4e \rightleftarrows 2H_2O\,, \tag{5.1}$$

by reduction with polyphenols and other organic substrates. Chemical and biochemical properties have been extensively reviewed [70–72]. Laccase contains four copper atoms. These represent the three common biological classes, i.e., T1, T2 and T3 and are all engaged in the four-electron reduction of O_2. Other observations are:

(a) The MW $\approx$ 70 kDa.
(b) No crystal structure is available but reference to the closely related ascorbate oxidase is frequently made. The T1 centre in this enzyme is separated by 12 Å from the structurally and electronically linked T2/T3 centre [73]. It is the site for the reducing substrate while the T3 centre

is the site for oxidation and O_2-binding. The two sites are connected by a conspicuous covalent facile electron tunnelling route. An interesting and highly important structural observation is that the bond length between the two copper atoms in the T3 centre of ascorbate oxidase increases by ≈1.5 Å on reduction of the T3 centre, i.e., the bond is virtually dissociated. This could impose drastic conformational changes, with parallel changes in the enzyme activity.

(c) Kinetic and spectroscopic properties have been investigated in detail. Long-range ET between the T1 and T2/T3 centres, and the accumulation of intermediates in the different oxidation states have been key issues. O_2-binding also induces drastic increases in the enzyme activity (cf. above) [70–72]. Much in these observations seems to carry over to the catalytic and non-catalytical *electrochemical* electron exchange of the laccases.

(d) Reports of anaerobic laccase voltammetry under specified conditions are sparse, but there are a number of studies of surface immobilized laccase catalysis of electrochemical O_2-reduction [60–62, 74].

Figures 5.4 and 5.5 summarize results of a recent study of *P. versicolor* laccase electrochemistry based on cyclic and rotating disk voltammetry [60]. Figure 5.4 shows unequivocally that this laccase is voltammetrically active and gives a kinetically controlled, unpromoted four-electron peak at edge-plane pyrolytic graphite. Electrochemical reduction of O_2 catalyzed by an immobilized laccase monolayer is close to reversible, and unrestricted by mass transport. The electrocatalysis follows, moreover, a Michaelis–Menten pattern (Fig. 5.5). Finally, there is a characteristic bell-shaped functional pH-profile with a pronounced maximum at pH ≈ 3.1.

The data strongly imply that intramolecular ET is a feature of both the overall catalytic and non-catalytic process. The four-centre nature of laccase, the large number of interactions, microscopic reduction potentials and rate constants, and the accompanying protonation equilibria precludes a detailed analysis here, such as for cyt c_4. In the steady-state limit of electrocatalytic O_2-reduction the rate determining step is thus most likely either one of the intramolecular ET steps, or liberation of a bound product water molecule. From the data values of these rate constants, the range 20–40 s^{-1} emerges. This is the same *range* but smaller than reported values for laccase activity in homogeneous solution. Suitably conducted, direct

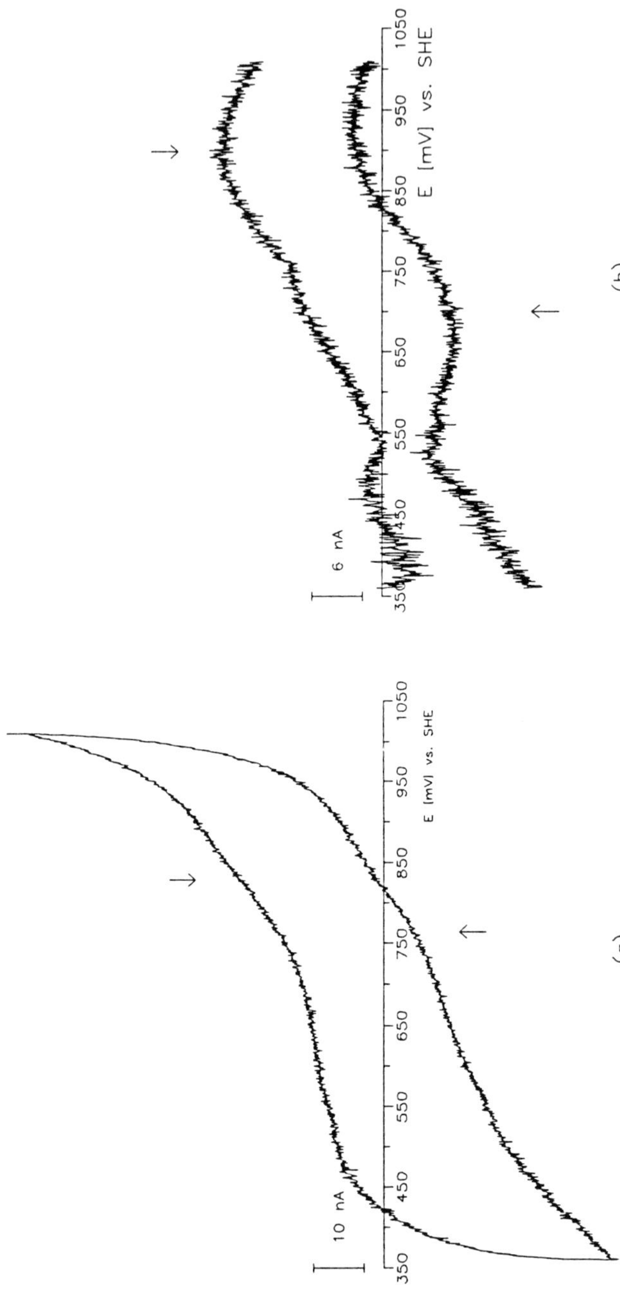

Fig. 5.4. Left: Anaerobic cyclic voltammogram of 2.5 pmol (≈ one monolayer) *P. versicolor* laccase on pyrolytic graphite electrode. pH = 3.31, ionic strength 0.1 M. Right: background corrected signal [60].

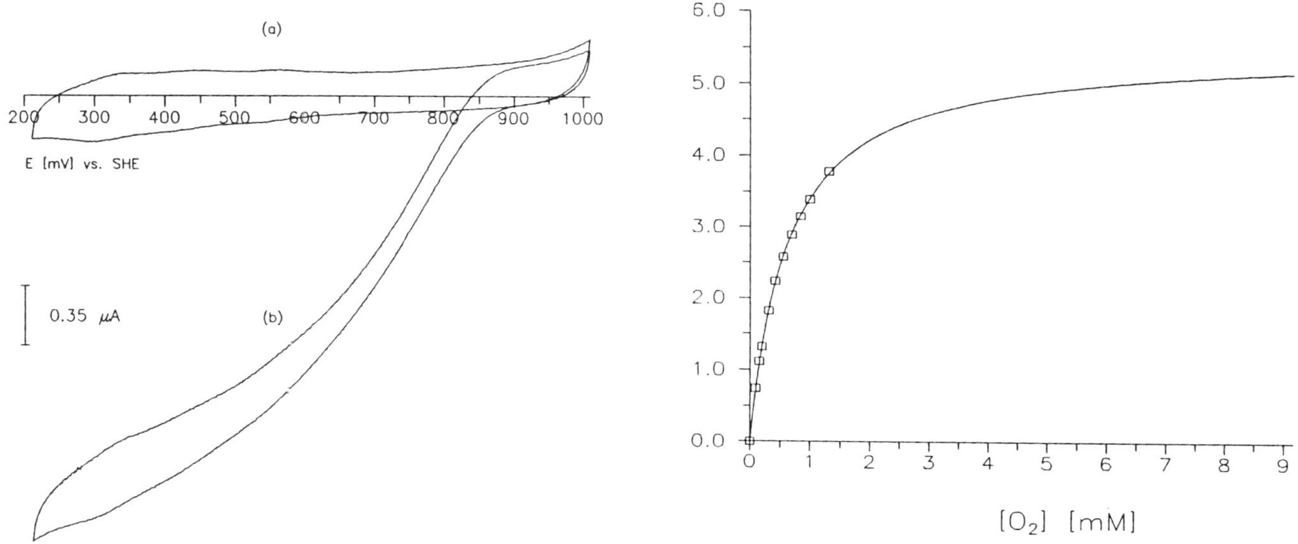

Fig. 5.5. Left: Catalytic voltammogram of dioxygen reduction at 17 pmol *P. versicolor* laccase on pyrolytic graphite. pH = 3.82, ionic strength 0.1 M. (a) no dioxygen; (b) 1.2 mM dioxygen. Right: Electrochemical Michaelis–Menten plot of data for variable dioxygen concentration [60].

and catalyzed voltammetry of multicentre metalloenzymes can be undertaken, therefore, to some degree of mechanistic detail.

5.4 *IN SITU* AND AFM OF SINGLE-METAL METALLOPROTEINS; HORSE HEART CYTOCHROME *C* AND *PSEUDOMONAS AERUGINOSA* AZURIN

STM and AFM offer new prospects for direct visualization of molecular and biomolecular entities in the resting states and in chemical action. Fundamental and technological perspectives of this were noted above. So were recognized fundamental, preparative and technological difficulties, associated with the configurational shift from the *ex situ* to the *in situ* mode. In this section, we overview some approaches to *in situ* STM and AFM imaging of the two single-metal redox metalloproteins horse heart cytochrome *c* and *P. aeruginosa* azurin. The approaches involve different kinds of immobilization and substrate, giving images at molecular resolution. The images are, however, robust to different degrees and give different surface adsorption patterns. We first discuss a few elements of *in situ* STM theory that also illustrate why metalloproteins offer attractive targets for long-range ET in the three-level *in situ* STM configuration.

5.4.1 The role of redox metal centres in the in situ *STM configurations*

In situ STM of metalloproteins with localized low-lying redox levels can be expected to follow ET patterns similar to metalloprotein ET in homogeneous solution and at electrochemical surfaces. The redox level is thus strongly coupled to the protein and solvent environment. A key notion is that the vacant local level (oxidized form) at equilibrium with the environmental nuclear motion is located well *above* the Fermi levels of both the substrate and tip, whereas, the occupied level (reduced form) at equilibrium is located well *below* the Fermi levels. Another central notion is that the local redox level at the transition metal centre is still much lower than environmental protein or solvent electronic levels. The redox level therefore constitutes a pronounced indentation in the tunnel barrier. This alone would strongly enhance tunnelling. Configurational *fluctuations* in the environment can, secondly take the redox level to such low values that temporary physical population occurs. This requires nuclear activation but can still be favourable due to the much shorter electron tunnel distances in such a two-step process. These views are illustrated in Fig. 5.6 [47, 48]. Environmental configurational fluctuations thus take the initially equilibrated vacant local level from positions well above the Fermi levels of both substrate and tip, down in energy and close to resonance with metallic electronic levels around the Fermi level of the negatively biased electrode (highest Fermi level) The first ET step is likely to occur around this region. Relaxation towards the equilibrium configuration of the occupied (reduced) redox level well *below* the Fermi levels continues subsequent to the first ET step. The STM process is completed by a second ET step where the electron is transferred from the occupied redox level to vacant metallic electronic levels around or above the Fermi level of the positively biased electrode (lower Fermi level), restoring the original molecular nuclear configuration. The potential surfaces corresponding to such a sequence are shown in Fig. 5.6b. Completion of the STM process by the second ET step can happen, however, in at least four physically different modes:

(1) Configurational relaxation to the intermediate equilibrium configuration of the occupied (reduced) redox level. Completion of the second step then requires renewed environmental activation, and the overall process can be viewed as two consecutive, fully equilibrated ET steps according with common formal consecutive schemes of chemical kinetics. This limit would prevail when the electronic coupling (tunnel distance) between the redox metal centre and the positively biased electrode is weak (long) so that the ET probability during intermediate state configurational relaxation is small [47, 48].
(2) When the electronic coupling (tunnel distance) is strong (short) the transition acquires a different character. The electron can proceed "directly" from a given metallic electronic donor, say substrate level to a resonating acceptor (tip) level via the local redox level in a nuclear configuration where the latter resonates with *both* the metallic donor and acceptor levels. Nuclear configurations corresponding to any local redox level position between the two Fermi levels contribute to this mechanism [49, 75].
(3) In the limit of strong interactions the overall process can follow a multitude of forward and reverse transitions between the redox level and either set of metallic electronic levels along the motion on the

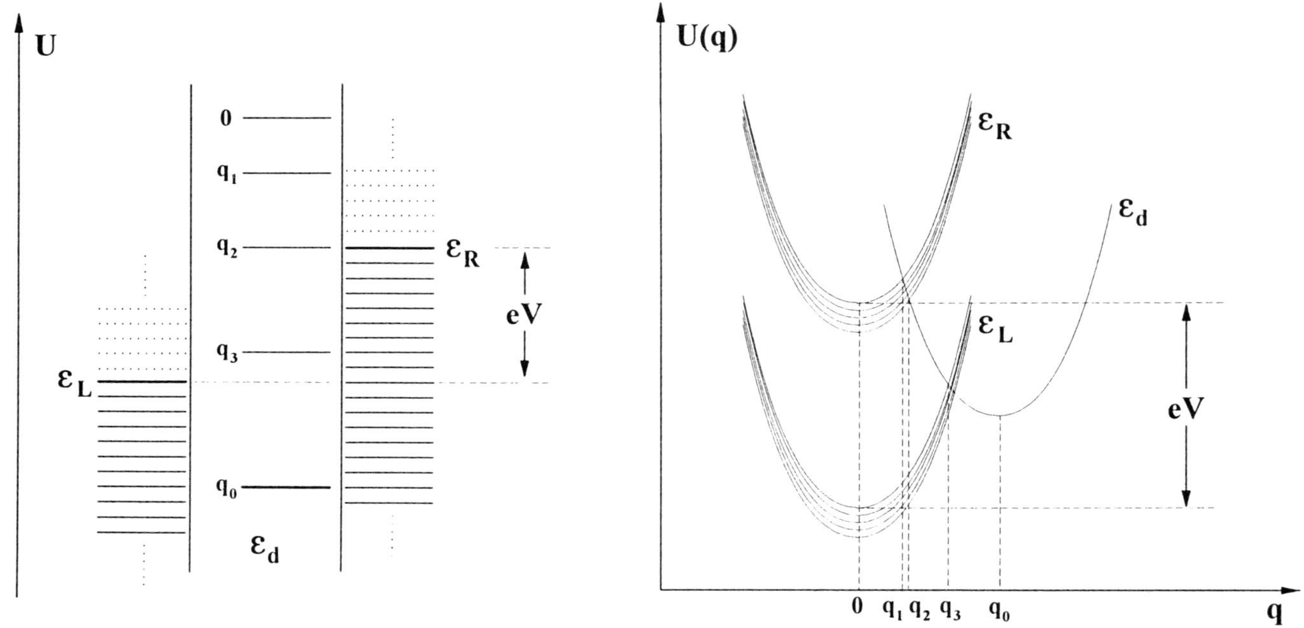

Fig. 5.6. Energy diagram of *in situ* three-level electronic-vibrational STM configuration. Left: The manifolds of electronic energy levels of the positively (left) and negatively (right) biased electrode. The Fermi levels are ϵ_L and ϵ_R, respectively. The discrete molecular level in the gap region assume different values depending on the fluctuating nuclear coordinate q. $q = 0$ corresponds to equilibrium in the initial, vacant state. $q = q_0$ corresponds to equilibrium in the final, occupied state. Resonance or coherent tunnelling can proceed via level positions corresponding to q-values between $q = 0$ and $q = q_0$. Right: Potential surface configurations corresponding to the level system to the left.

potential surfaces. This is the *adiabatic* limit where there is no explicit appearance of the electronic tunnel factors in the tunnel current.

(4) Intermediate strength electronic couplings (tunnel distances) give, finally, still another physically different mechanism. Nuclear relaxation again begins after the first ET step but the second step now follows *before* full intermediate state relaxation is reached. The two ET steps occur vibrationally *coherently* here, with the second transition in an activationless, strongly non-equilibrated mode.

Analogous considerations apply to hole transfer where an occupied redox level *below* the two Fermi levels is first thermally activated to values above the Fermi level of the positively biased electrode where ET to this electrode occurs. This is followed by ET from the negatively biased electrode to the, now vacant, redox level.

The observations illustrate that inelastic and thermally activated tunnel channels may apply to metalloproteins and large transition metal complexes. The channels hold perspectives for mapping protein structure, adsorption and electronic *function* at metallic surfaces. One observation regarding the latter is, for example that the two electrode potentials can be varied in parallel, relative to a common reference electrode potential, at fixed bias potential. This is equivalent to taking the local redox level up or down relative to the Fermi levels (Fig. 5.6a). If both electrode potentials are shifted negatively, and the redox level is empty (oxidized), then the current at first rises. It reaches a maximum, convoluted with the bias potential *between* the two Fermi levels, and then drops as further potential variation takes the redox level below the Fermi level of the positively biased electrode. The relation between such current-voltage patterns and other three-level processes, such as molecular resonance Raman scattering [76], has been discussed [38].

5.4.2 In situ *STM of horse heart cytochrome*

Cytochrome *c* is a single-heme protein, MW ≈ 12 kDa, fairly universal in respiratory ET chains [77]. The reduction potential is 265 mV (SHE, horse heart). ET reactions in homogeneous solution and promoter-induced reversible cyclic voltammetry are well characterized. Horse heart cyt *c* has been in recent focus in some of the first STM investigations of metalloproteins *in situ* [51, 52, 78]. Figure 5.7 shows patterns for non-covalently assembled cyt *c* on flame annealed gold [51]. Immobilization is at low ionic strength. No images can be observed at high ionic strength that are more representative of cyt *c* electrochemistry. At low ionic strength cyt *c* assembles in laterally extended aggregates that contain 100+ molecules. The aggregates exhibit dynamic features such as pinhole formation and annihilation. Smaller structures of the size of individual cyt *c* molecules can, however, sometimes be observed *between* the larger structures. These could well be individual *molecules* possibly separated from the aggregates by tip interference.

The balance between immobilization and aggregation is, however, fragile. Covalent linking gives more confined mobility with prospects for better resolution. Such conditions are, on the other hand, less suitable for electrochemical characterization. Cytochrome *c* can be attached to Au(111) by N-acetylcysteine chemisorption and subsequent carbo-diimide coupling. Polycrystalline platinum surfaces are broadly composed of large (up to several hundred nm) areas that are also almost atomically flat. Cytochrome *c* can be attached to these areas by silanization to surface oxides and binding via glutaric aldehyde. Cytochrome *c* is reported to retain redox activity towards xanthine oxidation in either case [79]. Figure 5.7 also shows a high-resolution image of cyt *c* immobilized covalently on polycrystalline platinum by this procedure [52]. The surface layer is quite different from both the bare surface and the surface exposed solely to the linker molecules. Spots in the size range 4–15 nm are scattered all over the surface, the broad topological features of which can be distinguished, however, *underneath* the adsorbates (not shown). The pattern is robust and independent of the electrode potential over several hundred mV across the formal reduction potential of cyt *c*. Similar kinds of observation are encountered when cyt *c* is covalently immobilized on Au(111). The smaller structures accord with the crystallographic size of cyt *c*. Observation of laterally larger structures, corresponding in size to a few molecules is most likely caused by limited resolution when the tip is close to the irregular dense surface layer. The regular lateral organization of azurin (cf. below) thus gives a more uniform tip contact and better lateral resolution. Altogether covalent immobilization of cyt *c* gives improved image quality compared with the spontaneously, non-covalently assembled layers. The covalent immobilization procedure is, finally general and expected to apply

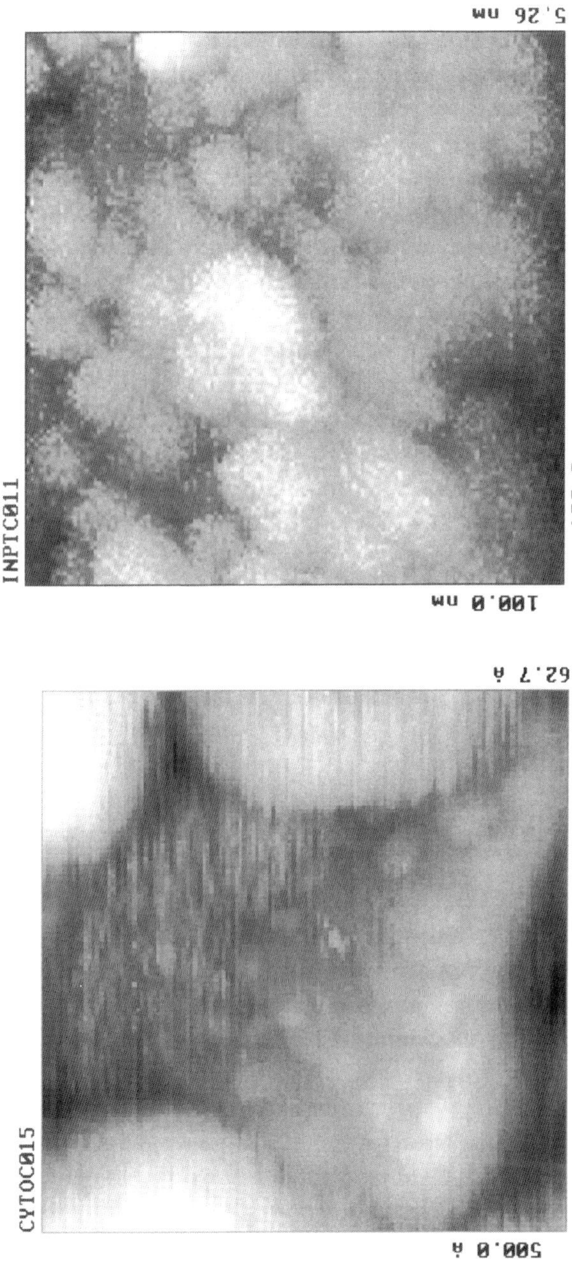

Fig. 5.7. Left: *In situ* STM image of non-covalently immobilized cyt *c* (low ionic strength). Small molecular-size structures between larger (≈100 mol) flat aggregates. Tunnel current 1.25 nA (other details in ref. [51]). Right: *In situ* STM image of cyt *c* covalently immobilized on polycrystalline platinum by silanization and glutaric aldehyde coupling. Tunnel current 1.78 nA. Substrate potential −500 mV (Au/AuO_x). Bias voltage 100 mV (other details in ref. [52]).

broadly to proteins, and perhaps to imaging of suitably chosen proteins in chemical action.

5.4.3 In situ AFM and STM of P. aeruginosa azurin on gold(111)

The bacterial single-copper protein azurin (*P. aeruginosa*) has a surface disulphide group (Cys3Cys26) opposite to the strongly asymmetrically located copper atom (Fig. 5.8) [80, 81]. This group is a suitable linker for attachment to gold and other chemically soft metals. The disulphide group and the copper centre are connected by a route for facile electron tunnelling through the protein such as shown by studies of pulse radiolysis induced intramolecular ET [82]. The azurins offer other properties suitable for STM/AFM imaging. They are robust, structurally and kinetically well characterized, and many mutants are available. *P. aeruginosa* azurin shows, finally unpromoted stable and reversible electrochemistry at pyrolytic graphite. Reversible electrochemistry at gold electrodes is also found but here the voltammograms decay within 10–20 min [85], as could be anticipated if there is a change from direct ET at the copper atom in freely mobile azurin to an adsorbed configuration, where ET is attenuated by longer range, although still facile tunnelling through the protein. XPS subsequent to exposure of the gold substrate to azurin solution has substantiated that approximately a monolayer of azurin is adsorbed.

Figure 5.9 shows contact mode AFM images of oven-annealed Au(111) in contact with aqueous ammonium acetate buffer after exposure to azurin solution [83]. New structures, quite different from the triangular steps of the pure surface appear in azurin-exposed samples. Their vertical dimension ($\approx$5 nm) accords with the crystallographic size while the lateral dimension is up to five times larger This kind of discrepancy is not uncommon and most likely caused by convolution with the tip for which the surface dimensions are much more critical than in STM [43]. Successive scanning pushes off the adsorbed azurin, exposing again the free surface. This underlines the strong effects of the tip-sample interactions when soft material is imaged by AFM. Preliminary investigations have in fact shown that resolution and stability are improved when azurin is imaged by AFM in the tapping and phase contrast modes.

Figure 5.10 shows recent high-resolution *in situ* STM images of azurin, recorded under full and independent potentiostatic control of substrate

Fig. 5.8. Three-dimensional structure of *P. aeruginosa* azurin. Coordinates from ref. [80] and Brookhaven Data Bank. Graphic representation by Molscript [66]. The copper atom is indicated by the large sphere at the top, the disulphide group by the two smaller, lightly shaded spheres at the bottom.

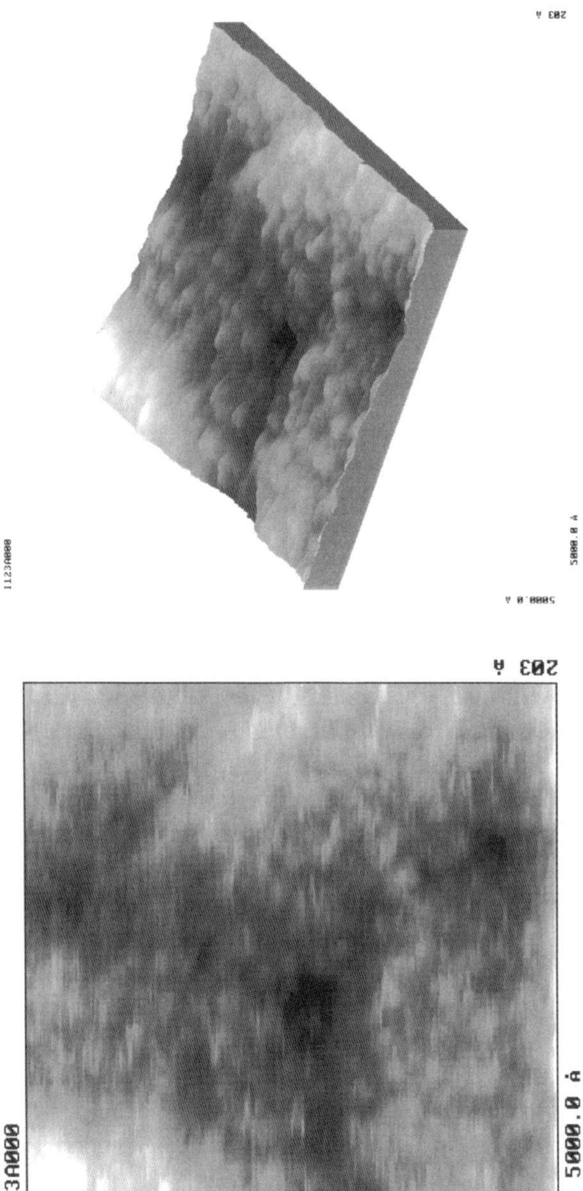

Fig. 5.9. Left: *In situ* (open circuit) AFM image of pre-adsorbed *P. aeruginosa* azurin on Au(111). 50 mM ammonium acetate, pH = 4.65 (isoelectric point). Right: Three-dimensional view of the same image (other details in ref. [83]).

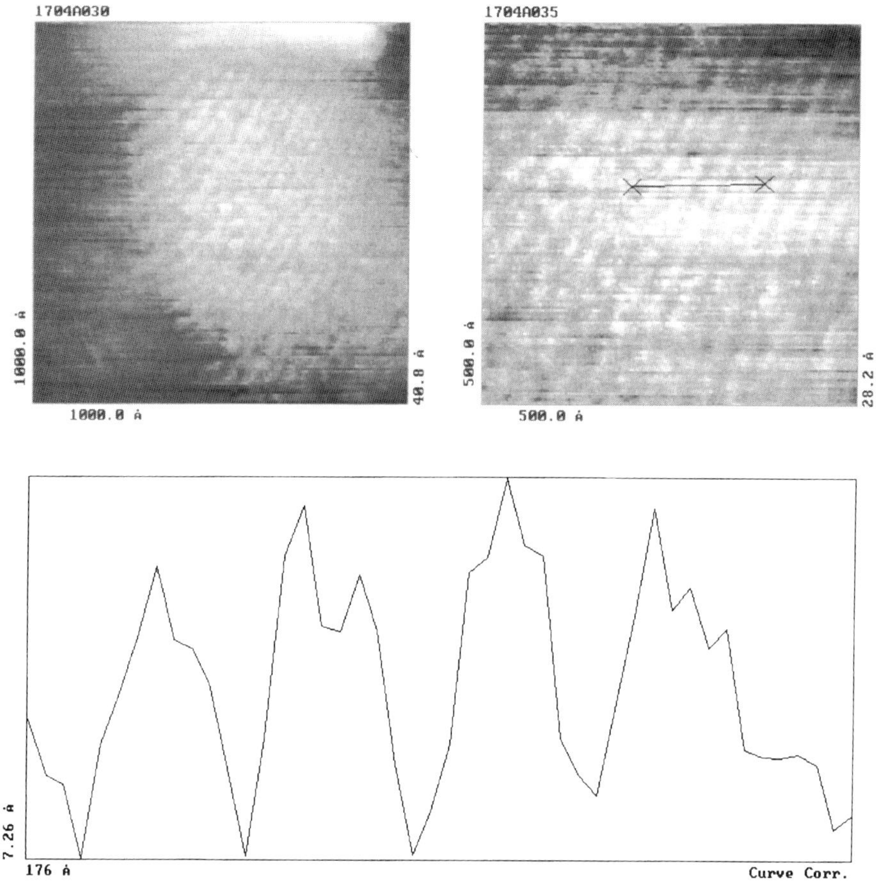

Fig. 5.10. Left: *In situ* STM image of pre-adsorbed *P. aeruginosa* azurin on Au(111). Tunnel current 2 nA, Bias voltage 50 mV. Substrate potential −400 mV (Au/AuO_x). other conditions as in Fig. 5.9 (further details in refs. [83] and [84]).

and tip. These images are the first of their kind and show several notable features [84]:

(a) The lateral resolution is much better than for AFM. The individual structures now accord well with the crystallographic size (Fig. 5.10, bottom). This substantiates that the disulphide group is a suitable structural element for gentle protein self-assembly on gold.

(b) The images are robust over a potential range of a few hundred mV around the reduction potential of azurin. No proper current-voltage relations are presently available.

(c) The images of highest resolution show a smaller central region of brighter contrast than the rest of the molecule. This submolecular feature is in line with expectations from the activated and resonance tunnel mechanisms discussed in Section 5.4.1, and indicative of the particular role of the low-lying redox level of the copper atom in enhancing the tunnel currents. It also parallels the tunnel patterns for adsorbed protoporphyrin IX recently reported [56].

5.5 CONCLUSIONS

Our short overview illustrates some prospects for investigation of metalloprotein dynamics at metal-solution interfaces. Cyclic voltammetry of small single metalloproteins is straightaway feasible. Reversible electrochemistry can be achieved but molecular detail such as adsorption patterns and precise promoter function remain elusive.

Investigations of multicentre electrochemical metalloprotein function including metallo*enzyme* function have also been brought to a level, where both direct and catalytic modes, and elements of molecular mechanisms can be addressed. The latter are, however, entangled by features such as composite electrochemistry, extremely complicated molecular interaction patterns when more than *two* metallic redox centres are involved, fragile surface enzyme preparations, and lack of structural surface characterization of the adsorbed metalloenzymes. In this respect, *two-centre* metalloproteins constitute interesting promising intermediates where the cooperativity between the metallic redox centres can be accurately addressed with molecular resolution within reach.

Our discussion of scanning probe approaches to metalloproteins at surfaces indicates, first that *in situ* STM and AFM mapping of metalloproteins with *molecular* (rather than atomic) resolution is definitely feasible. In addition, submolecular detail is possibly within reach. This is interesting as proteins constitute a class of "soft" materials. It is also comforting as aqueous solution is otherwise the natural functional medium for proteins.

Protein mobility and aggregation are inherent problems in the *in situ* STM and AFM characterization. This can, however, also be turned to the better and brought to open other new perspectives for mapping of molecular and collective protein dynamics such as phase transitions, heterophase fluctuations etc.

Twenty-first century nanotechnology aspires to endeavour into impressive multidisciplinarity where chemistry, physics, materials science etc., *as well as* biology and biotechnology are expected to interface. The observations and conclusions above hold intriguing perspectives in these respects.

ACKNOWLEDGEMENTS

Financial support from the Danish Natural Science and the Technical Science Research Councils, the Carlsberg Foundation, the Novo Nordisk Foundation, the Lundbeck Foundation and the P.A. Fisker's Foundation, is acknowledged.

REFERENCES

[1] H. Sigel and A. Sigel, eds, Metal Ions in Biological Systems, Vol. 27 (Marcel Dekker, New York, 1991).
[2] M.A. Fox, ed, Chem. Rev. 92 (1992) 365–490, Special Issue on Electron Transfer.
[3] L. Arnaut, ed, J. Photochem. Photobiol. A: Chem. 82 (1994) 1–232, special issue on photoinduced electron transfer reactions.
[4] For a list of many systems, see, G.W. Canters and C. Dennison, Biochimie 77 (1995) 506.
[5] H.B. Gray and J.R. Winkler, Chem. Rev. 92 (1992) 369.
[6] G. Iversen, Yu.I. Kharkats, A.M. Kuznetsov and J. Ulstrup, Adv. Chem. Phys. (in press).
[7] D.S. Bendall, ed, Protein Electron Transfer (Bios Scientific Publishers, Oxford, 1996).
[8] L.S. Conrad, J.-J. Karlsson and J. Ulstrup, Eur. J. Biochem. 231 (1995) 133.
[9] I.B. Coutinho and A.V. Xavier, Meth. Enzymol. 243 (1994) 119.
[10] O. Farver, S. Wherland and I. Pecht, J. Biol. Chem. 269 (1994) 22933.
[11] G. Tollin, J.K. Hurley, J.T. Hazzard and T.E. Meyer, Biophys. Chem. 48 (1993) 259.
[12] P. Kyritsis, A. Messerschmidt, R. Huber, G.A. Salmon and A.G. Sykes, J. Chem. Soc., Dalton Trans. (1993) 731.
[13] B.G. Malmström, Acc. Chem. Res. 26 (1993) 332.

[14] J.R. Winkler, B.G. Malmström and H.B. Gray, Biophys. Chem. 54 (1995) 199.
[15] T. Tsukihara, H. Aoyama, E. Yamashita, T. Tomizaki, H. Yamaguchi, K. Shinzawa-Itoh, R. Nakashima, R. Yaono and S. Yoshikawa, Science 269 (1995) 1069.
[16] J. Kim and D.C. Rees, Biochemistry 33 (1994) 389, and references therein.
[17] M.E. Michel-Beyerle and G.J. Small, eds, Chem. Phys. 197 (1995) 223–472, special issue on photosynthesis and the bacterial reaction centre.
[18] M.E. Michel-Beyerle, ed, The Reaction Centre of Photosynthetic Bacteria (Springer-Verlag, Berlin, 1996).
[19] G.A. Mines, M.J. Bjerrum, M.G. Hill, D.R. Casimiro, I.-J. Chang, JR. Winkler and H.B. Gray, J. Am. Chem. Soc. 118 (1996) 1961.
[20] H.B. Gray and J.R. Winkler, Ann. Rev. Biochem. 65 (1996) 537.
[21] N.H. Andersen, M. Hervàs, J.A. Navarro, M.A. De la Rosa and J. Ulstrup, Inorg. Chim. Acta 272 (1998) 109.
[22] I.V. Kurnikov and D.N. Beratan, J. Chem. Phys. 105 (1996) 9661.
[23] H.E.M. Christensen, L.S. Conrad, K.V. Mikkelsen, M.K. Nielsen and J. Ulstrup, Inorg. Chem. 29 (1990) 2808.
[24] P. Siddarth and R.A. Marcus, J. Phys. Chem. 97 (1993) 2400.
[25] J.M. Gruschus and A. Kuki, J. Phys. Chem. 97 (1993) 5581.
[26] A.A. Stuchebrukhov and R.A. Marcus, J. Phys. Chem. 99 (1995) 7581.
[27] A.M. Kuznetsov, J. Ulstrup and M.D. Vigdorovich, Chem. Phys. 176 (1993) 539.
[28] Yu.I. Kharkats, AM. Kuznetsov and J. Ulstrup, J. Phys. Chem. 99 (1995) 13545.
[29] H. Sumi and T. Kakitani, Chem. Phys. Lett. 252 (1996) 85.
[30] A.A. Kornyshev, A.M. Kuznetsov, J. Ulstrup and U. Stimming, J. Phys. Chem. 100 (1996) 11175, 11184.
[31] T.A. Horbett and J.L. Brash, eds, Proteins at Interfaces II, ACS Symp Ser. 602 (1995).
[32] F.A. Armstrong, Struct. Bond. 72 (1990) 138.
[33] G. Huo and H.A.O. Hill, Adv. Inorg. Chem. 36 (1991) 341.
[34] F.A. Armstrong, in Bioelectrochemistry of Biomacromolecules, eds, G. Lenaz and G. Milazzo (Birkhäuser, Basel, 1996) p. 205.
[35] A.A. Kornyshev, A.M. Kuznetsov and J. Ulstrup, J. Phys. Chem. 98 (1994) 3832.
[36] A.M. Kuznetsov, Charge Transfer in Physics, Chemistry and Biology (Gordon & Breach, Reading, UK, 1995).
[37] A.A. Kornyshev, A.M. Kuznetsov, G. Makov and M.D. Vigdorovich, J. Chem. Soc., Faraday Trans. 92 (1996) 3997, 4005.
[38] J.E.T. Andersen, A.A. Kornyshev, A.M. Kuznetsov, P. Møller and J. Ulstrup, Electrochim. Acta 42 (1997) 819.
[39] J.N. Butt, A. Sucheta, L.L. Martin, B. Shen, B.K. Burgess and F.A. Armstrong, J. Am. Chem. Soc. 115 (1993) 12587.
[40] J.-J. Karlsson, M.F. Nielsen, M.H. Thuesen and J. Ulstrup, J. Phys. Chem. 101 (1997) 2430.
[41] A. Gewirth and H. Siegenthaler, eds, Nanoscale Probes of the Solid/Liquid Interface (Kluwer, Dordrecht, 1995).
[42] O. Marti and M. Armrein, eds, SPM and SFM in Biology (Academic Press, San Diego, 1993).

[43] S.N. Magonov and M.H. Whangbo, Surface Analysis with STM and AFM (Verlag Chemie, Weinheim, 1996).
[44] B. Drake, C.B. Prater, A.L. Weisenkorn, S.A.C. Gold, T.R. Albrecht, C.F. Quate, D.S. Cannel, H.G. Hansma and P.K. Hansma, Science 243 (1989) 1586.
[45] M. Radmacher, M. Fritz, H.G. Hansma and P.K. Hansma, Science 265 (1994) 1577.
[46] A. Moyak, A. Nitzan and R. Kosloff, J. Chem. Phys. 104 (1996) 1549.
[47] A.M. Kuznetsov, P. Sommer-Larsen and J. Ulstrup, Surface Science 275 (1992) 52.
[48] A.M. Kuznetsov and J. Ulstrup, Surf Coat Technol. 67 (1994) 193.
[49] W. Schmickler, Surface Science 295 (1993) 43.
[50] J. Halbritter, in Electron and Ion Transfer in Condensed Media, eds, A.A. Kornyshev, M.P. Tosi and J. Ulstrup (World Scientific, Singapore, 1997), p. 278.
[51] J.E.T. Andersen, P. Møller, M.V. Pedersen and J. Ulstrup, Surface Science 325 (1995) 193.
[52] J.E.T. Andersen, K.G. Olesen, A.I. Danilov, C.E. Foverskov, P. Møller and J. Ulstrup, Bioelectrochem. Bioeng 44 (1997) 57.
[53] A.I. Danilov, Russ. Chem. Rev. 64 (1995) 767.
[54] S.M. Lindsay, N.J. Tao, J.A. de Rose, P.I. Oden, Y.L. Lyubchenko, R.E. Harrington and L.S. Shlyakhtenko, Biophys. J. 61 (1992) 1570.
[55] M. Kunitake, N. Batina and K. Itaya, Langmuir, 11 (1995) 2337.
[56] N.J. Tao, Phys. Rev. Lett. 76 (1996) 4066.
[57] T.A. Jung, R.R. Schlitter and J.K. Gimzewski, Nature 386 (1997) 696.
[58] See for example: J. Ruz, M.D.L. de Castro and M. Valcarcel, Analyst 112 (1987) 259.
[59] C.-W. Lee, H.B. Gray, F.C. Anson and B.G. Malmstrøm, J. Electroanal. Chem. 172 (1984) 289.
[60] M.H. Thuesen, O. Farver, B. Reinhammar and J. Ulstrup, Acta Chem. Scand. 52 (1998) 555.
[61] V.A. Bogdanovskaya and M.P. Tarasevich, Sov. Sci. Rev. B Chem. 17 (1992) 1.
[62] W.J. Albery, P.N. Bartlett and D.H. Craston, J. Electroanal. Chem. 194 (1985) 223.
[63] Y. Degani and A. Heller, J. Phys. Chem. 91 (1987) 1285.
[64] R. Wilson and A.P.F. Turner, Biosens. Bioelectron. 7 (1992) 165.
[65] A. Kadziola and S. Larsen, Structure 5 (1997) 203.
[66] P. Kraulis, J. Appl. Cryst. 24 (1991) 946.
[67] Z.-W. Chen, M. Koh, G. van Driessche, J.J. van Beeumen, R.G. Bartsch, T.E. Meyer, M.A. Cusanovich and F.S. Mathews, Science 266 (1994) 430.
[68] M. Nissum, J.-J. Karlsson, P.W. Jensen, J. Ulstrup and G. Smulevich, J. Biol. Inorg. Chem. 2 (1997) 302.
[69] I.B. Coutinho, J.-J. Karlsson, J. Ulstrup and A.V. Xavier (in preparation).
[70] J.A. Fee, Struct. Bond. 23 (1975) 1.
[71] B. Reinhammar, Chem. Scripta 25 (1985) 172.
[72] L. Morpurgo, Life Chem. Rept. 5 (1987) 277..
[73] A. Messerschmidt, R. Ladenstein, R. Huber, M. Bolognesi, L. Avigliano, A. Marchesini, R. Petruzelli, A. Rossi and A. Finazzi-Agró, J. Mol. Biol. 224 (1992) 179.
[74] A.I. Yaropolov, A.N. Kharybin, J. Emnéus, J. Marko-Varga and L. Gorton, Bioelectrochem. Bioeng. 40 (1996) 49.

[75] A.M. Kuznetsov and J. Ulstrup, Elektrokhimiya 31 (1995) 244.
[76] M.G. Zakaraya and J. Ulstrup, Chem. Phys. 135 (1989) 49.
[77] R.A. Scott and A.G. Mauk, Cytochrome *c*. A Multidisciplinary Approach (University Science Books, Sausalito, 1996).
[78] B. Zhang and E. Wang, J. Chem. Soc., Faraday Trans. 93 (1997) 327.
[79] C.J. McNeil, D. Atley and W.O. Ho, Biosens. Bioelectron. 10 (1995) 75.
[80] H. Nar, A. Messerschmidt and R. Huber, J. Mol. Biol. 221 (1991) 765.
[81] A.G. Sykes, Adv. Inorg. Chem. 36 (1991) 377.
[82] O. Farver, L.K. Skov, T. Pascher, B.G. Karlsson, M. Nordling, L.G. Lundberg, T. Vänngård and I. Pecht, Biochemistry 32 (1993) 7317.
[83] E.P. Friis, J.E.T Andersen, L.L. Madsen, N. Bonander, P. Møller and J. Ulstrup, Electrochim. Acta 43 (1998) 2889.
[84] E.P. Friis, J.E.T. Andersen, L.L. Madsen, P. Møller and J. Ulstrup, J. Electroanal. Chem. 431 (1997) 35.

Chapter 6

The Kinetics of the Partitioning of Compounds Between Octanol and Water, and its Relationship to the Movement of Molecules in Biological Systems

PETER R. FISK, MARTYN G. FORD and PETER WATSON

SUMMARY

The significance of the kinetics of partitioning in liquid–liquid systems, and experimental methods used are reviewed. Experiments are then reported which investigate the rate of partitioning of molecules from octanol to water, using the rotating diffusion cell technique. The objective of this work was to examine aspects of whether the rate of crossing interfaces is an important component in the movement of molecules within biological systems.

The results allow a comparison with the theoretical description of the rate of partitioning, which is largely obeyed for the set of 23 compounds studied. The overall rate of transfer is proportional to, and dominated by, the position of equilibrium above log P_{ow} (at pH 7) $\simeq$ 1.2 where P_{ow} is the octanol-water partition coefficient. In the intermediate range of lipophilicity the rate of resolvation at the molecule crosses from octanol to water becomes important in influencing the rate. However, these resolvation rate constants are not low enough for that process to become rate-determining under the conditions studied, but diffusion through the solvent phases was more significant.

1. INTRODUCTION

This overview sets out the rationale for, and potential benefits of, studies of the kinetics of partitioning of compounds between immiscible phases.

The use of the equilibrium constant P_{ow} (the octanol-water partition coefficient) as a descriptor of chemical processes in biological systems is well known; the study of kinetics is more general than that of the special case of equilibrium, but is more difficult. The emphasis here will be on pesticides and model compounds, but is relevant to other systems.

6.1.1 Importance of the rate of partitioning

Partitioning of compounds between a solvent and water has been studied extensively as a model for processes concerned with cuticular penetration and systemic movement with plants. There are many papers and reviews [1–6], almost all emphasising the utility of the octanol-water partition coefficient of a molecule, P_{ow}. The use of this property as part of a statistical approach to understanding structure–biological activity relationships, is well known [7–9]. However, even if a correlation of biological activity with log P_{ow} exists, it does not help us to know which physicochemical property the correlation is due to. Better understanding may be possible by closer consideration of the transport steps involved prior to a molecule arriving at an active site, such as cuticular and membrane permeation. Study of the kinetics of partitioning in a model system [10, 11] is one aspect of trying to understand the movement of compounds to active sites.

A further element in the investigation of the use of P_{ow} is that it seems reasonable to assume that the system "plant + pesticide" is not at thermodynamic equilibrium, especially considering that some crop protection agents act over short time scales. The literature of penetration of the plant cuticle, for example, indicates that deposits on leaves decline in quantity over (typically) several days. The compound then will be redistributed within the plant over a longer interval. During this time metabolism and removal to vacuoles will have been occurring. The effectiveness of a herbicide, for example, will depend upon how fast a compound arrives at an active site and whether a sufficient concentration accumulates for a biological effect to be achieved. These concepts are expressed in compartmental models widely presented in the literature.

In this overview, the direct measurement of the rate of a molecule crossing the octanol–water interface is presented as an appropriate extension to measurement of P_{ow}, a thermodynamic property. In addition to reviewing previous work, we include results from a wide variety of chemical structures. The reliable measurement of these rates is achieved with the rotating diffusion cell (RDC) method. The principle of the RDC is to impose a convection pattern on to both octanol and water phases, giving known mass transfer of the compound to and from the liquid–liquid interface.

The concept of measuring such rates is not new, particularly in the pharmaceutical field. Van de Waterbeemd [14] measured rates of transfer of various drugs from octanol to water and empirically related these rates to the partition coefficient. Similarly Brodin [15], using a different experimental method, obtained rates of transfer for another series of compounds between cyclohexane and water. The rotating diffusion cell has been introduced for similar purposes [16–18]. It is necessary to look into the broader background of liquid–liquid interfacial kinetics, in order to illustrate aspects of the issues under consideration. The subject has been reviewed in part by Noble [19].

There are many topics in chemistry that are studied in biphasic systems rather than in homogeneous solution. These include aromatic nitration [20], oxidation [21], polymerisation [22], phase transfer catalysis [23] and solvent extraction [24]. The partitioning of a molecule not undergoing chemical transformation is part of the general topic of kinetics of solvent extraction; at this point models of biological systems (such as octanol and water mixtures) meet with a topic of importance in physical chemistry and chemical engineering. In studies of the rate of solvent extraction processes, progress has been considerably aided by the introduction of the "rotating diffusion cell" [25], which has enabled detailed reaction models to be tested for the first time. The rotating diffusion cell (RDC) was developed by Albery and Hadgraft [16] to study the partitioning of compounds, and in particular, models of skin penetration [26].

In liquid–liquid systems, several qualities are needed in an experimental approach. Firstly, convenience for analysis of concentration *vs.* time, with reproducibility within and between laboratories. Secondly, the experimental design should lend itself to quantitative and mathematically based interpretation *via* soundly-based kinetic equations (as opposed to empirical observations that cannot be fully tested).

At any interface there will be boundary layers of liquid not subject to convection. Diffusion of materials to and from the interface will be determined by the thickness of such layers. In previous studies of solvent extraction and partitioning, a wide variety of techniques to set up the liquid–liquid interface have been employed. These are illustrated in Fig.

6.1 and compared in Table 6.1. The clear conclusion is that the RDC is the method of choice for this study of interfacial liquid–liquid kinetics. If reactions or processes are very fast (i.e. faster than diffusion) other methods are needed. It should be noted that Brodin [15] used the drop method and van de Waterbeemd [14] the Lewis cell, in their studies related to partitioning rate.

It therefore is a logical extension of the prior literature to propose the use of the RDC method to the understanding of partitioning properties of pesticides. By the same logic, Hadgraft [27] and Leahy [28, 29, 30] have studied systems of pharmaceutical interest. The work of Leahy is of particular importance and will be discussed below. These workers and the present author consider that a study of the rate of partitioning of a biologically active molecule can be more appropriate than use of log P_{ow} alone. The RDC enables such measurements to be made reliably; the theoretical description of its use is given in Section 6.2, with experimental detail in Section 6.3.

6.1.2 Objectives

The argument that kinetics are important has lead to the present study of *in vitro* partitioning rates. How are such data to be used? There are several stages, which are developing in parallel in the literature.

(1) Establish a model of the whole plant, and determine when rate of crossing cuticles or membranes is likely to be rate-limiting.
(2) Compare data from *in vivo* and *in vitro* experiments.
(3) Compare partitioning rate data to molecular features, e.g., is rate proportional to partition coefficient? Are there common structural features associated with rapid or slow transport?

The main objective of this work has been to gain a broad view of relevant structural types. Before detailed questions can be addressed, indication of the range of data exhibited and their reliability, was necessary. Therefore, the compounds chosen include the following:

(a) some known herbicides;
(b) sub-structural features of compounds in set (a), in order to see if factors affecting rate can be "built" via sub-structure;

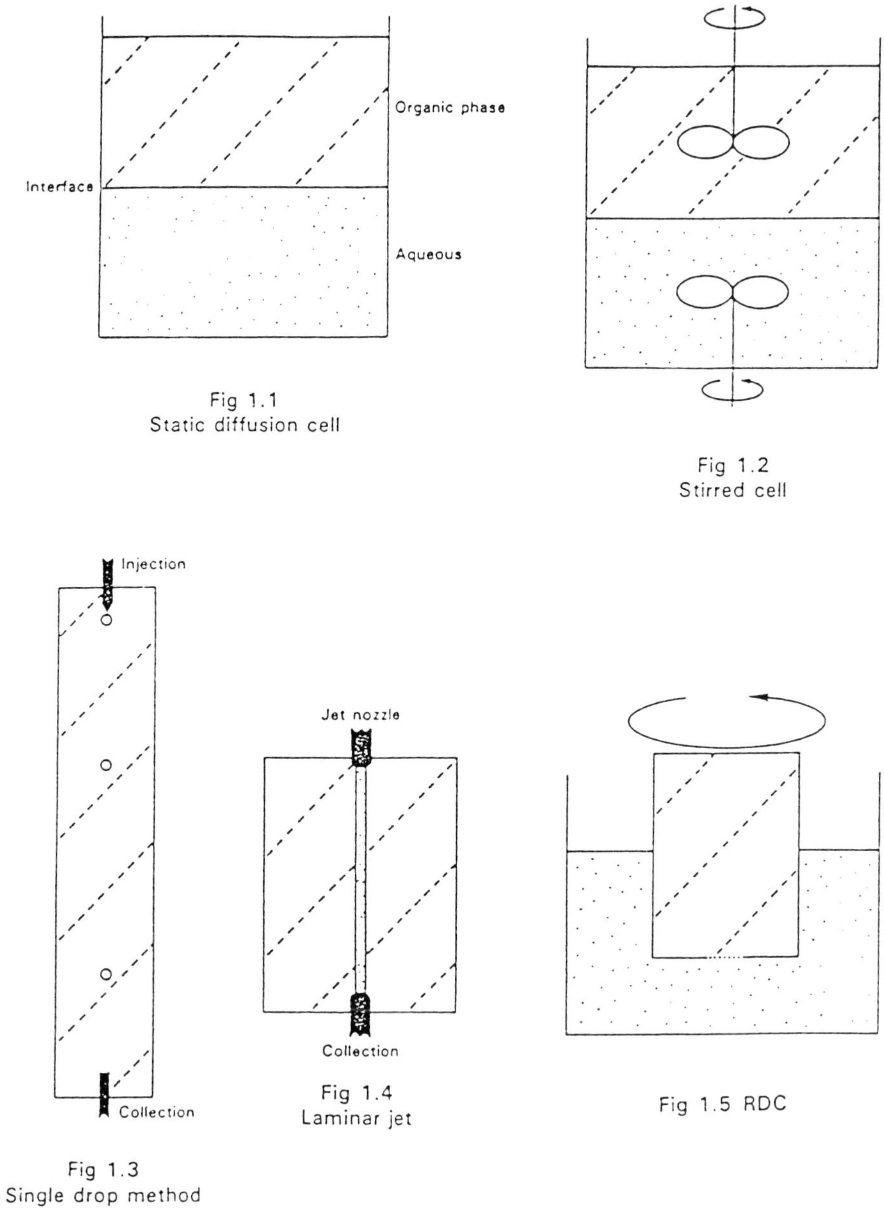

Fig. 6.1. Experimental methods to study rate of partitioning.

TABLE 6.1
Relative merits of techniques for study of kinetics of partitioning (see Fig. 6.1 for schematic)

Technique	Principle	Appraisal
Static cell	Two phases contained without agitation	Irreproducible, subject to ill defined and large diffusion layers, and stray vibrations of the apparatus.
Capillary cell	As for static cell, but use a capillary	Capillary prevents spurious agitation interfering, analytically difficult
Stirred (Lewis) cell	Impellers gently stir the two phases	Irreproducible, undefined boundaries
Slightly stirred	Agitate so as to produce a dispersion	Irreproducible between laboratories. Rapid equilibration can be useful. Unknown interfacial area
Laminar jet	Pass a jet of one phase through the other	Experimentally difficult especially if surfactants present. Thickness of the diffusion layers can be calculated
Single drop	Pass drops of one phase through the other	Useful for reliable relative rates but not amenable to thorough mathematical analysis because boundary films cannot be described
Rotating diffusion cell	Set up hydrodynamics of a rotating disc on both sides of a supported interface	Reproducible between laboratories. Known interfacial area. Mathematically straight-forward but with a limit on measurement of the fastest rates

Ref. [19] gives further information.

(c) a small series of nitro-methylene insecticides, to examine the effect of small structural change;
(d) "calibration" compounds, for which literature is available;
(e) compounds with functional groups not exhibited within (a) to (d).

6.2 ROTATING DIFFUSION CELL THEORY

The published literature is not comprehensive in respect of the theory behind the operation of the RDC, therefore a full description of it is given here. The conventional RDC and clamp-RDC are shown in Figs. 6.2 and 6.3. The latter is very similar in construction to that used by David Leahy [28, 29, 30]. The RDC can be used with a variety of procedures for establishment of one of two liquid–liquid interfaces.

6.2.1 Organic/organic/water system

The key principle of the RDC method is that the liquid–liquid interface is set up at the surface of a membrane filter, which is rotated in order to impose a known convection pattern on the two liquid phases. The set-up used most frequently has the filter filled with octanol, and octanol (initially containing the diffusant) in the inner compartment of the cell. The outer compartment contains water. Under these conditions we may represent a general concentration profile of the diffusant by Fig. 6.4. The system is termed "organic/organic/water" or o → /o/w, indicating that the compound moves from the inner compartment containing the organic solvent, though the supporting filter also containing organic solvent, through the interface into the outer compartment containing water. Other frequently used systems are w → /o/o, o/o/ ← w. The mathematical analysis follows the general approach described in [16].

The figure shows known diffusion layers and steady state concentration gradients. These are consequences of the thickness of the diffusion layers being known (calculable) and that the experiment is performed under steady-state conditions. This condition may break down if rates are very fast or slow or concentration in the donor phase is outside the range 0.5–100 mM (which does not apply to his work). At a particular rotation speed W ($1 < W < 8$ Hz for experimental reasons), the steady flux j (in mol m^{-2} s^{-1}) is given by the following,

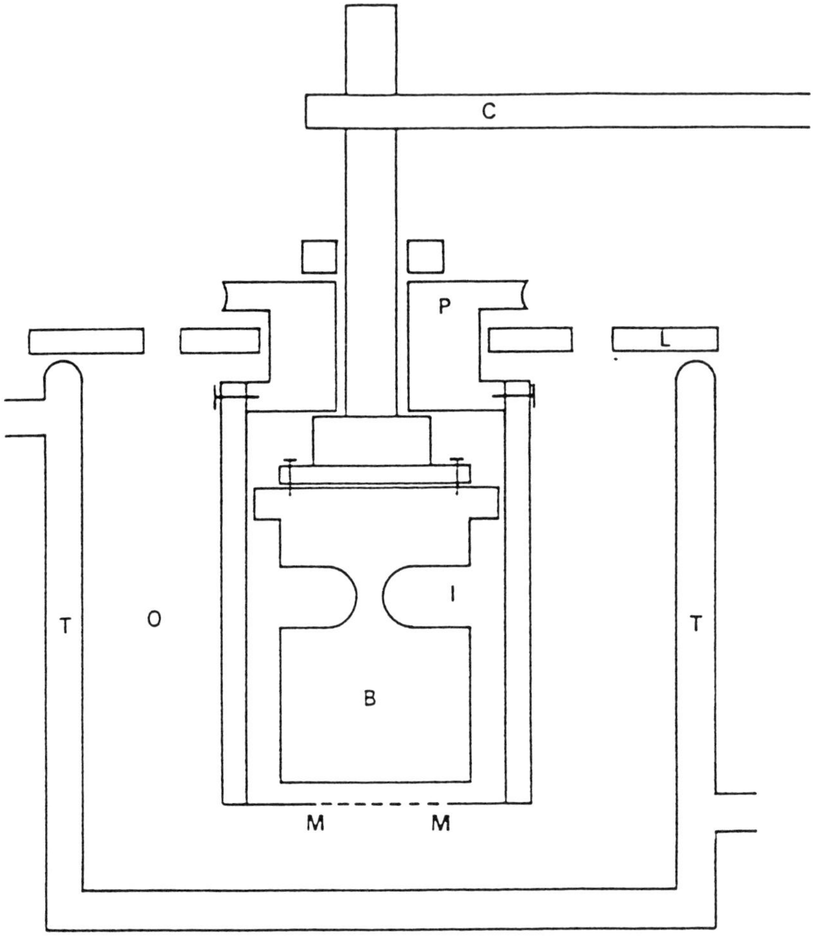

M = Membrane
B = Baffle
I = Inner compartment
O = Outer compartment
T = Thermostatting jacket
L = Lid
P = Pulley
C = Clamp

Fig. 6.2. Cross section of standard rotating diffusion cell.

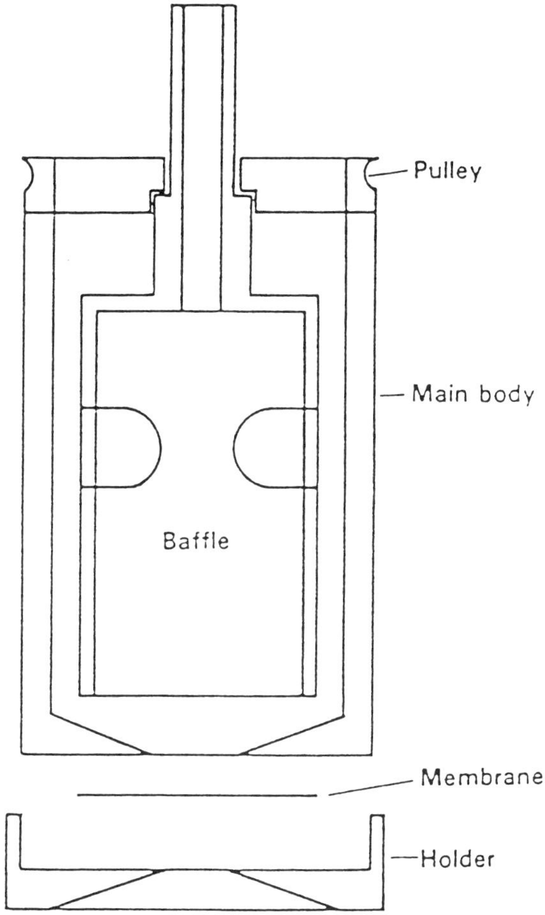

Fig. 6.3. Cross section of clamp rotating diffusion cell.

$$j = D_o(C_{\infty,org} - C_1)/Z_o, \tag{6.1}$$

where D_o is the diffusion coefficient in the organic solvent,

$$j = aD_o(C_1 - C_{o,org})/L, \tag{6.2}$$

where a is the porosity of the filter,

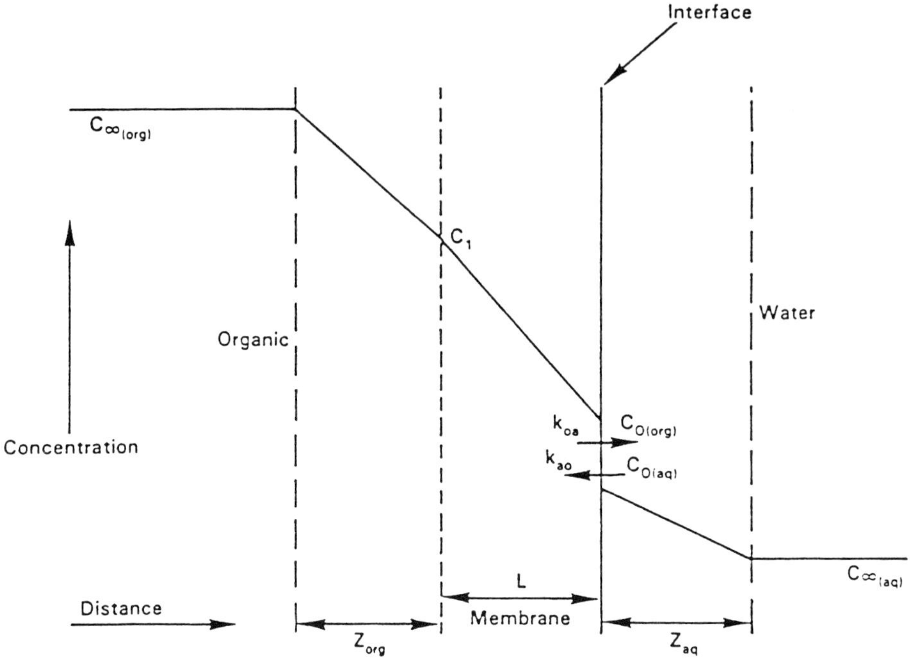

Fig. 6.4. Concentration profile in partitioning.

$$j = a(k_{oa}C_{o,org} - k_{ao}C_{o,aq}), \tag{6.3}$$

where k_{oa} is the rate constant for crossing the interface from organic to aqueous; k_{ao} is the reverse.

$$j = D_a(C_{o,aq} - C_{\infty,aq})/z_a, \tag{6.4}$$

where D_a is the diffusion coefficient in the aqueous solution.

Elimination gives

$$\frac{(c_{\infty,org} - Pc_{\infty,aq})}{j} = \frac{Pz_a}{D_a} + \frac{z_o}{D_0} + \frac{L}{aD_o} + \frac{1}{ak_{oa}}, \tag{6.5}$$

provided that $P_{ow} = k_{ao}/k_{ao}$ = the partition coefficient. It is not necessary to accept this equation as the only valid one. For example, a new theoretical description of the partitioning process, based on an interfacial adsorption model is given below. It has not yet been fully tested but will be referred to in the Discussion.

With the RDC, the concentrations at the surface $C_{o,org}$ and $C_{o,aq}$ are exactly described by the above equations because, in general, the thickness of the diffusion layer, z is given by

$$z = 0.634\nu^{1/6}D^{1/3}W^{1/2}, \tag{6.6}$$

where W is the rotation speed, ν the kinematic viscosity and D the diffusion coefficient of the diffusant under study; this is the "Levich equation" [33].

The term $P_{ow}C_{\infty,aq}$ accounts for the back reaction. During most experiments $C \gg P_{ow}C_{\infty,aq}$. We replace $(C_{\infty,org} - PC_{\infty,aq})$ by C' for clarity but the calculation is explicit throughout the calculations and the interpretation of data.

As a consequence of Equation (6.6), Equation (6.5) becomes:

$$\frac{c'}{j} = 0.643W^{-1/2}\left[\frac{\nu_{org}^{1/6}}{D_{aq}^{2/3}} + \frac{P_{ow}\nu_{aq}^{1/6}}{D_{aq}^{1/6}}\right] + \frac{L}{aD_{org}} + \frac{1}{ak_{oa}}, \tag{6.7}$$

which predicts a linear response of C'/j versus $W^{-1/2}$, the gradient of which depends upon known or calculable constants.

Leahy [30], van de Waterbeemd [14] and Albery [16] discussed the meaning of the rate constants, k_{oa}, k_{ao}; it is sufficient to state here that the primary process involved is the resolvation of the compound as it moves from the octanol to water environments, that is, the loss of any strongly associated octanol molecules and gain of water. Because of the mutual solubility of these solvents it seems reasonable to say that some solvent molecules may transfer along with the compound (which may not be true for other solvent systems where the mutual solubilities are lower). There may also be molecular conformation changes during this process of transfer, which would also be likely to affect solvation.

Equation (6.7) shows that: the relationship between flux and partition coefficient may be completely due to diffusion, as seen in the diffusion terms (those dependent on $W^{-1/2}$), in the case that the k_1 term is irrelevant. If the k_{oa} term is important relative to the diffusion terms, we may be able to learn about resolvation and interfacial processes that are of considerable relevance to *in vivo* phenomena. The assertions of some previous workers [15, 16], that the rate that compounds cross the interface is proportional to log P_{ow}, would seem to be based on a false premise, that they were measuring the resolvation rates; in fact they were probably measuring the diffusion characteristics of their experimental set up. Note that the RDC mathematics facilitates the discussion because of the separation of all the "resistances" into separate components.

The significance of the relative sizes of diffusion and kinetic terms will feature in the discussion.

6.2.2 Theory for other experimental procedures

Using the same terminology as before, except $c_{\infty,d}$ is the donor solution and $c_{\infty,ac}$ the acceptor, various relationships may be obtained. For transfer from water, through a solvent-impregnated membrane, into water

$$\text{w} \rightarrow /\text{o}/\text{w}: \quad \frac{(c_{\infty,d} - c_{\infty,ac})}{j} = \frac{2z_a}{D_a} + \frac{L}{aP_{ow}D_0} + \frac{2}{ak_{ao}}. \tag{6.8}$$

For transfer from water through a solvent-impregnated membrane, into solvent:

$$\text{w} \rightarrow /\text{o}/\text{o}: \quad \frac{(c_{\infty,d} - c_{\infty,ac}/P_{ow})}{j} = \frac{z_a}{D_a} + \frac{z_o}{P_{ow}D_o} + \frac{L}{aD_oP_{ow}} + \frac{1}{ak_{ao}}. \tag{6.9}$$

For transfer from solvent through a water-impregnated membrane, into water:

$$\text{o} \rightarrow /\text{w}/\text{w}: \quad \frac{(c_{\infty,d} - P_{ow}c_{\infty,ac})}{j} = \frac{P_{ow}z_a}{D_a} + \frac{z_o}{D_o} + \frac{P_{ow}L}{aD_a} + \frac{1}{ak_{oa}}. \tag{6.10}$$

For diffusion from an aqueous donor compartment, through a water-impregnated membrane, into water, and similarly for solvent:

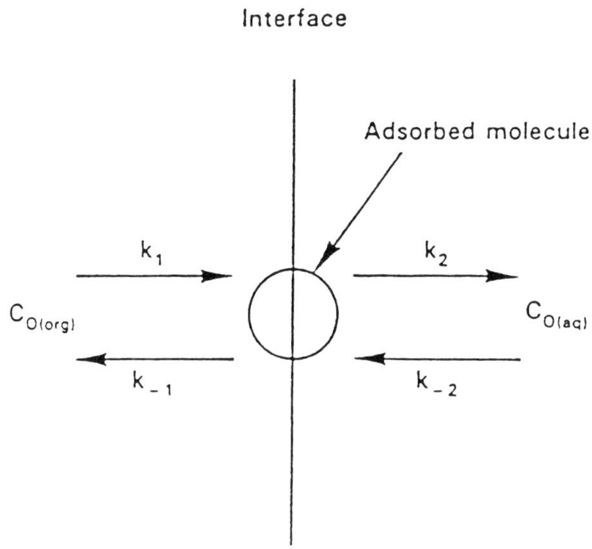

Fig. 6.5. Adsorption model for partitioning.

$$\mathrm{w} \rightarrow /\mathrm{w}/\mathrm{w}: \quad \frac{(c_{\infty,\mathrm{d}} - c_{\infty,\mathrm{ac}})}{j} = \frac{2z_\mathrm{a}}{D_\mathrm{a}} + \frac{L}{aD_\mathrm{a}}, \tag{6.11}$$

$$\mathrm{o} \rightarrow /\mathrm{o}/\mathrm{o}: \quad \frac{(c_{\infty,\mathrm{d}} - c_{\infty,\mathrm{ac}})}{j} = \frac{2z_\mathrm{o}}{D_\mathrm{o}} + \frac{L}{aD_\mathrm{o}}. \tag{6.12}$$

6.2.3. *Adsorption model for interfacial transfer*

Figure 6.3 shows the basis of the adsorption model, and the four rate constants used. For simplicity, we assume a Langmuir isotherm model for the occupation of the interface. If the fraction of interface occupied is χ then the adsorption from the octanol side is

$$j = k_1 c_\mathrm{o}(1 - \chi) - k_{-1}\chi, \tag{6.13}$$

and from the water side

$$j = k_2\chi - k_{-2}c_a(1 - \chi). \tag{6.14}$$

Elimination of χ gives

$$\frac{c_o - K'c_a}{j} = \frac{1}{k_1} + \frac{c_o}{k_2} + \frac{k_{-2}c_a}{k_1k_2} + \frac{k_{-1}}{k_1k_2}, \tag{6.15}$$

where

$$K_1 = \frac{k_{-2}k_1}{k_1k_2}. \tag{6.16}$$

Hence equation (6.15) has terms which can account for all types of rate limiting behaviour within this adsorption model. If we consider that the back transfer is unimportant, then

$$\frac{c'}{j} = \frac{1}{k_1} + \frac{c}{k_2} + \frac{X_D}{D} + \frac{L}{aD}. \tag{6.17}$$

Therefore, the intercept of the C'/j vs. $W^{-1/2}$ plot would be expected to show dependence on C' if this model pertains.

The "Langmuir" approach described here is very suitable for analysis of results where surfactants are used, which may adsorb at the liquid–liquid interface in competition with the diffusant.

6.3 EXPERIMENTAL USE OF THE RDC

6.3.1 *Chemicals*

Water was doubly distilled, octanol was HPLC grade (from Aldrich), and sodium hydroxide and potassium dihydrogen phosphate were used to make pH 7 buffer.

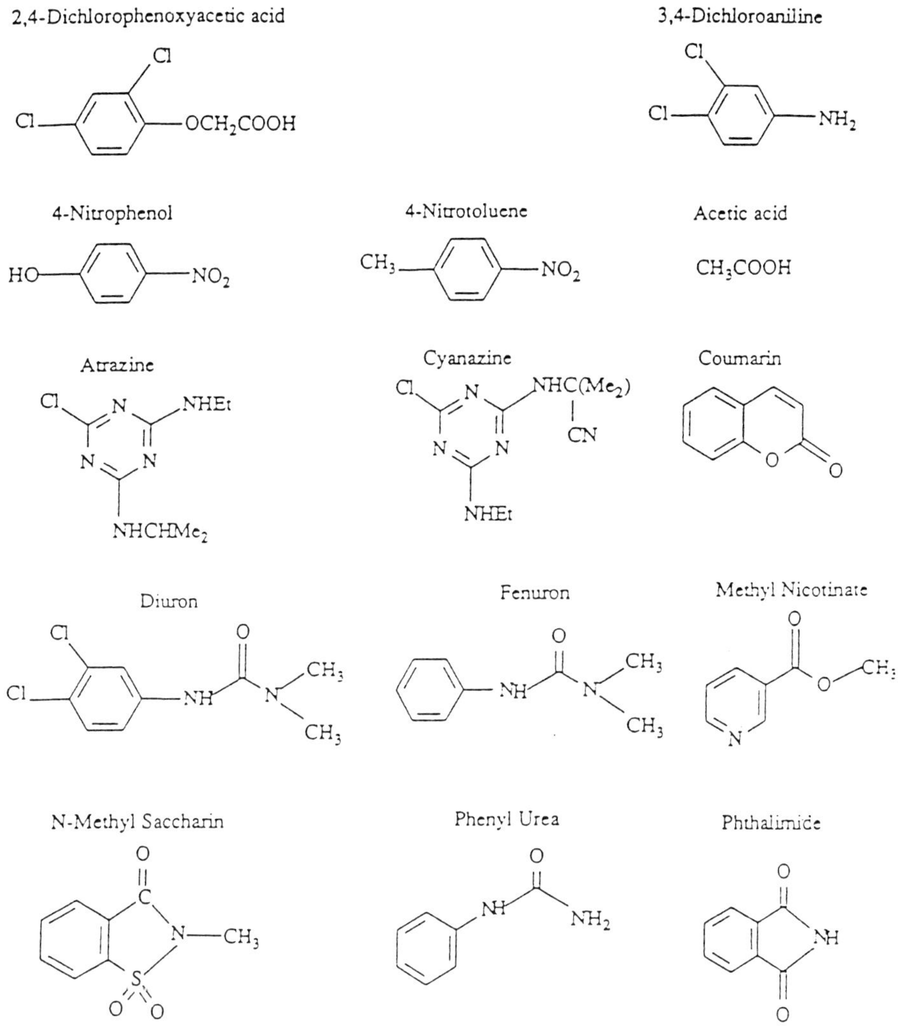

Fig. 6.6. Structures of the compounds studied.

The structures of the substances studied are shown in Fig. 6.6. "WL" compounds and fenuron, synthesized at Sittingbourne Research Centre were typically of 95% purity and used without further purification. Other compounds were obtained from standard commercial sources.

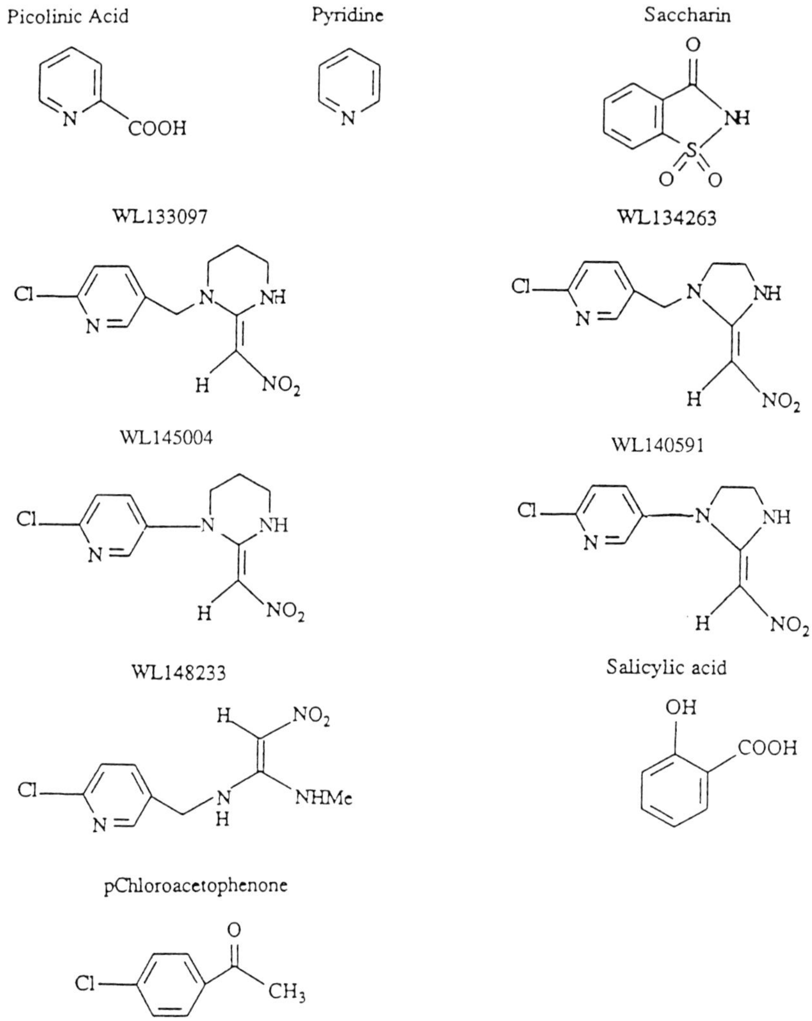

Fig. 6.6. Continued.

6.3.2 *Rotating diffusion cell*

The RDC used in the majority of this work is shown in Fig. 6.2, and is based on the design of Albery *et al.* [16]. The baffle in the interior is designed to minimise the influence of the walls of the vessel causing convection to be lateral rather than toward the disk. A further method of supporting an inert filter is to use a clamp design RDC (Fig. 6.3). The use is essentially the same except that particular care is needed to ensure bubbles are not trapped under the cell.

6.3.3 Measurement of flux

Two methods have been used, both using the ultra-violet (UV) absorption of the compound under test. In order to interpret absorbance vs. time data, extinction coefficients were obtained by conventional methods.

6.3.3.1 Continuous measurement

The contents of the outer compartment were recirculated through a Cecil double beam UV spectrophotometer by a peristaltic pump or an HPLC pump. The absorbance readings were recorded on an IBM PC via a Compuface interface. A program automatically changed the rotation speed every 25 min. The program analysed the gradient at each speed every 25 min by linear regression with an iterative rejection of outliners, and recorded the rotation speed and the absorbance mid-way through the particular interval.

The C'/j vs. $W^{-1/2}$ data were collected and analysed by linear regression.

6.3.3.2 Manual sampling

This was occasionally necessary if it was shown that the compound was absorbed by the pump tubing. The total experimental duration is similar to (a) but the rate was measured by sampling the outer compartment.

6.3.4 RDC procedure, $o \rightarrow /o/w$ system

The filters used were from Millipore (various pore sizes, typically 0.22 μm, with filters 0.15 mm in thickness). In order to render the outer part of the filter impermeable, the literature procedure [16] was used, except that no surfactants or glycerol were used in the process. Mutually saturated octanol and pH 7 buffer were prepared; all subsequent references to octanol or buffered water pertain to these mutually saturated phases.

Sixty cubic centimetres of pH 7 buffer were placed in the jacketed vessel thermostatted at 25 °C. Five cubic centimetres of octanol were pipetted into the interior of the RDC which was then quickly placed in the 60 cm^3 of buffer. This procedure ensures the formation of the liquid–liquid interface at the outer surface of the filter. The volume of octanol in the inner compartment was made up to 20 cm^3 and the RDC rotation was started, usually at $W^{-1/2} = 0.6$ (2.78 Hz). When it is established that the liquid–liquid interface was intact and the spectrophotometer reading was stable at zero; an aliquot of an octanol solution of the compound was added to the inner compartment and measurement of flux (as UV absorbance vs. time) began. Typically, a cycle of $W^{-1/2} = 0.6, 0.9, 0.5, 0.8, 1.0, 0.4$ was followed twice. This is at variance with the method of Leahy [28] who monitored the rate at lower concentration and at one speed only, thus requiring many experiments, all conducted much nearer equilibrium, and not in a steady state.

6.3.5 RDC procedure, w → /o/w system

The w → /o/w approach requires additional care in the setting up of two liquid–liquid interfaces. The difference from the o → /o/w procedure is that a drop of the organic phase was placed onto the Millipore membrane filter, so as to fill its pores. Any excess was wiped off with a tissue. The empty RDC is placed into the 60 cm^3 of buffer, and 20 cm^3 of water added to the inner compartment. Aqueous concentrate of diffusant was added to the inner compartment to initiate the experiment.

The w → /o/o experiment was set up exactly as for o → /o/w except that aqueous concentrate is added to the outer compartment to initiate the experiment.

6.3.6. Interpretation of flux data

The RDC arrangement used most frequently in this work was o → /o/w, and the outer with aqueous (termed o/o/w). The theory developed above leads to an Equation (6.8), describing the flux of molecules, which may be simplified as follows:

$$\frac{c'}{j} = K_1 W^{-1/2} + K_2 + \frac{1}{ak_{\mathrm{oa}}}, \tag{6.18}$$

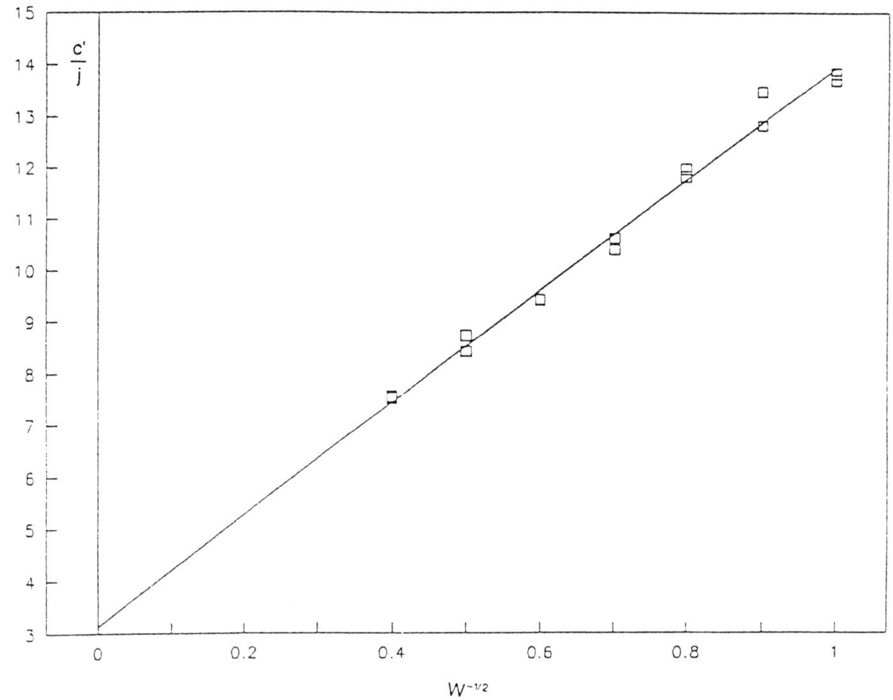

Fig. 6.7. Partitioning kinetics of cyanazine: standard plot of rotation speed dependence.

where K_2 is a constant describing diffusion within the membrane filter, K_1 is a constant describing diffusion to and from the membrane filter, and C', W, a and k_{oa} are as described above.

Figure 6.7 illustrates some data for cyanazine transfer from octanol to water. The gradient of this is K_1, and the intercept $K_2 + (1/a\ k_{oa})$. The results section analyses the data generated according to this method.

6.3.7 Diffusion coefficients

Diffusion coefficients are required to interpret all the data, but measured values are not available in most cases. The method of Wilke and Chang [33] represents the widely accepted standard for prediction of diffusion coefficients:

$$D = \frac{7.4 \times 10^{-12}(\phi_s M_s)^{1/2}}{\eta_s V_D^{0.6}}, \quad (6.19)$$

where D is the diffusion coefficient ($m^2\ s^{-1}$); ϕ_s is a constant characteristic of a solvent; M_s is the molar mass of the solvent; T is the absolute temperature; η_s is the viscosity of the solvent (cpoise); V_D is the LeBas molar volume ($cm^3\ mol^{-1}$) [35].

Aqueous diffusion coefficients may also be given by the following [34]:

$$D = \frac{13.3 \times 10^{-9}}{\eta_s^{1.14} V_B^{0.589}}, \quad (6.20)$$

where the symbols are as above.

Therefore, D_{aq} was obtained from equation (6.20) and D_{org} from equation (6.19) via

$$\frac{D_{org}}{D_{aq}} = \frac{(\phi_{org} M_{org})^{1/2} \eta_{aq}}{(\phi_{aq} M_{aq})^{1/2} \eta_{org}}. \quad (6.21)$$

During the course of the work, it was discovered that the Wilke and Chang equation (19) contains certain deficiencies; however, these are insignificant in the present context. Chapter 15 of this volume gives full information.

6.4 RESULTS

The results all pertain to transfer from octanol to water, i.e., o → /o/w.

6.4.1 Validation of the RDC technique by comparison of results obtained with different RDC and membrane types

Albery and Hadgraft [16] have taken the trouble to validate aspects of the procedures and mathematics of RDC work, particularly the applicability of the Levich equation and rotating disk hydrodynamics. The work reported here confirms and extends their conclusions, by explicit discussion of concentration dependence, and use of a wider range of substances than carboxylic acids and nicotinate esters. This contrasts with results from

TABLE 6.2
Investigation of the effect of membrane type on the interfacial transfer of methyl nicotinate

System	No. of runs	Gradient	Intercept	r^2	Comments
o → /o/w	5	0.878 ± 0.055	1.443 ± 0.044	87%	Conventional RDC
w → /o/w	4	0.233 ± 0.017	0.307 ± 0.014	82%	Conventional RDC
w → /o/w	1	0.257 ± 0.022	0.263 ± 0.030	85%	MF membrane in clamp cell
w → /o/w	1	0.182 ± 0.011	0.111 ± 0.008	96%	Fluoropore membrane in clamp cell
o → /o/w	1	0.873 ± 0.074	0.649 ± 0.053	91%	Fluoropore membrane in clamp cell
w → /o/w	1	0.251 ± 0.039	0.333 ± 0.028	84%	Durapore membrane in clamp cell
o → /o/w	1	1.109 ± 0.055	1.661 ± 0.040	97%	Durapore membrane in clamp cell
o → /w/w	1	0.963 ± 0.145	2.789 ± 0.150	77%	Durapore (hydrophilic) in clamp cell
w → /w/w	1	0.224 ± 0.038	0.439 ± 0.027	71%	Durapore (hydrophilic) in clamp cell

Leahy *et al.* [34] which cast doubt on the work of Albery and Hadgraft. Leahy states that no interfacial resistance was measurable (i.e., rate constants are high). However, the paper did not attempt to repeat the Albery and Hadgraft work completely. A clamp design using a PTFE membrane was employed, and much lower concentration of diffusant. Evidently the contradiction between the results from the two laboratories cannot be ignored since the whole validity of the method is being questioned. An earlier paper by Leahy [29] studied methyl nicotinate in a variety of solvent systems. For this reason, methyl nicotinate as a probe diffusant was employed in a series of experiments, full results from which are given below. The data for the Millipore membrane system are in good agreement with Albery *et al.* [16] except that a small dependence of the intercept (of the C'/j vs. $W^{-1/2}$ graphs) on C_{org} was found.

6.4.1.1 Results

All experiments reported in this section were performed at 15 °C in order to enhance stability of the interface. Table 6.2 gives the results.

Using these results and the constants listed below, calculations of k_{oa} and k_{ao} are possible.

6.4.1.2 Constants used

For methyl nicotinate, $P_{ow} = 4.6$ at 15 °C, 6.76 at 25 °C, $D_{aq} = 6.6 \times 10^{-10}\ m^2\ s^{-1}$ at 15 °C, and $D_{org} = 2.7 \times 10^{-10}\ m^2\ s^{-1}$ at 15 °C.

The membranes used were all from Millipore, with thickness and porosity as shown: MF Cellulose (150 μm, 0.75), Fluoropore (60 μm, 0.85) and Durapore (125 μm, 0.75). Viscosity values were obtained from previous papers [16] and standard literature sources, giving $\nu_{org}^{1/6} = 0.153$ at 15 °C, and $\nu_{aq}^{1/6} = 0.102$ at 15 °C.

6.4.1.3 Gradients

From equations (6.7) and (6.8), for the o → /o/w system, the predicted gradient = 0.951 $\mu m\ s^{-1}\ Hz^{1/2}$ and for w → /o/w, the predicted gradient = 0.173 $\mu m\ s^{-1}\ Hz^{1/2}$.

Agreement with experiment is acceptable (Table 6.2). Note that Guy *et al.* [17, 18] forced predicted gradients through the measured data, although there is no compelling reason to do so. This has not been done here.

6.4.1.4 Summary of k_{oa} and k_{ao}, methyl nicotinate, 15 °C

The gradients described above relate to diffusion; the point of contention being addressed is whether rate constants k_{oa} and k_{ao} are measurable. The results obtained in this study were:

Membrane	Experiment	$k_{ao}/\mu m\ s^{-1}$	$k_{oa}/\mu m\ s^{-1}$
MF, cellulose	w → /o/w	18	(3.7)
	o → /o/w	(9.3)	1.9
Fluoropore	w → /o/w	41	(3.1)
	o → /o/w	(15)	3.1
Durapore	w → /o/w	13	(2.7)
	o → /o/w	(6.3)	1.3
	o → /w/w	(6.8)	1.4

Figures in parentheses are calculated from the other rate constant and the P_{ow} value.

The RDC—clamped membrane configuration gave rise to estimates of

k_{oa} and k_{ao} generally in agreement for Durapore (polyvinylidene fluoride) and cellulose membranes, which in turn agreed with the conventional RDC, within experimental error. Results obtained with the Fluoropore (PFTE) membranes lead to faster rate constants being obtained. This appears not to be due to the clamp system itself. In experimental use the Fluoropore membranes tend to oscillate which makes them of no use for valid use in the RDC experiment. The results of Leahy *et al.* [30] are therefore in considerable doubt for experimental reasons, and the apparent contradictions between his and other [17] work are resolved.

6.4.2 Detailed description of generation of results for the o → /o/w system

This section starts with analysis of the results of one particular experiment in order to detail properly the process gone through for each. There follows a validation of the RDC mathematics (an extension of that done by the originators of the method [16]). Results for the o → /o/w programme are then summarised, including some unusual effects that have been observed. In this section, discussions of procedure and rationale will be made, in order to concentrate on deeper issues in Section 6.5.

The inner compartment concentration is calculated by mass balance. C' (which is $C_{inner} - P_{ow}C_{outer}$) is then used in a form appropriate to Equation (6.7), i.e., C'/j plotted against $W^{-1/2}$ and linear regression performed, as illustrated in Figure 6.7. This is the fundamental form of RDC results; hereafter the gradient and intercept of such analyses are termed plot gradient and plot intercept respectively.

6.4.2.1 Results for the o → /o/w system

A given compound was tested in more than one run in order to establish the repeatability. In some cases two were sufficient, in other more were necessary. The data from each run were combined and analysed by using a SAS routine [32]. The results are summarised in Table 6.3. The headings have the following meanings:

Conc. (mM)	= concentration of diffusant in the inner compartment.
Area (cm^2)	= nominal area of untreated membrane filter.
Grad.	= gradient of C'/j vs. $W^{-1/2}$ plot, termed plot gradient.

TABLE 6.3
RDC results for the 23 compound set

No.	Compound	Log P_{ow}	pKa	Log P'_{ow}	V	Calculated $\times 10^{10}$ D_{aq}	D_{oct}	Predicted plot gradient
1	2,4-Dichlorophenoxy acetic acid	2.81	2.87	−1.1	209.7	6.54	1.33	0.362
2	3,4-dichloroaniline	2.69		2.69	152	7.91	1.61	36.438
3	4-Nitrophenol	1.91	7.15	1.68	131.9	8.60	1.75	3.616
4	4-Nitrotoluene	2.37		2.37	146.7	8.08	1.64	17.360
5	Acetic acid	−0.17	4.76	−2.41	48.4	1.55	3.16	0.200
6	Atrazine	2.7		2.7	250.6	5.89	1.20	45.366
7	Coumarin	1.39		1.39	250.6	5.89	1.20	2.584
8	Cyanazine	2.22		2.22	262.5	5.73	1.17	15.557
9	Diuron	2.77		2.77	239.4	6.05	1.23	52.286
10	Fenuron	0.98		0.98	194	6.85	1.39	1.119
11	Methyl nicotinate	0.83		0.83	148.3	8.02	1.63	0.804
12	*M*-Methyl saccharin	1.45		1.45	179.4	7.17	1.46	2.552
13	Phenylurea	0.77		0.77	148.1	8.03	1.64	0.740
14	Phthalimide	1.15	8.3	1.13	137.2	8.40	1.71	1.253
15	Picolinic acid	−1.1		−2.35	128.3	8.74	1.78	0.293
16	Pyridine	0.65	5.25	0.65	93.1	1.06	2.15	0.530
17	Saccharin	0.91	1.31	−2.20	157.2	7.75	1.58	0.317
18	Salicylic acid	2.26	2.98	−1.22	142.2	8.23	1.67	0.309
19	WL134263	−0.1		−0.1	260.5	5.76	1.17	0.459
20	WL140591	−0.21		−0.21	238.3	6.07	1.24	0.427
21	WL145004	−0.42		−0.42	253.3	5.85	1.19	0.417
22	WL148233	0.13		0.13	253.5	5.85	1.19	0.504
23	*p*-Chloroacetophenone	2.32		2.32	161.3	7.64	1.56	16.094

Int. = intercept of C'/j vs. $W^{-1/2}$ plot, termed plot intercept.

$\log P_{ow}$, $\log P'_{ow}$ = log of the octanol–water partition coefficient for the neutral molecule and at pH 7 respectively (calculated from $\log P_{ow}$ and pK_a or measured) $\log P_{ow}$ values have been obtained from the Medchem MASTER 3.53 database.

V = Le Bas molar molecular ($cm^3\ mol^{-1}$) calculated[37].

D_{aq} = Diffusion coefficient in water ($m^2\ s^{-1}$).

D_{org} = Diffusion coefficient in octanol ($m^2\ s^{-1}$).

"Predicted plot gradient" = i.e., from equation (6.7).

Calculated L/aD_{org}	Individual Adj R^2	Common gradient			All regressed		j at 2.78 Hz	k_{oa}
		Adj R^2	Gradient	S Error	Adj R^2	Intercept		
1.50	93.7	93.9	0.886	0.084	14.5	0.248	0.148	0.28
1.24	96.8	95.3	39.18	1.72	79.3	1.428	0.040	7.16
1.14	90.4	90.8	5.13	0.32	89.8	2.235	0.188	1.22
1.22	85.3	86.2	54.45	5.26	86.9	−13.61	0.052	
0.63	64.5	65	0.886	0.172	24.4	1.583	0.473	1.4033
1.67	97.6	–				3.86	0.037	0.4455
1.67	75.4	73	6.73	0.62	66.9	1.124	0.194	
1.71	95.2	95.3	10.49	0.38	87.7	3.151	0.106	0.9274
1.62	65.5	65.8	131.6	11.8	64.2	−28.59	0.20	
1.43	80.1	80.7	1.86	0.19	72.8	2.728	0.260	1.0302
1.22	79.2	78.6	0.933	0.082	66.4	1.461	0.495	5.6256
1.37	82.3	83	1.843	0.13	82.6	1-147	0.444	
1.22	95.5	95.7	1.096	0.074	12.5	3.405	0.246	0.6110
1.17	87.9	88.4	0.769	0.063	63.7	1.34	0.555	7.8049
1.12	64.3	65.6	0.317	0.0562	38.7	2.804	0.334	0.7935
0.93	78.8	80.2	0.515	0.065	15.5	1.523	0.546	2.2501
1.27	74.1	73.3	0.219	0.0268	64.4	1.554	0.593	4.6414
1.19	2.2	8.7	0.391	0.41	0	2.658	0.346	0.9107
1.71	68.8	68.4	0.278	0.215	0	4.27	0.225	0.5199
1.62	50.6	52.8	0.725	0.294	4.8	4.78	0.192	0.4217
1.68	66.5	65.5	0.165	0.12	0	3.536	0.275	0.7175
1.68	44.3	29.1	1.883	0.47	20.6	4.115	0.191	0.5472
1.29	86.1	88.7	23.83	3.31	27.1	3.43	0.056	0.6219

L/aD_{org} = the term found in equation (6.7) for diffusion in the filter.

2.78 Hz j = Flux in μmol m^{-2} s^{-1} at 2.78 Hz, calculated from the fitted line.

The routine found a common plot gradient for all but one case (atrazine) which is considered in further detail below. In the majority of cases, the runs did not give a common plot intercept at the 95% confidence level. However, some estimate of the combined plot intercept is needed, and these are given.

Results for atrazine in Table 6.4 are based on the calculation of a common plot intercept. The overall r^2 for a common plot intercept is 96.7% compared to 97.6% for the free combined regression.

The finding of constant plot gradient for each compound (except one) means that attention must be paid to factors which do not affect the plot gradient in order to explain the plot intercept variation. Therefore, miscalculation of concentration, filter area or extinction coefficient are excluded. The variability in plot intercept may be due to several factors:

(1) The procedure for establishment of the interface is variable because of the possibility of small leaks from the inner compartment (major leaks are easily observed).
(2) Traces of impurity may adsorb at the liquid–liquid interface in a concentration-dependent manner.
(3) The membrane filters have a variability (in the porosity and thickness), although this is believed to be only of the order of 2–5% [16].

6.4.2.2 Atrazine

The results for atrazine are unusual compared to the rest of the set, in that a marked dependence of Plot gradient on concentration was found (Fig. 6.8); for this, plot gradient = 5.40 + 1.21 [Atrazine] and $r^2 = 0.84$, $n = 7$.

It is interesting that the plot gradient only approaches the theory prediction at high atrazine concentration. No explanation of this dependency is available.

An overview of the results illustrates some other surprising features. For diuron, *N*-methyl saccharin, coumarin, 3,4-dichloroaniline (probably) and 4-nitrotoleune, plot intercepts less than the anticipated L/aD_{org} were found, which should be the lowest allowed value for the plot intercept. These are largely high log P_{ow} compounds. In this case, the calculation $C' = C_{inner} - P_{ow}C_{outer}$ is particularly important. The flux was very small, and it is possible that true steady state fluxes were not established [32].

The problem also sheds some light on the Leahy data [30]. First, there is the problem of using a non-planar membrane in a clamp-RDC, as discussed above. Second, he was using an inner compartment concentration of 0.1 mM, and assuming steady state kinetics, which work in this laboratory has clearly shown do not apply under their conditions.

Finally, the inherent heterogeneity varies from compound to compound. This is only partly due to the compounds with larger extinction coefficients giving higher rates of change of absorbance and more accurately determined flux. This point is discussed further in the following section.

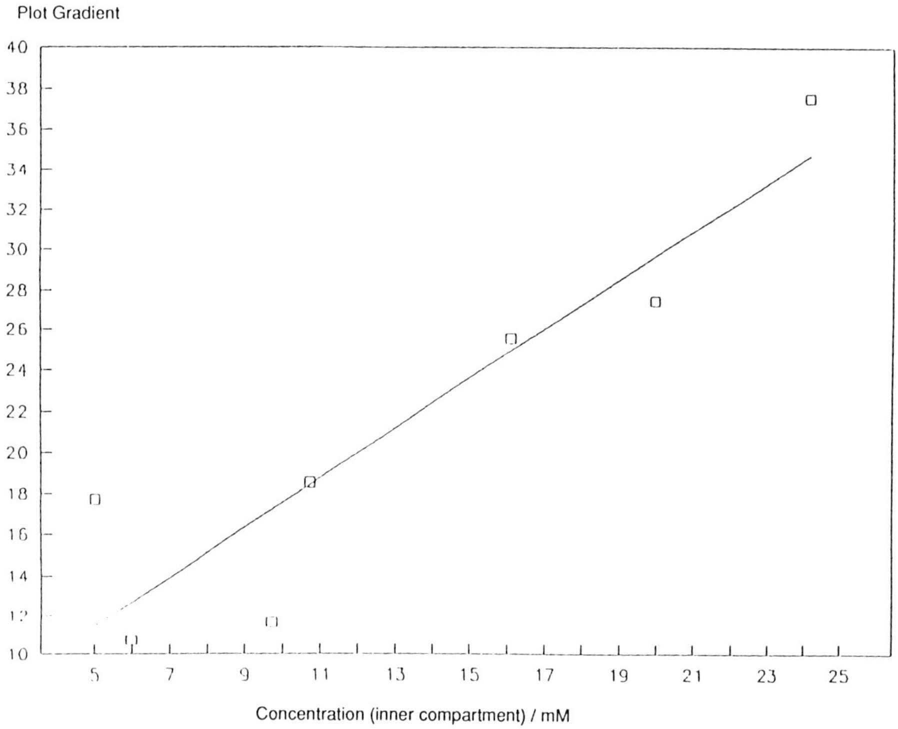

Fig. 6.8. Atrazine: concentration dependence of the plot gradient.

6.5 DISCUSSION

6.5.1 Comparison of measured and predicted plot gradients

The results of Table 6.4 are used in Fig. 6.9, which shows a comparison of the 23 compounds; a logarithmic scale is used to enable a useful overview. Linear regression ($n = 23$, $r^2 = 0.893$), gives

$$\log(\text{Predicted plot gradient}) = -0.077 + 0.887\,(\text{Measured plot gradient}). \quad (6.22)$$

The line on the graph is not that of the best fit but one of slope = 1 passing through the origin, i.e., the expected dependence based on theory. There

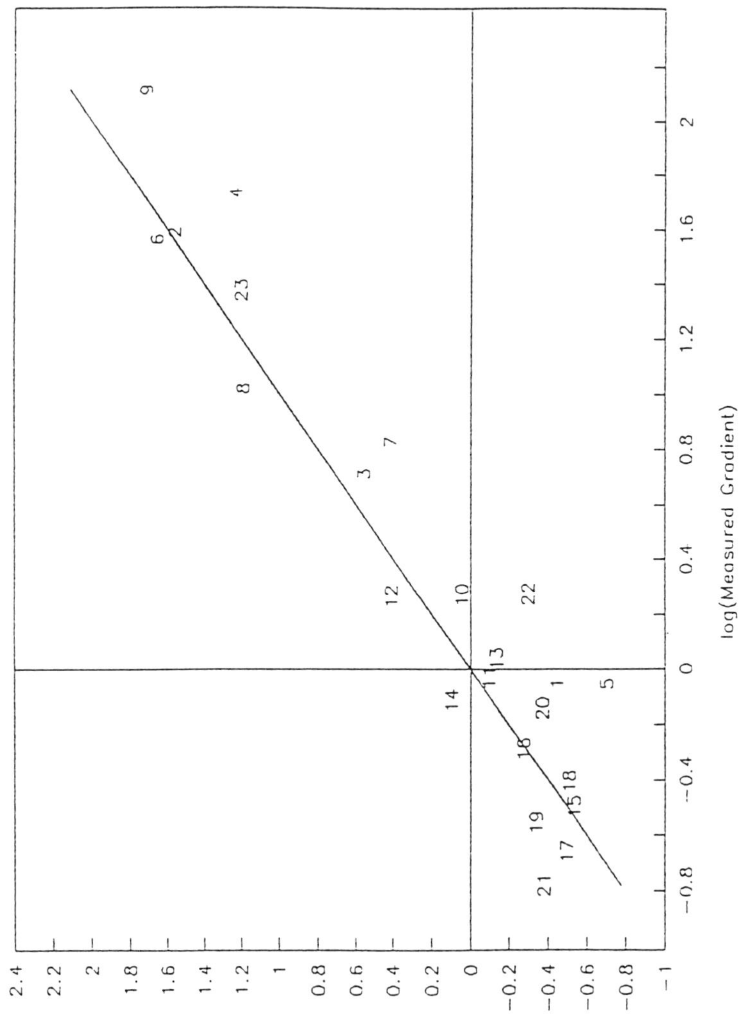

Fig. 6.9. Plot gradient: comparison of theory and measured values.

appears to be no systemic trend on the outliers. The deviations imply that in some cases measured plot gradient is higher than predicted: i.e. P_{ow} is higher, and/or D values are actually lower than those predicted.

Table 6.4 uses calculated diffusion coefficients to predict plot gradient. The plot gradient prediction would be more useful with measured D values, but factors greater than 2 between calculated and measured plot gradient are significant deviations from accepted theory, and the source of the difference is unlikely to be error in D values. Anomalies were found for two compounds with high log P_{ow} which also exhibit a negative plot intercept, and for two carboxylic acids (2,4-D and acetic acid).

It is important to note that the P_{ow} used is that for pH 7 (P'_{ow}). Consider 2,4-dichlorophenoxyacetic acid: the predicted plot gradient would be much higher if P_{ow} was used in equation (6.8) rather than P'_{ow}. This implies that the compound ionises from neutral to anion on or near the liquid–liquid interface (within the diffusion layer). Therefore, for ionisable compounds, the rate constants k_{oa} and k_{ao} must also include the ionisation process. Because proton transfer is fast it has not been explicitly described here.

6.5.2 Flux as a function of low P'_{ow}

Figure 6.10 shows the average flux at $W^{-1/2} = 0.6$ for the 23 compounds, based on the overall regression plot gradient and plot intercept (combined runs). The results fall into two regions: the first, log $P'_{ow} < 1.2$, where flux is largely independent of P'_{ow}; the second, where log $P'_{ow} > 1.2$ and flux drops rapidly with increasing log P_{ow}, as predicted by equation (6.8). The line shows the diffusion limited rates. Deviation below the line are due to the term in k_{oa}.

6.5.3 Comparison of rate constants (k_{oa})

Figure 6.11 also uses log P'_{ow} as a suitable variable on which to present the information. The vertical axis is the term $(ak_{oa})^{-1}$ obtained from the plot intercept of a C'/j vs. $W^{-1/2}$ graph, after subtraction of the calculated L/aD_{org} term. Although, according to equation (6.1), no values below 0 should be found (i.e., rate constants cannot be negative), inspection of Table 6.5 reveals some. Two high log P_{ow} compounds, 4-nitrotoluene and diuron are particularly pronounced, both having much higher plot gradients than predicted (both fall outside Figure 6.11). 4-chloroacetophenone is another high log P_{ow} (high plot gradient) example. However, overall

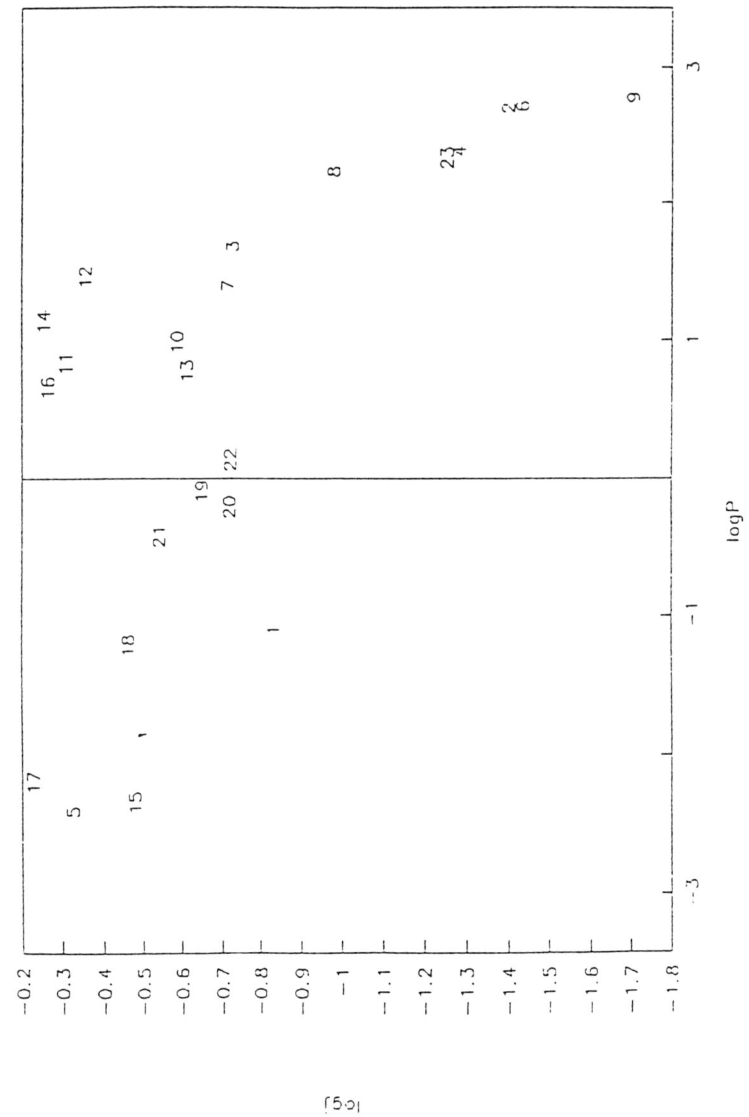

Fig. 6.10. 23 compound set: dependence of partioning rate on log P_{ow}.

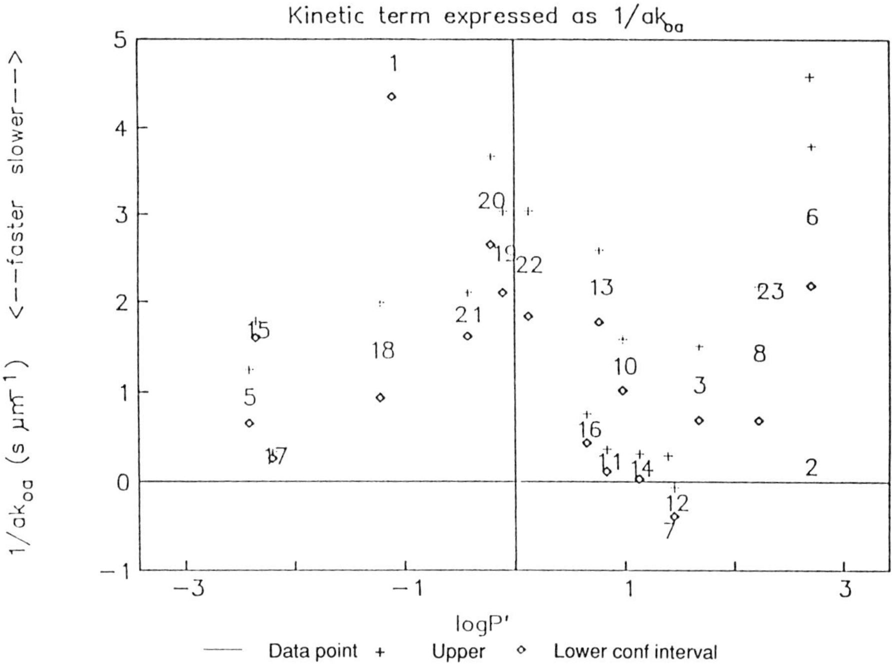

Fig. 6.11. 23 compound set: dependence of k_{oa} on log P_{ow}.

there is no evidence for log P'_{ow} dependence in Figure 6.10. Some observations are:

(a) The size of the 90% confidence limits shows that it may be difficult to get structural information from differences in rate constants for members of a closely related series (e.g., compounds 19–22).

(b) The $(ak_{ao})^{-1}$ values of the sub-set fenuron, diuron, 3,4-dichloroaniline and phenyl urea show a decreasing kinetic term (i.e., faster interfacial transfer) as polarity decreases: phenyl urea ($(ak_{ao})^{-1} = 2.18$), then fenuron, (1.29), diuron (nominally 0) and 3,4-dichloraniline (nominally 0).

Fenuron showed how steric hindrance of the polar group gives it faster movement through the interface than phenyl urea.

(c) The weak acids (carboxylic acids and 4-nitrophenol) generally showed slow rate constants for crossing from octanol to water.
(d) Reduction in accessibility of the polar group in saccharin by *N*-methylation increased the rate of interfacial transfer considerably.
(e) The set of nitroenamines (19–21), have slow rates of interfacial transfer, are highly dipolar and accessible to solvent water molecules.
(f) The triazines give slow rates of interfacial transfer. Although slightly weaker bases than anilines, the triazine rings exhibit dipolarity around the ring *N* atoms.

Therefore, empirically, the rates of interfacial transfer seem to be relating to the dipolar nature of parts of the molecule, and the accessibility to solvent of such groups. Molecules without dipolar regions move rapidly through the liquid–liquid interface. Those with strong dipoles move more slowly.

Albery *et al.* [16] have used Marcus' Theory [16] interfacial processes to calculate that an interfacial rate constant of 2 $\mu m\ s^{-1}$ constitutes a free energy barrier of 44 $kJ\ mol^{-1}$. The slowest rate here (2,4-*D*) gives 49 $kJ\ mol^{-1}$, and rates >100 $\mu m\ s^{-1}$ have barriers of 34 $kJ\ mol^{-1}$, still rather larger than that of diffusion itself (20 $kJ\ mol^{-1}$). However, the RDC method is not suitable for accurate measurement of $k > 20\ \mu m\ s^{-1}$.

The indication that the rate constants relate to the dipolar nature of the molecule moving from organic to aqueous phase agrees well with the assignment of these rate constants to resolvation processes at the interface.

6.5.4. Relationship of the results to biological data and compartmental models

It has been shown that the magnitude of the rate constant for crossing the octanol-water interface makes the energy barrier significantly larger than the diffusional barrier. It has also been shown that for compounds with log P'_{ow} less than ≈1.2, the overall rates are faster and the interfacial kinetics term more important. However, detailed development of a model would be needed to understand what the relative importance of diffusion and interfacial terms (such as cuticle or membrane permeation) are *in vivo*. No clear dependence of interfacial rate constants on log P_{ow} was seen, but the initial emphasis of such a study should be on the intermediate region of log P_{ow}. However, it is important to note that compounds with a rapid rate of transfer overall are considered to be mobile *in vivo*.

The authors have studied the effects of some surfactants on the rate of interfacial transfer; this work, which has not been published, has shown that the rate may be increased or decreased. The structure of the surfactant is important, and the effects apply both above and below the critical micelle concentration of the surfactant. The effect in certain cases was to impede transfer totally, which suggests that much more work is required to build up an understanding of the properties of biological systems, and of the effects of surfactants on them. Surfactants are commonly present in pesticide formulations.

6.6 CONCLUSIONS

The development and experimental use of the rotating diffusion cell method of measuring rates of interfacial transfer have been described. The set of 23 compounds studied shows that the rate of transfer from octanol to water depends upon both diffusion and the rate of crossing the liquid–liquid interface. There is no relationship between rate of interfacial transfer and log P_{ow}, but the overall rate depends upon P_{ow} above log $P_{ow} \sim 1.2$ because one of the diffusional rate terms contains P_{ow}. The measured ranges of overall rate and interfacial transfer rate constants cover two orders of magnitude only. Below log $P'_{ow} \simeq 1.2$, kinetics of interfacial transfer are generally as important as rate of diffusion, or more so. Above log $P'_{ow} \simeq 1.2$, the position of equilibrium controls the rate. The 90% confidence intervals on the interfacial rate constants are sufficiently wide that the RDC method will not distinguish close members of a homologous series. Observations of the structural effects upon rate of crossing the liquid–liquid interface support the literature view that the rate of resolvation is involved during interfacial transfer, which relates to dipolarity in the transferring molecule. Future work should concentrate on experiments that model aspects of the biological system more closely than octanol/water.

ACKNOWLEDGEMENTS

Thanks are due to S.F. Gallagher for the computer system controlling the RDC, S.A. Ballard for performing the majority of the experimental work, and S.E. Pack and J.A. Reading for help with the statistical analyses.

REFERENCES

[1] P.J. Quinn, The Molecular Biology of Cell Membranes (Macmillan, London, 1985).
[2] D.F. Sears, Biological Horizons in Surface Science (1973).
[3] L.V. Edgington, Ann. Rev. Phytopathol. 19 (1981) 107.
[4] G.G. Briggs and R.H. Bromilow, Chem. Soc. Spec. Pub. 29 (1976) 129.
[5] F.T. Lichtner, What's New Plant Phys. 15 (1984) 29.
[6] C.E. Price, in D.F. Cutler (ed.), The Plant Cuticle, (Academic, 1980, 237).
[7] Y.C. Martin, Quantitative Drug Design (Marcel Dekker, New York, 1978).
[8] ACS Symposium Series 225, Pesticide Synthesis through Rational Approaches, (ACS, 1984).
[9] ACS Symposium Series 112, Computer-Assisted Drug Design (ACS, 1979).
[10] D.E. Fenton, Chem. Soc. Rev. 6 (1977) 325.
[11] H.T. Tien, Bilayer Lipid Membranes (Dekker, New York, 1974).
[12] B.E. Cohen and A.D. Bangham, Nature (London), 236 (1972) 173.
[13] J.C. Dearden and M.S. Townend, see page 135 in ref. [4].
[14] H. van de Waterbeemd, P. van Bajel and A. Jansen, J. Pharm. Sci. 70 (1981) 1081.
[15] A. Brodin and A. Argen, Acta Pharm. Suec. 8 (1971) 609.
[16] W.J. Albery, J.F. Burke, E.B. Leffler, J. Hadgraft, J.C.S. Far. Trans. I. 72 (1976) 1618.
[17] R.H. Guy and D.H. Honda, Int. J. Pharm. 19 (1984) 129.
[18] R. Fleming, R.H. Guy and J. Hadgraft, J. Pharm. Sci. 72 (1983) 142.
[19] G.J. Hanna and R.D. Noble, Chem. Rev. 85 (1985) 583.
[20] J.W. Chapman, P.R. Cox and A.N. Strachan, Chem. Eng. Sci 29 (1974) 1247.
[21] G.A. Mansoori and A.J. Madden, A.I.Ch.E.J. 15 (1969) 245.
[22] Interfacial Synthesis, F. Millich and C.E. Carraher (eds.) (Dekker, New York, 1977).
[23] C.M. Starks and R.M. Owens, J. Amer. Chem. Soc. 95 (1973) 3613.
[24] I.L. Jenkins, Hydrometallurgy 4 (1979) 1.
[25] W.J. Albery, R.A. Choudhery and P.R. Fisk, Faraday Disc. Chem. Soc. 77 (1984) 53.
[26] J. Hadgraft and G. Ridout, J. Pharm. Pharm. 37 (1985) 75 pp.
[27] M. Ahmed, J. Hadgraft and I.W. Kellaway, Int. J. Pharm. 13 (1983) 227.
[28] D.E. Leahy and A.R. Wait, J. Pharm. Sci. 75 (1986) 1157.
[29] D.E. Leahy, in QSAR in D. Hadzi and B. Jerman-Blazic (eds.), QSAR in Drug Design and Toxicology (Elsevier, Amsterdam, 1987).
[30] D.E. Leahy, Int. J. Pharm. 50 (1989) 117.
[31] J. Hadgraft and R.H. Guy (personal communication).
[32] W.J. Albery, Electrode Kinetics (Clarendon, Oxford, 1975).
[33] C. Wilke and P. Chang, A.I.Ch.E.J. 1 (1955), 264.
[34] W. Hayduk and H. Laudie, A.I.Ch.E.J. 20 (1974) 611.
[35] W.J. Lyman, W.F. Reehl and D.H. Rosenblatt, Handbook of Chemical Property Estimation Methods, (McGraw-Hill, New York, 1982).

Chapter 7

Redox Mediated Whole Cell Biosensors for Toxicity Assessment and Environmental Protection

BARRY G.D. HAGGETT

7.1. INTRODUCTION

In the thirty years since they were first described, biosensors have created an enormous amount of academic and commercial interest. Nearly every conceivable combination of biological material and physicochemical transducer has been explored in more or less detail. Enzyme based devices have been particularly widely investigated while sensors that incorporate immuno materials are of increasing importance. However, there are substantial problems inherent in bringing together biological and other materials in a stable and reproducible manner. So, despite massive investments in time and money there have been a very limited number of commercially viable devices successfully launched into the marketplace. Of these, perhaps the most conspicuous has been the range of rapid glucose in blood tests for diabetic users to monitor and control their condition. The enzyme glucose oxidase (E.C. 1.1.3.4) is used to specifically target glucose, whilst a variety of techniques are used to transduce the chemical signal into an electronic signal—however, the use of redox (electron transfer) mediators is well established [1–6]. The mediator interacts with the enzyme cofactor:

$$\beta - \mathrm{D} - \text{glucose} + \text{FAD} + \mathrm{H_2O} \xrightarrow{\text{glucose oxidase}} \text{gluconic acid} + \text{FADH}, \tag{7.1}$$

$$\text{FADH}_2 + \text{M}_{\text{ox}} \rightarrow \text{FAD} + \text{M}_{\text{red}} + 2\text{H}^+. \tag{7.2}$$

FAD (flavine adenine dinucleotide) is located at the enzyme's active site (as). In oxidizing glucose, the cofactor is reduced to $FADH_2$, but the oxidized form of an appropriate mediator (M_{ox}) has a redox potential sufficient to re-oxidize the cofactor whilst itself being reduced. In its simplest embodiment, the reduced mediator (M_{red}) diffuses from the active site at the enzyme to an electrode surface poised at a potential sufficiently oxidizing to re-oxidize the mediator:

$$M_{red,as} \xrightarrow{\text{diffusion}} M_{red,electrode}, \tag{7.3}$$

$$M_{red,electrode} \xrightarrow{\text{electrode, } E>E^0} M_{ox,electrode}. \tag{7.4}$$

Diffusion of the reoxidized mediator back to the active site completes the recycling of the redox mediator.

Glucose in blood biosensors have been heralded as paradigms for other enzyme based biosensor developments, but they are in fact atypical. While there may be analytes other than glucose that are of similar or greater importance, there are few enzymes that are as stable, active and cheap as glucose oxidase and there are few applications with such a large user base where the results can be of such immediate use. These prosaic factors, rather than a lack of scientific or technical ingenuity, have restricted a wider and faster uptake of enzyme based biosensors in the commercial market. With this background, it may be thought perverse to pursue, for anything other than academic curiosity, the development of sensing devices that incorporate biological whole cells, since such materials bring problems over and above those of enzyme based devices. Nevertheless, as will be shown later, there are important applications that *require* the use of *living* biological materials and in some of these applications whole cell biosensing devices offer actual or potential benefits.

Whole cell biosensors are those devices that incorporate biological cells rather than cellular components such as enzymes, membrane fragments or spheroplasts. The biological material may be in the form of tissue slices or organelles [7–9] but the emphasis of this work is on the use of isolated cell cultures (bacteria, micoalgae, microfungi, invertebrate or mammalian cell cultures, etc.) or the use of consortia of cell populations (e.g., activated sludge). The number and variety of "natural" cell types is truly enormous—for example, the National Collection of Industrial and Marine Bacteria (NCIMB) contains thousands of strains and this is just one of many collections throughout the world [10]. Bearing in mind the possibility of genetically manipulating cell lines, the potential for exploitation is immense.

Cellular biosensors have been widely described [11–55]. In many cases, the cells have been used in a manner analogous to that of enzyme based devices simply because they contain substantial quantities of particular enzymes. There are, of course, advantages to this approach since the enzymes do not have to be isolated and so may be cheaper but also more active and more stable than the purified components. However, the reproducibility, speed of response and selectivity of the cell based devices will, in general, be less favorable than their enzyme based counterparts. This is because of the relatively large physical size of the cells, the presence of membranes that hinder diffusion and the presence of enzymes other than the one(s) of particular interest. Nevertheless, a range of approaches has been adopted to improve the selectivity and other characteristics of whole cell biosensor devices. These were reviewed by Racek [11] and include:

- selecting cells for which the substance to be determined is the only source of carbon and/or energy for the cells;
- use of genetically manipulated cells;
- permeabilization of the cell membrane;
- induction of enzyme synthesis;
- inhibition of enzymes of other metabolic pathways, etc.

Whole cell biosensors may be divided into two groups depending on whether or not it is necessary for the cellular membranes to be intact. If the cells are being used simply as a source of one or more bound enzymes then the integrity of the cellular membranes may be of little import whereas if the enzymes are not bound within the cells then the contiguity of the cellular membranes is of vital importance. The status of the cell membranes is of particular importance when it is necessary that the cells be "alive" (i.e., potentially or actually capable of cell division, etc.). This requirement imposes additional difficulties on the fabrication of devices. However, biosensors that contain intact whole cells have the potential to act as convenient surrogates for "traditional" applications of biological cells and systems. At the forefront are those applications orientated towards environmental protection, e.g., biochemical oxygen demand [32–40], rapid toxicity assessment [41–54] and biodegradation [55]. Whilst indirect measurements of these parameters are of interest, the parameters are essentially biological and direct measurements *require* living biological systems.

7.2 WHOLE CELL BIOSENSORS FOR ESTIMATION OF BIOCHEMICAL OXYGEN DEMAND

Biochemical oxygen demand (BOD) is a measure of the quantity of biological substrate materials within a water or wastewater. Each year, many millions of these measurements are made using the so-called BOD_5 test [56]. In short, the sample to be tested is seeded with a range of microflora and the oxygen content of the sample determined before and after a five-day incubation period. The change in oxygen concentration is indicative of the concentration of substrates, ranging from easily assimmilable sugars to complex polymers. However, interpretation of the results is complicated by a variety of factors including the chemical oxygen demand (oxygen uptake due to reaction of chemical species in solution) and the presence of toxic or inhibitory components within the solutions.

Aerobic respiration is a fundamental energy source to a wide range of microflora:

$$\text{substrates} + O_2 \xrightarrow{\text{respiration}} \text{products} + \text{energy}. \tag{7.5}$$

In closed systems, or in systems to which the flux of oxygen is restricted, respiration brings about changes in oxygen tension, which makes the process accessible to measurement by appropriate sensing systems [57].

$$O_2 \xrightarrow{\text{dissolved oxygen probe}} \text{current signal}. \tag{7.6}$$

In the classical BOD_5 test, a membrane covered dissolved oxygen (DO) probe is used to measure changes in the oxygen concentration brought about by respiration of the suspended microflora. Several workers have investigated the use of immobilized microorganisms at the surface of a dissolved oxygen probe as a means of developing rapid BOD tests.

Li *et al.* [37–39] described the use of the bacterial species *Bacillus subtilis* and *Bacillus licheniformis* entrapped between a polycarbonate membrane and a Teflon-covered DO probe. The differences between the steady-state signals before and after exposure to the test samples were used as a measure of the sample BOD levels. Riedel *et al.* [32–33] described the use of the yeast *Trichosporon cutaneum*, or both *T. cutaneum* and *B. subtilis*, sandwiched between a dialysis membrane and a polyethylene-covered DO probe. In these cases, the sensor response times were speeded up by measuring the initial rates of change of the signals. In this way, measurements could be made within 30 s rather than within 15–20 min for the steady-state approach [33].

These systems for rapid determination of BOD have two particular disadvantages. First, only easily assimilable substrates are measured since the complex substrates present in environmental samples are not metabolized in the short time scales of rapid measurement systems. Second, the sensor systems described each utilize one or two microbial species: in comparison with the diversity of microflora employed in the BOD_5 test, a narrow range of substrates is accessible to the sensor measurements. Taking these two factors together, there may be substantial differences between the biosensor BOD and the classical BOD_5 results. Nevertheless, the rapid sensor measurements offer considerable advantages to process control and may be particularly applicable to the monitoring of effluent streams whose compositions vary in loading (*g*/*l*) rather than in the chemical composition.

7.3 TOXICITY ASSESSMENT FOR ENVIRONMENTAL PROTECTION

In the context of environmental impact, toxicity is the resultant detrimental effect of an effluent on a biological system. The toxic effect of a particular chemical or effluent depends on a wide variety of factors, but foremost amongst these is the nature of the biological system. A given effluent may have a severe impact on one biological species or cell type while having no, or negligible, impact on another species. Other factors that effect the toxicity of a given effluent include physicochemical factors such as pH, redox potential, speciation, temperature, etc. Synergistic and antagonistic effects of other chemical species are also important. In consequence, a chemical analysis of an effluent or wastewater stream is insufficient to determine the likely toxic effects of the waste stream on a particular environment. However, where there has been regulation of industrial discharges to the environment it has, historically, been on the basis of restricting the effluent volume and maximum permissible chemical concentrations. More recently, there have been moves towards regulating effluent discharges to the environment on the basis of the effluent toxicity ("direct toxicity assessment").

In general, toxic substances exhibit inhibitory effects that are a function of the concentration of the material. Thus, by carrying out a number of such measurements, over a range of concentrations, it is possible to establish a pattern of responses (see Fig. 7.1). This information may be used to determine the toxicity of the test material: the concentration that brings about a 50% measured effect (the "EC50") is widely used as a measure of the material's toxicity. The pattern of responses may also be used to determine the required dilution of the material needed to reduce its toxicity to a suitably low level before discharge for further treatment or direct to the environment. A particularly important application of this approach is in the protection of wastewater treatment plants "sewage farms" which in

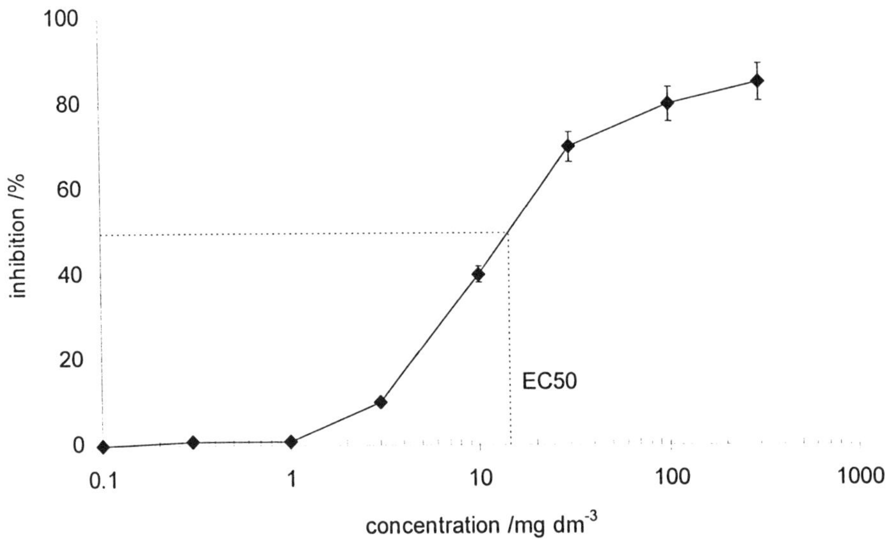

Fig. 7.1. Idealized dose-response relation between measured toxic effect, e.g., respiration inhibition, and concentration of test substance or effluent.

effect are complex bioreactors which degrade effluents and turn them into more acceptable outputs. However, a sudden influx of toxic material has the potential to irreversibly damage the microflora and render the treatment plant ineffective. This can have potentially devastating effects on the environment into which untreated effluent is discharged. Therefore there is a need to assess the toxicity of effluent streams before they reach the treatment plants. In these cases, what is important is the toxicity of the effluents towards the particular consortia of biological cells and organisms in the treatment plants—the range and concentrations of individual species will be different from one plant to another. Therefore, for example, it is common practice to remove samples of activated sludge from a treatment plant and to assess the toxicity of effluents on the fresh sludge (activated sludge respiration inhibition test [58], Fig. 7.2). However, an effluent may effect some species within the sludge more than others and the depression or annihilation of some components (e.g., the denitrifiers) of the total microflora can have a very adverse effect on the utility of the waste

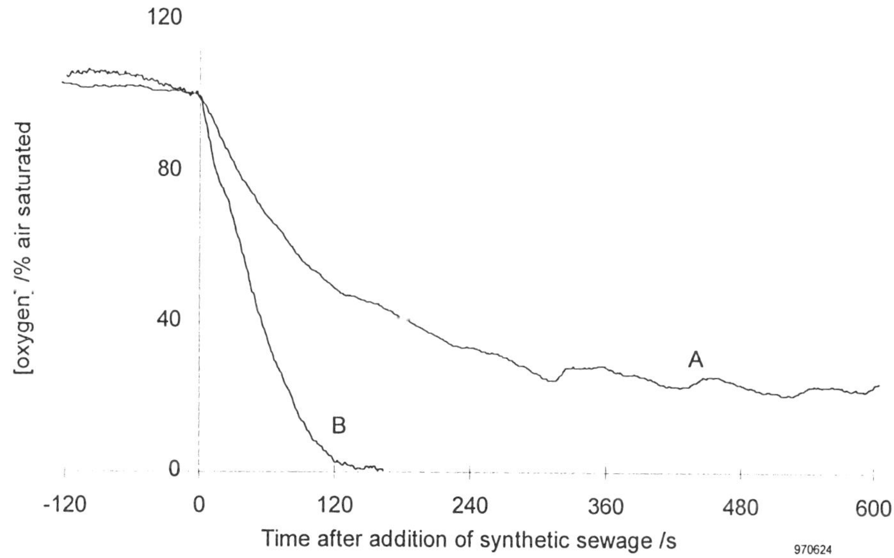

Fig. 7.2. Activated sludge respiration inhibition test. Oxygen uptake due to addition of synthetic sewage to activated sludge (A) with test effluent and (B) with tap water.

treatment plant. Thus, it is good practice to assess the effluent toxicity on a range of representative species.

7.4 TOXICITY ASSESSMENT FOR ENVIRONMENTAL PROTECTION

The trend towards direct toxicity assessment has significant ramifications both for regulators and for industry. Traditional methods of assessing toxic impact, such as algal cell growth or daphnia mortality, are very slow. Like the BOD_5 (*q.v.*), the endpoints for the tests may not be until after several days have elapsed: particularly for process control, the results may then be only of historical interest. Therefore, there has been considerable activity directed towards the development of rapid methods for direct toxicity assessment. Biosensors have featured amongst this activity and dissolved oxygen sensors are widely used to monitor perturbations in cellular respiration or photosynthesis. In these cases the immobilized cells are provided with an ample supply of respiratory substrates and exposed to the test materials. Toxicity may be determined on the basis of the measured perturbation in the oxygen flux:

$$\text{inhibition}(\%) = 100 \times \left\{1 - \frac{(\mathrm{d}[O_2]/\mathrm{d}t)_{\text{test}}}{(\mathrm{d}[O_2]/\mathrm{d}t)_{\text{control}}}\right\}. \qquad (7.7)$$

7.5 REDOX MEDIATED BIOSENSORS

Redox mediators provide an alternative to dissolved oxygen measurements for monitoring of respiration or photosynthesis. The mechanism is based on the reducing activity of metabolising biological whole cells:

$$M_{ox} + n\bar{e} \xrightarrow{\text{biological activity}} M_{red}, \qquad (7.8)$$

e.g.,

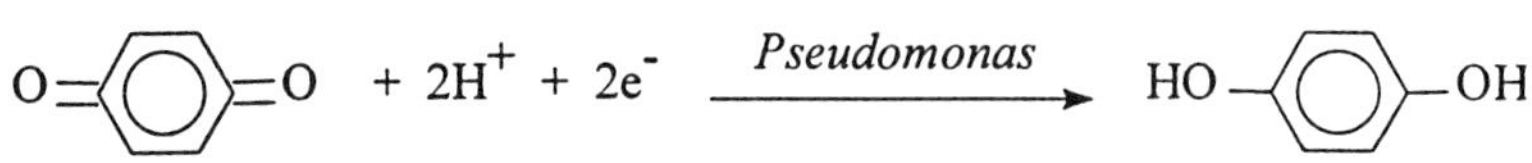

(7.9)

In common with enzyme based biosensors (equations (7.3) and (7.4)) the reduced mediator can be re-oxidized when it diffuses to an electrode poised at a suitably oxidising potential, e.g.,

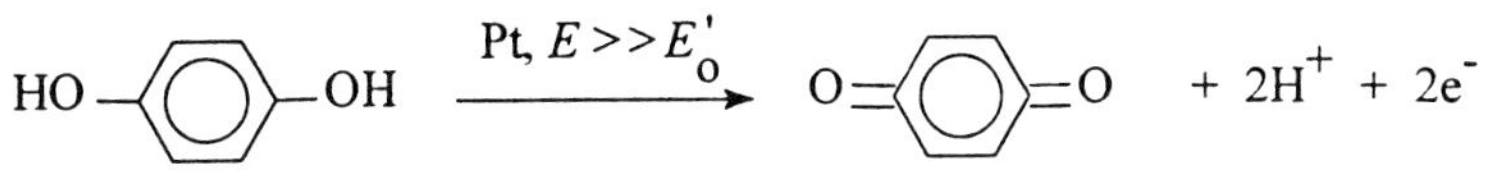

(7.10)

7.6 TIME DEPENDENCE OF MEASURED TOXIC EFFECT

A common problem of toxicity assessment is that the duration of exposure of the biological system to the test material can have a significant influence on the magnitude of the measured effect. For example, the mortality observed after 24 h may be significantly different from that observed after 72 h exposure to the same starting material. Thus, it is usual to see toxicity assessment results together with the associated exposure period. Methods for rapid toxicity assessment can also suffer from a temporal dependence, Fig. 7.3. In some cases, the maximum rate of change of the signal is used as a measure of the toxic effect; in other cases the toxic effect is determined from the change in signal over the exposure period.

An alternative approach, adopted by Albery *et al.* [59–61], is to determine the mechanism giving rise to the sensor response and to use this information together with the measured data at short times to calculate the final response. This was used for an electrochemical sensor system incorporating cytochrome oxidase where the steady-state responses of the measurement system were insufficiently fast for useful measurement of respiratory inhibitors such as cyanide, hydrogen sulphide, etc. By using mechanistic information, it was possible to successfully calculate the concentration in a test sample by real-time analysis of the sensor signals at short times after exposure to the test sample. The analysis could cope with the gradual loss of enzyme activity commonly found in these biosensor devices.

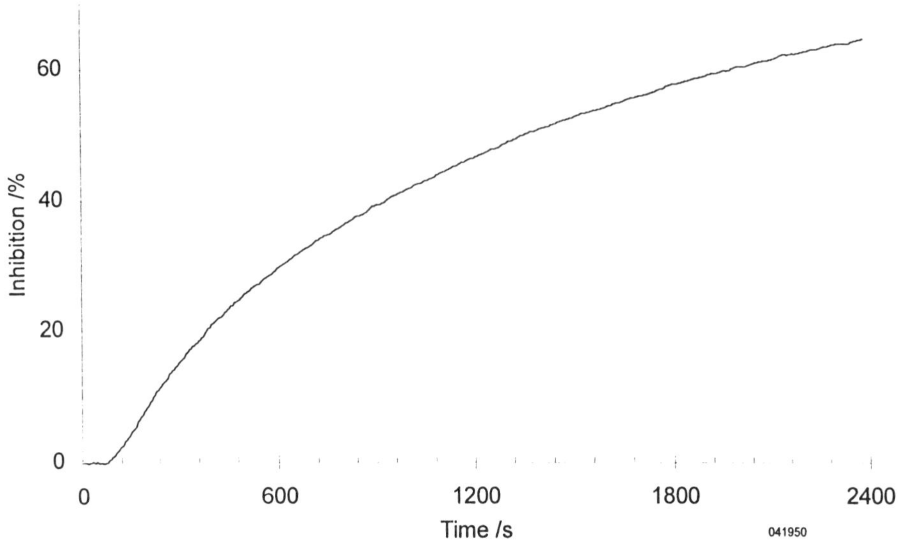

Fig. 7.3. Time dependence of inhibition calculated from *Escherichia coli* biosensor exposure to 3,5-dichlorophenol.

7.7 KINETIC MODELS OF WHOLE CELL BIOSENSORS

Despite their potential importance, there are few analytical models of whole cell biosensing devices—particularly when compared to the plethora of models describing enzyme based biosensors [62]. Although aspects of cellular biochemistry are similar to those of isolated enzymes [63], problems arise in modelling the physicochemistry of whole cells due to their complex nature: they are large (typically 0.2–10 μm); they may contain a variety of biological structures (membranes, organelles, etc.); they incorporate a diversity of biochemical pathways; and they may contain many types of active site.

With respect to whole cell biosensors utilising dissolved oxygen probes, Liang *et al.* reported a four-layer model to describe the signals from an oxygen probe used to transduce the respiratory activity of mouse leukaemia tumour cells [13]. This work was carried out with a view to using the biosensor to study the dynamic responses to various drugs. Finite difference methods were used to solve the material balances between the inner electrolyte and the Teflon-covered dissolved oxygen sensor, together with the gel-cell and mass transfer boundary layers. The oxygen uptake rate of the cells was assumed to follow Monod growth kinetics:

$$\frac{\partial[O_2]}{\partial t} = D_{O_2}\frac{\partial^2[O_2]}{\partial x^2} - \frac{k_T[O_2]}{k_{O_2} + [O_2]}\mathrm{N}, \tag{7.11}$$

where D_{O_2} is the diffusion coefficient of oxygen in the biological layer, k_T the turnover number, k_{O_2}, the respiration constant, and N the viable cell concentration.

The first term on the right hand side is simply a statement of Fick's second law of diffusion. The second term describes the rate of consumption of oxygen by the cells in terms of familiar Michaelis–Menten type enzyme kinetics. Parameter estimation gave good fits to measured data when studying the effects of step changes in dissolved oxygen concentration. However, the results of toxic challenge were not well simulated by the simple model that assumed that cell death was proportional to an arbitrary power of the toxicant concentration.

With respect to redox mediated whole cells, there are many reports of the interaction of such mediators with isolated fragments of respiratory and photosynthetic electron transport chains. Nevertheless, the interactions of the same mediators with **intact** cells have not been described in detail. In consequence, biosensors that use redox mediators, to monitor whole cells, generate amperometric signals that are difficult to describe analytically. The following text develops an earlier model [64] that described redox mediated responses of microbial cells physically entrapped between a working electrode surface and a membrane with capillary pores. In order to simplify the mathematics, the original model was very simplistic—in particular, it assumed that the biological cells were localized at one edge of the active layer. That simplification is not made in this present paper so as to more adequately describe the structure of the biological layer, and the effects of microbial perturbations.

7.8 NOTATION

θ	Fractional coverage of pores ($0 \leqslant \theta \leqslant 1$)
A	Area (m^2)
$c(x, t)$	Reduced mediator concentration at distance x and time t (mol m^{-3})
D	Diffusion coefficient of the reduced mediator in aqueous solution ($m^2\ s^{-1}$)
D_f	Effective diffusion coefficient of the reduced mediator in the biological layer ($m^2\ s^{-1}$)
$\bar{f}$	Laplace transform of variable $f(t)$ which is a function of time and other parameters
F	Faraday constant (C mol^{-1})
$I(t)$	Current at time t (A)
I_{max}	Maximum current available due to the microbial turnover of mediator (A)
$j(x, t)$	Flux of reduced mediator normal to, and away from, the electrode surface (mol $m^{-2}\ s^{-1}$)
k	First-order rate constant (s^{-1})
k_T	Turnover number of mediator at the microbial cell (s^{-1})
l	Thickness of the biological layer (m)
m	Thickness of the membrane containing capillary pores (m)
n	The number of electrons in the stoichiometric equation describing the redox reaction of the mediator
n_m	Number of moles of microorganisms per unit area (mol m^{-2})
N	Amount of reduced mediator in the microbial layer (mol)
q	Smallest microbial dimension, e.g., radius for cocci, diameter for bacilli (m)
s	Positive real number that is sufficiently large to ensure convergence of the Laplace integral (s^{-1})
t	Time, set to be zero at the start of the perturbation (s)
x	Distance normal to the plane of the electrode (m)
x^*	Distance normal to the plane of the electrode at which the maximum in the reduced mediator concentration occurs (m)

The superscripts 0 and ∞ indicate, respectively, the value of the parameter at the start of a perturbation, and the limiting value of the parameter—the value the parameter would eventually attain if the system was otherwise undisturbed. Other symbols are defined in the text when they are first used.

7.9 PHYSICAL MODEL

In this model, the biocatalyst is entrapped in a thin layer of solution between a working electrode and a membrane with capillary pores [64]. The electrode is poised at a potential sufficiently positive to ensure that the surface concentration of reduced mediator is negligible. In addition, it is assumed that: (i) the microbial activity is constant in the absence of external perturbations; (ii) the concentration of reduced mediator in the external medium is negligibly different from zero; (iii) the microbial cells are point sources of reduced mediator homogeneously distributed throughout the biological layer; and (iv) the concentration gradients are linear.

The first assumption may be a good approximation over a short time scale (minutes to hours, depending on the cells and the system configuration). The second approximation is a good one—providing that the sensor is placed either in a flow cell, so that reaction products are continuously washed away from the front face of the device, or in a large volume of stirred solution, so that the reduced mediator is significantly diluted. The third and fourth assumptions greatly simplify the mathematics and allow closed form solutions to be generated.

In contrast to the earlier model [64], the present model recognizes that the cells are distributed throughout the biological layer rather than being confined to the interface with the porous membrane. In consequence, a maximum in the concentration distribution may occur within the biological layer, Fig. 7.4(a). In addition, cognisance is given to the existence of the barrier presented by the immobilized cells, which impedes the diffusion of mediator through the biological layer.

7.9.1 Effective diffusion coefficient, D_f

Dependent upon the nature of the immobilized cells and the lipophilicity of the mediator molecule, the reduced mediator may be formed at redox active sites on the outside of the plasma membrane or at sites within the

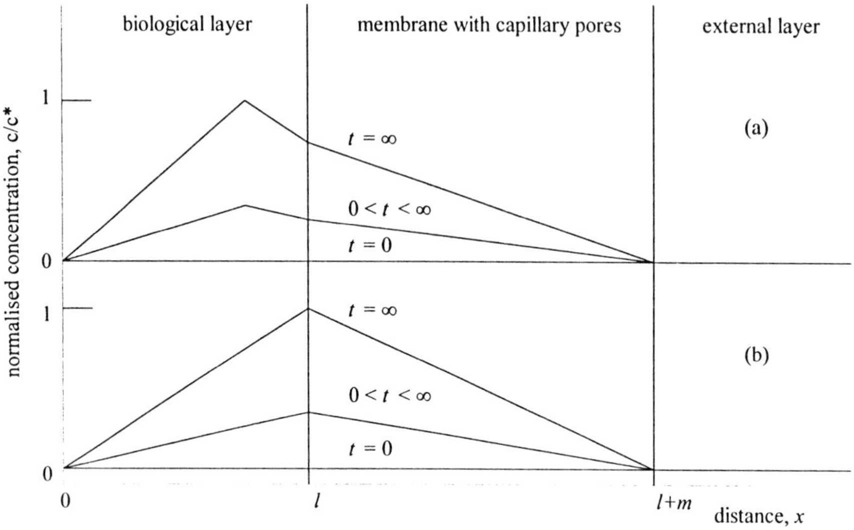

Fig. 7.4. Schematic diagram showing initial, transient and steady-state concentrationsof reduced mediator in the layers adjacent to the working electrode surface. For comparison, the redox active sites are shown in two distributions: (a) distributed throughout the biological layer (this paper); (b) localized at the member interface [64]. The effect of distributing the active sites throughout the biological layer is to introduce a maximum in the concentration profile across the layer.

cytoplasm of the cell, Fig. 7.5. In either case the biological layer would present a significant barrier to the diffusion of reduced mediator:

- molecules of reduced mediator must move through, or between the cells, to get to either boundary of the biological layer;
- redox mediator formed at sites within cells must move through the cellular membranes before they can diffuse further.

Consequently, the ratio of the diffusion coefficient in bulk aqueous solution and the effective diffusion coefficient in the biological layer is likely to be large (i.e., $D/D_f \gg 1$). The actual magnitude will depend on the size, shape, membrane structures, coverage, and packing arrangement of the immobilized cells.

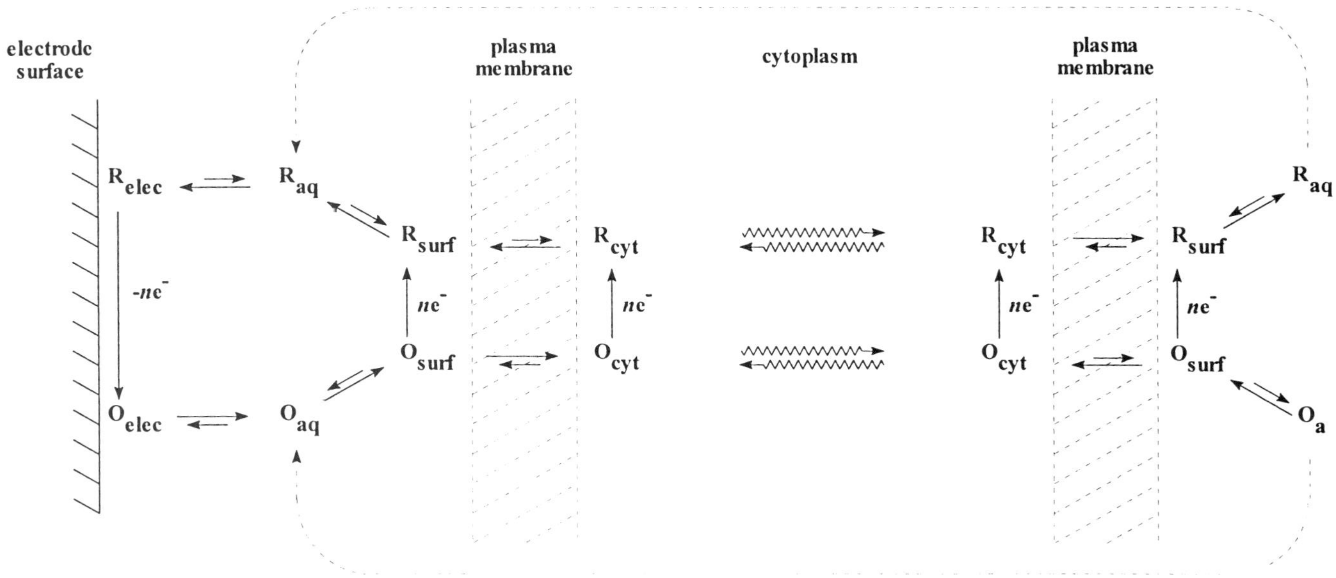

Fig. 7.5. Schematic representation of some of the redox mediator processes at a whole cell biosensor. Lipohilic mediators may be reduced at redox active sites in the plasma membrane or at sites within the cytoplasm or both processes may occur—depending on the cell type and the mediator. Lipophobic mediators can only be reduced at sites on the outside edge of the plasma membrane. The oxidized form of the mediator, O, may be present in excess, but much of the reduced form, R, may need to diffuse between packed cells (dotted arrows) or through the cytoplasm (squiggly arrows). The subscripts aq, cyt, elec, and surf represent mediator in the aqueous phase, within the cytoplasm, at the electrode surface, and at the plasma membrane-aqueous interface, respectively.

7.9.2 Mathematical model

The initial and boundary conditions are defined by the following equations:

$$c(x, 0) = 0, \tag{7.12}$$

$$c(0, t) = 0, \tag{7.13}$$

$$D_f \frac{\mathrm{d}c(l, t)_b}{\mathrm{d}x} = \theta D \frac{\mathrm{d}c(l, t)_m}{\mathrm{d}x}, \tag{7.14}$$

$$c(l, t)_b = c(l, t)_m, \tag{7.15}$$

and

$$c(l + m, t) = 0, \tag{7.16}$$

where the subscripts b and m indicate the biological and membrane layers, respectively.

From Fick's first law, the fluxes to the electrode surface, and through the membrane layer are given by, respectively:

$$j(0, t) = -D_f \frac{c(x^*, t)}{x^*}, \tag{7.17}$$

and

$$j(l, t) = \theta D \frac{c(l, t)}{m}. \tag{7.18}$$

The position, x^*, of the maximum in the concentration profile is given by:

$$x^*/l = j(0, t)/[j(0, t) - j(l, t)]. \tag{7.19}$$

From geometric considerations, the maximum concentration is given by:

$$c(x^*, t) = \left\{\frac{l - x^*}{D_f} + \frac{m}{\theta D}\right\} j(l, t). \tag{7.20}$$

Similarly, the amount of reduced mediator in the biological layer, given by:

$$N = A\{lc(x^*, t) + (l - x^*)c(l, t)\}/2. \tag{7.21}$$

The time-dependent accumulation of reduced mediator is given by the difference between its biological turnover and the fluxes to the electrode and into the membrane:

$$\mathrm{d}N/\mathrm{d}t = A[k_T n_m u(t) + j(0, t) - j(l, t)], \tag{7.22}$$

where $u(t)$ is the unit step function:

$$u(t) = \begin{cases} 0 & t < 0 \\ \frac{1}{2} & t = 0. \\ 1 & t > 0 \end{cases} \tag{7.23}$$

The turnover number, k_T, reflects the redox activity of the immobilized cells. The resulting electrode current is given by Faraday's law:

$$I(t) = -nFAj(0, t). \tag{7.24}$$

Equations (7.12)–(7.24) apply to both the transient and steady-states.

7.10 SENSOR RESPONSE

The method of Laplace transforms can be used to solve equations (7.17)–(7.24), Appendix 1. Re-arrangement of (7.A14) yields:

$$I(t) = I_{\max} u(t) \left\{ 1 - \frac{\theta l D}{2\theta l D + m D_f} \right\} [1 - \exp(-k_E t)], \tag{7.25}$$

where

$$I_{\max} = nFAk_T n_m. \tag{7.26}$$

The maximum "available" current, I_{max}, is determined by the density of cells on the electrode surface and by the activity of those cells. However, not all of this current is seen in the external circuit, since some of the reduced mediator is lost through the membrane layer.

7.10.1 Steady-state current

The variable quantity within the braces of equation (7.25) describes the loss of reduced mediator from the biological layer and its consequent effect on the steady-state current. It indicates that the smallest current is obtained in the absence of a membrane layer—in which case there are equal fluxes of reduced mediator to the working electrode and into the external medium—and the current is one-half of the maximum value. The largest steady-state current is obtained with a biological layer that is negligibly thin. However, microbial whole cells are of a significant size (typically 0.2–10 μm), so there may be an appreciable loss of reduced mediator. The steady-state current is insensitive to the membrane porosity (θ), but it is sensitive to the thicknesses of the biological and membrane layers (l and m) and to the ratio of the diffusion coefficients (D/D_f) in the two layers, Fig. 7.6.

7.10.2 Transient current

The bracketed term in equation (7.25) describes the current transient as an exponential function of time. The rate constant, k_E, is a characteristic of the sensor—it describes the fastest rate of change that the output signal is capable of achieving. The constant is a function of the sensor geometry and the diffusion coefficients. From (7.A7):

$$k_E = \frac{2D_f[2\theta lD + mD_f]^2}{l^2[(\theta lD + mD_f)^2 + \theta lmDD_f]}. \tag{7.27}$$

The transient $1 - \exp(-k_E t)$ is relatively insensitive to the membrane properties (θ and m), but sensitive to the thickness of the biological layer (l) and to the ratio of the diffusion coefficients (D/D_f) in the two layers, Fig. 7.7.

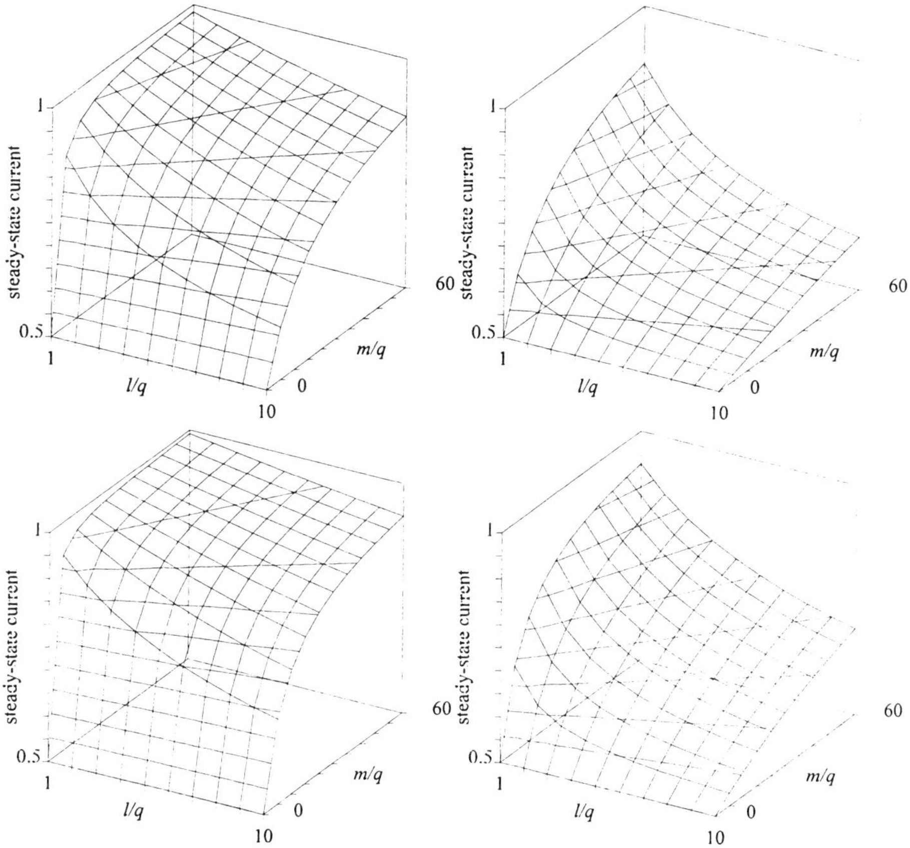

Fig. 7.6. Effect of variation in the thicknesses of the microbiol layer, l, and the membrane layer, m, on the steady-state sensor current, which was normalized by the maximum attainable value I_{max}. The range of the l/q axis takes cognisance of the fact that the thickness of the biological layer cannot be less than the smallest microbiol dimension, q. The data indicate that the current is insensitive to the membrane porosity ($\theta = 1$, top; $\theta = 0.5$, bottom), but sensitive to the diffusion barrier presented by the biological material ($D/D_f = 1$, left; $D/D_f = 10$, right).

7.11 PERTURBATIONS

The step-function in equation (7.22) describes an effectively instantaneous perturbation of biological redox activity—such as might be ob-

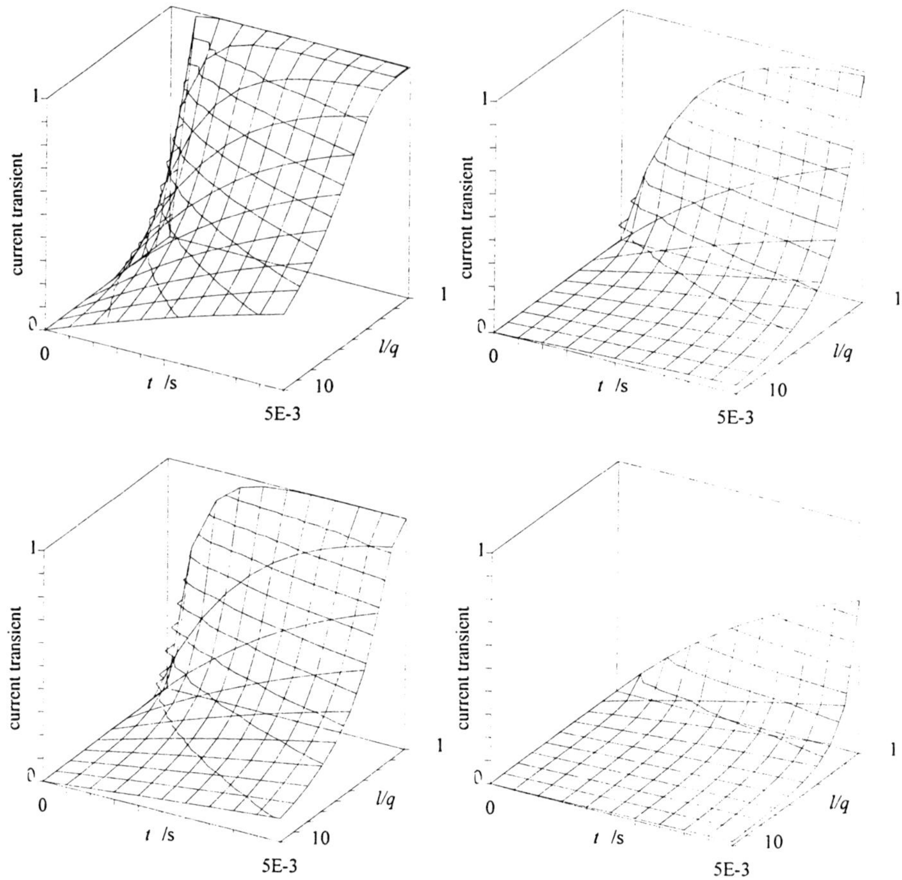

Fig. 7.7. Transient current response, $1 - \exp(-k_E t)$, as a function of time, and the biological layer thickness, l/q. The data indicate that the response is relatively insensitive to the thickness of the porous membrane ($m/q = 0$, top; $m/q = 60$, bottom), but sensitive to the diffusion barrier presented by the biological layer ($D/D_f = 1$, left; $D/D_f = 10$, right). $D = 10^{-5}$ cm^2 s^{-1}; $\theta = 0.5$.

tained by rapid stimulation of photosynthetic redox centres [65]. A more general relationship between the sensor signal and the microbial activity can be obtained by using an exponential function to describe perturbations of redox activity:

$$k_T = [(k_T^o - k_T^\infty)\exp(-k_Bt) + k_T^\infty]u(t) + k_T^0[1 - u(t)] \qquad 0 < k_B < \infty. \tag{7.28}$$

The modified sensor response can be obtained by using Laplace transformation of equations (7.17)–(7.22) with k_T substituted using equation (7.28) to yield (Appendix 2):

$$I(t) = I_{\max}\left\{1 - \frac{\theta lD}{2\theta lD + mD_f}\right\}\psi, \tag{7.29}$$

where

$$I_{\max} = nFAn_m. \tag{7.30}$$

The result in equation (7.29) is comparable to (7.25)—the only significant difference being that the shape of the transient responses is described by the multiple exponential function ψ rather than the simple exponential $1 - \exp(-k_Et)$. In addition, the turnover rate (k_T), that previously appeared in $I_{\max}$, is now contained within ψ. The perturbation, ψ, can be calculated from (7.A23), by setting appropriate values for ψ^0, k_T^0 and k_T^∞. It is satisfactory that the calculated response approaches the simple exponential when the biological perturbation is rapid (i.e., $k_B \gg k_E$), Figure 7.8.

7.12 CONCLUSION

There are practical difficulties in producing redox mediated whole cell biosensors that are reproducible within and between batches [53]. Models such as the one here can provide insights into the influence of various physical factors on the sensor outputs. Perturbations such as those described by the function ψ may be used to describe both stimulation and inhibition of these biosensors which are of importance in rapid toxicity assessment.

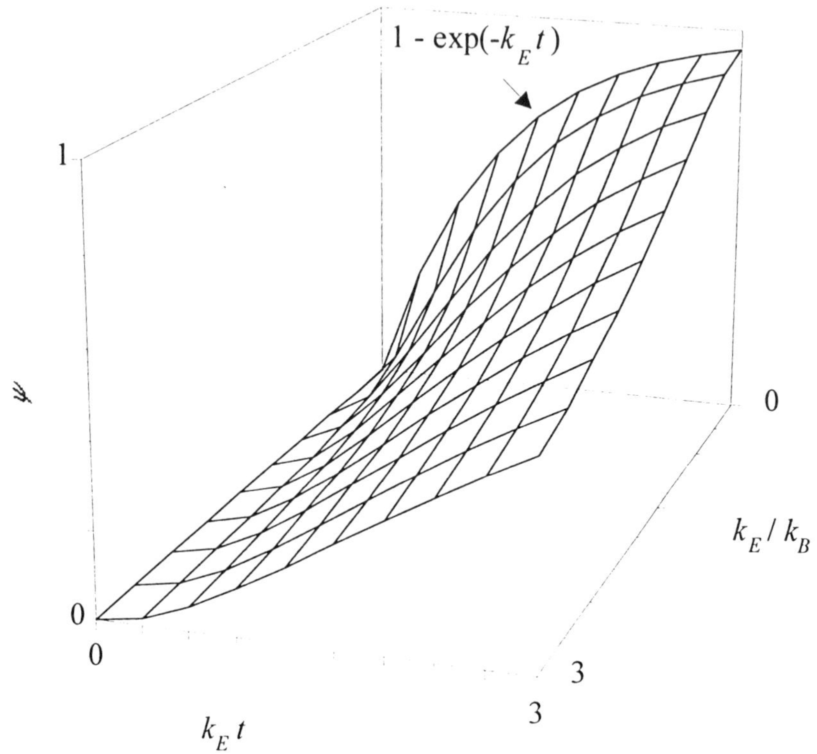

Fig. 7.8. 3-D plot of ψ as a function of dimensionless $(k_E t)$ and as a function of the perturbation rate constant (k_B). The quantity k_E is characteristic of the sensor. The plot shows that for $k_B \gg k_E$ the measured response is effectively exponential.

ACKNOWLEDGEMENTS

Grateful thanks to Prof. P. N. Bartlett for suggesting that this type of model be investigated.

APPENDIX 7.1

Equations (7.6), (7.8) and (7.9) can be manipulated to yield the variables $j(0, t)$ and $c(x^*, t)$ in terms of the variable $j(l, t)$:

$$j(0, t) = -\frac{\epsilon}{l} j(l, t), \tag{7.A.1}$$

$$c(x^*, t) = \frac{\epsilon^2 j(l, t)}{D_f(\epsilon + l)}, \tag{7.A.2}$$

where

$$\epsilon = l + mD_f l/(\theta D), \tag{7.A.3}$$

and the parameter x^* is found to depend on the diffusion coefficients and the sensor geometry:

$$x^* = \epsilon l/(\epsilon + l). \tag{7.A.4}$$

The results (7.A.2) and (7.A.4) can be used together with (7.7) and (7.10) to yield:

$$N = \frac{Al}{2}\left\{\frac{\epsilon(\epsilon + l) - l^2}{D_f(\epsilon + l)}\right\} j(l, t), \tag{7.A.5}$$

or

$$N = (\epsilon + l)\frac{Aj(l, t)}{k_E l}, \tag{7.A.6}$$

where

$$k_E = \frac{2D_f(\epsilon + l)^2}{l^2\{\epsilon(\epsilon + l) - l^2\}}. \tag{7.A.7}$$

It is straightforward to write down the complementary equations of (7.A1) and (7.A.6) in Laplace space:

$$\bar{j}(0, s) = -\frac{\epsilon}{l}\bar{j}(l, s), \tag{7.A.8}$$

and

$$\bar{N} = (\epsilon + l)\frac{A\bar{j}(l, s)}{k_E l}. \tag{7.A.9}$$

The complementary equation for (7.11) is given by:

$$s\bar{N} - N^0 = A\left[\frac{k_T n_m}{s} + \bar{j}(0, \mathrm{s}) - \bar{j}(l, s)\right]. \tag{7.A.10}$$

Assuming that the initial concentration of reduced mediator is negligible, then $\bar{N}$ and $\bar{j}(0, s)$ can be eliminated from the simultaneous equations (7.A.8)–(7.A.10) to yield:

$$\bar{j}(l, s) = \frac{lk_T n_m}{(\epsilon + l)(s/k_E + 1)s}, \tag{7.A.11}$$

which can be inverted to yield:

$$j(l, t) = \frac{lk_T n_m}{\epsilon + l}\{1 - \exp(-k_E t)\}. \tag{7.A.12}$$

The other time-dependent parameters can be found by substituting this result into (7.A1)–(7.A3), and (7.13), e.g.,

$$j(0, t) = -\frac{\epsilon k_T n_m}{\epsilon + l}\{1 - \exp(-k_E t)\}, \tag{7.A.13}$$

and

$$I(t) = nFAk_T n_m \frac{\epsilon}{\epsilon + l}\{1 - \exp(-k_E t)\}. \tag{7.A.14}$$

APPENDIX 7.2

The Laplace transform of (7.11), modified by substitution from (7.17), is given by:

$$s\bar{N} - N^0 = A\left\{\frac{k_T^0 - k_T^\infty}{k_B + s} n_m + \frac{k_T^\infty}{s} n_m + \bar{j}(0, s) - \bar{j}(l, s)\right\}. \quad (7.A.15)$$

Substitution from (7.A.8), (7.A.9) yields:

$$\bar{j}(l, s) = \frac{l n_m}{(\epsilon + l)(1 + s/k_E)}\left\{\frac{N^0}{A n_m} + \frac{k_T^0 - k_T^\infty}{k_b + s} + \frac{k_T^\infty}{s}\right\}, \quad (7.A.16)$$

which can be inverted to yield:

$$j(l, t) = \frac{l n_m}{\epsilon + l}\psi, \quad (7.A.17)$$

where

$$\psi = \frac{k_E N^0}{A n_m}(1 - \alpha) + \frac{k_T^0 - k_T^\infty}{k_B/k_E - 1}[(1 - \alpha) - (1 - \alpha)^{k_B/k_E}] + \alpha k_T^\infty, \quad (7.A.18)$$

and

$$0 \leqslant \alpha \equiv 1 - \exp(-k_E t) \leqslant 1. \quad (7.A.19)$$

The other time-dependent parameters can be found by substituting (7.A.17) into (7.A.1)–(7.A.3), and (7.13). In particular:

$$j(0, t) = -\frac{\epsilon n_m}{\epsilon + l}\psi, \quad (7.A.20)$$

$$I(t) = nFA\frac{\epsilon n_m}{\epsilon + l}\psi, \quad (7.A.21)$$

and

$$N = A n_m \psi / k_E. \quad (7.A.22)$$

The result (7.A.22) can be substituted back into (7.A.18) to yield:

$$\psi = (1 - \alpha)\psi^0 + \frac{k_T^0 - k_T^\infty}{k_B/k_E - 1}[(1 - \alpha) - (1 - \alpha)^{k_B/k_E}] + \alpha k_T^\infty. \qquad (7.A.23)$$

REFERENCES

[1] A.E.G. Cass, G. Davis, G.D. Francis, A.H.O. Hill, W.J. Aston, J. Higgins, E.V. Plotkin, L.D.L. Scott and A.P.F. Turner, Anal. Chem. 56 (1984) 667.

[2] C.G.J. Kopal, M.C. Feiters, R.J.M. Nolte, B. de Ruiter and R.B.M. Schasfoort, Biosens. Bioelectron. 7 (1992) 461.

[3] J.F. Cassidy, C. Clinton, W. Breen, R. Foster and E. O'Donoghue, Analyst 118 (1993) 415.

[4] N. Martens, A. Hindle and E.A.H. Hall, Biosens. Bioelectron. 10 (1995) 393.

[5] S.M. Zakeeruddin, M. Grtzel and D.M. Fraser, Biosens. Bioelectron. 11 (1996) 305.

[6] W.R. Everett and G.A. Rechnitz, Anal. Chem. 70 (1998) 807.

[7] M.A. Arnold and G.A. Rechnitz, in A.P.F. Turner, I. Karube and G.S. Wilson (eds), Biosensors: Fundamentals and Applications, Oxford University Press, Oxford (1987) 35.

[8] R.M. Buch, T.Q. Barker and G.A. Rechnitz, Anal. Chim. Acta 243 (1991) 157.

[9] B. Cheun, H. Endo, T. Hayashi, Y. Nagashima and E. Watanabe, Biosens. Bioelectron. 11 (1996) 1185.

[10] Catalogue of strains, NCIMB, Aberdeen, Scotland (1994)

[11] J. Racek, Cell-based Biosensors (Technomatic Publishing, Lancaster PA, 1995).

[12] R.R. Walters, B.E. Moriarty and R.P. Buck, Anal. Chem. 52 (1980) 1680.

[13] S.B. Liang, X.-M. Li and H.Y. Wang, Biotechnology Progress 2 (1986) 187.

[14] K. Riedel, R. Renneberg and P. Liebs, Bioelectrochem. Bioenergetics 19 (1988) 137.

[15] K. Riedel, A.V. Naumov, V.G. Grishenkov, A.M. Boronin, H.J. Stein, F. Scheller and H-G. Mueller, Appl. Microbiol. Biotechnol. 31 (1989) 502.

[16] K. Riedel, J. Huth, M. Kuehn and P. Liebs, J. Chem. Tech. Biotechnol. 47 (1990) 109.

[17] S Gamati, J.H.T. Luong and A. Mulchandani, Biosens. Bioelectron. 6 (1991) 125.

[18] M. Hoshi, Y. Sasamoto, M. Nonaka, K. Toyama and E. Watanabe, Biosens. Bioelectron. 6 (1991) 15.

[19] N.J. Richardson, S Gardner and D.M. Rawson, J. Appl. Bact. 70 (1991) 422.

[20] K. Riedel, Bioelectrochem. Bioenergetics 25 (1991) 19.

[21] K. Riedel, A.V. Naumov, A.M. Boronin, L.A. Golovleva, H.J. Stein and F. Scheller, Appl. Microbiol. Biotechnol. 35 (1991) 559.

[22] A. Heitzer, K. Malachowsky, J.E. Thonnard, P.R. Bienowski, D.C. White and G.S. Sayler, Appl. Env. Microbiol. 60 (1994) 1487.

[23] T. Kalab and P. Skladal, Electroanalysis 6 (1994) 1004.

[24] D. van der Lelie, P. Corbisier, W. Baeyens, S. Wuertz, L. Diels and M. Mergeay, Res. Microbiol. 145 (1994) 67.

[25] H-M. Tan, S-P. Cheong and T-C. Tan, Biosens. Bioelectron. 9 (1994) 1.

[26] U. Wollenberger, B. Neumann, K. Riedel, F.W. Scheller, Fresenius, J Anal. Chem. 348 (1994) 563.
[27] B.G.D. Haggett, J.G. Rogerson and D.M. Rawson, Bioelectrochem. Bioenergetics 37 (1995) 77.
[28] J. Peter, W.Hutter, W. Stöllnberger and W. Hampel, Biosens. Bioelectron. 11 (1996) 1215.
[29] O.V. Ignatov, S.M. Rogatcheva, S.V. Kozulin and N.A. Khorkina, Biosens. Bioelectron. 12 (1997) 105.
[30] A.N. Reshetilov, P.V. Iliasov, M.V. Donova, D.V. Dovbnya, A.M. Boronin, T.D. Leathers and R.V. Greene, Biosens. Bioelectron. 12 (1997) 241.
[31] M. Naessens and C. Tran-Minh, Biosens. Bioelectron., 13 (1998) 341.
[32] K. Riedel, R. Renneberg, M. Kuhn and F. Scheller, Appl. Microbiol. Biotechnol. 28 (1988) 316.
[33] K. Riedel, K-P. Lange, H-J. Stein, M. Kuhn, P. Ott and F. Scheller, Wat. Res. 24 (1990) 883.
[34] Y-R. Li and J. Chu, Appl. Biochem. Biotech. 28/29 (1991) 855.
[35] T.C. Tan, F. Li, K.G. Neoh and Y.K. Lee, Sensors & Actuators B 8 (1992) 167.
[36] T.C. Tan, F. Li and K.G. Neoh, Sensors & Actuators B 10 (1993) 137.
[37] F. Li, T.C. Tan and Y.K. Lee, Biosens. Bioelectron. 9 (1994) 197.
[38] F. Li and T.C. Tan, Biosens. Bioelectron. 9 (1994) 315.
[39] F. Li and T.C. Tan, Biosens. Bioelectron. 9 (1994) 445.
[40] K. Riedel, Application Report Bio Nr 202. Dr Bruno Lange, GmbH, Berlin (1994).
[41] E.J. Dorward and B.G. Barisas, Environ. Sci. Technol. 18 (1984) 967.
[42] I. Campanella, A.M. Paoletti and G. Tranchida, Chimicaoggi, March (1987) 61.
[43] D.M. Rawson and A.J. Willmer, Toxicity Assessment 2 (1987) 325.
[44] I. Karube, K. Hiramato, M. Kawarai and K. Sode, Membrane 14 (1989) 311.
[45] D.M. Rawson, A.J. Willmer and A.P.F. Turner, Biosensors 4 (1989) 299.
[46] W.C. Gaisford, N.J. Richardson, B.G.D. Haggett, and D.M. Rawson, Biochem. Soc. Trans. 19 (1991) 15.
[47] P-D. Hansen, Acta hydrochim. hydrobiol. 20 (1992) 92.
[48] S. Lee, K. Sode, K. Nakanishi, J-L. Marty, E. Tamiya and I. Karube, Biosens. Bioelectron. 7 (1992) 273.
[49] Z. Kong, P.A. Vanrolleghem and W. Verstraete, Biosens. Bioelectron. 8 (1993) 49.
[50] P. Pandard and D.M. Rawson, Environmental Toxicology 8 (1993) 323.
[51] P. Pandard, P. Vasseur and D.M. Rawson, Wat. Res. 27 (1993) 427.
[52] W. Bains, Biosens. Bioelectron. 9 (1994) 111.
[53] M.E. Polak, D.M. Rawson and B.G.D. Haggett, Biosens. Bioelectron. 11 (1996) 1253.
[54] J.G. Rogerson, A. Atkinson, M.R. Evans and D.M. Rawson. In Toxic Impacts of Wastes on the Aquatic Environment, J.R. Tapp, J.R. Wharfe and S.M. Hunt (eds.), Royal Society of Chemistry, Cambridge, 1996, 74.
[55] T. Henrysson and B. Mattiasson, Biodegradation 4 (1993) 101.
[56] 5 Day biochemical oxygen demand (BOD5), 2nd edition, HMSO, London (1989).
[57] M.L. Hitchman. Measurement of dissolved oxygen. Wiley-Interscience, 1978.
[58] Organisation for Economic Cooperation and Development (OECD), Guideline for testing chemicals, No. 209, (1984).

[59] W.J.Albery, A.E.G. Cass and Z.X. Shu, Biosens. Bioelectron. 5 (1990) 367.
[60] W.J. Albery, A.E.G. Cass and Z.X. Shu, Biosens. Bioelectron. 5 (1990) 379.
[61] W.J. Albery, A.E.G. Cass, B.P. Mangold and Z.X. Shu, Biosens. Bioelectron. 5 (1990) 397.
[62] P.N. Bartlett, and K.F.E. Pratt, Biosens. Bioelectron. 8 (1993) 451.
[63] T. Ikeda, T. Kurosaki, K. Takayama, K. Kano and K. Miki, Anal. Chem. 68 (1996) 192.
[64] B.G.D. Haggett, Analyst 119 (1994) 197.
[65] D.O. Hall and K.K. Rao, Photosynthesis, 5th edition, Cambridge University Press, 1994.

Chapter 8

Photoelectrochemical Kinetics at Semiconductor Electrodes

LAURENCE M. PETER

8.1 INTRODUCTION

The foundations of semiconductor electrochemistry were laid more than forty years ago by Gerischer, Pleskov and others, but initially semiconductor photo-electrochemistry remained a rather academic topic. An excellent overview of work up to 1980 has been given by Morrison [1]. Interest in potential applications of photoelectrochemistry was triggered in the 1980s by suggestions that semiconductor/electrolyte junctions could be used for solar energy conversion. Development of this concept led to the development of regenerative photovoltaic cells based on redox electrolytes, and the 1980s saw a rapid expansion of activity in semiconductor photoelectrochemistry driven primarily by the oil crisis [2–4]. Progress during this period is reflected in the impressive conversion efficiencies achieved for liquid junction photovoltaic cells, but fundamental problems with stability [5], and ultimately with cost, have so far prevented practical large scale application of either regenerative photoelectrochemical cells or photoelectrolysis systems. Failure to develop low cost electrolyte-based systems to challenge established photovoltaic technologies led inevitably to a hiatus in interest (and funding) in photoelectrochemistry at the end of the last decade. However, interest in the practical potential of photoelectrochemical solar energy conversion was revived following the development of efficient dye sensitized nanocrystalline TiO_2 cells by Grätzel and co-workers [6–11] in the early 1990s. As a consequence, the photoelectrochemical properties of nanocrystalline semiconductor electrodes have become the subject of widespread research activity.

Light driven electron transfer is central to many important processes. Photosynthesis and colour photography both involve light stimulated charge carrier separation, and mundane but practically important processes such as the photodegradation of paint layers containing TiO_2 are also essentially photoelectrochemical in nature. In principle, photoelectrochemistry also includes photogalvanic reactions, which involve light absorption by solution species rather than by the electrode, and attempts have been made to design solar cells based on photogalvanic effects [12, 13]. In this chapter, attention is focused on photoelectrochemical systems in which light absorption takes place either in a semiconducting solid or in a monolayer of dye adsorbed on the surface of a wide bandgap semiconductor like TiO_2. Some aspects of the dynamic behaviour of photoelectrochemical systems have been reviewed previously [14]; the present chapter gives an indication of progress made in the last seven or eight years.

8.2 THE SEMICONDUCTOR/ELECTROLYTE JUNCTION

The bulk electronic properties of extrinsic semiconductors are largely determined by the level of doping that is used to make the materials *n*-type or *p*-type. For non-degenerate semiconductors, the electron concentration in the conduction band and the hole concentration in the valence band are related to the Fermi energy E_F and to the effective densities of states in the conduction and valence bands (N_C and N_V respectively) by

$$n = N_C\,\mathrm{e}^{-(E_C-E_F)/k_BT}, \qquad p = N_V\,\mathrm{e}^{-(E_F-E_V)/k_BT}. \tag{8.1}$$

It can be seen that the product of the hole and electron concentrations is given by

$$np = N_C N_V\,\mathrm{e}^{-(E_C-E_V)/k_BT}. \tag{8.2}$$

The Fermi energy is usually included in the band diagrams as shown in Fig. 8.1.

In the dark, the junction between an extrinsic (doped) semiconductor and a redox electrolyte behaves as a diode because only one type of charge carrier (electrons for *n*-type and holes for *p*-type) is available to take part in electron transfer reactions. The potential distribution across the semiconductor/electrolyte interface differs substantially from that across

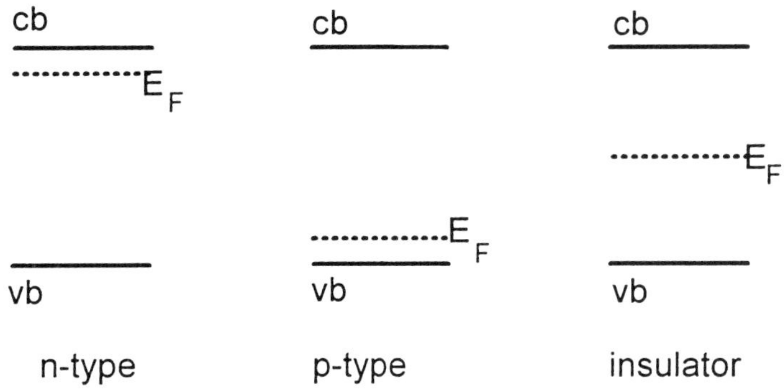

Fig. 8.1. Band diagrams for semiconductors and insulators showing position of the Fermi energy relative to the valence and conduction bands.

the metal electrolyte/interface. At a defined potential referred to as the flatband potential U_{FB}, there is no excess charge in the semiconductor and hence no potential drop in the solid. If the potential of an n-type electrode, for example, is made more positive than U_{FB}, electrons are withdrawn from the surface leaving positively charged ionised donor species as a space charge region. The dimensions of the space charge region are nearly always much greater than those of the electrical double layer in the electrolyte (the Helmholtz layer). Solution of Poissons equation shows that the width, W, of the depletion layer is related to the potential drop between the bulk and the surface of the semiconductor, $\Delta\phi_{SC}$:

$$W = \left(\frac{2\Delta\phi_{SC}}{\epsilon\epsilon_0 q N_d}\right)^{1/2} = W_0(\Delta\phi_{SC})^{1/2}, \tag{8.3a}$$

where

$$W_0 = \left(\frac{2}{\epsilon\epsilon_0 q N_d}\right)^{1/2}. \tag{8.3b}$$

Here ϵ is the relative permittivity of the solid and N_d is the donor density. W_0 ranges from 10^{-4} to 10^{-6} cm for doping densities from 10^{15} to 10^{19} cm^{-3}.

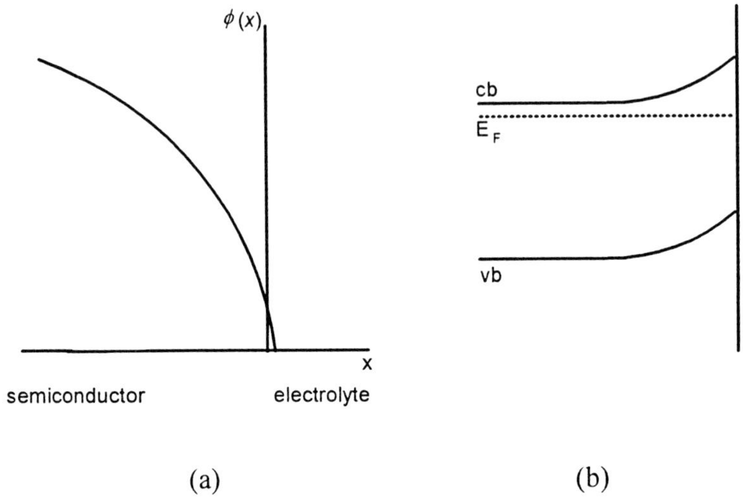

Fig. 8.2. (a) Potential distribution across the semiconductor/electrolyte junction under depletion conditions. (b) Band bending corresponding to (a).

The potential distribution across the interface under depletion conditions is illustrated in Fig. 8.2. It can be seen that the majority of the potential drop is located in the solid under depletion conditions, and consequently changes in electrode potential have little effect on the potential drop across the Helmholtz layer, $\Delta\phi_H$. The existence of a parabolic potential drop across the space charge region means that the potential energy of electrons depends on position. This "band bending" is shown in Fig. 8.2.

Since, under depletion conditions, the potential drop is mostly located in the space charge region unless the doping density is higher than 10^{18} cm^{-3}, $\Delta\phi_{SC}$ is given to a good approximation by the difference between the applied potential and the flat band potential, $(U - U_{FB})$. The storage of charge in the semiconductor electrolyte junction can be described in terms of the space charge capacitance $C_{SC} = \mathrm{d}Q_{sc}/\mathrm{d}\Delta\phi_{SC}$. The potential dependence of C_{SC} is commonly used to determine the flatband potential of semiconductors from the Mott Schottky equation [1].

$$C_{SC}^{-2} = \frac{2}{\epsilon\epsilon_0 q N_d}\left(U - U_{FB} - \frac{k_B T}{q}\right). \tag{8.4}$$

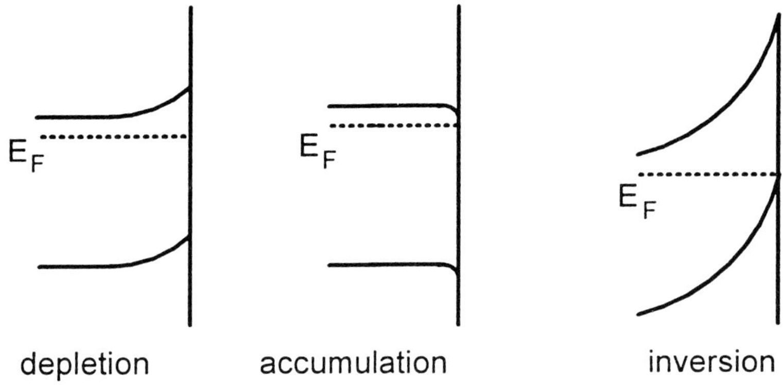

Fig. 8.3. Band bending and Fermi energy positions for depletion, accumulation and inversion conditions (*n*-type semiconductor).

In general, C_{SC} is orders of magnitude smaller than the Helmholtz capacitance, reflecting the fact that the major apart of any change in electrode potential appears across the space charge region.

Under accumulation conditions, mobile carriers accumulate at the surface, and when their density is sufficiently high, the Fermi level enters the majority carrier band. Under these degenerate conditions, the semiconductor behaves essentially as a metal. If the potential is changed in the opposite direction, the band bending can become sufficiently large that the Fermi level enters the minority carrier band. For example, an *n*-type electrode will in principle become *p*-type at the surface. However, this inversion condition is usually not observed at room temperature, at least for wide bandgap semiconductors, because the rate of hole generation by thermal excitation is inadequate to maintain equilibrium conditions if carriers are consumed in interfacial reactions. The band bending under depletion, accumulation and inversion conditions is illustrated in Fig. 8.3 for the case of an *n*-type semiconductor/electrolyte junction.

8.3 ELECTRON TRANSFER AT SEMICONDUCTOR ELECTRODES IN THE DARK

Before considering the kinetics of photoelectrochemical reactions, it is useful to compare electron transfer at metal electrodes with electron transfer at n-type semiconductor electrodes in the dark. Heterogeneous electron transfer reactions at metal electrodes have been studied extensively, and theoretical treatments within the framework provided by Marcus, Gerischer and others can be found in the literature [15–17]. Electron exchange between the metal and a redox system transfer takes place close to the Fermi energy, and the electron density in the metal is usually considered to be constant, so that the electron transfer is characterised in terms of a first order heterogeneous rate constant (cm s^{-1}). Changes in electrode potential appear across the Helmholtz region, affecting the activation energy for the electrode process and giving rise to an exponential dependence of the rate constant on potential (the Butler Volmer equation). The importance of the electronic density of states is illustrated by studies of electrode kinetics at semi-metallic highly oriented pyrolytic graphite (HOPG), which has a considerably lower electron density than that of metals. It has been observed that the electron transfer rate constants for a range of redox systems are lower at HOPG by 3–5 orders of magnitude compared with metals [18]. Theoretical considerations suggest that the upper limit for the first order heterogeneous rate constant at metal electrodes is of the order of 10^4 cm s^{-1}. It is, however, extraordinarily difficult to measure such high rate constants experimentally due the limitations imposed by mass transport of reactant to the electrode surface. This limitation and other experimental factors such as the RC time constant of the electrochemical cell usually restrict the experimentally accessible upper limit of rate constants to 10 cm s^{-1} or less.

It is important to note that the description of electron transfer kinetics is different in the case of semiconductor electrodes. For an n-type semiconductor electrode in the dark, the rate of electron transfer depends not only on the concentration of redox species in the solution but also on the potential dependent density of electrons in the semiconductor. Under depletion conditions, most of the potential drop is located in the solid, so that to a good approximation the activation energy for electron transfer is independent of potential. Electron transfer at semiconductor electrodes is therefore characterised in terms of a second order heterogeneous rate constant with units cm^4 s^{-1}.

Electron transfer rate constants of outer sphere redox reactions can be measured relatively easily at n-type semiconductor electrodes. This is because electrons are withdrawn from the surface under depletion conditions, so that their concentration is lower than in the bulk. Under ideal conditions, the electron concentration $n_{x=0}$ at the surface varies exponentially with the potential, U, according to the Boltzmann equation

$$n_{x=0} = n_{\text{bulk}}\, e^{-q(U-U_{FB})/k_BT}. \tag{8.5}$$

For an n-type semiconductor at 300 K which has $n_{\text{bulk}} = 10^{16}\,\text{cm}^{-3}$, $n_{x=0}$ is reduced to $10^{14}\,\text{cm}^{-3}$ at a potential 120 mV positive of the flatband potential. The electron transfer rate is therefore expected to be six to seven orders of magnitude lower than on a metal electrode where the electron density is of the order of $10^{21}\,\text{cm}^{-3}$. High values of the electron transfer rate constant can therefore be determined directly from steady state current-voltage characteristics by appropriate selection of the potential range. Transient and periodic methods can also be used. The experimental access to high values of the electron transfer rate constant provided by semiconductor electrodes has recently led to controversy about the upper limit to the value of rate constants. Lewis and co-workers [19–22] have argued that the upper limit of the second order heterogeneous rate constants of outer sphere redox reactions at semiconductor electrodes is around $10^{-17}\,\text{cm}^4\,\text{s}^{-1}$, and their measurements on carefully prepared single crystal surfaces of n-Si and n-InP support this conclusion. However, rate constants as high as $10^{-10}\,\text{cm}^4\,\text{s}^{-1}$ have been derived by Memming, Nozik and co-workers [23–25] from measurements on n-GaAs and n-GaInP. To place this value in context, it can be translated into an equivalent rate constant for a metal electrode, where for an electron density of $10^{21}\,\text{cm}^{-3}$, the corresponding rate constant would be $10^{11}\,\text{cm s}^{-1}$. This is more than seven orders of magnitude higher than the generally accepted upper limit of $10^4\,\text{cm s}^{-1}$.

8.4 ELECTRON TRANSFER AT ILLUMINATED SEMICONDUCTOR ELECTRODES

When a photon with energy exceeding that of the bandgap is absorbed in a semiconductor, an electron is promoted from the valence band to the conduction band, creating an electron hole pair. In an n-type semiconductor, the electron is the majority carrier and the hole is the minority carrier: the converse is true for a p-type semiconductor. The minority carrier is not a stable species; it will eventually recombine by an electron returning from the conduction band to the valence band. Since, by definition, majority carriers are in large excess, the recombination process in the bulk (uncharged) region of the semiconductor is pseudo-first-order, and the recombination of minority carriers in the neutral region is characterised by the lifetime, τ_p or τ_n for electrons or holes respectively. By contrast with the situation in the dark, where only majority carrier reactions can take place at any appreciable rate, it is generally photogenerated minority carriers that take part in photoelectrochemical reactions. Formulation of the kinetics of photoelectrochemical reactions is complicated by the fact that the rates depend on the concentration of minority carriers at the interface, and this is not known *a priori*. It depends on the potential, the illumination intensity, the concentration of redox species and the rate constant for interfacial electron transfer.

The generation and collection of photogenerated minority carriers can be illustrated for the case of an *n*-type semiconductor, where holes generated by absorption of light in the bulk of the semiconductor are collected at the interface by diffusion and migration. Under depletion conditions, electrons are withdrawn from the surface, leaving a space charge region due to ionised electron donors in the crystal lattice. The electric field in the space charge region will drive holes to the surface, where they can take part in electron transfer reactions. Holes generated deeper in the bulk of the semiconductor can reach the edge of the space charge region by diffusion or they may be lost by recombination with electrons. As shown in Fig. 8.4, the generation/collection problem is defined in terms of three characteristic lengths: the width of the space charge region $W = (2\Delta\phi_{SC}/\epsilon\epsilon_0 qN_d)^{1/2}$, the penetration depth of the light $1/\alpha$ and the diffusion length of holes $L_p = (D_p\tau_p)^{1/2}$. Here $\Delta\phi_{SC}$ is the potential drop across the space charge region, ϵ is the relative permittivity of the solid, N_d is the doping density, α is the absorption coefficient, τ_p is the hole lifetime and D_p is the hole diffusion coefficient.

The solution of this boundary value problem was given originally by Gärtner [26] for the solid state junction. It is assumed that holes generated within the space charge region are all collected at the contact and that holes generated in the neutral region are swept rapidly to the surface when they reach the edge of the space charge region, so that their concentration at $x = W$ approaches zero. The flux, g, of holes at $x = 0$ and therefore the

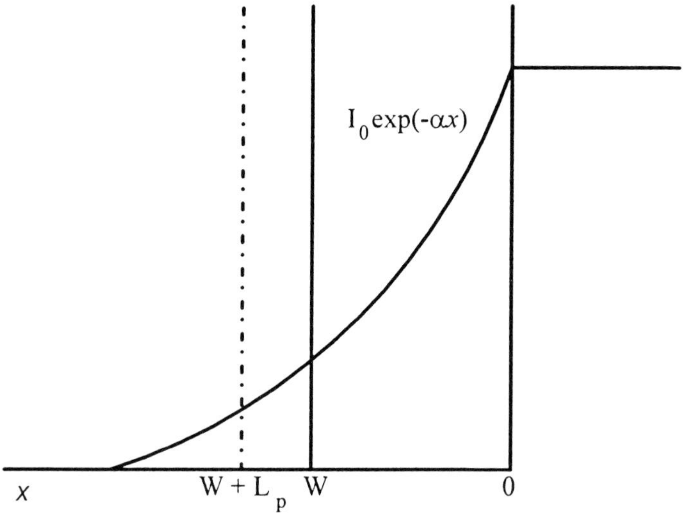

Fig. 8.4. Profile of light intensity at the semiconductor electrolyte junction. W is the width of the depletion layer and L_p is the hole diffusion length. The penetration depth of the light is defined as $1/\alpha$.

measured photocurrent are linearly dependent on I_0, the incident photon flux (corrected for reflection):

$$j_{\text{photo}} = qg = qI_0\left(1 - \frac{e^{-\alpha W}}{1 + \alpha L_p}\right). \tag{8.6}$$

The Gärtner equation (equation (8.6)) provides a satisfactory description of the photocurrent voltage curves measured at many semiconductor electrodes, but it contains no information about electron transfer kinetics because its derivation involves an *a priori* assumption about the boundary conditions.

Under steady state conditions, the concentration of holes at the surface is determined by the rate of their arrival from the bulk and on the rate of their removal by electron transfer and surface recombination. It is usually assumed that the potential distribution across the semiconductor/electrolyte junction is not affected by illumination under potentiostatic conditions, so that W is independent of I_o. In fact, W may change under illumination due to the release of carriers from traps or due to charge accumulation at the interface. It is instructive to estimate the steady state concentration of holes at the surface for the case of an illuminated n-type semiconductor in contact with a redox couple that has an electron transfer rate constant at the theoretical upper limit for outer sphere reactions of $10^{-17}\,\mathrm{cm^4\,s^{-1}}$. Electron transfer is restricted to a small reaction volume defined by the tunnelling distance δ, and the steady state concentration of holes in this region can be calculated from the flux of holes into the surface, g, by noting that

$$\delta\frac{\mathrm{d}p_{x=0}}{\mathrm{d}t} = g - k_p p_{x=0}[\mathrm{Red}] = \frac{\mathrm{d}p_s}{\mathrm{d}t}, \tag{8.7}$$

so that

$$p_{x=0} = \frac{g}{k_p[\mathrm{Red}]}. \tag{8.8}$$

For $\alpha = 10^4\,\mathrm{cm^{-1}}$, $W = 10^{-5}\,\mathrm{cm}$, $\tau_p = 10^{-7}\mathrm{s}$, $D_p = 10\,\mathrm{cm^2\,s^{-1}}$ and $I_0 = 10^{15}\,\mathrm{cm^{-2}s^{-1}}$, the flux of holes is $g = 9.2\times10^{14}\,\mathrm{cm^{-2}\,s^{-1}}$. If $[\mathrm{Red}] = 10^{19}\,\mathrm{cm^{-3}}$ (1.6×10^{-2} mol dm^{-3}) and $k_p = 10^{-17}\,\mathrm{cm^4\,s^{-1}}$, the surface hole concentration $p_{x=0}$ is $9.2\times10^{14}/(10^{19}\times(10^{-17}) = 9.2\times10^{12}\,\mathrm{cm^{-3}}$. The value of $p_{x=0}$ is much smaller than the charge density due to ionised donors in the space charge region ($10^{15}-10^{18}\,\mathrm{cm^{-3}}$). For $\delta = 10^{-7}$ cm, it corresponds to $p_s \cong 10^6\,\mathrm{cm^{-2}}$. However, if, as is usually the case, the rate constant for the redox reaction is smaller than the upper limit of $10^{-17}\,\mathrm{cm^4\,s^{-1}}$, the concentration of holes at the surface will be correspondingly larger. The accumulation and decay of photogenerated charge has been detected using microwave reflectivity measurements [27, 28], and this offers an exciting new approach to the study of photoelectrochemical kinetics. The accumulation of charge will tend to increase the potential drop across the Helmholtz layer and to reduce the potential drop across the semiconductor. The photoinduced change in band bending in the semiconductor can complicate the analysis of recombination kinetics, and for this reason, low light intensities should be used wherever possible in studies of photoelectrochemical kinetics in order to ensure that the constant band bending approximation is valid.

8.5 MULTISTEP PHOTOELECTROCHEMICAL REACTIONS

Many photoelectrochemical reactions of interest involve multistep electron transfer reactions. Examples that have been studied are hydrogen evolution on InP [29] and Si [30], oxygen evolution on TiO_2 [31] and photodecomposition of CdS [32]. These reactions involve the formation of charged surface bound intermediates, and the accumulation of surface charge modified the potential distribution across the interface. For example, the photodecomposition of *n*-CdS is believed to occur by the route

$$CdS + h^+ \rightarrow Cd^+ + S^0, \tag{8.9a}$$

$$Cd^+ + h^+ \rightarrow Cd^{2+}. \tag{8.9b}$$

The Cd^+ state corresponds to a surface trapped hole, and its concentration under steady state conditions will depend on illumination intensity and the rate constants for the two steps.

The photoanodic dissolution of *n*-silicon in acidic fluoride media provides an example of the complexity of multistep photoelectrochemical reactions [33, 34]. The reaction requires the transfer of four electrons, but it is clear that not all of the steps involve photogenerated holes because the photocurrent quantum efficiency is between 2 and 4. The explanation of the high quantum efficiencies is that the initial hole capture step can be followed by a series of steps in which intermediates with low electron affinity inject electrons into the conduction band. These intermediates can be assigned nominal oxidation states as shown in the following scheme.

$$Si + h^+ \rightarrow Si(I), \tag{8.10a}$$

$$Si(I) \rightarrow Si(II) + e^-, \tag{8.10b}$$

$$Si(II) \rightarrow Si(III) + e^-, \tag{8.10c}$$

$$Si(III) \rightarrow Si(IV) + e^-. \tag{8.10d}$$

In fact, each of the steps in equation (8.10) competes with an equivalent step in which a hole is captured instead of an electron injected. The consequence of this kind of reaction mechanism is that the absorption of a single photon can result in the passage of more than one electron across the interface (in the limiting case, the scheme in equation (8.10) can result in the passage of 4 electrons per absorbed photon). The phenomenon is referred to as photocurrent multiplication. It has also been observed for the two electron oxidation of secondary alcohols and other species such as formic acid on illuminated *n*-type semiconductors [1, 35] and for the reduction of oxygen and hydrogen peroxide on *p*-type semiconductors [1, 36]. In these cases a radical intermediate is formed by reaction of the majority carrier with the solution species. In the case of formic acid oxidation at *n*-TiO_2 or *n*-CdS, for example, $HCO_2^{\bullet}$ formed in the first step is able to inject an electron into the conduction band to give photocurrent doubling. Similarly in the case of oxygen reduction at *p*-GaP, $HO_2^{\bullet}$ formed by electron capture is able to inject a hole into the valence band, again giving photocurrent doubling.

8.6 SURFACE RECOMBINATION

In many cases, the experimental photocurrent-voltage characteristics of semiconductor electrodes deviate from the form predicted by the Gärtner equation. The deviation is particularly noticeable close to the flatband potential, where the photocurrent is much lower than predicted by equation (8.6). The loss of photocurrent is attributed to the recombination of electrons and holes via surface energy levels that lie in the bandgap. These energy levels, which are referred to as surface states, may result from mechanical or chemical treatment of the surface, or they may result from reactions taking place under illumination. An interesting example of a photogenerated surface state is hydrogen in *p*-GaP [38]. Illumination of *p*-GaP in acid solutions results in the reduction of protons to hydrogen. Some of the hydrogen evidently diffuses into the lattice as atomic hydrogen, where as H^+ it can act as a recombination centre by capturing alternately electrons and holes. The centres are located close to the surface and intercept electrons as they migrate towards the interface. These "near surface" recombination centres lead to substantial reduction in photocurrent at low band bending. At higher band bending, the states are filled with electrons so that they no longer act as recombination centres.

Detailed treatments of surface recombination have been given elsewhere [39, 40]. From the point of view of photoelectrochemical kinetics, surface recombination might appear simply as an uninteresting loss mechanism. However, as shown in the next section, the existence of surface recombination opens the way to the determination of rate constants for photoelectrochemical reactions involving minority carrier electron transfer.

The rate of recombination depends on the density of surface states and their occupation with electrons. In the case of an *n*-type semiconductor, the competition between electron transfer involving the reduced component of a redox couple and surface recombination can be described by the simplified reaction scheme

$$h^+ + \mathrm{Red} \rightarrow \mathrm{Ox} \tag{8.11a}$$

$$X + h^+ \rightarrow X^+ \tag{8.11b}$$

$$X^+ + e^- \rightarrow X \tag{8.11c}$$

Here X is a surface state that is occupied by an electron and is therefore able to capture a hole. X^+ is the vacant state created that is now able to accept an electron from the conduction band. The surface concentration of X will depend on the total surface density of surface states and their electron occupancy. The rate constants for hole and electron capture, β_p and β_n can be defined as the products of the thermal velocities, v_p and v_n and the capture cross sections σ_p and σ_n of X and X^+.

The scheme in equation (8.11) is not complete, because a hole that has been trapped at a surface state can also react with the reduced component of a redox couple. This can be written in terms of X^+ as

$$X^+ + \mathrm{Red} \rightarrow \mathrm{Ox} \tag{8.11d}$$

In principle, all of the steps in equation (8.11) must be reversible in order to satisfy the equilibrium condition in the dark. However, it is often convenient to regard them as irreversible under illumination in order to simplify the description of electron transfer kinetics.

8.7 COMPETITION BETWEEN RECOMBINATION AND ELECTRON TRANSFER (THE STEADY STATE CASE)

This section considers the competition between electron transfer and surface recombination under steady state conditions, and shows that no kinetic information can be obtained from the measured photocurrent.

In the presence of surface recombination, the minority carrier density at the surface is determined by the rate of their arrival from the collection region $W + L_P$ and the rate of their removal by the routes illustrated in Fig. 8.5. The concentration of holes accumulating at the surface can be expressed in terms of the equivalent surface concentration, p_s (cm^{-2}) = ($p_{x=0}$ since this allows a convenient formulation of the kinetic equations. Further simplification is achieved by considering the concentrations of redox species and majority carriers to be time invariant. This simplified scheme is illustrated for the case of an n-type semiconductor in Fig. 8.5.

Consideration of the rates of arrival, charge transfer, trapping and recombination of minority carriers leads to expressions for the time dependent surface concentrations (cm^{-2}) of free $p_{s,\mathrm{free}}$ and trapped $p_{s,\mathrm{trap}}$ holes:

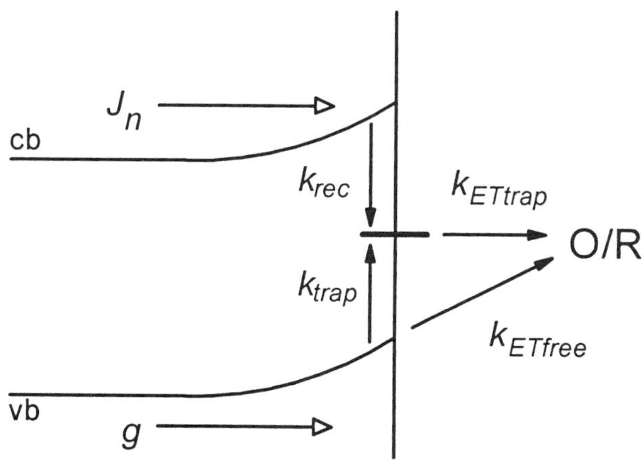

Fig. 8.5. Fluxes of holes (g) and electrons (J_n) into the interface. The possible reaction routes for holes are direct or indirect electron transfer to redox species and recombination.

$$\frac{\mathrm{d}p_{s,\mathrm{free}}}{\mathrm{d}t} = g - k_{ET\mathrm{free}}[\mathrm{Red}]p_{s,\mathrm{free}} - k_{\mathrm{trap}}p_{s,\mathrm{free}}, \tag{8.12}$$

and

$$\frac{\mathrm{f}p_{s,\mathrm{trap}}}{\mathrm{d}t} = k_{\mathrm{trap}}N_{\mathrm{trap}}p_{s,\mathrm{free}} - k_{ET\mathrm{trap}}p_{s,\mathrm{trap}}[\mathrm{Red}] - k_r n_{x=0}p_{s,\mathrm{trap}}. \tag{8.13}$$

Here g is the Gärtner flux of holes into the surface generated by illumination, [Red] is the bulk concentration of reduced species in solution and $n_{x=0}$ is the electron density at $x = 0$. The rate constants, k_{ET}, for electron transfer involving free and trapped holes, and the recombination rate constant, k_r, are shown in Fig. 8.5. They have units $\mathrm{cm}^3\ \mathrm{s}^{-1}$ and can be converted to conventional heterogeneous rate constants (units $\mathrm{cm}^4\ \mathrm{s}^{-1}$) by multiplying by δ.

Two limiting cases of the scheme shown in Fig. 8.5 have been considered by Peat and Peter [41]. In one case, hole transfer to redox species occurs entirely via the valence band and trapped holes can only recombine. This limit is appropriate for so-called "near-surface" states, which arise from defects close to the surface that interact strongly with the conduction and valence bands but not with electrolyte energy levels. This is the case, for example, for hydrogen in p-GaP, which was discussed in Section 8.6 [38]. In the second case, holes are trapped so rapidly at surface states (reaction (8.9b)) that electron transfer to the valence band can be neglected, but the resulting vacant surface state X^+ can accept an electron from a redox species as well as from the conduction band (recombination). In terms of the scheme this mean that $k_{\mathrm{trap}}N_{\mathrm{trap}} \gg k_{ET1}[\mathrm{Red}]$).

For the present discussion, it is assumed that the second case applies, i.e., that electron transfer and recombination proceed via X/X^+. In this case, the surface hole concentration is given by

$$\frac{\mathrm{d}p_s}{\mathrm{d}t} = g - k_{ET2}[\mathrm{Red}]p_s - k_r n_{x=0}p_s. \tag{8.14}$$

Here $p_{s,\mathrm{trap}}$ has been replaced by p_s, since essentially all holes are trapped. If [Red] is considered constant (the pseudo first order situation), $k_{ET2}[\mathrm{Red}]$ can be replaced simply by k_{tr} (units s^{-1}). Similarly, if the density of electrons is assumed to be constant, i.e., there is no change in band bending under illumination, $k_r\ n_{x=0}$ can be replaced by k_{rec} (units s^{-1}). Equation (8.14) then becomes

$$\frac{dp_s}{dt} = g - k_{\text{tr}}p_s - k_{\text{rec}}p_s, \tag{8.15}$$

where p_s is the density (cm^{-2}) of trapped holes at the surface.

The surface concentration of electrons depends on the potential drop (band bending) in the semiconductor, and in the absence of complications due to surface state charging (Fermi level pinning), it is given by (cf. equation (8.5))

$$\frac{n_s}{\delta} = n_{\text{bulk}}\, e^{-q[(U-U_{fb})/k_BT]}. \tag{8.16}$$

Consequently the first order rate constant for surface recombination $k_{\text{rec}} = k'_{\text{rec}}n_s$ is expected to depend exponentially on potential:

$$k_{\text{rec}} = k^0_{\text{rec}}\, e^{-q[(U-U_{fb})/k_BT]}. \tag{8.17}$$

The recombination rate constant $k^0_{\text{rec}}(U = U_{FB})$ can be expressed as the product of the doping density N_d, the thermal velocity v_n of electrons and the electron capture cross section σ_n of X^+:

$$k^0_{\text{rec}} = N_d\sigma_n v_n. \tag{8.18}$$

For $N_d = 10^{17}\ \text{cm}^{-3}$, $\sigma_n = 10^{-16}\ \text{cm}^2$, $v_n = 10^5\ \text{cm s}^{-1}$, $k^0_{\text{rec}} = 106\ \text{s}^{-1}$.

Under conditions of steady state illumination, $dp_s/dt = 0$ in equation (8.15), so that the surface hole concentration is given by the ratio

$$p_s = \frac{g}{k_{\text{tr}} + k_{\text{rec}}}. \tag{8.19}$$

The photocurrent density, j_{photo}, is therefore

$$j_{\text{photo}} = qk_{\text{tr}}p_s = qg\frac{k_{\text{tr}}}{k_{\text{tr}} + k_{\text{rec}}}, \tag{8.20}$$

where q is the elementary charge. The dimensionless ratio $k_{\text{tr}}/(k_{\text{tr}} + k_{\text{rec}})$ represents the fraction of minority carriers that undergo interfacial electron transfer. This simple treatment suffices to show that the rate constants for electron transfer and recombination cannot be separated by steady state measurements.

8.8 SEPARATING THE RATE CONSTANTS FOR RECOMBINATION AND CHARGE TRANSFER (THE NON-STEADY-STATE CASE)

This section shows that in order to separate k_{tr} and k_{rec} it is necessary to carry out non-steady-state measurements. A simple example of a non steady-state measurement is switching the illumination on and off. The photogenerated flux of holes, g, towards the surface feeds into three processes: surface charge storage, interfacial electron transfer and recombination. The magnitudes of the corresponding components of the total current density can be written in terms of the surface hole charge Q_s:

$$j_{\text{ch}} = \frac{dQ_s}{dt}, \tag{8.21a}$$

$$j_{\text{tr}} = k_{\text{tr}}Q_s, \tag{8.21b}$$

$$j_{\text{rec}} = k_{\text{rec}}Q_s. \tag{8.21c}$$

The measured photocurrent response is the sum of the charging and transfer components or alternatively the difference between the photogenerated hole current density qg and the recombination current density j_{rec}

$$i_{\text{photo}} = j_{\text{ch}} + j_{\text{tr}} = qg - j_{\text{rec}}. \tag{8.22}$$

The time dependent behaviour is therefore described by the differential equation

$$qg(t) = \frac{dQ_s(t)}{dt} + (k_{tr} + k_{rec})Q_{s(t)}. \tag{8.23}$$

In the non steady-state experiment, the illumination is switched on at $t = 0$, defining the initial condition $Q_s(0) = 0$. The transfer, recombination and charging components of the time dependent photocurrent are

$$j_{tr}(t) = \frac{qgk_{tr}[1 - e^{-(k_{tr}+k_{rec})t}]}{k_{tr} + k_{rec}}, \tag{8.24a}$$

$$j_{rec}(t) = \frac{qgk_{rec}[1 - e^{-(k_{tr}+k_{rec})t}]}{k_{tr} + k_{rec}}, \tag{8.24b}$$

$$j_{ch}(t) = qg\, e^{-(k_{tr}+k_{rec})t}. \tag{8.24c}$$

The total time dependent photocurrent for $t \geqslant 0$ can be expressed in the dimensionless form

$$\Phi(t) = \frac{j_{photo}(t)}{qg} = \frac{(k_{tr} + k_{rec}\, e^{-(k_{tr}+k_{rec})t})}{k_{tr} + k_{rec}}. \tag{8.25}$$

Figure 8.6 illustrates the time dependence of the photocurrent and its relationship to the charging and transfer components.

At $t = 0$, $Q_s = 0$ and the measured photocurrent is entirely due to charging. At longer times the surface charge tends towards its steady state value, and the charging current falls to zero. The current due to charge transfer is proportional to Q_s, so it is zero at $t = 0$ and increases with time towards its final steady state value. Since the recombination term also depends on Q_s, the fraction of charge carriers at the surface that are transferred is determined by the ratio $k_{tr}/(k_{tr} + k_{rec})$ for all values of t.

The key point of this discussion is that it shows that analysis of the transient photocurrent response allows separation of k_{tr} and k_{rec}. The ratio $j_{photo}(\infty)/j_{photo}(0)$ is given by $k_{tr}/(k_{tr} + k_{rec})$ and a plot of $\ln(j_{photo}(t) - j_{photo}$(/infinity)) vs. t has a slope of $-(k_{tr} + k_{rec})$. It is important to note that the analysis is only possible if recombination occurs. For $k_{rec} \to 0$, the photocurrent simply follows the illumination step, and $\Phi(t) = 1$.

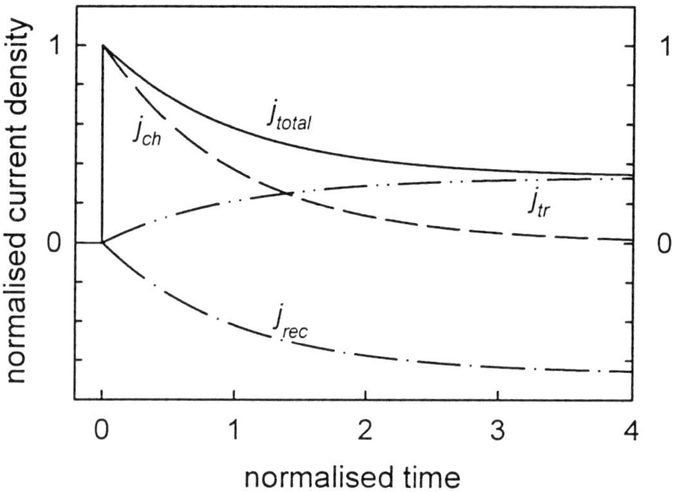

Fig. 8.6. Components of the current response of a n-type semiconductor electrode to an illumination step. j_{ch} is the charging current, j_{tr} is the current due to interfacial electron transfer and j_{rec} is the current due to electrons recombining with holes via surface states. The total current, given by the sum of j_{ch} and j_{tr}, is equal to $qg - j_{rec}$, where g is the flux of minority carriers given by the Gärtner equation. The dimensionless normalised time axis is $(k_{tr} + k_{rec})t$. The dimensionless normalised current axis is j/qg.

8.9 INTENSITY MODULATED PHOTOCURRENT SPECTROSCOPY

The light step method is only one of several possible non steady-state perturbations that can be used to investigate the kinetics of photoelectrochemical processes. A more powerful method involves sinusoidal modulation of the illumination intensity. This method is known as Intensity Modulated Photocurrent Spectroscopy (IMPS), and it has been used extensively to investigate a variety of photoelectrochemical reactions at semiconductor electrodes [42–72] The experimental arrangement is illustrated in Fig. 8.7. The light source can be a Light Emitting Diode (LED) or a laser beam modulated by an acousto-optic device. LEDs are available with emission wavelengths ranging from the blue to near infrared, and their output can be modulated over a wide frequency range by controlling the current with a frequency response analyser. The use of a laser with an acousto-optic modulator allows access to the ultraviolet. However, the acoustic delay in the modulator introduces a frequency dependent phase lag into the

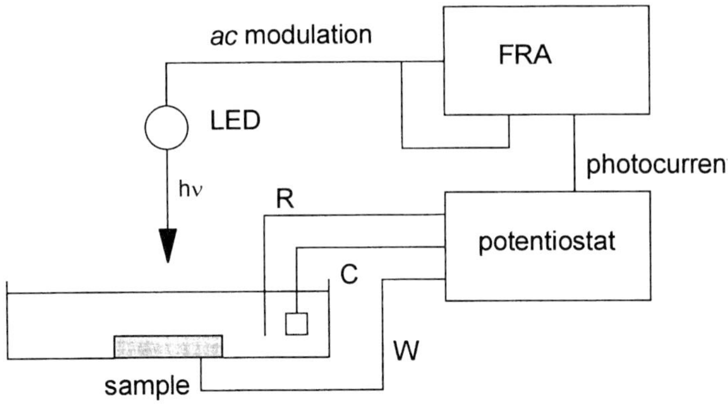

Fig. 8.7. Experimental arrangement for intensity modulated photocurrent spectroscopy (IMPS). FRA: frequency response analyser. LED: light emitting diode.

modulation, so that a beam-splitter and fast photodiode are necessary to provide the reference signal for the frequency response analyser. In order to allow linear analysis of the response, the modulation can be a small fraction of the total intensity from the LED or laser, or alternatively a second source can be used to provide the dc component of the illumination.

The illumination intensity consists of a dc and an ac component.

$$I(t) = I_{dc} + I_{ac}\sin(\omega t), \tag{8.26}$$

where $\omega = 2\pi f$. Generally I_{ac} is chosen to be a small fraction of I_{dc} to ensure linearity in more complex systems. The Gärtner flux of minority carriers into the surface can also be resolved into corresponding dc and ac components:

$$g(t) = g_{dc} + g_{dc}\sin\omega t. \tag{8.27}$$

The accumulated surface charge $Q_s(t)$ oscillates periodically, and the time dependent charging, transfer and recombination terms are replaced by the sums of their steady state and periodic equivalents.

$$Q_s(t) = \frac{qg_{\rm dc}}{k_{\rm tr} + k_{\rm rec}} + \frac{qg_{\rm ac}(k_{\rm tr} + k_{\rm rec})}{(k_{\rm tr} + k_{\rm rec})^2 + \omega^2} \sin \omega t - \frac{qg_{\rm ac}\omega}{(k_{\rm tr} + k_{\rm rec})^2 + \omega^2} \cos \omega t, \tag{8.28}$$

$$j_{\rm photo} = \frac{qg_{\rm dc}k_{\rm tr}}{k_{\rm tr} + k_{\rm rec}} + \frac{qg_{\rm ac}(k_{\rm tr}(k_{\rm tr} + k_{\rm rec}) + \omega^2)}{(k_{\rm tr} + k_{\rm rec})^2 + \omega^2} \sin \omega t - \frac{qg_{\rm ac}\omega k_{\rm rec}}{(k_{\rm rec} + k_{\rm tr})^2 + \omega^2} \cos \omega t. \tag{8.29}$$

Equation (8.29) can be conveniently expressed in the dimensionless complex form

$$\Phi(\omega) = \frac{j_{\rm photo}(\omega)}{qg_{\rm ac}} = \frac{k_{\rm tr} + i\omega}{k_{\rm rec} + k_{\rm tr} + i\omega}. \tag{8.30}$$

The corresponding real and imaginary components of the IMPS response are

$$\mathrm{RE}\left(\frac{j_{\rm photo}}{qg_{\rm ac}}\right) = \frac{(k_{\rm tr} + k_{\rm rec})k_{\rm tr} + \omega^2}{(k_{\rm tr} + k_{\rm rec})^2 + \omega^2}, \tag{8.31a}$$

and

$$\mathrm{Im}\left(\frac{j_{\rm photo}}{qg_{\rm ac}}\right) = \frac{k_{\rm rec}\omega}{(k_{\rm tr} + k_{\rm rec})^2 + \omega^2}. \tag{8.31b}$$

Figure 8.8 illustrates the semicircular complex plane plot predicted by equation (8.31).

The rate constants for electron transfer and recombination are readily separated because in the limit $(\omega \to 0)$, equation (8.31) tends to $k_{\rm tr}/(k_{\rm tr} + k_{\rm rec})$, and the maximum of the semicircle occurs when $\omega = 2\pi f = k_{\rm t} + k_{\rm rec}$. In the absence of RC attenuation effects, the high frequency intercept of the IMPS plot $\Phi(\omega \to \infty)$ is unity, so that the ac photocurrent can be used to determine $g_{\rm ac}$, the Gärtner flux of minority carriers. Measurements of $g_{\rm ac}$ as a function of potential (band bending) can be used to determine the minority carrier lifetime and absorption coefficient [46]. The main advantage of using the IMPS data rather than dc measurements of

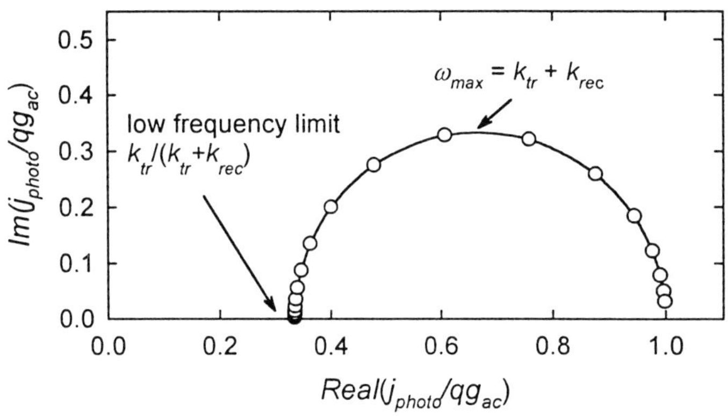

Fig. 8.8. Analysis of normalised IMPS response in the complex plane used to derive values of k_{tr} and k_{rec}.

the photocurrent is that the high frequency limit of the IMPS effectively "freezes out" the effects of surface recombination which are responsible for substantial deviations of the dc photocurrent from the Gärtner equation.

An example of the analysis of IMPS data based on the competition between electron transfer and recombination is provided by a detailed study of hydrogen evolution on illuminated *p*-InP in acid solution [29]. Since the photogenerated minority carriers are electrons in the case of *p*-InP, they are driven to the surface under depletion conditions, where they can reduce protons to hydrogen. Figure 8.9 is a set of experimental IMPS responses measured for *p*-InP in 1.0 mol dm^{-3} H_2SO_4 at different potentials. The measurements were performed using a small ac modulation of the illumination intensity superimposed on a larger steady component.

It can be seen that diameters of the IMPS semicircles decrease as the potential becomes more negative, and the band bending increases. The high frequency intercept, which remains almost constant, corresponds to the electron current density in the absence of recombination (qg_{ac}). The low frequency intercept, which moves away from the origin as the band bending is increased, corresponds to the steady state current. The sum of the rate constants ($k_{tr} + k_{rec}$) is derived from the frequencies, ω_{max} at which the maxima in the semicircles occur. The ration $k_{tr}/(k_{tr} + k_{rec})$ is obtained from the low and high frequency intercepts after normalisation by dividing by the high frequency intercept to obtain j_{photo}/qg_{ac}.

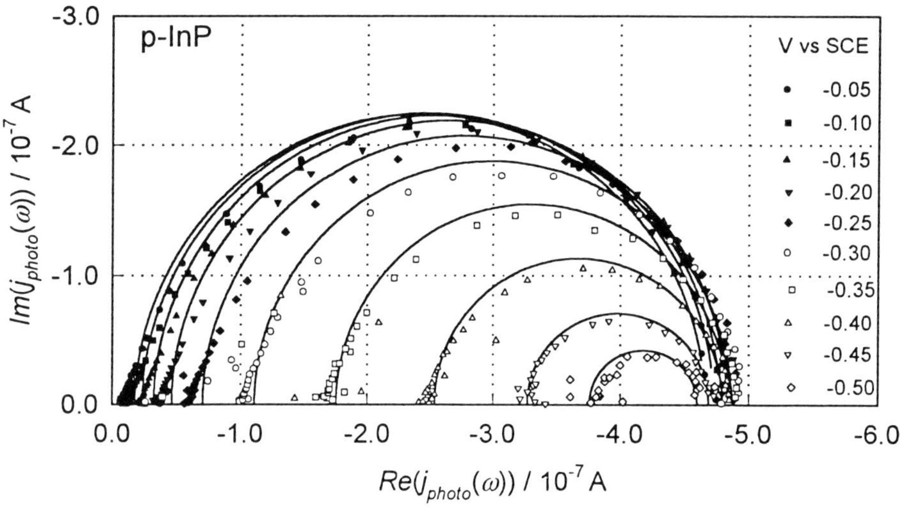

Fig. 8.9. IMPS plots recorded at different potentials for p-InP in 1.0 mol dm^{-3} H_2SO_4. Note that the diameters of the semicircles decrease as the band bending increases and recombination becomes less effective. In the saturation photocurrent region, the IMPS response contracts to a point.

Figure 8.10 shows how the derived values of k_{tr} and k_{rec} depend on electrode potential. The ratio $k_{tr}/(k_{tr} + k_{rec})$ is also shown; it represents the fraction of photogenerated electrons that react to produce hydrogen rather than recombining.

Closer examination of the IMPS data reveals that the p-InP system does not follow the behaviour predicted by the simple one electron transfer model. Figure 8.10 shows that, contrary to expectations, the rate constant for electron transfer, k_{tr}, is not independent of potential. The behaviour of the rate constant for recombination, k_{rec}, is also not ideal. It is expected that k_{rec} will decrease by an order of magnitude for every 59 mV increase in band bending as the potential is made more negative and the hole concentration falls (cf. equation (8.17)). In fact, however, Fig. 8.10 shows that k_{rec} decreases by less than two orders of magnitude for a potential change of 0.5 V. This non-ideal behaviour suggests that the surface hole concentration changes much more slowly with electrode potential than predicted by equation (8.17) for an ideal junction. It appears likely that

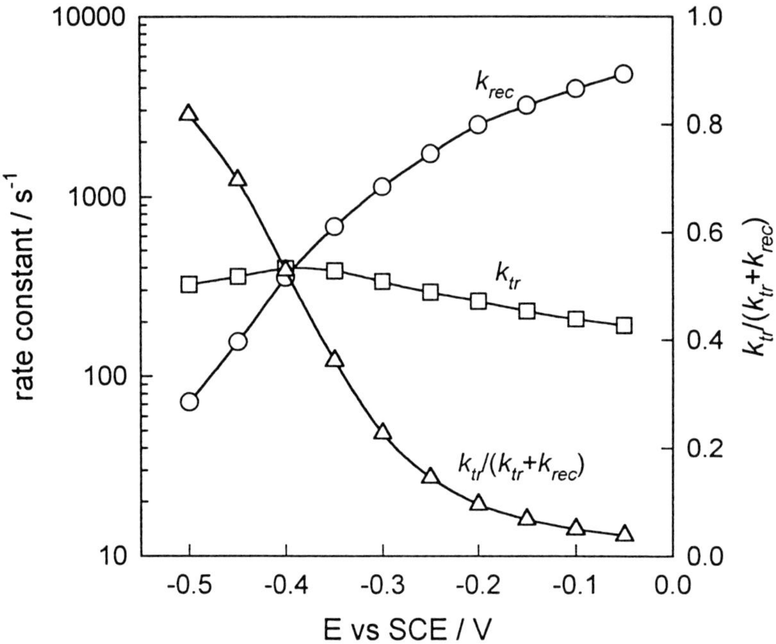

Fig. 8.10. Variation of k_{tr} and k_{rec} for hydrogen evolution on illuminated *p*-InP. The ratio $k_{tr}/(k_{tr} + k_{rec})$ represents the fraction of photogenerated electrons that are transferred across the interface. Note that k_{tr} appears to depend weakly on potential. The non-ideal variation of k_{rec} with potential is interpreted as evidence for partial Fermi level pinning.

not all of the change in applied potential appears across the space charge region of the semiconductor; instead some fraction must appear across the Helmholtz layer as the result of the surface charging. This phenomenon is common in semiconductor electrolyte junctions and is often referred to as Fermi level pinning. The observed potential dependence of k_{tr} is less easily explained, but as shown in Section 8.11, it may arise from the fact that the photoelectrochemical reduction of H^+ to H_2 is a two electron process.

8.10 THE INFLUENCE OF THE *RC* TIME CONSTANT, τ_{cell}

The time resolution of electrochemical measurements is usually limited by the *RC* time constant, τ_{cell}, of the electrochemical cell, which arises

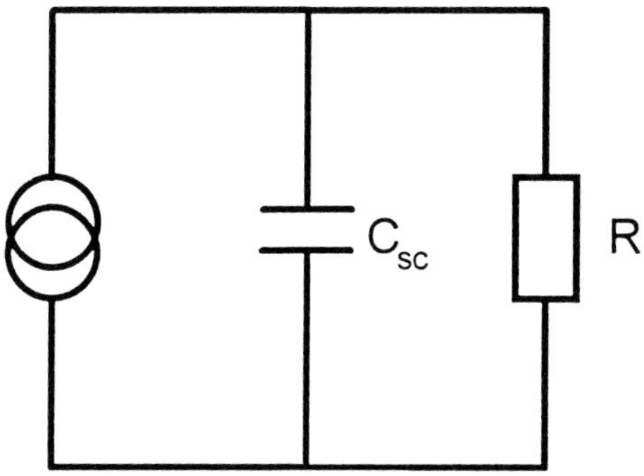

Fig. 8.11. Equivalent circuit for a photoelectrochemical cell under potentiostatic conditions. The externally measured photocurrent is less than the generated current at high frequencies because the space charge capacitance shunts the cell resistance.

from the combination of the series resistance of the electrolyte with the double layer capacitance of the electrode. The frequency range in which the simple analysis of IMPS data is applicable is also limited by the τ_{cell}. This effect of τ_{cell} on the IMPS response was originally treated by Li and Peter [44] who assumed that the space charge capacitance is much smaller than that of the Helmholtz layer and showed that it gives rise to an additional semicircle in the lower complex plane IMPS plot. The effect of τ_{cell} on the IMPS response arises from the fact that the internally generated photocurrent is attenuated by the space charge capacitance which effectively appears in parallel with the contact and electrolyte resistance as shown in Fig. 8.11.

The frequency dependent attenuation factor $A(\omega)$ for this parallel combination is given by

$$A(\omega) = \frac{1}{1 + i\omega RC} = \frac{1}{1 + i\omega\tau_{cell}}, \tag{8.32}$$

so that the observed modulated photocurrent response is obtained by multiplying equation (8.31) by equation (8.32). The result of this *RC* attenuation is seen in the complex plane IMPS plots as an additional semicircle in the lower complex plane, giving a total IMPS response of the

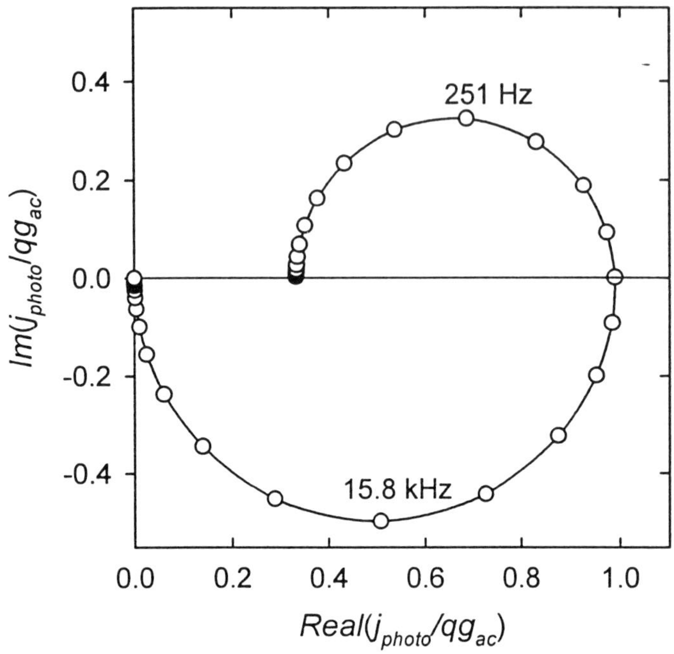

Fig. 8.12. IMPS plot showing the effect of $\tau_{\rm cell}$. Calculated for $R = 100\ \Omega$, $C_{\rm SC} = 100$ nF, $C_{\rm H} = 10\ \mu$F, $k_{\rm tr} = 500\ {\rm s}^{-1}$, $k_{\rm rec} = 103\ {\rm s}^{-1}$. The upper semicircle arises from surface recombination and the lower from RC attenuation.

type illustrated in Fig. 8.12. It is important to note that if the time constants for recombination and RC attenuation are closer than those chosen in this example, the IMPS plot crosses the real axis at a point less than unity.

Li and Peter assumed that the space charge capacitance is much smaller than the capacitance of the Helmholtz layer: the more generalised theory of Ponomarev and Peter considers the case where the space charge capacitance and Helmholtz capacitances are of comparable magnitude [60]. The attenuated IMPS response is then given by

$$\frac{j_{\rm photo}}{qg_{\rm ac}} = \frac{k_{\rm tr} + i\omega(C/C_{\rm sc})}{k_{\rm tr} + k_{\rm rec} + i\omega}\left(\frac{1}{1 + i\omega RC}\right), \tag{8.33}$$

where the effective capacitance C is defined as

$$C = \frac{C_{sc}C_H}{C_{sc} + C_H}. \tag{8.34}$$

The upper limit of rate constants that can be accessed by IMPS is normally determined by τ_{cell}. Typical values of C_{sc} for a moderately doped semiconductor under depletion conditions are of the order of 10^{-7} F cm^{-2}. If care is taken to minimise the sum of the solution and contact resistances below 10 Ω cm^2, τ_{cell} will be less than 10^{-6} s. For highly doped semiconductors or for samples with poor ohmic contacts, larger *RC* time constants will restrict the range of rate constants that can be determined by IMPS.

If the series resistance is deliberately increased by adding an external resistor, then $\omega RC > 1$ in equation (8.33), and the measured response is effectively an ac photovoltage $U_{photo}(\omega)$. If the modulation of the light source is sufficiently small to guarantee that the photovoltage is of the order of mV, k_{rec} can be treated as constant to a first approximation, and in this case the modulated photovoltage is given by [60]

$$\omega U_{photo}(\omega) = \frac{qg_{ac}}{C}\left[\frac{k_{tr} + i\omega(C/C_{sc})}{k_{tr} + k_{rec} + i\omega}\right]. \tag{8.35}$$

The complex plane plot of $\omega U_{photo}(\omega)$ predicted by equation (8.35) is a semicircle which for $(RC)^{-1} \ll \omega \ll (k_{tr} + k_{rec})$ has a low frequency intercept given by

$$\omega U_{photo}(\omega)|_{\omega \to 0} = \frac{qg_{ac}}{C}\frac{k_{tr}}{k_{tr} + k_{rec}}, \tag{8.36}$$

and a high frequency intercept

$$\omega U_{photo}(\omega)|_{\omega \to \infty} = \frac{qg_{ac}}{C_{sc}}. \tag{8.37}$$

The maximum of the semicircle appears at $\omega_{max} = k_{rec} + k_{tr}$. It follows that frequency response analysis of the small signal photopotential response should give access not only to k_{tr} and k_{rec} but also to C_{sc} and C_H. The possibilities offered by intensity modulated photopotential measurements remain largely unexplored, although Kamieniecki used the method to determine space charge capacitances as long ago as 1982 [73, 74]. Since the input impedances of most voltage amplifiers are at least $10^8\ \Omega$ and typical space charge capacitances are in excess of 10^{-8} F cm^{-2}, the *RC* time constant for a typical electrochemical cell is of the order of 1 sec or more.

8.11 IMPS ANALYSIS OF MULTISTEP PHOTOELECTROCHEMICAL REACTIONS WITH ADSORBED INTERMEDIATES

The behaviour of *p*-InP discussed in Section 8.9 was judged to be non-ideal on the basis of a simplified theory for one electron transfer reactions. However, the reduction of protons to hydrogen by photogenerated electrons must be a two step process. There are two possible mechanisms:

$$h\nu \rightarrow h^+_{vb} + e^-_{cb}, \tag{8.38a}$$

$$H^+ + e^-_{cb} \rightarrow H^\bullet, \tag{8.38b}$$

followed either by

$$H^+ + H^\bullet + e^-_{cb} \rightarrow H_2 \qquad \text{mechanism I}, \tag{8.38c}$$

or by

$$H^\bullet + H^\bullet \rightarrow H_2 \qquad \text{mechanism II}. \tag{8.38d}$$

Both mechanisms involve a hydrogen intermediate which must be adsorbed at the surface, modifying the potential distribution across the interface and hence the band bending. This intermediate may also be responsible for surface recombination by the reaction

$$H^\bullet_{ads} + h^+ \rightarrow H^+. \tag{8.38e}$$

The intermediate can therefore be considered as a photogenerated surface state. The formation of adsorbed intermediates is a common feature of multi-electron transfer reactions. Other examples are encountered in the photodecomposition of compound semiconductors, for example the photoanodic decomposition of *n*-CdS (cf., equation (8.9)).

Multistep photoelectrochemical mechanisms introduce several complications into the kinetic analysis. Accumulation of the photogenerated intermediate results in the build up of surface charge, which reduces the band bending. For an IMPS measurement, the surface charge will have not only a steady state component but also a modulated component because modulation of the surface intermediate concentration results in a corresponding modulation of the potential (i.e., an ac photovoltage is generated). This ac photovoltage in turn modulates the surface concentration of majority carriers and hence the rate of surface recombination. Finally the ac photovoltage also results in a charging current associated with charging of the space charge and Helmholtz capacitances. These effects, which lie outside the scope of this chapter, have been discussed by de Wit *et al.* [53] and by Peter *et al.* [71].

Peter *et al.* [71] have given a detailed analysis of mechanisms I and II. It can be shown that apparent rate constants k_{tr} and k_{rec} derived from IMPS data are in fact functions of the rate constants for all the steps in the reaction. For both mechanisms, the analysis shows that k_{tr} is no longer potential independent because it contains a term associated with recombination via the intermediate. Similarly k_{rec} is determined by the all three rate constants, and no longer follows a simple Boltzmann dependence on potential. Figures 8.13a and 8.13b illustrate the quite complicated potential dependence of the apparent rate constants k_{tr} and k_{rec} predicted for mechanism I when dynamic surface charging is taken into account.

Recently Fermin *et al.* [31] have used IMPS to study the photo-evolution of oxygen on *n*-TiO_2 and have concluded that the response is consistent with a Case II type mechanism involving $OH^{\bullet}_{ads}$ intermediates. The same work showed that the $OH^{\bullet}_{ads}$ intermediate is also involved in the photo-oxidation of formic acid.

Oskam *et al.* [66] have used IMPS to investigate the role of surface states at the *n*-Si(111)/NH_4F interface. In this case, the redox reaction is simpler, and appears not to involve holes trapped at surface states. This is probably due to the presence of a surface oxide layer. Electron transfer is evidently exceptionally slow in this case, since these authors observed a modulated photocurrent even at potentials far from the flatband potential where recombination is expected to be negligible. Accumulation of holes modifies the potential drop across the Helmholtz (and presumably also surface oxide region), leading to a capacitive charging current. This effect has also been treated in more detail by Peter *et al.* [71].

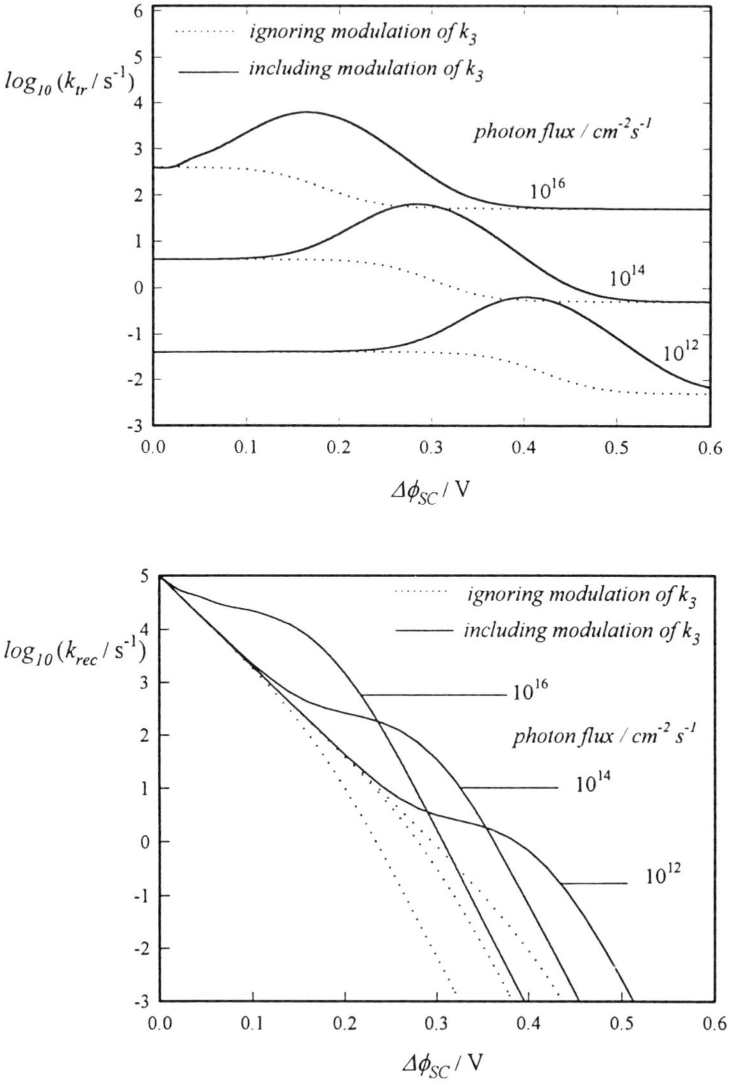

Fig. 8.13 (a) Potential dependence of the phenomenological rate constant k_{tr} derived for mechanism I. Note that for a two step mechanism, this rate constant contains terms associated with surface recombination, so that it is not the true rate constant for electron transfer. The influence of the modulation of potential due to surface charging is shown. (b) Potential dependence of the phenomenological rate constant k_{tr} for case *I*. The influence of the dynamic modulation of surface potential by accumulated reaction intermediates is shown.

8.12 STUDIES OF PHOTOCURRENT MULTIPLICATION BY IMPS

Examples of photocurrent multiplication [1] include the photo-oxidation of formic acid and of secondary alcohols at *n*-type semiconductors, and the photoreduction of oxygen at *p*-type semiconductors. The mechanisms involve majority carrier injection by a photogenerated intermediate, and IMPS has been used to determine the rate constant for these processes.

The first system exhibiting photocurrent multiplication to be studied by IMPS was the photoreduction of oxygen to H_2O_2 at *p*-GaP [36, 37]. The oxidation of formic acid at *n*-CdS was then characterised by the same method [35]. Further examples of current doubling that have been studied by IMPS include the photo-oxidation of formic acid at TiO_2 [31] and the photoanodic oxidation of *n*-Si [33, 34].

The oxidation of formic acid to CO_2 is a two step reaction which involves the following steps

$$h\nu \rightarrow h^+ + e \tag{8.39a}$$

$$HCOOH + h^+ \rightarrow COOH^\bullet + H^+ \tag{8.39b}$$

$$COOH^\bullet + h^+ \rightarrow CO_2 + H^+ \tag{8.39c}$$

$$COOH^\bullet \rightarrow CO_2 + H^+ + e^- \tag{8.39d}$$

Reaction (8.39d) is responsible for photocurrent doubling since it results in the two electron oxidation of formic acid to CO_2 for the absorption of only one photon. The electron injection step competes with the hole capture reaction, (8.39c), and as a result the photocurrent quantum efficiency depends on illumination intensity. At high intensities, the supply of photogenerated holes to the surface favours reaction (8.39c), and the quantum efficiency is 1. At low light intensities, electron injection becomes predominant, and the quantum efficiency tends towards 2.

IMPS is a powerful technique for the study of photocurrent multiplication because it allows deconvolution of the minority and majority carrier contributions to the total photocurrent. The component of the photocurrent flux due to injection of majority carriers lags behind the in-phase component associated with the flux of photogenerated minority carriers. The time delay corresponds to the first order lifetime of the injecting intermediate, and the injection component is attenuated progressively as the modulation frequency is increased. The phase lag and attenuation give rise to a characteristic IMPS plot which is a semicircle in the lower quadrant of the complex plane given by

$$\frac{j_{\text{photo}}}{qg_{\text{ac}}} = 1 + \frac{1}{1 + i\omega\tau_{\text{inj}}}. \tag{8.40}$$

The low frequency intercept corresponds to a quantum efficiency of 2, and the high frequency intercept tends towards a quantum efficiency of 1. The minimum of the semicircle occurs at an $\omega\tau = 1$, i.e., $\omega = k_{\text{inj}}$, the first order rate constant for majority carrier injection [36, 37].

A more detailed treatment takes recombination and *RC* attenuation into account [35], and Fig. 8.14 illustrates potential dependence of the IMPS response expected in this case.

Figure 8.15 illustrates the experimental IMPS response for formic acid oxidation on CdS reported by Herrasti and Peter [35]. As predicted, the IMPS response in the photocurrent saturation region exhibits a semicircle in the lower quadrant. At lower potentials, recombination is more favoured, and the IMPS response becomes a semicircle in the upper quadrant. The rate constant for electron injection derived from ω_{min} is $6.0 \times 10^4\ \text{s}^{-1}$. This is a surprisingly low value since the HCO_2/CO_2 energy level is expected to be well above the conduction band of CdS. The low constant probably reflects the high activation energy for the conversion of the non-linear HCO_2 radical to linear CO_2.

It is clear from Fig. 8.15 that it is not possible to measure the IMPS response at sufficiently high frequencies to observe the limit where the quantum efficiency tends towards unity (i.e., where $\omega \gg k_{\text{inj}}$). The limitations arise in this case from the dynamic response of the potentiostat. In other cases, attenuation due to the *RC* time constant of the system may obscure the injection semicircle. The upper limit to the majority carrier injection rate constants that can be obtained by IMPS is around $10^5\ \text{s}^{-1}$.

IMPS has been used to investigate the involvement of electron injection in the photodissolution of *n*-Si [33, 34]. In the case of the photodissolution *n*-Si in fluoride solutions, the quantum efficiency varies from 4 at low light intensities to 2 at high light intensities. A quantum efficiency of 4 corresponds to a mechanism in which the capture of one photogenerated hole is followed by the injection of three electrons into the conduction

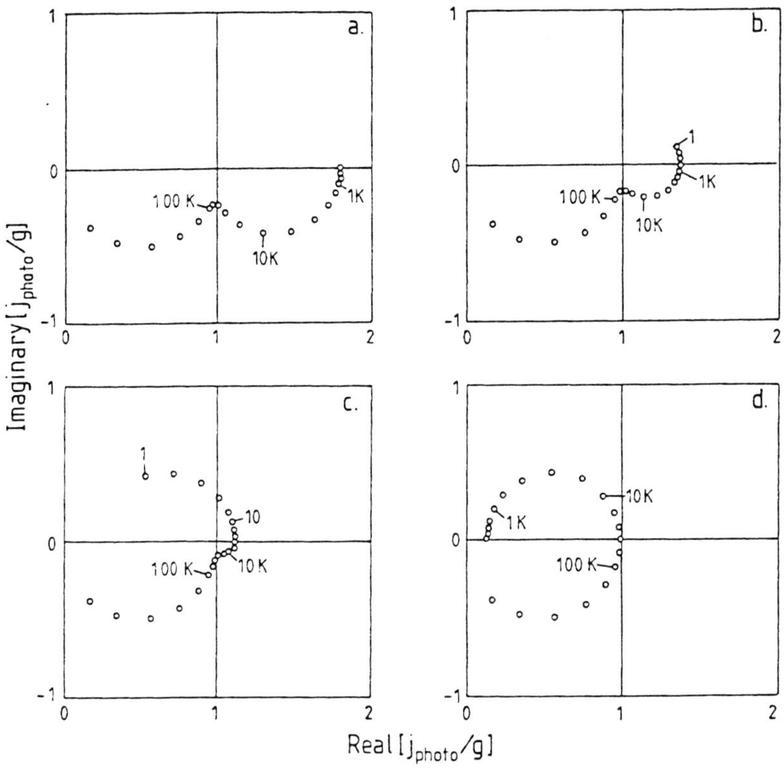

Fig. 8.14. Theoretical plots showing competition between recombination and current doubling. Calculated for a surface state density of 5×10^{12} cm^{-2}. The surface state is located 0.3 eV below the bulk Fermi level. Donor density 1.5×10^{16} cm^{-3},. k_{inj} 5×10^{4} s^{-1}. τ_{cell} = 2×10^{-7} s. Band bending values (a) 0.40 eV, (b) 0.35 eV, (c) 0.30 eV, (d) 0.2 eV. Note the transition in the IMPS response from current from current doubling control at 0.4 eV to recombination control at 0.2 eV.

band by dissolution intermediates. Peter *et al.* [34] have given a detailed treatment of both the steady state and periodic response expected for a mechanism involving surface bound silicon intermediates with nominal valence states between (I) and (III). The general reaction scheme shown in Fig. 8.16 considers competition between the capture of photogenerated holes and the injection of electrons by intermediates. Surface recombination occurring in parallel *via* surface states was also considered (a more

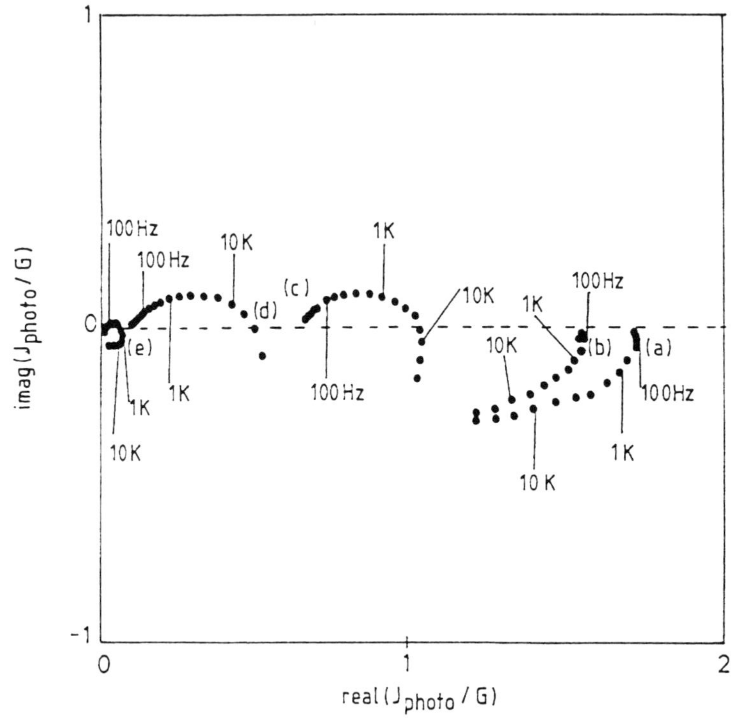

Fig. 8.15. Experimental IMPS for formic acid oxidation on *n*-CdS.. Potentials vs. RHE: (a) 0.8 V, (b) 0.0 V, (c) −0.2 V, (d) −0.4 V, (d) −0.6 V. At high band bending, the response is characteristic of current doubling. The value of k_{inj} derived from ω_{min} is $6 \times 10^4\ s^{-1}$.

realistic approach might be to consider recombination reactions that involve the photogenerated intermediates).

If $k_a > k_b > k_c$, three semicircles are expected to appear in the lower quadrant of the IMPS plot with ω_{min} values are equal to k_a, k_b and k_c (Fig. 8.17). Each semicircle has a diameter corresponding to a quantum efficiency of one. If $k_a < k_b > k_c$, two semicircles are seen, one with a diameter twice that of the other and with ω_{min} values equal to k_a and k_c. If $k_a < k_b < k_c$, only one semicircle is seen with $\omega_{min} = k_a$.

Figure 8.18 illustrates the fitting used to obtain the rate constants for the injection processes from the experimental IMPS response for *n*-Si in NH_4F. The flattening of the semicircles is attributed to surface heterogeneity, and this effect was modelled by assuming a normal distribution of

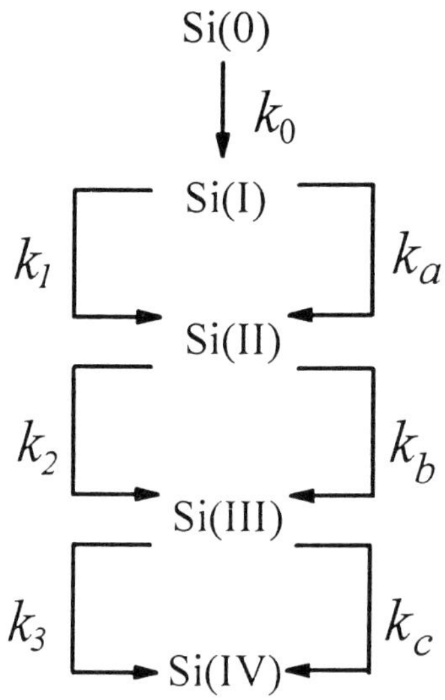

Fig. 8.16. Reaction scheme for photoanodic dissolution of silicon in low intensity limit illustrating the competition between hole capture steps (rate constants k_0 to k_3) and electron injection steps (rate constants k_a to k_c). The nominal valence states of the silicon intermediates are indicated. The final product Si(IV) is the soluble hexafluorosilicate species.

activation energies for electron injection with a standard deviation of the order of k_BT. The rate constants obtained from the fit are $k_a = 2 \times 10^4\,s^{-1}$, $k_b = 500\,s^{-1}$ and $k_c = 0.5\,s^{-1}$. The value of k_c is so low that electron injection only competes with hole capture in the third step at very low light intensities.

An estimate of the energy levels of the injecting states relative to the conduction band can be made by assuming that the rate constants for electron injection are given by

$$k_{\mathrm{inj}} = \nu\, e^{-E_a/k_BT} \tag{8.41}$$

where ν is a pre-exponential factor (of the order of $10^{12}\,s^{-1}$) and the activation energy E_a is determined by the energy difference between the injecting state and the conduction band of the silicon. The activation energies obtained from the values of k_{inj} are 0.44, 0.54 and 0.69 eV.

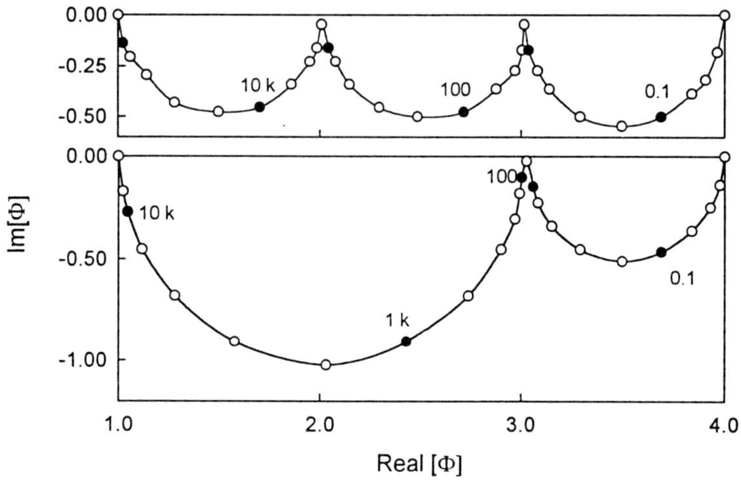

Fig. 8.17. Theoretical IMPS plots calculated for the scheme shown in Fig. 8.16. The upper plot is for the case where $k_a > k_b > k_c$. The values of these rate constants are given by the corresponding $\omega_{\min}$ values. The lower figure.shows the IMPS response expected for $k_a < k_b > k_c$. In this case the two higher frequency semicircle condense into one determined by k_a.

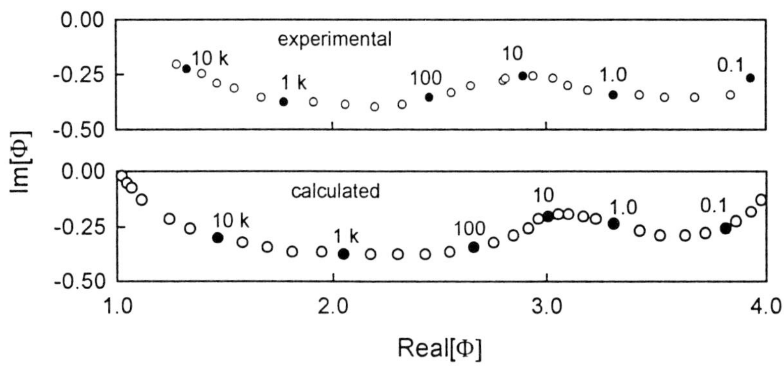

Fig. 8.18. The upper plot shows the experimental IMPS response measured at very low light intensities for the photodissolution of *n*-Si in 1.0 NH_4F at pH 4.5. The lower plot is the calculated best fit. Note that the photocurrent efficiency varies from 4 at low frequencies to 1 at high frequencies as expected for the scheme shown in Fig. 8.16.

8.13 PHOTOELECTROCHEMICAL IMPEDANCE SPECTROSCOPY

In an IMPS experiment the light intensity is modulated to produce an ac photocurrent that is analysed to obtain kinetic information. An alternative approach is to modulate the electrode potential while keeping the illumination intensity constant. This method is called PhotoElectrochemical Impedance Spectroscopy (PEIS).

In a PEIS experiment, the flux of minority carriers generated by illumination is constant to a good approximation, and if the condition $1/\alpha < d_{SC} + L$ is fulfilled, then $g \approx I(0)$. The ac potential perturbs the density of majority carriers at the surface and therefore modulates k_{rec} about a mean (dc) value, giving rise to an ac photocurrent. The photoelectrochemical admittance, ($Y_{PEIS} = 1/Z_{PEIS}$), is the ratio of the ac component of the total photocurrent to the ac voltage. For competition between recombination and electron transfer, the photoelectrochemical admittance is given by [29]

$$Y_{PEIS} = i\omega C_{sc} + \frac{qI(0)}{kT}\frac{k_{rec}(k_{tr} + i\omega)}{(k_{tr} + k_{rec})(k_{tr} + k_{rec} + i\omega)}. \tag{8.42}$$

The complex plane plot of Y_{PEIS} exhibits two semicircles, and it can be fitted using the series/parallel circuit shown in Fig. 8.19.

The R and C values for the low frequency semicircle, which depend on the dc illumination, $I(0)$, are given by

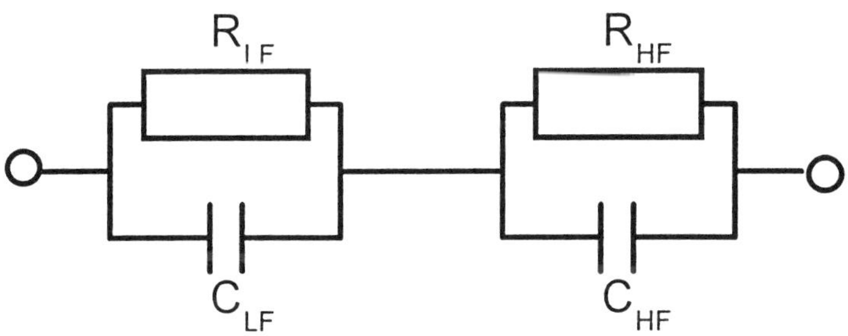

Fig. 8.19. Series/parallel equivalent circuit that can be used to model PEIS response (see text for definition of components).

$$R_{LF} = \left(\frac{kT}{q^2 I(0)}\right)\frac{k_{tr} + k_{rec}}{k_{tr}}, \tag{8.43a}$$

and

$$C_{LH} = \left(\frac{q^2 I(0)}{kT}\right)\frac{1}{k_{tr} + k_{rec}}. \tag{8.43b}$$

The frequency of the maximum of the low-frequency semicircle, $\omega_{max}(LF)$ is equal to k_{tr} and is independent of light intensity.

The R and C values for the high frequency semicircle are

$$R_{HF} = \left(\frac{kT}{q^2 I(0)}\right)\frac{k_{tr} + k_{rec}}{k_{rec}}, \tag{8.44a}$$

$$C_{HF} = C_{SC}. \tag{8.44b}$$

ω_{max} for the high frequency semicircle is depends linearly on light intensity:

$$\omega_{max}(HF) = \left(\frac{q^2 I(0)}{kTC_{SC}}\right)\frac{k_{rec}}{k_{tr} + k_{rec}}. \tag{8.44c}$$

It is important to note that the physical significance of the RC components only emerges from an analysis of the kinetics. Attempts to interpret these elements without relating them to the kinetics and mechanism of the photoelectrochemical process will inevitably result in conceptual difficulties.

Figure 8.20 is an example of a typical PEIS response measured for *p*-InP in 1.0 mol dm^{-3} H_2SO_4 [29]. The size of the two semicircles depends on potential because the band bending determines the mean or dc value of k_{rec}. The values of k_{rec} and k_{tr} derived from PEIS measurements on *p*-InP agree well with those derived from IMPS.

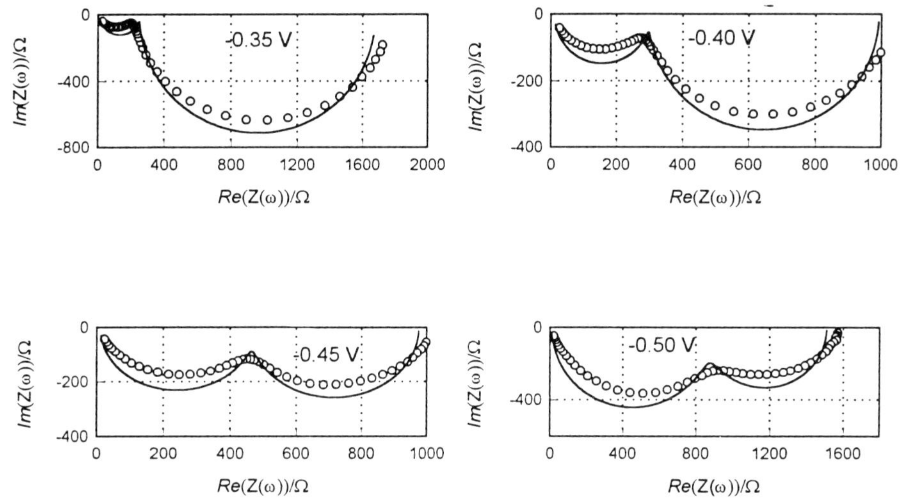

Fig. 8.20. Experimental PEIS plots for p-InP in 1.0 mol dm^{-3} H_2SO_4 at the potentials shown. Experimental data shown as open circles. Lines, best fit to the theory.

8.14 LIGHT MODULATED MICROWAVE MEASUREMENTS

Light and potential modulated microwave reflectivity measurements offer a novel approach to the study of the semiconductor electrolyte interface. Perturbation of the density of electrons and holes in a semiconductor influences the conductivity and hence the imaginary component of the dielectric constant at microwave frequencies. For small perturbations, the change ΔR_m in microwave reflectivity depends linearly on the change in conductivity [27, 28, 75]. The application of frequency response analysis to light modulated microwave reflectance is relatively new [30]. Although the technique is analogous to IMPS, it provides additional information.

Figure 8.21 shows the experimental arrangement for light modulated microwave measurements. The semiconductor sample terminates an X-band microwave waveguide. The front of the wafer is in contact with an electrolyte solution, and a modulated light source (for example a light emitting diode) illuminates the sample. The microwave radiation incident from the other side probes the conductivity of the sample. The light induced changes in microwave reflectivity are a few parts per million, and an ac coupled low noise preamplifier is needed to amplify the signal so

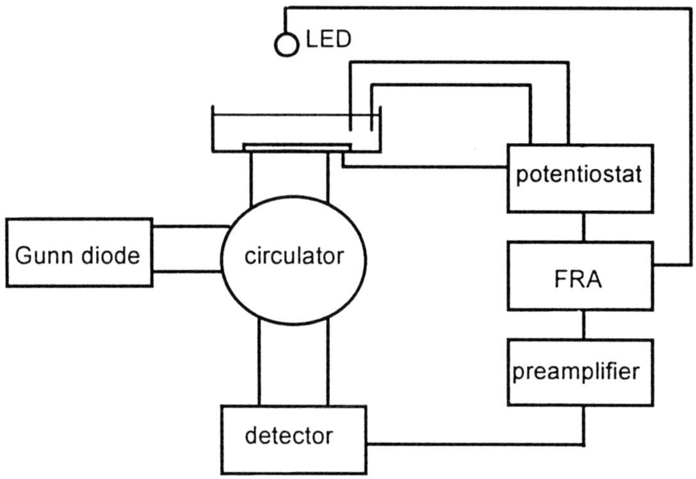

Fig. 8.21. Experimental arrangement for frequency resolved measurements of light induced microwave reflectivity.

that it can be detected by the frequency response analyser. The magnitude and phase of the harmonic component of the modulated microwave reflectivity are measured relative to the illumination.

For the very small perturbations involved in the measurements, the light modulated microwave reflectivity change, ΔR_m is a linear function of the ac component of the photogenerated charge that accumulates at the semiconductor/electrolyte interface

$$\Delta R_M(t) = SQ_s(t), \tag{8.45}$$

where S is a sensitivity factor.

The flux of minority carriers into the surface is linked to surface charging, electron transfer and recombination, so that

$$qg_{\mathrm{ac}}(t) = \frac{\mathrm{d}Q_s(t)}{\mathrm{d}t} + (k_{\mathrm{tr}} + k_{\mathrm{rec}})Q_s(t). \tag{8.46}$$

The photocurrent in the external circuit is

$$j_{\text{photo}}(t) = qg_{\text{ac}}(t) - k_{\text{rec}}Q_s(t). \tag{8.47}$$

If the illumination is switched on at $t = 0$ (i.e., the initial condition is $Q_s = 0$), the transient microwave response is given by

$$\Delta R_M(t) = SQ_s(t) = Sqg\frac{(1 - \mathrm{e}^{-(k_{\text{tr}}+k_{\text{rec}})t})}{k_{\text{tr}} + k_{\text{rec}}}. \tag{8.48}$$

Examination of equation (8.48) shows that the transient microwave response still contains information when there is no surface recombination. By contrast, if $k_{\text{rec}} = 0$, the photocurrent simply follows the illumination step and contains no information about the rate of charge transfer at the interface. If $k_{\text{rec}} = 0$, equation (8.48) becomes

$$\Delta R_M(t) = Sqg\frac{(1 - \mathrm{e}^{-k_{\text{tr}}t})}{k_{\text{tr}}}. \tag{8.49}$$

This time dependence of the microwave reflectivity predicted by equation (8.48) is contrasted in Fig. 8.22 with the time dependent photocurrent for the case where $k_{\text{rec}} = 0$. The comparison shows that, unlike IMPS and PEIS, light modulated microwave reflectivity measurements still provide kinetic information at high band bending where recombination is negligible and the steady state photocurrent is described by the Gärtner equation.

The microwave response to sinusoidal modulation of the light intensity is given by [30]

$$\text{Re}(\Delta R_M) = Sqg_{\text{ac}}\frac{k_{\text{tr}} + k_{\text{rec}}}{(k_{\text{tr}} + k_{\text{rec}})^2 + \omega^2}, \tag{8.50a}$$

and

$$\text{Im}(\Delta R_M) = -Sg_{\text{ac}}\frac{\omega}{(k_{\text{tr}} + k_{\text{rec}})^2 + \omega^2}. \tag{8.50b}$$

The microwave response is therefore a semicircle in the lower complex plane with a high frequency intercept of zero and a low frequency intercept given by

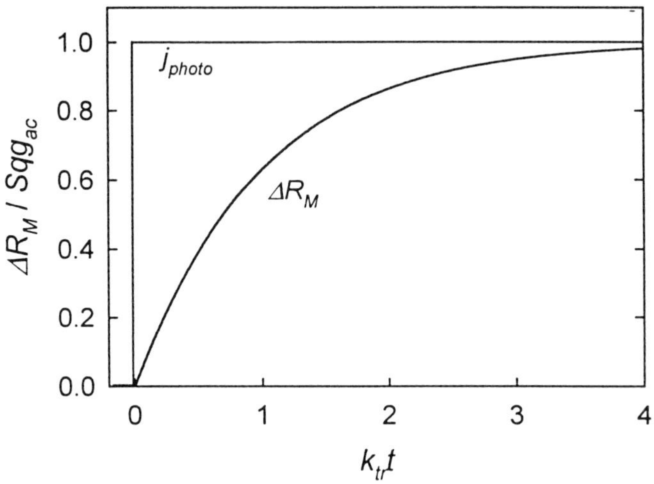

Fig. 8.22. Comparison of the time dependence of the photocurrent step and of the slower increase in microwave reflectivity predicted for a light step in the case where recombination is negligible. Note that no kinetic information can be derived from j_{photo}, whereas the time dependence of the microwave reflectivity provides access to the value of k_{tr} even in the potential region where recombination is negligible.

$$\text{Re}(\Delta R_M)_{\omega \to 0} = \frac{Sqg_{\text{ac}}}{k_{\text{tr}} + k_{\text{rec}}}. \tag{8.51}$$

The frequency of the minimum in the semicircle is equal to the sum of the rate constants for charge transfer and recombination: $\omega_{\text{min}} = k_{\text{tr}} + k_{\text{rec}}$. Since it is possible to measure k_{tr} at high band bending where k_{rec} is zero, the rate constants can be separated if it is assumed that k_{tr} is independent of potential.

IMPS and light modulated microwave reflectivity have been used to study the photo-evolution of hydrogen evolution on oxide free *p*-Si in acidic fluoride solution at low light intensities [30]. The set of IMPS plots in Fig. 8.23 can be compared with those for *p*-InP shown in Fig. 8.9. It can be seen that the response tends towards a point on the real axis as the potential becomes more negative and k_{rec} becomes negligible. Clearly all information about k_{tr} is lost in this region. The behaviour of the frequency resolved microwave is quite different. As Fig. 8.24 shows, a semicircular microwave response is still observed at $-1.0\,\text{V}$, which is well into the

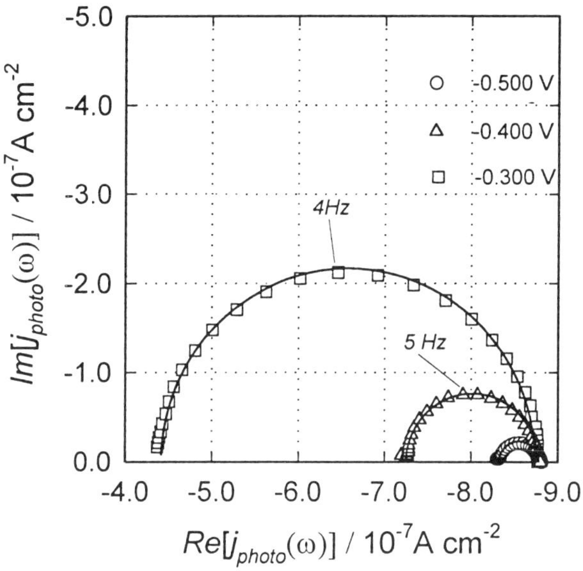

Fig. 8.23. Set of IMPS plots for *p*-Si in 1.0 mol dm^{-3} HF. Note that the plots condense to a point as the band bending is increased, so that no kinetic information can be derived at potentials more negative than −0.5 V.

saturation photocurrent region. Since recombination is negligible at this potential, the pseudo-first-order rate constant for charge transfer can be obtained directly from $\omega_{min} = 23\ s^{-1}$. At lower band bending, where recombination takes place, the minimum frequency is determined by the sum of the charge transfer and recombination rate constants.

The values of k_{tr} and k_{rec} derived from IMPS and from the microwave measurements agree to better than 5% [30]. Frequency resolved microwave techniques evidently have much to offer in the study of photoelectrochemical kinetics, and they should find more widespread application in future.

8.15 PHOTOELECTROCHEMICAL KINETICS IN NANOCRYSTALLINE SYSTEMS

Porous and nanocrystalline semiconductor electrodes have unusual properties because the semiconductor/electrolyte interphase is three-dimensional. The Helmholtz electrical double layer can extend throughout the

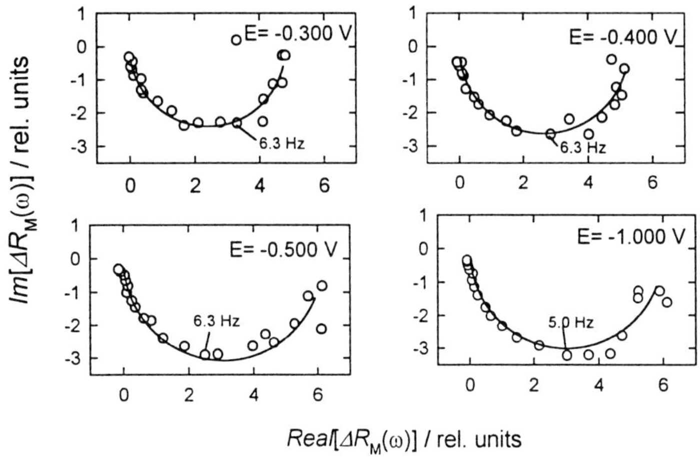

Fig. 8.24. Set of light modulated microwave reflectivity plots for p-Si in 1.0 mol dm^{-3} HF. Contrast these plots with the IMPS plots in Fig. 8.23. It is clear that k_{tr} can be derived from the microwave response even at -1.0 V, where the quantum efficiency is unity and the IMPS plot has condensed to a point.

large internal surface area of a nanoporous network, and in consequence, nanoporous semiconductor electrodes behave as "supercapacitors" under accumulation conditions. Their internal surface area can be estimated experimentally by measuring the interfacial capacitance under electron accumulation conditions, where the capacitance is determined primarily by the Helmholtz-layer, as shown for porous silicon [76].

The large surface/volume ratio characteristic of nanostructured electrodes is exploited in the Grätzel dye-sensitized photovoltaic cell [6–11], in which light is absorbed by a monolayer of sensitizer dye adsorbed on the surface of nanocrystalline TiO_2. The order of magnitude of the surface area/volume ratio can be estimated by assuming that a nanocrystalline electrode consists of hexagonally close packed spheres of radius r (fractional volume occupancy 0.73). The ratio of surface area to total volume $A/V = 0.73\ 4\pi r^2/(4/3)\pi r^3 = 2.2/r$. For $r = 15$ nm, the A/V ratio is of the order of 10^6, and for a 1 cm^2 film of 10 μm thickness, this A/V ratio corresponds to an internal surface area of 1000 cm^2. A monolayer of dye adsorbed on this high internal surface area will absorb almost all of the incident light over a wide wavelength range, whereas at a conventional flat electrode, a monolayer of sensitizer dye absorbs less than 0.1% of incident radiation.

Charge transport in nanocrystalline electrodes is clearly strongly influenced by the inter-penetration of the solid and liquid phases. If electron hole pairs are generated by band to band excitation, it is usually observed that one type of carrier is transferred to the solution, while the other is transported to the substrate contact. In the case of the dye sensitized nanocrystalline systems, an electron is injected into the conduction band from the photoexcited dye and is then transported to the substrate. The dye is regenerated by reaction of its oxidised state with a supersensitiser such as I^- as shown in Fig. 8.25.

Södergren *et al.* [77] have considered generation, diffusion and back reaction of electrons with I_3^- ions in the dye-sensitized nanocrystalline TiO_2 cell. The time dependence of the electron density, $n(x, t)$, is given by

$$\frac{\partial n(x, t)}{\partial t} = \eta \alpha I_0 e^{-\alpha x} + D \frac{\partial^2 n(x, t)}{\partial x^2} - \frac{n(x, t) - n_0}{\tau}, \qquad (8.52)$$

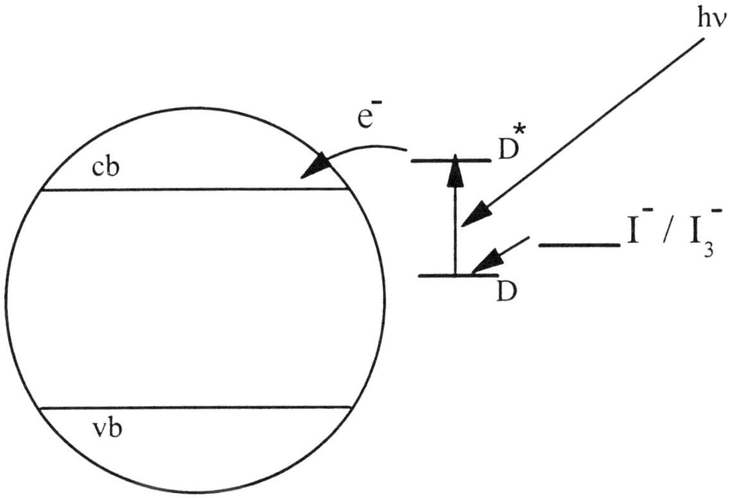

Fig. 8.25. Dye sensitisation of TiO_2. The photoexcited dye injects an electron into the conduction band of TiO_2. The oxidised dye is regenerated by electron donation from I^-.

where D is the diffusion coefficient of electrons, n_0 is the electron density in the dark, τ is the pseudo-first order lifetime of electrons determined by back reaction with tri-iodide ions in the electrolyte, α is the effective absorption (cm^{-1}) coefficient of the dye sensitized TiO_2 (determined by the absorption cross-section and surface coverage of the dye), I_0 is the incident photon flux corrected for reflection losses and η is the net efficiency of electron injection.

Södergren *et al.* obtained the steady state solution of equation (8.52) by assuming that D and τ are constant and $\eta = 1$. Under these conditions, the photocurrent due to photoinjected electrons is independent of voltage, and the steady state photocurrent is given by

$$j_{\text{photo}}\big|_{t\to\infty} = \frac{qI_0\alpha L\left(-\sinh\left(\frac{d}{L}\right)\right) - \alpha L\exp(-\alpha d) + \alpha L\cosh\left(\frac{d}{L}\right)}{(\alpha^2L^2 - 1)\cosh\left(\frac{d}{L}\right)}, \tag{8.53}$$

where d is the film thickness and $L = (D\tau)^{1/2}$ is the electron diffusion length.

Transient and periodic photocurrent measurements have revealed that the diffusion coefficient of electrons in nanocrystalline TiO_2 is several orders of magnitude smaller than in the bulk crystalline material [65, 72, 78]. During their travel through a nanoporous system, photogenerated electrons may be trapped temporarily in surface energy levels. Local polarisation of the solvent molecules by charge carriers may also lead to surface localisation (polaron formation), reducing surface mobility. Since the surface area/volume of nanocrystalline electrodes is very large, such effects are likely to slow down electronic transport.

The photocurrent responses of nanocrystalline electrodes to stepped or pulsed illumination exhibit features on rather slow timescales. This is illustrated, for example, by Fig. 8.26, which is a set of photocurrent transients reported by Solbrand *et al.* [78] for band-band excitation at 308 nm of nanocrystalline TiO_2 films of differing thicknesses permeated by 0.7 mol dm^{-3} $LiClO_4$ in ethanol. The 30 ns excimer laser pulse was incident from the solution side, and since the penetration depth of the light was much smaller than the film thickness, electron-hole pairs were effectively

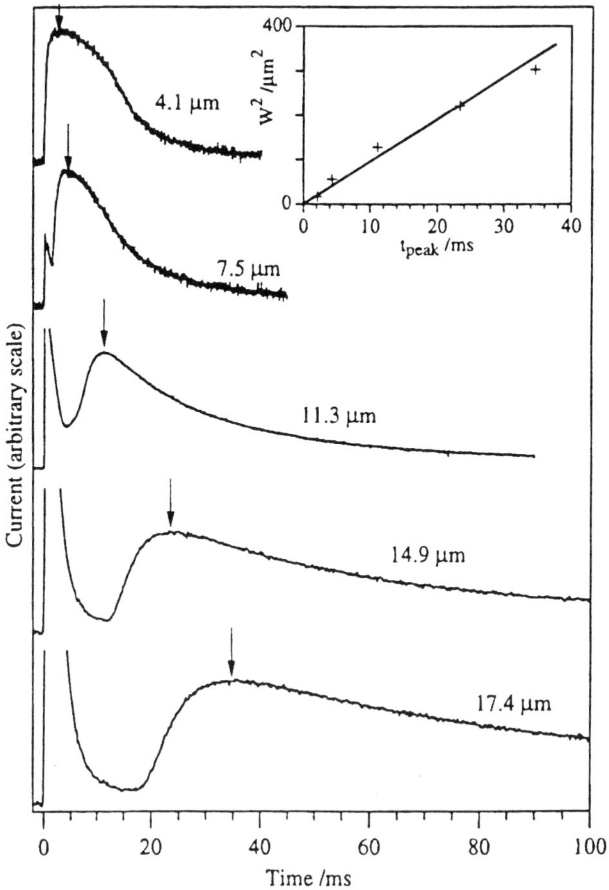

Fig. 8.26. Experimental photocurrent transients for pulsed excimer laser excitation of nanocrystalline TiO_2 electrodes of differing thicknesses taken from Ref. [78]. Illumination from the electrolyte side (200 mJ, 30 ns, λ 308 nm). Electrolyte 0.7 mol dm^{-3} $LiClO_4$ in ethanol. The insert shows that time t_{peak} at which the current peak occurs depends on the square of the film thickness (W), as expected for diffusion controlled electron transport.

generated in a thin sheet at the front surface of the film. The photocurrent peaks on a timescale of milliseconds, and the insert shows that the time at which this peak appears varies approximately with the square of the film thickness as expected for diffusion controlled electron transport. Solbrand *et al.* have interpreted these photocurrent transients in terms of an electron diffusion model that ignores trapping effects. Initially electrostatic repulsion between electrons will force some of the electrons towards the rear contact giving rise to an exponentially decaying current, $j_{\text{photo}}(t)$ that can be approximated by an RC equivalent circuit.

$$j_{RC}(t) = \frac{Q'}{\tau_{RC}} \mathrm{e}^{-(t/\tau_{\text{cell}})}, \tag{8.54}$$

where Q' is the charge density induced at the back contact and τ_{cell} is the RC time constant of the cell. After the RC charging pulse, electron-electron repulsion terms become negligible due to effective screening by the electrolyte, and electron transport is diffusion controlled. The initial concentration distribution is approximated by a delta function since the penetration depth of the illumination is small compared with the film thickness. The diffusion controlled current at the back contact is found to be

$$j_{\text{diff}} = \frac{qd\Delta n}{2\sqrt{\pi D t^{3/2}}} \mathrm{e}^{-(\mathrm{d}^2/4Dt)}. \tag{8.55}$$

The total photocurrent is given by the sum of equations (8.54) and (8.55).

Figure 8.27 illustrates the theoretical electron density profiles and photocurrent transients calculated by Solbrand *et al.* The transients exhibit a maximum at a time $t_{\text{peak}} = d^2/6D$. The inset in Fig. 8.26 shows that a plot of t_{peak} vs. d^2 is linear as predicted (the authors use W rather than d to denote the film thickness), and the slope of the plot gives a value of $1.5 \times 10^{-5}\,\text{cm}^{-2}\,\text{s}^{-1}$ for the electron diffusion coefficient.

Dye sensitized nanocrystalline have also been studied by IMPS [65, 72]. Interpretation of photocurrents generated by large amplitude perturbations is complicated by the fact that it is not possible to treat D and τ as constant if the electron density is changing by orders of magnitude during the perturbation. By contrast, analytical non-steady-state solutions of the generation collection equation can be obtained for small amplitude perturbation conditions where D and τ can be treated as constants. In practice, small amplitude means that the amplitude of the light pulse or modulation are small compared to the dc illumination level. An example of an experimental IMPS response for a dye sensitized TiO_2 cell measured under these

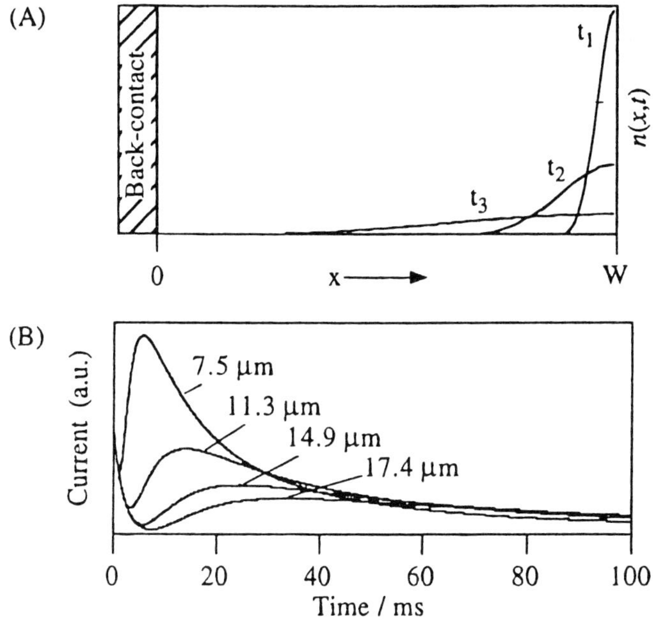

Fig. 8.27. Illustration of electron density profiles at different times $t_1 < t_2 < t_3$ following pulsed excitation from the electrolyte side. W is the film thickness. The lower part of the figure.shows transients calculated for different film thicknesses. Taken from Ref. [78].

conditions is shown in Fig. 8.28 [72]. The response is clearly quite different to the IMPS semicircle in the upper complex plane that is usually seen for normal semiconductor electrodes. The fact that the IMPS response for the nanocrystalline electrode is largely located in the lower complex plane indicates that the photocurrent lags behind the illumination due to the transit time required for electrons to reach the back contact. This is confirmed by light step measurements which show that the photocurrent rises slowly after the light is switched on [65].

Dlocik *et al.* [72] have taken a general approach to the generation collection problem which considers voltage dependent rate constants for electron extraction/injection at the TiO_2/substrate interface. The diffusion controlled limit is obtained by making the rate constant for electron extraction sufficiently large. For illumination from the electrolyte side, the ac photocurrent conversion efficiency $\Phi(\omega) = j_{photo}/qI_0$ is given by

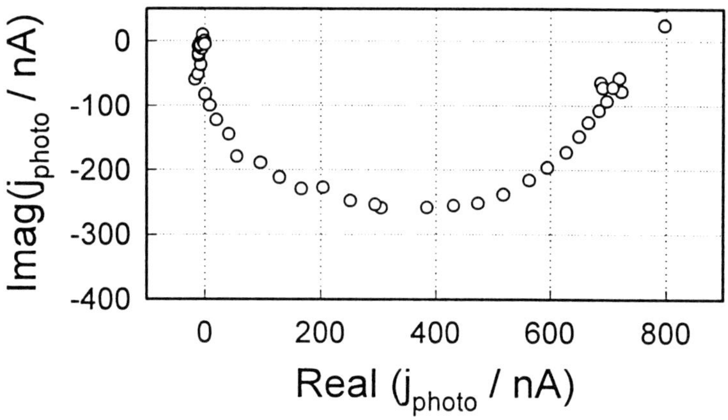

Fig. 8.28. Experimental IMPS response for a dye sensitized TiO_2 cell (illumination from the substrate side, $\lambda = 514$ nm). dc current 6.14 mA. Note that the modulated photocurrent is orders of magnitude lower than the dc photocurrent.

$$\Phi(\omega) = \frac{\alpha}{\alpha + \gamma} \cdot \frac{e^{(\gamma-\alpha)d} - e^{-(\ +\gamma)d} + 2\alpha \dfrac{e^{(\gamma-\alpha)d} - 1}{\gamma - \alpha}}{e^{\gamma d} + e^{-\gamma d}}. \tag{8.56a}$$

For illumination from the substrate side,

$$\Phi(\omega) = \frac{\alpha}{\alpha + \gamma} \cdot \frac{e^{\gamma d} - e^{-\gamma d} + 2\alpha \dfrac{e^{-\alpha d} - e^{-\gamma d}}{\gamma - \alpha}}{e^{\gamma d} + e^{-\gamma d}}. \tag{8.56b}$$

Here α is the effective absorption coefficient, d is the film thickness and γ is given by

$$\gamma = \sqrt{\left(\frac{1}{D\tau} + \frac{i\omega}{D}\right)}. \tag{8.56c}$$

The dc solutions obtained by setting $\omega = 0$ in equation (8.56) are identical with those derived by Södergren *et al.* [77].

The influences on the IMPS response of the illumination geometry and absorption coefficient are illustrated in Figs. 8.29 and 8.30. Two cases are considered corresponding to interband excitation (Fig. 8.29 small penetration depth of the light) and to dye sensitization (Fig. 8.30 large penetration depth of the light). In each case, the calculated IMPS responses are shown for illumination from the electrolyte side and from the substrate side. It can be seen that the imaginary component of the IMPS response is negative over most of the frequency range. This contrasts with the IMPS response for a minority carrier semiconductor/electrolyte junction, which is located in the upper complex plane. The difference arises from the fact that the IMPS response of the nanocrystalline system is determined by the delay or phase lag associated with carrier transport, whereas the IMPS response of a normal semiconductor electrolyte junction appears to "lead" the illumination due to the contribution of majority carriers involved in surface recombination (the charge of these majority carriers has the opposite sign and therefore their response appears as a component 180° out of phase from the minority carrier current).

If the penetration depth of the light is much smaller than the film thickness, illumination of the nanocrystalline electrode from the electrolyte side is expected to give a characteristic spiral in the high frequency IMPS response (Fig. 8.29). This spiral is typical for a constant time lag (i.e., frequency dependent phase shift), and it arises simply from the transit time required for carriers to move from the front face to the rear contact. By contrast, if the electrode is illuminated through the substrate, electrons are generated close to the contact and the transit time is much smaller (Fig. 8.29). This is reflected in the high value of $\omega_{\min}$.

The difference between the two illumination geometries is much less marked when the penetration depth of the light is comparable with the film thickness, as is usually the case for dye sensitized cells. As Fig. 8.30 shows, the characteristic frequency is quite similar in both cases since carriers are generated throughout the whole of the film.

The IMPS response is dominated by transport effects provided that the lifetime, τ_n, of injected electrons is sufficiently long that all electrons reach the back contact ($\tau_D < \tau_n$). In this case, the dc value of Φ is determined by the fraction of incident light that is absorbed, and the characteristic frequency is determined by the mean electron transit time τ_D. The effect on the IMPS response of reducing τ_n is illustrated in Fig. 8.31. The dc

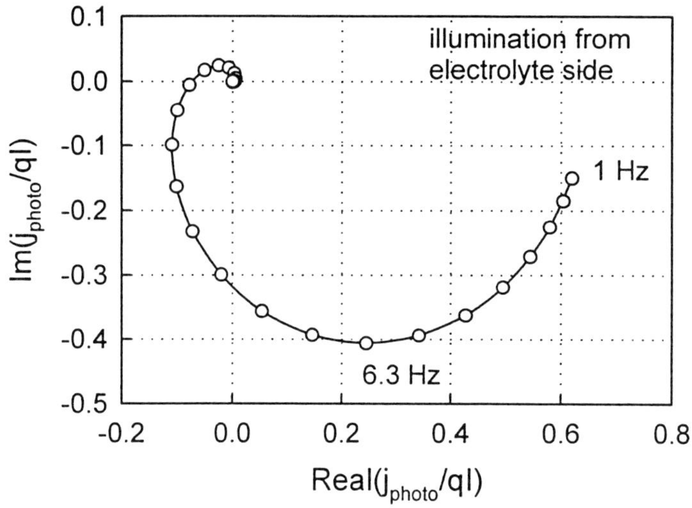

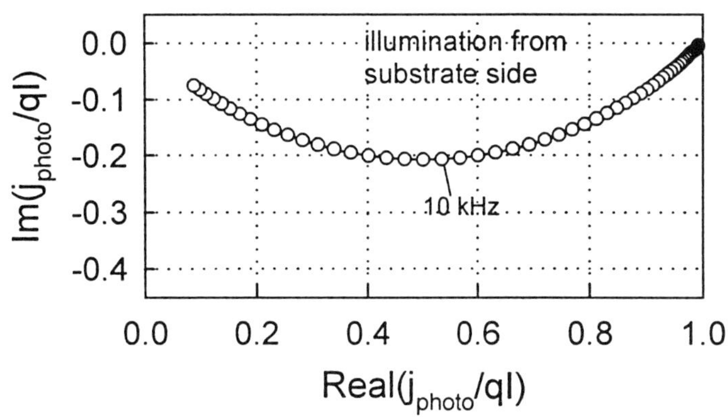

Fig. 8.29. (a) Diffusion controlled IMPS response calculated for illumination from the electrolyte side for small penetration depth. $\alpha = 10^5\ \text{cm}^{-1}$, $d = 10\ \mu\text{m}$, $D = 10^{-5}\ \text{cm}^2\ \text{s}^{-1}$, $\tau = 0.1$ s. The value of $\omega_{\min}$ is equal to the inverse of the transit time for electrons to reach the substrate. Note how the response spirals into the origin at high frequencies. (b) IMPS response calculated for the same values as in (a), except that illumination is from the substrate side. In this case electrons have to diffuse only a very short distance, and this is reflected in the much higher value of $\omega_{\min}$.

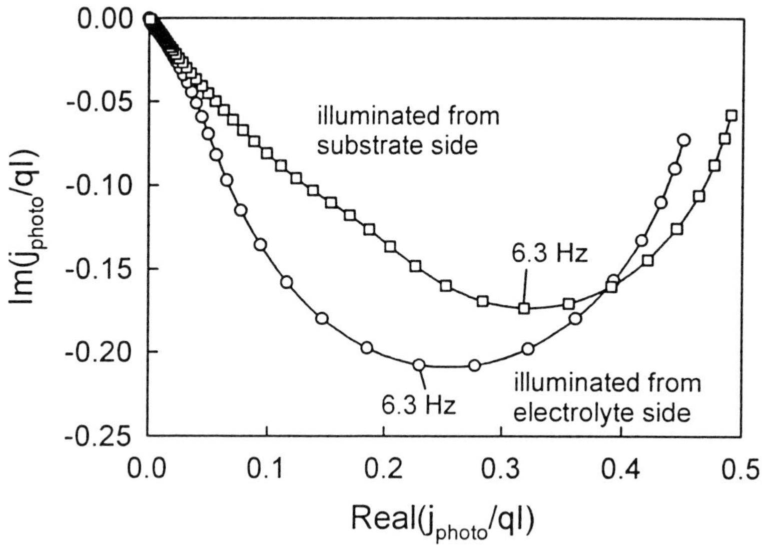

Fig. 8.30. IMPS responses calculated for the case where the penetration depth of the exciting light is comparable with the film thickness as is usually the case for dye sensitized cells. $\alpha = 10^3\ \mathrm{cm}^{-1}$, $D = 10^{-5}\ \mathrm{cm}^2\ \mathrm{s}^{-1}$, $d = 10\ \mu\mathrm{m}$, $\tau = 0.1$ s. Direction of illumination as shown. Note that the ω_{min} values for the two directions of illumination are similar because the penetration depth of the light is of the same order of magnitude as the film thickness.

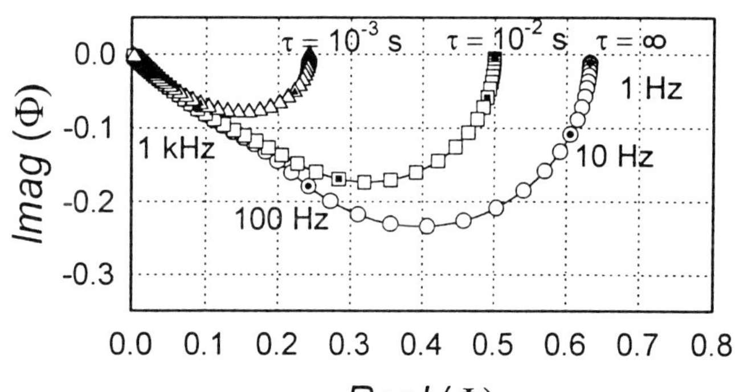

Fig. 8.31. Calculated IMPS responses showing the influence of the electron recombination lifetime τ. Film thickness $12\ \mu\mathrm{m}$, $D = 10^{-4}\ \mathrm{cm}^2\ \mathrm{s}^{-1}$, $\alpha = 10^3\ \mathrm{cm}^{-1}$. Illumination from the substrate side. Note that the characteristic frequency tends towards $1/\tau$ as τ becomes smaller.

photocurrent conversion efficiency depends on the ratio between the electron diffusion length $L = (D\tau_n)^{1/2}$ and the film thickness d, and in the limit that $L < d$, ω_{min} tends to $1/\tau_n$.

The IMPS response of nanocrystalline electrodes can be strongly influenced by the RC time constant τ_{cell}. The capacitance of nanocrystalline TiO_2 electrodes is strongly dependent on potential, and under accumulation conditions capacitance values in the mF cm^{-2} range are common. RC attenuation alters the shape of the IMPS response, and in the limit $\tau_{cell} > \tau_d$, the IMPS plot becomes semicircular and is dominated by τ_{cell}. Further details are given in ref. [72].

IMPS data are usually fitted in the complex plane, but Bode plots of the magnitude and phase of the IMPS response offer a more sensitive way of deconvoluting RC attenuation effects from the kinetics of carrier transport and recombination. An example of the comparison between theory and experiment in Bode plot format is shown in Fig. 8.32.

The survey in this section, which describes progress in a rapidly developing field, demonstrates that many problems remain to be solved. It is clear, however, that time and frequency resolved photocurrent measurements hold the key to understanding the behaviour of these novel electrode/electrolyte systems.

8.13 CONCLUSIONS

Photoelectrochemistry remains a challenging and fertile area of research. Although theoretical and experimental developments have made the kinetics and mechanisms of light induced reactions more accessible to quantitative characterisation, little is known in general about the structure and composition of the interfacial region and the role that these play in electrode reactions at illuminated semiconductors. More experimental work is needed on chemically stable semiconductors using simple one electron redox systems. A range of experimental tools is now available, and it should be possible to make further progress towards understanding electrode processes at illuminated semiconductor electrodes. Many of the reactions that have been studied so far are rather complicated. To take an example: the kinetics of photoelectrochemical hydrogen evolution have been studied on p-InP and on p-Si, but in neither case is the adsorption site for the hydrogen intermediate known. The case of p-Si in acid fluoride

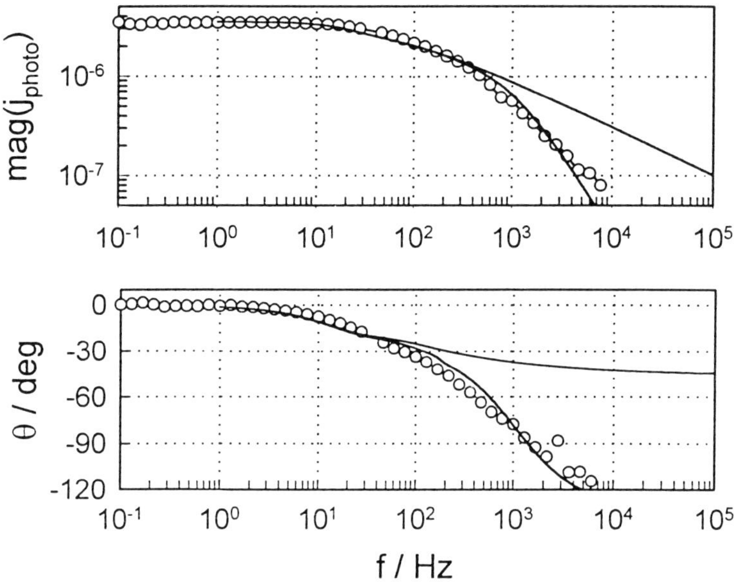

Fig. 8.32. Bode plot showing experimental data and best and fit to the short circuit IMPS response of a dye sensitized cell. Dc photocurrent 6.3 mA. The upper solid line in each plot shows the fitted response in the absence of *RC* attenuation. Note the limiting high frequency phase shift is 45°, which is characteristic of diffusion control. The lower line illustrates the improved fit obtained by including the influence of τ_{cell}. $R = 10\,\Omega$, $C = 5 \times 10^{-5}$ F.

solution is particularly interesting, since the surface is hydrogen terminated—each silicon atom in the surface is bound either to one hydrogen in the case of the (111) surface or to two hydrogen atoms in the case of the (100) surface. The capture of the first electron by H^+ is expected to lead to an adsorbed hydrogen intermediate, but the identity of possible adsorption sites remains obscure. This situation is reminiscent of the high coverage of adsorbed hydrogen on platinum that is reached before hydrogen evolution becomes thermodynamically feasible (underpotential adsorption). Hydrogen evolution on illuminated *p*-Si also illustrates the difficulty in distinguishing experimentally between Heyrovsky and Tafel mechanisms. Although the potential dependences of k_{tr} and k_{rec} have been predicted for the two cases (Case I and Case II: cf. equation (8.38)), surface charging and the diffusion of hydrogen into the semiconductor lattice introduce additional complications.

Nanocrystalline systems display a number of unusual features that are not fully understood at present. In particular, further work is needed to clarify the relationship between carrier transport, trapping, inter-particle tunnelling and electron-electrolyte interactions in three dimensional nanoporous systems. The photocurrent response of nanocrystalline electrodes is nonlinear, and the measured properties such as electron lifetime and diffusion coefficient are intensity dependent quantities. Intensity dependent trap occupation may provide an explanation for this behaviour, and methods for distinguishing between trapped and mobile electrons, for example optically, are needed. Most models of electron transport make *a priori* assumptions that diffusion dominates because the internal electric fields are small. However, field assisted electron transport may also contribute to the measured photocurrent response, and this question needs to be addressed in future work.

References

[1] For a general introduction see S.R. Morrison, Electrochemistry at Semiconductor and Oxidized Metal Electrodes (Plenum, New York, 1980).
[2] A. Heller, Acc. Chem. Res. 14 (1981) 154.
[3] R. Memming, Topics in Current Chemistry 153 (1988) 79.
[4] Y.V. Pleskov, Solar Energy Conversion—A photoelectrochemical approach (Springer, Berlin, 1990).
[5] H. Gerischer, Faraday Disc. Chem. Soc. 70 (1980) 137.
[6] B. O'Regan and M. Grätzel, Nature (London) 353 (1991) 737.
[7] M. Grätzel, Coord. Chem. Rev. 111 (1991) 167.
[8] M.K. Nazeeruddin, A. Kay, I. Rodicio, R. Huphrey-Baker, E. Muller, P. Liska, N. Vlachopoulos and M. Grätzel, J. Am. Chem. Soc. 115 (1993) 6382.
[9] M. Grätzel and K. Kalayasundaram, Current Science 66 (1994) 706.
[10] Kay and M. Grätzel, Solar Energy Mat. and Solar Cells 44 (1996) 99.
[11] T. Gerfin, M.Grätzel and L. Walder, Prog. Inorg. Chem. 44 (1997) 345.
[12] W.J. Albery, Acc. Chem. Res. 15 (1982) 142.
[13] W.J. Albery and P.N. Bartlett, J. Electroanal. Chem. 131 (1982) 145.
[14] L. M. Peter, Chem. Rev. 90 (1990) 753.
[15] W. Schmickler, Interfacial Electrochemistry, Chapters 8 and 19 (Oxford University Press, Oxford, 1996).
[16] R.J. Dwayne Miller, G.L. McLendon, A.J. Nozik, W. Schmickler and F. Willig, Surface Electron Transfer Processes, Chapter 4 (VCH, Weinheim, 1995).
[17] A.M. Kuznetsov, Charge Transfer in Physics, Chemistry and Biology, Chapter 8 (Gordon and Breach, Amsterdam, 1995).

[18] K.R. Kneten and R.L. McCreery, Anal. Chem. 64 (1992) 2518.
[19] G.A.Shreve and N.S. Lewis, J. Electrochem. Soc. 142 (1995) 112.
[20] K.E. Pomykal, A.M. Fajardo and N.S. Lewis, J. Phys. Chem. 99 (1995) 8302.
[21] K.E. Pomykal, A.M. Fajardo and N.S. Lewis, J. Phys. Chem. 100 (1996) 3652
[22] K.E. Pomykal and N.S. Lewis, J. Phys. Chem. 101 (1997) 2476.
[23] I. Uhlendorf, R. Reinekekoch and R. Memming, J. Phys. Chem. 100 (1996) 4930.
[24] R. Memming, Solid State Ionics 94 (1997) 131.
[25] A. Meier, S.S. Kocha, M.C. Hanna, A.J. Nozik, K. Siemoneit, R. Reinekekoch and R. Memming, J. Phys. Chem. 101 (1997) 7038.
[26] W. W. Gärtner, Phys. Rev. 116 (1959) 84.
[27] G. Schlichthörl and H. Tributsch, Electrochim. Acta 37 (1992) 919.
[28] H. Tributsch, G. Schlichthörl and L. Elstner, Electrochim. Acta 38 (1993) 141.
[29] E.A. Ponomarev and L.M. Peter, J. Electroanal. Chem. 397 (1995) 45.
[30] G. Schlichthörl, E.A. Ponomarev and L.M. Peter, J. Electrochem. Soc. 142 (1995) 3062.
[31] D.J. Fermn, E.A. Ponomarev and L.M. Peter, in: Photoelectrochemistry. Proc. Symp. Photoelectrochemistry. p. 62. eds. K. Rajeshwar, L.M Peter, A. Fujishima, D. Meissner and M. Tomkiewicz (The Electrochemical Society Pennington, N.J., 1997).
[32] P. Herrasti and L.M. Peter, J. Electroanal. Chem. 334 (1992) 133.
[33] H-J. Lewerenz, J. Stumper and L.M. Peter, Phys. Rev. Lett. 61 (1988) 1989.
[34] L.M. Peter, A.N. Borazio, H.J. Lewerenz and J. Stumper, J. Electroanal. Chem. 290 (1990) 229.
[35] P. Herrasti and L.M. Peter, J. Electroanal. Chem. 305 (1991) 241.
[36] R. Peat and L.M. Peter, Electrochim. Acta 31 (1986) 731.
[37] R. Peat and L.M. Peter, J. Electroanal. Chem. 209 (1986) 307.
[38] J. Li, R. Peat, and L. M. Peter, J. Electroanal. Chem. 165 (1984) 41.
[39] L.M. Peter, in: Electrochemistry, Specialist Periodical Report. p. 66. ed. D. Pletcher (Royal Society of Chemistry, London, 1984).
[40] L.M. Peter and D. Vanmaekelbergh, in: Advances in Electrochemical Science and Engineering, Vol. 6. eds. R.C. Alkire and D.M. Kolb (VCH Weinheim, in press).
[41] R. Peat and L.M. Peter, J. Electroanal. Chem. 228 (1987) 351.
[42] W.J. Albery and P.N. Bartlett, J. Electrochem. Soc. 129 (1982) 2254.
[43] J. Li and L.M. Peter, J. Electroanal. Chem. 193 (1985) 27.
[44] J. Li and L.M Peter, J. Electroanal. Chem. 199 (1986) 1.
[45] R. Peat and L.M. Peter, Ber. Bunsenges. Phys. Chem. 91 (1987) 382.
[46] R. Peat and L.M. Peter, Appl. Phys. Lett. 51 (1987) 328.
[47] L.M. Peter, J. Li, R. Peat, H.J. Lewerenz and J. Stumper, Electrochim. Acta 35 (1990) 1657.
[48] B.P. Minks, N.A.M. Verhaegh, J.J. Kelly, J. Stumper and L.M. Peter, J. Electroanal. Chem. 303 (1991) 277.
[49] J.J. Kelly, B.P. Minks, N.A.M. Verhaegh, J. Stumper and L.M. Peter, Electrochim. Acta 37 (1992) 877.
[50] R.S. Hutton and L.M. Peter, J. Electroanal. Chem. 332 (1992) 315.
[51] J. Schefold, J. Electroanal. Chem. 341 (1992) 111.

[52] P.C. Searson, D.D. MacDonald and L.M. Peter, J. Electrochem. Soc. 139 (1992) 2538.
[53] A.R. de Wit, D. Vanmaekelbergh and J.J. Kelly, J. Electrochem. Soc. 139 (1992) 2508.
[54] D. Vanmaekelbergh, A.R. de Wit and F. Cardon, J. Appl. Phys. 73 (1993) 5049.
[55] B.H. Erné, D. Vanmaekelbergh and I.E. Vermeir, Electrochim. Acta 38 (1993) 2559.
[56] I.E. Vermeir, W.P. Gomes, B.H. Erné and D. Vanmaekelbergh, Electrochim. Acta 38 (1993) 2659.
[57] P.C. Searson and D.D. Macdonald, Electrochim. Acta 38 (1993) 1913.
[58] A. Goossens and D.D Macdonald, J. Electroanal. Chem. 352 (1993) 65.
[59] A. Goossens and D.D. Macdonald, Electrochim. Acta 38 (1993) 1965.
[60] E.A. Ponomarev and L.M. Peter, J. electroanal. Chem. 396 (1995) 209.
[61] J. Schefold, J. electroanal. Chem. 394 (1995) 35.
[62] G.H. Schoenmakers, D. Vanmaekelbergh and J.J. Kelly, J. Phys. Chem. 100 (1996) 3215.
[63] D. Vanmaekelbergh, F. Iranzo Marn and J. van de Lagemaat, Ber. Bunsenges. Phys. Chem. 100 (1996) 616.
[64] P.E. de Jongh and D. Vanmaekelbergh, Phys. Rev. Lett 77 (1996) 3427.
[65] F. Cao, G. Oskam, G.J. Meyer and P.C. Searson, J. Phys. Chem 100 (1996) 17021.
[66] G. Oskam, J.C. Schmidt and P.C. Searson, J. Electrochem. Soc. 142 (1996) 2538.
[67] E.A. Meulenkamp and A.R. deWit, Electrochim. Acta 41 (1996) 109.
[68] D. Vanmaekelbergh. P.E. de Jongh and J. van de Lagemaat, ref. 31, p. 72.
[69] P.E de Jongh and D. Vanmaekelbergh, J. Phys. Chem. B 101 (1997) 2716.
[70] G.H. Schoenmakers, D. Vanmaekelbergh and J.J. Kelly, J. Chem. Soc. Faraday Trans. 93 (1997) 1127.
[71] L.M. Peter, E.A. Ponomarev and D.J. Fermn, J. Electroanal. Chem. 427 (1997) 79.
[72] L. Dloczik, O. Ileperuma, I. Lauermann, L.M. Peter, E.A. Ponomarev, G. Redmond, N.J. Shaw and I. Uhlendorf, J. Phys, Chem. B 101 (1997) 10281.
[73] E. Kamieniecki, J. Vac. Sci. Technol., 20 (1982) 811.
[74] E. Kamieniecki, J. Appl. Phys. 54 (1983) 6481.
[75] M. Kunst and G. Beck, J. Appl. Phys. 63 (1988) 1093.
[76] L.M. Peter, D.J. Riley and R.I. Wielgosz. Appl. Phys. Lett. 66 (1995) 2355.
[77] S. Södergren, A. Hagfeldt, J. Olsson and S.-E. Lindquist, J. Phys. Chem. 98 (1994) 5552.
[78] A. Solbrand, H. Linström, H. Rensmo, A. Hagfeldt, S-E. Lindquist and S. Södergren, J. Phys. Chem. B 101 (1997) 2514.

Chapter 9

Kinetic Modeling of Electron Transfer Processes in Colloidal Semiconductor Photocatalysis

COLIN BOXALL

9.0. PREFACE

This review concentrates on John Albery's work in the field of colloidal semiconductor photoelectrochemistry. John's major contributions to this area, as in so many others, have been through his astounding facility for generating useful asymptotic solutions for highly complex kinetic models of electrochemical systems. So as to put John's work in colloidal photoelectrochemistry into context, Sections 9.1–9.3 of this chapter provide a review of the more salient kinetic models of semiconductor photocatalysis developed over the last 20 years or so. Section 9.4 then concentrates on the Alberian view and presents, for the first time, John's model of the chronoamperometric behaviour of colloidal CdS.

1. INTRODUCTION

This review is a discussion of the kinetic modelling of the photoelectrochemistry of colloidal semiconductor systems. This area is currently attracting significant attention from the scientific community due to the applications of colloidal semiconductors within two rapidly advancing research fronts: heterogeneous photocatalysis and nanocrystalline particle technology.

9.1.1 Heterogeneous photocatalysis

Colloidal semiconductor particles have been found to act as heterogeneous photocatalysts in a number of environmentally important reactions.

Materials such as TiO_2, ZnO, α-Fe_2O_3, SnO_2, WO_3, ZnS and CdS and have been used in laboratory-scale pollution abatement systems, reducing both organic (esp. halocarbons [see e.g., 1–10] and benzene derivatives [see, e.g., 11–20]) and inorganic (e.g., cyanide, sulphite, sulphide, nitrites etc. [21–25]) impurities/toxins to chemically useful and/or biologically harmless compounds. Colloidal TiO_2 and CdS can act as both CO_2 [26–28] and nitrogen [29] fixatives, in the latter case generating ammonium and nitrate ions, while airborne particulate oxide semiconductors photocatalytically contribute to acid rain generation [23]. Further, semiconductor photocatalysis has been found to have application in the killing of human pathogens [30–32], for the photodynamic treatment of cancer cells [33–36], for the control of (especially noxious) odours [37, 38] and for the clean up of oil spills [39–43].

This has led to an increased interest in the elucidation of the mechanistic details of such reactions and the photocatalytic properties of particulate semiconductors for the purposes of either trying to improve process efficiency prior to engineering scale-up or, in the case of acid rain production, deliberate inhibition. Indeed, as has been noted by Ollis and Al-Ekabi [44] and more recently reiterated by Hoffmann *et al.* [45], the average publication rate over the last 10 years in the areas of water, air and wastewater treatment alone exceeds 200 papers per year. The publication of authoritative review articles on colloidal semiconductors over the last decade has been correspondingly prolific [44–68] and the interested reader is referred to them for information beyond the scope of this review.

9.1.2 Nanocrystalline systems

One of the most attractive features of colloidal semiconductor systems is the ability to control the mean particle size and size distribution by judicious choice of experimental conditions (such as reactant concentration, mixing regimen, reaction temperature, type of stabilizer, solvent composition, pH) during particle synthesis. Over the last decade and a half, innovative chemical [69], colloid chemical [69–72] and electrochemical [73–75] methods have been developed for the preparation of relatively monodispersed ultrasmall semiconductor particles. Such particles (typically <10 nm across [50, 59, 60]) are found to exhibit quantum effects when the particle radius becomes smaller than the Bohr radius of the first exciton state. Under this condition, the wave functions associated with photogenerated charge carriers within the particle (*vide infra*) are subject to extreme spatial confinement within the three dimensional potential well described by the particle surface. As a result of this confinement, many important photochemical properties of the semiconductor, such as its optical absorption edge, optical absorption profile and luminescence emission wavelength, intensity and duration, are liable to significant alteration [60, 76, 77].

These nanometer sized particles, also known as nanocrystallites, quantum dots or Q-state particles, are currently the subject of substantial international research effort, especially in regard to the following areas: (i) photocatalysis, wherein it has been predicted that photogenerated electrons in Q-state semiconductor particles are long-lived with a high reducing power [77], so enhancing the photocatalytic activity of semiconductors for redox reactions induced by photogenerated electrons and holes; (ii) the conversion of light to electricity, wherein the fractal nature of deposited nanocrystalline semiconductor films, in conjunction with charge-transfer sensitization dyes (*vide infra*), has been exploited to yield a AM 1.5 solar-to-electric conversion efficiency of 10%, approaching the performance of polycrystalline silicon photovoltaic cells [78]; (iii) self assembling systems, wherein CdS nanocrystallites self assemble on thiol functionalised metal surfaces, with the potential for formation of closely packed CdS cluster monolayers which may be important for molecular electronics applications [79]; and (iv) non-linear optics (especially resonant and non-resonant third order non-linear processes) [58, 60, 76, 77, 80–88], with potential applications as optical switching and frequency conversion elements in the telecommunications and information processing industries. As in the case of heterogeneous photocatalysis, the field of Q-state semiconductor particles has been the topic of several extensive and authoritative reviews over the last ten years [see e.g., 60, 69, 77, 89–94], to which the interested reader is again referred for information beyond the scope of this review.

9.1.3 The structures of this review

This review will proceed as follows. After a description of the primary processes common to all colloidal semiconductor applications, the thermodynamic requirements of those processes will be discussed, with special emphasis on those relevant to photocatalysis. This will be followed by a short section on how the colloidal state affects the semiconductor band structure and photogenerated charge carrier recombination dynamics, leading into a longer section on the dynamics of interfacial charge transfer including a discussion of the quantum efficiency of charge transfer, static quenching and Langmuir–Hinshelwood kinetics, dynamic quenching and disperse kinetics, and the factors affecting them. Review and discussion will be restricted to general models wherever possible although, for illustration purposes, the latter part of this section of the review will be concerned with those models developed specifically to describe the photocatalytic behaviour of the most studied colloidal semiconductor materials—TiO_2 and CdS. The review will end with a discussion of some electrochemical techniques (and their attendant kinetic models) developed to characterise the photoelectrochemistry of colloidal semiconductors.

9.1.4 Notation used in this review

The following sections are in part concerned with the thermodynamics of interfacial charge carrier transfer in colloidal semiconductors, a subject that necessarily involves discussion of band edge energies, semiconductor and electrolyte Fermi energies, the redox potential of the electrolyte and the electrostatic potential distribution at the semiconductor-electrolyte interface. The symbol E is popularly used to represent energy in the semiconductor physics literature and redox potential in the electrochemical literature. Unfortunately, this has led to some confusion in the semiconductor electrochemistry literature wherein both quantities are used simultaneously (and sometimes interchangeably within the same communication) to describe the thermodynamic behaviour of electrons. Thus, to avoid confusion, this review will use the following convention, based on that used by Morrison [95]. The symbol q will refer to the magnitude but not the sign of the charge on an electron (1.60219×10^{-19} C). E will be taken to represent an energy and at all times be expressed in eV (electron volts)*. ϕ will be taken to represent the local electrostatic potential and,

---

* Owing to the peculiar nature of the relationship between the Joule and the electron volt, E may also be expressed in Joules. For example a photon with an energy of 5.12×10^{-19} J ($\equiv$ a photon wavelength of 388 nm) is required to excite one electron from the valence band to the conduction band of TiO_2. The energy of the photon, $E_{h\nu}$, may be expressed as:

$$\begin{aligned} E_{h\nu} &= 5.12 \times 10^{-19}\,\text{J} \\ &= 3.2 \times 1.60219 \times 10^{-19}\,\text{J} \\ &= 3.2 \times (1.60219 \times 10^{-19}\,\text{C})\,\text{V} \\ &= 3.2\,\text{eV}. \end{aligned}$$

Therefore, the unitary interconversion between the Joule and the electron volt occurs *within* unless otherwise stated, will at all times be expressed in volts relative to vacuum. V will represent a voltage or measured potential difference. The only exception to this convention will be E^0 which will be taken to represent the standard electrode potential of a redox pair. As such it symbolises a voltage rather than an energy and is so used to conform with common usage. Consequently, its use results in some queer-looking equations.

9.2 PRIMARY PROCESSES ON COLLOIDAL SEMICONDUCTORS

9.2.1 Light harvesting by semiconductor band gap excitation

The electronic structure of semiconductors is usually comprised of a filled valence band and an empty conduction band which are energetically separated by an inter-band gap, E_g. All of the reactions described above involve the initial absorption of photons by the semiconductor and, for ultra-band gap photon energies, the subsequent generation of electron-hole pairs within the material lattice (e^-_{CB}, h^+_{VB}), as shown in Fig. 9.1a and equation (9.1). The entire process is thought to occur in <1 fs [96].

$$\text{semiconductor} \xrightarrow{h\nu} \text{semiconductor}\ (e^-_{CB}, h^+_{VB}). \tag{9.1}$$

Semiconductor particles typically contain a high density of lattice defect sites (T_e, T_h), most of which are concentrated at the particle surface. The nature of these defect sites depends strongly on the constituent material of the particle and the method of particle synthesis [61]. Thus, charge carriers photogenerated in accordance with equation (9.1) may subsequently either recombine directly, re-emitting the absorbed energy as heat (Δ) or light ($h\nu$),

$$\text{semiconductor}\ (e^-_{CB}, h^+_{VB}) \rightarrow \text{semiconductor} + h\nu/\Delta, \tag{9.2}$$

or become entrapped in metastable material lattice defect states,

---

the symbol $E_{h\nu}$ and, necessarily, $E_{h\nu}$ may be considered to be expressed in either J or eV. A useful rule of thumb is $1\ \text{eV} \equiv 1.60219 \times 10^{-19}$ J.

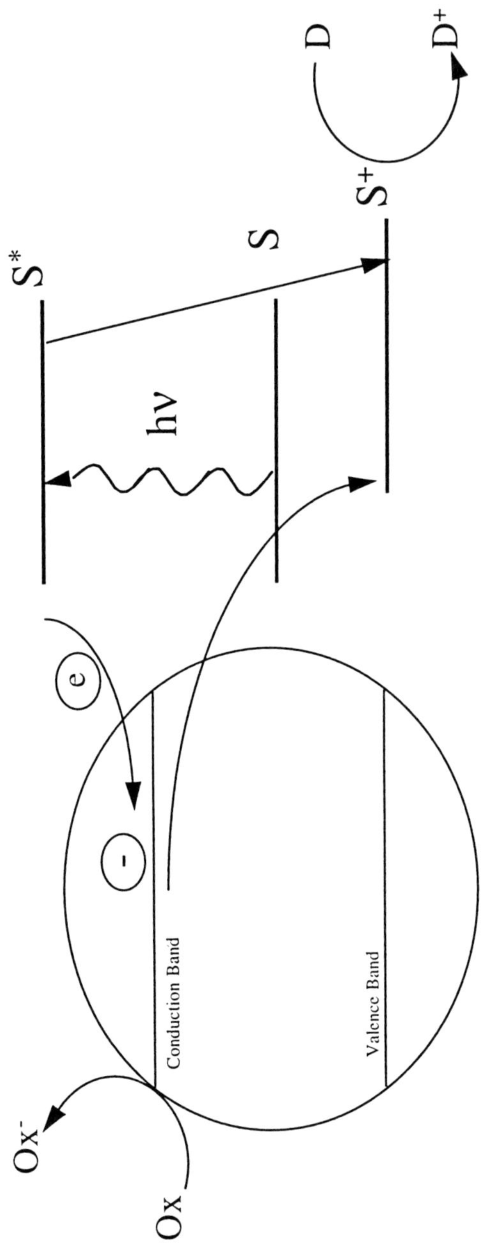

Fig. 9.1. (a) Semiconductor particle charge transfer processes induced by direct band gap excitation.

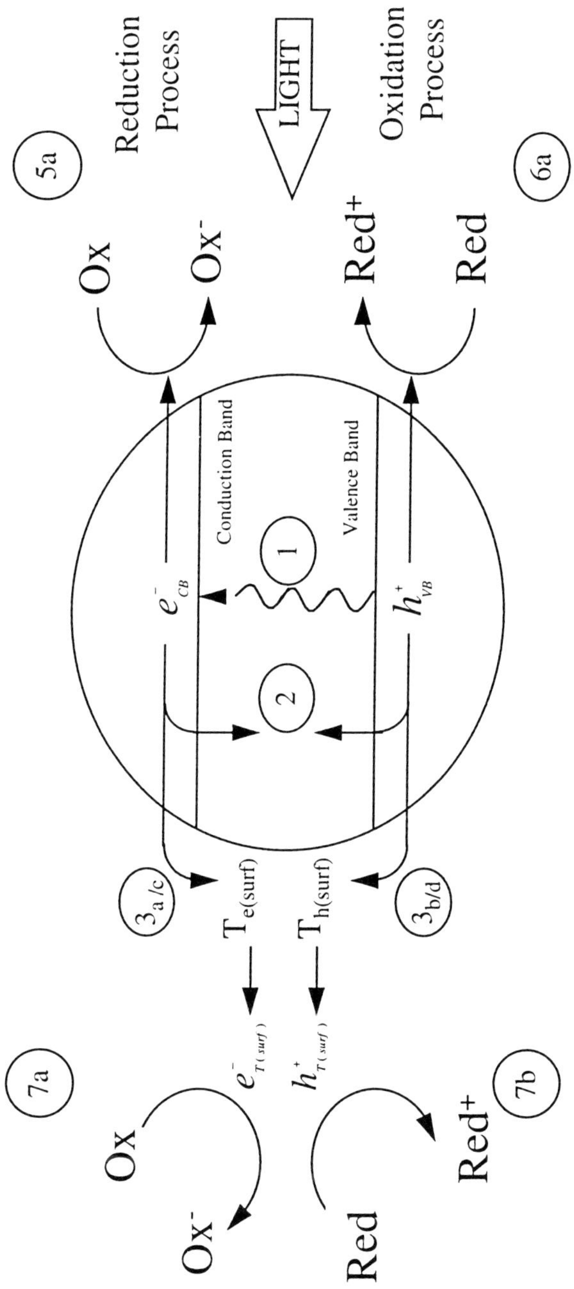

Fig. 9.1. (b) Charge transfer processes induced by charge injection from the excited state of an adsorbed molecule into the conduction band of a semiconductor particle.

$$e^-_{\mathrm{CB}} + T_e \rightarrow e^-_T, \tag{9.3a}$$

$$h^+_{\mathrm{VB}} + T_h \rightarrow h^+_T, \tag{9.3b}$$

or migrate to the surface of the semiconductor where they can participate in an electrochemical reaction, subject to the reaction being thermodynamically allowed, or become entrapped in metastable surface states in surface reactions analogous to equations (9.3a) and (9.3b):

$$e^-_{\mathrm{CB}} + T_{e,\mathrm{surf}} \rightarrow e^-_{T,\mathrm{surf}}, \tag{9.3c}$$

$$h^+_{\mathrm{VB}} + T_{h,\mathrm{surf}} \rightarrow h^+_{T,\mathrm{surf}}. \tag{9.3d}$$

Femtosecond pump-probe flash photolysis measurements on Q-state semiconductor nanoclusters indicate that typical electron trapping times are less than 100 fs for CdS, wherein the electrons are thought to reside at the particle surface [96–99] and less than 200 fs for TiO_2 [96], wherein the electrons are thought to reside in both shallow Ti^{4+} surface traps (within 25–50 meV of the conduction band edge [100, 101]), as suggested by EPR studies on TiO_2 [102–104] and deep traps associated with either interstitial Ti(IV) sites [105] or oxygen vacancies [106] within the particle lattice; the relative amounts in each type of trap has been found to be determined by the laser flash intensity [102]. The exact nature of the electrons trapped on the surface of CdS particles is open to debate considering the large number of surface states or defect sites available in those particles [86, 103,104,107–110], although it is well established that they can undergo subsequent chemical reaction via the reduction of Cd^{2+} sites [91].

Rothenberger *et al.* report that hole trapping on TiO_2 particles requires on average 250 ns [103]. In a more recent communication, Lepore *et al.* report a hole trapping time of 1 ns [111]. The former value was calculated from a stochastic kinetic model of the observed behaviour of trapped photogenerated electrons. The latter was measured from the rise time of a directly observed, photogenerated absorbance in the visible spectrum (λ_{max} = 420 nm) of laser-illuminated TiO_2 particles, attributed to the behaviour of trapped holes. Differences notwithstanding, both the Rothenberger and Lepore values indicate that the trapping of holes is considerably slower than the trapping of electrons. The holes are thought to be trapped as surface-bound $OH^{\bullet}$ species derived from interfacial transfer of photogenerated h^+_{VB} to >Ti-OH groups at the particle surface [111–113].

Duonghong *et al.* report that hole trapping on CdS particles requires an average of 0.3 ns [114]. This was measured by determining the lifetime of a photogenerated strong red luminescence ($\lambda_{max} = 700$ nm) from the particles by a single photon counting technique and is attributed to a dominant valence band hole capture process involving sulphur vacancies, V_S, within the CdS lattice. Kamat and co-workers [115, 116] have studied the time-resolved transient absorption spectra recorded after 355 nm laser pulse excitation of colloidal CdS. They observed a growth in the absorption at wavelengths >460 nm which they attribute to the formation of $S^{\bullet -}$ from lattice S^{2-} ions at the CdS particle surface, this assignment being in agreement with pulse radiolysis experiments performed by Baral *et al.* [109]. Kamat *et al.* determined a quantum yield of 0.77 for $(S^{\bullet -})_{surf}$ generation, indicating that a substantial majority of photogenerated holes may be trapped at the particle surface. This is by no means surprising as CdS particles readily photocorrode [91], the valence band holes being thermodynamically capable of oxidising S^{2-} species to SO_4^{2-} ions [117]. As such, valence band hole capture by surface S^{2-} ions may be regarded as the first step in the photoanodic corrosion of CdS. Pseudo-first order kinetic analysis indicates that the rate constant for the formation of $(S^{\bullet -})_{surf}$ is of the order $5 \times 10^8\ s^{-1}$ with a characteristic time of ca. 2 ns, approximately 10 times slower than hole capture by sulphur vacancies within the CdS lattice.

Trapped charge carriers (e_T^-, h_T^+) can further participate in radiative and non-radiative recombination processes:

$$e_T^- + h_{VB}^+ \rightarrow h\nu/\Delta, \tag{9.4a}$$

$$h_T^+ + e_{CB}^- \rightarrow h\nu/\Delta, \tag{9.4b}$$

$$h_T^+ + e_T^- \rightarrow h\nu/\Delta. \tag{9.4c}$$

Femtosecond flash photolysis studies on Q-state CdS [107] indicate that reaction (4a) proceeds via two recombination processes: a 50 ps decay at low excitation intensities, postulated to correspond to geminate e^-h^+ recombination, and a faster 2 ps decay at higher flash fluences, corresponding to non-geminate or possibly three body Auger charge carrier recombination. Other studies by Nosaka and Fox [118] indicate that the second order rate coefficient for electron-hole recombination within CdS particles is of the order $9 \times 10^{-17}\ m^3\ s^{-1}$.

Femtosecond studies on Q-state TiO_2 particles indicate that electrons residing in the shallow Ti^{4+} surface traps have lifetimes of <50 ps, while those in deep interstitial interior traps have lifetimes of >200 ns [102] before they recombine with free or trapped photogenerated holes. These results are in broad agreement with earlier studies by Rothenberger *et al.* on TiO_2 particles with a mean radius of 6 nm [103] which report trapped electron lifetimes of <200 ps and >30 ns at high and low flash fluences, respectively, and more recent studies by Serpone *et al.* on TiO_2 particles with mean radii in the range 2–27 nm [106] which report lifetimes of ~70 ps and >20 ns for electrons residing in shallow and deep traps respectively.

In the absence of any deliberately added solution phase charge scavenger, the majority of trapped holes on TiO_2 particles recombine with free conduction band or trapped photogenerated electrons; however, a small fraction of the (>Ti-OH$^{\bullet}$)+ sites do react to form thermodynamically stable >TiOOH or >TiOOTi< sites (*vide infra*) [119]. Measurement of the rate constants for these processes by photoelectrophoresis has already been reviewed elsewhere [47].

As shown in Fig. 9.1a, the electrochemistry that the conduction band electrons may participate in at the semiconductor surface can take the form of reaction with reducible solution species or reductive decomposition of the constituent material lattice of the particles:

$$e_{CB}^- + \text{Ox} \rightarrow \text{Ox}^-, \tag{9.5a}$$

$$e_{CB}^- + \text{M}_m\text{X}_x \rightarrow \text{lattice reduction products}. \tag{9.5b}$$

Similarly, the photogenerated valence band holes may also oxidise solution species or cause oxidative decomposition of the particle material lattice.

$$h_{VB}^+ + \text{Red} \rightarrow \text{Red}^+, \tag{9.6a}$$

$$h_{VB}^+ + \text{M}_m\text{X}_x \rightarrow \text{lattice oxidation products}. \tag{9.6b}$$

Charge carriers trapped in surface states may also react with reducible/oxidisable species in solution:

$$e_{T,\text{surf}}^- + \text{Ox} \rightarrow \text{Ox}^-, \tag{9.7a}$$

$$h_{T,\text{surf}}^+ + \text{Red} \rightarrow \text{Red}^+. \tag{9.7b}$$

Therefore, there are two general types of transfer of photogenerated charge carriers from the particle band structure to acceptor species in solution—that mediated by surface states, and that involving direct interfacial charge carrier transfer from the conduction band/valence band to the reductant/oxidant. These processes may be further classified as being of two general types: (i) charge transfer to an electron donor/acceptor adsorbed onto the particle surface (so-called *static quenching*); or (ii) charge transfer to an electron donor/acceptor within the electrical double layer surrounding the particles (so-called *dynamic quenching*). These will be discussed in greater detail below.

9.2.2 Light harvesting by sensitized semiconductor particles

An alternative mechanism by which colloidal particulate semiconductor systems may harvest light is shown in Fig. 9.1b. In this configuration, a sensitizer molecule, usually a photoactive redox couple or dye with a high extinction coefficient in the visible, e.g., eosin [120], cresyl violet [121], rhodamine B [122], a phthalocyanine [123] or ferrocyanide and its analogues [124], is attached to the particle surface. Illumination results in electron injection from a photoexcited state of the sensitizer into the conduction band of the semiconductor particle support, thus effecting charge separation with light of less energy than band gap. This has implications both in solar energy conversion strategies and imaging science. However, this review is concerned with photocatalytic applications of particulate semiconductors and the majority of such applications that have been studied to date operate through band gap excitation. Thus, it is to semiconductor particulate systems that harvest photon energy through this mode of light capture that this chapter will be restricted. Readers interested in dye-sensitised colloidal semiconductor systems are referred to the many excellent reviews that already exist on the subject [see e.g., 51, 61].

9.3 PHOTOGENERATED CHARGE TRANSFER IN COLLOIDAL SEMICONDUCTORS

The overwhelming majority of colloidal semiconductors studied to date are metal oxides and chalcogenides with the greatest interest being shown in the chemically durable titanium dioxide. In turn, most of the research into the applications of semiconductor photocatalysis for waste and contaminant abatement has concentrated on the remediation of organic compounds, although the destruction of inorganic species/pollutants and biological entities is attracting increased attention (*vide supra*). That these processes involve photogenerated charge carriers whose reactions are intrinsically electrochemical in nature facilitates the use of electrochemical models in the design of efficient photocatalytic particle systems [114]. Indeed, as will be seen later in this review, the kinetics of solution species consumption and/or particle decomposition may be detected by a change in the electrochemical response of a semiconductor particle system upon illumination.

In the previous section, we described in detail the primary processes that give rise to photogenerated electrochemically active (and potentially electrochemically useful) charge carriers on illuminated semiconductor particles (reactions (9.1)–(9.4)). Thus, in this section, we will consider, from a predominantly electrochemical viewpoint, the thermodynamics and kinetics associated with the interfacial charge transfer processes that occur post-charge carrier generation (reactions (9.5)–(9.7)).

A complete treatment of interfacial charge transfer in colloidal semiconductor systems with band gap excitation should consider the following factors:

(1) the energetics of the conduction band of the semiconductor;
(2) the redox potential of the charge carrier acceptor;
(3) competition with recombination;
(4) the ratio of the radius of the particles to the Debye length of the particle constituent material;
(5) the adsorption of the charge carrier acceptor (or inert species) at the particle surface;
(6) the intensity of the incident light;
(7) charge carrier trapping and the participation of surface states; and
(8) the surface charge of the particles.

The thermodynamic requirements for efficient photoelectrolysis may be described in isolation from the kinetic requirements if factors (3)–(8) are ignored and this simplest case scenario will be considered in the next section. The remainder of Section 9.3 will discuss the modifications that must be made to the simple thermodynamic model when factors (3)–(8) are introduced in turn.

9.3.1 Thermodynamics of photoinduced charge transfer in semiconductor particles

The Fermi energy or Fermi level, E_f, of a solid is that energy at which the probability of electronic energy level occupancy is exactly 0.5. Chemically, the Fermi energy corresponds to the electrochemical potential of electrons in the solid. At equilibrium, all electronically conducting materials in contact have the same Fermi energy.

It can be shown that the Fermi energy of an electrolyte containing a redox pair is equal to the Nernst energy level associated with the redox pair [95, 125], i.e., for the general redox couple

$$\mathrm{Ox} + \mathrm{ne}^- \leftrightarrow \mathrm{Red}, \tag{9.8}$$

the associated Nernst energy level $E(\mathrm{Ox/Red})$ is related to the electrolyte Fermi energy, $E_{f,\mathrm{el}}$, by:

$$E(\mathrm{Ox/Red}) = E_{f,\mathrm{el}} = -q\left(E^0_{\mathrm{Ox/Red}} + \frac{RT}{nF}\ln\left(\frac{a_{\mathrm{Ox}}}{a_{\mathrm{Red}}}\right)\right) + \mathrm{C}, \tag{9.9}$$

where C is the difference between this electrochemical scale and the physical vacuum scale and all other symbols have their usual meanings. As stated in Section 9.1.4, all energies are expressed in eV. Thus, the energetic requirement for spontaneous electron transfer from a solid to a redox active species in an electrolyte is

$$E_f > E(\mathrm{Ox/Red}). \tag{9.10}$$

The requirement is reversed for spontaneous electron transfer from electrolyte to solid. If the solid in question is a semiconductor, the location of the Fermi energy is given by either of the following expressions:

$$E_f = E_{\mathrm{CB}} + kT\ln\left(\frac{n}{N_{\mathrm{CB}}}\right), \tag{9.11a}$$

$$E_f = E_{\mathrm{VB}} - kT \ln\left(\frac{p}{N_{\mathrm{VB}}}\right), \tag{9.11b}$$

where E_{CB}, N_{CB} and n are the conduction band edge energy, the density of states in the conduction band and the density of electrons in the conduction band respectively, and E_{VB}, N_{VB} and p are the corresponding quantities for holes in the valence band. Thus, from equations (9.11) it can be seen that in the dark, for an n-type semiconductor, E_f lies very close to E_{CB}, for a p-type, E_f lies close to E_{VB} and for an intrinsic semiconductor, E_f lies in the centre of the band gap. Nevertheless, no matter the type of semiconductor, the condition described by equation (9.10) still holds for spontaneous electron transfer between an electrolyte containing a redox active species and a semiconductor in the dark.

In order to make electrolyte Nernst energies comparable with energies of charge carriers in semiconductors under non-equilibrium conditions, the concept of the quasi-Fermi energy has been developed [95]. For example, if a sample is illuminated, a quasi-Fermi energy for photogenerated electrons, ${}_nE_f$, and a quasi-Fermi energy for photogenerated holes, ${}_pE_f$, are used to describe the free energies of electrons and holes in the conduction and valence bands respectively. However, as pointed out by Salvador [126], the use of this statistical concept is questionable in situations involving the absence of equilibrium between electrons and holes; ${}_pE_f$ is a function of the concentration of holes at the semiconductor surface, which in turn depends in a complicated fashion on the illumination intensity, the electron-hole recombination rates and the rate of charge transfer to the electrolyte [127]. The utility of the non-equilibrium quasi-Fermi energy concept when applied to photoreactions is, therefore, doubtful. Instead, this article will follow Salvador's recommendation and use the simple concept of individual energies of photogenerated electrons and holes and assume that after thermalization within larger semiconductor particles, those energies reach levels near E_{CB} and E_{VB} respectively. The situation with respect to the smaller, Q-state particles must take into account a photo-induced blue shift in the semiconductor absorption edge; this will be discussed in more detail in Section 9.4.1.2.

In the absence of kinetic complications, the interfacial charge transfer reactions that occur during photoelectrolytic processes at illuminated semiconductor particles can be schematically represented by reactions (9.5a) and (9.6a):

$$e^-_{CB} + Ox \rightarrow Ox^-, \tag{9.5a}$$

$$h^+_{VB} + Red \rightarrow Red^+. \tag{9.6a}$$

Thus, the two energetic requirements for reactions (9.5a) and (9.6a) to take place spontaneously are:

$$E_{CB} > E(Ox/Ox^-), \tag{9.12a}$$

$$E_{VB} < E(Red^+/Red), \tag{9.12b}$$

i.e., the energy of the photogenerated conduction band electrons must be above the Nernst energy level associated with the Ox/Ox^- redox system and the energy of the photogenerated holes must be below the Nernst energy level associated with the Red^+/Red redox system. As the semiconductor band gap energy is given by:

$$E_g = E_{CB} - E_{VB}, \tag{9.13}$$

then, in the absence of any kinetic considerations and their consequent overpotentials, equations (9.12a) and (9.12b) allow us to derive the condition that, for spontaneous photoelectrolysis to occur, the minimum semiconductor band gap should be:

$$E_g = E(Ox/Ox^-) - E(Red^+/Red). \tag{9.14}$$

Figure 9.2 shows the bulk band gaps and band edge positions for a range of semiconductor materials where the semiconductor is in contact with aqueous electrolytes of pH 1. As has been indicated by Hagfeldt and Grätzel [50], the outer ordinate in Fig. 9.2 represents the internal energy relative to vacuum and not the free energy. The free energy of an electron-hole pair is smaller than the energy of the band gap due to the mobile charge carriers having significant translational entropy within the conduction and valence bands.

It has been shown that the relative energies of the semiconductor valence band edge and the Nernst energy level associated with a redox pair can

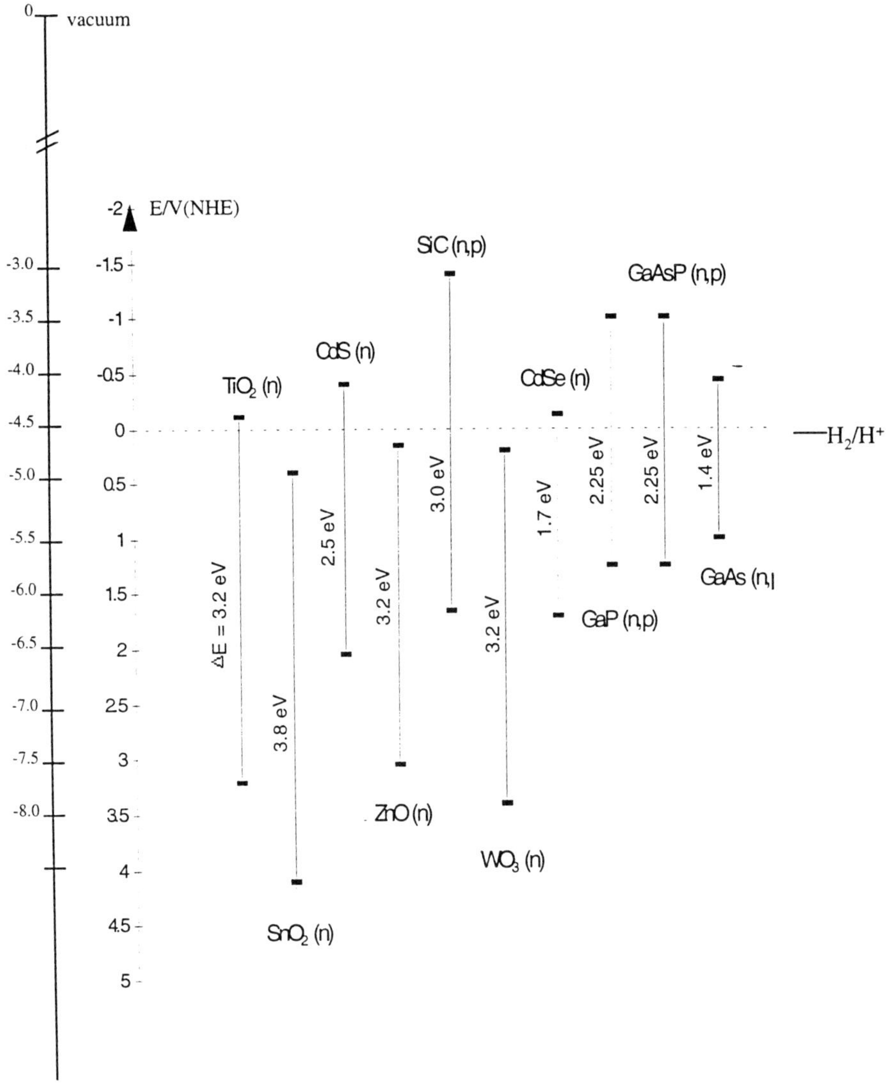

Fig. 9.2. Position of the band edges at the surface of a range of semiconductors in contact with aqueous electrolyte of pH 1.

affect photocatalytic oxidation efficiency. For instance, the efficiency of halide oxidation at TiO_2 follows the sequence $I^- > Br^- > Cl^-$, correlating with the Nernst energy level sequence $E(I_2/I^-) > E(Br_2/Br^-) > E(Cl_2/Cl^-)$ [128, 129]. Consistent with this result is the observation that decreasing the solution pH, and so shifting the TiO_2 valence band to more positive potentials, results in an enhancement in the quantum efficiency of halide oxidation.

9.3.2 The space charge layer within semiconductor particles

When mobile charge carriers are transferred into or out of the surface of a bulk semiconductor, a region of charge, or a "space charge layer" is produced within the semiconductor adjacent to that surface. There are several mechanisms by which this charge transfer and concomitant space charge layer generation may occur [95, 130]. Almost inevitably, a semiconductor in a vacuum will develop a space charge layer adjacent to the surface due to charge carrier capture from within the semiconductor bulk by surface states. If the semiconductor is brought into contact with a liquid or solution, the charge in the space charge layer may change due to:

(1) interaction of the surface states with the liquid/solution, resulting in a change in the charge within the states; and/or
(2) charge transfer between the bulk semiconductor or the surface states with redox active species in the liquid/solution.

These two effects will occur if the semiconductor is under conditions of open circuit. Further variation in the charge stored within the space charge layer may occur if

(3) a second electrode is dipped into the liquid/solution within which the semiconductor is immersed, and a voltage applied between the semiconductor and second electrode different from the voltage at open circuit. The capacitance associated with the space charge layer will then charge up in accordance with

$$dQ = CdV$$

and a charge Q will be added or subtracted from that which obtains within the space charge layer at open circuit.

In many respects, unitary semiconductor particles may be considered to be analogous to bulk semiconductor electrodes at open circuit: thus, effect (3) finds little utility in the systems described in this review. As the space charge layer arises from the local depletion or accumulation of one type of charge carrier within a region adjacent to the surface of the semiconductor, it necessarily has associated with it an intrinsic electric field. As result of the presence of this field, the conduction and valence bands within the space charge layer are bent. If, for simplicity of explanation, we ignore charge carrier exchange between the semiconductor bulk and surface states, and consider only interfacial faradaic charge transfer between the semiconductor and redox active species in solution, four different situations may be envisaged pertaining to band bending. These are shown in Fig. 9.3 for an *n*-type semiconductor (such as TiO_2 or CdS) in contact with an electrolyte containing a suitable redox pair and may be described as follows.

(1) If charge carriers which have the same sign as the semiconductor majority carrier are injected into the semiconductor, then an accumulation layer results at the semiconductor surface (Fig. 9.3a).
(2) If there is no space charge layer, the semiconductor electrode is said to be at its flat band potential (Fig. 9.3b).
(3) If majority carriers are extracted into solution from the semiconductor in moderate amounts, a depletion layer—sometimes also termed an exhaustion layer—forms (Fig. 9.3c). The surface region is depleted of majority carriers, and the minority carriers (valence band holes in the case of an *n*-type semiconductor) are not present, so the surface region is depleted (exhausted) of both forms of mobile carrier. Thus, the excess space charge within this layer is given by the ionised donor states and, unless the extent of band bending is very small, the surface region is essentially insulating.
(4) If majority carriers are extracted into solution from the semiconductor in excessive amounts, an inversion layer forms (Fig. 9.3d) wherein the majority carriers are so depleted that their concentration falls below the intrinsic level. If it is assumed that electronic equilibrium is maintained, the local concentration of minority carriers in the space charge

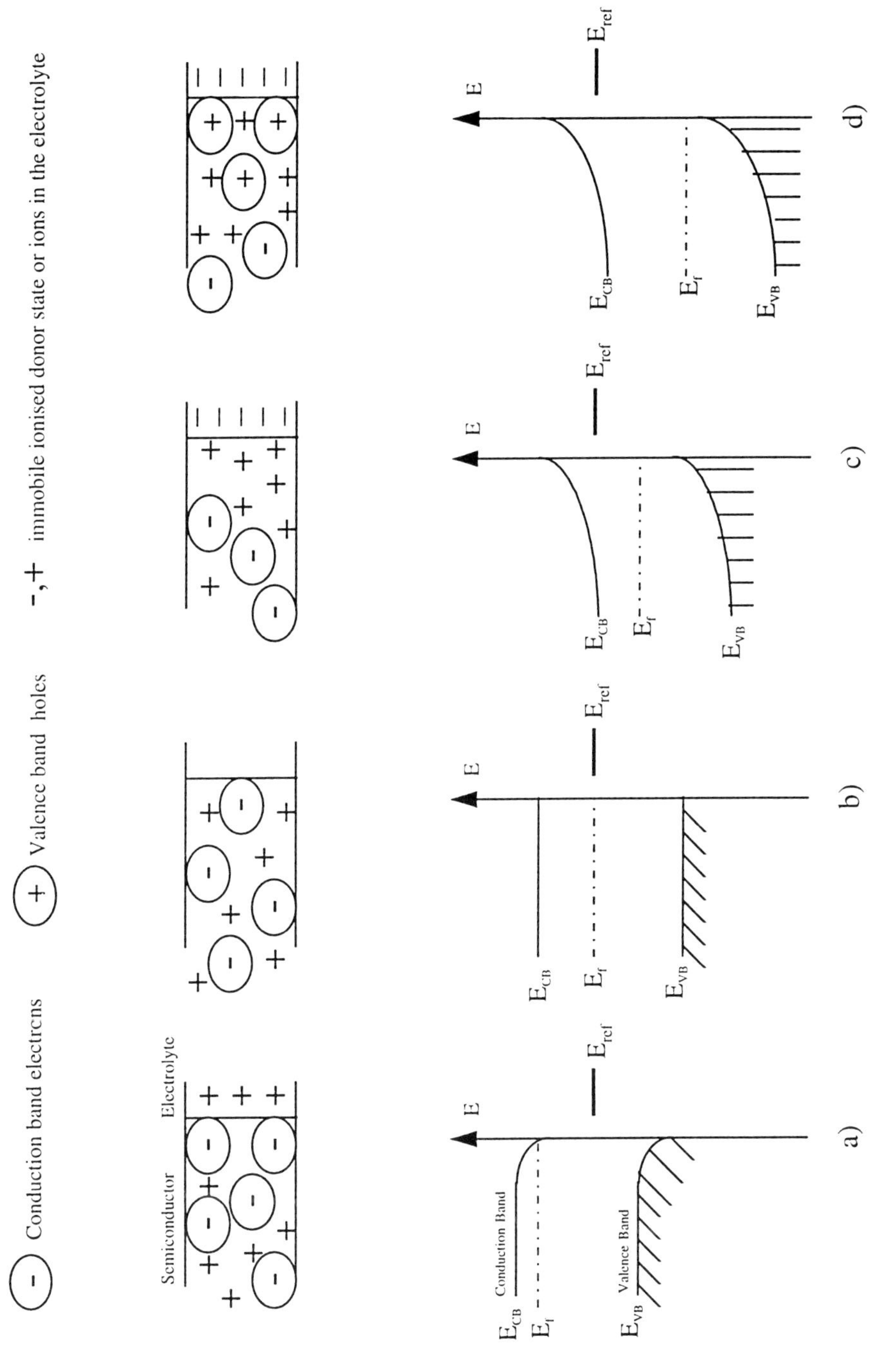

Fig. 9.3. Band bending and space charge layer formation at an *n*-type semiconductor-electrolyte interface: (a) accumulation layer, (b) at flat band, (c) depletion layer (d) inversion layer.

layer exceeds that of the majority carriers, i.e., majority carriers have also been extracted from the minority carrier band. Consequently, the Fermi level is closer to the valence than the conduction band and, in the example of an n-type semiconductor shown in Fig. 9.3, may be considered to be p-type within the space charge layer and n-type in the bulk.

(5) A fifth type of space charge layer, the deep depletion layer, may be formed under non-equilibrium conditions at the semiconductor surface when a high voltage is applied such that an inversion layer should form, but either (a) minority carriers are not available to accumulate at the surface in the time allotted or (b) the minority carriers are consumed in an electrochemical reaction as soon as they reach the surface. Such a space charge layer is unlikely to form within semiconductor electrodes at open circuit and is included here solely for completeness.

The potential distribution, and hence the extent of the band bending, within the space charge layer of a planar macroscopic electrode may be obtained by solution of the one-dimensional Poisson–Boltzmann equation [95]. However, since the particles may be assumed to have spherical geometry, the Poisson–Boltzmann for a sphere must be solved. This has been done by Albery and Bartlett [131] in a treatment that was recently extended by Liver and Nitzan [125]. For an n-type semiconductor particle of radius r_0, the Poisson–Boltzmann equation for the case of spherical symmetry takes the form:

$$\frac{1}{r^2}\frac{\partial}{\partial r}\left(r^2\frac{\partial \phi}{\pi r}\right) = -\frac{qn_b}{\epsilon_0\epsilon_r}\left(1 - \exp\left(\frac{q(\phi - \phi_b)}{kT}\right)\right), \tag{9.15}$$

where r is the distance from the centre of the particle and n_b and ϕ_b are, respectively, the local concentration of electrons and local potential that would obtain in the interior of a macroscopic ($r_0 \to \infty$) bulk semiconductor. If it is assumed that intrinsic carrier densities are negligible relative to donor originated ones, then $n_b = N_D$, the number of donor (dopant) atoms m^{-3}. If it further assumed that the particle has a negative surface potential large enough so that the particle is entirely depleted of its majority carriers, equation (9.15) becomes:

$$\frac{1}{r^2}\frac{\partial}{\partial r}\left(r^2\frac{\partial \phi}{\partial r}\right) = -\frac{qN_D}{\epsilon_0\epsilon_r}. \tag{9.16}$$

Integration of equation (9.16) using the boundary conditions that $\phi = \phi_w$ and $(\partial\phi/\partial r) = 0$ at $r = (r_0 - w)$, where w = the width of the space charge layer, yields:

$$(\phi - \phi_w) = \frac{kT}{6q}\left(\frac{r - (r_0 - w)}{L_D}\right)^2\left(1 + \frac{2(r_0 - w)}{r}\right), \tag{9.17}$$

where $L_D = (\epsilon_r\epsilon_0 kT/q^2 N_D)^{1/2}$ is the semiconductor Debye length and $(\phi - \phi_w)$ is the potential drop within the space charge layer. Two limiting cases of equation (9.17) may be identified. For large particles, where the width of the depletion layer is small compared to the radius of the particle, $r \approx r_0$ throughout the depletion layer and equation (9.17) simplifies to

$$(\phi_{ro} - \phi_w) = \frac{kT}{2q}\left(\frac{w}{L_D}\right)^2, \tag{9.18}$$

where ϕ_{ro} is the electrical potential at the particle surface (i.e., at $r = r_0$). This is the same result as for a planar electrode. For small colloidal particles $r_0 \approx 10^{-8}$ m, the radius of the particle may be smaller than the width of the depletion layer so that $r_0 = w$ with $(\partial\phi/\partial r) = 0$ at $r = 0$. Equation (9.17) then simplifies to

$$(\phi_{\mathrm{ro}} - \phi_w) = (\phi_{\mathrm{ro}} - \phi_0) = \frac{kT}{6q}\left(\frac{r_0}{L_D}\right)^2, \tag{9.19}$$

where ϕ_0 is the electrical potential at the centre of the particle. Under these conditions, the whole of the particle is depleted of majority carriers and there is no field free region. The situations described by equations

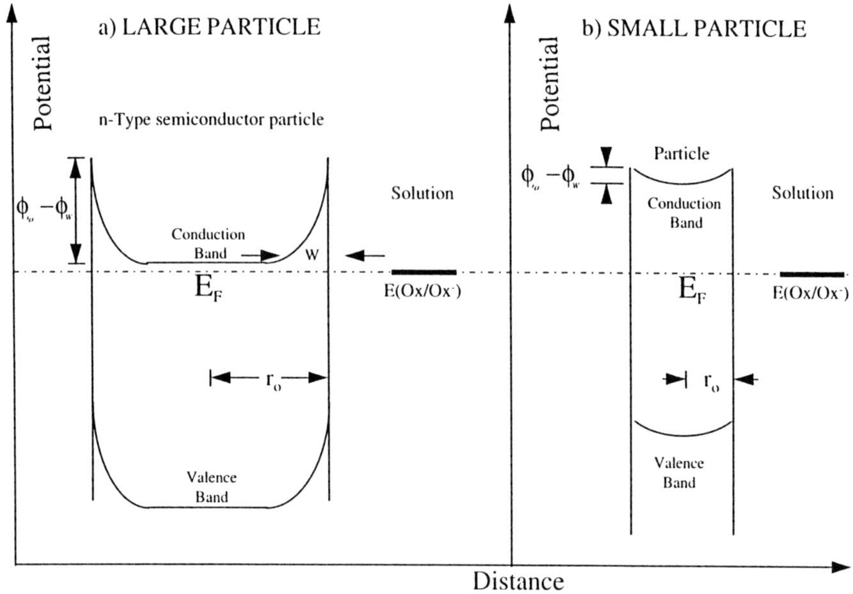

Fig. 9.4. Comparison of the band bending, space charge layer formation and Fermi levels (E_F) for a large particle when $r \approx r_0$ throughout the depletion layer and equation (9.18) applies, and for a small particle when $r \approx w$ and equation (9.19) applies. The semiconductor particles are considered to be in thermodynamic equilibrium with a redox pair of Nernst energy $E(\text{Ox/Ox}^-)$.

(9.18) and (9.19) are illustrated in Fig. 9.4 wherein an n-type semiconductor particle is considered to be in equilibrium with a redox pair of Nernst energy $E(\text{Ox/Ox}^-)$.

These results have important ramifications for photo-induced interfacial charge transfer reactions on semiconductor particles. It is apparent from equation (9.19) that the electric field in colloidal semiconductors is usually small and high doping densities are required if a significant potential difference is to be generated between the surface and centre of the particle. Grätzel and co-workers have suggested that if the majority carriers are depleted from a particle in solution in the dark and the particles are too small to develop a space charge layer, then the electrical potential difference resulting from the transfer of charge from the semiconductor to the solution has to drop within the Helmholtz layer if contributions from the diffuse layer are neglected [48, 50]. A consequence of this argument is that the position of the band edges of the semiconductor will shift. However, a recent study of the electrical potential distribution at the semiconductor particle-electrolyte interface, wherein the particles are considered to be depleted of majority carriers and the potential distribution on the electrolyte side of the interface is explicitly calculated [125], indicates that, even for very small particles where $r_0 \approx 10^{-8}$ m and is of the order of L_D, the electrical potential difference $(\phi_0 - \phi_{ro}) \approx (\phi_0 - \phi_{el})$ where ϕ_{el} is the potential in the bulk of the electrolyte. In other words, no matter the size of the particles, under conditions of majority carrier depletion the main part of the interfacial potential drop lies in the space charge layer of the semiconductor, where its magnitude is asymptotically determined by equation (19). As a result, it can be seen that under the dark conditions described by Grätzel *et al.*, the position of the band edges of the semiconductor particle do not shift or, if they do, it is not to any significant extent.

9.3.3 The behaviour of photogenerated carriers in the space charge layer

It can be seen from the previous section that the ratio of the radius of the particle to the Debye length of the constituent material of the particle has direct bearing on the (energetic) depth of the space charge layer and hence affects the competition between photogenerated charge carrier separation and recombination kinetics. This is because the local electrostatic field associated with a space charge layer at a bulk semiconductor-electrolyte interface serves to enhance the separation of photogenerated electron-hole pairs within the semiconductor. In the case of an *n*-type material under conditions of majority carrier depletion, the direction of the resultant field is such that holes generated within the space charge layer migrate to the semiconductor surface, while the corresponding electrons are swept back into the semiconductor bulk. Charge carrier pairs photogenerated within the field free bulk of the semiconductor may also experience field induced separation if the lifetime of the pair is long enough such that it may diffuse into the space charge layer.

As discussed above, the band bending in semiconductor particles may be small and in this situation photogenerated charge carriers may either recombine or separate *via* diffusion. In the case of the latter, the charge carriers may ultimately diffuse to the particle surface where they participate in the reactions described in Section 9.2.1. Calculations of photogenerated charge carrier intra-particle dynamics based on a random walk model [129] have revealed that, for semiconductor particles of colloidal dimensions, the transit time of electrons/holes from the particle interior to the surface, τ_d, is given by

$$\tau_d = \frac{r_0^2}{\pi^2 D_{\mathrm{rs}}}, \tag{9.20}$$

where D_{rs} is the diffusion constant for the redox charge carrier species within the particle and is given by

$$D_{\mathrm{rs}} = \frac{\mu k T}{q}, \tag{9.21}$$

where μ is the mobility of the charge carrier. For the majority of colloidal semiconductors, τ_d equates at most to a few tens of picoseconds [47]. For a conduction band electron in a TiO_2 particles of radius 10 nm and carrier mobility $0.5 \times 10^{-4}\,\mathrm{m}^2\,\mathrm{Vs}^{-1}$, the average transit time is 8 ps which is of the same order as the charge carrier recombination time. If the doping density of the semiconductor or the particle radius is large enough to support a space charge layer, the transit time of the minority carriers to the particle surface is further reduced as the Debye length must be substituted for the particle radius in equation (9.20) [132].

Thus, it may be seen that, by reducing the particle radius, it is possible to obtain systems where transit from the particle interior to the surface occurs more rapidly than recombination, implying that quantum efficiencies for photoredox reaction of near unity are feasible. However, the achieving of such high quantum efficiencies depends very much upon the rapid removal of one or both types of charge carrier upon their arrival at the semiconductor surface, underlining the importance of the interfacial charge-transfer kinetics. This is the subject of the next section.

9.3.4 The kinetics of photoinduced interfacial charge transfer in semiconductor particles

Kinetic analysis of photoinduced interfacial charge transfer processes in colloidal semiconductor systems is frequently complex. Indeed, it is difficult to derive a generic kinetic expression from a completely generalised model of photocatalytic systems. Nonetheless, it is informative at this point to review briefly some aspects of the fundamental theories of charge transfer between a redox active solution species and a semiconductor with a view to how those ideas might subsequently be applied to colloidal semiconductor systems.

During this discussion we will refer only to conduction band processes (i.e., exchanges of electrons between species in solution and the conduction band) as the valence band processes are analogous in principle. Several approaches have been adopted by various workers [133–136], each of which is based on a number of assumptions, some of which may be invalid. The overwhelming majority of these treatments are based on the fluctuating energy level model which was first applied to heterogeneous electrochemical systems in a non-quantum mechanical analysis by Marcus [137–139]. Gerischer's treatment [134, 136, 140] of the fluctuating energy level model is a convenient generalisation which states that the flux of electrons, j_e, from the conduction band to the acceptor species in solution is given by

$$j_e = \int_{-\infty}^{\infty} \nu(E) D_{\text{ox}}(E)_{\text{sc}}^{\text{occup}}(E)\, \mathrm{d}E-, \qquad (9.22)$$

where $D_{\text{ox}}(E)$ is a function describing the density of vacant electron acceptor levels per unit energy in the electrolyte volume adjacent to the semiconductor surface, $D_{\text{sc}}^{\text{occup}}(E)$ is a function describing the density of occupied electronic donor levels per unit energy in the semiconductor and $\nu(E)$ is the tunnelling collision frequency factor. It assumed in the derivation of equation (9.22) that electron transfer obeys the Franck–Condon principle and is therefore adiabatic, and that electron transfer is most likely to occur when the electron acceptor species is at the Helmholtz layer of the electrolyte. For particulate systems, it is also necessary to assume that the thermally-induced fluctuations in the solvent cage of the particle have no effect on particle electron energetics. Whilst this is reasonable enough for larger particles, such an assumption may break down in the case of Q-state particles; consequently, some level of caution should be exercised when interpreting the kinetics of interfacial charge transfer in nanocrystalline systems in terms of the conventional fluctuating energy level model.

Marcus has recently returned to this problem [133] and, by analogy with the problem of donor-acceptor electron transfer at the interface between two immiscible liquids, has derived the following expression for k_{ct}, the heterogeneous rate constant for electron transfer from a semiconductor to a species in a contacting electrolyte:

$$k_{\mathrm{ct}} = \kappa\nu\left(\frac{2\pi(r_A + r_e)}{\beta^3}\right)\exp\left(\frac{-(\Delta G^{0\prime} + \lambda_A)^2}{4kT\lambda_A}\right), \tag{9.23}$$

where r_A is the radius of the acceptor ion, r_e is the effective radius of the electron in the semiconductor, λ_A is the reorganisation energy of the solvent, β represents the attenuation of the electron coupling in a direction perpendicular to the electrode surface, ν is some relevant frequency for molecular motion (most appropriately the vibrational frequency of the molecule in the solvent sheath) and κ is a Landau–Zener transmission coefficient formulated to account for the relative adiabacity/non-adiabacity of the electron transfer process. $\Delta G^{0\prime}$ is the energy difference between the Nernst energy associated with the acceptor redox couple in the electrolyte under standard conditions and the conduction band energy ($\Delta G^{0\prime} = E(\mathrm{Ox/Ox^-}) - E_{\mathrm{CB}}$) and as such corresponds to the additional activation energy needed for adiabatic interfacial charge transfer.

As mentioned in Section 9.2.1, interfacial reaction of photogenerated electrons or holes with a suitable solution species may be of two general types: (i) charge transfer to an electron donor/acceptor within the electrical double layer surrounding the particles, referred to as *dynamic quenching*; or (ii) charge transfer to an electron donor/acceptor adsorbed onto the particle surface, referred to as *static quenching*. We will now consider each in turn in light of the concepts developed above.

9.3.4.1 Photoinduced interfacial charge transfer with dynamic quenching

Dynamic electron transfer is observed when the adsorption between the charge carrier acceptor species and the semiconductor particle is weak. As such, charge carrier acceptor molecules have to diffuse from the homogeneous solution bulk to the surface of the semiconductor particle in order to accept the photogenerated electrons/holes. As described in Section 9.2.1, photogenerated carriers typically recombine on the sub-nanosecond time scale; thus, it is necessary to accumulate one of the charge carriers within the particle if efficient charge transfer is to take place. This may be achieved by using sacrificial charge scavengers to capture one type of carrier (e.g., alcohols or certain halogens for holes; oxygen or hydrogen peroxide for electrons), leaving its opposite number to accumulate within/on the particles at surface traps. Thus, in addition to the thermodynamic driving force, the kinetics of dynamic charge transfer will also be controlled by factors such as particle size (and its attendant polydispersity) and particle surface charge.

Dynamic charge transfer of photogenerated electrons has been observed by many authors with probably the most significant reports including the article by Bahnemann *et al.* [113] on the photocatalytic reduction of tetranitromethane and the work by Brown and Darwent [141] and Duonghong *et al.* [114] on the reduction of MV^{2+} by colloidal TiO_2. In the same paper, Duonghong *et al.* report the observation of dynamic photogenerated hole transfer from colloidal CdS particles to solution phase HS^- ions. A more recent study by Bedja *et al.* [142] reports that the reduction of several solution phase dyes (e.g., thionine, methylene blue, cresyl violet) by photoexcited colloidal WO_3 occurs with quantum efficiencies of 0.5 and at rates close to diffusion control in the presence of the sacrificial hole scavenger, oxalic acid.

Broadly speaking, dynamic electron transfer involves two steps. The first step is the diffusion controlled formation of an encounter complex between the electron/hole acceptor molecule and the particle. The second step is the electrochemical interfacial charge-transfer, which may be characterized by a rate constant k_{ct}. Albery *et al.* [143] and Grätzel *et al.* [129] have independently arrived at the same result, relating the observed effective bimolecular rate constant for hole/electron acceptor oxidation/reduction, k_{obs} ($m^3\,mol^{-1}\,s^{-1}$) to reactant diffusion and k_{ct}.

$$\frac{1}{k_{obs}} = \frac{1}{4\pi R^2 L}\left(\frac{1}{k_{ct}} + \frac{R}{D}\right), \tag{9.24}$$

where R and D are, depending on the treatment adopted, either the sums of the radii and diffusion coefficients respectively of the particle and the charge carrier acceptor OR the radius of the particle and the diffusion coefficient of the acceptor. It is worth noting that these definitions of R and D are equivalent in any case when the radius of the particle is much greater than that of the acceptor molecule. In essence, Grätzel derives his result by solving Ficks second law of diffusion (equation (9.25)) for a charge carrier acceptor within the diffusion field of a spherical particle photoelectrode:

$$D\left(\frac{\partial^2 c}{\partial r^2} + \frac{2}{r}\frac{\partial c}{\partial r}\right) = 0, \tag{9.25}$$

where c is the concentration of the charge carrier acceptor at a distance r from the centre of the particle photoelectrode. Albery's treatment considers the similar case of a relay compound, Ox_R, transferring electrons that have been homogeneously photogenerated in the solution bulk to a colloidal Pt collector particle according to the simplified reaction scheme:

$Ox_R \xrightarrow{g} Ox_R^-$ Homogeneous generation of reduced relay in solution bulk with rate g (9.26a)

$Ox_R^- \xrightarrow{k_b} Ox_R$ Homogeneous back reaction of reduced relay in solution bulk with rate constant k_b (9.26b)

$Ox_R^- \xrightarrow{k_{ct}} Ox_R$ Heterogeneous electron transfer at particle surface with rate constant k_{ct} (9.26c)

Albery solves the corresponding reaction-diffusion equation for this scheme to produce equation (9.24):

$$D\left(\frac{\partial^2 c}{\partial r^2} + \frac{2}{r}\frac{\partial c}{\partial r}\right) + g - k_b c = 0, \tag{9.27}$$

where, in this instance, c refers to the concentration of the reduced relay, Ox_R^-, at a distance r from the centre of the collector particle. Equation (9.24) has two limiting asymptotic forms. The first case arises when interfacial charge-transfer is much slower than diffusion ($k_{ct} \ll D/R$) and therefore rate determining:

$$k_{obs} = 4\pi R^2 L k_{ct}. \tag{9.28}$$

The second case arises when diffusion is much slower than charge transfer ($k_{ct} \gg D/R$) and therefore rate determining:

$$k_{obs} = 4\pi RLD. \tag{9.29}$$

However, due to several complicating factors, these expressions are not as generally applicable to real systems as may first appear. The first complicating factor to be considered is that of adsorption or chemical fixation of the charge carrier acceptor onto the particle surface. This will be considered in detail in Section 9.3.4.2.

The second complicating factor is that, unless steps are taken to prepare a monodisperse sample, colloidal dispersions frequently exhibit a distribution of particle sizes. Equation (9.24) shows that k_{obs} is particle size dependent and that consequently particles with different radii should have different rate constants, the larger particles reacting faster than the smaller. Such effects have indeed been observed and Albery *et al.* [144] have proposed a general model based on a Gaussian distribution of the logarithm of the first order rate constant, k_x, about some mean $\langle kx \rangle$ given by the general expression:

$$k_x = \langle k_x \rangle \exp(\gamma x), \tag{9.30}$$

where γ describes the spread of the Gaussian distribution and the concentration of reactant, $c_{kx,0}$, with rate constant k_x at time $t = 0$ is given by

$$c_{kx,0} = c_{\langle kx \rangle,0} \exp(-x^2), \tag{9.31}$$

where $c_{\langle kx \rangle,0}$ is the concentration of reactant with the mean rate constant $\langle k_x \rangle$ at time $t = 0$. Thus, for a first order process, the concentration of reactant, c_{kx}, with rate constant k_x at any time t is given by

$$c_{kx} = c_{\langle kx \rangle,0} \exp(-x^2) \exp(-k_x t) = c_{\langle kx \rangle,0} \exp(-x^2) \exp[-\langle k_x \rangle t \exp(\gamma x)], \tag{9.32}$$

and the total reactant concentration, c, at any time t is given by

$$c = c_{\langle kx\rangle,0} \int_{-\infty}^{\infty} \exp(-x^2) \exp[-\langle k_x\rangle t \exp(\gamma x)] \, \mathrm{d}x-, \tag{9.33}$$

or, with respect to the total reactant concentration, c_0, at time $t = 0$:

$$\frac{c}{c_0} = \frac{\int_{-\infty}^{\infty} \exp(-x^2) \exp[-\langle k_x\rangle t \exp(\gamma x)] \, \mathrm{d}x-}{\int_{-\infty}^{\infty} \exp(-x^2) \, \mathrm{d}x-}. \tag{9.34}$$

Equation (9.34) may be simplified by noting that

$$\int_{-\infty}^{\infty} \exp(-x^2) \, \mathrm{d}x = \pi^{1/2}. \tag{9.35}$$

It is also worth noting that when $\gamma = 0$, corresponding to no dispersion, equation (9.34) reduces to the simple first order exponential decay

$$\frac{c}{c_0} = \exp[-kt]. \tag{9.36}$$

In the case of colloidal semiconductors whose interfacial charge transfer kinetics are consistent with those assumed during the derivation of equation (9.28), the rate of charge transfer from the particles to the charge carrier acceptor in solution is proportional to the surface area of the particles. Albery *et al.* assume that, within the systems that they studied, the particle radii conform to a Gaussian distribution. Thus, in analogy to equations (9.30) and (9.31):

$$r_x = \langle r_x\rangle \exp(\rho x), \tag{9.37}$$

$$c_{rx} = c_{\langle rx\rangle} \exp(-x^2), \tag{9.38}$$

where c_{rx} is the concentration of particles with radius r_x. Substitution of equation (9.37) in equation (9.28) gives:

$$k_{rx} = k_{\langle rx \rangle} \exp(2\rho x), \tag{9.39}$$

where ρ describes the width of the radial dispersion and k_{rx} and $k_{\langle rx \rangle}$ are the pseudo-first order electron transfer rate constants for particles of radius r_x and $\langle rx \rangle$ respectively. This equation has the same form as equation (9.30). Hence the same result as that obtained in equation (9.34) will hold with:

$$\gamma = 2\rho. \tag{9.40}$$

For colloidal semiconductor systems, Albery *et al.* observed good agreement between the value of the radial dispersion obtained from dynamic light scattering and the value found from application of the above kinetic analysis to flash photolysis experiments [144]. It should be remembered that this disperse kinetics model can only be applied to the decay of heterogeneous species under unimolecular or pseudo-first order conditions and that for colloidal semiconductors it may only be applied to dispersions whose particle radii conform to equation (37), i.e., a log normal distribution. However, other authors [145] have recently refined the model so that assumptions about the particle size distribution may be avoided in the kinetic data analysis.

The third complicating factor arises from the particle charge. Apart from under exceptional conditions, the surface of a solid immersed in a liquid will, by one or more of a variety of mechanisms, acquire a net charge (the nature of these mechanisms has been reviewed by this author elsewhere [47]). The kinetic models of interfacial electron-transfer reactions described in equations (9.24)–(9.29) ignore Coulombic effects such as electrostatic migration and only consider the approach of the particle and acceptor in terms of diffusion. Equations (9.24)–(9.29) would therefore only be valid under conditions where the particle is uncharged (exceptional and difficult to achieve, *vide infra*) or the acceptor is uncharged or the ionic strength is high. Obviously, the latter is an undesirable condition to impose on any colloidal dispersions which derive their kinetic stability with respect to aggregation solely from inter-particle electrostatic repulsion. Thus, in general, equations (9.24)–(9.29) must be corrected to account for electrostatic attraction/repulsion and, to facilitate this, the magnitude of the particle charge quantified.

The latter is most readily achieved by conducting electrophoretic measurements on illuminated semiconductor particles and has been reviewed by this author elsewhere [47]. The former has been investigated by Brown and Darwent [141] who demonstrated that the rate of reduction of methyl viologen, MV^{2+}, by the conduction band electrons of colloidal TiO_2 is significantly affected by the particle charge and the ionic strength of the sol. Further, they semi-quantify this effect by empirically deriving an equation relating k_{ct} to pH and ionic strength from the control these two quantities are known to exercise over the particle surface charge/potential. According to Brown and Darwent, in the case of the reduction of MV^{2+} by the conduction band electrons of colloidal TiO_2,

$$\log_{10} k_{ct} = \log_{10} k_{ct,0} + \alpha'\text{pH} - \chi \frac{(\text{p.z.z.p.})}{\sqrt{I}}, \tag{9.41}$$

where $k_{ct,0}$ is the rate constant for charge transfer at pH 0 when there is no electrostatic effect, α' is a constant closely related to the Tafel transfer coefficient, the p.z.z.p. is the point of zero zeta potential, I is the ionic strength and χ is a constant which is found to be approximately equal to 0.04 $\text{mol}^{1/2}\,\text{dm}^{-3/2}$ (or 1.26 $\text{mol}^{1/2}\,\text{m}^{-3/2}$) for the reaction described. Therefore, it can be seen from equation (9.41) that the control of the particle surface charge by manipulation of the pH and solution ionic strength could lead to improvement in efficiency of photogenerated charge transfer processes on colloidal semiconductors.

9.3.4.2 Photoinduced interfacial charge transfer with static quenching

Static processes dominate when the charge acceptor is strongly adsorbed on the semiconductor surface. Photogenerated charge carrier transfer occurs rapidly and directly competes with charge recombination. This is especially true if the charge carrier acceptor species is chemisorbed onto the semiconductor surface since the associated electron transfer can happen directly without the necessity of a tunnelling process [126]. Under these circumstances, the net effect is that of simplification of equation (9.24) to equation (28) as diffusion effects may be eliminated and interfacial charge transfer kinetics take sole control [48, 50]. k_{ct} may then be expressed as

$$k_{ct} = \frac{\delta}{\tau_{ct}}, \tag{9.42}$$

where δ is the reaction layer thickness and τ_{ct} is the average time taken for a charge carrier to tunnel across the interface.

The chemical and electrostatic forces that operate to bring a charge carrier acceptor into contact with a photoactive semiconductor surface include [45]: inner-sphere ligand substitution for metal ions and conventional (in)organic ligands, van der Waals forces, dipole-dipole interactions, induced dipole–dipole interactions, hydrogen bonding, outer-sphere complexation, ion exchange, surface organic partitioning, hemimicelle formation and sorbate hydrophobicity. Most of these interactions yield Langmuirian sorption isotherms (*vide infra*). The hydrophobicity, which serves to drive hydrophobic organic compounds out of the bulk solution phase in aqueous systems and onto the semiconductor surface increases with the size of the nonpolar part of the molecule. The free energy of adsorption associated with the hydrophobicity of linear aliphatic organics is given by

$$\Delta G_{\text{hydrophobicity}} = m\Delta G_{\text{—CH}_2\text{—}},$$

where m is the number of CH_2 groups in the carbon chain and $\Delta G_{\text{—CH}_2\text{—}}$ is the hydrophobic contribution made by each CH_2 group to the net driving force for the substrate into the electrical double layer around the particle. A charged organic substrate may be attracted to the surface by both electrostatic and hydrophobic means, a combined effect which may be quantified by:

$$\frac{[\text{substrate}]_{\text{surface}}}{[\text{substrate}]_{\text{solution}}} = \exp\left(\frac{-(\Delta G_{\text{electrostatic}} + \Delta G_{\text{hydrophobicity}})}{RT}\right)$$

$$= \exp\left(\frac{-(zF\psi_0 + \Delta G_{\text{hydrophobicity}})}{RT}\right).$$

Static electron transfer from photoexcited particles to adsorbed substrates has been observed for a wide range of semiconductor and organic materials respectively. For example, the photoinduced reduction of zwitterionic viologen (ZV) by CdS conduction band electrons is found by time resolved transient absorption spectra to occur with a risetime for $ZV^{\bullet -}$ formation of less than 20 ps [116], consistent with a rate constant for interfacial electron transfer of $>5 \times 10^{10}\,s^{-1}$. Strong adsorption of the acceptor onto the particle surface is essential for such ultrafast charge transfer to take place.

Static hole transfer has also been observed from photoexcited semiconductor particles to a wide range of adsorbed (esp. organic) substrates. For example, in their seminal paper, Duonghong *et al.* [114] found that the photocatalysed oxidation of SCN^- to $(SCN)_2^{\bullet -}$ was completed during the interrogating laser pulse of a flash photolysis experiment due to the strong adsorption of SCN^- at the TiO_2 particle surface. Surface hydroxyl group mediated hole transfer to adsorbed organic species is also found to be the actuating process during semiconductor photocatalysis driven organic pollutant oxidation reactions (*vide infra*).

However, a strongly adsorbed product may inject the transferred charge back into the semiconductor. The importance of such back reaction suppression for improving the quantum efficiency of charge transfer for both static and dynamic processes has been demonstrated by Curran and Lamouche [132]. The phenomenon has been directly investigated by Kamat *et al.* [116] who report that while a fraction of $ZV^{\bullet -}$ produced by ZV reduction with photogenerated conduction band electrons on TiO_2 diffuses away from the particle surface, a major fraction of the anion radical decayed within 500 ps of its formation as a result of its reaction with photogenerated trapped holes. Thus, both the adsorption equilibrium and the quick removal of the product from the particle surface are crucial in determining the overall efficiency of charge transfer. Consequently, Langmuir–Hinshelwood (LH) kinetics have been found to play an important role in the analysis of rate of reaction data obtained from photocolloid sensitised redox processes [1–3, 112, 146–150].

The LH model assumes that (i) at equilibrium, the number of adsorption sites is fixed; (ii) only one substrate may adsorb at each surface site; (iii) the enthalpy of adsorption is identical for each site and independent of surface coverage; (iv) there are no lateral interactions between adsorbed molecules; (v) the rate of adsorption of the initial substrate is faster than the rate of any following chemical reactions; and (vi) no irreversible blocking of active sites by binding of product occurs (i.e., the sorption of the substrate is a rapid equilibrium process in both the forward and the reverse directions). Using these assumptions, it can be shown that the fractional coverage, θ_{sub}, of the particle surface by the substrate is given by:

$$\theta_{sub} = \frac{K_{sub,d}[S]_0}{1 + K_{sub,d}[S]_0}, \qquad (9.43)$$

where $[S]_0$ is the initial solution concentration of the substrate and $K_{sub,d}$ is the apparent adsorption equilibrium constant of the substrate obtained from a classical Langmuir sorption isotherm recorded in the dark. The initial rate of disappearance of the substrate, R_0, can then be written as a single component LH kinetic rate expression:

$$R_0 = -\left(\frac{d[S]}{dt}\right)_0 = k'_{obs}\theta_{sub} = \frac{k'_{obs}K_{sub,d}[S]_0}{1 + K_{sub,d}[S]_0}, \tag{9.44}$$

where k'_{obs} is the apparent pseudo-first order rate constant occurring at the active site on the photocatalyst surface. It is found that k'_{obs} is proportional to I^{φ}_{abs}, where I_{abs} is the rate of light absorption and φ is a power term which is equal to $\frac{1}{2}$ or 1 at high or low light intensities respectively. The linearity of $1/R_0$ versus $1/[S]_0$ plots tests the validity of the LH model and many photocatalytic systems yield good straight lines in such plots [1–3, 112, 146–149]. In a more sophisticated treatment of the Langmuir–Hinshelwood model, the rate of the general photocatalysed reaction

$$\text{Ox} + \text{Red} \xrightarrow{\text{photocatalyst}} \text{products} \tag{9.45}$$

can be expressed in general terms for both the oxidant (e.g., O_2) and the reductant (e.g., an organic targeted for photodegradation) as follows:

$$R_0 = -\left(\frac{d[\text{Ox}]}{dt}\right)_0 = -\left(\frac{d[\text{Red}]}{dt}\right)_0 = k''_{obs}\theta_{\text{Ox}}\theta_{\text{Red}}. \tag{9.46}$$

where k''_{obs} is the apparent second order rate constant for reaction (9.45) on the photocatalyst surface and [Ox], [Red] and θ_{Ox}, θ_{Red} are analogous to $[S]$ and θ_{sub} respectively. This treatment is subject to the further assumptions that the rate determining step of the reaction involves both species present in a monolayer at the solid-liquid interface and that Ox and Red non-competitively adsorb at different surface sites. A commonly studied example of such a system is the photocatalytic degradation of an organic pollutant substrate by illuminated colloidal TiO_2 where Ox = O_2 and Red = the organic substrate [see e.g., 1–5, 45, 56, 112]. O_2 appears to non-competitively adsorb at Ti(III) sites, while hydroxyl radicals (produced by hole-driven oxidation of OH^-/H_2O species) and organic substrates are believed to be exclusively adsorbed at Ti(IV)-lattice oxygen sites. Application of equations (9.43) and (9.46) to this system yields

$$R_0 = \frac{\gamma_k I^{\varphi}_{\mathrm{obs}} K_{\mathrm{O_2,d}}[O_2]_0 K_{\mathrm{sub,d}}[S]_0}{(1 + K_{\mathrm{O_2,d}}[\mathrm{O_2}]_0)(1 + K_{\mathrm{sub,d}}[S]_0)}, \tag{9.47}$$

where R_0 is still the initial rate of disappearance of the organic substrate and γ_k is a proportionality constant. Equation (9.47) reduces to equation (9.44) if it is assumed, say, that atmospheric oxygen acts as the source of O_2 and that, consequently, $[O_2]$ is invariant with time.

The Langmuir–Hinshelwood treatment has also been modified to account for the more complicated conditions presented by competitive adsorption between two or more species for a single adsorption site such as might occur within multicomponent systems or with intermediates formed during multi-step oxidation processes [3, 151–153]. This modification yields the following expression for θ_{sub}:

$$\theta_{\mathrm{sub}} = \frac{K_{\mathrm{sub,d}}[S]_0}{1 + K_{\mathrm{sub,d}}[S]_0 + \sum_{i=1}^{n} K_{\mathrm{Si}}[S_i]}, \tag{9.48}$$

which may be substituted into equation (9.46) to give:

$$R_0 = \frac{\gamma_k I^{\varphi}_{\mathrm{abs}} K_{\mathrm{O_2,d}}[\mathrm{O_2}]_0 K_{\mathrm{sub,d}}[S]_0}{(1 + K_{\mathrm{O_2,d}}[\mathrm{O_2}]_0)(1 + K_{\mathrm{sub,d}}[S]_0 + \sum_{i=1}^{n} K_{\mathrm{Si}}[S_i])}, \tag{9.49}$$

where K_{Si} and $[S_i]$ are the adsorption equilibrium constant and the concentration respectively of the reaction intermediates or other competitive species and n is the number of those species present [151].

Cunningham *et al.* [1, 2, 149, 154] and other workers [3, 155–158] have measured classical Langmuir adsorption isotherms on TiO_2 for a range of organic substrates in the dark and have made the following observations: (i) strongly adsorbed organic substrates exhibit self-inhibition of photocatalysed substrate removal either via blockage of sites for the photocatalytic reaction on the TiO_2 surface or through their action as locations for surface recombination of photogenerated holes and electrons; and (ii) $K_{sub,d}$ values measured from dark Langmuir isotherm measurements on weakly adsorbing pollutants are significantly smaller than the values obtained from $1/R_0$ versus $1/[S]_0$ plots, indicating the possible existence of photoinduced adsorption processes.

Unfortunately, the observation of a good linear fit in a plot of $1/R_0$ versus $1/c_0$ for a photocatalyst/substrate system cannot be taken as solid proof of preadsorption and does not necessarily imply that the photocatalytic reaction occurs completely on the particle surface. Let us return to our example of an organic pollutant being photocatalytically degraded by illuminated colloidal TiO_2 in the presence of O_2. A key study by Turchi and Ollis [112] on possible mechanisms involving $OH^{\bullet}$ radical attack on organic substrates during substrate destruction by illuminated colloidal TiO_2 shows that at least four different reaction mechanisms yield a rate equation of the form:

$$R_t = -\left(\frac{\mathrm{d}[S]}{\mathrm{d}t}\right)_t = \frac{k'_{\mathrm{obs}}\kappa_s[S]_t}{1 + \kappa_s[S]_t + \sum_{i=1}^{n} \kappa_{\mathrm{Si}}[S_i]_t}, \tag{9.50}$$

where t is the time after the start of the irradiation. Although of the same empirical form as equation (49) under condition of constant $[O_2]$, the fundamental meanings of the constants κ_s and κ_{Si} are different for each of the four reaction mechanisms proposed by Turchi and Ollis, which may explain why LH expressions can adequately account for the observed kinetics of a photocatalytic system without having values for the "adsorption constant" (determined from $1/R_0$ versus $1/[S]_0$ plots) that correlate with adsorption data (determined from classical Langmuir isotherm measurements). Interestingly, k'_{obs} adopts the same form for each of the four mechanisms, i.e., it is reaction independent, and this prediction is supported by a detailed experimental studied conducted by Turchi and Ollis [112]. Due to the applicability of equation (9.50) to at least four photocatalytic mechanisms, it is worthwhile considering the Turchi and Ollis model in more detail.

The sequence of elementary steps which Turchi and Ollis suggest describe the initial photocatalytic oxidation mechanism is shown in Table

TABLE 9.1
The photocatalytic reaction scheme of Turchi and Ollis [112]

Primary step	Primary reaction	Reaction number[a]
Excitation	$TiO_2 + h\nu \rightarrow e^- + h^+$	(T1) ≡ (eqn. (1)
Adsorption	$O_L^{2-} + Ti(IV) + H_2O \leftrightarrow O_LH^- + Ti(IV)\text{-}OH^-$	(T2a)[b] ≡ (none)
	$Ti(IV) + H_2O \leftrightarrow (Ti(IV)\text{-}H_2O$	(T2b) ≡ (none)
	$\text{site} + S \leftrightarrow S_{ads}$	(T3) ≡ (none)
	$OH^\bullet + Ti(IV) \leftrightarrow Ti(IV) \int OH^\bullet$	(T4) ≡ none
Recombination	$e^- + h^+ \rightarrow \text{heat}$	(T5) ≡ eqn. (2)
Trapping	$Ti(IV)\text{-}OH^- + h^+ \leftrightarrow Ti(IV) \int OH^\bullet$	(T6a) ≡ eqn. (3d)
	$Ti(IV)\text{-}H_2O + h^+ \leftrightarrow Ti(IV) \int OH^\bullet + H^+$	(T6b) ≡ eqn. (3d)
	S_{ads} *or* $S_{i,ads} + h^+ \leftrightarrow S^+_{ads}$ or $S^+_{i,ads}$	(T7) ≡ eqn. (6a)
	$Ti(IV) + e^- \leftrightarrow Ti(III)$	(T8a) ≡ eqn. (3c)
	$Ti(III) + O_2 \leftrightarrow Ti(IV)\text{-}O_2^{\bullet-}$	(T8b) ≡ eqn. (3c)
Hydroxyl attack		
Case I	$Ti(IV) \int OH^\bullet + S_{ads} \rightarrow Ti(IV) + S_{1,ads}$	(T9) ≡ eqn. (7b)
Case II	$OH^\bullet + S_{ads} \rightarrow S_{1,ads}$	(T10) ≡ none
Case III	$Ti(IV) \int OH^\bullet + S \rightarrow Ti(IV) + S_1$	(T11) ≡ eqn. (7b)
Case IV	$OH^\bullet + S \rightarrow S_1$	(T12) ≡ none
Reactions of other radicals	$e^- + Ti(IV)\text{-}O_2^{\bullet-} + 2(H^+) \leftrightarrow Ti(IV)(H_2O_2)$	(T13) ≡ none
	$Ti(IV)\text{-}O_2^{\bullet-} + (H^+) \leftrightarrow Ti(IV)(HO_2^\bullet)$	(T14) ≡ none
	$(H_2O_2) + (OH^\bullet) \leftrightarrow (HO_2^{\bullet-}) + (H_2O)$	(T15) ≡ none

Species in parentheses may be adsorbed or in the aqueous phase. In the treatment in the text, k_{T1} refers to the rate constant for reaction (T1), K_{T2}, k_{T2} and k_{-T2} and refer to the equilibrium constant and rate constants for the forward and back reactions respectively of the process described by reactions (T2a/T2b). The equilibrium and rate constants for reactions (T3) to (T15) are defined in a similar manner.

[a] Where applicable, correspondences are shown between the reactions of Turchi and Ollis's scheme and those listed in Section 9.2.1.

[b] Involves the displacement of a proton from H_2O to one lattice oxygen, O_L^{2-}.

9.1. So as to avoid confusion if the reader should wish to consult Turchi and Ollis's original text, we have retained their system of reaction identification ((T1), (T2), (T3) etc.) and most of their nomenclature, e.g., Ti(IV) refers to a Ti(IV) surface site, $Ti(IV)\int OH^\bullet$ refers to a hydroxyl radical that is still closely associated with a Ti(IV) site due to their close proximity immediately after $OH^\bullet$ formation. S refers to the original organic substrate while S_1 refers to the first reaction intermediate. We have also indicated correspondences (where correspondences exist) between the elementary processes suggested by Turchi and Ollis and those listed in Section 9.2.1 of this chapter. In light of the fact that no photomineralization of the organic substrate occurs in the absence of oxygen, Mills *et al.* [46] have modified Turchi and Ollis's original treatment so as to consider the role of O_2 in determining the efficiency of electron trapping on the particles. Below, we indicate the nature of this modification and incorporate it into the kinetic expressions presented.

As may be seen from Table 9.1, the kinetic model is based on photolytically generated (by reaction (T6)) hydroxyl radical attack of the organic substrate. Four cases of hydroxyl radical attack on the original organic substrate are considered:

(I) Reaction occurs while both species are adsorbed (Langmuir–Hinshelwood kinetics).
(II) A nonbound radical reacts with an adsorbed organic molecule (Eley–Rideal kinetics).
(III) An adsorbed radical reacts with a free organic molecule arriving at the catalyst surface (Eley–Rideal kinetics).
(IV) Reaction occurs between two free species in the fluid phase.

Obviously, similar distinctions may be drawn between the various modes of $OH^{\bullet}$ attack on the intermediate species generated by reactions (T9)–(T12). However, so as to condense the system into a representative rate form of practical use, it is assumed that the rate constants for the reaction of $S_{i,\mathrm{ads}}$ with Ti(IV) $\int OH^{\bullet}$ and $OH^{\bullet}$ are the same, as are the rate constants for the reaction of S_i with Ti(IV) $\int OH^{\bullet}$ and $OH^{\bullet}$. This necessitates the consideration of only three further processes which, with their attendant rate and equilibrium constants may be expressed as:

$$S_i + \text{site} \xleftrightarrow{K_{\mathrm{Si}}} S_{i,\mathrm{ads}} \tag{9.51a}$$

$$(\text{Ti(IV)} \int OH^{\bullet} \text{ or } OH^{\bullet}) + S_i \xrightarrow{k_{\mathrm{Si}}} S_{i+1} \tag{9.51b}$$

$$(\text{Ti(IV)} \int OH^{\bullet} \text{ or } OH^{\bullet}) + S_{i,\mathrm{ads}} \xrightarrow{k_{\mathrm{Si,ads}}} S_{i+1,\mathrm{ads}}. \tag{9.51c}$$

All other rate and equilibrium constants in the following treatment are defined in the notes accompanying Table 9.1.

The treatment starts with an application of the steady state approximation to the concentration of the holes:

$$\frac{\mathrm{d}[h^+]}{\mathrm{d}t} = k_{\mathrm{T1}}\phi I_{\mathrm{abs}}A_C - k_{\mathrm{T5}}[h^+][e^-]V_P - k'_{\mathrm{T6}}[h^+]A_S + k_{-\mathrm{T6}}[\mathrm{Ti(IV)} \int \mathrm{OH^\bullet}]A_S \approx 0, \tag{9.52}$$

where it is assumed that reaction (T2) lies far to the right hand side (not unreasonable in an aqueous system) so that [Ti(IV)-(H_2O/$OH^\bullet$)] may be taken as constant and that k_{T6}[Ti(IV)-(H_2O/$OH^\bullet$)] may be written as k'_{T6}. I_{abs} is the absorbed light intensity, ϕ is the quantum yield for reaction (T1), A_C is the area of a particle normal to the illumination, A_S is the surface area of a particle and V_P is the volume of a particle. If the semiconductor intrinsic carrier density is low, because the photo-generation rates of electrons and holes are equal, it may be assumed that $[h^+] = [e^-]$ and that:

$$k_{\mathrm{T1}}\phi I_{\mathrm{abs}}A_C = k_{\mathrm{T5}}[h^+]^2V_P + k'_{\mathrm{T6}}[h^+]A_S - k_{-\mathrm{T6}}[\mathrm{Ti(IV)} \int \mathrm{OH^\bullet}]A_S. \tag{9.53}$$

Under conditions of high I_{abs}, $[h^+]$ will be high and consequently

$$k_{\mathrm{T5}}[h^+]^2V_P \gg k'_{\mathrm{T6}}[h^+]A_S - k_{-\mathrm{T6}}[\mathrm{Ti(IV)} \int \mathrm{OH^\bullet}]A_S.$$

Under these conditions, equation (9.53) simplifies to:

$$[h^+] = \left(\frac{k_{\mathrm{T1}}\phi I_{\mathrm{abs}}A_C}{k_{\mathrm{T5}}V_P}\right)^{1/2}. \tag{9.54}$$

Under conditions of low I_{abs}, $[h^+]$ will be correspondingly low and consequently:

$$k'_{\mathrm{T6}}[h^+]A_S - k_{-\mathrm{T6}}[\mathrm{Ti(IV)} \int \mathrm{OH^\bullet}]A_S \gg k_{\mathrm{T5}}[h^+]^2V_P.$$

Further, as the reverse process of reaction (T6) is merely one path for the deactivation of photogenerated $OH^\bullet$, it is clear that $k_{\mathrm{T1}}\phi I_{\mathrm{abs}}A_C > k_{-\mathrm{T6}}[\mathrm{Ti(IV)} \int \mathrm{OH^\bullet}]A_S$ and that, under these conditions, equation (9.53) simplifies to:

$$[h^+] = \left(\frac{k_{T1}\phi I_{abs}A_C}{k'_{T6}A_S}\right). \tag{9.55}$$

Equations (9.54) and (9.55) suggest that the order with respect to light intensity of the rate of organic substrate destruction will change from 1 to $\frac{1}{2}$ as the light intensity is increased from low to high levels. Such a result has indeed been observed experimentally [159] and was alluded to earlier in this section in reference to equation (44).

No photooxidation of the original organic substrate occurs in the absence of O_2. Therefore the efficiency of electron trapping, δ_T, is a function of the dissolved oxygen concentration and is given by:

$$\delta_T = \frac{[\mathrm{Ti(III)}] + [\mathrm{Ti(IV)} - O_2^{\bullet-}]}{[e^-] + [\mathrm{Ti(III)}] + [\mathrm{Ti(IV)} - O_2^{\bullet-}]},$$

which, if it is assumed that $K_{T8b}[O_2] \gg 1$, gives:

$$\delta_T = \frac{K_{O_2,d}[O_2]}{1 + K_{O_2,d}[O_2]}, \tag{9.56}$$

where $K_{O_2,d} = K_{T8a}[\mathrm{Ti(IV)}]K_{T8b}$. Similar treatments of $[\mathrm{Ti(IV)} \int \mathrm{OH}^\bullet]$ and $[\mathrm{OH}^\bullet]$ in accordance, where applicable, with each of the kinetic models described by Cases (I)–(IV) yields four rate expressions, each with the same general form as equation (50):

$$-\frac{d[S]}{dt} = \frac{k'_{obs}\kappa_s[S]}{1 + \kappa_s[S] + \sum_{i=1}^{n} \kappa_{Si}[S_i]}, \tag{9.50}$$

k'_{obs} adopts the same form for each of the four mechanisms and is given by

$$[h^+] = \alpha\delta_T k_{T1}\phi I_{abs}A_C \text{ at low light intensities, and} \tag{9.57a}$$

$$[h^+] = \alpha\delta_T A_S k'_{T6}\left(\frac{k_{T1}\phi I_{abs}A_C}{k_{T5}V_P}\right)^{1/2} \text{ at high light intensities,} \tag{9.57b}$$

where δ_T is defined by equation (9.56) and α is a proportionality constant introduced to account for possible radical formation through the reduction of O_2 by conduction band electrons via equations (T13)–(T15), and which may take any value between 1 and 2, depending on the efficiency of those processes. κ_s and κ_{Si} adopt the following forms for each case.

Case I: the reaction between adsorbed $OH^{\bullet}$ and adsorbed substrate:

$$\kappa_s = \frac{k_{T9}K_{T3}[\text{site}]A_S}{k_{-T6}}; \qquad \kappa_{Si} = \frac{k_{Si,ads}K_{Si,ads}[\text{site}]A_S}{k_{-T6}}. \tag{9.58a}$$

Case II: the reaction between free solution phase $OH^{\bullet}$ and adsorbed substrate:

$$\kappa_s = \frac{k_{T10}K_{T3}[\text{site}]}{k_{-T6}K_{T4}[\text{Ti(IV)}]}; \qquad \kappa_{Si} = \frac{k_{Si,ads}K_{Si,ads}[\text{site}]}{k_{-T6}K_{T4}[\text{Ti(IV)}]}. \tag{9.58b}$$

Case III: the reaction between adsorbed $OH^{\bullet}$ and free solution phase substrate:

$$\kappa_s = \frac{k_{T11}}{k_{-T6}}; \qquad \kappa_{Si} = \frac{k_{Si}}{k_{-T6}}. \tag{9.58c}$$

Case IV: the reaction between free solution phase $OH^{\bullet}$ and free solution phase substrate:

$$\kappa_s = \frac{k_{T12}}{k_{-T6}K_{T4}[\text{Ti(IV)}]A_S}; \qquad \kappa_{Si} = \frac{k_{Si}}{k_{-T6}K_{T4}[\text{Ti(IV)}]A_S}. \tag{9.58d}$$

In their review, Mills *et al.* [46] suggest that, given the mounting evidence that the photogenerated hydroxyl radical remains bound to the semiconductor surface during organic substrate destruction by photocatalysis [56, 160], it is unlikely that cases II and IV ever obtain in any practical situation. Nevertheless, at time of writing, it is still experimentally impossible to discriminate between cases I and III.

The Turchi and Ollis treatment finds greatest applicability at time $t = 0$ and tells us little about the evolution of $[S]$ as a function of t without

TABLE 9.2
The photocatalytic reaction scheme of Minero [161]

Primary step	Primary reaction	Reaction number[a]
Excitation	$TiO_2 + h\nu \rightarrow e^- + h^+$	(T1)
Adsorption	site $+ S \leftrightarrow S_{ads}$	(T3)
Recombination	$e^- + h^+ \rightarrow$ heat	(T5)
Trapping	$H_2O + h^+ \rightarrow Ti(IV) \int OH^{\bullet}$	(T6)
	$(O_2)_{ads} + e^- \rightarrow (O_2^{\bullet -})_{ads}$	(T8)
Hydroxyl attack (Case I)	h^+ or $Ti(IV) \int OH^{\bullet} + S_{ads} \rightarrow S_{1,ads}$	(T7/9)

Where no direct correspondences exist between the mechanisms of Minero and Turchi & Ollis, the closest match has been chosen and then suitably modified.

recourse to integration of equation (50). A recent paper by Minero [161] indicates how such an integration can be accomplished for a much simplified kinetic model of Turchi and Ollis's case I under conditions where the rate of free carrier recombination is much slower than the rate of electron entrapment. Table 9.2 describes Minero's model, recast in terms consistent with Turchi and Ollis's nomenclature, which yields the following expression for the photocatalytic destruction of the organic substrate S:

$$-\frac{d[S]}{dt} = \frac{(1 + K_3[S])^2(k_{T1}I_{abs}\phi A_c)(k_{T8}A_S[O_2]_{ads})(k_{T7/9}[S])}{k_{T5}(k_{T1}I_{abs}\phi A_c)(1 + K_3[S]) + (k_{T8}A_S[O_2]_{ads})(k_{T7/9}A_S[S])K_3[\text{site}]_T}, \quad (9.59)$$

where $[\text{site}]_T$ is the total number of available adsorption sites for the organic substrate on a particle surface. Minero identifies two limiting forms for equation (9.59) and solves them to give $[S]$ as a function of t.

In **Minero's Case 1**, the concentration of the substrate is small and/or the adsorption of the substrate on the catalyst is small. Such a situation might be expected to obtain on a hydrophilic metal oxide surface and under these conditions, equation (9.59) reduces to

$$-\frac{\mathrm{d}[S]}{\mathrm{d}t}=\frac{(k_{\mathrm{T1}}I_{\mathrm{abs}}\phi A_c)(k_{\mathrm{T8}}A_S[\mathrm{O_2}]_{\mathrm{ads}})(k_{\mathrm{T7/9}}A_S[S])}{k_{\mathrm{T5}}(k_{\mathrm{T1}}I_{\mathrm{abs}}\phi A_c)+(k_{\mathrm{T8}}A_S[\mathrm{O_2}]_{\mathrm{ads}})(k_{\mathrm{T7/9}}A_S[S])K_3[\mathrm{site}]_T}. \tag{9.60}$$

Whilst equation (9.60) may be solved to generate kinetic working curves, Minero identifies two subsidiary limiting forms of this equation that yield simple $[S](t)$ functions that are of practical use. In **Minero's Case 1a**, it is assumed that the rate of oxidation is small and/or I_{abs} is high. Under these conditions, equation (9.60) reduces to

$$-\frac{\mathrm{d}[S]}{\mathrm{d}t}=\frac{(k_{\mathrm{T8}}A_S[\mathrm{O_2}]_{\mathrm{ads}})(k_{\mathrm{T7/9}}A_S[S])}{k_{\mathrm{T5}}}, \tag{9.61}$$

which, if it is assumed that $[\mathrm{O_2}]_{\mathrm{ads}}$ is invariant with time, integrates to give a simple first order expression in $[S]$—an exponential decay of $[S]$ with time. In **Minero's Case 1b**, it is assumed that the rate of oxidation is high and/or I_{abs} is very low. Under these conditions, equation (9.60) reduces to:

$$-\frac{\mathrm{d}[S]}{\mathrm{d}t}=\frac{(k_{\mathrm{T1}}I_{\mathrm{abs}}\phi A_c)}{K_3[\mathrm{site}]_T}. \tag{9.62}$$

Integration gives a zero order expression in $[S]$—a linear decay of $[S]$ with time. Thus, the complete solution of equation (9.60) would be expected to produce $[S]$ versus t plots of intermediate behaviour between a straight line and an exponential and this is indeed what is observed [161].

In **Minero's Case 2**, the concentration of the substrate is high or it is very hydrophobic, although it should be borne in mind that such a scenario may result in the oxidised substrate behaving as an extrinsic recombination centre. A direct consequence of this is that Minero's model is no longer applicable as his simplified kinetic scheme did not consider this reaction. Thus, the utility of the resultant expressions is limited to mainly very hydrophobic or poor water soluble compounds at low concentrations. Under these conditions, equation (9.59) simplifies to

$$-\frac{\mathrm{d}[S]}{\mathrm{d}t}=\frac{(k_{\mathrm{T1}}I_{\mathrm{abs}}\phi A_c)(k_{\mathrm{T8}}A_S[\mathrm{O_2}]_{\mathrm{ads}})(k_{\mathrm{T7/9}}A_S)(K_3[S]^2)}{k_{\mathrm{T5}}(k_{\mathrm{T1}}I_{\mathrm{abs}}\phi A_c)+(k_{\mathrm{T8}}A_S[\mathrm{O_2}]_{\mathrm{ads}})(k_{\mathrm{T7/9}}A_S)[\mathrm{site}]_T}, \quad (9.63)$$

which may be easily integrated to give the hyperbolic expression

$$[S]_t=\frac{[S]_0}{1+k_{\mathrm{obs}}[S]_0 t}, \quad (9.64)$$

where

$$k_{\mathrm{obs}}=\frac{(k_{\mathrm{T1}}I_{\mathrm{abs}}\phi A_c)(k_{\mathrm{T8}}A_S[\mathrm{O_2}]_{\mathrm{ads}})(k_{\mathrm{T7/9}}A_S)K_3}{k_{\mathrm{T5}}(k_{\mathrm{T1}}I_{\mathrm{abs}}\phi A_c)+(k_{\mathrm{T8}}A_S[\mathrm{O_2}]_{\mathrm{ads}})(k_{\mathrm{T7/9}}A_S)[\mathrm{site}]_T}. \quad (9.65)$$

Discrimination between Minero's Cases 1a, 1b and 2 can be achieved by simply carrying out experiments as a function of $[S]_0$ [161].

Extensive studies by a number of workers on the kinetics of organic substrate destruction through a range of static and dynamic semiconductor photocatalytic processes have identified several key factors as being important in determining overall reaction efficiency. Whilst properly the province of reviews dedicated to pollutant destruction by photocatalysis [44–46, 52, 54, 56, 57, 61, 65, 68] it is worth briefly summarising those factors here. As well as the adsorption/desorption equilibria in the dark and light discussed above, factors found to play a role include: the particle size (*vide supra*), the material crystal phase, the concentrations of the catalyst, substrate and sacrificial agent (usually O_2), surface condition/modification (especially with respect to surface states and deliberately adsorbed photosensitising dyes), the light intensity and the pH and ionic strength of the solution. A wide range of experimental techniques have been employed in determining in these factors. In the next section we discuss some of those techniques, with special emphasis on electrochemical methods for characterising colloidal semiconductors.

9.4 SOME ELECTROCHEMICAL TECHNIQUES FOR CHARACTERIZING COLLOIDAL SEMICONDUCTORS

Since aqueous dispersions of colloidal semiconductors are transparent (light scattering depending on particle size), optical methods may be used for reactant, reactant intermediate and product analysis, enabling reaction kinetics to be determined, mechanisms to be elucidated and the probing of some catalyst properties *in situ*. Photochemically speaking, time resolved transient fluorescence and absorption spectroscopies (including UV-visible, Raman, epr and diffuse relectance techniques) over a wide range of time scales down to the femtosecond domain, have proven especially useful, as have time resolved microwave conductivity studies and quantum yield measurements conducted with a range of ion, liquid and gas phase chromatographies. Direct electrochemical measurements on colloidal semiconductors usually involves precipitation of the particles onto a supporting electrode. The resultant assembly may then be interrogated by any of the existing battery of photoelectrochemical techniques available for the study of bulk semiconductor electrodes, e.g., steady state photocurrent methods, measurement of the photocurrent in response to chopped illumination, laser induced photocurrent transients, photoelectrochemical ac impedance, intensity modulated photocurrent spectroscopy etc. Many of these techniques were reviewed by Peter in 1990 [162].

Unfortunately, the to-electrode precipitation required for conventional (photo)electrochemical measurements on colloidal semiconductors necessarily perturbs the (assumed) spherical diffusion fields and surface adsorption equilibria that obtain at particles in the free solution state, phenomena which are instrumental in determining the dynamic and static charge transfer characteristics of the semiconductor. Consequently, there is a requirement for photoelectrochemical techniques capable of *in situ*, non-perturbative investigations of the mechanistic details and catalytic properties of colloidal semiconductors in solution conditions typical of their intended ultimate application. Two such techniques are photoelectrophoresis and the Optical Rotating Disc Electrode (ORDE, developed by Albery *et al.*). As mentioned above, the former technique has already been reviewed by this author elsewhere [47]. Thus, the remainder of this review will concentrate on measurements that can be made with the latter

9.4.1 The optical rotating disc electrode (ORDE)

When combining photochemistry and electrochemistry there are many advantages in using a semitransparent optical electrode. If light is shone through solution at the electrode then the solution has to be transparent enough to allow passage of light to the electrode. This means that only a small fraction of light can be absorbed close to the electrode surface. High absorption close to the electrode surface is desirable though, so as to maximise the current obtained from any photogenerated intermediates with short lifetimes. When light is shone through an electrode, the solution can contain much higher concentrations of absorbing species so that a large part of the light can be absorbed in the diffusion layer of the electrode. Such a semitransparent electrode [163] may be fabricated either by the sputtering of a thin film of metal (usually gold or platinum) or the pyrolytic deposition of a layer of quasi-metallic semiconductor (*n*-type SnO_2) on to either a quartz microscope slide or the polished end of a quartz rod light guide. The use of a quartz rod light guide as the electrode support has a further advantage in that the end of the rod forms a disc electrode which may be rotated. The reason for rotating the electrode is that one controls the supply of reactant (or the precursor of a photogenerated reactant) to the electrode surface; a genuine steady state can be established and the rotation speed is a useful experimental variable.

Thus, it can be seen that the ORDE is a powerful tool for the interrogation of both the thermodynamics and kinetics of dynamic charge transfer from the particles to the electrode surface. Further, the introduction to the system of a solution phase charge carrier acceptor to act as a charge carrier shuttle between the particles and the electrode will facilitate the investigation of static and dynamic charge transfers between the particles and a wide range of charge carrier acceptors.

Let us consider the electrode kinetics associated with charge transfer from an *n*-type semiconductor particle to an electrode. As indicated by Albery *et al.* [164], the crucial difference between the electrochemistry of a colloidal particle and an ordinary electrochemically active solution phase species is the number of electrons transferred from the particle to the electrode may be large and will depend upon the potential of the electrode. Fig. 9.5 shows the model for an encounter of a particle with an electrode used by Albery and co-workers. k_D is the mass-transfer coefficient for the transport of the particles to the electrode surface. In the simplest case, wherein it is assumed that the lifetime of the transferable electrons (majority carriers of thermal or photonic origin) is greater than the time taken by a particle to traverse the ORDE diffusion layer, this is given by

$$k_D = \frac{D_P}{X_D}, \tag{9.66}$$

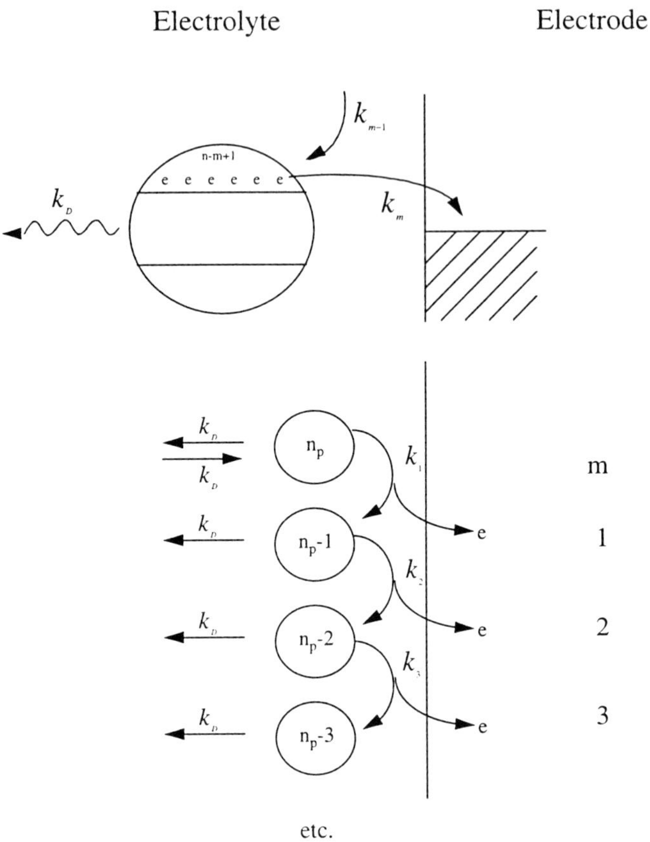

Fig. 9.5. Kinetic scheme for the successive transfer of electrons from a colloidal particle to a macroscopic electrode.

where D_P is the diffusion coefficient of the particle and X_D is the diffusion layer thickness of a rotating electrode and is defined in the appendix. The flux of particles arriving at the electrode surface is given by:

$$j_0 = k_D c_P, \tag{9.67}$$

where c_P is the total concentration of the particles. Albery *et al.* assume that each particle has n electrons (majority carriers, generated by thermal or photonic means) in the conduction band and that the rate constant for electron transfer from particle to electrode is proportional to the number of electrons in the conduction band:

$$k_m = k_I(n - m + 1), \tag{9.68}$$

where m is the number of the electron transfer in the cascade sequence of Fig. 9.5 and k_I describes the transfer of the last electron from the particle conduction band at that particular electrode potential. They further assume that the potential distribution between the particle and the electrode and within the particle are not perturbed by the successive electron transfers, assumptions supported by the results obtained by Albery and Bartlett [131] and Liver and Nitzan [125] and described in Section 9.3.2.

Summing the transfer of electrons for each partition with these assumptions in mind, Albery *et al.* obtained for the total flux of electrons j_e:

$$\frac{j_e}{j_0} = \frac{n}{\left(1 + \frac{k_D}{k_I}\right)}, \tag{9.69}$$

an equation which has the same form as the current voltage curve for a single electron transfer. This is because, in Albery and co-worker's model, each electron has the chance $k_I/(k_I + k_D)$ of being transferred at the electrode and as long as k_I is independent of the number of electrons in the particle, it makes no difference whether the electrons all come together in one big particle, or separately, each on its own little ion. In the light of this result, Albery *et al.* concluded that, for this model, the Tafel relation and the Koutecky–Levich equation should hold. The limiting current i_L for large overpotentials is given by

$$i_L = nFAk_Dc_P, \tag{9.70}$$

and under these conditions, all the conduction band electrons are removed and the current is limited by the transport of the particles to the electrode surface.

To fully utilise the result expressed by equation (9.70), we must now consider the solution of the convective diffusion equation for a system of n-type semiconductor particles as they diffuse towards an ORDE. Whilst

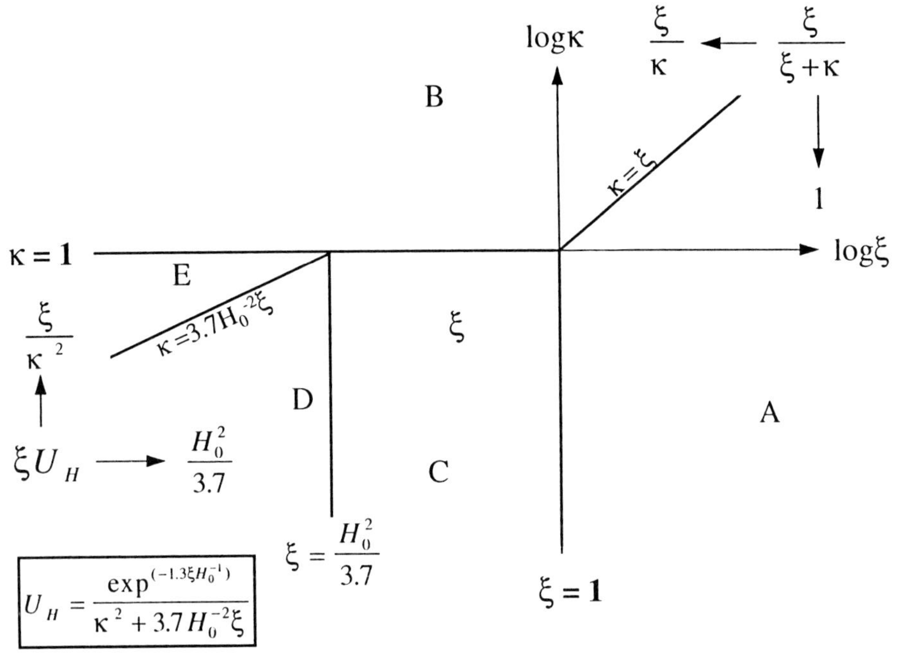

Fig. 9.6. Approximate solutions for the photoelectrochemical collection efficiency $N_{h\nu}$ assuming that $\phi = 1$. The different approximations hold for different values of ξ and κ where ξ ($= X_D/X_\epsilon$) compares the thickness of the diffusion layer to the distance over which the light is absorbed and κ ($=X_D/X_k$) compares the thickness of the diffusion layer to the distance a photogenerated species diffuses before decomposing. The parameter H_0 ($= X_D/X_k$) compares the thickness of the diffusion layer to the thickness of the hydrodynamic layer.

the theory relating to the use of conventional Rotating Disc Electrodes (RDEs) and Rotating Ring Disc Electrodes (RRDEs) is well documented and easily accessible to the interested reader [165], the theoretical treatment of the ORDE for a general homogeneous photochemical system is not and, therefore, some explanation is needed. This can be found in Appendix A.1; Fig. 9.6 shows a case diagram illustrating the approximate solutions obtained therein for $N_{h\nu}$, the photoelectrochemical collection efficiency (defined in Appendix 9.A.1) describing the recovery of photogenerated species at the ORDE surface under conditions of mass transport control. Appendix 9.A.1 also contains a discussion of the solutions for $N_{h\nu}$.

The convective diffusion equation for a particle of concentration c_m and which contains m electrons in the conduction band and surface states is given by:

$$D_p \frac{\partial^2 c_m}{\partial x^2} + v_x \frac{\partial c_m}{\partial x} + k_g c_{m-1} - kmc_m + k(m+1)c_{m+1} - k_g c_m = 0, \qquad (9.71)$$

It is assumed that all electrons transfers from the particle conduction band and surface states to the electrode take place under conditions where the current is mass transport controlled. The first order rate constant k_g describes electron promotion either by thermal or photonic processes, and the rate constant k describes the loss of the electrons from the conduction band or surface states by a process which is first order in electron concentration. The validity of this assumption will be discussed later. There will be an equation similar to equation (71) for each value of m. If each equation is multiplied by its value of m and the engendered set of equations summed, it is possible to obtain the simple result that:

$$D_p \frac{\partial^2 c_e}{\partial x^2} + v_x \frac{\partial c_e}{\partial x} + k_g c_P - kc_e = 0, \qquad (9.72)$$

where c_e is the total concentration of electrons in the conduction band and surface states, and is given by:

$$c_e = \sum_{m=0} mc_m.$$

This result shows that the transport of the particles may be treated by the usual convective diffusion equation, using c_e as the variable and supports the analysis of the system in terms of the cascade model; simple Tafel and Levich relations will hold for the system of particles. Thus, in the dark

$$k_g c_P = k_T c_P$$

where k_T is the rate constant for the thermal generation of electrons in the conduction band and surface states in the dark. The result expressed by equation (9.72) also implies that, from equation (9.70), the mass transport limited current in the dark, $(i_D)_L$, is given by

$$(i_D)_L = nFAk_D c_P = \frac{nFAD_P c_P}{X_D} = \frac{nFAD_P c_P}{0.643 D_P^{1/3} \nu^{1/6} \omega^{-1/2}}. \tag{9.73}$$

Under conditions of irradiation

$$k_g c_P = k_T c_P + I_o \phi \epsilon_P c_P \exp(-\epsilon_P \epsilon_P x),$$

where ϵ_P is the natural extinction coefficient of the particles at the wavelength of light used. If it is assumed that the rate of production of photonically generated electrons is greater than the rate of production of thermally generated electrons under conditions of high photon flux close to the electrode surface (i.e., within the diffusion layer) then

$$k_g c_P = I_o \phi \epsilon_P \epsilon_P \exp(-\epsilon_P c_P x).$$

Therefore, equation (9.72) may be rewritten as:

$$D_p \frac{\partial^2 c_e}{\partial x^2} + v_x \frac{\partial c_e}{\partial x} + I_0 \phi \epsilon_P c_P \exp(-\epsilon_P c_P x) - kc_e = 0. \tag{9.74}$$

Comparison of equations (9.A.4) and (9.74) shows that all results deduced for the ORDE for a general homogeneous photochemical system can be applied to a heterogeneous colloidal system.

Of particular application in colloidal systems are the results that obtain in regions B and C of Fig. 9.6. Movement between these regions may be accomplished by varying κ, the ratio between X_D the diffusion layer thickness, and X_k the kinetic length or reaction layer thickness. In experiments on colloidal CdS described below, X_D and X_k may be of the same order of size; therefore, variation of the rotation speed of the electrode, and thus X_D, may move the system under study from one region of the case diagram to another. Thus, in such a situation, the general form of the photoelectrochemical collection efficiency in regions B and C must be used, and this is given by:

$$N_{h\nu} = \frac{1}{X_\epsilon}\left[\frac{\phi}{X_D^{-1} + X_k^{-1}}\right]. \tag{9.75}$$

Thus, when X_k is much more than X_D (region C):

$$N_{h\nu} = \frac{\phi X_D}{X_\epsilon},$$

and when X_k is much less than X_D (region B)

$$N_{h\nu} = \frac{\phi X_k}{X_\epsilon}.$$

Therefore, by using equations (9.A.7) and (9.74), the following result for the mass transport limited photocurrent, $(i_{h\nu})_L$ in case regions B and C of Fig. 9.6 may be obtained:

$$\begin{aligned}(i_{h\nu})_L &= \frac{\phi AFI_o}{X_\epsilon}\frac{X_D X_k}{X_D + X_k} \\ &= \frac{\phi AFI_o(0.643\nu^{1/6}D_P^{1/3}\epsilon_P c_P \omega^{-1/2})(D_P/k)}{(0.643D_P^{1/3}\nu^{1/6}\omega^{-1/2}) + (D_P/k)}.\end{aligned} \tag{9.76}$$

Equations (9.73) and (9.76) were derived under conditions of mass transfer control. Equation (9.69) describes the Tafel behaviour for majority carriers whose lifetimes are assumed to be greater than the time taken by a particle to traverse the ORDE diffusion layer, i.e., in the case of an n-type particle, those electrons generated by thermal means. However, the model for the current-voltage behaviour associated with electrons generated by photonic means is slightly different to that which pertains for thermally generated charge carriers. The general model is still based on the idea of sequential electron transfer presented in Fig. 9.5. However, the transport of the particles to the ORDE is described by the mass transfer coefficient k_{PH}, where

$$k_{PH} = D_p\frac{X_D + X_k}{X_D X_k},$$

k_{PH} is used instead of the analogous and more straightforward k_D of equation (9.66) because, for the particle photocurrent-voltage behaviour, it is no longer applicable to assume that the photogenerated carrier lifetime is greater than the time taken by a particle to traverse the electrode diffusion layer. Thus, at lower electrode rotation speeds and, therefore, larger values of X_D, the diffusion layer thickness may become greater than the kinetic length, X_k. X_k is then the transport controlling distance and k_{PH} must be written as

$$k_{PH} = D_p/X_k.$$

At higher rotation speeds, X_D is less than X_k and k_{PH} can be written as

$$k_{PH} = D_p/X_D,$$

where D_p/X_D can be identified with k_D from equation (9.66).

Using k_{PH} as the mass transfer coefficient, the flux of particles arriving at the electrode surface is then given by:

$$j_{0,PH} = k_{PH}c_P. \tag{9.77}$$

Use of this equation in place of equation (9.67) in the mathematical treatment of sequential cascade model of Fig. 9.5 produces the following expression for the total flux of particle ferried, photogenerated electrons, $j_{e,PH}$, to the surface of the ORDE:

$$\frac{j_{e,PH}}{j_{0,PH}} = \frac{n_{PH}}{\left(1 + \dfrac{k_{PH}}{k_I}\right)}, \tag{9.78}$$

where n_{PH} is mean number of photogenerated transferable electrons per particle. Equation (9.78) has exactly the same form as equation (9.69), the current-voltage curve for a single electron transfer in the dark because, again, each electron has the chance $k_I/(k_I + k_{PH})$ of being transferred. Hence, according to the cascade model, it can be concluded that all of the commonly used electrochemical relationships should hold for photogenerated species, including the Tafel relationship, which may be derived from equation (9.78):

$$\frac{j_{0,PH}n_{PH}}{j_{e,PH}} = \frac{j_{0,PH}n_{PH}FA}{j_{e,PH}FA} = \frac{(i_{h\nu})_L}{i_{h\nu}} = 1 + \frac{k_{PH}}{k_I}.$$

Now,

$$k_I = k_i^0 \exp\left(\frac{\beta F(V - E_{CB}^0)}{RT}\right),$$

where, in accordance with the symbol convention used in this review, V refers to the potential of the electrode and E_{CB}^0 is the standard redox potential associated with the conduction band edge. β is the anodic transfer coefficient and k_I^0 is the standard heterogeneous rate constant for electron transfer from the particle conduction band. Therefore,

$$V = E_{CB}^0 + \frac{RT}{\beta F}\ln\left(\frac{k_{PH}}{k_I^0}\right) - \frac{RT}{\beta F}\ln\left(\frac{(i_{h\nu})_L}{i_{h\nu}} - 1\right). \tag{9.79}$$

9.4.1.1 Steady state dark electrochemistry of colloidal CdS

Albery *et al.* tested the validity of equations (9.69) and (9.73) on 8 nm radii CdS particles in pH 7 solution [166], made according to a recipe published by Grätzel [114] with one modification: H_2S was reacted with $Cd(ClO_4)_2$ rather than $Cd(NO_3)_2$ under N_2 (to avoid O_2 inclusion in the particle lattice). Point by point voltammograms recorded in the dark indicated an onset voltage of ca. 0.2 V versus SCE for an oxidation reaction, which was interpreted as harvesting of thermally generated majority carriers from the particle conduction band. The current then rose slowly with potential, reaching a plateau at about 0.6 V. The nearly pH-invariant flat band potential of bulk CdS is approximately −0.9 V versus SCE [167] indicating that an overpotential of about 1 V is required to drive the electron-transfer to the macroscopic electrode. A plot of $(i_D)_L$ versus $\omega^{1/2}$ yields a straight line in accordance with the Levich expression given by equation (9.73), from which the number of electrons per particle is found to 4. Tafel analysis of the current-voltage curve produced a straight line, predicted by equation (9.69), from which a transfer coefficient value of 0.34 was obtained. This result, in conjunction with the apparent overpotential needed to drive electron transfer, indicated that the electron transfer reaction needed to be driven hard by the electrode potential.

Albery *et al.* suggest that solvent shell reorganisation may not be relevant to electron transfer between two massive entities, implying that the charge transfer between particle and electrode should have been rapid. They further suggest that the observed irreversibility must therefore arise from the mutual repulsion of the particle and electrode double layers.

A repeated study of this work [168] led to results that qualitatively matched, but quantitatively slightly varied from, those reported. CdS particles were again made according to the modified Grätzel recipe. The radii of the resultant particles were measured by dynamic light scattering and found to be in the range 5–12 nm, the lower end of this range corresponding to the upper particle radius at which Q-state behaviour may be expected for CdS [48, 50]. Point by point voltammograms (Fig. 9.7) indicated an onset voltage of ca. −0.7 V versus SCE for the aforementioned oxidation wave, an overpotential of about 0.2 V with respect to the flat band potential of bulk CdS.

The current of the wave then rose slowly with potential, reaching a

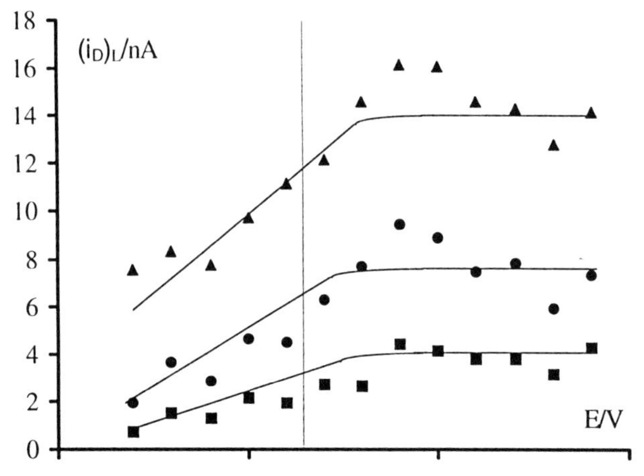

Fig. 9.7. Dark current as a function of applied electrode potential at a range of rotation speeds for a 5 mol m^{-3} colloidal suspension of CdS: (■), 4 s^{-1}, (●), 16 s^{-1}, (Δ), 36 s^{-1}. Electrode radius = 2 × 10^{-3} m.

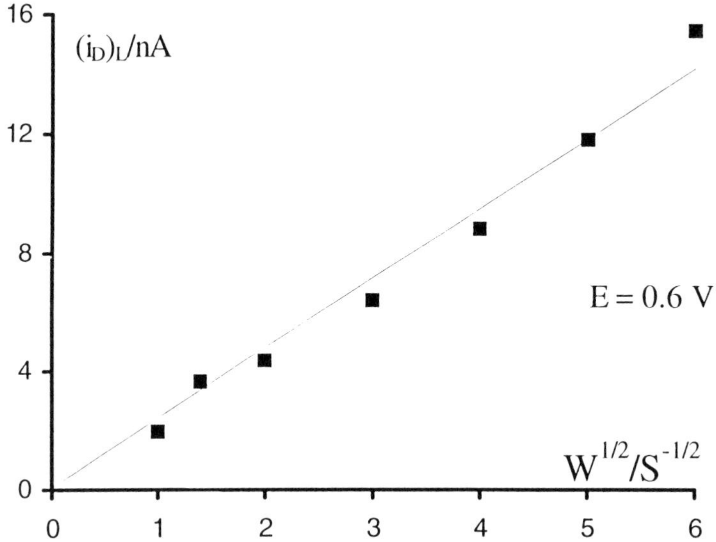

Fig. 9.8. Typical plot of mass transport limited dark current $(i_D)_L$ versus $\omega^{-1/2}$ for a $5\ \text{mol m}^{-3}$ colloidal suspension of CdS. Electrode radius $= 2 \times 10^{-3}$ m.

plateau at about 0.3 V. Plots of $(i_D)_L$ versus $\omega^{1/2}$ (e.g., Fig. 9.8) yield straight lines, again in accordance with the Levich expression given by equation (9.73), from which the number of electrons per particle was found to be, variously, 2 to 27 depending on the mean particle size. Tafel analysis of the current-voltage curve produced a straight line, as predicted by equation (9.69), from which a transfer coefficient value of 0.13 was obtained, again indicative of irreversible electrode kinetics.

Importantly, one further observation was recorded. Equation (9.69) indicates that, in analogy to equations (9.78) and (9.79), for a reaction with irreversible electrode kinetics:

$$\frac{nj_0}{j_e} = \frac{i_L}{i} = \left(1 + \frac{k_D}{k_I^0 \exp\left(\frac{\beta F(V - E_{CB}^0)}{RT}\right)}\right). \tag{9.80}$$

Rearrangement of equation (9.80) reveals that the half-wave potential,

$V_{1/2}$, for the removal of thermally (or photonically) generated electrons from the particles is given by:

$$V_{1/2} = E^0_{CB} + \frac{RT}{\beta F}\ln\left(\frac{k_D}{k_I^0}\right) = E^0_{CB} + \frac{RT}{\beta F}\ln\left(\frac{1.555 D_P^{2/3}\nu^{-1/6}}{k_I^0}\right) + \frac{RT}{2\beta F}\ln(\omega). \tag{9.81}$$

Thus, the half wave potential for electron transfer from the particles to the electrode should shift anodically with increasing electrode rotation speed if the reaction occurs with irreversible electrode kinetics. However, it is found that $V_{1/2}$ is invariant with rotation speed [168], implying that the particles have reversible electrode kinetics, a result in conflict with the observation of a reaction overpotential of 0.2 V and a transfer coefficient of 0.13 for this set of data.

These contradictions may be reconciled in two ways. The first explanation involves invoking the participation of surface states with reversible electrode kinetics in mediating the electron transfer process. Energetically, the postulated surface states must reside between the band edges of the particles. Thus, electron transfer occurs reversibly from an increasing number of surface states as the electrode is scanned anodically from the oxidation wave onset potential until, when the observed plateau on the voltammetric wave is reached, all the states are accessible to the electrode and the rate of charge transfer is governed by the transport of the particles to the electrode surface. In this model, the observed "transfer coefficient" is then a measure of the compound function describing the energy distribution and relative populations of the particle surface states. This model is supported by the observation that the Levich relationship is obeyed over the studied potential range of the wave (−0.5 to 0.8 V versus SCE) implying that $k_I \gg k_D$ and the possession of reversible electron transfer kinetics for the majority of surface states accessible at any potential within that range. This explanation is plausible, especially considering the particle surface inhomogeneity and the associated wide range of surface state or defect site energies available, as has been demonstrated by experimental studies of photogenerated charge-carrier trapping at the semiconductor solid-liquid interface [86, 103, 104, 107–110]. That surface states are seen to participate in the repeated study but not in Albery and co-workers original paper may be explained as follows: The number and nature of surface/defect sites in/on colloidal semiconductors is strongly dependent on the method of chemical synthesis (see Section 9.2.1). By studying the photoluminescent properties of CdS particles, Ramsden and Grätzel have shown that the number of particle defect sites decreases with increasing sol aging time [169]. All colloidal CdS samples of the repeated study were used in experiments conducted no longer than 1 week after particle synthesis; as such, they would have been possessed of a relatively high concentration of defect sites. The history of the samples used in the original study by Albery *et al.* is unknown, although it may be that their experiments were performed on aged suspensions with consequently comparatively low concentrations of defect sites. This is by no means an unreasonable conclusion in light of the above.

A second explanation is suggested by the work of Heyrovsky and Jirkovsky on the electrochemistry of Q-state colloidal SnO_2, TiO_2 and TiO_2/Fe_2O_3 suspensions at renewed mercury drop electrodes [170]. The band gap of Q-state particles is known to increase with decreasing particle size, usually corresponding to simultaneous cathodic and anodic shifts in the conduction and valence band edges respectively [48, 50]. Thus, the onset of the wave corresponding to electron transfer from the conduction bands of an assembly of Q-state monoradial particles to the electrode would be expected to shift in a cathodic direction as the particle radius is decreased. For a polydisperse suspension, this necessarily implies the existence of a range of particle-to-electrode electron transfer onset potentials, the smaller particles starting to participate in the charge transfer reaction at potentials cathodic of the larger particles. The observed polarographic wave is then a compound entity comprised of contributions made by individual groups of particles of equal size, the relative size of each contribution being determined by the particle size distribution. Thus, if it is assumed, in accordance with the suggestion made by Albery *et al.* [164] (*vide supra*) that all particles are possessed of reversible electrode kinetics, the manner in which, for example, the observed anodic current increases with potential for electron transfer from Q-state or near Q-state CdS particles to the electrode is a reflection of the particle size distribution, with $E_{1/2}$ approximately denoting the centre of the potential region in which the particles of a given size distribution undergo the electrode process.

As mentioned above, the CdS particle samples examined in the repeated study were polydisperse with mean radii in the range 5–12 nm. The lower end of this range corresponds to the approximate upper radius at which Q-state behaviour obtains in colloidal CdS, allowing us to speculate that some fraction of the CdS particles in any one sample examined in the initial Alberian or subsequent repeated study may exhibit Q-state behaviour with a consequent dispersion in particle-to-electrode electron transfer onset potentials. Simultaneously, this again allows us to account for both the slopes of the reported polarographic waves (and the corresponding low values for the "transfer coefficients") and the quantitative discrepancies observed between those results obtained by Albery and co-workers [166] and those obtained in the repeated study. A range of particle samples, all of which possess some Q-state character, and which have slightly different mean radii and size distributions will exhibit a range of "transfer coefficient" values (determined by the polydispersity of the sample) and different polarographic wave onset potentials (the cathodic limit of the wave being determined by the electrode kinetics of the smallest particle in each sample). The particles studied by Albery and co. would seem to be characterised by a narrower size distribution than those examined in the repeated study, allowing us to predict that the polarographic wave derived in the repeated study should have both a smaller anodic "transfer coefficient"/less steep current-voltage curve (the sample is of greater polydispersity) and more cathodic onset potential (the sample contains particles of smaller radii). This is indeed what is observed experimentally.

9.4.1.2 Steady state photoelectrochemical behaviour of colloidal CdS

For the purposes of the studies reported here, the photocurrent was taken to be the total current recorded at the ORDE from an illuminated colloidal dispersion of CdS minus the current recorded under identical condition from the same dispersion in the dark. In both studies, the photocurrents generated by CdS particles illuminated at the ORDE exhibited a wavelength dependence (action spectrum) identical to the absorption spectrum of colloidal and bulk CdS [166,168], unambiguously indicating that the observed photocurrent is due entirely to ultra-band gap photoexcited conduction band electrons. However, it should be noted that, unless stated otherwise (e.g. the action spectrum experiments), the particle suspensions of both studies were usually irradiated with white light from a 250 W quartz iodine projector lamp to maximise the photocurrents observed.

Albery *et al.* [164, 166] tested the validity of simplified versions of equations (9.76) and (9.78) by studying 8 nm CdS particles, prepared as described in Section 9.4.1.1, at an illuminated ORDE. They found that the photocurrent-voltage curve had a similar shape to the dark current-voltage curve, from which they concluded that all conduction band electrons have the same electrode kinetics. The rising part of the photocurrent-voltage curve was plotted according to the Tafel relationship and was found to give similar values for the electrochemical rate constant and transfer coefficient as the corresponding dark current data. Further, the limiting photocurrent was found to be inversely proportional to the square root of the electrode rotation speed, indicating that the photogenerated electrons are kinetically stable in their passage across the diffusion layer. Using equation (9.82),

$$(i_{h\nu})_L = \frac{\phi AFI_o X_D}{X_\epsilon} = \phi AFI_o(0.643\nu^{1/6} D_P^{1/3} \epsilon_P c_P \omega^{-1/2}), \qquad (9.82)$$

which is derived from equation (9.76) under conditions where $X_k > X_D$, Albery *et al.* were able to obtain a value of the quantum efficiency for detectable charge carrier generation, ϕ, of 2×10^{-3} from the slope of a graph of $(i_{h\nu})_L$ vs. $\omega^{-1/2}$. Such a low value of ϕ was taken as indicating that, in the absence of a suitable scavenger for photogenerated holes, the main process for photogenerated electrons is intra-particle electron-hole recombination. Albery *et al.* posit that those electrons that survive recombination to reach the electrode do so because corresponding holes have been lost to a particle corrosion reaction or an adventitious, trace electron donor. By comparing the observed dark and photocurrents, they estimate that a typical number of electrons transferred per particle under illuminated conditions is about 20.

Again, a repeated study of this work [168] on 5–12 nm radii CdS particles (*vide supra*) led to results that qualitatively matched, but quantitatively varied from, those previously reported by Albery *et al.*. Observed point by point photocurrent-voltage curves are presented in Fig. 9.9.

A number of important points arise from these plots. The first is that the observed photocurrent of the repeated study increases with applied potential at a faster rate than the associated dark current. This may be attributed to the electrons being transferred from the conduction band or one type of shallow electron trap at or near the particle surface (typical electron trapping times are <100 fs for CdS, see Section 9.2.1 above) [96–

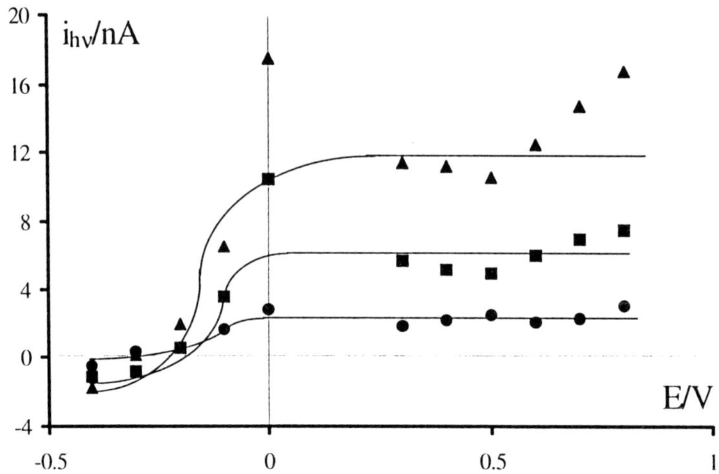

Fig. 9.9. Photocurrent as a function of applied electrode potential at a range of rotation speeds for a 5 mol m^{-3} colloidal suspension of CdS: (●), 1 s^{-1}, (■), 4 s^{-1}, (Δ), 36 s^{-1}. Electrode radius = 2×10^{-3} m.

99]. If the data of Fig. 9.9 is plotted according to the appropriate Tafel relationship of equation (9.79), a good straight line with a β of 0.45 is obtained, in broad agreement with the results of Albery *et al.* Once again, with the diminished relevance of solvent reorganisation to such an electron transfer event, we would expect the charge transfer itself to occur quickly and reversibly. A β of 0.45 indicates that this is not the case and that there still exists some barrier to electron transfer the nature of which is uncertain but which Albery *et al.* attribute to mutual double layer repulsion. Heyrovsky and Jirkovsky [170] attribute the observed value of β to any one particle sample being comprised of particles of a range of radii which consequently exhibit a "transfer coefficient" value and polarographic wave onset potential determined by the polydispersity and mean particle size of the sample.

The absence of a current signal derived from the oxidation of particle surface states in the photocurrent voltage curve (Fig. 9.9) compared to that observed in the dark current voltammogram (Fig. 9.7) may be attributed to the photogenerated valence band hole-driven oxidation of those states precluding their participation in a particle-to-electrode electron transfer. As described in Section 9.2.1, the particle surface states may be surface

S^{2-} ions, sulphur vacancies (V_S) or some other surface defect. Thus, particle illumination is essentially mimicking the role of the aging process eluded to by Ramsden and Grätzel [169] in that it eliminates the participation of the surface states in the particle-to-electrode electron transfer event.

It should be noted that the Heyrovsky and Jirkovsky model does not by itself account for the change in the value of β observed when the particles are illuminated. Thus, it would seem that, depending upon the age of the sample, surface states/defect sites may play a major role in determining the form of the observed current-voltage curve in the dark; however, this behaviour may itself be superimposed over an underlying current behaviour of the form described by Heyrovsky and Jirkovsky [170]. Upon illumination, photogenerated valence band holes oxidise the surface states, precluding their participation in a particle-to-electrode electron transfer, so allowing pure Alberian behaviour (described by the cascade model (Fig. 9.5) modified to account for mutual double layer repulsion) or Heyrovsky and Jirkovsky behaviour to be observed. Differentiation between these two effects could be easily achieved by the study of monodiperse CdS particles. This will be the subject of later work.

The second point to note is that the onset of the wave has moved by about 0.4 V to the more anodic potential of −0.3 V vs SCE, which is itself about 0.6 V more positive than the band edge of macroscopic CdS [171]. This could be due to the accumulation of holes at the surface caused by the photogenerated oxidation of particle surface states/hole traps (*vide supra*). Such a concentration of positive charge at the particle surface might lead to an unpinning of the band edges and their subsequently shifting to more anodic potentials. Such a phenomenon has been described by Bard for both macroscopic electrode and particulate semiconductor systems [172, 173]. However, early work by Albery *et al.* [84] and Henglein *et al.* [85], indicates that, under high intensity illumination, a photoinduced blue shift in the semiconductor absorption edge may be seen in the transient absorption spectrum. A later and more sophisticated treatment of the data by Bard [174] invokes the concept of a "Burstein shift" [175–180] to explain this observation. Bard suggests that photoinduced high electron concentrations in the semiconductor particles raise the Fermi level into the conduction band, so increasing the energy necessary to optically excite electrons from the valence band to the conduction band to a value larger than the material band gap. A consequence of this effect is that the potential of the Fermi level of electrons within the particles (which may be identified with the conduction band edge in n-type CdS particles) can be considerably more negative than that predicted from E_{CB} of the bulk semiconductor. This has been found to be so in experimental studies which show that the potential of the Fermi level shifts cathodically as the photon flux increases [181–184].

Whilst this may initially appear to be in opposition to the results of the optical rotating disc electrode study on colloidal CdS (Fig. 9.9), this may be readily explained by consideration of the relatively low illumination intensities used in the ORDE experiments, and the high surface state concentrations typical of the samples employed therein. The former precludes the generation of a "Burstein shift" while the latter, with a quantum yield of 0.77 for $(S^{\bullet -})_{surf}$ generation from S^{2-} ions at the CdS particle surface [115, 116], provides a highly efficient mechanism for positive charge accumulation at the particle surface.

An alternative explanation for the positive shift in the current-voltage curve onset potential upon illumination is that an overpotential exists for electron transfer directly out of the conduction band, which would be in keeping with the observation that β has a value of 0.45. However, the origin of such an overpotential is not readily explained and requires further study.

Fig. 9.10 shows a plot of the mass transport limited photocurrent, $(i_{h\nu})_L$, against $\omega^{-1/2}$. In agreement with the data of Albery *et al.* [166], a good straight line is obtained indicating that $(i_{h\nu})_L$ obeys equation (76) and that, consequently, X_k is more than X_D at all rotation speeds. Thus, the photogenerated electrons which contribute to the photocurrent are kinetically stable in their passage across the diffusion layer. By use of the data of Fig. 9.10 and equation (9.76), a value for the quantum efficiency for detectable charge carrier generation, ϕ, may be calculated; this is discussed in greater detail in the next section.

An estimate of n_{PH}, the number of photogenerated electrons per particle transferred to the electrode, may be obtained by comparing the size of the photocurrent with that of the dark current. From n_{PH} a value of the photogenerated carrier density may be calculated. This is found to be ca. 1.6×10^{26} m^{-3}, which is about 50 times greater than the value in the dark, indicating that the particles may indeed assume some metallic character and that any variation in the "dark majority carrier population" is negligible compared with the number of photogenerated charge carriers.

The photogenerated carrier density, in conjunction with the value of

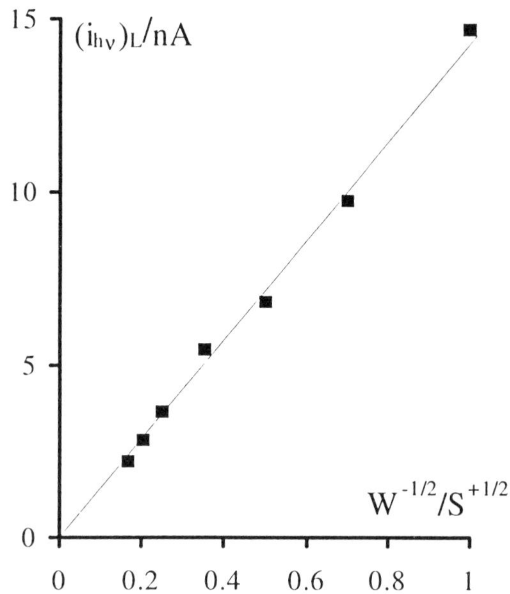

Fig. 9.10. Typical plot of mass transport limited photocurrent $(i_{h\nu})_L$ versus $\omega^{-1/2}$ for a 5 mol m^{-3} colloidal suspension of CdS. Electrode radius = 2×10^{-3} m.

9×10^{-17} m^3 s^{-1} recorded for the second order rate coefficient for electron-hole recombination within CdS particles by Nosaka and Fox [118] (see Section 9.2.1), allows us to calculate a characteristic lifetime for photo-generated charge carriers in CdS. This equates to 70 ps, in agreement with a value of 50 ps recorded by femtosecond flash photolysis for the slower, geminate electron-hole recombination process that obtains at lower excitation intensities [107] (again, see Section 9.2.1).

Thus, it can be seen that a study of the steady state photoelectrochemistry of colloidal semiconductors with the ORDE can provide information relating to the energy distribution of the particle surface states, the photo-generated carrier density and the quantum efficiency of carrier generation. The next section describes how to obtain information pertaining to intra-particle charge carrier dynamics from a study of the behaviour of transient photocurrents at the ORDE.

9.4.1.3 Transient photoelectrochemical behaviour of colloidal CdS
The experiments described in this section are performed by recording light-on transient photocurrents from aqueous dispersions of 2–12 nm radii CdS particles (prepared as above) at a stationary optical rotating disc electrode. However, to be able to interpret the results from these experiments, it was first necessary to model the time-dependent behaviour of the mass transport limited photocurrent at the ORDE.

9.4.1.3.1 Theory of the transient photocurrent behaviour at the ORDE. The stationary optical disc electrode is assumed to be uniformly illuminated by parallel light which is switched on at time $t = 0$, and which produces a measurable concentration of photogenerated electrons on the particles denoted by c_e^*. The differential equation for the generation and transport of these electrons to the electrode surface, with concurrent homogeneous back reaction is set up with the following assumptions.

(1) Electron transfer kinetics are extremely rapid so that the photocurrent is under either mass transport or photochemical kinetic control.
(2) The light makes only a small perturbation to the population of particles in the ground state, allowing us to assume that the "dark" current does not change significantly upon illumination. It also means that the colloidal dispersion does not bleach and that, throughout the experiment, the light intensity away from the electrode surface conforms to the Beer–Lambert law.
(3) All photogenerated electrons that are kinetically stable during their journey across the diffusion layer undergo electron transfer at the electrode surface, effectively regenerating particles in their ground state.
(4) The homogeneous back reaction of the photogenerated electrons is pseudo-first order with respect to c_e^* with a rate coefficient k_0.

Assumption number 4 can be justified as follows. Electrons that survive to reach the electrode do so because their corresponding holes are no longer available for the recombination reaction. While the hole reaction responsible cannot be identified for certain, in light of the information presented in Sections 9.4.1.2 and 9.2.1, it is reasonable to assume that it involves rapid hole capture (on the 0.2–2 ns timescale) by either sulphur vacancies or surface S^{2-} ions. At higher ODRE rotation speeds, the photogenerated electrons then survive in the particles for long enough to cross the electrode diffusion layer as recombination between the electrons (which themselves reside in electron traps, having been captured within 100 fs of their photogeneration) and trapped holes is kinetically slow.

As it takes a 5 nm particle typically about 1 second to cross the diffusion layer of an electrode rotating at 16 revolutions s^{-1}, the characteristic time for the reaction between photogenerated electrons and trapped holes must be more than 1 second. Valence band hole capture by sulphur vacancies/surface S^{2-} ions occurs on the nanosecond timescale, i.e., within 0.2–2 nanoseconds of a photogenerated electron recombining with a hole residing in an occupied hole trap, another hole is trapped by the same surface state. Thus, trapped holes are regenerated on the ns timescale while the reaction of photogenerated electrons with those holes occurs on the higher millisecond/second timescale. Photogenerated electrons therefore see an effectively constant concentration of holes trapped in surface states, allowing us to say that the recombination/back reaction of electrons with trapped holes is a pseudo-first order process that is first order with respect to c_e^*, i.e.:

$$k_e[h_T]c_e^* = k_0 c_e^*, \tag{9.83}$$

where k_e is the second order rate coefficient for the reaction between photogenerated electrons and trapped holes (at sulphur vacancies or surface S^{2-} ions) of concentration $[h_T]$.

Using assumptions 1–4 and the procedures outlined in Section 9.4.1, the time-dependent differential equation for semiconductor particles at the ORDE is then given by:

$$\frac{\partial c_e^*}{\partial t} = D_p\left(\frac{\partial^2 c_e^*}{\partial x^2}\right)_{t=t} - k_o c_e^* + \phi I_o \epsilon_p c_p \exp(-\epsilon_p c_p x). \tag{9.84}$$

Using the dimensionless variables

$$\tau = k_0 t, \tag{9.85}$$

$$X_k = \sqrt{\frac{D_p}{k_o}}, \tag{9.86}$$

$$\chi = x/X_k, \tag{9.87}$$

$$X_\epsilon = (\epsilon_p c_p)^{-1}, \tag{9.88}$$

$$\gamma = X_k/X_\epsilon, \tag{9.89}$$

$$U_\tau = \frac{k_o c_e^* X_\epsilon}{\phi I_o}, \tag{9.90}$$

the equation can be normalised to give:

$$\frac{\partial U_\tau}{\partial \tau} = \frac{\partial^2 U_\tau}{\partial \chi^2} - U_\tau + \exp(-\gamma\chi). \tag{9.91}$$

Since all electrons arriving at the electrode undergo electron transfer, the boundary condition at the surface at all t is:

$$c_e^* = 0 \quad \text{at } x = 0,$$

and since the light only penetrates a finite distance into solution;

$$c_e^* \to 0 \quad \text{as } x \to \infty \text{ for all } t.$$

Also, since at time $t = 0$ no electrons have been photogenerated:

$$c_e^* = 0 \quad \text{at } t = 0 \text{ for all } x.$$

The boundary conditions transform to give:

$$U_\tau = 0 \quad \text{at } \chi = 0 \text{ for all } \tau, \tag{9.92}$$

$$U_\tau \to 0 \quad \text{as } \chi \to \infty \text{ for all } \tau, \tag{9.93}$$

$$U_\tau = 0 \quad \text{at } \tau = 0 \text{ for all } \chi. \tag{9.94}$$

We wish to know the flux of electrons reaching the electrode at time t, that is

$$j_{e,t} = D_p\left(\frac{\partial c_e^*}{\partial x}\right)_{x=0} = \gamma\phi I_0\left(\frac{\partial U_\tau}{\partial \chi}\right)_{\chi=0}. \tag{9.95}$$

Laplace transformation of equation (9.33) gives:

$$\frac{\partial^2 \bar{U}_\tau}{\partial \chi^2} - (s+1)\bar{U}_\tau + \frac{1}{s}\exp(-\gamma\chi) = 0, \tag{9.96}$$

where the Laplace variable, s, is assumed to be real. The solution to this equation is:

$$\bar{U}_\tau = -\frac{(e^{-\chi\sqrt{s+1}} - e^{-\gamma\chi})}{s(s+1-\gamma^2)}. \tag{9.97}$$

Therefore,

$$\left(\frac{\partial \bar{U}_\tau}{\partial \chi}\right)_{\chi=0} = \frac{1}{s(\sqrt{s+1}+\gamma)}. \tag{9.98}$$

Inverse Laplace transformation of equation (9.98) gives the complete solution [185]:

$$\left(\frac{\partial U_\tau}{\partial \chi}\right)_{\chi=0} = \frac{1}{\gamma^2-1}(\gamma + \mathrm{erf}(-\sqrt{\tau}) - \gamma\, \mathrm{e}^{(\gamma^2-1)\tau}\, \mathrm{erfc}(\gamma\sqrt{\tau})). \tag{9.99}$$

This is a somewhat opaque and over-complicated result. Thus, to facilitate the extraction of useful kinetic information from experimental data, we return to equation (9.98) in order to identify simpler, asymptotic solutions. These are illustrated in the case diagram of Fig. 9.11. At the bottom of the diagram, where s is small and τ is therefore large, we obtain the asymptotic steady state solution:

$$\left(\frac{\partial \bar{U}_\tau}{\partial \chi}\right)_{\chi=0} = \frac{1}{s(1+\gamma)},$$

which, upon inverse transformation, gives:

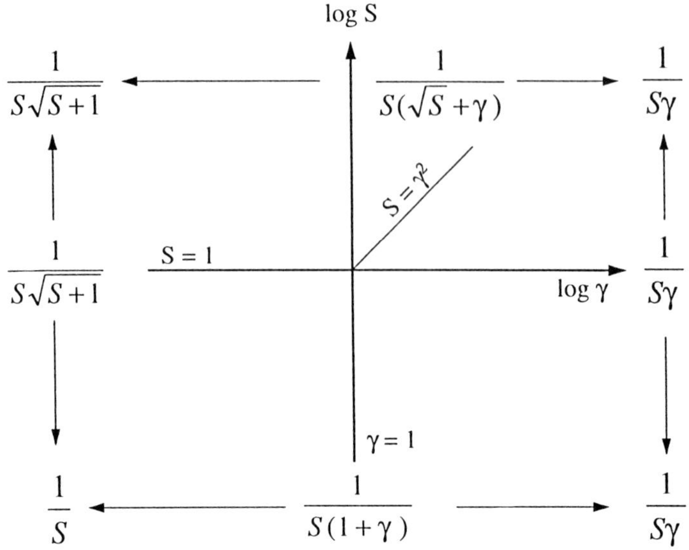

Fig. 9.11. Case diagram for the approximate solutions of equation (9.99). Different approximations hold for different values of the Laplace variable, s, and γ where $\gamma(=X_k/X_\epsilon)$ compares the distance a photogenerated electron diffuses through solution on a particle before participating in a loss reaction with the distance over which the light is absorbed.

$$\left(\frac{\partial U_\tau}{\partial \chi}\right)_{\chi=0} = \frac{1}{1+\gamma}.$$

Therefore, the steady state flux is given by:

$$j_e = \frac{\phi I_o \gamma}{1+\gamma} \quad \text{as } t \to \infty. \tag{9.100}$$

Now, if γ is very much less than 1, i.e., $X_k \ll X_\epsilon$ then:

$$j_e = \phi I_o \gamma = \frac{\phi I_o X_k}{X_\epsilon}. \tag{9.101}$$

Under these conditions, only those electrons photogenerated with a distance X_k of the electrode may be detected at the electrode surface. Those

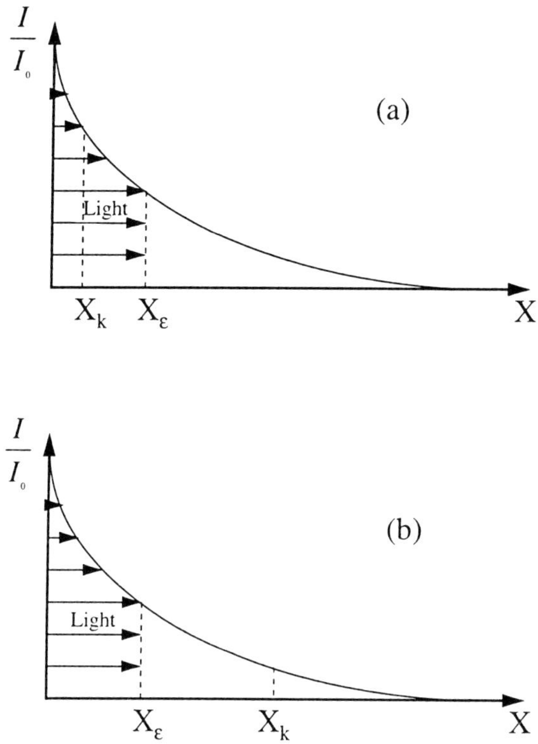

Fig. 9.12. (a) Beer–Lambert profile perpendicular to the surface of the ORDE when $X_k < X_\epsilon$; (b) Beer–Lambert profile perpendicular to the surface of the ORDE when $X_\epsilon < X_k$.

electrons generated by light absorbed at $x > X_k$ do not survive long enough to reach the electrode. This is illustrated in Fig. 9.12a. If, however, (is much greater than 1 (i.e., $X_k \gg X_\epsilon$) then

$$j_e = \phi I_o.$$

In this case, as X_k is greater than X_ϵ, all the light absorbed by the solution photogenerates electrons which survive long enough to reach the electrode before they undergo the back/loss reaction. This is illustrated in Fig. 9.12b. Thus, there is a difference between the form of the light-on transient when $\gamma < 1$, and when $\gamma > 1$. When γ is less than 1:

$$\left(\frac{\partial \bar{U}_\tau}{\partial \chi}\right)_{\chi=0} = \frac{1}{s\sqrt{s+1}} \quad \text{at small } \tau \text{, and}$$

$$\left(\frac{\partial \bar{U}_\tau}{\partial \chi}\right)_{\chi=0} = \frac{1}{s} \quad \text{at large } \tau .$$

When γ is more than 1:

$$\left(\frac{\partial \bar{U}_\tau}{\partial \chi}\right)_{\chi=0} = \frac{1}{s\sqrt{s}+\gamma} \quad \text{at small } \tau \text{, and}$$

$$\left(\frac{\partial \bar{U}_\tau}{\partial \chi}\right)_{\chi=0} = \frac{1}{\gamma s} \quad \text{at large } \tau .$$

Thus, by reverse transform, when $\gamma < 1$:

$$\left(\frac{\partial U_\tau}{\partial \chi}\right)_{\chi=0} = \operatorname{erf}\sqrt{\tau}, \tag{9.103}$$

and when $\gamma > 1$:

$$\left(\frac{\partial U_\tau}{\partial \chi}\right)_{\chi=0} = \frac{1}{\gamma}\left(1 - e^{\gamma^2\tau}\operatorname{erfc}\sqrt{\gamma^2\tau}\right). \tag{9.104}$$

Equations (9.103) and (9.104) may be obtained from equation (9.99) by making γ very much less than one, and very much more than one in turn. Combining equations (9.95), (9.101) and (9.103), and remembering that the current at time t is given by:

$$i_t = FAj_{e,t}, \tag{9.105}$$

it can be shown that, when $\gamma < 1$:

$$\frac{(i_{h\nu})_t}{(i_{h\nu})_\infty} = \operatorname{erfc}\sqrt{\tau}, \tag{9.106}$$

and, by combining equations (9.95), (9.102) and (9.104), it can be shown that, when $\gamma > 1$:

$$\frac{(i_{h\nu})_t}{(i_{h\nu})_\infty} = 1 - e^{\gamma^2\tau}\,\mathrm{erfc}\sqrt{\gamma^2\tau} = 1 - e^{\tau'}\,\mathrm{erfc}\sqrt{\tau'}, \tag{9.107}$$

where $\tau' = \gamma^2\tau$. In the next section, these asymptotic theoretical results will be used to derive kinetic parameters from the experimentally observed behaviour of the light-on transient of CdS particles at the ORDE.

9.4.1.3.2 Transient photocurrent experiments at the ORDE. Fig. 9.13 shows a typical light-on current-time transient for colloidal CdS recorded at an ORDE. It must now be determined which of the two cases described above ($\gamma < 1$ or $\gamma > 1$) this curve fits. This is most easily accomplished by plotting $(i_{h\nu})t/(i_{h\nu})_\infty$ against $\ln(\tau/\tau_{t,1/2})$, for equations (9.106), (9.107) and the experimental data, where τ is the relevant time variable (which may be τ in equation (9.106), τ' in equation (9.107) or t for the experimental data) and $\tau_{1/2}$ is the value of τ for which $(i_{h\nu})_t/(i_{h\nu})_\infty$, $= 0.5$. As $(i_{h\nu})_t/(i_{h\nu})_\infty$ is plotted as a function of normalized time in all three cases, the experi-

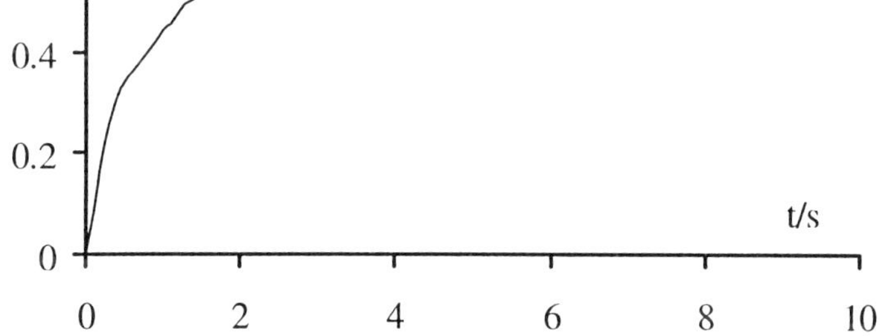

Fig. 9.13. Typical light-on current time transient for colloidal CdS ([CdS] = 5 mol m^{-3}) at a stationary optical disc electrode.

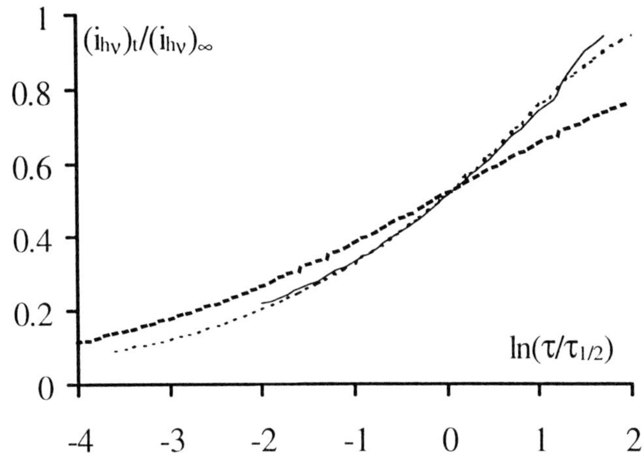

Fig. 9.14. Matching the experimental data of Fig. 9.13 to theoretical curves by means of $(i_{h\nu})_t/(i_{h\nu})_\infty$ versus $\ln(\tau/\tau_{1/2})$ plots. ———— represents the experimental data;------------ represents the $1 - \exp(\sqrt{\tau'})\,\mathrm{erfc}\sqrt{\tau'}$ versus $\ln(\tau'/\tau'_{1/2})$ plot; ·············· represents the $\mathrm{erf}\sqrt{\tau}$ versus $\ln(\tau/\tau_{1/2})$ plots.

mental data will overlay the theoretical case by which it is most appropriately described. A graph of a typical set of experimental data and the two theoretical cases is shown in Fig. 9.14, from which can be seen that the experimental data most closely fits the plot of $\mathrm{erf}\sqrt{\tau}$ against $\ln(\tau/\tau_{1/2})$. Thus, the light-on transient follows the equation:

$$\frac{(i_{h\nu})_t}{(i_{h\nu})_\infty} = \mathrm{erf}\sqrt{k_o t}. \tag{9.108}$$

Experimental values of $(i_{h\nu})_t/(i_{h\nu})_\infty$ are larger than predicted theoretical values at high τ, which is attributed to the deposition of particles on the electrode surface. As the electrode is stationary and there exists no convection to sweep the deposited particles away from the electrode surface, this residual current persists after illumination is stopped, distorting the form of the observed light-off current-time transient. Consequently, theoretical analysis of the transient was not attempted. The rate constant k_o may be obtained from $(i_{h\nu})_\infty$ and the rotation speed dependence of the photocurrent. From equations (9.101) and (9.76), it can be seen that:

TABLE 9.3
Experimentally observed values of k_0 and (for a range of CdS colloids ([CdS] = 5 mol m^{-3})

Sample	k_0/s^{-1} (from equation (9.109))	k_0/s^{-1} (from equation (9.108))	ϕ
A	0.15	0.10	0.011
B	0.58	0.12	0.018
C	0.02	0.05	0.005
D	0.06	0.07	0.009
E	0.16	0.13	0.014
F	0.64	0.18	0.022
G	0.06	0.03	0.007

Average $k_0 = 0.168 \pm 0.052\ s^{-1}$.
Average $\phi = 0.012 \pm 0.002$.

$$\frac{(i_{h\nu})_\infty}{(i_{h\nu})_L} = 1 + 1.555 D_p^{1/6} \nu^{-1/6} \omega^{1/2} k_0^{-1/2}. \tag{9.109}$$

Thus, k_0 may be readily determined from a plot of $(i_{h\nu})_\infty/(i_{h\nu})_L$ against $\omega^{1/2}$. k_o may also be found by the curve fitting of the experimental data, as plotted in Fig. 9.14, to equation (9.108). Table 9.3 shows some typical experimental values of k_o, obtained by both methods. Fair agreement is observed between the two values of k_o for any one sample, with the average rate coefficient being $0.168 \pm 0.052\ s^{-1}$. This corresponds to a photogenerated electron lifetime of 14.2 ± 3.6 s. Albery *et al.* [84] report photogenerated electron lifetimes of more than 10 s when CdS colloids are illuminated in the presence of a hole scavenger such as cysteine. In this case, holes in the valence band are captured by cysteine and the photoinduced electrons in the conduction band or shallow surface traps can only relax through slow electron-transfer processes. Such a situation is directly analogous to that which obtains at colloidal CdS particles that are characterised by a large number of surface states/defect sites wherein the surface states mimic the role of the cysteine molecule. Thus, the good agreement between the photogenerated electron lifetime reported by Albery *et al.* and that derived from transient photocurrent measurements on CdS using the ORDE is not unexpected. Further, these lifetimes are in keeping with the prediction that photogenerated electrons in Q-state semiconductor particles are long lived and have a high reducing power [77].

TABLE 9.4
Beer–Lambert lengths for a typical colloidal dispersion of CdS at a range of illumination wavelengths ([CdS] = 5 mol m^{-3})

Wavelength of light, λ/nm	Beer–Lambert length, X_ϵ/m
540	3.2×10^{-2}
520	1.3×10^{-2}
500	2.8×10^{-3}
450	1.2×10^{-3}
400	8.7×10^{-4}
350	6.3×10^{-4}
300	4.2×10^{-4}

Equation (9.86) allows calculation of X_k, the distance diffused by a particle over which photogenerated electrons are kinetically stable. Using a value of D_p of 3.2×10^{-11} m^2 s^{-1}, this is found to be approximately 1.38×10^{-15} m. Table 9.4 shows the Beer–Lambert lengths, X_ϵ, for a typical 5 mol m^{-3} suspension of colloidal CdS at a range of illumination wavelengths. It can be seen that $X_\epsilon \gg X_k$ for all ultra-band gap energy photons ($\lambda < 533$ nm) in the white light output of the projector lamp used to illuminate the particles. Thus, γ is less than one, supporting the conclusion that equation (9.106) provides the best fit for the experimental data (*vide supra*).

Using the experimental values of k_o in Table 9.3, and equation (9.76), we may now calculate values of ϕ, the quantum efficiency for the photogeneration within the particles of electrons detectable by the ORDE. Calculated values of ϕ are shown in Table 9.3, the average value being 0.012 ± 0.002. ϕ is given by

$$\phi = \frac{\left\{\sum_i k_i[T_h]_i[h_{VB}^+]\right\}}{\left\{\sum_i k_i[T_h]_i[h_{VB}^+]\right\} + \left\{k_{\mathrm{R1}}[e_{CB}^-][h_{VB}^+] + k_{\mathrm{R2}}[e_T^-][h_{VB}^+]\right\}},$$

where k_i is the second order rate coefficient for the trapping of valence band holes in hole trap type i of concentration $[T_h]_i$, k_{R1} is the second order rate coefficient for the direct recombination of valence band holes with conduction band electrons and k_{R2} is the second order rate coefficient for the indirect recombination of valence band holes with photogenerated electrons residing in shallow traps at the particle surface. Thus, it can be seen that the dominant processes in this expression are the direct and indirect photogenerated electron-valence band hole recombinations, and not valence band hole loss to hole traps.

In conclusion then, we can say that the cascade model proposed (see Fig. 9.5) is supported experimentally. However, the model must be modified to take into account (i) the effect of inter-band surface states/defect sites on the form of the current voltage curve and (ii) the effect of mutual double layer repulsion (the Albery theory [166]) and/or particle sample polydispersity (the Heyrovsky and Jirkovsky theory [170]) on the observed value of β. A value of k_o of $0.168 \pm 0.052\, s^{-1}$ supports the assumption that the process being observed is not direct photogenerated electron-valence band hole recombination (which other workers have shown takes place on the sub-nanosecond timescale [107]), but the loss of electrons left on the particles after some valence band hole filling reaction has occurred. This reaction is the filling of the holes by electrons originating from inter-band surface states and defect sites. The trapped holes then recombine with the photogenerated electrons in a pseudo-first order process with a rate constant of $0.168 \pm 0.052\, s^{-1}$, in good agreement with a value of >10 s previously report by Albery *et al.* [84].

A similar model has been advanced by Brown and Rajapakse [186], again based on the work done by Albery *et al.* [84]. They also suggest that electron-hole recombination can be effectively slowed by the presence of interband hole traps, some of which can be attributed to sulphur vacancies, in agreement with conclusions made by Ramsden and Grätzel [169]. Work must be done to refine this model and to identify the nature of the inter-band hole traps. Nonetheless, the pseudo-first order rate constant for electron loss to them is $0.168 \pm 0.052\, s^{-1}$ a value which has far-reaching implications for the design of photocatalytic systems employing colloidal CdS in photoreduction reactions.

It is clear that the observed photocurrent transient is the composite of two effects; diffusion of the particles to the electrode and the pseudo-first order decay of the photo-generated electrons. In a subsequent communication, this technique will be taken one step further and used to obtain rate constants for dynamic and static electron transfer from the particles to a redox active species deliberately introduced into the system.

9.5 CONCLUDING REMARKS

Since interest was first prompted by Fujishima and Honda's 1972 discovery that water could be split by an illuminated TiO_2 electrode [187], applications of semiconductor photocatalysis have proliferated finding greatest utility in the fields of pollution remediation and non-linear optics, as described in Section 9.1. The summary of the kinetic models for semiconductor photocatalysis presented in Section 9.3 indicates not only how much is already known about these systems, but also how much still remains to be understood. Whilst the primary processes are well characterised (see Section 9.2) and reaction products are relatively straightforward to identify (see Sections 9.3 and 9.4), efforts still need to be made to convincingly kinetically model the reaction sequences that bridge these two extremes. Information obtained from studies to date indicate that combinations of photochemical and surface electrochemical techniques offer the greatest opportunity for fully understanding these systems.

ACKNOWLEDGEMENTS

The author wishes to thank Gwenaelle Le Gurun for useful discussions and aid with drawing the diagrams, Diane Tempest and Betty Valentine for help with producing the first draft, Peter K.J. Robertson for helpful discussions and proof-reading the manuscript, and Geoff Kelsall for finishing what WJA started.

DEDICATION

I first met John Albery during the early spring of 1983 when he (very) informally interviewed me for a Ph.D. position within his research group at Imperial College. At the time, John's group was some 26 strong and active at the leading edges of a number of electrochemical research fronts. It was only when I started working for John in the autumn of 1983 that I realised he was also on the board of several London theatres, a playwright, the chair of the SERC Chemistry Committee and a bon viveur of quite staggering proportions. During my time with him, he sloughed off some of these responsibilities (except the last) only to take on others, often in greater number such was the demands for his talents. He also entered new areas of research, becoming involved in photocatalysis, electrochemical spectroscopy and fuel cells, to name but a few.

I left John's group in 1986. In 1988, John moved back to Oxford to become Master of University and, except for meeting him at conferences, I lost touch with him. I have fond memories, both scientific and personal, of my times with John (I can still remember the dryness of mouth and weakness of knee that would accompany the delivery of the results of Johns latest weekend with Stegun and Abramowitz—often left mysteriously on your desk on a Monday morning in screed form, and written in a hand that most general practitioners would have been proud of). It is as a mark of appreciation for those times that I take great pleasure in dedicating this article to him on the event of his 60th birthday.

APPENDICES

A.1 The optical rotating disc electrode (ORDE)

In this appendix we review the general theoretical aspects of the semi-transparent Optical Rotating Disc Electrode. The electrode is assumed to be uniformly illuminated by parallel light passing through the electrode and into the solution where it generates, from species A, a photoproduct B, with a quantum efficiency ϕ. The reaction is:

$$A \xrightarrow{h\nu,\phi} B, \tag{9.A.1}$$

$$B \xrightarrow{k_{-2}} A. \tag{9.A.2}$$

The differential equation describing the convective diffusion of B is set up with the following assumptions:

(1) The light makes only a small perturbation to the concentration of A, thus not altering the dark current value significantly. This also means that the solution does not bleach and that the light has a Beer–Lambert profile given by

$$I = I_o \exp(-\epsilon_\lambda[A]x), \tag{9.A.3}$$

where I_o is the flux of light at the electrode surface in Einsteins $m^{-2}\ s^{-1}$, and I is the flux of light at a distance x from the electrode.

(2) The back reaction is (pseudo-) first-order, with rate constant k_{-2}.

(3) All B arriving at the electrode is converted into A.

With these assumptions, the basic convective diffusion equation for B is

$$D_B \frac{\partial^2[B]}{\partial x^2} + v_x \frac{\partial[B]}{\partial x} + \phi I_o \epsilon_\lambda[A] \exp(-\epsilon_\lambda[A]x) - k_{-2}[\mathrm{B}] = 0, \qquad (9.A.4)$$

where D_B is the diffusion coefficient of B in $m^2\ s^{-1}$ and v_x is the velocity of flow towards the electrode ($m\ s^{-1}$). This system has four characteristic lengths, all in metres:

the thickness of the hydrodynamic layer $\quad X_H = \left(\frac{\nu}{2\pi\omega}\right)^{1/2},$

the thickness of the diffusion layer $\quad X_D = 0.643\omega^{-1/2}\nu^{1/6}D_B^{1/3},$

the thickness of the reaction layer $\quad X_k = \left(\frac{D_B}{k_{-2}}\right)^{1/2},$

the thickness of the absorbance layer $\quad X_\epsilon = \frac{1}{\epsilon_\lambda[A]},$

where ω is the rotation speed of the electrode in s^{-1}, ν is the kinematic viscosity of the solvent in $m^2\ s^{-1}$. Recasting equation (9.A.4) in terms of the dimensionless variables

$H_0 = X_D/X_H,$

$\xi = X_D/X_\epsilon,$

$\kappa = X_D/X_k,$

$m' = [B]D_B/\phi\epsilon_\lambda X_D I_0$, the normalized concentration,

$\chi = x/X_D$, the normalized distance normal to the electrode gives:

$$\frac{\partial^2 m'}{\partial \chi^2} + \left(\frac{v_x X_D}{D_B}\right)\frac{\partial m'}{\partial \chi} + \exp(-\xi\chi) - \kappa^2 m' = 0. \tag{9.A.5}$$

Since the electrode destroys all B reaching it, the boundary condition at the surface of the ORDE is

$$m' = 0 \quad \text{at } \chi = 0,$$

and, since the light only penetrates a finite distance into the solution,

$$\text{as } \chi \to \infty,\ m' \to 0.$$

We wish to know the flux of B reaching the electrode given by

$$j'_B = D_B\left(\frac{\partial[B]}{\partial x}\right)_{x=0} = \phi\xi I_0\left(\frac{\partial m'}{\partial \chi}\right)_{\chi=0}. \tag{9.A.6}$$

In solving equation (9.A.5), it is useful to define a photoelectrochemical collection efficiency, $N_{h\nu}$, to describe the recovery of photogenerated species at the electrode:

$$N_{h\nu} = \frac{j'_B}{I_0} = \phi\xi\left(\frac{\partial m'}{\partial \chi}\right)_{\chi=0}. \tag{9.A.7}$$

Fig. 9.6 shows a case diagram illustrating the approximate solutions for $N_{h\nu}$ derived by Albery *et al.* [163] from equations (9.A.5) and (9.A.7), assuming that ϕ has a value of one. We now discuss each of these solutions. In case A, the collection efficiency is unity, as the light is absorbed close to the electrode ($\xi > 1$) and all the photogenerated material reaches the electrode. In case B, the kinetics of the decomposition are rapid enough to destroy the material in its passage across the diffusion layer, and material that reaches the electrode has to be generated close to the electrode. In cases C, D and E the kinetics are slow enough to allow material generated outside the diffusion layer to reach the electrode. In case C, the light does not penetrate far outside the diffusion layer, and roughly equal amounts of the material reaching the electrode come from inside and outside the diffusion layer. In case D, the light penetrates outside the hydrodynamic layer and the collection efficiency becomes constant, most of the material generated in the solution flowing towards the electrode being lost by radial convection; only a small constant fraction reaches the electrode. In case E, the kinetics of the back reaction are rapid enough for the concentration of photogenerated species in the stream to be determined by the back reaction rather than by the convective loss.

The variation of the current with rotation speed for each case is as follows. For cases A and B, there is no rotation speed dependence of the photocurrent, since all the material reaching the electrode originates within the diffusion layer. In case D, increasing the rotation speed increases both convective dilution and transport. These two effects balance, and again there is no net variation of photocurrent with rotation speed. In case C, the photocurrent decreases as rotation speed increases, as it varies with $\omega^{-1/2}$. This curious effect arises because as the rotation speed increases the diffusion layer becomes narrower and there is less space for species to be generated without being swept away by radial convection. For case E, the current increases with rotation speed. Here, the more rapid transport carries the species to the electrode before it decomposes.

REFERENCES

[1] J. Cunningham and P. Sedlk, J. Photochem. Photophys. A: Chem. 77 (1994) 255.

[2] J.Cunningham and G. Al-Sayyed, J. Chem. Soc. Faraday Trans. 1 86 (1990) 3935.

[3] A. Mills, S. Morris and R. Davies, J. Photochem, Photobiol. A: Chem. 70 (1993) 183.

[4] A. Mills and S. Morris, J. Photochem, Photobiol. A: Chem. 71 (1993) 75.

[5] A. Mills and S. Morris, J. Photochem, Photobiol. A: Chem. 71 (1993) 285.

[6] U. Stafford, K.A. Gray and P.V.Kamat, J. Phys. Chem. 98 (1994) 6343.

[7] J.-C. D'Oliveira, C. Guillard, C. Maillard and P. Pichat, J. Environ. Health A28 (1993) 941.

[8] E. Pelizzetti, V. Carlin, C. Minero and M. Grätzel, N. J. Chem. 15 (1991) 351.

[9] E. Pelizzetti, M. Borgarello, C. Minero, E. Pramaura, E. Borgarello and N. Serpone, Chemosphere 17 (1988) 499.

[10] P.C. Zhang, R.J. Scrudato, J.J. Pagano and R.N. Roberts, Chemosphere 26 (1993) 1213.

[11] A. Vidal, J. Herrero, M. Romero, B. Sanchez and M. Sanchez, J. Photochem, Photobiol. A: Chem. 79 (1994) 213.

[12] S. Das, M. Muneer, K.R. Gopidas, J. Photochem, Photobiol. A: Chem. 77 (1994) 83.
[13] R.W. Matthews, J. Chem. Soc., Faraday Trans. 1, 80 (1984) 457.
[14] R.W. Matthews, Water Res. 20 (1986) 569.
[15] C. Richard and P. Boule, N. J. Chem. 18 (1994) 547.
[16] R.W. Matthews and S.R. McEvoy, Solar Energy 49 (1992) 507.
[17] R.W. Matthews and R. McEvoy, J. Photochem. Photobiol. A: Chem. 64 (1992) 231.
[18] R. Terzian, N. Serpone, C. Minero, E. Pelizzetti and H. Hidaka, J. Photochem. Photobiol. A: Chem. 55 (1990) 243.
[19] R. Terzian, N. Serpone, C. Minero, and E. Pelizzetti, J. Catal. 128 (1991) 352.
[20] L. Amalric, C. Guillard, N. Serpone and P. Pichat, J. Environ. Sci. Health, A 28 (1993) 1393.
[21] S.N. Frank and A.J. Bard, J. Phys. Chem. 81 (1977) 1484.
[22] S.N. Frank and A.J. Bard, J. Am. Chem. Soc. 99 (1977) 303.
[23] Y. Hori, A. Bandoh and A. Nakatsu, J. Electrochem. Soc. 137 (1990) 1155.
[24] M.M. Kosanic and A.S. Topalov, Int. J. Hydrogen Energy 45 (1990) 319.
[25] C.H. Pollema, J.L. Hendrix, E.B. Milosavljevic, L. Solujic and J.H. Nelson, J. Photochem. Photobiol. A: Chem. 66 (1992) 235
[26] B.R. Eggins, P.K.J. Robertson, J.H. Stewart and E. Woods, J. Chem. Soc., Chem.-Commun. 1993, 349.
[27] B.R. Eggins, J.T.S. Irvine and J. Grimshaw, Solar Energy 45 (1990) 27.
[28] B.R. Eggins, J.T.S. Irvine, E.P. Murphy and J. Grimshaw, J. Chem. Soc., Chem.-Commun.1988, 1123.
[29] G.K.-C. Low, S.R. McEvoy and R.W. Matthews, Environ. Sci. Technol. 25 (1991) 460.
[30] P.K.J. Robertson, L.A. Lawton, B. Münch and J. Rouzade, J. Chem. Soc., Chem.-Commun.1997, 393.
[31] J.C. Ireland, P. Klostermann, E.W. Rice and R.M. Clark, Appl. Environ. Microbiol. 59 (1993) 1668.
[32] P. Zhang, R.J. Scrudato and G. Germano, Chemosphere 28 (1994) 607.
[33] R. Cai, K. Hashimoto, K. Itoh, Y. Kubota and A. Fujishima, Bull. Chem. Soc. Jpn. 64 (1991) 1268.
[34] R. Cai, K. Hashimoto, Y. Kubota and A. Fujishima, Chem. Lett. 1992, 427.
[35] R.X. Cai, Y. Kubota, T. Shuin, H. Sakai, K. Hashimoto and A. Fujishima, Cancer Res. 52 (1992) 2346.
[36] A. Fujishima, R.X. Cai, J. Otsuki, K. Hashimoto, K. Itoh, T. Yamashita and Y. Kubota, Electrochimica Acta 38 (1993) 153.H.
[37] H.W. Read, X. Fu, L.A. Clark, M.A. Anderson and T. Jarosch, J. Soil. Contam. 5 (1996) 187.
[38] K.Suzuki in Photocatalytic Purification and Treatment of Water and Air, D.F. Ollis and H. Al-Ekabi (eds.), Elsevier, Amsterdam, 1993.
[39] N.B. Jackson, C.M. Wang, Z. Luo, J. Schwitzgebel, J.G. Ekerdt, J.R. Brock and A. Heller, J. Electrochem. Soc. 138 (1991) 3660.
[40] H. Gerisher and A. Heller, J. Electrochem. Soc. 139 (1992) 113.

[41] A. Heller and J.R. Brock, in: Aquatic and Surface Photochemistry, eds G.R. Helz, R.G. Zepp and D.G. Crosby (CRC Press Inc., Boca Raton, Florida, 1993).
[42] M. Nair, Z.H. Luo and A. Heller, Ind. Eng. Chem. Res. 32 (1993) 2318.
[43] R.J. Berry and M.R. Mueller, Microchem. J. 50 (1994) 28.
[44] D.F.Ollis and H.Al-Ekabi, in: Photocatalytic Purification and Treatment of Water and Air, eds D.F.Ollis and H.Al-Ekabi (Elsevier, Amsterdam, 1993).
[45] M.R. Hoffman, S.T. Martin, W. Choi and D. Bahnemann, Chem. Rev. 95 (1995) 69.
[46] A. Mills, R.H. Davies and D. Worsley, Chem. Soc. Rev. 22 (1993) 417.
[47] C. Boxall, Chem. Soc. Rev. 23 (1994) 137.
[48] M. Grätzel, in: Electrochemistry in Colloids and Dispersions, eds R.A. Mackay and J. Texter (VCH Publishers Inc., New York, 1992).
[49] N.Serpone, D.Lawless, R.Terzian and D.Meisel, in: Electrochemistry in Colloids and Dispersions, eds R.A. Mackay and J. Texter (VCH Publishers Inc., New York, 1992).
[50] A. Hagfeldt and M. Grätzel, Chem. Rev. 95 (1995) 49.
[51] P.V. Kamat, Prog. Reaction Kinetics 19 (1994) 277.
[52] M.M. Halmann, Photodegradation of Water Pollutants (CRC Press Inc., Boca Raton, Florida, 1996).
[53] N. Nageswara Rao and P. Natarajan, Curr. Sci. 66 (1994) 742.
[54] O. Legrini, E. Oliveros and A.M. Braun, Chem. Rev. 93 (1993) 671.
[55] U. Stafford, K.A. Gray and P.V. Kamat, Heterogen. Chem. Rev. 3 (1996) 77.
[56] M.A. Fox and M.T. Dulay, Chem. Rev. 93 (1993) 341.
[57] A.L. Linsebigler, G. Lu and J.T. Yates Jr., Chem. Rev. 95 (1995) 735.
[58] Y. Wang, Acc. Chem. Res. 24 (1991) 133.
[59] A. Henglein, Ber. Bunsenges. Phys. Chem. 99 (1995) 903.
[60] A. Henglein, Chem. Rev. 89 (1989) 1861.
[61] P. Kamat, Chem. Rev. 93 (1993) 267.
[62] D.W. Bahnemann, Isr. J. Chem. 33 (1993) 115.
[63] P. Pichat, Catal. Today 19 (1994) 313.
[64] E. Pelizzetti, N. Serpone (eds.), Homogeneous and Heterogeneous Photocatalysis (D. Reidel Publishing Co., Dordrecht, 1986).
[65] D. Bahnemann, J. Cunningham, M.A. Fox, E. Pelizzetti and N. Serpone, in: Aquatic and Surface Photochemistry, eds G.R. Helz, R.G. Zepp and D.G. Crosby (CRC Press Inc., Boca Raton, Florida, 1993).
[66] N.L. Lewis, Chem. Rev. 93 (1993) 2693.
[67] M.Grätzel, in: Heterogeneous Photochemical Electron Transfer (CRC Press Inc., Boca Raton, Florida, 1989).
[68] N. Serpone and E. Pelizzetti (eds.) Photocatalysis: Fundamentals and Applications (John Wiley & Sons, New York, 1989).
[69] H. Weller, Adv. Mater. 5 (1993) 88.
[70] J.H. Fendler, Membrane-Mimetic Approach to Advanced Materials, Advances in Polymer Science Series Vol. 113 (Springer-Verlag, Berlin 1994).
[71] J.H. Fendler and F.C. Meldrum, Adv. Mater. 7 (1995) 607.
[72] Y. Tian, T. Newton, N.A. Kotov, D.M. Guldi and J.H. Fendler, J. Phys. Chem. 100 (1996) 8927.

[73] K. Rajeshwar, Adv. Mater. 4 (1992) 23.
[74] R.C. DeMattei and R.S. Feigelson, in: Electrochemistry of Semiconductors and Electronics, eds J. McHardy and F. Ludwig (Noyes Publications, Park Ridge, New Jersey, 1992).
[75] G. Hodes, Isr. J. Chem. 33 (1993) 95
[76] M.L. Steigerwald and L.E. Brus, Acc. Chem. Res. 23 (1990) 183.
[77] Y. Wang and N. Herron, J. Phys. Chem. 95 (1991) 525.
[78] M.K. Nazeeruddin, A. Kay, I. Rodicio, R. Humphrey-Baker, E. Müller, P. Liska, N. Vlachopoulos and M.Grätzel, J. Am. Chem. Soc. 115 (1993) 6382.
[79] A. Ulman, Adv. Mater. 5 (1993) 55
[80] C. Kormann, D.W. Bahnemann and M.R. Hoffman, J. Phys. Chem. 92 (1988) 5196.
[81] E.F. Hilenski, P.A. Lucas and Y. Wang, J. Phys. Chem. 89 (1988) 3435.
[82] Y. Wang, N. Herron, W. Mahler and A. Suma, J. Opt. Soc. Am. B6 (1989) 808.
[83] L.-T. Cheng, N. Herron and Y. Wang, J. Appl. Phys. 66 (1989) 3417.
[84] W.J. Albery, G.T. Brown, J.R. Darwent and E. Saievar-Iranizad, J. Chem. Soc. Faraday Trans. 1 81 (1985) 1999.
[85] A. Henglein, A. Kumar, E. Janata and A. Heller, Chem. Phys. Lett. 132 (1986) 133.
[86] M. Haase, A. Heller and A. Henglein, J. Phys. Chem. 92 (1988) 4706.
[87] C. Liu and A.J. Bard, J. Phys. Chem. 93 (1989) 3232.
[88] C. Liu and A.J. Bard, J. Phys. Chem. 93 (1989) 7749.
[89] H. Weller, Agnew. Chem. 105 (1993) 43.
[90] H. Weller, Agnew. Chem. Int. Ed. Engl. 32 (1993) 41.
[91] A. Henglein, Pure Appl. Chem. 56 (1984) 1215.
[92] L.E. Brus, IEEE J.Quantum Electron. 22 (1986) 1909.
[93] L.E. Brus, J. Phys. Chem. 86 (1986) 2555.
[94] A. Henglein, Top. Curr. Chem. 143 (1988) 115.
[95] S.R. Morrison, Electrochemistry at Semiconductor and Oxidised Metal Electrodes (Plenum Press, New York, 1980).
[96] D.E. Skinner, D.P. Colombo, Jr., J.J. Cavaleri and R.M. Bowman, J. Phys. Chem. 99 (1995) 7853.
[97] J.Z. Zhang, R.H. O'Neil, T.W. Roberti, J.L. McGowen and J.E. Evans, Chem. Phys. Lett. 218 (1994) 479.
[98] J.Z. Zhang, R.H. O'Neil and T.W. Roberti, Appl. Phys. Lett. 64 (1994) 1989.
[99] J.E. Evans, K.W. Springer and J.Z. Zhang, J. Chem. Phys. 101 (1994) 6222.
[100] S.T. Martin, H. Herrmann, W. Choi and M.R. Hoffmann, J. Chem. Soc., Faraday Trans. 90 (1994) 3315.
[101] S.T. Martin, H. Herrmann and M.R. Hoffmann, J. Chem. Soc., Faraday Trans. 90 (1994) 3323.
[102] D.P. Colombo, Jr., K.A. Roussel, J. Saeh., D.E. Skinner, J.J. Cavaleri and R.M. Bowman, Chem. Phys. Lett. 232 (1995) 207.
[103] G. Rothenberger, J. Moser, M. Grätzel, N. Serpone and D.K. Sharma, J. Am. Chem.Soc. 107 (1985) 8054.
[104] R.F. Howe and M. Grätzel, J. Phys. Chem. 89 (1985) 4495.
[105] D.P. Colombo, Jr. and R.M. Bowman, J. Phys. Chem. 99 (1995) 11752.

[106] N. Serpone, D. Lawless, R. Khairutdinov and E. Pelizzetti, J. Phys. Chem. 99 (1995) 16655.
[107] J.Z. Zhang, R.H. O'Neil and T.W. Roberti, J. Phys. Chem. 98 (1994) 3859.
[108] Z. Alfassi, D. Bahnemann and A. Henglein, J. Phys. Chem. 86 (1982) 4656.
[109] S. Baral, A. Fojtik, H. Weller and A. Henglein, J. Am. Chem. Soc. 108 (1986) 375.
[110] M.G. Bawendi, W.L. Wilson, L. Rothberg, P.J. Carroll, M.L. Steigerwald and L.E. Brus, Phys. Rev. Lett. 65 (1990) 16.
[111] G.P. Lepore, C.H. Langford, J. Víchová and A.Vlcek, Jr., J. Photochem. Photobiol. A: Chem. 75 (1993) 67.
[112] C.S. Turchi and D.F. Ollis, J. Catal. 122 (1990) 178.
[113] D.A. Bahnemann, A. Henglein, J. Lilie and L. Spanhel, J. Phys. Chem. 88 (1984) 709.
[114] D.Duonghong, J. Ramsden and M. Grätzel, J. Am. Chem. Soc. 104 (1982) 2977.
[115] P.V. Kamat, K.R. Gopidas and N.M. Dimitrijevic, Mol. Cryst. Liq. Cryst. 183 (1990) 439.
[116] P.V. Kamat, T.W. Ebbesen, N.M. Dimitrijevic and A.J. Nozik, Chem. Phys. Lett. 157 (1989) 384.
[117] C. Boxall and G.H. Kelsall, J. Chem. Soc., Faraday Trans. 87 (1991) 3537.
[118] Y. Nosaka and M.A. Fox, J. Phys. Chem. 90 (1986) 6521.
[119] C. Boxall and G.H. Kelsall, J. Chem. Soc., Faraday Trans. 87 (1991) 3547.
[120] J. Moser and M. Grätzel, J. Am. Chem. Soc. 106 (1984) 6557.
[121] R.L. Crackel and W.S. Sturve, Chem. Phys. Lett. 120 91985) 473.
[122] Y. Liang, P.F. Moy, J.A. Poole and A.M. Ponte Goncalves, J. Phys. Chem. 88 (1984) 2451.
[123] F.R.F. Fan and A.J. Bard, J. Am. Chem. Soc. 101 (1979) 6139.
[124] E. Vrachnou, N. Vlachopoulos and M. Grätzel, in: Photocatalytic Production of Energy-Rich Compounds: Energy from Biomass 2, eds G. Grassi and D.O. Hall (Elsevier, London, 1988).
[125] N. Liver and A. Nitzan, J. Phys. Chem. 96 (1992) 3366.
[126] P. Salvador, New J. Chem. 12 (1988) 35
[127] W. Lorenz, C. Aegester and M. Handschuh, J. Electroanal. Chem. 200 (1986) 83.
[128] D.A. Bahnemann, A. Henglein and L. Spanhel, Faraday Discuss. Chem. Soc. 78 (1984) 151.
[129] M. Grätzel and A.J. Frank, J. Phys. Chem. 86 (1982) 2964.
[130] Y.V. Pleskov and Y.Y. Gurevich, Semiconductor Photoelectrochemistry, Consultants Bureau, London, 1986.
[131] W.J. Albery and P.N. Bartlett, J. Electrochem. Soc. 131 (1984) 315.
[132] J.S. Curran and D. Lamouche, J. Phys. Chem. 87 (1983) 5405.
[133] R.A. Marcus, J. Phys. Chem. 94 (1990) 1050.
[134] H. Gerischer, Pure Appl. Chem. 52 (1980) 2649.
[135] Page 124 in ref. 130.
[136] Page 97 in ref. 95.
[137] R.A. Marcus, J. Chem. Phys. 24 (1956) 966.
[138] R.A. Marcus, Can. J. Chem. 37 (1959) 155.

[139] R.A. Marcus, J. Chem. Phys. 43 (1965) 679.
[140] H. Gerisher, D.M. Kolb and J.K. Sass, Adv. Phys. 27 (1978) 437.
[141] G.T. Brown and J.R. Darwent, J. Chem. Soc., Chem. Commun. 1985, 93.
[142] I.Bedja, S. Hotchandani and P.V. Kamat, J. Phys. Chem. 97 (1993) 11064.
[143] W.J. Albery and P.N. Bartlett, J. Electroanal. Chem. 139 (1982) 57.
[144] W.J. Albery, P.N. Bartlett, C.P. Wilde and J.R. Darwent, J. Am. Chem. Soc. 107 (1985) 1854.
[145] H. Bogaert, P. Douglas and P. Van der Meeren, J. Chem. Soc., Faraday Trans. 88 (1992) 3467.
[146] B.Jenny and P.Pichat, Langmuir, 7 (1991) 947.
[147] R.W. Matthews, J. Chem. Soc., Faraday Trans. 1 85 (1989) 1291.
[148] R.W. Matthews, J. Catal. 111 (1988) 264.
[149] J. Cunningham and S. Srijaranai, J. Photochem. Photobiol. A: Chem. 58 (1991) 361.
[150] I. Langmuir, Trans. Faraday Soc. 17 (1921) 621.
[151] H. Al-Ekabi, N. Serpone, E. Pelizzetti, C. Minero, M.A. Fox and R.B. Draper, Langmuir, 5 (1989) 250.
[152] C.S. Turchi and D.F. Ollis, J. Catal. 119 (1989) 483.
[153] D.F. Ollis and C.S. Turchi, Environ. Prog. 9 (1990) 229.
[154] J. Cunningham, G. Al-Sayyed and S. Srijaranai, in: Aquatic and Surface Photochemistry, eds G.R. Helz, R.G. Zepp and D.G. Crosby (CRC Press Inc., Boca Raton, Florida, 1993).
[155] R.W. Matthews, in: Photochemical Conversion and Storage of Solar Energy, eds E. Pelizzetti and M. Schiavello (Kluwer, Dordrecht, 1991.
[156] G. Al-Sayyed, J.-C. D'Oliveira and P. Pichat, J. Photochem. Photobiol.:A Chem. 58 (1991) 99.
[157] R. Terzian, N. Serpone, C. Minero and E. Pelizzetti, J. Catal. 128 (1991) 352.
[158] E. Pelizzetti, E. Praumauro, C. Minero, N. Serpone and E. Borgarello, in: Photocatalysis and Environment, ed M. Sciavello, NATO ASI Ser.C237 (Kluwer, Dordrecht, 1988).
[159] T.A. Egerton and C.J. King, J. Oil Col. Chem. Assoc. 62 (1979) 386.
[160] D.F. Ollis, E. Pelizzetti and N. Serpone, in: Photocatalysis: Fundamentals and Applications, eds N.Serpone and E.Pelizzetti (Wiley Interscience, New York, 1989).
[161] C. Minero, Solar Energy Mat.Solar Cells 38 (1995) 421.
[162] L.M. Peter, Chem. Rev. 90 (1990) 753.
[163] W.J. Albery, M.D. Archer and R.G. Egdell, J. Electroanal. Chem. 82 (1977) 199.
[164] W.J. Albery, P.N. Bartlett and J.D. Porter, J. Electrochem. Soc. 131 (1984) 2892.
[165] W.J. Albery and M.L. Hitchman, Ring-Disc Electrodes (Clarendon Press, Oxford, 1971).
[166] W.J. Albery, P.N. Bartlett and J.D. Porter, J. Electrochem. Soc. 131 (1984) 2896.
[167] N. Buhler, K. Meier and J.F. Reber, J. Phys. Chem. 88 (1984) 3261.
[168] C. Boxall, Ph.D. Thesis, University of London, 1987.
[169] J.J. Ramsden and M. Grätzel, J. Chem. Soc., Faraday Trans. 1 80 (1984) 919.
[170] M. Heyrovsky and J. Jirkovsky, Langmuir 11 (1995) 4288.
[171] D. Lawless, S. Kapoor and D. Meisel, J. Phys. Chem. 99 (1995) 10329.

[172] A.J. Bard, A.B. Bocarsly, F.-R.F. Fan, E.G. Walton and M.S. Wrighton, J. Am. Chem. Soc. 102 (1980) 3671.
[173] O. Enea and A.J. Bard, J. Phys. Chem. 90 (1986) 301.
[174] C.Y. Liu and A.J. Bard, J. Phys. Chem. 93 (1989) 3232.
[175] B.E. Burstein, Phys. Rev. 93 (1954) 632.
[176] T.S. Moss, Proc. Phys. Soc. (London) B 76 (1954) 775.
[177] L. Banyai and S. Koch, Phys. Rev. Lett. 57 (1986) 2722.
[178] T. Moss, Phys. Status Solidi 101 (1980) 555.
[179] P.D. Dapkus, N. Holonyak, R.D. Burnham and D.L. Keune, Appl. Phys. Lett. 16 (1980) 93.
[180] K. Seeger, Semiconductor Physics, p. 342 (Springer-Verlag, New York, 1973).
[181] W.W. Dunn, Y. Aikawa and A.J. Bard, J. Electrochem. Soc. 128 (1981) 222.
[182] M.D. Ward and A.J. Bard, J. Phys. Chem. 86 (1982) 3599.
[183] J.K. Leland and A.J. Bard, J. Phys. Chem. 91 (1987) 5076.
[184] J.K. Leland and A.J. Bard, J. Phys. Chem. 91 (1987) 5083.
[185] M. Abramowitz and I.A. Stegun, Handbook of Mathematical Functions (Dover Publications, New York, 1970).
[186] G.T. Brown and R.M.G. Rajapakse, Personal Communication.
[187] A. Fujishima and K. Honda, Nature 37 (1972) 238.

Chapter 10

Hydrodynamic Modulation Methods in Electrochemistry

DAVID E WILLIAMS and JULIE V MACPHERSON

LIST OF SYMBOLS

ac	alternating current
CSFM	channel stopped flow method
CE	chemical reaction preceding electrochemical discharge
dc	direct current
EHD	electrohydrodynamic impedance
FIA	flow injection analysis
HECI	hydroelectrochemical impedance
HM	hydrodynamic modulation
HMRDE	hydrodynamically modulated rotating disc electrode
LC	liquid chromatography
PF	pulsed flow
PRV	pulsed rotation voltammetry
RDE	rotating disc electrode
RFMRE	radial flow microring electrode
RVC	reticulated vitreous carbon
SECM	scanning electrochemical microscope
SF	stopped flow
SRV	stopped rotation voltammetry
TPM	tip position modulation
a	vibration amplitude
a_o	amplitude of the velocity oscillation
A	electrode area
c	concentration of species
c_b	concentration of species in bulk solution

c_o	concentration of species at the surface of the electrode
C_{dl}	double layer capacitance
d	diameter of wire or disc electrode
d_s	site diameter
d_{secm}	tip/substrate separation
D	diffusion coefficient of species
D_f	diffusion coefficient of species in a film
f	frequency (in Hz)
F	Faradays constant
g	force per unit volume exerted on an element of a fluid by gravity
h	half the nozzle/substrate separation in a RFMRE geometry
H	transfer function relating current modulation to modulation of velocity gradient at the interface
$H_0(p)$	transfer function for ideal mass transport control in the HMRDE
$\bar{\imath}$	mean current density
$\tilde{\imath}_D$	AC current density generator or induced current density modulation
$\tilde{\imath}_e$	observed modulated current density in the external circuit
i_{Levich}	limiting current density at a rotating disc electrode
i_{lim}	limiting current density
$\bar{\imath}_{\text{lim}}$	mean limiting current density
$\tilde{\imath}_{\text{lim}}$	modulated limiting current density
I	current
I_{lim}	limiting current
$\bar{I}_{\text{lim}}$	mean limiting current
I_{min}	limiting current at a microwire in the absence of electrode motion
j	$\sqrt{-1}$
k_m	mass-transfer coefficient
l	length of electrode
L	characteristic length of a tubular electrode
n	number of electrons transferred per redox event
p	dimensionless variable $(=\omega/\Omega_o)$
p_D	dimensionless diffusional relaxation time
p	pressure
r	radius of wire or disc electrode

r_o	tube electrode radius
Re	Reynolds number (characteristic solution velocity × characteristic length/kinematic viscosity)
R_F	faradaic resistance for an electrochemical reaction
R_S	electrolyte resistance between a reference electrode and a working electrode
Sc	Schmidt number ($=\nu/D$)
Sh	Sherwood number
t	time
u	solution velocity
$\mathbf{u}$	solution velocity vector
$\bar{u}$	mean solution velocity
$\tilde{u}$	modulated velocity
u_o	centreline velocity in tubular flow
u_r	ratio of δ_D to δ_{AC}
U_o	solution velocity at the centre of a channel
V	applied electrode potential
V_f	volume flow rate
W	frequency of rotation (Hz)
$w_{\mathrm{rf\,mre}}$	circumference of ring electrode
x	distance along the flow direction from the upstream edge of an electrode or plate
x_e	entry length in a channel flow cell
$x_{\mathrm{rf\,mre}}$	ring electrode thickness
y	distance normal to the electrode or plate
Z_{AC}	AC impedance measured at constant Ω_o
Z_D	Warburg diffusional impedance
$(Z_{\mathrm{EHD}})_I$	EHD impedance with galvanostatic regulation
$(Z_{\mathrm{EHD}})_V$	EHD impedance with potentiostatic regulation
$(Z_{\mathrm{EHD}})_{V,\,\mathrm{lim}}$	potentiostatic EHD impedance determined at the diffusion-limited current
$\bar{\alpha}$	gradient of velocity tangential to the wall with respect to distance normal to the wall
δ_{AC}	characteristic penetration depth into the solution phase of the concentration wave during AC impedence
δ_D	concentration boundary layer
δ_f	film thickness
δ_{hm}	distance over which oscillatory motions are damped out

δ_H	hydrodynamic boundary layer
δ_{le}	concentration boundary layer calculated invoking the Lévêque approximation
ϕ	phase angle of the AC impedance
θ	phase angle of the EHD impedance
ρ_L	density of the fluid
σ	Warburg coefficient
τ_D	characteristic time taken for diffusional relaxation
τ_f	characteristic time for diffusion through a surface film
τ_H	characteristic time taken for hydrodynamic relaxation
τ_t	time between pulses in a square wave motion
μ	viscosity
ν	kinematic viscosity
ω	angular frequency of the velocity or rotation modulation
ω_D	characteristic relaxation frequency for diffusion across the concentration boundary layer
Ω	angular frequency of rotation (rad s^{-1})
$\tilde{\Omega}$	modulated rotation speed
Ω_o	central value for the rotation speed about which a sinusoidal modulation is applied

10.1 INTRODUCTION

Hydrodynamic modulation (HM) methods [1, 2] refer to the modulation of motion of an electrode or the flow of solution past an electrode. Examples include: the rotating disc electrode (RDE) with modulation of the rotation frequency by a fixed amount about a central frequency; different geometry vibrating electrodes; and various methods involving interruption or periodic variation of solution flow. HM techniques are extremely useful in electrochemistry because by extracting the alternating current (ac) component it is possible to separate mass transport-limited processes from those which are not, the so-called background processes. These can include solvent decomposition, double layer charging, irreversible surface reactions and transformations of surface layers [3].

To put HM methodology into the more general context of developments in techniques for electroanalysis, for a modulated macroelectrode the significant diminution in the charging current to the desired current signal contribution is comparable to that observed by decreasing the characteristic dimension of an electrode in quiescent solution from macro- to micro-. Further advantages accrue when the modulation methods are applied in conjunction with ultramicroelectrodes (UMEs), as will be shown later.

Two of the major goals in electrochemical analytical science are (i) lowering the detection limits of electroactive species and (ii) extending the potential window of a given solvent in order to increase the range of electrolytically detectable species. Using conventional electrochemical techniques at solid electrodes, the diffusion-controlled current from electroactive species present can be masked by solvent electrolysis or, for species present at low concentrations, background processes in general. Since HM with lock-in detection can reject these residual currents from the diffusion-controlled component of the current signal, with an improved potential range compared to direct current (dc) voltammetry, it represents an extremely useful analytical technique.

One important and often neglected aspect of HM methodology is the ability to resolve subtle effects on mass transport reflecting details of reaction mechanism. Mass transport to an electrode and therefore HM is sensitive to the geometry of the electrode. This is due to the interaction between two important boundary layers: the concentration boundary layer, δ_D, the zone of the solution near the electrode within which the concentrations of species change as a consequence of the interfacial electrode reaction; and the hydrodynamic boundary layer, δ_H, the region of the solution near to the electrode surface within which the solution velocity changes rapidly. Modulation of the electrode motion or of the flow causes a modulation of δ_H which has a characteristic length scale dependent on the modulation frequency. The effect on the measured current depends on the ratio of this characteristic length to the dimension of the concentration boundary layer, which is in turn dependent on details of the electrode geometry and of the reaction mechanism. This is the basis of the use of hydrodynamic modulation to explore details of reaction mechanism. The difficulty with the method is that, although the idea is simple in principle, the mathematical analysis is difficult, so that interpretation of the results is not necessarily straightforward.

10.2 CLASSIFICATION OF METHODS

HM methods can be classified in three ways either (1) according to their objective (analysis or exploration of reaction mechanism), (2) the nature of the modulation (flow versus electrode motion) or (3) the method by which the modulation is performed. HM methods have been widely applied to the development of chemical analysis, as will be highlighted in this review. Applications of various forms of HM to enhance yield and selectivity in electrochemical synthesis, are not covered here. In general, for analysis a single modulation frequency is employed, whereas for exploration of reaction mechanisms, the variation of current signal with modulation frequency is typically the primary objective. This latter method is termed electrohydrodynamic impedance (EHD) [1] and it is one of a general class of transfer function measurements, in which the response of the system to some perturbation is considered [4, 5]. The AC impedance (response of current to perturbation of potential) is the best known of these. Methods combining different modulation techniques have also been described (e.g., Ref. [6]).

In the following, a theoretical background is given with a view to providing a firm qualitative understanding of the factors which affect the hydrodynamically modulated signal. Reference should be made to the original literature cited for a more extensive treatment.

10.3 THEORETICAL BACKGROUND

10.3.1 Introduction: Some basic principles of hydrodynamics for electrochemistry

Modulation of the relative velocity of the solution or of the electrode causes a change in the concentration gradient of reagents in the diffusion layer near the electrode. This results in modulation of the concentration at the electrode (galvanostatic control) or concentration gradient normal to the electrode (potentiostatic control), and hence modulation of the potential, for galvanostatic control, or of the current, for potentiostatic control. Analysis of the resultant electrochemical modulation may be used either to characterise aspects of electrode surface processes or of the mass transport and hydrodynamic regime. There is no general description applicable to all cases; rather, each case must be treated individually.

The concentration distribution of species in the diffusion layer, δ_D, obeys the convective-diffusion equation:

$$\frac{\partial c}{\partial t} = D\nabla^2 c - \mathbf{u} \cdot \nabla c, \tag{10.1}$$

where $\mathbf{u}$ denotes the solution velocity vector, t is the time, D is the diffusion coefficient and c the concentration of the species. The velocity profile is obtained by solution of the continuity equation, equation (10.2), which is a statement of incompressibility of the fluid, and the Navier–Stokes equation, equation (10.3).

$$\nabla \cdot \mathbf{u} = 0 \tag{10.2}$$

$$\rho_L \frac{\mathrm{d}\mathbf{u}}{\mathrm{d}t} = -\nabla P + \mu\nabla^2\mathbf{u} + \mathbf{g}. \tag{10.3}$$

Equation (10.3) represents Newtons second law of motion, where the left hand side of the equation denotes the product mass × acceleration, ρ_L is the density of the fluid, and the right hand side represents the net force exerted on a volume element of the fluid, P is the pressure, μ the viscosity and g the force per unit volume exerted on an element of the fluid by gravity. The term $\mu\nabla^2\mathbf{u}$ represents frictional forces.

Equation (10.3) is generally simplified for electrochemical problems by assuming the absence of a pressure gradient and neglecting the effect of gravitational forces and hence can be written as equation (10.4):

$$\frac{\mathrm{d}\mathbf{u}}{\mathrm{d}t} = \nu\nabla^2\mathbf{u}, \tag{10.4}$$

where $\nu = \mu/\rho_L$ is termed the kinematic viscosity. Solution of these hydrodynamic equations requires modelling of the system under the appropriate co-ordinates and boundary conditions and as such will be sensitive to the geometry of the electrode and the flow conditions utilised.

The equations are often written in terms of dimensionless variables which will become evident later. The two most important for this review are the Reynolds number, Re and the Schmidt number, Sc. The Reynolds number describes the flow characteristics of the system. It is directly proportional to some characteristic solution velocity and length (normally associated with the electrode, but for flow in a channel it is related to the half height of the channel) and is normalised with respect to the kinematic viscosity of the solution. At low solution velocities Re is low and the fluid flow profile is smooth and steady, described as laminar. As the velocity increases, Re increases, with the flow becoming increasingly more chaotic and turbulent. The Schmidt number is characteristic of the solution properties, where $Sc = (\nu/D)$.

For HM a perturbation is introduced to the velocity vector in equation (10.4) and the corresponding change in concentration or concentration gradient calculated from equation (10.1). Finally the resultant perturbation in potential (for galvanostatic control or constant concentration gradient at electrode surface) or current (potentiostatic control) is evaluated by relating the current density to concentration or concentration gradient and integrating over the electrode surface. The following discussion attempts to elicit some general principles and give results from the literature for particular cases. The application of these ideas is developed in more detail in Sections 10.4 and 10.5.

10.3.2 Boundary layers and distance scales

At an electrode surface, the velocity of the solution with respect to the surface will be zero and the concentration of electroactive species will be different from that in the bulk of the solution (zero at the limiting current). The behaviour of an electrochemical system under the influence of a steady flow is therefore governed by two boundary layers: δ_H within which the solution velocity increases until it reaches its limiting value far from the surface; and δ_D, within which the reagent concentration increases until it reaches its bulk solution value. In the case of a modulated flow or modulated motion of the electrode, there is a third distance scale which must be considered, δ_{hm}, the boundary layer over which the modulation is damped out.

When a fluid flow of constant velocity, u, impinges parallel to the edge of a plate, the boundary condition is such that the fluid velocity is zero on the surface of the plate. This results in the formation of a hydrodynamic boundary layer in which the flow velocity parallel to the surface varies with distance normal to the surface. The hydrodynamic boundary layer thickness increases with distance, x, from the upstream edge of the plate as given by equation (10.5) [7]:

$$\delta_H \approx 1.7\sqrt{\frac{x\nu}{u}}. \tag{10.5}$$

When the flow is constrained to a channel bounded on both faces, then at a sufficient distance from the entry point, x_e, the two hydrodynamic boundary layers associated with each wall overlap and the fluid velocity varies parabolically with position across the channel.

The development of δ_D is easiest illustrated for the case where the electrode operates under diffusion-controlled conditions, i.e., at the limiting current. When the flow impinges on the upstream edge of an electrode embedded in a wall the concentration should be reduced from a constant value before the electrode to zero at the edge. The concentration boundary layer will increase in thickness with position along the electrode, resulting in a non-uniform current density distribution across the electrode. In the absence of limitations imposed by the solution resistance the current density at the leading edge approaches infinity. As a result, due to these extremely high mass transport rates, the condition $c = 0$ at the upstream edge of the electrode might not be obtained.

When the concentration boundary layer is sufficiently thin the mass transport problem can be solved under the approximation that the solution velocity within the concentration boundary layer varies linearly with distance away from the surface. This is called the Lévêque approximation [8, 9] and is satisfactory under conditions where convection is efficient compared with diffusion. More accurate treatments of mass transfer taking account of the full velocity profile can be obtained numerically [10, 11] but the Lévêque approximation has been shown to be valid for most practical electrodes and solution velocities. Using the Lévêque approximation, the local value of the concentration boundary layer thickness, δ_{le}, (determined by equating the calculated flux to the flux that would be obtained according to a Nernstian diffusion layer approximation; that is with a linear variation of concentration across the boundary layer) is given by equation (10.6) [12].

$$\delta_{le} = 1.8575\left(\frac{Dx}{\bar{\alpha}}\right)^{1/3}, \tag{10.6}$$

where $\bar{\alpha} = (\partial u/\partial y)_{y=0}$ denotes the gradient of the velocity tangential to the electrode with respect to distance normal to the electrode. Averaged over the whole electrode, of length, l,

$$\delta_{le} = 1.2383\left(\frac{Dl}{\bar{\alpha}}\right)^{1/3}, \tag{10.7}$$

provided that the electrode is small enough so that $\bar{\alpha}$ does not vary significantly along its length. For parallel flow impinging on a plate at a distance x from the upstream edge [7],

$$\bar{\alpha} = 0.332\sqrt{u^3/x\nu}. \tag{10.8}$$

For fully-developed flow in a channel [8],

$$\delta_{le} = 1.4925\left(\frac{Dhl}{U_0}\right)^{1/3}, \tag{10.9}$$

where U_o is the solution velocity at the centre of the channel and $2h$ is the channel height. Figure 10.1 compares δ_H and δ_{le} for flow of aqueous solution at 10 cm s^{-1} incident on the edge of a plate containing an embedded plane electrode.

Although in general the calculation of δ_{hm} for modulation is difficult, a simplistic understanding can be obtained through consideration of the flow pattern above an oscillating infinite plate [13]. The velocity profile $u(y, t)$ parallel to the wall, at a distance y above the wall, is a lagged, damped simple-harmonic motion:

$$u(y, t) = a_o \exp(-\eta)\cos(\omega t - \eta), \tag{10.10}$$

where ω is the angular frequency of the velocity modulation, a_o is the amplitude of the velocity oscillation and $\eta = y\sqrt{\omega/2\nu}$. Two fluid layers, a distance $\bar{y} = 2\pi\sqrt{2\nu/\omega}$ apart, oscillate in phase. This distance can be regarded as being similar to the wavelength of the motion. The fluid layer which is carried by the wall, δ_{hm}, has a thickness of the order of $\delta_{hm} \approx \sqrt{\nu/\omega}$, within which the amplitude of the motion drops to 13% of the value at the wall.

Equations describing the motion and consequently the variation in con-

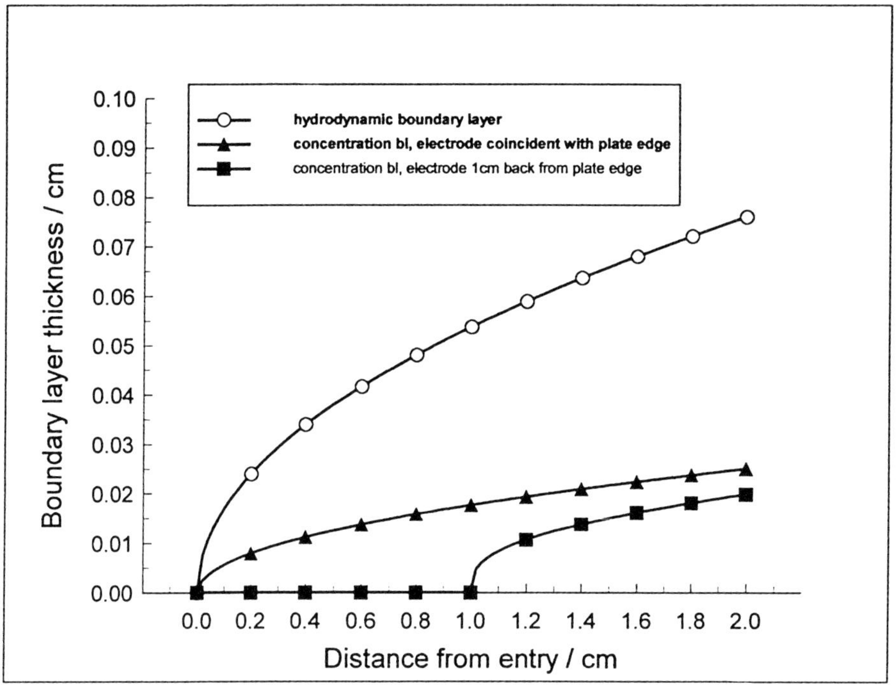

Fig. 10.1. Comparison of hydrodynamic and concentration boundary layer thicknesses. An aqueous flow ($\nu = 1.0 \times 10^{-2}\ \mathrm{cm^2\,s^{-1}}$ for water, $D = 1 \times 10^{-5}\ \mathrm{cm^2\,s^{-1}}$) at $10\ \mathrm{cm\,s^{-1}}$ is incident on the edge of a plate with an embedded plane electrode. The electrode leading edge is either coincident with the plate edge or set 1 cm back. The concentration boundary layer thickness is calculated using the Lévêque approximation.

centration gradient can be separated into a steady (equation (10.8)) and an unsteady part (equation (10.10)). Hence equations (10.8) and (10.10) can be used to assess the effect of an oscillation imposed upon a stationary flow. The frame of reference is simply changed so that the electrode is considered to move rather than the solution and then the two motions added. Figures 10.2(a) and 10.2(b) compare δ_{hm} and δ_{le} calculated for an electrode embedded in a wall, as a function of x and ω. Evidently δ_{hm} is always significantly greater than δ_{le} except at positions far from the leading edge of the electrode, or at high modulation frequencies. Later it will be shown that this effect is evident in the relationship between electrode size and the frequency dependence of the modulated current.

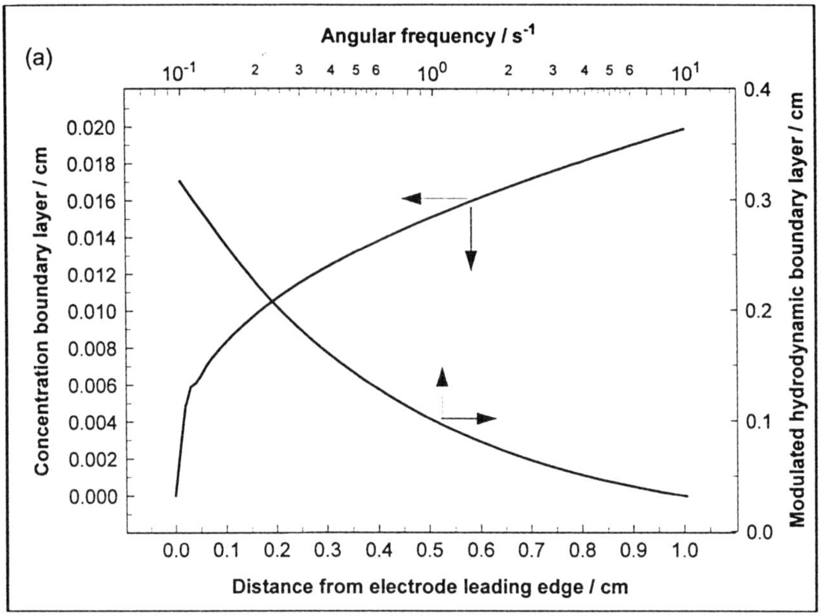

Fig. 10.2(a).

10.3.3 Relaxation times for flow and for diffusion

In response to a step change in the flow, the velocity distribution near a wall relaxes with a characteristic time scale, τ_H, where $\tau_H \approx \delta_H^2/4\nu$ (Ref. 13, p. 91), whereas the characteristic time scale for diffusional relaxation, τ_D, is $\tau_D \approx \delta_D^2/2D$. Figure 10.3 demonstrates that for an electrode embedded in a wall, and hence by extension to a channel, under typical aqueous electrolyte solution conditions, the hydrodynamic relaxation time is much less than the diffusion relaxation time and therefore the experimentally observed relaxation time is that for diffusion across the concentration boundary layer. An important exception is at the extreme leading edge of the electrode. Thus there might be special effects observable for sufficiently small electrodes. For a RDE, the ratio between hydrodynamic and concentration boundary layer thicknesses is $\delta_H/\delta_D \approx 2\mathrm{Sc}^{1/3}$. For typical species in water, $D \approx 1 \times 10^{-5}\ \mathrm{cm^2\,s^{-1}}$ and $\nu \approx 1 \times 10^{-2}\ \mathrm{cm^2\,s^{-1}}$, so $\delta_D \approx 0.05\ \delta_H$, and $\tau_D/\tau_H \approx 5$ again demonstrating that the relaxation time is determined

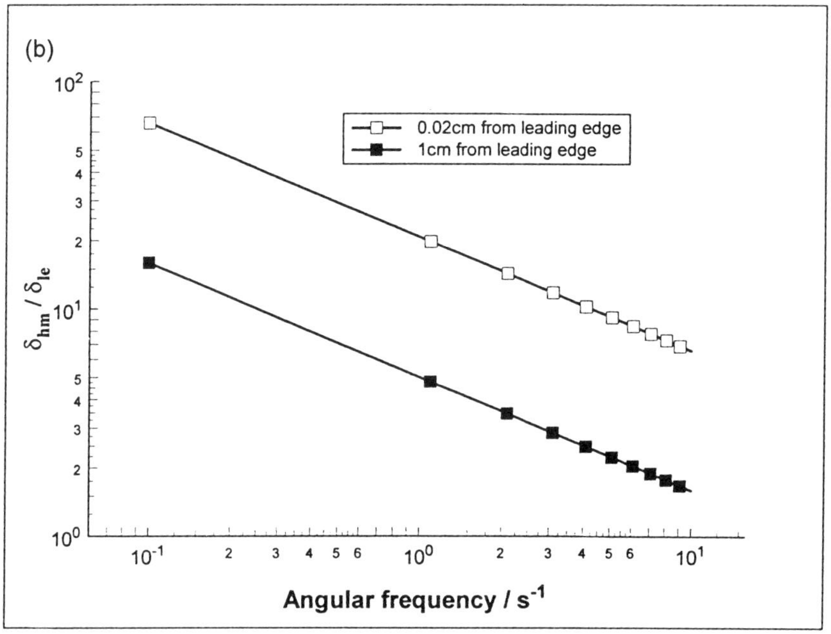

Fig. 10.2. Comparison of concentration boundary layer thickness, δ_D, developed over an electrode embedded in a plate and set 1 cm back from the leading edge of the plate (aqueous flow, $\nu = 1 \times 10^{-2}\,\mathrm{cm^2\,s^{-1}}$, $D = 1 \times 10^{-5}\,\mathrm{cm^2\,s^{-1}}$, free stream flow velocity $10\,\mathrm{cm\,s^{-1}}$) with the thickness of the hydrodynamically modulated layer over an infinite oscillating plate, δ_{hm}, angular frequency of oscillation ω. (a) boundary layers as a function of position on the electrode and angular frequency of oscillation; (b) ratio δ_{hm}/δ_D for two different positions on the electrode, near and far from the leading edge.

by diffusion across the concentration boundary layer. Experimental confirmation of these timescales is given later in Sections 10.4 and 10.5.

As the boundary layer thickness is fixed by the hydrodynamics the relaxation time will be governed primarily by the diffusion coefficient of the solution species. This parameter can be determined very accurately using either sinusoidal flow modulation over a range of modulation frequencies [1, 14], (Sections 10.4.1.1 and 10.5.3.1) or by stepping the fluid flow rate (and electrode rotation rate) and following the current relaxation (Section 10.4.3.1) [15, 16]. In an effort to reduce diffusional relaxation times, workers have attempted to decrease δ_D by either adopting thin layer geometries (Section 10.4.3.2) and/or by employing very small electrodes,

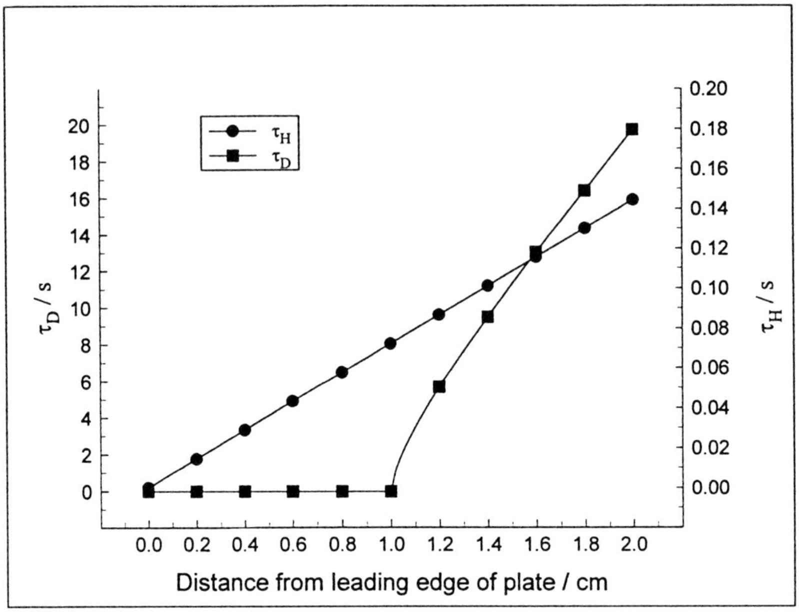

Fig. 10.3. Relaxation times for concentration boundary layer, τ_D and hydrodynamic boundary layer, τ_H, calculated for an aqueous flow ($\nu = 1.0 \times 10^{-2}\ \mathrm{cm^2\ s^{-1}}$, $D = 1 \times 10^{-5}\ \mathrm{cm^2\ s^{-1}}$) at $10\ \mathrm{cm\ s^{-1}}$ impinging on the edge of a plate with an embedded electrode set 1 cm back from the plate edge.

in order to increase mass transfer rates (Sections 10.4.2.5, 10.4.2.6 and 10.4.3.3).

10.3.4 Effects of electrode size

Figure 10.2(a), for solution impinging on the upstream edge of an electrode embedded in a wall, shows the importance of electrode size in hydrodynamic modulation measurements. The modulation of current density at the leading edge of the electrode is much greater than at the downstream edge. As the measured effect is averaged over the whole electrode it can be seen to be strongly dependent upon the electrode size. If the electrode becomes sufficiently small, then the frequency response of the current to the imposed perturbation is significantly changed (see Section 10.3.6.1 below).

10.3.5 Equivalent circuit description

Figure 10.4 shows a typical equivalent circuit description for an electrochemical system. To consider the case of potentiostatic control, the hydrodynamic modulation is modelled [17] as an alternating current density generator in parallel with the Warburg diffusional impedance, Z_D. Because the system is under potentiostatic control, there is no resultant alternating potential difference between the two terminals (A, B in Figure 10.4) of the circuit, hence these are modelled as being connected. Consequently, the alternating current density observed in the external circuit between terminals A and B, $\tilde{\imath}_e$, can be related to the induced current modulation, $\tilde{\imath}_D$, which is itself given by equations such as equation (10.14) or equation (10.19) below (notation in Figure 10.4):

$$\tilde{\imath}_e = \tilde{\imath}_D \frac{Z_D}{R_S + (1 + j\omega C_{dl})(Z_D + R_F)}, \tag{10.11}$$

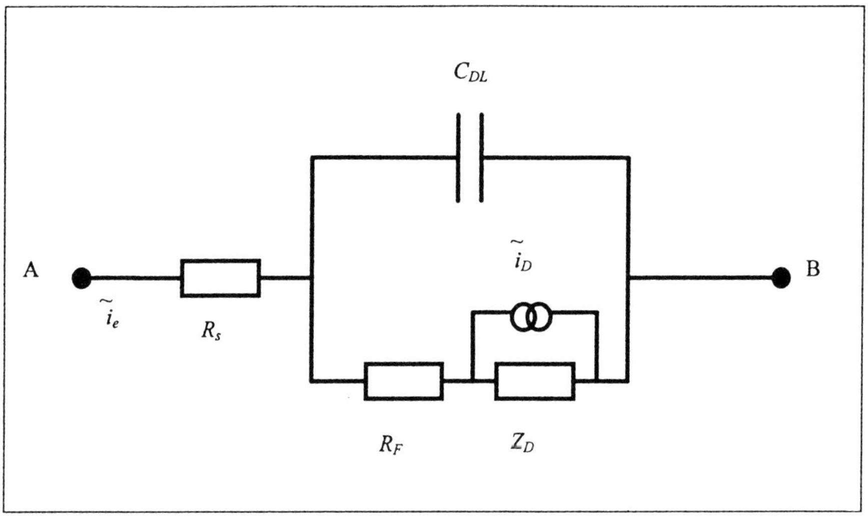

Fig. 10.4. Equivalent circuit for the electrode with hydrodynamic modulation: R_s denotes the solution resistance, C_{DL} the double layer capacitance, R_F the faradaic resistance and Z_D the impedance contributed by diffusional transport of the reactants to the surface. Hydrodynamic modulation is represented by the modulated current, $\tilde{\imath}_D$ and the resultant current in the external circuit represented by $\tilde{\imath}_e$.

where R_S is the electrolyte resistance between the reference electrode and working electrode, C_{dl} is the double layer capacitance and R_F is the faradaic resistance for an electrochemical reaction. Equation (10.11) thus relates the observed current modulation to the applied flow modulation [1, 18, 19]. Equation (10.11) demonstrates that the modulated flow signal is largest when $Z_D \gg R_F$, the condition pertaining at the limiting current.

10.3.6 Particular cases

10.3.6.1 Laminar pulsatile flow in a tube

Flow in a tube is in one direction, parallel to the electrode surface, (taken as the x-direction). The time-dependent convective-diffusion equation for this geometry is given by equation (10.12). Mass transport to the surface of the electrode is thus determined both by the gradient perpendicular to the surface of the tangential flow, $\partial u_x/\partial y$ and the concentration gradient perpendicular to the surface:

$$\frac{\partial c}{\partial t} = D\left(\frac{\partial^2 c}{\partial y^2}\right) - u_x(y)\,\frac{\partial c}{\partial x}. \tag{10.12}$$

The concentration gradient parallel to the surface is only important for small electrodes and low volume flow rates, as will be discussed later.

There have been a number of approaches to the study of transport across a wall bounding a pulsating flow. The response of the limiting current density, i_{lim}, at a tubular electrode in the presence of pulsations was obtained numerically by Mizushina *et al.* [20, 21] and confirmed analytically by Patel *et al.* [22]. By invoking the Lévêque approximation [8, 9], equation (10.13);

$$u_x = \frac{2u_o}{r_o}, \tag{10.13}$$

where u_o is the centreline velocity in tubular flow and r_o is the tube electrode radius, the modulated current density, over a wide range of mean flow rates and modulation frequencies, can be represented by equation (10.14):

$$\frac{\tilde{i}_{\text{lim}}}{\tilde{u}} = \frac{\bar{i}_{\text{lim}}}{\bar{u}} \cdot H\left(\frac{\omega}{\omega_D}\right), \tag{10.14}$$

where $\tilde{i}_{\text{lim}}$ is the modulated limiting current density, $\bar{i}_{\text{lim}}$ is the mean limiting current, $\tilde{u}$ is the modulated velocity and $\bar{u}$ is the mean solution velocity. H is a transfer function relating current modulation to modulation of the velocity gradient at the interface and ω_D is the characteristic diffusion relaxation frequency, $\omega_D = (\tau_D L^{2/3})^{-1}$ where the characteristic length $L = (l/2r_o \text{ReSc})$. The relaxation time for diffusion across the concentration boundary layer, in this case, is equal to r_o^2/D. Here $\text{Re} = 2r_o\bar{u}/\nu$.

The variation of H with ω/ω_D is shown in Figure 10.5. A noticeable roll-

(a)

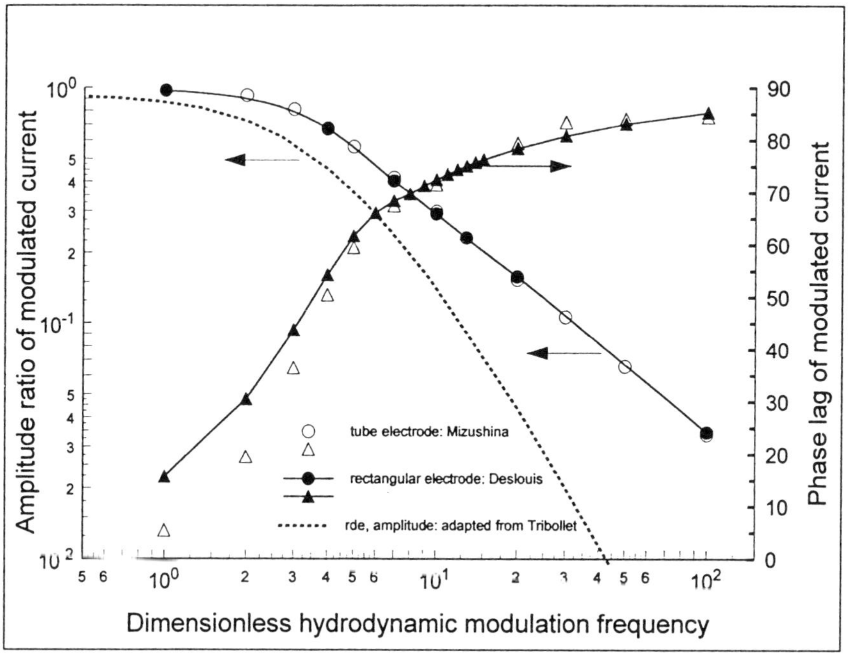

Fig. 10.5(a). Transfer function, H, relating modulated current response to modulated flow rate for tube electrode (Reference [20]), rectangular electrode embedded in a wall (Reference [12]) and modulated RDE (amplitude only). The dimensionless modulation frequency (ω/ω_D see text) is the ratio of the time scale for diffusion across the concentration boundary layer to the timescale for modulation of the hydrodynamics.

(b)

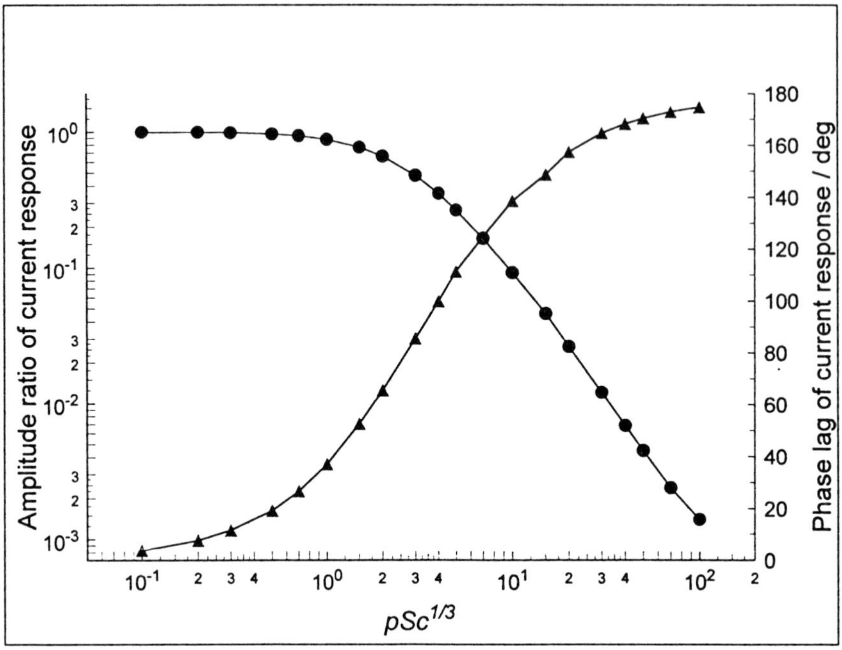

Fig. 10.5(b). (b) Transfer function $H(p)$, where $p = \omega/\Omega_o$, linking modulated current response to modulation of rotation frequency of a RDE, plotted for Sc = 1000 (Reference [30]). For the modulated RDE the axes in Fig. 10.5(a) and 10.5(b) are related by $\omega/\omega_D = 1.296\ p\mathrm{Sc}^{1/3}$.

off in response and corresponding phase lag occurs when $\omega \approx \omega_D$. When $H \approx 0.5$, ω ca. $5\omega_D$. At high frequency, the amplitude decreases as ω^{-1}. Importantly, over a range of mean flow velocities, the results for the modulated current will scale onto a common curve. Additionally ω_D scales with the dimension of the electrode because δ_D, averaged over the whole electrode length, increases with increasing length, see Figure 10.1.

If the electrode were partially blocked, or behaved as an array of UMEs, then the characteristic length for the scaling would be different and would depend upon whether the concentration boundary layers for the different active areas interacted. If the electrode could be represented as consisting of two or more independent active areas, then two or more characteristic relaxations would be observed in the response, and the length scale could be deduced from the dependence of the measured H values on Re. Additionally, if the electrode were wholely or partially covered with a permeable film within which transport was by diffusion alone, then the modulation of solution velocity would not extend to the electrode surface, and H for different Re would not superimpose in the manner dictated by equation (10.14).

Deslouis, Gil and Tribollet [12] have given a solution for a rectangular or circular electrode embedded in a wall, calculating the modulation in current density caused by a modulation of the tangential velocity gradient normal to the wall. The electrode was considered small enough so that variation of this velocity gradient with position over the electrode was negligible. Usually, this would mean that the electrode is small in comparison with the hydrodynamic boundary layer thickness. The solution is analogous in form to equation (10.14), with a similar roll-off frequency and identical high-frequency behaviour, where the characteristic diffusion relaxation frequency is now defined as; $\omega_D = (l^2/D\bar{\alpha}^2)^{-1/3}$. As before, ω_D is approximately the reciprocal of the time scale for diffusion across the concentration boundary layer: $\omega_D \approx 2D/\delta_{le}^2$, with the thickness δ_{le}, given by equation (10.6). The relationship between the velocity gradient, $\bar{\alpha}$, and the flow velocity depends upon the geometry. The transfer functions for different electrode geometries are compared in Figure 10.5.

The effects of electrode size are evident in the results given in Reference [12]. Although the overall current modulation decays as ω^{-1} at high frequency, the local modulated current density from a small element of the electrode decays at sufficiently high frequencies as $(\omega/\omega_D)^{-3/2}$. At the leading edge of the electrode, ω/ω_D may be small whereas at the trailing edge it may be large. Clearly, as the electrode becomes smaller, the roll-off frequency for relaxation of the current in response to the modulated flow increases, and, with a sufficiently small electrode, the high frequency behaviour will change from $(\omega/\omega_D)^{-1}$ to $(\omega/\omega_D)^{-3/2}$. Results for a circular electrode embedded in a wall were very similar.

The derivation leading to equation (10.14) assumes that mass transport to the electrode by diffusion occurs only normal to the surface. This is not true at the edge of the electrode, where axial diffusional rates can be quite high. The effect becomes increasingly evident when the electrode behaviour is dominated by the edges. For a channel microband electrode this is evident with short length electrodes and low solution velocities [23–25]. For an UME (i.e., an electrode where at least one dimension is comparable to δ_D) implanted in the wall of a flow channel, when the mean flow velocity is small enough, the transfer function relating current fluctuation to flow velocity modulation should be different. This is because in addition to planar diffusion there is likely to be a significant contribution of radial diffusion, to the edges of the electrode [26].

10.3.6.2 Rotating disc electrode

Miller and Bruckenstein [27, 28] introduced the hydrodynamically modulated rotating disc electrode (HMRDE) in 1974. The steady-state current density at a rotating disc electrode is well-defined, given by the Levich equation (equation (10.15)):

$$i_{\mathrm{Levich}} = 0.62nFD^{2/3}\Omega_o^{1/2}\nu^{-1/6}(c_b - c_o), \tag{10.15}$$

where n is the number of electrons transferred per redox event, F is Faradays constant, Ω_o is the angular frequency of rotation (rad s^{-1}) and c_b and c_o are the bulk and surface concentrations of electroactive species. Sinusoidal modulation of the rotation frequency about a fixed central value, Ω_o, with a peak to peak amplitude, $\Delta\Omega$, such that

$$\Omega = \Omega_o + \Delta\Omega \sin(\omega t), \tag{10.16}$$

gives a current modulation:

$$\delta i_{\mathrm{Levich}} = 0.31nFD^{2/3}\nu^{-1/6}(c_b - c_o)\Omega_o^{-1/2}\,\Delta\Omega \sin(\omega t). \tag{10.17}$$

This analysis holds in practice provided that the modulation amplitude is small enough ($\Delta\Omega$ is typically 0.01–0.1 Ω_o) and the modulation frequency low enough.

Formal development of HMRDE, leading to the general method of EHD spectroscopy notably pioneered by Delouis and Tribollet [1], was given by Tokuda *et al.* [29], Albery *et al.* [15], Deslouis *et al.* [18], Tribollet and Newman [30] and Deslouis and Tribollet [31], who have also summarised and reviewed the mathematical and experimental details of the method and its applications [1, 19].

For the rotating disc electrode, a uniformly accessible electrode, the concentration only varies in the direction perpendicular to the electrode hence the time-dependent convection-diffusion equation is given by equation (10.18):

$$\frac{\partial c}{\partial t} = D\frac{\partial^2 c}{\partial y^2} - u_y\frac{\partial c}{\partial y}. \qquad (10.18)$$

Modulation of the rotation frequency about Ω_o gives a response of the velocity component normal to the electrode, which can be obtained by a series expansion in the dimensionless variable $p\text{Sc}^{1/3}$ where $p = (\omega/\Omega_o)$. This leads to an expression for the variation of the concentration, or concentration gradient, at the interface, and hence, by way of consideration of the equivalent circuit of the interface, to an expression for the modulated potential or current in response to flow modulation.

Another way of expressing the oscillatory behaviour is by starting from the steady-state Levich behaviour (equation (10.15)) where $i_{\text{Levich}} \propto \Omega_o^{1/2}$, write that, for small variations, $\delta i_{\text{Levich}} \propto \delta\Omega(\Omega_o^{-1/2})$ so that $\delta i/i = -(1/2)$ $\delta\Omega/\Omega_o$ and then generalise for the modulated response:

$$\frac{\tilde{\imath}}{\tilde{\Omega}} = 0.5\,\frac{\bar{\imath}}{\bar{\Omega}_o}H_0(p), \qquad (10.19)$$

where $H_0(p)$ is a transfer function for ideal mass transport control at the RDE, and in the low frequency limit, $H_0(p) = 1$. For operation at the limiting current, at high frequency, $H_0(p)$ decays as p^{-2}, with a roll-off frequency $p \approx 3$ for $H_0 \approx 0.5$. H_0 is a product of two transfer functions: one, A_0, expresses the response assuming that the fluid responds without lag to the motion of the electrode, whilst the other describes the response of the fluid itself [31]:

$$A_0(\omega)_{\omega\to\infty} = 3.06[(\omega/\Omega_o)\text{Sc}^{1/3}]^{-3/2}. \qquad (10.20)$$

This is the same asymptotic modulation frequency dependence as for a small element of a channel electrode, and arises because the concentration boundary layer is the same thickness over the whole of the rotating disc.

The variation of $H_0(p)$ is shown in Fig. 10.5(b). A point immediately obvious is that the limiting value of the phase lag at high frequency is 180°, which contrasts with the behaviour of the channel electrodes for which the limiting phase lag is 90° (Fig. 10.5(a)). Writing, as above, $\omega_D = 2D/\delta_D^2$, where the concentration boundary layer thickness is given by the Levich formula: $\delta_D = 1.61D^{1/3}\Omega_0^{-1/2}\nu^{1/6}$ then $\omega/\omega_D = 1.296p\text{Sc}^{1/3}$ and a direct comparison of the variation of the amplitude ratio for the disc and channel electrodes can be made, as shown in Fig. 10.5(a). The response of the RDE rolls off at a somewhat lower relative frequency, and falls more rapidly than does that for the channel electrode. The reason both for the different phase shift and for the different amplitude variation lies in the response of the various components of the fluid velocity to modulation of the rotation rate.

10.3.6.3 Perturbation imposed upon stationary solution or electrode: vibrating electrodes

Section 10.4.2 gives a detailed description of the behaviour and electroanalytical use of vibrating wire and band electrodes. In this section, some general remarks are given to link this discussion with the information presented above.

If the oscillation of motion of fluid with respect to an electrode is imposed upon a stationary system and the perturbation is not small with respect to the steady-state then new effects can appear. For example, the flow induced by the motion can have both an oscillating and a steady component (the latter due to the interaction between viscous and inertial effects in the boundary layer) and the oscillating part may have components which are harmonics of the imposed oscillation. In some cases the stationary component of the induced flow can, indeed, dominate the oscillating component. These effects are particularly seen for vibrating electrodes.

A simple oscillating flow is found for the case of an infinite horizontal oscillating plane, as described in Section 10.3.2. For practical systems, where the critical dimension describing the flow (e.g., the radius of the oscillating electrode) is finite, a complex current waveform in response to a sinusoidal perturbation is often found. The current can be expressed as a Fourier series of harmonics of the oscillation frequency [32]. The enhancement of the DC signal is the zeroth order term. The same effects occur when the perturbation is applied to a steady flow, but only if the amplitude of the perturbation is large, for example, large-amplitude hydrodynamic modulation voltammetry on a rotating disc electrode as used by Schwartz [32].

The behaviour of a vibrating microband electrode [33] (Fig. 10.6) also illustrates these effects. The use of these devices in electroanalysis is described comprehensively in Section 10.4.2.2. Typical microband current (I)–potential (V) characteristics for the oxidation of 5.5×10^{-3} mol dm^{-3}

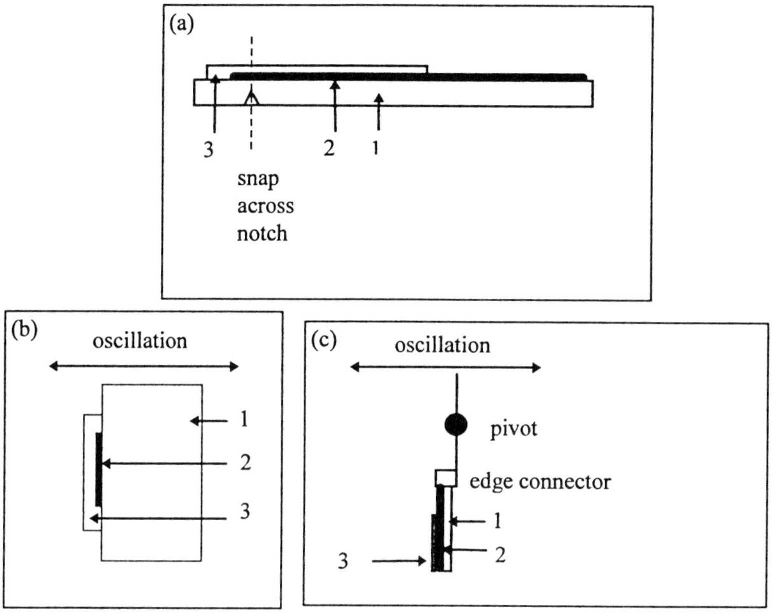

Fig. 10.6. Schematic diagrams of a microband electrode prepared by screen-printing gold onto an alumina substrate, over-printing with an insulator and then snapping to expose a fresh line electrode (Reference [33]). Substrate (1) 500 μm thick; gold (2) 10 μm thick; insulator (3) 20 μm thick. (a) Cross section showing the different layers; (b) cross-section of the exposed surface at the snap line; (c) scheme of oscillation.

$Fe(CN)_6^{4-}$ at a static and a vibrating electrode (oscillating parallel to its short axis) are shown in Fig. 10.7. As can be seen, the effect of the vibration is to induce both an increase in the stationary current and a modulation. Typical current waveforms and their Fourier decomposition are shown in Fig. 10.8.

The periodic asymmetry of the signal, (a small and a large peak which regularly alternate with the period of the motion—Fig. 10.8(a) is due to the geometry of the fabricated microband electrodes, as shown in Fig. 10.6. On one side of the electrode there is only a 20–40 μm thick layer of insulator while on the other this layer is 500 μm thick (the ceramic substrate). It has already been stated in Section 10.3.2 and shown in Fig. 10.1 that for flow impinging parallel to a wall in which an electrode is embedded, δ_H over the electrode will differ depending on the distance of the electrode

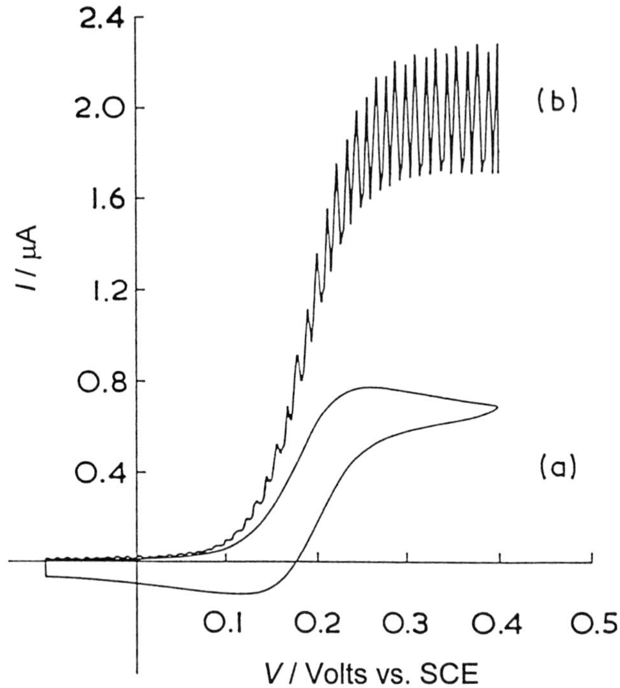

Fig. 10.7. Microband current (I)–potential (V) characteristics for the oxidation of 5.5×10^{-3} mol dm^{-3} $Fe(CN)_6^{4-}$ in phosphate buffer at pH 6.8, recorded at a potential scan rate of 0.1 V s^{-1}. A SCE served as the reference electrode and a Ag wire as the counter. (a) stationary; (b) in sinusoidal motion, amplitude 2 mm, frequency 10 Hz. The electrode width is 10 µm and length 2 mm (after Reference [33]).

from the leading edge. If δ_H over the electrode differs then the resulting electrochemical signal will differ too. To a first approximation the effect of the vibrating electrode is to induce solution flow first one way across the electrode and then the other. For both directions the electrode is at a different distance from each of the two leading edges. The small peak (in Fig. 10.8(a)) is interpreted as the longer entry length (ceramic substrate) leading and the large as the smaller entry length (insulator) leading the motion.

The velocity in a sinusoidal oscillation is proportional to the amplitude, a, and the frequency, f, of the oscillation. For vibrating electrodes, the simple interpretation relating the instantaneous current to the instan-

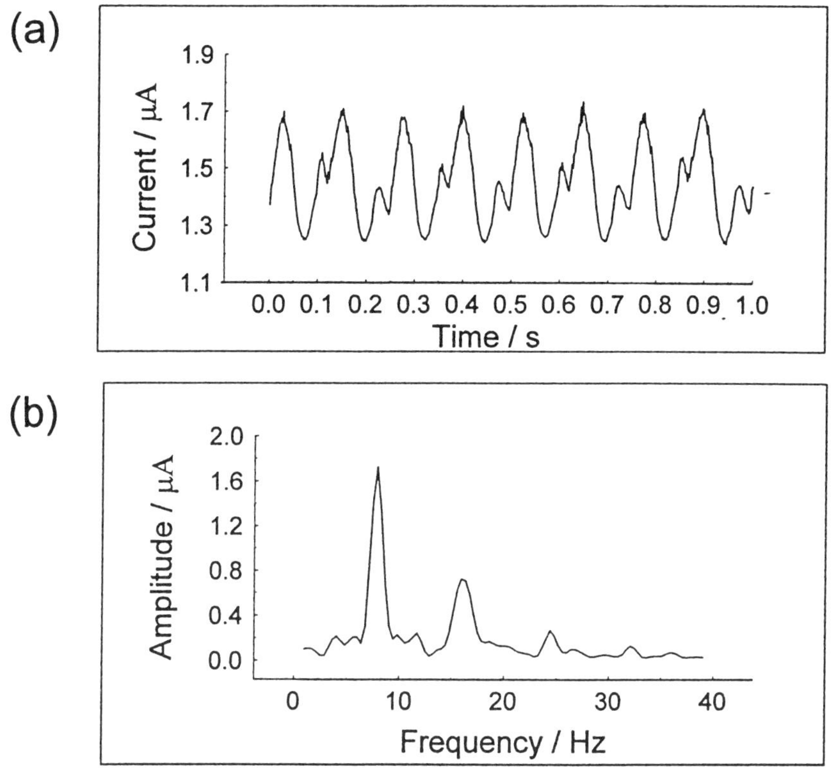

Fig. 10.8. (a) Current waveform for the oxidation of 5.5×10^{-3} mol dm^{-3} $Fe(CN)_6^{4-}$ in phosphate buffer at pH 6.8 and (b) its frequency components: microband electrode width 10 μm and length 2 mm in sinusoidal oscillation at 8.1 Hz, amplitude 2 mm (after Reference [33]).

taneous velocity of the solution relative to the electrode does not, in general, work. This is due to the effect of the damping of the oscillating motion away from the surface. In fact for the vibrating band electrode, the observed mean and maximum current is independent of the amplitude of the vibration and empirically, a dependence on $f^{1/2}$ is found. In contrast, for a vibrating microwire electrode, when the diameter of the wire is sufficiently small such that δ_H is sufficiently thin, then the current response fits the expected sinusoidal oscillatory behaviour [34], but this is a special case. Under these conditions it is possible to achieve very high modulation frequencies without attenuation of the current signal. This results in larger signals, shorter analysis times and greater levels of noise rejection (Section 10.4.2.1).

A rather simple interpretation of the behaviour of vibrating electrodes can be obtained by considering the response to a square-wave motion, to which a sinusoid rather crudely approximates [33]. Here, it is considered that the concentration boundary layer is periodically renewed by the instantaneous rapid motion and that in the intervals between the square-wave steps the solution is at rest. This is a reasonable approximation for most practical purposes because the hydrodynamic boundary layer relaxation time is short, (Section 10.3.3). In this simple model, the waveform would instantaneously rise to a limit during the motion, decaying as a function of $t^{-1/2}$ during the static phase. This decay rate will obviously be dependent on the size and geometry of the electrode: wire, microwire, band or microband. If the delay time between steps were τ_t, then the mean current would vary as $(1/\tau_t)\int_0^{\tau_t} t^{-1/2}\mathrm{d}t$, i.e., as $\tau_t^{-1/2}$ or as $f^{1/2}$.

This discussion emphasises a major difference between a vibrating electrode in a stationary solution and a stationary electrode in a steady flow upon which a modulation is imposed. In the former case the signal increases with increasing modulation frequency whereas in the latter case it decreases. This is a consequence of the nonlinear behaviour of the vibrating electrode in quiescent solution.

10.4 APPLICATION TO ELECTROANALYSIS

10.4.1 Frequency modulated rotating disc electrodes

10.4.1.1 Hydrodynamically modulated rotating disc electrode

Miller and Bruckenstein [27] noted that in the in-phase HMRDE signal, contributions from charging currents, solvent decomposition, surface transformations or irreversible reaction components were negligible. Using this technique they demonstrated selective discrimination of the mass transport-controlled component of the current signal for a variety of electrode materials (gold, amalgamated gold, silver, platinum, and graphite) and electroactive species (sulphide, nitrobenzene, quinhydrone, Tl^{+}, Hg^{2+} and Fe^{2+}–Fe^{3+}). In some cases concentration detection limits were as low as

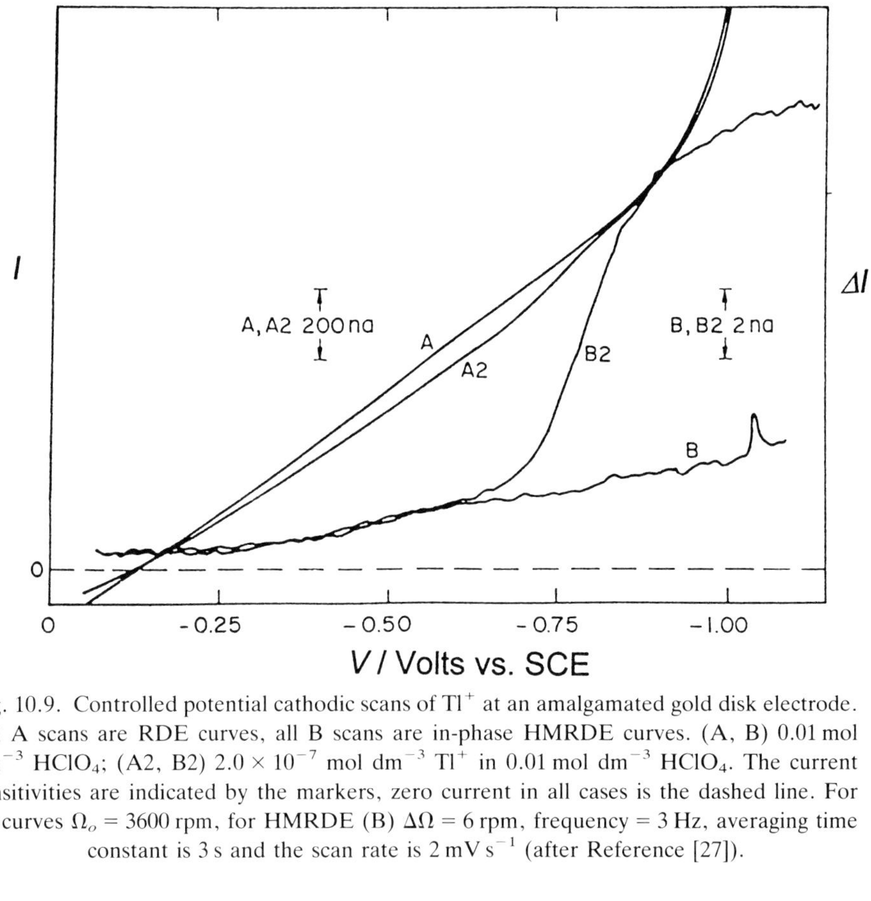

Fig. 10.9. Controlled potential cathodic scans of Tl^+ at an amalgamated gold disk electrode. All A scans are RDE curves, all B scans are in-phase HMRDE curves. (A, B) 0.01 mol dm^{-3} $HClO_4$; (A2, B2) 2.0×10^{-7} mol dm^{-3} Tl^+ in 0.01 mol dm^{-3} $HClO_4$. The current sensitivities are indicated by the markers, zero current in all cases is the dashed line. For all curves $\Omega_o = 3600$ rpm, for HMRDE (B) $\Delta\Omega = 6$ rpm, frequency = 3 Hz, averaging time constant is 3 s and the scan rate is 2 mV s^{-1} (after Reference [27]).

5×10^{-8} mol dm^{-3}. Figure 10.9 shows the RDE and in-phase HMRDE response for the detection of 2.0×10^{-7} mol dm^{-3} Tl^+ in 0.01 mol dm^{-3} $HClO_4$ at an amalgamated gold electrode. It can be seen that the HMRDE dramatically reduces background currents, making the amperometric detection of trace Tl^+ in a potential range that is dominated by residual currents and solvent electrolysis possible.

Rosamilia and Miller [35] presented instrumentation developments allowing rapid potential scans to be made at the HMRDE. They demonstrated a linear concentration calibration in the range 2–10 ×

10^{-6} mol dm^{-3} for dopamine on a carbon RDE with scan rates up to 80 mV s^{-1} and 15 Hz modulation of rotation rate.

A detailed examination of the mass transport effects of the HMRDE has been made. At low rotation speeds and for small amplitude modulations (as defined in Section 10.3.6.2) the response of the current is found to agree exactly with that predicted by the steady-state Levich theory (equations (10.15)–(10.17)) [27, 36, 37]. Theoretical and experimental application of the HMRDE, under these conditions, to cases where the electrode reaction rate constant was comparable to the mass-transfer coefficient has also been made [36]. At higher rotation speeds and/or larger amplitude modulations, the observed current response deviated from the expected Levich behaviour.

The first results deriving the frequency response for the HMRDE were given by Tokuda *et al.* [29]. This treatment also represented the first approach to a formal theory of EHD (Section 10.5). The theoretical results were experimentally confirmed over a wide range of Schmidt numbers (230–2100) and p values for the reduction of Fe^{2+} to Fe^{3+} in 1.0 mol dm^{-3} H_2SO_4 in the temperature range 25–85 °C. Although this treatment is mathematically more complicated than the Levich approach, it does allow measurements to be quantitatively interpreted at higher frequencies and therefore under shorter times with easier signal processing. As the frequency response of the sinusoidal HM is dependent only on p and Sc, it is possible to obtain the diffusion coefficient of an electroactive species independent of knowledge of its concentration or the number of electrons transferred, by monitoring relaxation phenomena as a function of Ω_o after an initial perturbation has been applied. This technique has been used to establish the diffusion coefficient of O_2 in 1.0 mol dm^{-3} KOH [14] and verify the number of electrons transferred for the reduction of C_{60} and C_{70} in dichlorobenzene [38]. The theory of HMRDE at high rotation speed has been further developed [39, 40]. Unfortunately with small amplitude modulation, the resulting modulated electrochemical signal is only a small proportion of the mean current, resulting in possible signal to noise limitations. This problem has been addressed both experimentally and theoretically, by considering the response of the RDE to a large amplitude modulation [32, 41–43].

10.4.1.2 Stopped rotation voltammetry

One special case of HMRDE is stopped-rotation voltammetry (SRV) introduced by Wang [44]. As the name suggests, the rotation is switched on (e.g., 1600 rpm for 3 s) and off (e.g., for periods of 25 s) while maintaining a constant applied potential. The major advantage of this technique is exploitation of all the rotation-dependent analytical signal, i.e., 100% modulation, although lengthy "off" times must be employed to achieve complete relaxation of the concentration boundary layer after the step (see Section 10.3.3). ΔI is measured at discrete applied potentials and the full current-voltage characteristics plotted in a point-wise fashion. In general, the modulation frequency is limited to low values due to the inertia and response time of the motor electrode assembly. SRV has been used to detect sub-micromolar levels of ascorbic acid in phosphate buffer at a glassy carbon electrode [44]. A detection limit of 4×10^{-8} mol dm^{-3} was estimated. Unfortunately data acquisition times are long (ca. 20 mins), increasing the possibility of contamination of the electrode surface with impurities. To address this point Wang and Freiha introduced scanning potential SRV [45] where the potential of the electrode was linearly scanned while the rotation was switched on and off. This reduced experimental times to ca. 3 mins. Detection levels of 7×10^{-8} mol dm^{-3} for dopamine at a glassy-carbon disk electrode were achieved.

Amperometric detection of the end-point of a titration is a long-standing technique in analytical chemistry [46]. It is applied particularly to titrations with iodine. Detection of the difference in current with the solution stirred and then stationary is a very simple method of increasing the sensitivity and reliability of detection of the end point. Wang developed this method using SRV, or by simply interrupted stirring using a magnetic stirrer. One example is the determination of ascorbic acid by an iodine titration [47]. As a result of modulating the current response by switching a rotating disc on and off, the signal due to oxidation of iodide was separated from the signals due to the irreversible oxidation of ascorbic acid and the drifting background imposed by use of a solid electrode. The detection limit for the titration was ca. 1×10^{-6} mol dm^{-3} ascorbic acid. Each titrant addition corresponded to a decrease in the ascorbic acid concentration of 100×10^{-9} mol dm^{-3}. The concentration of ascorbic acid in a mixture of 5×10^{-6} mol dm^{-3} of the acid with 5×10^{-6} mol dm^{-3} dopamine (both have oxidation potentials ca. +0.35 V versus a Ag/AgCl reference at a glassy carbon electrode) was also determined using SRV in conjunction with titration of ascorbic acid with iodine.

10.4.1.3 Pulsed rotation voltammetry

A variant of SRV is pulsed rotation voltammetry (PRV) [48–50]. Here, the rotation rate of a RDE is switched between two well separated speeds (e.g., 500 and 1500 rpm) at a rate less than $0.1\ s^{-1}$. This results in a greater measurable steady-state current difference than for HMRDE leading to increased sensitivity. The voltammetric behaviour is recorded as for SRV. Using PRV at a glassy carbon electrode, Blaedel and Engstrom [48] reported an experimental detection limit of 1×10^{-8} mol dm^{-3} for $Fe(CN)_6^{4-}$ in a phosphate buffer. PRV has additionally been used for kinetic studies of different heterogeneous redox systems (see also Section 10.5.3.1), such as Fe^{2+}/Fe^{3+}, riboflavin/flavin co-enzymes and NAD^+/NADH [2]. Further increases in the measured current difference have been achieved by employing a conductive porous material as the disk electrode to increase electrode surface area [49], and further still by incorporating this electrode in an electrochemical flow-through cell [50].

10.4.2 Vibrating electrodes

Vibration of an electrode is an experimentally simple way of producing hydrodynamic modulation. Whilst the advantage is experimental simplicity, the disadvantage is that the hydrodynamic regime is difficult to describe accurately (see Section 10.3.6.3).

10.4.2.1 Wire electrode and microcylinder electrode

The use of vibrating wire electrodes for electroanalysis was introduced by Harris and Lindsey [51–54]. One particular objective was reliable end-point detection in iodimetric titrations. Subsequent authors showed that these simple devices offered very low detection limits in amperometric measurements and a very wide linear dynamic range. Pratt and Johnson [55] quote a detection limit of 5×10^{-10} mol dm^{-3} and a range of 6.3 decades for I^- in a flow-injection system with a vibrating wire amperometric detector. Schuette and McCreery [34], using a vibrating microcylinder electrode at high frequency (80–160 Hz), achieved a detection limit approaching 1×10^{-8} mol dm^{-3} for ferrocene and excellent rejection of the background signal at a Pt electrode in aqueous solution. Further developments to this technique were made using a double modulation method, where the potential and vibration of the electrode were modulated simultaneously [6]. The current change was then recorded at the sum frequency. Fourier transforms of the raw data produces peak-shaped voltammograms, improving detection limits and sensitivity [56]. Using this technique a detection limit for 1,1′-Bis(hydroxymethyl)ferrocene in water at pH 7 of 4×10^{-7} mol dm^{-3} was quoted [6].

Comparison of the work of different authors demonstrates that the results obtained from a vibrating wire electrode are very sensitive to the geometry of the electrode. This is because the instantaneous velocity of the electrode relative to the solution depends upon both the frequency and amplitude of the vibration whilst the details of the flow pattern depend upon the radius of the wire and upon the ratio of wire radius to vibration amplitude. If the wire diameter is small enough, then the current follows the wire velocity. If the wire diameter is large, then the modulation of current caused by the oscillating wire velocity is greatly diminished, though there is an enhancement of the mean current. If the wire diameter is small enough, then the flow across the surface remains laminar up to high vibration frequencies. If the wire diameter is large, then there is a transition to turbulent flow at low frequencies.

Schuette and McCreery [34] demonstrated that with decreasing wire diameter there was a significant increase in current enhancement and modulation depth. This approached 100% modulation for a wire of diameter, $d = 25$ μm vibrated at 160 Hz. They showed that in these circumstances, for low Re numbers, the limiting current strictly followed the wire velocity and used [6] an empirical power-law correlation of mass-transfer coefficient to flow velocity: $I_{\text{lim}} = I_{\text{min}}(1 + \Delta I|\cos(\omega.t)|^{s})$ with $s \approx 0.7$. They also noted that the frequency and amplitude dependence of the mean current, and the modulation depth, was linked to whether the flow was strictly laminar or not. Flow modelling indicated that for $\text{Re} > 5$ where $\text{Re} = \bar{u}\, d/v$, there was separation of the boundary layer at the wire surface, when $a/d \gg 1$. For $\text{Re} > 40$ the flow pattern became very irregular. Under these circumstances, a direct relation between velocity and current should be lost, and they indeed showed that the modulation depth decreased steeply with increase of wire diameter, down to ~10% for 0.8 mm diameter wire.

The above studies supported the findings of Lindsay [54], who for a 0.5 mm diameter electrode vibrated with 3 mm amplitude, showed a weak linear dependence of mean current on amplitude and frequency, for sufficiently low values of these parameters. The current eventually became independent of a and f at sufficiently high values. The maximum augmentation of the current due to vibration was ca. 15%. With similar electrode dimensions and vibration amplitudes, Pratt and Johnson [57]

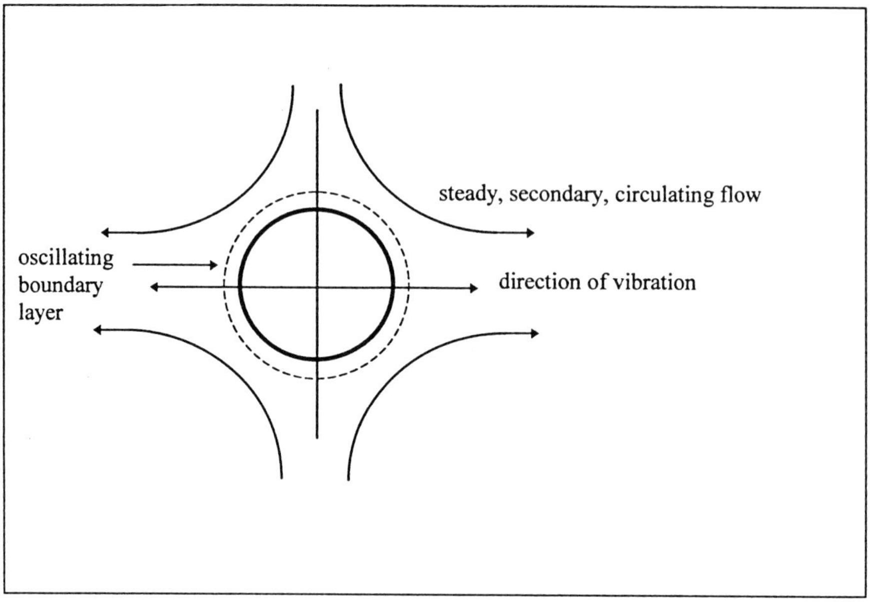

Fig. 10.10. Schematic of the flow pattern around a vibrating wire (cross section).

showed $\overline{i_{\rm lim}} \propto \mathrm{Re}^{m}$, with m typically 0.62, but varying both with the range of vibration frequencies employed and with interchange of nominally identical electrodes.

For sufficiently large electrodes with a small vibration amplitude, $a/d \ll 1$, a solution of the hydrodynamic problem is possible [58, 59]. As well as the periodic flow pattern, a steady secondary flow is induced as a consequence of the interaction of viscous and inertial effects in the boundary layer [13] as shown in Fig. 10.10. It is this flow which causes the enhancement of mass-transfer. The theory developed by Schlichting [13] and Jameson [58] applies when the time of oscillation, ω^{-1} is small in comparison with the time taken for a species to diffuse across the hydrodynamic boundary layer (thickness $\delta_H \approx (\nu/\omega)^{1/2}$; diffusion timescale δ_H^2/D), i.e., when $\nu/D \gg 1$. Re needs to be sufficiently high for the calculation to converge but sufficiently low such that the flow does not become turbulent. Experiment shows that, for large diameter wires (radius, r, $\sim$1 cm), the condition is $\mathrm{Re} \leqslant 2000$. The solution: $\mathrm{Sh} = 0.746\mathrm{Re}^{1/2}\ \mathrm{Sc}^{1/3}(a/r)^{1/6}$, where Sh (the Sherwood number) $= k_m r/D$ and k_m is the mass-transfer coefficient, was compared with experiment for the dissolution of rods of benzoic acid [58]. The experimental results were slightly high in comparison with the theory, but the expected correlation held over the range Re = 2–2000, $a/r = 0.1–0.3$.

The theory was refined by Dumarque and Humeau [59], who gave the expected mean current response for an electrochemical experiment at the limiting current as:

$$\mathrm{Sh} = \frac{k_m \, . \, r}{D} = \frac{\overline{I_{\mathrm{lim}}}}{(2\pi \, . \, DLcnF)} = 0.5166\mathrm{Re}^{1/2}\,\mathrm{Sc}^{1/3}\left(\frac{a}{r}\right)^{1/6},$$

$$\overline{I_{\mathrm{lim}}} = 2.580lrnFcD^{1/2}\mathrm{Sc}^{-1/6}\left(\frac{a}{r}\right)^{2/3} f^{1/2},$$

where l is the active length of the electrode. The theory described experimental results only for $\mathrm{Re}^{1/2}\ \mathrm{Sc}^{1/3}(a/r)^{1/6} < 150$ (in practical terms $a/r \leqslant 0.06$, $f < 4$ Hz and Re < 2000). At larger values there was an enhancement of the mass-transfer coefficient above that predicted.

10.4.2.2 Microband electrodes

Vibrating microband electrodes (introduced in Section 10.3.6.3) prepared by screen printing have been used by Williams *et al.* [33] for the detection of chlorine in water. No prior calibration was necessary and a detection limit of approximately 0.3×10^{-6} mol dm^{-3} was determined with linearity into the 1×10^{-3} mol dm^{-3} range. The microband electrode construction procedure is advantageous to electroanalysis due to the low cost, ease of preparation and repeatability of bulk manufacture [33]. Reagents could also be printed onto the device which were rapidly dissolved and mixed into the test solution by the vibration. The resulting microband electrodes when vibrated showed the signal enhancement typical of vibrated wire electrodes (see Section 10.4.2.1), the modulation depth and signal enhancement being smaller than those characteristic of a vibrated microwire but larger than those from larger diameter wires.

10.4.2.3 Multiple microband electrode

Multiple microband electrodes have been used to implement, in a simple and inexpensive way, generator-collector methods of electrochemical titration [33]. This is exemplified by the determination of ascorbic acid and thiamine using electrogenerated Fe^{3+}. The microband electrodes were prepared in a similar manner to that discussed above in Section 10.4.2.2.

The basic experimental idea is adapted from the earlier work of Albery *et al.* [60], who used rotating ring-disc electrodes to perform generator-collector titration experiments, as shown in Fig. 10.11. Either the diminution of collector current upon addition of analyte, or the value of the generator current required to produce a given (low) collector current can be used as the analytical signal. The method is analogous to a titration, with the generator current being the analogue of the volume of titrant added. The method relies on a balance of diffusional fluxes rather than of molar amounts. When the flux of the electrogenerated titrant away from the surface matches and then exceeds the flux of analyte towards the surface, a signal will appear on the collector electrode.

If a stationary multiple microband electrode is used, then the collector current is rather sensitive to adventitious vibrations. If the electrode assembly is vibrated parallel to the inter-electrode gap, then although the collection efficiency is reduced the collector current is now insensitive to such random vibrations (of a non-modulatory nature). Repeatable, reliable titration using electrogenerated reagents has been demonstrated in this way [33].

10.4.2.4 Disc electrode

Podesta *et al.* [61] showed for a horizontal disc electrode, under longitudinal vibration, the current was of a complex form with the amplitude of the ac component amounting to only 1.5% of that of the average current. The major contribution was an enhancement of the current caused by a steady circulating flow, established as a consequence of the oscillation. This is similar to the effects seen with a vibrating wire, shown in Fig. 10.10. With disc diameters in the range 0.3–1.5 cm and an oscillation amplitude and frequency of <0.11 cm and 50 Hz respectively, the oscillating flow pattern was constrained to a cylinder defined by the area of the disc surface and the oscillation amplitude. At larger amplitude, turbulent motion set in. Defining the average Reynolds number in terms of the rms velocity, $\mathrm{Re} = \omega ad/\nu\sqrt{2}$, it was shown that the correlation $\mathrm{Sh} = (d\overline{I_{\mathrm{lim}}}/AnFcD) = 0.07\mathrm{Re}^{3/4}\,\mathrm{Sc}^{1/2}$ was valid over the range Re = 10 to 10^3 and $\mathrm{Sc} = 10^3$–10^5.

The current to planar electrodes (1.1 mm square) oriented vertically and

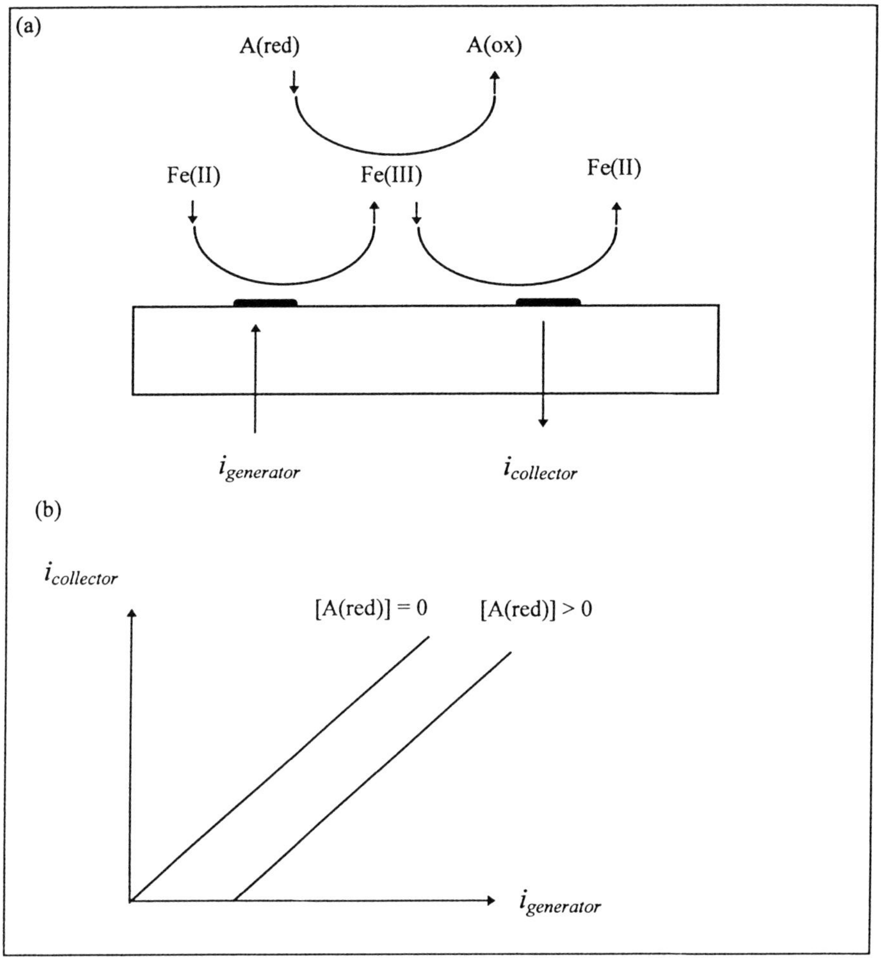

Fig. 10.11. (a) Generator-collector scheme for electrochemical titration. The generator electrode is galvanostatically controlled so the titrant, Fe(III) given as an example, is formed at the generator at a fixed rate. The collector is controlled potentiostatically to reduce Fe(III) at a diffusion-limited rate. The collector current is therefore determined by the flux of Fe(III) across the gap between the electrodes. The titrant reacts with the analyte, denoted A(red), so the flux across the gap is decreased and hence the collector current is decreased. As the concentration of the analyte is increased, the generator current required to give a current at the collector increases. (b) Schematic dependence of $i_{collector}$ on $i_{generator}$. The line relating the two is displaced as the analyte concentration increases, the displacement being approximately proportional to the analyte concentration.

vibrated along the surface normal was also not significantly modulated: instead, a uniform circulating flow appeared to be set up, which served to increase the limiting current [33]. Speculations as to the reason for this effect included: firstly, that the electrode was aligned vertically, facing horizontally, so the circulation could have been caused by the effect of the force due to gravity; and secondly that the motion was not strictly normal to the surface, being instead along a slight arc and hence having a small component parallel to the surface.

10.4.2.5 Ultramicrodisc electrode

In a potentiostatic step experiment on an ultramicrodisc electrode, the current attains a limiting value in a time of the order r^2/D. In addition to this time scale becoming shorter with decreasing r, the mass transport rate of species to the electrode increases. As a consequence, one motivation for the development of devices using UMEs has been the resulting insensitivity of the limiting current to fluctuations of solution flow rate [62]. Hence UMEs have found widespread use as electrochemical detectors in fluid flow experiments, such as amperometric liquid chromatography (e.g., Ref. [63]) and flow injection analysis (e.g., Ref. [64]). Incidentally, this implies that the current should also be insensitive to vibration.

This latter statement is only true if the time scale for the motion is long compared to the time taken for the electrode to reach a mass transport controlled steady-state. Vibration effects upon the current will be observed when the time scale of the vibration (ω^{-1}) is of the order r^2/D or less; i.e., $\omega \geqslant D/r^2$. This is 10 Hz for a 10 μm radius disc or 1000 Hz for a 1 μm radius disc electrode, assuming a typical aqueous solution diffusion coefficient of 1×10^{-5} cm^2 s^{-1}. Vibration as a means of reducing fluid-flow artefacts on the amperometric measurement of oxygen concentration using a 100 μm disc electrode (a large UME) was described by Towe [65]. The electrode was vibrated using a piezo-electric actuator with a typical vibration frequency and amplitude of 600 Hz and 1 mm respectively. Towe demonstrated an enhancement and modulation of the signal.

10.4.2.6 Ultramicroelectrode tip position modulation

As a consequence of the extremely high mass transport rates attainable at UMEs, HM effects at vibrating UMEs are normally small. However when the electrode is in close proximity to a surface these effects can be greatly enhanced. Although the modulation of the signal is not hydrodynamic in origin, UME tip position modulation (TPM) is included in this review for completeness.

The scanning electrochemical microscope (SECM) [66, 67] is a scanned probe microscope, whose response is dependent on mass transport between an UME and a sample, which in turn is governed by chemical reactions occurring at the sample. The most common mode of operation of the SECM is the feedback mode [68]. Here the UME is used to generate a redox mediator in its oxidised or reduced form, at a diffusion-controlled rate, with the UME in close proximity to the surface of a substrate (of the order of the electrode radius or less). For an insulating sample, diffusion of species to the surface is simply hindered, the degree of hindrance increasing as the tip/substrate separation, d_{secm}, is decreased, hence the limiting tip current decreases. In contrast for a conducting surface it is possible to regenerate the original form of the redox mediator at the surface providing an additional flux of species to the UME tip. In this case as d_{secm} decreases, the tip current increases. These principles are illustrated schematically in Fig. 10.12. Mass transport in the feedback mode has been extensively treated [68, 69].

In essence, the feedback current flowing as an UME tip is scanned across a sample surface yields information on the surface topography and redox activity of the substrate. In order to discriminate between these two substrate effects, which both contribute to the limiting current, it is often necessary to work with either a well-defined surface (to obtain reactivity data) or one of known uniform activity (to obtain topographical information). The introduction of SECM tip position modulation [70] goes part way to addressing this problem, by separating topographical and reactivity effects and also improves the resolution capabilities of the technique as an imaging device. In SECM TPM the tip is moved sinusoidally in the y direction, normal to the tip and sample, so that its distance from the substrate is modulated by $\pm a \sin 2\pi ft$, about d_{secm}. Typical values of f and a in the range 40–160 Hz and $(0.01–1.0)r$ respectively, were employed, where r is either 1 or 4 μm.

As the tip position is modulated, the tip current alters accordingly. For a negative change in y (move closer to the substrate) the tip current increases over a conductor but decreases over an insulator with the two currents occurring 180° out of phase. By detecting the in-phase component of the modulated signal it is possible to identify exclusively the redox activity of the surface, with the added benefits of noise reduction (due to

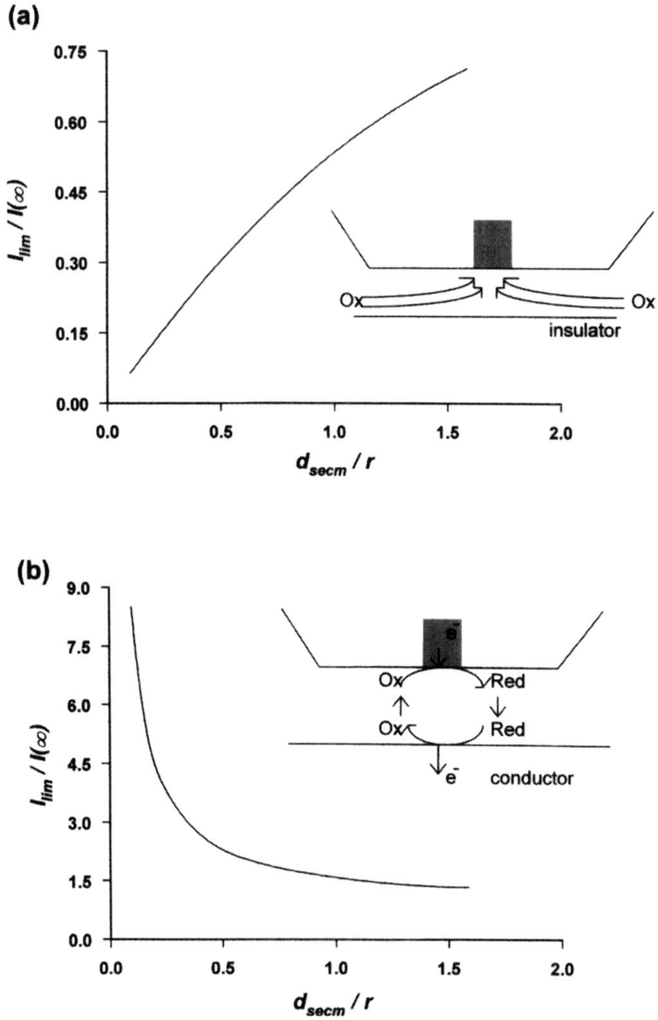

Fig. 10.12. General principles of the SECM feedback mode. The UME, normally a disk electrode of radius r, is used to generate a redox mediator in its oxidised or reduced form (a reduction process is shown here) at a diffusion-controlled rate. As the UME approaches an insulating surface (a) diffusion of Ox to the electrode simply becomes hindered and the recorded limiting current is less than the steady-state value measured when the electrode is placed far from the surface, in the bulk of the solution, $I(\infty)$. This effect becomes more pronounced as the tip/substrate separation, d_{secm}, is decreased. As the UME approaches a conducting surface (b) the original form of the redox mediator (Ox) can be regenerated at the substrate establishing a feedback cycle and an additional flux of material to the electrode. In this case I_{lim} will be greater than $I(\infty)$.

the rejection of low frequency noise) and increased sensitivity. This is especially evident against a redox inert surface, where the in-phase component of the modulated current now increases with decreasing d_{secm} as a result of the removal of the dc offset. SECM TPM has been used to image an interdigitated array electrode, consisting of 3 μm wide Pt bands separated by 5 μm wide SiO_2 bands [70]. Greatly improved resolution was evident compared with conventional SECM operating in the feedback mode. Typical images of a section of the interdigitated array electrode are shown in Fig. 10.13. Figures 10.13(a) and 10.13(b) are plotted as gray scale with the smaller (a) or more negative (b) current shown as a darker shade. It is clearly evident that TPM enhances the quality of the topographical image considerably, especially in insulating areas. Presentation of the TPM image can be improved further, as shown in Fig. 10.13(c), by using the absolute value of the TPM in-phase current with separate gray or colour scales for the insulating and conducting regions.

10.4.3 Flow modulation methods

Electroanalytical detection in flowing streams is particularly important as it offers, amongst other advantages, the opportunity for multiple sample analysis, high signal to noise ratios, small sample volumes and low concentration detection limits. One drawback is the presence of the notorious background signal, which can limit electrochemical measurement to a specific potential window or concentration level. Methods involving interruption or pulsation of flow in a channel or through a tube with electrochemical detection have attempted to overcome this problem.

10.4.3.1 Stopped channel or tubular flow

In 1971 Blaedel and Boyer [71] introduced the stopped flow (SF) tubular electrode in an effort to decrease the concentration detection limit in their steady flow tubular electrode. At 0.1×10^{-6} mol dm^{-3} concentrations of $Fe(CN)_6^{3-}$ the analytical signal was only 10% of the background current. By measuring the difference between a high flow rate (ca. 3 ml min^{-1}) and zero flow rate they immediately obtained improved sensitivity for the tubular electrode. Figure 10.14 is a schematic showing typical diffusion limited current-time tracings for a blank solution and a sample solution containing an electroactive mediator under SF conditions. In this situation

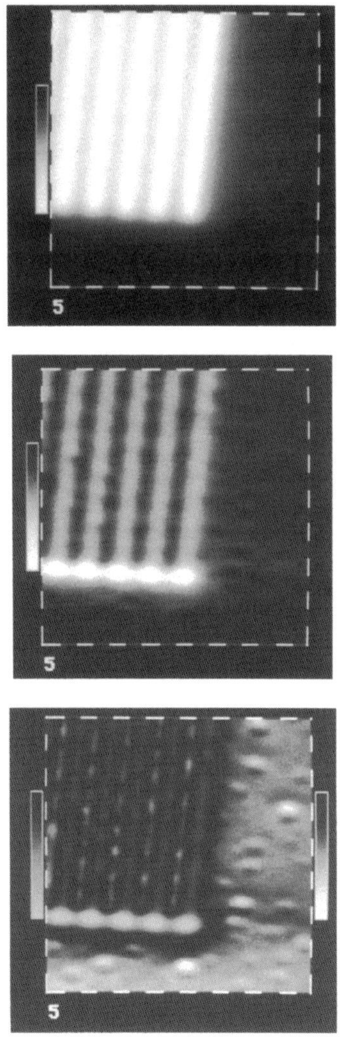

Fig. 10.13. dc SECM and TPM images of a 80 μm × 80 μm portion of an interdigitated array substrate. Conditions: mediator, 1.5×10^{-3} mol dm^{-3} $Ru(NH_3)_6^{3+}$ in pH 4.0 buffer; tip electrode, 1 μm radius Pt; $d_{secm} \approx 0.9$ μm; $f = 160$ Hz, $a/r = 0.1$; scan speed, 10 μm s^{-1}. (A) SECM image, scale is 200–850 pA, $I(\infty) = 470$ pA. (B) TPM SECM image, scale −7 to 15 pA_{rms}. (C) Absolute value TPM SECM image, left scale = 0–15 pA_{rms}; right scale = 0–7 pA_{rms} (after Reference [70]).

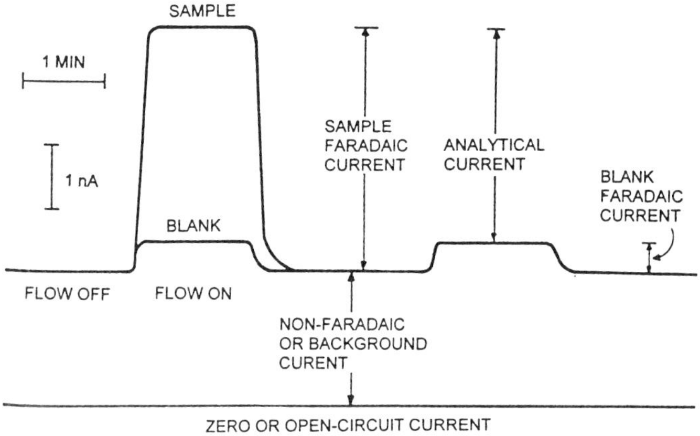

Fig. 10.14. Typical steady-state current-time curves for a blank solution and sample solution under stopped flow conditions. The annotations give the definitions of the various currents which can be measured by the stopped flow technique (after Reference [71]).

the signal of interest is the analytical current, which is taken as being proportional to the concentration of the electroactive species of interest.

This technique is analogous to SRV, described in Section 10.4.1.2, and also offers minimum current attenuation as all the steady-state analytical signal is exploited for the measurement. With a tubular Pt electrode typically 2 mm diameter and 4 mm long, concentrations in the range 0–1×10^{-6} mol dm^{-3} $Fe(CN)_6^{4-}$ were determined with a standard deviation of $\pm 0.01 \times 10^{-6}$ mol dm^{-3}. By employing a plug of reticulated vitreous carbon (RVC) disks as the working electrode [72], offering a high ratio of porous electrode surface to solution volume, it was possible to further increase the sensitivity of the technique. Given the reduced background currents resulting from measuring the current difference, and the increased working electrode signal from utilising RVC, a detection limit of ca. 1×10^{-9} mol dm^{-3} was achieved for $Fe(CN)_6^{4-}$.

Using an alternative geometry Evans *et al.* [16] developed the channel stopped flow method (CSFM). This technique, to date, has been used to measure solution diffusion coefficients (independent of knowledge of the concentration of the electroactive species) and crystal dissolution kinetics. The channel flow cell consisted of a rectangular electrode, typical dimensions 2.5 mm long and 6.25 mm wide, situated in a rectangular duct, 10 mm wide and 0.25–1.0 mm high. The electrode was placed a suitable distance from the solution inlet such that fluid flow over it was fully-developed and Poiseuille in nature. After arresting the flow the ensuing current-time profile was measured, for the diffusion-controlled electrolysis of an electroactive mediator. As the diffusion layer relaxed the chronoamperometric profile gave information on the value of D for the electroactive species of interest. The transient decay time was found to decrease with an increase in the volume flow rate, V_f, and D and a decrease in the height of the channel flow cell, all as predicted by theoretical modelling. A typical match between experiment and theory for the oxidation of $Fe(CN)_6^{4-}$ in a 0.5 mm high cell with an initial volume flow rate of 0.197 $cm^3\,s^{-1}$ is shown in Fig. 10.15.

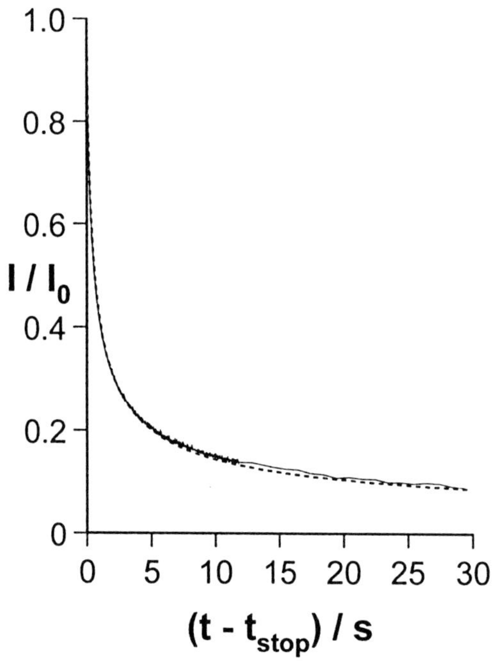

Fig. 10.15. Experimental (—) and theoretical (--) chronoamperometric response for the diffusion-limited oxidation of 2×10^{-3} mol dm^{-3} $Fe(CN)_6^{4-}$ in 0.1 mol dm^{-3} KCl at a rectangular electrode, 2.5 mm long and 6.25 mm wide, in a 0.5 mm high channel flow cell under channel stopped flow conditions. The initial volume flow rate of the solution was 0.197 $cm^3\,s^{-1}$, which gave a limiting current at the channel electrode, defined as I_o. At time, t_{stop}, solution flow was retarded (Evans *et al.*, in preparation). The theoretical data has been simulated assuming $D_{Fe(CN)_6^{4-}} = 6.5 \times 10^{-6}\ cm^2\,s^{-1}$.

10.4.3.2 Pulsed channel or tubular flow

The pulsed flow (PF) technique at the tubular electrode was introduced by Blaedel and Iverson [73]. The sensitivity of the electrode, over that achieved with steady flow, was increased by monitoring the change in current as the solution flow rate was switched between two well separated values (e.g., 9 ml min^{-1} and 0.5 ml min^{-1}), in a manner similar to PRV. ΔI in this case is given by equation (10.21). The time between flow rate pulses was of the order 10–30 s.

$$\Delta I = I_{\lim,hi} - I_{\lim,lo} = 2.01nF(Drl)^{2/3}c_b(V_{f,hi}^{1/3} - V_{f,lo}^{1/3}). \qquad (10.21)$$

The lower the minimum V_f or higher the maximum V_f the greater the analytical signal, although the sensitivity of PF will never be as high as SF due to a nonzero $V_{f,lo}$. Blaedel and Inverson found that for a tubular electrode 1.4 mm in diameter and 2.5 mm long, the limit of detection was 0.05×10^{-6} mol dm^{-3} $Fe(CN)_6^{3-}$, with a linear calibration of ΔI versus concentration up to 4×10^{-6} mol dm^{-3}, the highest concentration tested.

Unfortunately the long cycling times employed in these experiments, in order for the system to reach steady-state, precluded their use as detection systems in FIA or electrochemical LC. In an effort to address this problem and reduce this relaxation time, different thin-layer cells were designed which decreased the relaxation distance of the concentration boundary layer when the flow was pulsed [74, 75]. Figure 10.16 illustrates this point by showing the current response at a wide (3 mm diameter and 4 mm long) and thin (0.2 mm diameter and 1 mm long) channel tubular electrode under identical pulsed flow conditions. Using this thin layer approach, rapid PF techniques with response times of ca. 0.5 s have been developed [74]. At higher pulse rates >0.5 Hz the observed current difference decreases from the expected value, as predicted by the theory given in Section 10.3.6.1. At pulse rates >2 Hz, ΔI tends to zero [74]. Nevertheless, the sensitivity of these systems is very high, for example, using a small porous carbon flow-through electrode a detection level for $Fe(CN)_6^{3-}$ of 2×10^{-6} A/2×10^{-6} mol dm^{-3} was obtained [74].

10.4.3.3 Modulated jet flow: The radial flow microring electrode

In the radial flow microring electrode (RFMRE) [76] geometry, solution flows from a capillary (typical nozzle diameter 100 μm), which is positioned very close to a planar substrate using micropositioners. The ring electrode,

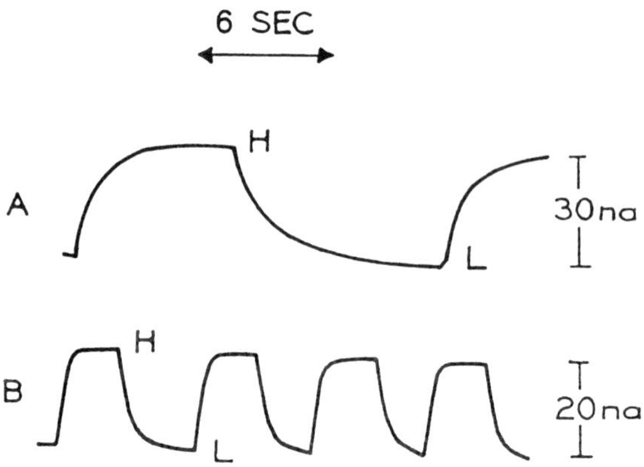

Fig. 10.16. Pulsed flow current response for the oxidation of 3×10^{-6} mol dm^{-3} $Fe(CN)_6^{4-}$ at 0.4 V versus SSCE in 0.1 mol dm^{-3} KCl and phosphate buffer, pH 7.4. (A) Wide channel tubular electrode (3 mm diameter, 4 mm long). (B) Thin channel tubular electrode (0.2 mm diameter, 1 mm long). H and L represent flow rates of 6 and 3 ml min^{-1} respectively (after Reference [74]).

which has a typical ring thickness 0.1–0.4 μm is positioned around the edge of the capillary and electrochemically insulated using an epoxy resin. As fluid leaves the capillary it is forced into the nozzle/substrate gap (typical separation of 5–40 μm) and flows radially past the ring electrode. This experimental arrangement is depicted schematically in Fig. 10.17(a). The mass transport characteristics of the RFMRE have been fully analysed [76], demonstrating that this device is effectively analogous to a microband channel electrode [8]. Under these conditions the mass transport-limited current to the electrode in the absence of diffusional edge effects is given by equation (10.22).

$$I_{\lim} = 0.925nFc_b\left(\frac{Dx_{rfmre}w_{rfmre}}{h}\right)^{2/3} V_f^{1/3}, \tag{10.22}$$

where $2h$ is the distance between the substrate and the nozzle, x_{rfmre} is the ring thickness and w_{rfmre} is the circumference of the ring electrode.

Equation (10.22) demonstrates that in the RFMRE geometry, modulation of the current, for a given electrode, can be achieved in two ways: variation of V_f or h. For experimental simplicity h was modulated, in a

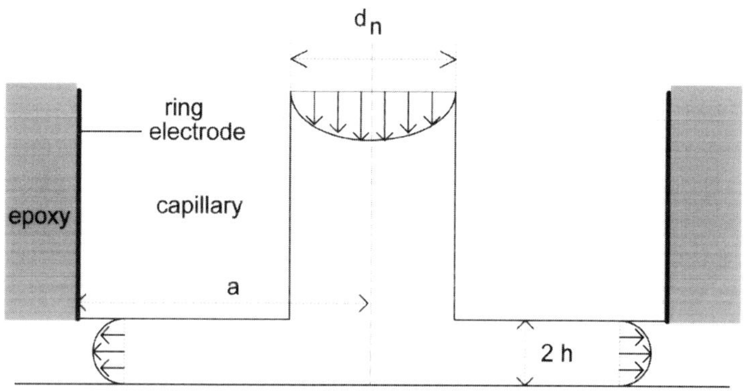

Fig. 10.17(a).

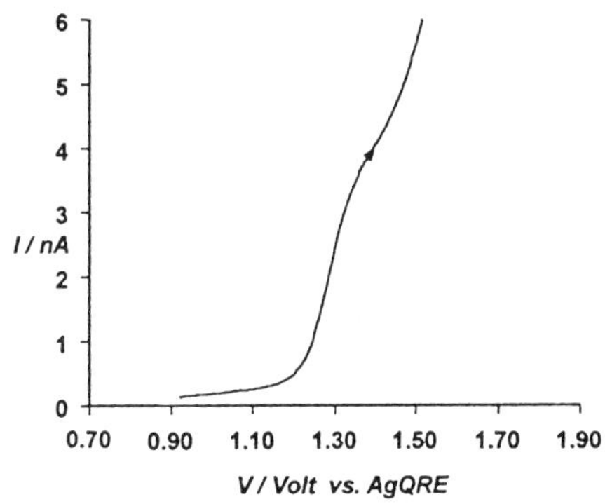

Fig. 10.17(b).

manner similar to that described for SECM in Section 10.4.2.6. For a range of modulation amplitudes (1 μm < a < 6 μm) the current difference was well-defined [77] in terms of steady-state Levich theory, equation (10.23).

$$\Delta I_{\mathrm{lim}} = 0.925nFc_b(Dx_{rfmre}w_{rfmre})^{2/3}V_f^{1/3}(h_{hi}^{-2/3} - h_{lo}^{-2/3}). \quad (10.23)$$

In addition since extremely high mass transport rates are achievable at small h values, in excess of 2 cm s^{-1} [76], concentration boundary layer

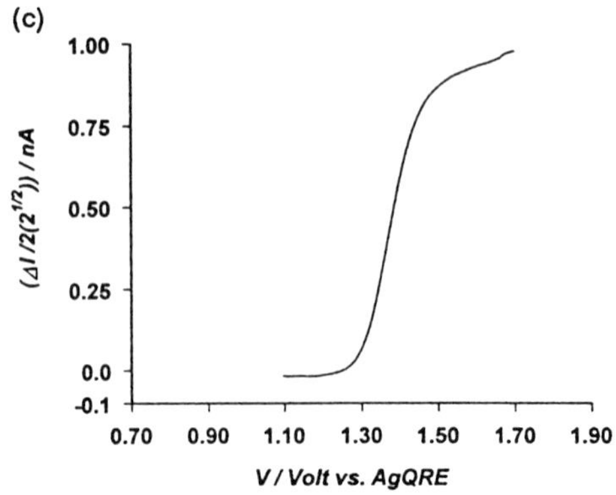

Fig. 10.17. (a) Schematic cross-section of the RFMRE illustrating the fluid velocity profile. The ring electrode is placed on the outer side of the nozzle and is exposed at the end. The epoxy coating provides electrical insulation and completes the formation of a thin channel with the glass substrate through which solution radially flows. A (b) RFMRE and (c) corresponding in-phase HMRFMRE current-voltage curve for the diffusion-controlled oxidation of 1.0×10^{-5} mol dm^{-3} $Ru(bipy)_3^{2+}$ in 0.1 mol dm^{-3} KNO_3, versus a Ag quasi reference electrode (AgQRE). The RFMRE is characterised by an internal nozzle diameter = 100 μm, inner radius of ring electrode = 71 μm, $x_{rf\,mre}$ = 0.28 μm, V_f = 1.67×10^{-2} cm^3 s^{-1}, h = 15 μm (b) and h = 15 μm ± 3.8 μm (c). (Macpherson and Unwin, submitted).

relaxation times are very small. For example for a RFMRE system characterised by an internal nozzle diameter = 88 μm, inner radius of ring electrode = 66 μm, $V_f = 1.67 \times 10^{-2}$ cm^3 s^{-1}, $x_{rf\,mre}$ = 0.14 μm and Re = 20, relaxation times of the order of 1×10^{-5} s and less are achieved, ultimately leading to short analysis times. These relaxation times are spectacularly shorter than those obtained at conventional channel and RDEs. This would be an advantageous feature for the device were it to be employed as an electrochemical detector in a flow analysis system. Figure 10.17 shows (b) a typical RFMRE current-voltage curve and (c) the corresponding in-phase I–V signal, for the oxidation of 1×10^{-5} mol dm^{-3} $Ru(bipy)_3^{2+}$ in 0.1 mol dm^{-3} KNO_3. Solvent decomposition mostly masks the limiting current for oxidation of this species at the platinum electrode in the steady flow system (Fig. 10.17(b), whereas Fig. 10.17(c) demonstrates excellent rejection of the background signal.

10.4.3.4 Ultrasonic modulation

Dewald and Peterson [78] implemented a form of hydrodynamic modulation using pulsed ultrasonic agitation. The ultrasonic probe (20 kHz, up to 54 W) was centred 1 cm above a working electrode (3 mm diameter glassy carbon) and pulsed on and off. The working electrode was operated at the limiting current for the species to be detected and the difference in current between the states with sound on and off recorded. The current decay time after the pulse was ca. 1 s, whilst the pulse width was 0.2 s and cycle time typically 1 s. The stirring was caused by bubble collapse inducing micro-jets which impact on the electrode surface [79, 80]. The detection limit for reversibly electroactive species was ca. 0.1×10^{-7} mol dm^{-3}, with a relative standard deviation, caused by fluctuations in the stirring, of 1–4% of the limiting current. The sensitivity was 50 nA/1×10^{-6} mol dm^{-3}.

10.5 ELECTROHYDRODYNAMIC IMPEDANCE STUDIES

10.5.1 General remarks

In these studies, a modulation of the transport rate is imposed upon a steady-state rate. As noted in Section 10.3, the ideas can be generalised through the concept of a transfer function linking fluctuations in current to fluctuations in the velocity gradient normal to the electrode. There are two distinct themes in the literature: one is to impose a flow with known fluctuation characteristics in order to deduce information about electrochemical processes occurring at or near the interface, this being the focus of the present review; the other is to use the variations in limiting current to deduce the characteristics of the flow, with an emphasis on analysis of the fluctuations in current to deduce characteristics of turbulent flow [81–85].

In EHD impedance studies, the relaxation times for the different transport processes are obtained by variation of the modulation frequency of the flow. The utility of the technique relies on being able to separate out individual transport relaxation times. This is possible because, with EHD, each relaxation time will have a different functional dependence upon the perturbation. The theory and methodology were first developed for sinusoidal modulation of the flow velocity in a tube [20, 22]. The results can also be directly applied to a rectangular channel [8]. Experimentally, modulating flow in a tube imposed difficulties which limited the range of modulation frequencies obtainable. Patel *et al.* [22] employed pumps to alter the level of solution in a header tank whereas Mizushina *et al.* [20] used a pair of bellows pumps to change the flow characteristics. In the latter case as well as the fundamental frequency of flow modulation, higher harmonics were present [20]. As a result, Fourier transformation and filtering of perturbation (determined via the pressure drop across the test section) and response was used in order to obtain the transfer function at the fundamental frequency.

Small-amplitude modulation of the angular velocity of a RDE [1] (refer to Section 10.3.6.2) offers a relatively wide range of modulation frequencies with experimental simplicity and the additional advantage of a uniform concentration boundary layer across the whole electrode. Consequently, the theory and methodology have been fully developed for this particular technique. In fact, the terminology "electrohydrodynamic impedance" was developed specifically for the RDE but is now used in a wider context. With this particular definition, the EHD impedance is the transfer function relating the oscillating current or potential to the oscillating rotation rate and is a product of two functions, one relating oscillating current to oscillating velocity gradient and the other (referred to as a hydrodynamic transfer function) relating the oscillating velocity gradient to the oscillating rotation frequency. Deslouis *et al.* [86] introduced an ingenious "cone and plate" arrangement: an electrode is embedded in the face of a cone which rotates with its apex in close proximity to a plate, as in a cone-and-plate viscometer. The result is a flow which is very similar to that for a channel electrode, with the advantage of the ease of modulation conferred by use of a rotating system. A limitation of the methodology using the RDE is the noise introduced by the rotating contacts. This particularly influences results at high modulation frequency, where although the amplitude of the transfer function is small, particular information of chemical significance may be found.

An alternative approach is to impose random fluctuations upon the flow and then use Fourier analysis of the resultant current or potential fluctuations to obtain the transfer function. MartemYanov and Grafov [17, 87, 88] used a particularly simple method employing a paddle stirrer to generate a turbulent flow, with a uniform spectrum of fluctuations over the frequency range (<100 Hz). This methodology was termed "hydroelectrochemical impedance (HECI)". They introduced the idea of a fluctuating current generator, as in Fig. 10.4 [17, 89]. In this case, the power spectral density of the current fluctuations is measured. It depends upon the spectrum of velocity fluctuations both normal and parallel to the interface: the result is the same as that described in Section 10.3, namely a fluctuating current caused by fluctuations in the concentration boundary layer thickness. The turbulent flow ensures a uniformly accessible electrode surface, and the transfer function [87], $H(\omega)$, relating the fluctuating current to the fluctuating velocity, has a high-frequency asymptotic behaviour $|H(\omega)| \sim (\omega/\omega_D)^{-3/2}$, with $\omega_D = 2D/\delta_D^2$. This is identical behaviour to the HMRDE. The advantage of the method is its simplicity, although the steady flow conditions of HECI are not so well controlled. This is in contrast to the HMRDE and, as will be seen in Section 10.5.3.3, there are electrochemical systems where the surface state is dependent upon the steady flow on which the probe perturbation is imposed. In this context, it is perhaps surprising that the use of turbulence promoters in a flow channel, such as a rod or a mesh across the flow, has not been explored as a means to induce the required fluctuations.

The relaxation time for diffusion has the form $\delta_D^2/2D$. In most cases for a flat, clean electrode surface δ_D is the diffusion layer thickness determined by the steady-state transport rate. In other cases it could relate to the thickness of some layer covering the electrode, which could be fixed in thickness, or could vary according to the current density, the dimension of the electrode or of a single active site upon the electrode, the spacing between active sites, or the radius of curvature of a protrusion from the surface. The diffusion coefficient will be determined by the characteristics of the medium immediately above the electrode. This may be simply the same as the bulk solution, or different. For example, the solution locally could be more viscous, varying in viscosity with distance away from the electrode, as a consequence of dissolution from the electrode, or the characteristics of the medium immediately above the electrode could be completely different from those of the bulk solution, as a consequence of the deposition of some surface layer. EHD modulation separates out these individual transport relaxation times with a different functional dependence upon the perturbation than is provided by other modulation techniques such as the AC impedance method. By changing the potential or current and steady-state transport rate, and comparing the consequent changes in steady-state behaviour, AC impedance and EHD response, phenomena involving rather complex changes in the electrode surface and in the region immediately above it can be resolved [90–92].

Of importance for any experimental technique which is to be used to fit some complex reaction model is the way in which experimental errors influence the result [93]. The error structure for the EHD method utilising the RDE has been analysed in detail by Orazem *et al.* [94]. These authors showed that information could reliably and accurately be extracted even at high modulation frequencies (up to 20 Hz). In principle the determination of Sc should only require data at low modulation frequency. They demonstrated that extraction of accurate values for Sc required data which had been recorded over a relatively wide frequency range and had been weighted according to a reliable model for the errors. They also showed that the EHD ($\tilde{\imath} - \tilde{\Omega}$) response could be fitted empirically to the form:

$$H(p) = \sum_k \frac{H_{0,k}}{1 + jp(\tau_k \Omega_o)},$$

where $H_{0,k}$ denotes the low frequency limit for component k of the response, having relaxation time τ_k. Typically, two or three components were required for the fit. The advantage of this approach is that the Kramers–Kronig transform [5] can be used to verify whether the system is evolving with time during the measurement. Orazem *et al.* [94] used this method to show that the reduction of $Fe(CN)_6^{3-}$ to $Fe(CN)_6^{4-}$, in KCl at a platinum electrode, was affected by the formation of a layer blocking part of the electrode surface. The deduction of partial blocking of the interface was made from the observation of a high-frequency element in the EHD response. This will be described in more detail later.

10.5.2 Relationship between AC impedance and EHD impedance

Deslouis *et al.* [95] stated the general functional relation which must exist between the different impedances measured on an electrochemical system. For the current flowing at a RDE, at constant temperature, there exists an implicit functional relation:

$$f(I, V, \Omega) = 0. \tag{10.24}$$

The state of the electrochemical system is controlled by imposing a given Ω and polarisation, which is in turn fixed by using either a galvanostat (I controlled) or a potentiostat (V controlled). The partial derivatives of the three variables are related:

$$\left(\frac{\partial \Omega}{\partial I}\right)_V = -\left(\frac{\partial \Omega}{\partial V}\right)_I\left(\frac{\partial V}{\partial I}\right)_\Omega. \tag{10.25}$$

Therefore, the impedances are related:

$$Z_{AC} = -(Z_{EHD})_V/(Z_{EHD})_I, \tag{10.26}$$

where Z_{AC} denotes the AC impedance measured at constant Ω_o, $(Z_{EHD})_I$ denotes the EHD impedance with galvanostatic regulation and $(Z_{EHD})_V$ denotes the EHD impedance with potentiostatic regulation. A simple redox reaction might be represented by an equivalent circuit as in Fig. 10.4. The EHD and AC impedances are then further related:

$$(Z_{EHD})_V = (Z_{EHD})_{V,\,\mathrm{lim}} \cdot Z_D/Z_{AC}, \tag{10.27}$$

where $(Z_{EHD})_{V,\,\mathrm{lim}}$ denotes the diffusion-limited potentiostatic EHD impedance.

In addition to the formal relationship between the different impedances, stated above, it is useful to develop a qualitative understanding of the way in which the different methods probe the interface. It is then possible to appreciate that in some circumstances the information from the AC and EHD impedance measurements is very similar, whilst in other cases it is subtly, but significantly, different.

In the AC impedance method, a modulation of potential causes a modulation of concentration at the interface. As a consequence there is a modulation of the concentration gradient and hence of the current density. The modulation of concentration decays with distance away from the interface. Thus there is a characteristic penetration depth into the solution phase of the concentration wave, δ_{AC}, where $\delta_{AC} = (2D/\omega)^{1/2}$. The time scale for chemical processes, either in the solution or at the interface, probed by the perturbation, is simply $1/\omega$. If the concentration boundary layer has a fixed thickness, then the expression for the mass-transfer impedance has the following form [96, 97]:

$$Z_{D,\text{real}} = \sigma\omega^{-1/2}\frac{\sinh(2u_r) + \sin(2u_r)}{\cosh(2u_r) + \cos(2u_r)},$$

$$Z_{D,\text{imagl}} = \sigma\omega^{-1/2}\frac{\sinh(2u_r) - \sin(2u_r)}{\cosh(2u_r) + \cos(2u_r)}, \tag{10.28}$$

where σ denotes the Warburg coefficient and $u_r = \delta_D/\delta_{AC}$. These expressions are approximate, because experimentally δ_D is normally not fixed. Tabulations of the convective-diffusion impedance, obtained by numerical solution of the unsteady Navier–Stokes equation, have been given by Tribollet and Newman [30] for the RDE and by Deslouis *et al.* [98] for the channel electrode using the Lévêque approximation. From equation (10.28), the phase angle of the AC impedance, $\phi = \arctan(Z_{D,\text{imag}}/Z_{D,\text{real}})$, goes through a maximum when $u_r \approx 1.964$. Hence, δ_D may be determined. This is the key to the use of the AC impedance method to deduce information concerning the nature of the interface, i.e., does the interface have localised reactive sites or is it uniformly reactive; is it covered by some porous layer or is it excessively rough? Finally, it is self-evident that the AC impedance method does not apply to processes at the diffusion-limit since, there, modulation of electrode potential causes no change in surface concentration.

In the EHD impedance method, modulation of the flow velocity causes a modulation of the velocity gradient at the interface which, in turn, causes a modulation in the concentration boundary layer thickness. As demonstrated previously in Section 10.3.3 and Fig. 10.3 the experiment shows a relaxation time determined solely by the time for diffusion across the concentration boundary layer. Although there is a characteristic penetration depth, δ_{hm}, of the velocity oscillation above the surface, and at sufficiently high modulation frequencies this is smaller than the concentration boundary layer thickness, any information associated with the variation of δ_{hm} with ω is generally lost, unless the solution is very viscous. The reason is simply that, at sufficiently high modulation frequencies, the amplitude of the transfer function between flow modulation and current density is small. So, in contrast to the AC impedance experiment, the depth into the solution probed by the EHD experiment is not a function of the modulation frequency (except for very viscous solutions), but is determined by the steady flow condition upon which the modulation is superimposed. Given, typically, $\delta_D \approx 1 \times 10^{-3}$ cm (for a RDE at ca. 10 Hz), only slow chemical processes (relaxation times greater than ca. 0.1 s) within the boundary layer would significantly affect the EHD impedance. Hence the EHD method has a temporal restriction on the range of *chemical* processes that can be addressed. However, the method is particularly useful at the mass-transport limit, where the AC impedance method cannot be applied, and it does provide information directly about changes in the boundary layer. The temporal limit on EHD is lifted in the case of the RFMRE treated in Section 10.4.3.3, although the possibilities offered for study of chemical relaxation in the hydrodynamically modulated boundary layer have not yet been explored. EHD and AC impedance effects vary differently as a function of electrode size or active area. It is these characteristics that have determined the use of the EHD method.

10.5.3 Unique information from EHD impedance

10.5.3.1 Electrochemical parameter determination

Analysis of an electrochemical system under diffusion-limited conditions gives a product of values: for example of concentration and diffusion coefficient or of concentration, diffusion coefficient and viscosity. Analysis of the electrical relaxation time, for example by measurement of AC impedance, may give the parameters in a different combination. If EHD impedance is also measured then concentration, diffusion coefficient and solution viscosity may all be obtained, independent of one another. One example of the utility of such a procedure is the analysis of the mechanism of copper electropolishing in concentrated phosphoric acid [99]. Measurement of the EHD impedance gave accurate values of Sc and showed that these were very high (3×10^6 at ambient temperature). Using both the AC impedance results to obtain the Warburg coefficient (which varies as $1/(D^{1/2}c_b)$) and Levich analysis of the variation of limiting current with rotation frequency (equation (10.15)) it was possible to obtain an estimate of the bulk concentration of the active reagent. The value was in agreement with the known concentration of water in phosphoric acid, indicating that water was indeed the reactive species controlling the polishing.

Use of the chronoamperometric response following a step change in rotation rate, as a general method for determination of the diffusion coefficient of an electroactive solution species without knowledge of its concentration, was introduced for the RDE by Albery *et al.* [15] and refined later by Blauch and Anson [42]. The explanation for the success of this procedure is again that the relaxation time for the current is dominated by the relaxation of the concentration boundary layer. If the angular velocity is increased in a step to a final value of $\Omega_f/\text{rad s}^{-1}$, the concentration boundary layer thickness changes to:

$$\delta_{D,\text{Levich}} = 1.61 D^{1/3} \Omega_f^{-1/2} \nu^{1/6}, \tag{10.29}$$

and the relaxation time for the current following the change of boundary layer thickness is approximately $\delta^2_{D,\text{Levich}}/2D$. For $\Omega_f = 60$ rad s^{-1} (ca. 10 Hz), the relaxation time for $D = 1 \times 10^{-5}$ cm^2 s^{-1} is approximately 0.25 s.

Albery *et al.* [15] derived an accurate solution and then gave a particularly simple method, accurate to a few percent, for evaluation of D from experimental data: if the rotation rate is increased in a step to a final value of W_f/Hz, then the time t_q for the current to attain a fraction q of the final, steady-state value is given by the following, where the values of the proportionality constant B_q are tabulated below:

$$\text{Sc}^{1/3} \approx B_q(W_f t_q - 0.04) - 1 \tag{10.30}$$

q	0.3	0.5	0.7
B_q	20.6	12.9	8.7

For example, for $W_f = 10$ Hz and $q = 0.7$, $t_q = 0.13$ s, which may be compared with the approximate estimate of 0.25 s derived above.

Blaedel and Engstrom [48] noted that for a quasi-reversible process the current could be simply expressed in terms of the rate constant and mass-transport coefficient. Application of a square wave step in the rotation rate of a RDE (i.e., PRV, see Section 10.4.1.3) resulted in modulation of the diffusion-limited current and hence modulation of the mass-transfer coefficient. By solving the appropriate quadratic equation it was possible to derive a value for the heterogeneous rate constant for the electrochemical cathodic, k_f, or anodic, k_b, process of interest. Values for the standard heterogeneous rate constant and transfer coefficient were subsequently obtained from the intercept and slope, respectively, of ln k_f or ln k_b versus $V-V^{0'}$ plots, where $V^{0'}$ is the formal standard potential. Importantly using PRV, these values could be derived in the presence of parallel electrode surface reactions. Rate constants and transfer coefficients for the oxidation of $Fe(CN)_6^{4-}$ and the reduction of $Fe(CN)_6^{3-}$ on a glassy carbon electrode were obtained in this way.

A more sophisticated application of the concept of separating mass-transport controlled processes from those due to reaction of the surface itself, was made by Austin *et al.* [100]. These authors, using square wave modulation of the rotation frequency of a RDE, showed that the useful anodic limit for study of mass transfer-limited process could be extended by at least 350 mV above the practical limit for steady-state voltammetry. As a result they were able to investigate electrocatalysis of anodic reactions by surface oxides on noble metals, specifically the formation of IO_3^- from I^-(5×10^{-3} mol dm^{-3}) at +1.45 V versus SCE on a platinum electrode in 0.5 mol dm^{-3} H_2SO_4 [99]. Although the anodic reaction occurs simultaneously with oxygen evolution, using the HMRDE only the I^- transport-limited oxidation current was modulated. They demonstrated that the formation of IO_3^- was initiated by the growth of PtOH on the electrode surface. A potential step sequence was also used to form the oxide and then freeze its state on the surface, whilst the mass-transport modulation revealed the consequent effect upon the IO_3^- formation.

10.5.3.2 Effects of electrode geometry and heterogeneity

Two effects influence the modulation response when the electrode becomes sufficiently small, both of which are a consequence of the influence of the edges of the electrode. Firstly, for an electrode whose dimensions become comparable to the thickness of the concentration boundary layer, so-called UME behaviour is observed. This means that contributions to the diffusive flux along the coordinates parallel to the electrode surface become significant, resulting in a decrease in the size of δ_D. Secondly, as was shown in Section 10.3, in a flow channel, the concentration boundary layer is much thinner at the leading edge of the electrode. Hence as the electrode is made smaller, the average thickness of δ_D decreases, and the frequency response of the current to the imposed perturbation significantly changes.

As discussed in Section 10.3.6.1, the characteristic frequency, ω_D, which determines the roll-off frequency for the current response, increases with decreasing electrode length. If the electrode comprised a set of active areas whose spacing and spatial arrangement was such that the concentration boundary layers did not overlap then the roll-off frequency would be significantly higher than expected from the overall electrode dimensions. In contrast, if the concentration boundary layers did overlap, the behaviour would approach that expected for a uniform electrode.

In a channel or tube under pulsatile flow, nodes in the consequent oscillating flux should occur at positions along the tube which are integer multiples of the distance that the fluid travels during each oscillating cycle [101]. Such nodes are evident in all appropriate theoretical treatments [12, 20]. On a large electrode, the effects are averaged out, but Mizushina *et al.* [20] showed that these nodes could be detected with a sufficiently small probe, the effect showing up most noticeably in the phase angle of the response. Thus, there exists the potential to deduce the spatial location, along an electrode, of isolated active areas. This possibility, and its potential application, has yet to be realised experimentally.

On a RDE, if the electrode is not uniformly active, then the individual sites can behave as if they were isolated UMEs embedded in the wall of a channel. Except, in this case, the solution velocity parallel to the wall varies with radial position of the electrode. For an electrode with an array of active sites, sufficiently separated, two relaxations are expected, as illustrated in Fig. 10.18. At sufficiently low modulation frequencies, the distance scale, δ_{hm}, of the perturbation of the hydrodynamic boundary layer is much larger than the size and spacing of the sites, so the electrode behaves as a uniform surface. A deviation is observed when δ_{hm} becomes comparable to the spacing between the sites. Under these conditions, the surface responds to the perturbation as a set of isolated electrodes.

At higher modulation frequencies, as was noted in Section 10.3.6, for an isolated UME the roll-off frequency depends upon the electrode size. The modulus of the transfer function relating current response to perturbation of the velocity gradient decreases ultimately as p^{-1} [102]. For a RDE, at high modulation frequency ($p > 2$) the velocity gradient modulation decreases as $p^{-0.5}$ [31, 103], so the net effect for an UME embedded in a rotating disc, at high modulation frequency, is an asymptotic decrease as $p^{-1.5}$. This limit has not been observed experimentally or in numerical calculations over the experimentally accessible range of p: a high-frequency behaviour varying approximately as $p^{-0.8}$ is typical.

For a RDE, the roll-off frequency depends upon the disc radius, and

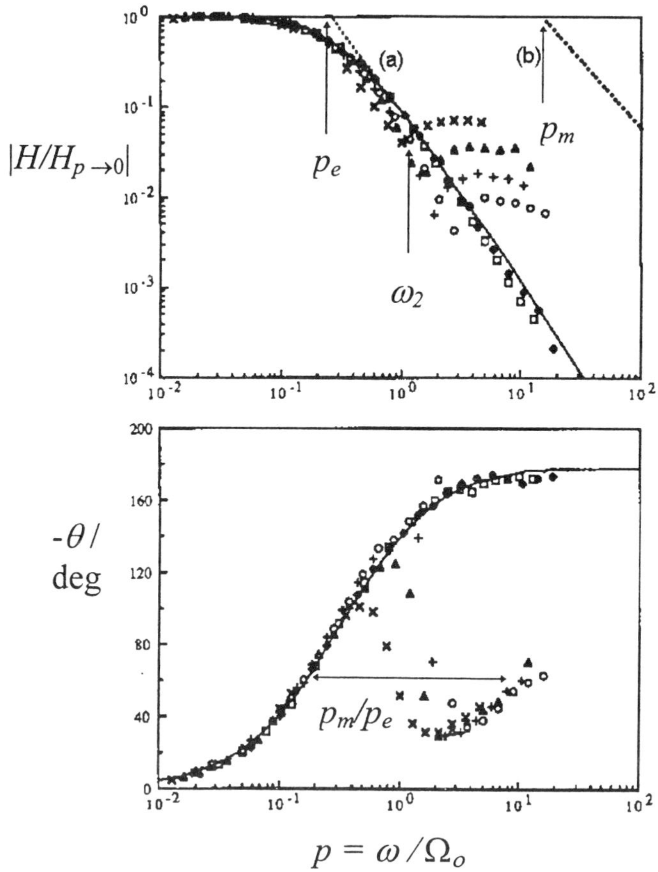

Fig. 10.18. Effects of surface roughness on EHD impedance (amplitude ratio, $|H(p)/H(p \to 0)|$, and phase lag, θ, against scaled frequency, p and comparison with the behaviour of a uniform disc—asymptotic line marked (a)—and an array of UMEs—asymptotic line marked (b). The frequency shift is deduced from the displacement between the two sections of the phase angle diagram where the data superimpose for different Ω_o. The modulation frequency, ω_2, at which the data deviate from that of a uniform electrode, is related to the amplitude of the surface roughness or the spacing between the elements of the UME array. Data from Reference [121], for $Fe(CN)_6^{3-}$ reduction on smooth Pt at □: 120 rpm; ◆: 240 rpm, and on a rough, Pt-coated silver electrode (roughness scale 5 μm, disc diameter 6 mm) at O: 120 rpm; +: 240 rpm; Δ: 500 rpm and ×: 1000 rpm.

the high-frequency asymptote of the transfer function relating current response to perturbation of the velocity gradient decreases as $p^{-1.5}$. Taking into account the decrease of velocity gradient modulation as p increases, the net effect is an asymptotic decay of current modulation as p^{-2} (Fig. 10.5). Typically, experimentally, a decay approximately as $p^{-1.8}$ is found over the accessible range. Figure 10.18 shows how both types of behaviour may be observed upon a RDE carrying an array of separated active sites. The frequency of first deviation from the behaviour of a uniform disc indicates the average site spacing, and the second roll-off frequency indicates the average site size.

Caprani *et al.* [104], defining the cut-off frequency as the intersection of the low and high-frequency asymptotes, as indicated on Fig. 10.18, have given an approximate method to deduce the size of active sites on a partially blocked electrode from the ratio of the two cut-off frequencies:

$$p_m/p_e \approx 2(r/d_s)^{2/3}, \tag{10.31}$$

where r denotes the disc radius, d_s the site diameter and p_m and p_e are as defined on Fig. 10.18. Strictly, r is the radial position of the active site but the disc radius is used as an approximation to average effects over the surface. The analysis was verified using an experimental model comprising a rotating disc electrode (5 mm radius) with an ultramicroelectrode (100 μm diameter) embedded 40 μm from the the edge of the disc. Results with an electrode fabricated from a grid confirmed the general features outlined above. However, the calculated site size was only indicative of that actually present, reflecting the approximations used in the analysis. A better agreement with equation (10.31) was obtained with a model surface prepared by photolithography, having a regular array of active sites of the same size [105]. The method seems very sensitive to partial blocking of the interface (see e.g., Ref. [94]). It has been used to track deposition of calcium carbonate on an electrode surface [106].

10.5.3.3 Effects of surface films

The effect of a surface film which partially blocks an electrode has been considered above (Section 10.5.2.1). This section considers the effect of a film which is permeable to reagents but which covers entirely the electrode

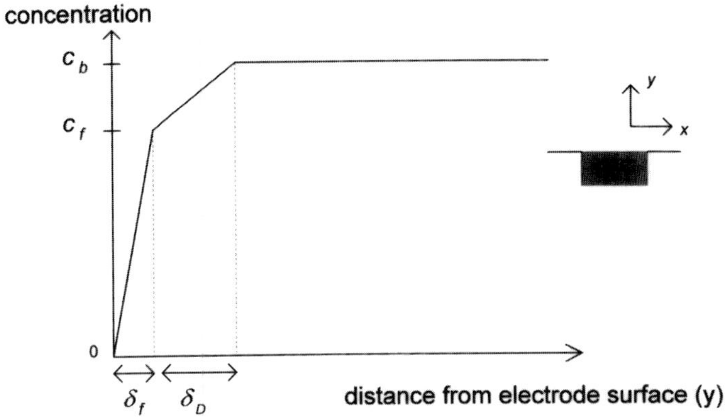

Fig. 10.19. Schematic of the concentration profile of a reagent which is consumed at the limiting current density at the surface of an electrode, covered entirely by a permeable film.

surface. Figure 10.19 shows a schematic of the concentration profile of a reagent being consumed at the limiting current density at such a surface. There are two diffusion time constants for this system: one, τ_f, corresponding to diffusion through the surface film, thickness δ_f, $\tau_f = \delta_f^2/2D_f$ and the other, $\tau_D = \delta_D^2/2D$, corresponding to the convective mass transport to the surface, across the Nernst diffusion layer, thickness δ_D.

On a RDE, in the absence of a surface layer, the EHD impedance is a function of a single dimensionless frequency, $p\mathrm{Sc}^{1/3}$. This means that if the viscosity of the medium directly above the surface of the electrode and the diffusion coefficient of the species of interest are independent of position away from the electrode, then the EHD impedance measured at different rotation frequencies reduces to a common curve when plotted as a function of p. In other words, there is a characteristic dimensionless diffusional relaxation time for the system, p_D, strictly $(p\mathrm{Sc}^{1/3})_D$, which is independent of the disc rotation frequency. However, if ν or D vary with position (for example, as a consequence of the formation of a viscous boundary layer or the presence of a surface film), then, except under particular circumstances described below, reduction of the measured parameters to a common curve is not possible. Under these conditions p_D is dependent upon the disc rotation frequency. The variation of the EHD impedance with Ω_o as a function of p is therefore the diagnostic for spatial variations near the electrode surface, and the means by which the properties of such surface layers may be analysed.

As it is well known that the EHD response is determined by the time scale for diffusion across the concentration boundary layer a qualitative appreciation and understanding of the surface film effects can thus be achieved. For a clean electrode surface δ_D is determined solely by the flow characteristics and the EHD response can be expressed in terms of dimensionless parameters of the flow, regardless of the particular electrode configuration and system geometry (e.g., disc, channel, tube) employed. Hence, for all flow conditions, the response will reduce to a common curve when expressed in terms of these parameters. However, when there is a porous layer covering the electrode, whose thickness does not change with changing flow conditions, the concentration boundary layer is composed of two parts (as in Fig. 10.19), δ_f—flow rate independent and δ_D—flow rate dependent, resulting in two impedances in series. In this case the measured response will not reduce to a common curve for all flow conditions when expressed in terms of the dimensionless parameters of the flow.

Modulation of the current density occurs as a result of modulation of the concentration boundary layer thickness caused by modulation of the flow (taken simply for convenience and generality of expression as modulation of Reynolds number), $\Delta\mathrm{Re}$. Under steady-state conditions and in the absence of a film over the surface, $i_{\lim} = nFDc_b/\delta_D$ and, in the limit of modulation frequency, $\omega \to 0$;

$$(Z_{\mathrm{EHD},0})_{\text{no film}} = \Delta i_{\lim}/\Delta\mathrm{Re} = (nFDc_b/\delta_D^2)(\Delta(\delta_D)/\Delta\mathrm{Re}).$$

In the presence of a film on the surface,

$$i_{\lim} = nFc_B \frac{D_f}{\delta_f}\left(\frac{1}{1+\dfrac{D_f\delta_D}{D\delta_f}}\right), \tag{10.32}$$

and the effect of flow modulation, in the limit of modulation frequency, $\omega \to 0$, is:

$$(Z_{\mathrm{EHD},0})_{\mathrm{film}} = \frac{\Delta i_{\mathrm{lim}}}{\Delta(\mathrm{Re})} = nFc_b \frac{D_f}{\delta_f^2}\left(\frac{D_f/D}{(1 + D_f\delta_D/D\delta_f)^2}\right)\left(\frac{\Delta\delta_D}{\Delta(\mathrm{Re})}\right). \tag{10.33}$$

Then the ratio of the EHD impedances in the low frequency limit is:

$$\frac{(Z_{\mathrm{EHD},0})_{\mathrm{film}}}{(Z_{\mathrm{EHD},0})_{\mathrm{no\,film}}} = \frac{\delta_D^2}{\delta_f^2}\left(\frac{D_f^2}{(1 + D_f\delta_D/\delta_f)^2}\right). \tag{10.34}$$

The presence of the film decreases the amplitude of the response to the flow perturbation. If the film develops over time, then the film thickness can be calculated, knowing δ_D from the characteristics of the flow. Furthermore, the relaxation time for the composite system is the sum of the two diffusional relaxation times:

$$\tau \approx \frac{\delta_D^2}{2D} + \frac{\delta_f^2}{2D_f}. \tag{10.35}$$

The discussion in Section 10.3.6 showed that in the absence of a film on the surface, the transfer function between current response and flow modulation had a characteristic roll-off frequency designating the point where the response starts to markedly decrease with increasing frequency. From equation (10.35), in the presence of a surface layer, the roll-off frequency moves to lower frequency:

$$\frac{\omega_{D,\mathrm{no\,film}}}{\omega_{D,\mathrm{film}}} \approx 1 + \frac{\delta_f^2}{2D_f}\,\omega_{D,\mathrm{no\,film}}. \tag{10.36}$$

Since the roll-off frequency in the absence of a surface layer changes in a known way with change of the stationary flow conditions, the variation of roll-off frequency with flow should give the ratio δ_f^2/D_f.

As an alternative to EHD measurements, the presence of a surface film of constant thickness can be detected using Koutecky–Levich analysis. From equation (10.32),

$$\frac{1}{i_{\mathrm{lim}}} = \frac{\delta_f}{nFc_bD_f} + \frac{\delta_D}{nFc_bD} = \frac{\delta_f}{nFc_bD_f} + \frac{1}{0.62nFD^{2/3}\Omega_o^{1/2}\nu^{-1/6}c_b}, \tag{10.37}$$

thus the ratio δ_f/D_f may be obtained from the intercept of the plot of i_{lim}^{-1} against $\Omega_o^{-1/2}$. The EHD approach is more elegant than the above, as it is much more sensitive and the measurements made to calculate δ_f^2/D_f are more accurate than the extrapolation procedure involved in a Koutecky–Levich plot. However, utilising both techniques does enable individual values for δ_f and D_f to be obtained. As an example, L'Hostis *et al.* [107] presented EHD and steady-state measurements for characterisation of the transport properties in a gel layer covering a RDE. Analysis of the results was made by fitting the full EHD impedance at different rotation rates. First, Sc for the electroactive species in the solution was obtained by fitting the EHD response in the absence of a gel layer over the electrode. Then, the steady-state limiting current for the coated electrode was obtained as a function of the rotation rate, to derive the ratio δ_f/D_f (equation (10.37)). Finally, the EHD impedance of the coated electrode was fitted [108], to obtain the ratio δ_f^2/D_f. Hence the values of δ_f and D_f could be individually calculated. Deslouis has applied this method to the study of a conducting polymer deposited on an electrode surface [109].

In contrast, Khomchenko *et al.* [88] have used the approach where turbulent fluctuations are imposed upon a flow, to detect the formation of a surface layer in the electroreduction of *N*-methyl- and *N*-butyl-pyridinium. At high frequencies, the spectral power density of current fluctuations decreased as ω^{-3}, as expected for a uniformly accessible surface subjected to a spectrum of velocity fluctuations uniform across the observed frequency range. When a film was deposited, the low-frequency limiting value of the current fluctations decreased, as expected from equation (10.34) and the roll-off frequency decreased as predicted by equation (10.36). Because the high-frequency asymptotic behaviour was unchanged, it could be assumed that the film was deposited uniformly over the electrode, at least on a distance scale larger than the scale length associated with the fluctuations of the concentration boundary layer (of the order of 10 μm at the highest frequencies).

The EHD method with a RDE has been applied to the characterisation of porous layers of corrosion products formed on carbon steel [110], for the characterisation of salt films formed on copper [90, 111, 112] and iron [113], and for biofilms developed in natural seawater [114]. Corrosion inhibition films formed by an organic surfactant acting on the surface of pure iron have been characterised in this way, too [115]. An effect of a mechanical action of the oscillating flow upon the surface film was deduced; diffusion across the inhibitor layer was extremely slow. Iron dissolution in sulphuric acid [116, 117] illustrates the effect of formation of a liquid layer of high viscosity at the interface, developed as a consequence of the high salt concentration formed there during the dissolution.

Subtle effects arise when the surface layer is modulated by the oscillating flow [116]. For example in response to the flow, the layer thickness may change or even deform. In the high-rate dissolution of copper in HCl [92], a salt film of CuCl is deposited onto the surface. The film is formed at the electrode-film interface, limited by transport of Cl^-, and dissolved at the same rate at the film-solution interface, limited by transport of $CuCl_2^-$. Because the film thickness varies as $\Omega_o^{-1/2}$, the presence of a film was only evident from a series of Koutecky–Levich plots where the slope decreased with increasing chloride concentration. In addition, given this dependence of Ω_o on film thickness, the EHD diagrams did reduce onto the same curve when plotted against p. Fortunately, at low frequency another relaxation effect was observed which was attributed to transport through the surface layer. The EHD method clearly detected the presence of the salt film. The transition between film-free and film-covered surface as potential or electrode rotation rate were varied was abrupt, and a boundary line could be defined on a V–Ω_o diagram. With this information, the complex AC impedance diagrams could be understood and fitted, and a kinetic scheme describing the electrode process derived [92].

In the high-rate dissolution of iron in sulfuric acid [116, 118], a surface layer is formed in which the viscosity of the solution varies with distance out from the surface. Again Koutecky–Levich plots passing through the origin, but with slopes diminishing with increasing sulfate concentration, were obtained and the EHD results at different Ω_o were reduced to a common curve as a function of p. There was no low-frequency relaxation, so the formation of a salt film could not explain the results. The low-frequency behaviour was consistent with a very high Schmidt number ($Sc \approx 15\,000$, implying high viscosity or low diffusion coefficient), but the high-frequency asymptote decreased more rapidly than expected, being more consistent with the more typical value $Sc \approx 1000$. However, the results could be modelled [116, 118], by making the assumption that within the concentration boundary layer the viscosity varied linearly with concentration.

10.5.3.4 Effects of surface roughness

There are two ways in which the effects of surface roughness might show up in an EHD experiment. Firstly, protrusions on the surface could act as ultramicroelectrodes, and the current flow could be predominantly to these sites. This being the case, behaviour reminiscent of that of a partially blocked surface might be obtained. Secondly, if the distance scale for the roughness were greater than the perturbation distance for the concentration boundary layer (δ_{hm} in our notation; where, for the modulated rotating disc, $\delta_{hm} \approx (D/\omega)^{1/2}$) then the zone of solution protected from the flow by the surface roughness could be detectable as the equivalent of a surface film whose thickness was not modulated by the flow.

Experimental confirmations of the expected effects, and use of the method to follow the evolution of surface roughness, have been obtained for Te [119], Cu [120], and Ag [121] electrodeposits. Because the measurements take some time, with a consequent evolution of the surface during them, for the studies of the metal surfaces the electrodes were removed from the plating bath and covered with a sputtered layer of platinum or an electrodeposited layer of nickel. EHD diagrams for $Fe(CN)_6^{3-}$ reduction in alkaline solution were then obtained on the modified surfaces. A typical example is the characterisation of the surface roughness of Cu electrodeposits [120]. On the Ni-coated surface in $Fe(CN)_6^{3-}$ solution, there was a small non-zero intercept in a Koutecky–Levich plot, indicative of a trapped surface layer. At low modulation frequency ($p < 0.1$) the behaviour was the same as that of a smooth electrode, with the data reduced to a common curve as a function of p for the various values of Ω_o. At higher modulation frequency, the observed transfer function deviated from that expected for a smooth surface: it decreased less rapidly and the curves for different Ω_o were not superimposed as a function of p. The value of ω when the deviation occurred gave an indication of δ_{hm} ($\delta_{hm} \approx 10\ \mu m$) which was an estimate of the average height of the surface roughness. No clearly defined asymptotic behaviour associated with an isolated site was observed. Instead, the frequency ratio was calculated from the variation of phase angle with p, as the ratio of frequencies giving the same phase angle in the low and high frequency regimes. Application of equation (10.31) then estimated the average dimension of the active sites as 30 μm. The results were consistent with the directly observed surface morphology. The same procedure was applied to characterise silver electrodeposits [121], with average crystallite size 3 μm (from microscopy) determined as 5–7 μm by EHD measurements. Figure 10.18 illustrates the effects.

10.5.3.5 Reaction kinetic studies

Although in general the use of the EHD method to address reaction mechanisms seems limited because the accessible time range is fairly long, there are circumstances where useful information can be derived. The presence of a chemical reaction preceding electrochemical discharge (CE process) gives a non-zero intercept in a Koutecky–Levich plot, though the effect may be small and difficult to detect. Given the general form:

$$(i_{\text{lim}})^{-1} = a + b\Omega_o^{-1/2}, \tag{10.38}$$

the low-frequency limit of the EHD response would be

$$Z_{\text{EHD},\omega\to 0} = \frac{\Delta i_{\text{lim}}}{\Delta\Omega} = b \cdot i_{\text{lim}}^2 \Omega_o^{-3/2}. \tag{10.39}$$

Vandeputte *et al.* [122] used both of these equations to derive more accurate values of the constants than could be obtained from the Koutecky–Levich analysis alone, and hence derived rate constants for the predissociation of the thiosulfate complex involved in the electrodeposition of silver from thiosulfate solutions.

In a slightly difference approach, Bruckenstein and Swathirajan [123], used the sinusoidally modulated RDE response over a range of electrode potentials, modulation amplitudes and frequencies to estimate the size of the reaction layer in the mass transport controlled reduction of dilute strong acids. The diffusion-controlled reduction of H^+ occurs simultaneously with the surface controlled reduction of H_2O. This latter process produces OH^- ions which react with the incoming H^+ in the reaction layer. As this layer is so thin, about two orders of magnitude smaller than δ_D, distortion of the limiting cathodic current is negligible with conventional RDE. In contrast, as the HMRDE can separate transport-controlled processes from those that are surface controlled, the HMRDE exclusively detects the true limiting current characteristics for H^+ reduction. At high overpotential, where significant H_2O reduction occurs, the diffusion-limited response for H^+ decreases with increasing electrode potential. For the CE dissociation of weak acids, contributions from OH^- reacting with the acid can cause deviations from Koutecky–Levich behaviour. Such effects can thus be minimised by using the above approach.

10.6 SUMMARY

This review has attempted to put hydrodynamic modulation methods for electroanalysis and for the study of electrochemical reactions into context with other electrochemical techniques. HM is particularly useful for the extension of detection limits in analysis and for the detection of heterogeneity on electrode surfaces. The timescale addressable using HM methodology is limited by the time taken for diffusion across the concentration boundary layer, typically >0.1 s for conventional RDE and channel electrode geometries. This has meant a restriction on the application of HM to deduce fast reaction mechanisms. New methodologies, employing smaller electrodes and thin layer geometries look to lift this restraint.

ACKNOWLEDGEMENTS

JVM would like to thank the EPSRC for support (GR/K97011). We both thank Prof. Patrick Unwin for reading this review; his comments were greatly appreciated.

References

[1] C. Deslouis and B. Tribollet, Flow Modulation Techniques in Electrochemistry, in: *Advances in (eds.) C. W. Tobias and H. Gerischer Electrochemical Science and Engineering*, Vol. 2, p. 205 (VCH Weinhein, New York, 1991).
[2] J. Wang, Talanta 28 (1981) 369.
[3] R. N. Adams, Electrochemistry at Solid Electrodes (Marcel Dekker: New York, 1969).
[4] C. Gabrielli and B. Tribollet, J. Electrochem. Soc. 141 (1994) 1147.
[5] D. D. Macdonald, E. Sikora and G. Englehart, Electrochim. Acta. 43 (1998) 87.
[6] S. A. Schuette and R. L. McCreery, Anal. Chem. 59 (1987) 2692.

[7] J. E. John and W. L. Haberman, Introduction to Fluid Mechanics (Prentice Hall, New Jersey, 1971, p 175).

[8] P. R. Unwin and R. G. Compton in: (eds.) R. G. Compton and A. Hamnett, Comprehensive Chemical Kinetics, vol. 29 (Elsevier, Amsterdam, 1989, p. 173–296 and references therein).

[9] M. A. Lévêque, Ann. Mines 13 (1928) 283.

[10] R. G. Compton and G. M. Stearn, J. Chem. Soc. Faraday Trans. 84 (1988) 4359.

[11] N. P. C. Stevens and A. C. Fisher, J. Phys. Chem. B 101 (1997) 8259.

[12] C. Deslouis, O. Gil and B. Tribollet, J. Fluid Mech. 215 (1990) 85.

[13] H. Schlichting, Boundary Layer Theory, 7th edn (McGraw-Hill, New York, 1979).

[14] J. M. Rosamilia and S. Bruckenstein, Anal. Chem. 56 (1984) 2410.

[15] W. J. Albery, A. R. Hillman and S. Bruckenstein, J. Electroanal. Chem. 100 (1979) 687.

[16] N. J. Evans, R. D. Martin and P. R. Unwin, paper in preparation.

[17] S. A. MartemYanov and B. M. Grafov, Soviet Electrochemistry 24 (1988) 94.

[18] C. Deslouis, C. Gabrielli, Ph. Sainte-Rose Fanchine and B. Tribollet, J. Electrochem. Soc. 129 (1982) 107.

[19] C. Deslouis and B. Tribollet, Electrochim. Acta. 35 (1990) 1637.

[20] T. Mizushina, T. Maruyama, S. Ide and Y. Mizukami, J. Chem. Eng. Japan 6 (1973) 152.

[21] T. Maruyama and T. Mizushina, A.I.Ch.E Journal 21 (1975) 1035.

[22] R. D. Patel, J. J. McFeeley and K. R. Jolls, A.I.Ch.E Journal 21 (1975) 259.

[23] J. A. Alden and R. G. Compton, J. Electroanal. Chem. 404 (1996) 27.

[24] K. Aoki, K. Tokuda and H. Matsuda, J. Electroanal. Chem. 217 (1987) 33.

[25] J. Newman, J. Electroanal. Chem. 6 (1973) 279.

[26] Y. Saito, Rev. Polarogr. 15 (1967) 177.

[27] B. Miller and S. Bruckenstein, Anal. Chem. 46 (1974) 2026.

[28] S. Bruckenstein and B. Miller, Acc. Chem. Res. 10 (1977) 54.

[29] K. Tokuda, S. Bruckenstein and B. Miller, J. Electrochem. Soc. 122 (1975) 1316.

[30] B. Tribollet and J. Newman, J. Electrochem. Soc. 130 (1983) 2016.

[31] C. Deslouis and B. Tribollet, J. Electroanal. Chem. 185 (1985) 171.

[32] D. T. Schwartz, J. Electrochem. Soc. 140 (1993) 452 and references therein.

[33] D. E. Williams, J. Ellis, A. Colville, S. J. Dennison, G. Laguillo and J. Larsen, J. Electroanal. Chem. 432 (1997) 159.

[34] S. A. Schuette and R. L. McCreery, Anal. Chem. 58 (1986) 1778.

[35] J. M. Rosamilia and B. Miller, J. Electroanal. Chem. 160 (1984) 131.

[36] B. Miller and S. Bruckenstein, J. Electrochem. Soc. 121 (1974) 1558.

[37] B. Miller, M. I. Bellavance and S. Bruckenstein, Anal. Chem. 44 (1972) 1983.

[38] B. Miller, J. M. Rosamilia, G. Dabbagh, A. J. Muller and R. C. Haddon, J. Electrochem. Soc. 139 (1992) 1941.

[39] K. Tokuda and S. Bruckenstein, J. Electrochem. Soc. 126 (1979) 431.

[40] Y. Kunzaki and S. Bruckenstein, J. Electrochem. Soc. 126 (1979) 437.

[41] D. T. Schwartz, J. Electrochem. Soc. 135 (1988) 31C.

[42] D. N. Blauch and F. C. Anson, J. Electroanal. Chem. 259 (1989) 1.

[43] J. M. Rosamilia and B. Miller, Anal. Chem. 55 (1983) 1142.
[44] J. Wang, Anal. Chem. 53 (1981) 1528.
[45] J. Wang and B. A. Freiha, Analyst 108 (1983) 685.
[46] A. I. Vogel, A Text-Book of Quantitative Inorganic Analysis Including Elementary Instrumental Analysis (Longmans, London, 1961).
[47] J. Wang and B. A. Freiha, Anal.Chem. 54 (1982) 12.
[48] W. J. Blaedel and R. C. Engstrom, Anal. Chem. 50 (1978) 476.
[49] W. J. Blaedel and J. Wang, Anal. Chem. 52 (1980) 1697.
[50] W. J. Blaedel and J. Wang, Anal. Chim. Acta 116 (1980) 315.
[51] E. D. Harris and A. J. Lindsey, Analyst 76 (1951) 647.
[52] E. D. Harris and A. J. Lindsey, Analyst 76 (1951) 650.
[53] E. D. Harris and A. J. Lindsey, Nature 162 (1948) 413.
[54] A. J. Lindsey, J. Phys. Chem. 56 (1952) 439.
[55] K. W. Pratt and D. C. Johnson, Anal. Chim. Acta 148 (1983) 87.
[56] B. Miller and J. M. Rosamilia, Anal. Chem. 55 (1983) 1281.
[57] K. W. Pratt and D. C. Johnson, Electrochim. Acta 27 (1982) 1013.
[58] G. J. Jameson, Chem. Eng. Sci. 19 (1964) 793.
[59] P. Dumargue and P. Humeau, Electrochim. Acta 27 (1982) 115.
[60] W. J. Albery, P. N. Bartlett, A. E. G. Cass, D. H. Craston and B. D. G. Haggett, J. Chem. Soc. Faraday Trans. 82 (1986) 1033.
[61] J. J. Podesta, G. F. Paus and A. J. Arvia, Electrochim. Acta 19 (1974) 583.
[62] W. L. Caudill, J. O. Howell and R. M. Wightman, Anal. Chem. 54 (1982) 2532.
[63] S. B. Khoo, H. Gunasingham, K. P. Ang and B. T. Tay, J. Electroanal. Chem. 216 (1987) 115.
[64] J. W. Bixler and A. M. Bond, Anal. Chem. 58 (1986) 2859.
[65] B. C. Towe, IEEE Trans. Biomed. Eng. 34 (1987) 657.
[66] P. R. Unwin and J. V. Macpherson, Chem. Ind. 25 (1995) 874.
[67] A. J. Bard, F.-R. Fan, D. T. Pierce, P. R. Unwin, D. O. Wipf and F. Zhou, Science 254 (1991) 68.
[68] J. Kwak and A. J. Bard, Anal. Chem. 61 (1989) 1221.
[69] M. V. Mirkin, F.-R. Fan and A. J. Bard, J. Electroanal. Chem. 328 (1992) 47.
[70] D. O. Wipf and A. J. Bard, Anal. Chem. 64 (1992) 1362.
[71] W. J. Blaedel and S. L. Boyer, Anal. Chem. 43 (1971) 1538.
[72] W. J. Blaedel and J. Wang, Anal. Chem. 51 (1979) 799.
[73] W. J. Blaedel and D. G. Iverson, Anal. Chem. 49 (1977) 1563.
[74] W. J. Blaedel and Z. Yim, Anal. Chem. 52 (1980) 564.
[75] W. J. Blaedel and J. Wang, Anal. Chem. 53 (1981) 78.
[76] J. V. Macpherson and P. R. Unwin, Anal. Chem. 70 (1998) 2914.
[77] J. V. Macpherson and P. R. Unwin, submitted.
[78] H. D. Dewald and B. A. Peterson, Anal. Chem. 62 (1990) 779.
[79] P. R. Birkin and S. Silva-Martinez, J. Electroanal. Chem. 416 (1996) 127.
[80] R. G. Compton, J. C. Eklund, F. Marken, T. O. Rebbitt, R. P. Akkermans and D. N. Waller, Electrochim. Acta 42 (1997) 2919.
[81] T. J. Hanratty, J. Appl. Electrochem. 21 (1991) 1038.

[82] C. Deslouis, O. Gil and B. Tribollet, Electrochim. Acta. 38 (1993) 1847.
[83] A. A. Vansteenhoven and F. J. H. M. Vandebeucken, J. Fluid Mechanics 231 (1991) 599.
[84] D. M. Wang and J. M. Tarbell, Int. J. Heat and Mass Transfer 36 (1993) 4341.
[85] M. Grafov, S. A. Martemyanov, N. V. Petrovskii and L. A. Sokolov, Soviet Electrochemistry 25 (1989) 482.
[86] C. Deslouis, O. Gil and B. Tribollet, Electrochim. Acta, 38 (1993) 1847.
[87] S. A. Martemyanov and B. M. Grafov, Soviet Electrochemistry 24 (1988) 344.
[88] T. N. Khomchenko, L. N. Nekrasov, Kh. Almualla, S. A. Martemyanov and B. M. Grafov, Russian J. Electrochem. 33 (1997) 26.
[89] S. A. Martemyanov and B. M. Grafov, Soviet Electrochemistry 24 (1988) 1052.
[90] O. E. Barcia, O. R. Mattos, N. Pebere and B. Tribollet, Electrochim. Acta 41 (1996) 1385.
[91] O. E. Barcia, O. R. Mattos, N. Pebere and B. Tribollet, J. Electrochem. Soc. 140 (1993) 2825.
[92] E. D'Elia, O. E. Barcia, O. R. Mattos, N. Pébère and B. Tribollet, J. Electrochem. Soc. 143 (1996) 961.
[93] M. E. Orazem, P. Agarwal, A. N. Jansen, P. T. Wojcik and L. H. Garcia-Rubio, Electrochim. Acta 38 (1993) 1903.
[94] M. E. Orazem, P. Agarwal, C. Deslouis and B. Tribollet, J. Electrochem. Soc. 143 (1996) 948.
[95] C. Deslouis, I. Epelboin, C. Gabrielli, Ph. Saint-Rose Fanchine and B. Tribollet, J. Electroanal. Chem. 107 (1980) 193.
[96] R. Greef, R. Peat, L. M. Peter, D. Pletcher and J. Robinson, Instrumental Methods in Electrochemistry (Ellis Horwood, Chichester, 1985, p. 274).
[97] M. Sluyters-Reybach and J. H. Reybach, in: Electroanalytical Chemistry, Volume 4, edn. A. J. Bard (Marcel Dekker, New York, 1970).
[98] C. Deslouis, B. Tribollet and M. A. Vorotyntsev, J. Electrochem. Soc. 138 (1991) 2651.
[99] R. Vidal and A. C. West, J. Electrochem. Soc. 142 (1995) 2682 and 2689.
[100] D. S. Austin, D. C. Johnson, T. G. Hines and E. T. Berti, Anal. Chem. 55 (1983) 2222.
[101] R. Siegel and M. Perlmutter, Trans. A.S.M.E. Ser. C 84 (1962) 111.
[102] E. F. Skurygin, S. A. Martemyanov, M. A. Vorotyntsev and B. M. Grafov, Soviet Electrochemistry, 25 (1989) 685.
[103] V. P. Sharma, Acta. Mech. 32 (1979) 19.
[104] A. Caprani, C. Deslouis, S. Robin and B. Tribollet, J. Electroanal. Chem. 238 (1987) 67.
[105] C. R. S. Silva, O. E. Barcia, O. R. Mattos and C. Deslouis, J. Electroanal. Chem. 365 (1994) 133.
[106] C. Deslouis, C. Gabrielli, M. Keddam, A. Khalil, R. Rosset, B. Tribollet and M. Zidoune, Electrochim. Acta 42 (1997) 1219.
[107] E. L'Hostis, C. Deslouis, B. Tribollet and D. Festy, Electrochim. Acta 41 (1996) 1393.

[108] C. Deslouis, B. Tribollet, M. Duprat and F. Moran, J. Electrochem. Soc. 134 (1987) 2496.
[109] C. Deslouis, Soviet Electrochemistry 29 (1993) 36.
[110] L. Bousselmi, C. Fiaud, B. Tribollet and E. Triki, Corros. Science 39 (1997) 1711.
[111] O. E. Barcia, O. R. Mattos, N. Pebere and B. Tribollet, J. Electrochem. Soc. 140 (1993) 2825.
[112] E. Delia, O. E. Barcia, O. R. Mattos, N. Pebere and B. Tribollet, J. Electrochem. Soc. 143 (1996) 961.
[113] A. C. West, R. D. Grimm, D. Landolt, C. Deslouis and B. Tribollet, J. Electroanal. Chem. 330 (1992) 693.
[114] R. Vidal and A. C. West, Electrochim. Acta. 41 (1996) 2417.
[115] C. Deslouis, M. C. Lafont, N. Pebere and D. You, Corros. Science 34 (1993) 1567.
[116] O. E. Barcia, O. R. Mattos, N. Pebere and B. Tribollet, Electrochim. Acta 41 (1996) 1385.
[117] O. R. Mattos, J. Chim. Phys. Physico-Chimie Biol. 93 (1996) 777.
[118] O. E. Barcia, O. R. Mattos and B. Tribollet, J. Electrochem. Soc. 139 (1992) 446.
[119] C. Deslouis, G. Maurin, N. Pebere and B. Tribollet, Electrochim. Acta 34 (1989) 1229.
[120] C. Deslouis, N. Tabti and B. Tribollet, J. Appl. Electrochem. 27 (1997) 109.
[121] S. Vandeputte, B. Tribollet, A. Hubin and J. Vereecken, Electrochim. Acta 39 (1994) 2729.
[122] S. Vandeputte, A. Hubin and J. Vereecken, Electrochim. Acta 42 (1997) 3429.
[123] S. Swathirajam and S. Bruckenstein, J. Electroanal. Chem. 163 (1984) 77.

Chapter 11

AC Impedance Spectroscopy of Polymer Films—An Overview

A.R. MOUNT and M.T. ROBERTSON

11.1 INTRODUCTION

Small amplitude AC impedance studies are invaluable to the characterisation of redox active polymer modified electrodes. These coats show profound changes in their cyclic voltammetric (CV) response on redox switching between their oxidised and reduced forms. As the coats typically have of the order of 1–5 M concentrations of redox active centres and associated counterions, present to preserve electroneutrality, the CV technique, which involves the complete oxidation and reduction of the film, therefore involves very large changes in these concentrations. This induces large changes in the amount of solvent in the layer, and the coat morphology, as the structure relaxes to attain the thermodynamically most stable state. Also, large fields can be developed within the film. The CV response is therefore often a complicated convolution of these processes, and the relative contributions of each process is often hard to discern.

In the small amplitude AC technique, a sinusoidal AC voltage perturbation of a few millivolts is applied on top of a constant DC potential. This DC potential is applied to the polymer modified electrode for several minutes to a few hours, in order to ensure that the coat has reached its thermodynamically most stable form at this potential. The small perturbation induces relatively small changes to the film concentration and relatively small fields, which greatly simplifies mathematical treatment. By applying sinusoids of varying frequency, the charging and discharging process can be studied for charged species which move at different speeds. Furthermore, by repeating this procedure at a wide variety of DC potentials, the response as a function of redox composition can be probed.

This therefore allows the relative contribution of the different processes occurring during redox reaction to be quantified.

In order to achieve this, analytical methods for producing physical parameters from this technique have developed by Albery and Mount [1–5]. This work uses a transmission line to model the response of the film to AC perturbation.

11.2 AC IMPEDANCE THEORY—THE TRANSMISSION LINE

In the electrical analogue of the actual transmission line, used to model the movement of charge down an electrical power line, the current is simply driven through the line by the voltage, or potential difference (in other words by migration). However, this is not the case for a film on an electrode such as an ICA trimer film (*vide infra*). If we consider a situation where the concentration of oxidised and reduced species and the potential are related by the Nernst equation throughout the layer, then the coat is at equilibrium and no current will pass, even though there may be variation in the potential throughout the layer. In the layer, we have positively charged trimer species and negatively charged counterions (as well as some positively charged counterions if the coat is mainly reduced). The flux of counterions in the pores is given by the Nernst–Planck equation

$$j = D\{\mathrm{d}c/\mathrm{d}z \pm c(F/RT)\,\mathrm{d}E/\mathrm{d}z\}, \tag{11.1}$$

where D is the diffusion coefficient of the ion, c is the concentration, z is the distance through the pore and E is the potential in the pore. The first term is therefore diffusional motion, and the second is migrational motion, and both charge transport methods may be important. The $\pm$ term indicates the effect of changing the sign of charge on the ion. We [1] have defined a "driving potential", E_D, which can be considered to be the equivalent of the electrical potential for the transmission line circuit of capacitors and resistances; this is the driving force of current through the pore resistance. For perchlorate ion, this is

$$E_D = E - (RT/F)\ln(c/s), \tag{11.2}$$

where c is the concentration of the ion in the pore, s is the concentration of ions in the electrolyte and E_D, is the driving potential, which we call the Donnan potential, as this equation was first produced by Donnan [6, 7] to describe the variation of ionic concentrations in membranes with potential. Here, it can be seen that a gradient in E_D produces the balance between the gradient in potential and the gradient in concentration, reflecting the true driving force for ion movement.

We now turn to the movement of charge between the polymer centres. In our particular case, the redox reaction for the trimer species we consider below is

$$(\mathrm{InCOOH})_3^+ + e^- \Leftrightarrow (\mathrm{InCOOH})_3, \qquad (11.3)$$

and we have ensured by our polymerisation procedure that the coat consists almost exclusively of free trimers in order to simplify the theoretical treatment (*vide infra*). It is often the case, however, that electron transfer between neighbouring sites involves the transfer of one electron, and so this approach is widely applicable to other films. (It is satisfying that for our chosen system electron paramagnetic resonance (epr) and superconducting quantum interference device (SQUID) measurements confirm that one radical spin is produced for every electron removed, in agreement with equation (11.3).) Most layers can be considered to contain fixed sites, and thus the total concentration of redox sites in the layer at any point remains constant. We assume that charge transfer occurs by electrons hopping between oxidised and reduced redox centres. By considering this electron hopping either by the kinetics of the hopping process or by using irreversible thermodynamics we have shown that the flux for this process can be considered to be [1]

$$j = D\{\mathrm{d}b/\mathrm{d}z + [ab/(a+b)](F/RT)(\mathrm{d}E/\mathrm{d}z)\}, \qquad (11.4)$$

where a and b are the concentrations of the reduced species (the reduced trimer centre, Tr, in our case, or more generally, the reduced centre) and the oxidised species (the oxidised trimer species, Tr^+ in our case) respectively and E is the potential at the redox centre. The first term is again the diffusion term, but now there is a cross term multiplying the second migration term. This is because maximum electron transfer will occur when there is maximum probability of exchange, i.e., when there is most likelihood of Tr having Tr^+ as a neighbour, or when $a = b$. Again, there is a driving potential for this process [1] given by

$$E_N = E + (RT/F)\ln(b/a). \tag{11.5}$$

This is termed a Nernst potential, as the expression is simply the Nernst equation for the one electron redox couple. Again, this driving potential tells us about the combination of the diffusional and migrational terms.

In AC impedance, where a small amplitude potential perturbation is applied, the driving potentials can be linearised, giving

$$\Delta E_D = \Delta E - (RT/F)\Delta c,$$

and

$$\Delta E_N = \Delta E + (RT/F)[1/a + 1/b]\Delta b,$$

where Δb and Δc are the induced changes in concentration due to the potential perturbation. These driving potentials are the potentials which drive current through the pores and the trimers. Each can be represented mathematically by a resistance. For the trimers, the resistance will be a minimum when Tr and Tr^+ have equal concentrations, as redox hopping will be most efficient when the nearest neighbour is most likely to be of the opposite oxidation state. The resistance is given by [1]

$$R_P = (RT/F^2)L/(AD_P)[1/a + 1/b], \tag{11.6}$$

where A is the electrode area, D_P is the diffusion coefficient of the trimer species and L is the thickness of the coat. For the pores, there are two cases; in the first, the concentration of Tr^+ is large and Donnan exclusion ensures that only counterions are present in the coat. In this case the pore resistance is given by [1]

$$R_X = (RT/F^2)\{L/(AD_Xc)\} = (RT/F^2)\{L/(AD_Xb)\}, \tag{11.7}$$

where D_X is the diffusion coefficient for the counterion, X, in the pores and b is the concentration of counterions (as the concentration must equal the concentration of positive ions to ensure electoneutrality). If the oxidised trimer is neutral and the reduced trimer is negatively charged, then the counterion would be M^+ and the concentration in equation (11.7) would be a rather than b. The second case is when the concentration of Tr^+ is negligibly small, and both counterions, X^-, and coions, M^+, from the electrolyte are present. Then

$$R_X = (RT/F^2)(L/A)\{1/D_X c + 1/D_M c)$$
$$= (RT/F^2)(L/A)\{(D_X + D_M)/D_x D_M\}1/c, \tag{11.8}$$

where D_M is the diffusion coefficient of the coion and c, the concentration of ions in the pores, would be equal to the solution concentration of electrolyte, s, neglecting activity effects. We have considered the transition between these two cases in previous papers [3, 4].

The redox reaction of the coat requires the storage of charge, and therefore a capacitance. It is the capacitance which ensures electroneutrality in the system and it is given by [1]

$$1/C_\Sigma = 1/C + 1/C_N + 1/C_D, \tag{11.9}$$

with the total capacitance, C_Σ, having three contributions. C_N is the Nernst capacitance, which tells us about the amount of charge stored in the trimers resulting from the potential difference between E_N and E, the potential at the redox centres

$$C_N = (ALF^2/RT)(1/a + 1/b)^{-1}, \tag{11.10}$$

C_D is the Donnan capacitance, which tells us about the amount of charge in the pores accumulated by the potential difference between E_D and E, the potential in the pores. When only counterion is present, C_D will be given by

$$C_D = (ALF^2/RT)b, \tag{11.11}$$

for counteranions or

$$C_D = (ALF^2/RT)a. \tag{11.12}$$

for countercations present for a negatively charged layer. For the other limiting case when both ions are present,

$$C_D = (ALF^2/RT)2s, \qquad (11.13)$$

neglecting activity effects. The third capacitance, C, is the extended double-layer capacitance between the trimers and the aqueous pore. We have already shown [3] that for a film such as polyvinylferrocene this capacitance is very large as the trimer/pore interfacial area throughout the film is very large and the pore contains a large amount of electrolyte; we therefore often have an electrode-like double layer extending over a very large area. This therefore allows the $1/C$ term to be approximated to zero in equation (11.9) when compared with the other two terms, in a large number of cases.

The final part of the transmission line circuit is the charge transfer elements at the electrode/film interface and at the film/solution interface. In the former case, at low AC frequency electrons are transferred from the electrode to the trimer centres. We have shown [5] that this process is controlled by the Nernst driving potential, and that the interfacial resistance is given by

$$R_{K,P} = (RT/AF^2)[a + b]^{-1}\{k'^{-1}_{ox} + k'^{-1}_{red}\}, \qquad (11.14)$$

where k'_{ox} and k'_{red} are the electrochemical rate constants for the oxidation and reduction reactions in equation (11.3) and a and b are the concentrations of oxidised and reduced trimer at the electrode surface. However, at high frequency, charging and discharging of the electrode double layer can occur through the potential [5]. At the latter interface, at low frequency the reaction involved is injection of ions, controlled by the Donnan driving potential [5]. This results in the kinetic resistance

$$R_{K,X} = (RT/AF^2)[s]^{-1}k'^{-1}_{ion}, \qquad (11.15)$$

where k'_{ion} is the interfacial rate constant for ion transfer. Again, at high frequency perturbation, charging of the film/solution double layer can occur, controlled by the potential [5]. We are therefore left with the equivalent transmission line circuit given in Fig. 11.1(a). In this circuit, the capacitances between R_P and R_X control the charging of R_P and R_X and produce the necessary "driving potentials", E_D and E_N. The potential in the layer at any point is given by the potential of the middle (dotted) line. Figure 11.1(b) shows the equivalent circuit when C is not negligible in equation (11.9). Now the potential difference between each trimer centre and its counterion is significant, as we have an extended double layer between the trimer stacks and the pores. We therefore have the capacitance C in the circuit, and the potential at the trimer and in the pore at any point in the film is given by E and E' respectively, as shown.

We have already shown the power of this approach, in that values for these parameters can be obtained by iteratively fitting AC data to this theoretical model [5]. Here we extend and further test this approach by obtaining parameters for ICA trimer films. We have shown [5] that there can be several different cases with a transmission line circuit with one kinetic barrier at one interface. This more general circuit has two kinetic steps, one at each interface, which further complicates matters. However, one simplification for our system is that the frequency of charging of the CR circuits for the kinetic steps are nearly always well separated from the frequency at which the charging of the transmission line occurs. In this case the impedance response is simply an addition of the response of the CR circuits for the kinetic steps (semicircles in the Nyquist plot of $-Z_{\text{im}}$ vs. Z_{re}, where Z_{re} and Z_{im} are the real and imaginary components of the impedance) and the response of the transmission line (a 45° line at high frequency tending to a vertical capacitive line at low frequency). In these cases, the semicircular features are given by [5]

$$Z_{\text{SC},?} = (R_{\text{K},?} - i\omega C_{\text{DL},?} R_{\text{K},?}^2)/[1 + (\omega C_{\text{DL},?} R_{\text{K},?})^2], \qquad (11.16)$$

where ω is the frequency in radians s^{-1}, and $R_{\text{K},?}$ and $C_{\text{DL},?}$ are the charge transfer resistance and double layer capacitance of the appropriate interface ($? = P$ or X as appropriate). The impedance of the resistive rails in the circuit (the transmission line) is given by [1]

$$\begin{aligned} Z_{\text{TR}} = {} & (1/R_X + 1/R_P)^{-1} + (1/R_X + 1/R_P)^{-1} f \\ & + \{(R_X + R_P)/2 + (1/R_X + 1/R_P)^{-1}\} g, \end{aligned} \qquad (11.17)$$

where the functions f and g are given by

$$\begin{aligned} f = {} & \{\sinh\theta\cos\theta - \cosh\theta\sin\theta - i(\sinh\theta\cos\theta + \cosh\theta\sin\theta)\}/ \\ & \theta[\sinh^2\theta\cos^2\theta + \cosh^2\theta\sin^2\theta], \end{aligned} \qquad (11.18)$$

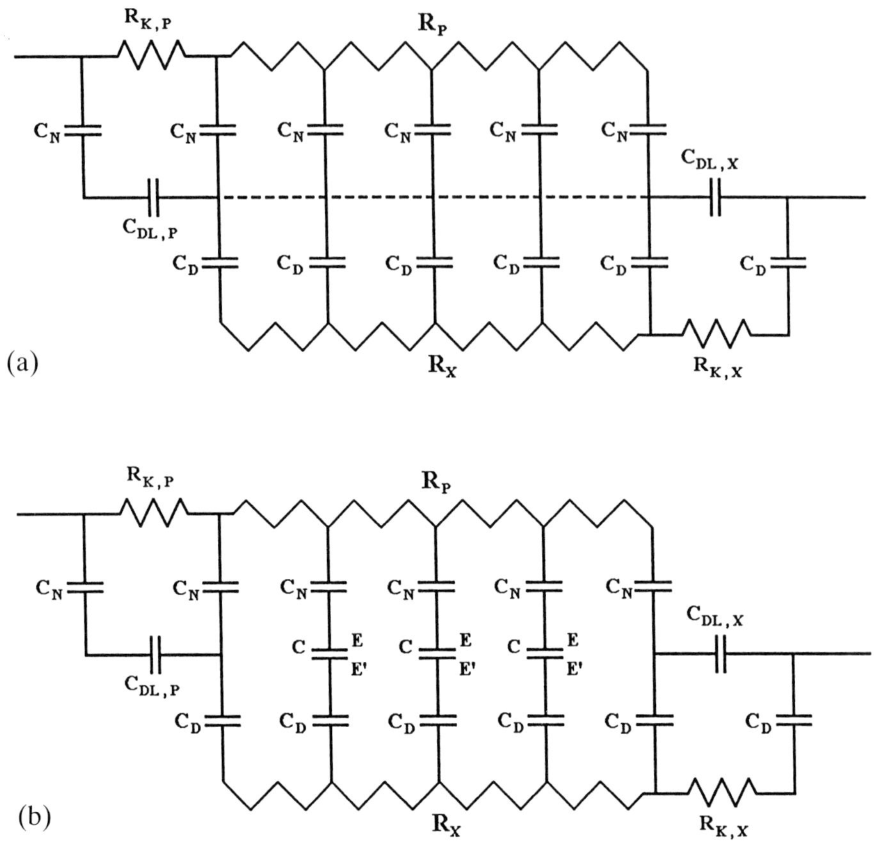

Fig. 11.1. The transmission line circuit used to model these data. The left hand end of the transmission line is at the electrode/film interface. The right hand end is at film/electrolyte interface. The extended resistances, R_P and R_X, correspond to the resistance to motion of electrons between trimer centres and ions through the pores respectively. (a) The potential in the central line of the diagram is the potential within the film, and the connecting capacitors modify this potential to produce the driving potentials to drive current through the resistors. The CR kinetic circuit elements for the interfacial process can be seen at each end of the transmission line. (b) The modified circuit when the capacitance, C in equation (9) is not negligible. The potential at the trimer and in the pores is given by E and E' respectively, as shown.

and

$$g = \{\sinh\theta\cosh\theta - \cos\theta\sin\theta - i(\sinh\theta\cosh\theta + \cos\theta\sin\theta)\}/ \theta[\sinh^2\theta\cos^2\theta + \cosh^2\theta\sin^2\theta]. \quad (11.19)$$

Therefore, experimental fits are made to the combined transmission line circuit

$$Z = Z_{el} + Z_{SC,P} + Z_{SC,X} + Z_{TR}, \quad (11.20)$$

where Z_{el} is the real impedance (resistance) of the electrolyte solution.

11.3 INDOLE-5-CARBOXYLIC ACID FILMS

We have previously considered [8, 9] the redox behaviour and film structure of indole-5-carboxylic acid trimer (ICA trimer) films formed by the electrooxidation of indole-5-carboxylic acid (ICA, Fig. 11.2). We have shown that electrooxidation involves the production and deposition of an asymmetric cyclic trimer (Fig. 11.3) onto the electrode surface

$$\mathrm{InCOOH} \rightarrow (\mathrm{InCOOH})_3^+ + 7e^- + 6\mathrm{H}^+. \quad (11.23)$$

We propose that this trimer forms from coupling of the 3,3′-dimer radical cation and a monomer radical cation. (The electron density in the monomer radical cation is largest at the 3-position. The electron density in the dimer radical cation is largest in the 2-position.) The trimer is initially formed near the electrode surface, where it deposits onto the electrode. Further growth then occurs by adsorption and cyclisation on the surface

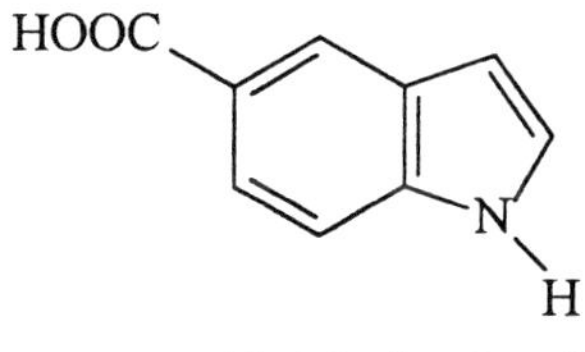

Fig. 11.2. The structure of indole-5-carboxylic acid (ICA).

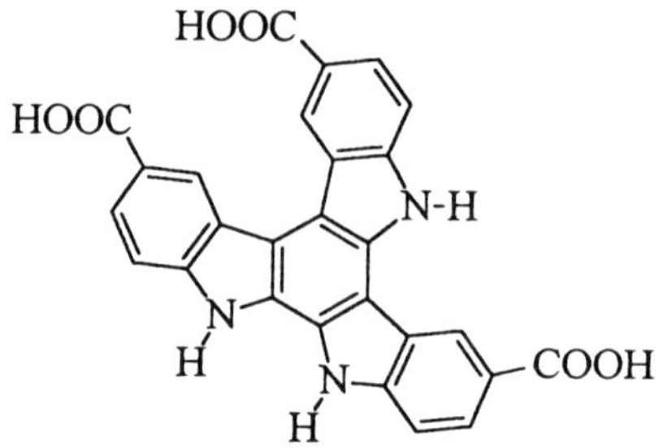

Fig. 11.3. The structure of the asymmetric cyclic trimer produced by the electrooxidation of ICA.

of the previously deposited film [8, 10]. This further growth is also in competition with a linking reaction between the cyclic trimers on the surface of the film to make a polymer of linked trimers [8]

$$n(\mathrm{InCOOH})_3^+ \rightarrow \{-(\mathrm{InCOOH})_3^+-\}_n + 2e^- + 2\mathrm{H}^+. \qquad (11.24)$$

Ring-disc studies have confirmed [11] that for ICA between 7 and 9 electrons are passed and between 6 and 8 protons are evolved in the production of these films. This shows that the trimer is deposited in its oxidised form, with the incorporation of an anion from the electrolyte in order to preserve electroneutrality.

ICA films have previously been studied for their potential use as mediating layers in biosensors [12], and as fast proton conductors [13]. In this work, performed in aqueous electrolyte, protons were found to be incorporated and expelled from the film during redox cycling. However, deprotonation is not observed during film formation in acetonitrile (*vide infra*). We would therefore like to know whether protons are evolved during redox cycling in acetonitrile-based background electrolyte. We will use this system to demonstrate the importance of the transmission line approach in analysing the factors which affect the film redox behaviour, as well as gaining important information on the redox reaction of ICA films.

11.4 EXPERIMENTAL

All chemicals were of AnalaR grade or equivalent and were used as received unless otherwise stated. Electrochemical control of the rotating-disc electrode (RDE) studies was by a modular potentiostat (Oxford Electrodes), with the data being collected on an *x-y-t* chart recorder (Bryans Instruments). An Oxford Electrodes motor controller was used for electrode rotation control. AC impedance data were collected on a Sycopel AC impedance analyser. The amplitude of the sinusoidal ac perturbation to the dc potential used in this work was 10 mV. The counter electrode was a 2 cm^2 platinum gauze, the working electrode was a platinum RDE (Oxford Electrodes, area 0.387 cm^2) and the reference electrode was made in-house and consisted of a silver wire dipping into a solution of silver perchlorate (0.01 M) in background electrolyte. All potentials are reported with respect to this electrode, which has a potential of +0.437 V with respect to the saturated calomel electrode (SCE). The background electrolyte solution used was 0.1 M anhydrous lithium perchlorate (Aldrich, dried in vacuo before use) in acetonitrile (Fisons, dried distilled) unless otherwise stated. ICA was recrystallised from water and dried at 130°C in vacuo for several days before use.

ICA trimer films were produced by the electropolymerisation of 100 mM ICA in background electrolyte at +1.46 V at the RDE, rotating at 4 Hz. Under these conditions [8, 9], electrooxidation and film production occurs under steady-state current conditions at close to 100% current efficiency, and the film is expected to consist almost entirely of free trimer. The electrooxidation reaction was terminated after passing a charge of 173 mC. From equation (11.1) and (11.23) we calculate that this corresponds to 25 mC of redox charge or approximately 250 nmoles of deposited trimer.

Iterative fits to the AC data were performed using SigmaPlot, a plotting program developed by Jandel Scientific. This procedure has been described in detail elsewhere [5], but it involves iteratively fitting with non-linear regression the real impedance, imaginary impedance and frequency data to a linear combination of two kinetic resistance CR circuits and a transmission line, given by equations (11.16)–(11.20). The physical parameters determined by iteration are then used to generate calculated values of real and imaginary impedance at each frequency. The "goodness of fit" is then determined by plotting the calculated and experimental values against each other; in each case for these data a gradient of 1 ± 0.01 was obtained with an intercept of 0 ± 0.01.

11.5 RESULTS AND ANALYSIS

Figure 11.4 presents typical linear sweep voltammetry (LSV) data for the ICA trimer film. It is clear from these data that profound changes occur in the redox behaviour of the ICA trimer film on cycling. The freshly prepared film shows a square voltammogram, indicating that the coat has a large pseudocapacitance, with little evidence of redox peaks (Fig.

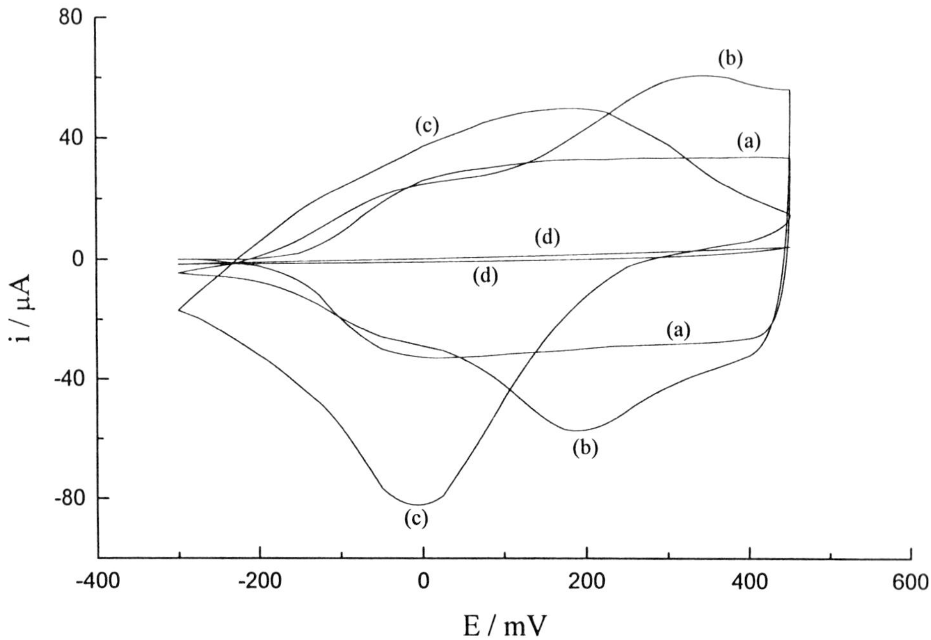

Fig. 11.4. Typical linear sweep voltammogram (LSV) data for the ICA film in background electrolyte LSV data is recorded (a) initially after film formation (b) after cycling and repeated holding first at a potential corresponding to complete oxidation (1 hour) and then at a potential for complete reduction of the film (1 hour) for an overall period of 2 days (c) same as (b) but for 5 days (d) in 0.1 M $NaClO_4$ rather than 0.1 M $LiClO_4$. The voltage sweep rate was 2.5 mV/s^{-1} in all cases.

11.4(a)). On cycling, peaks start to appear (Fig. 11.4(b)), and after several days of repeated oxidation and reduction, redox peaks dominate the voltammogram (Fig. 11.4(c)). That these redox peaks involve cation insertion and ejection can be demonstrated by immersing this conditioned film in 0.1 M $NaClO_4$. The resulting LSV data (Fig. 11.4(d)) show an almost complete loss of redox activity.

These LSV experiments demonstrate the change in the response on the film on cycling, but the overall current measured can be determined by a great many factors; therefore in order to deconvolute this response into the relative contributions of electron and ion injection at the interfaces and electronic and ionic motion in the film, we have performed small amplitude AC impedance studies at a variety of dc potentials which span the range of the LSV study.

Figure 11.5 shows a series of Nyquist plots for the freshly prepared

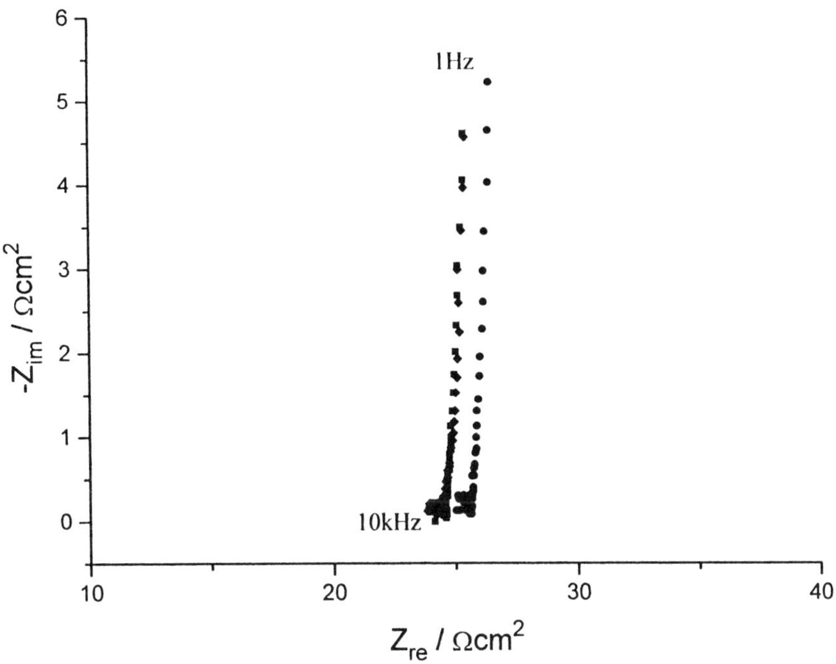

Fig. 11.5. Typical Nyquist plots of Z_{im} vs. Z_{re} for a freshly prepared coat at: (●) −200 mV, (□) −100 mV, (◆) 0 mV. The frequencies of the highest and lowest points in the plots are marked, along with the dc potential at which these data were collected.

coat. It is clear that the plots show a vertical line corresponding to a capacitance, offset on the real axis at a value corresponding the electrolyte resistance. These data indicate that the as-formed film is highly conducting, so that its ionic and electronic resistance within the film is negligible compared to the electrolyte resistance. This is to be expected, as during electrooxidation and film formation, a constant current is observed at the RDE despite the fact that with time the layer is growing to be several thousand layers thick. A significant film resistance would cause this current to drop with time as the film thickness increases. The value of the capacitance, C_{Σ}, obtained is 32 ± 2 mF, and is insensitive to dc potential. This value is much larger than a normal interfacial double layer capacitance, which is typically of the order of 10 s of microFarads, and its insensitivity to potential is the reason why the LSV data shows a square pseudocapacitive envelope (Fig. 11.4(a)). From equation (11.9), we would expect three terms in C_{Σ}, with the smallest capacitance dominating. This behaviour is indicative that C_{Σ} is being dominated by C, the distributed double layer capacitance between trimer centres and counterions, as the values of both C_N and C_D would be expected to show a potential dependence. Thus C_N and C_D appear both to be extremely large, and charge transport in the layer is occurring mainly by migration, with the trimer centres behaving as a distributed metal.

This behaviour is consistent with our thoughts on the structure of the newly-formed layer. We have proposed that the electrooxidation reaction involves the adsorption and cyclisation of monomer radical cations on the film that has already grown. It is most likely that this will result in these flat aromatic trimers forming stacks along which efficient π-electronic communication will occur, with associated anions in the channels between stacks which can also move extremely rapidly. The trimer centres and anion stacks will be separated by a relatively long distance with no solvent between them, which is probably the origin of the capacitance, C. Some preliminary evidence for this has recently been provided [14], with X-ray crystallography of an oxidised indole trimer crystal showing this structure.

In contrast, Fig. 11.6 shows a typical Nyquist plot for the layer after switching between the oxidised and reduced forms in background electrolyte for several days (Fig. 11.4(c)). A pronounced semicircular region, Warburg 45° line and vertical capacitive region can clearly be seen. We have fitted these data to the transmission line circuit (Fig. 11.1). The value of C_{Σ} obtained is found to vary with dc potential (Fig. 11.7) and with the

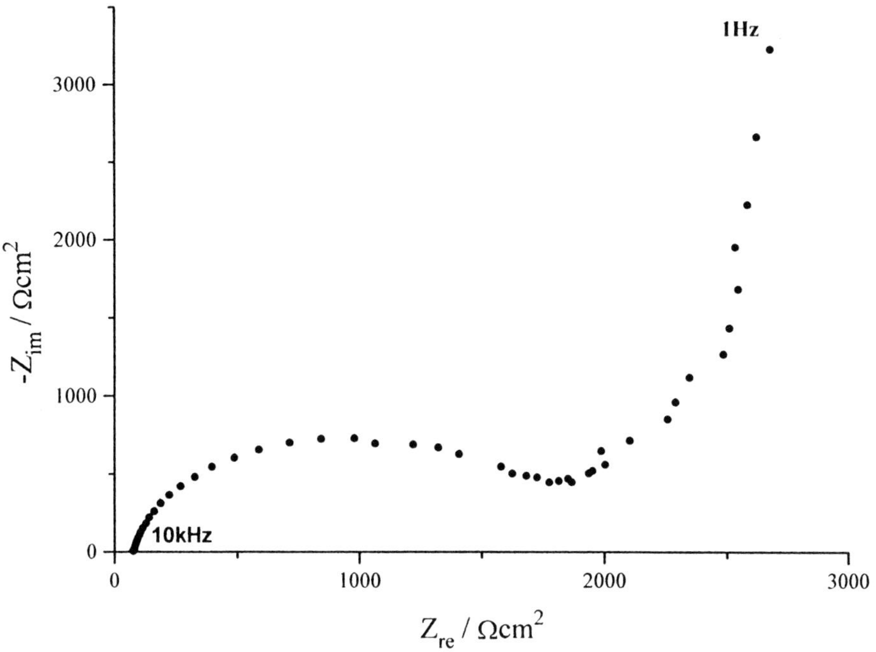

Fig. 11.6. Typical Nyquist plot for the cycled coat (as in Fig. 11.4(c)) at a dc potential of 33 mV.

concentration of background electrolyte. From equation (11.6), at each potential we can determine a value for R_P, which will be proportional to $(1/a + 1/b)$ for the film. Then using this proportionality and equations (11.9), (11.10) and (11.13), and assuming that the value of C is large enough so that $1/C$ can be assuming to approach zero, we can iteratively fit to these expressions, adjusting the values of D_P and L. The result is shown for the data at 0.04 M electrolyte by the theoretical line on Fig. 11.7. It is clear from these data that the theoretical fit is in good agreement with the experimental data. Also shown is the fit for 0.08 M and 0.10 M electrolyte. It can clearly be seen that the fit of theory to experiment is good, except for the highest capacitance values found near the half-oxidised, half-reduced state of the film at 0.10 M. Here the observed capacitance is found to be slightly lower than the calculated value, which may be because for these data the value of C cannot be assumed to be so large

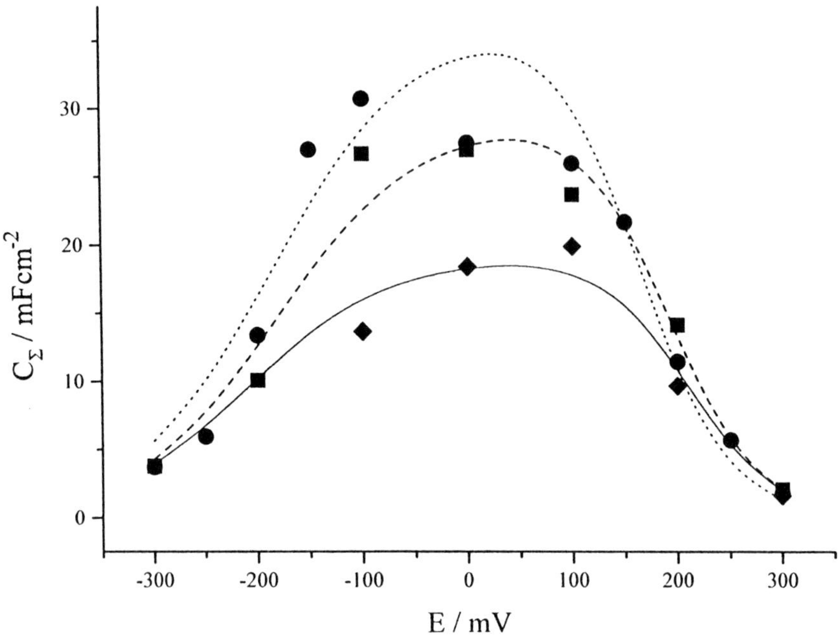

Fig. 11.7. The values of C_Σ obtained from the AC impedance data at various dc potentials (E) for the electrolyte concentrations of (●) 0.10 M, (□) 0.08 M,(◆) 0.04 M. The theoretical lines through these data (- - -) 0.10 M (— —) 0.08 M (———) 0.04 M, are those calculated from equations (11.6), (11.9), (11.10) and (11.13), assuming that C is sufficiently large that it is insignificant in equation (11.9).

that it can be ignored in equation (11.9). The good fit of the data to the experimental results confirms that the transmission line model (Fig. 11.1) is a good model for the redox behaviour of this layer, and the fact that the value of C_D varies according to equation (11.13) indicates that there is little Donnan exclusion in this film. It is therefore clear that the C_D and C_N terms in equation (11.9) have increased in importance compared to the C term. Turning to the kinetic resistances at each interface, these are plotted in Fig. 11.8. It can be seen that one resistance goes through a minimum at intermediate potentials. This is the barrier at the electrode/film interface, whose resistance should be given by equation (11.14). From this equation and the theoretical dependence of the oxidation and reduction

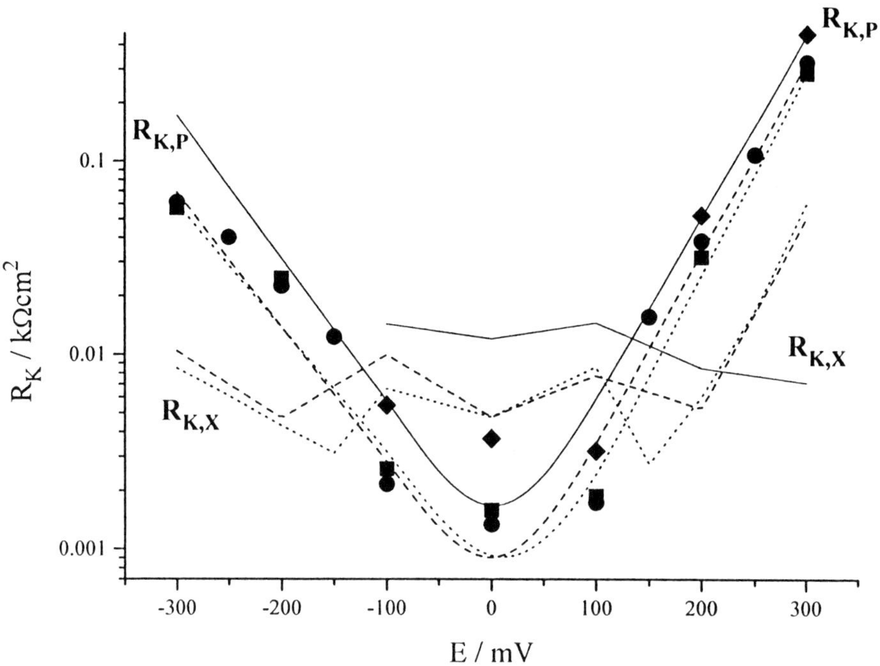

Fig. 11.8. The values of $R_{K,X}$ (lines) and $R_{K,P}$ (points) obtained from the AC impedance data at various dc potentials (E) and electrolyte concentrations of: (●) 0.10 M, (□) 0.08 M, (◆) 0.04 M. Also shown are the theoretical lines through the data points for $R_{K,P}$ at (- - -) 0.10 M, (— —) 0.08 M, (———) 0.04 M, calculated from equation (11.25). The more horizontal lines joining the data points for $R_{K,X}$ also follow the above convention for these electrolyte concentrations.

surface rate constants on potential for a one-electron reaction, the following expression can be derived

$$R_{K,P} = R_0[\exp\{-\alpha_{ox}F(E - E^\theta)/RT\} + \exp\{(1 - \alpha_{ox})F(E - E^\theta)/RT\}, \quad (11.25)$$

where R_0 is one half of the kinetic resistance at $E = E^\theta$ and α_{ox} is the transfer coefficient for oxidation. Also shown on Fig. 11.8 are the fits for the three sets of data corresponding to the three electrolyte concentrations. It is clear from these plots that the agreement between theory and experiment is very good, and from these data we obtain a value for α_{ox} of

0.38 ± 0.02 and for E^{θ} of 0.040 ± 0.010 V. It is also evident that the position and shape of the "bowl" is independent of the electrolyte concentration, which confirms that there is little Donnan potential at the outside of the film. If there were a significant Donnan potential, this would vary with the electrolyte concentration, which would vary the fraction of the overall potential across the film and change the position of the "bowl" on the potential axis.

Also shown in Fig. 11.8 are values for the resistance of the other CR kinetic term, which we can attribute to the ionic charge transfer resistance, $R_{K,X}$. This resistance can arise from the desolvation of ions as they enter the layer from electrolyte solution in order to maintain electroneutrality, but in this case, since the LSV experiments indicate that Na^+, a less well solvated ion, has a greater kinetic resistance, (Fig. 11.4(d)) we believe the barrier is insertion of the ion to create a pore, and that the smaller desolvated ion (Li^+) requires the creation of a smaller pore which results in a smaller resistance. In contrast to AC experiments performed after Fig. 11.4(b), which show the kinetic resistance to increase with increasing potential, we find this approximately constant value for $R_{K,X}$, which is again consistent with the lack of a significant Donnan potential at the outside of the film (the presence of which would be expected to alter $R_{K,X}$ through changing k'_{ion} in equation (11.15). The increase in the observed value of $R_{K,X}$ with potential observed after Fig. 11.4(b) further strengthens the view that it is lithium cation which is being inserted at these potentials, as for an anion $R_{K,X}$ would fall with increasing potential [5].

It is reassuring for the model that there is also an approximate doubling of the measured values of $R_{K,X}$ as s is deceased from 0.08 M to 0.04 M, in line with equation (11.15). This is to be expected, as increasing the concentration of ions in the electrolyte will increase the kinetics of ion insertion into the film.

11.6 INTERPRETATION OF TRANSMISSION LINE RESULTS FOR ICA

It is clear from these data that an ICA trimer film undergoes profound changes in structure and redox kinetics on redox cycling and holding the coat in first the reduced then the oxidised form for long periods of time. Initially, we would expect, and our results are consistent with, a structure where the trimer is deposited in π-conjugated stacks of oxidised trimer, with associated perchlorate ions between the stacks to preserve electroneutrality. On reducing this film and holding the film in the reduced form, we would expect the reaction

$$(\mathrm{InCOOH})_3^+\mathrm{ClO}_4^-(s) + \mathrm{e}^- \rightarrow (\mathrm{InCOOH})_3(s) + \mathrm{ClO}_4^-(\mathrm{aq}), \quad (11.26)$$

to occur. The film would then be expected to be much more compact and lyophobic as it will consist of neutral aromatic molecules. This may well lead to a change in coat morphology, the egress of solvent and an increase in film density.

On reoxidation ingress of anion into this compact layer would be hard. We have already shown that the carboxylic acid groups are delocalised into the trimer redox system [11]. We would therefore expect that field assisted dissociation of the carboxylic acid moieties would occur

$$(\mathrm{InCOOH})_3 \rightarrow (\mathrm{In}_3(\mathrm{COOH})_2(\mathrm{COO}))(s) + \mathrm{H}^+(\mathrm{aq}) + e^- \quad (11.27)$$

producing a self-charge compensating indole trimer film. (We have observed these proton fluxes using a ring electrode downstream from the RDE). These protons are then lost into the electrolyte solution, and subsequent rereduction of the film will result in the ingress of lithium cation

$$(\mathrm{In}_3(\mathrm{COOH})_2(\mathrm{COO}))(s) + \mathrm{Li}^+(\mathrm{aq}) + e^- \rightarrow (\mathrm{In}_3(\mathrm{COOH})_2(\mathrm{COOLi}))(s) \quad (11.28)$$

and any further redox cycling will result in the incorporation and expulsion of lithium ion (the forward and back reactions in equation (11.28)). As this reaction proceeds, pores in the film open up, the total charge in the film increases, and solvent will be reincorporated. This leads to a change in coat morphology, reducing the π-electronic interaction between stacks and increasing the electron transport resistance between trimer centres, whilst decreasing the ionic resistance by opening up the pores to lithium ions, increasing the value of C and making it much less important in determining the value of C_Σ for the film. Our AC data for this coat indicates that there is little Donnan exclusion in this film, and that the pores carry a similar concentration of electrolyte as the outside solution. This suggests that for the reduced form of the coat in equation (11.19), the lithium is strongly co-ordinated to the trimer centre, possibly as an ion-pair, thereby reducing the field in the pores of the film, and allowing equal concentrations of anions and cations to be present in the pores.

11.7 CONCLUSIONS

AC measurements on these ICA films clearly show the importance of this technique in monitoring slow changes in coat morphology and kinetics over several days. The AC data have been interpreted by using Transmission line theory with modified Nernst and Donnan potentials driving the current through the polymer strands and the ionic pores respectively. Freshly deposited layers show a simple capacitance indicative of fast switching of the layer. On cycling, the AC spectra change and a charge transfer resistance appears, indicating a resistance to lithium ion insertion. On continued cycling, the rate determining step in the redox reaction becomes charge transfer between the platinum electrode and the polymer at the electrode/polymer interface and the barrier to lithium insertion decreases in importance. Electronic motion through the polymer is consistent with fast electron hopping between trimer centres. This behaviour is explained in term of the dissociation of a carboxylic acid groups on oxidation and the insertion of lithium ions on reduction. The slow changes in the kinetics and mechanism of the coat redox reaction are attributed to the changes in coat structure as the composition of the layer changes. That this approach has been successful in modelling all these different cases, with widely differing charge transport mechanisms, demonstrates that the analytical Transmission line theory developed by Albery and Mount is widely applicable to interpreting the results of these, and by extension of other modified electrode films.

ACKNOWLEDGEMENTS

MTR wishes to thank the University of Edinburgh for funding. ARM thanks Professor W. John Albery FRS for his major contribution to the partnership which developed the Transmission line theory used in this work.

REFERENCES

[1] W.J. Albery, C.M. Elliott and A.R. Mount, *J. Electroanal. Chem.* 288 (1990) 15.
[2] W.J. Albery and A.R. Mount, *J. Chem. Soc. Faraday Trans.* 89 (1993) 2487.
[3] W.J. Albery and A.R. Mount, *J. Chem. Soc. Faraday Trans.* 90 (1994) 1115.
[4] W.J. Albery and A.R. Mount, *J. Electroanal. Chem.* 388 (1995) 1.
[5] W.J. Albery and A.R. Mount, Electroactive Polymer Electrochemistry Part 1: Fundamentals, Plenum, New York (1994) 443.
[6] F.G. Donnan, *Z. Elektrochem* 17 (1911) 572.
[7] F.G. Donnan, *J. Membrane Sci.* 100 (1995) 45.
[8] J.G. Mackintosh and A.R. Mount, *J. Chem. Soc. Faraday Trans.* 90 (1994) 1121.
[9] J.G. Mackintosh, C.R. Redpath, A.C. Jones, P.R.R. Langridge-Smith, D. Reed and A.R. Mount, *J. Electroanal. Chem.* 375 (1994) 63.
[10] J.G. Mackintosh, S.J. Wright, P.R.R. Langridge-Smith and A.R. Mount, *J. Chem. Soc. Faraday Trans.* 92 (1996) 4109.
[11] P. Jennings, A.C. Jones, A.R. Mount and A.D. Thomson, *J. Chem. Soc. Faraday Trans.* 93 (1997) 3791.
[12] P.N. Bartlett and J. Farrington, *J. Electroanal. Chem.* 261 (1989), 471.
[13] P.N. Bartlett and J. Farrington, *Bull. Electrochem.* 8 (1992) 208.
[14] V. Bocchi, A. Colombo, W. Porzio, *Synth. Metals* 80 (1996) 309.

Chapter 12

Kinetic Applications of the Electrochemical Quartz Crystal Microbalance (EQCM)

ERNESTO J. CALVO[1,2] and R. A. ETCHENIQUE

12.1 INTRODUCTION

Since the early work of Kanazawa [1] and Bruckenstein in 1985 [2], quartz crystal resonators have been used for more than 12 years in contact with liquids to assess changes in mass during electrochemical surface processes. Extensive use of the electrochemical quartz crystal microbalance (EQCM) has been done in the study of electrode processes with change of mass simultaneous to charge transfer.

In this paper we present a brief topical account of the kinetic possibilities and limitations of the EQCM. Extensive review chapters with theoretical aspects of the quartz crystal micro-balance have been written by Buttry and Ward [3–5] and the reader is referred to them for a more detailed discussion.

12.2 THE QUARTZ CRYSTAL MICROBALANCE

The quartz crystal micro-balance (QCM), the most extensively studied shear mode AT-cut quartz resonator, is comprised of a thin slice of quartz single crystal with two metal electrodes deposited on both faces of the crystal. These excitation electrodes generate a transverse shear wave across the thickness of the crystal that propagates into the film immobilized onto the crystal surface. When the over-layer is non-rigidly coupled to the

[1]In honour to Professor W.J. Albery on occasion of his 60th. anniversary.

[2]Permanent Research Staff of Conicet (Argentina).

quartz crystal the response also depends on the viscoelastic properties of the deposited layer [6]. In the particular case of the EQCM, one of the quartz crystal faces coated with gold is immersed in liquid electrolyte and connected to an electrochemical cell with the gold coated quartz acting as the working electrode.

Sauerbrey [7] in 1959 related the change in resonance frequency of a piezo-electric quartz crystal with the mass deposited onto or removed from the crystal surface. This approach has been used to perform micro-gravimetric measurements in the gas phase like metal evaporation. For ($\Delta f \ll f_o$) the Sauerbrey equation states that:

$$\Delta f = -\frac{2f_o^2}{\sqrt{\mu_Q \rho_Q}} \frac{\Delta m}{A}, \tag{12.1}$$

with Δf the measured frequency shift, f_o the resonant frequency of the quartz crystal, Δm the mass change, A the piezoelectrically active area, $\rho_Q = 2.648\ \mathrm{g\ cm^{-3}}$, the quartz density and $\mu_Q = 2.947 \times 10^{11}\ \mathrm{dyn\ cm^{-2}}$, the shear modulus of AT-cut quartz. The sensitivity factor increases with the resonant frequency of the resonator which corresponds to a thinner quartz crystal; for example a 10 MHz AT-cut quartz crystal 0.017 cm thick has a nominal sensitivity of $0.226\ \mathrm{Hz\ cm^2\ ng^{-1}}$.

The Sauerbrey equation assumes a rigid film with density and transverse velocity of the acoustic wave identical to those of the quartz crystal (equal acoustic impedance of the quartz and the over-layer). It also assumes that the deposit is uniform while the sensitivity of the QCM is non uniform across the radial direction of the resonant quartz crystal, with maximum sensitivity at the crystal centre [8].

Piezoelectric acoustic wave devices also respond to small changes in mass at surfaces immersed in (viscous) liquids [9]. The resonance frequency of AT-cut quartz resonators immersed in liquids depends on the liquid density and viscosity. The transverse shear wave which penetrates into the viscoelastic deposit and into the liquid is damped due to energy dissipation associated with the viscosity of the medium (film or liquid) at the acoustic frequencies.

In addition to mass changes at the quartz crystal surface and, liquid density and viscosity the resonant frequency can be affected by several other factors such as: the liquid conductivity [10], the hydrostatic pressure difference across both crystal surfaces and lateral stress at the overlayer deposited onto the quartz crystal [11], the surface roughness [12], longitudinal waves in the liquid [13] and viscoelastic properties and volume change of films immobilised at the quartz crystal [14].

12.3 THE ELECTROCHEMICAL QUARTZ CRYSTAL MICROBALANCE

The electrochemical quartz crystal microbalance (EQCM) has emerged as a very powerful "in situ" technique to complement electrochemical experiments [3–5]. Nomura and Okuhara [15] first used the quartz crystal microbalance (QCM) to detect mass changes at a metal coated quartz resonator immersed in electrolyte during electrochemical experiments.

The use of EQCM permits the kinetic examination of processes that involve events at the monolayer level and thin films which would be otherwise silent to electro-chemical methods, i.e., fluxes of solvent, neutral molecules or ion pairs. Bruckenstein weighed one adsorbed oxygen monolayer (20 ng) on the gold surface during the simultaneous recording of the current-voltage curve [16]. Partial electron transfer, i.e., electrosorption valency, during electrosorption phenomena could be detected by simultaneous measurement of the electrical charge in the electrochemical experiments and the EQCM frequency change during the adsorption of ionic species [17].

The technique has been extended to other electrochemical phenomena such as electro-deposition and dissolution of metal films, hydrogen absorption in metals, under-potential deposition, electrolyte adsorption, and ion and solvent exchange in redox polymer layers and electrochemically driven self-assembly, etc. [5].

The EQCM has been most commonly used simultaneously to quasi-steady state techniques like slow scan cyclic voltammetry. In this way mass changes during electrolysis can be obtained from $\Delta f(\Delta m/A)$ vs. potential curves, while $\Delta f(\Delta m/A)$ vs. charge density curves allow evaluation of the number of Faraday exchanged per mole of electro-active species by use of Faraday's law of electrolysis.

12.4 TRANSIENT STUDIES WITH THE EQCM

Potential or current step transients seem to be more appropriate for kinetic studies since the initial and boundary conditions of the experiment are better defined unlike linear scan or cyclic voltammetry where time and potential are convoluted. The time resolution of the EQCM is limited in this case by the measurement of the resonant frequency. There are different methods to measure the crystal resonance frequency. In the simplest approach, the Miller oscillator or similar circuit tuned to one of the crystal resonance frequencies may be used and the frequency can be measured directly with a frequency meter [18]. This simple experimental device can be easily built, but has a poor resolution which is inversely proportional to the measurement time; for instance for an accuracy of 1 Hz, a gate time of 1 second is needed, and for 0.1 Hz the measurement lasts as long as 10 seconds minimum to achieve the same accuracy. An advantage of the Miller oscillator is that the crystal electrode is grounded and can be used as the working electrode with a hard ground potentiostat with no conflict between the high ac circuit and the dc electrochemical circuit.

Modern frequency meters allow fast measurements with good resolution, i.e., 60 ms (Philips PM6654) or 100 ms (HP5384) for 0.1 Hz resolution. The ultimate resolution of the QCM is limited, however, to miliseconds by the time required in equilibrating the crystal with the deposited mass, which is determined by the operating frequency and the quality factor of the QCM [3].

In order to increase the resolution, a heterodyne circuit may be used, which consists of two oscillators: the quartz crystal microbalance and a second reference crystal. The frequency difference in the kilohertz range can be measured with a universal counter and the same resolution and gate time discussed above are valid; however if the period of the differential wave is measured, the accuracy can be highly increased and a 1 Hz resolution is easily attainable for a 1 ms gate time. This approach has been used by Sheng-li Chen *et al.* [19] to measure the kinetics of silver oxide formation on polycrystalline silver. These authors reported a resolution of 0.1 Hz in 1 ms which corresponds to 0.44 ng cm^{-2} in the mass sensitivity.

Bruckenstein [2] introduced a digital oscillator which comprised two TTL inverters with a positive feedback through the crystal. The circuit is very stable and can be interfaced directly with TTL or CMOS logic obtaining the difference frequency by digital subtraction. A limitation of this circuit, however, is that the working electrode is not grounded.

With the circuits described above, the resonance frequency can be directly measured as a digital signal useful to interface the EQCM experiment with a digital computer. On the other hand, when an analogue output signal proportional to the resonance frequency is needed, a frequency-to-voltage converter (FVC) can be added to the differential frequency output. This approach was first used by Bruckenstein [2] and more recently by Gabrieli *et al.*, who employed two Miller oscillators in parallel for the EQCM and the reference crystal, with the frequency-to-voltage converter connected to the oscillator output to perform electro-gravimetry of a Cu/Cu^{2+} electrode reaction in the 0.1 to 10 Hz range [20]. The voltage signal proportional to the resonance frequency shift of the EQCM was then input to a frequency response analyser (FRA) using as reference signal the electrode perturbation potential.

12.5 KINETIC APPLICATIONS

There are several of examples of kinetic studies in which the EQCM has been combined with other nonelecrochemical techniques in order to enhance the information of the electrochemical process under study.

Grzegorzewski and Heusler introduced a very nice experiment by combining the EQCM to the rotating disc electrode [21]. One of the electrodes of the EQCM was simultaneously a rotating disc electrode at which electrochemical reactions proceeded under controlled hydrodynamic conditions in the diffusion-convective steady state. A 10 MHz AT-cut quartz crystal was used as the base of a stainless steel hollow rotating cylinder that contained the oscillator electronics and the electrical contact was achieved through three mercury rotary contacts to ground, counter electrode and to the stabilised power source. Transients of the frequency shift and current under well defined mass transport conditions were recorded during current pulses at a MnO_2 electrode. The time dependence of hydrogen ion (j_{H^+}), Mn(II) ions ($j_{Mn(II)}$) and electron fluxes (current density) were simultaneously recorded.

MnO_2 is a ternary system and the statistically independent transfer of oxide ion (O^{2-}) and Mn^{2+} ions can take place at the oxide-water interface in addition to hydrogen ions and water. The respective component currents could be separated by combination of the rotating disc EQCM with a rotating ring disc electrode (RRDE): While pulsing the MnO_2 disc current, the ring measured hydrogen ions and the EQCM monitored changes in mass. These experiments showed that Mn^{2+} ions are transferred without water and that O^{2-} ions are transferred from the oxide surface to the electrolyte with several water molecules.

Another example of a similar rotating disc electrode with an EQCM has been reported by Ritchie *et al.* [22] for the study of gold and copper dissolution during the electroreduction of oxygen in alkaline cyanide solutions.

Gabrielli and Perrot [23] carried out "in situ" mass measurements in well-defined flowing electrolyte with an electrochemical quartz crystal microbalance (EQCM) adapted to a submerged impinging jet cell (wall tube configuration). The authors employed this new device for the study of nickel electrodeposition and evaluation of the cathodic efficiency. Under the conditions of their experiment (nozzle diameter $d = 7$ mm; disc electrode diameter $d_e = 5$ mm and nozzle-to-electrode distance $H = 2d$), the current that flows at the electrode increases with the square root of flow rate (0–10 $cm^3 s^{-1}$). It should be noted that this approach is much simpler to implement than the rotating EQCM, while keeping control of the convective-diffusion conditions.

By combination of ellipsometry and EQCM Gottesfel [24] studied the nucleation and growth of polyaniline (PANI). The mass deposited was obtained by the frequency change while the thickness resulted from the ellipsometry measurements. Therefore the variations of mass and PANI film density were followed in real time, which would otherwise be impossible with either individual technique alone.

Uosaki *et al.* combined EQCM and UV-visible absorption spectroscopy to study redox reactions of viologens [25]. Optically transparent Au coated AT-cut quartz crystals placed in the optical path of a commercial uv-visible spectrometer recording absorbance simultaneously with the oscillating frequency. Reduction of heptaviologen and methyl-viologen to the radical cation as shown by absorption at 550 nm was studied by cyclic voltammetry, recording Δf and ΔA vs. potential. Reversing the potential the radical cation was reoxidised to a dication. The large decrease of frequency during the cathodic process was ascribed to deposition of the radical cation returning to the initial frequency after dissolution of the deposit with decrease of 550 nm absorbance and a stripping voltammetric peak. Linear plots of Δf and ΔA vs. the reduction charge yielded one Faraday per mole of heptaviologen reduced, i.e., 433 vs. 434 g mol^{-1}, the molar mass of heptaviologen radical cation monobromide indicating that no solvent uptake was operative. Also the authors reported an extinction coefficient, $\epsilon = 1.3\,M^{-1}\,cm^{-1}$, from the slope of ΔA vs. charge plots for a one-electron reaction. However, much less change in frequency was found for the methylviologen reaction, i.e., 5 Hz vs. 3390 Hz due to a lighter radical cation and less deposit, while ΔA reflects the formation of the soluble radical cation close to the electrode.

Yao *et al.* [26] combined spectro-electrochemistry at grazing incidence angle with EQCM to study copper deposition from ammonial solutions. They also reported the use of a bifurcated fibre optic reflection spectroelectrochemical cell combined with an EQCM for the study of poly-naphthylamine film growth on platinum from acetonitrile solutions.

Photochemically induced changes in Cu_2O layers have been reported with simultaneous recording of the current transients and EQCM microgravimetry. The mass of the copper electrode decreased under illumination and increased in the dark under cathodic polarization [27].

Since micro-gravimetry with the EQCM lacks specificity only the difference of cation and anion fluxes can be obtained by microgravimetry and therefore an independent measurement of specific ions is needed. Scanning electrochemical microscopy (SECM) coupled with a quartz crystal microbalance with independent potential control of the tip and substrate has been recently done by Cliffel and Bard [28]. In this experiment generation at the substrate (EQCM crystal) working electrode and collection at the tip of an ultramicroelectrode (UNE) that was approached perpendicular to the EQCM crystal was employed with measurement of Δf. Hillier and Ward [8] had previously used a scanning microelectrode to map the mass sensitivity across the surface of the QCM crystal. Reflection of longitudinal waves at the UME tip limits these experiments due to oscillations.

Another possibility to independently measure ionic fluxes is to combine the EQCM with real time beam probe deflection technique to identify the exchanged ions [29, 30]. Ionic fluxes across the electrode-electrolyte interface result in changes of the solution refraction index adjacent to the electrode surface and a transient deflection of a laser beam is measured.

Kautek and co-workers [31] have combined "in situ" atomic force microscopy (AFM) and EQCM to study the specific adsorption and phase change processes at the silver/halide interface.

Bourkane, Gabrielli and Keddam reported an interesting kinetic study of electrode processes by ac-quartz electrogravimetry [32–36]. Dynamic information on mass changes was obtained by sinusoidal perturbation of the electrolysis current: Electrochemical impedance spectroscopy (EIS) and mass-voltage transfer functions were obtained during the deposition and dissolution of copper in sulfuric acid [32, 33]. This technique is able to reveal the kinetics of adsorbed reaction intermediates and complements the information supplied by EIS. The authors obtained the transfer function by vectorial comparison of the frequency-to-voltage converter output signal and the perturbing galvanostatic signal using a frequency response analyser [34].

Electrogravimetric transfer functions $\Delta m/\Delta E(\omega)$ and $\Delta m/\Delta q(\omega)$ associated with the electrochemical impedance, $\Delta E/\Delta I(\omega)$ were determined for the anodic dissolution of Fe in sulfuric acid. This gave the unique possibility of discriminating between the change in mass due to metal dissolution and the relaxation of the film thickness on the basis of their individual frequency responses [35, 36].

Mass transport of cations, anions and solvent was also extensively studied during redox processes in polymer film electrodes [37]. Hillman and Bruckestein have extensively studied poly(vinylferrocene) films (PVF) in different electrolytes under permselectivity and non-permselectivity conditions. Hillman [38] has also reported chrono-amperometric transient experiments to distinguish global equilibrium from rate limiting transfer of electrons, ions and neutral species during film redox conversion. The total mass change observed by the EQCM is the sum of the contributions of ions, solvent (S) and salt (CA) exchanged at the film-electrolyte boundary. If only anions (A) are exchanged, then:

$$\Delta M = \Delta M_A + \Delta M_S + \Delta M_{CA}. \tag{12.2}$$

The ionic flux is correlated to the charge (Q) through the electroneutrality condition and Faraday's law by ($m_A Q/F$) where m_A is the anion molar mass. Hillman further defined [39]:

$$\Phi = \Delta M_A - (m_A Q/F) = \Delta M_S + \Delta M_{CA}. \tag{12.3}$$

Under conditions of permselectivity, i.e., when the electrolyte concentration is much less than the redox concentration in the film, ϕ represents the mass of solvent exchanged.

Transient experiments with the EQCM have shown that for PVF in aqueous $NaClO_4$ exchange of neutral species is rate determining, with water entry into the reduced film very slow (D_{H_2O} (PVF) $\approx 10^{-10}\,cm^2\,s^{-1}$). At higher electrolyte concentration ΔM_{CA} is no longer negligible and slow exchange of solvent and salt is also observed.

Swelling phenomena due to solvent uptake by the films may lead to volume and viscoelastic changes, which limits the information that can be obtained on rigid mass as will be discussed below.

Ward [40, 41] studied bimolecular reactions between PVF films at open circuit and solution species by following mass changes with the QCM. Thus PVF oxidation by I_2/I^- and PVF^+ reduction by either $Fe(CN)_6^{4-}$ or $Ru(NH_3)^{2+}$ resulted in mass changes due to the ingress of I_3^- and expulsion of counterions in the latter two reduction reactions respectively.

Inzelt [42] described interesting periodic oscillations of the EQCM frequency during open-circuit copper dissolution in sulfuric $CuSO_4$ solutions. Below 45 mM cupric ion concentration the frequency time evolution showed a monotonous increase due to the copper mass loss. However above this critical concentration of copper ions in solution an oscillatory frequency behaviour was observed while the amplitude of the open circuit potential oscillation was less than 0.5 mV. The large-amplitude frequency oscillation lasted 1500 s and the period and amplitude of oscillation changed with time. Also, the frequency oscillation was sensitive to stirring in the solution so that mass transport rate determines at least some of the processes responsible for the frequency oscillation. During the anodic copper dissolution the following processes can take place:

$$Cu = Cu^{2+} + 2e, \tag{12.4}$$

and the disproportionation reaction at the copper surface:

$$2Cu^+ = Cu^{2+} + Cu. \tag{12.5}$$

Nonlinear feedback would produce multiple steady states and support oscillatory behavior. The increase of the analytical Cu(II) ion concentration leads to a mass decrease due to enhanced dissolution as shown by the continuous increase in frequency superimposed to the oscillations. Thus Cu(II) may act as cathodic depolariser in a non-uniform surface and the local decrease of its surface concentration (as reflected by the Nernst potential increase) may lead to the formation of Cu(I) which can further dis-proportionate to readjust the new steady-state diffusion layer. Beyond the complexity of the electrode kinetics and diffusion, the build up of viscosity and density gradients in the solution adjacent to the copper plated quartz crystal oscillator may be responsible for the oscillating phenomena.

The study of passive iron films has prompted to extensive investigations of γ-FeOOH layers. Oxidation of Fe^{2+} ions in neutral or alkaline solutions leads to the formation of γ-FeOOH as shown by Stimming [43] who reported that the time dependence of the anodic current for the formation of γ-FeOOH on gold is diffusion controlled by the ferrous ions through the Fe(III) oxyhydroxide layer under potentiostatic control. More recently Takahashi and co-workers [44] studied the anodic deposition reaction of ferrous ions on gold by "in situ" EQCM in solutions containing different anions. The time evolution of the anodic current and the Sauerbrey mass gain from the frequency shift was reported. From the analysis of the mass-charge relationship, the authors reported the apparent molecular weight of the deposited films with a strong dependence of the electrolyte anions.

In our hands, EQCM studies of this system have confirmed previous reports of γ-FeOOH deposition kinetics and the chemical reaction of ferrous ions with this film after a current interruption step. Figure 12.1 depicts the simultaneous transients of anodic current (Fig. 12.1(b)) and frequency shift (Δf) when a potential step (Fig. 12.1(c)) is applied from a potential where there is no reaction on gold to a potential where diffusion controlled oxidation of ferrous ions takes place. The current transient shown in Fig. 12.1(b) can be described by a diffusion process since a linear dependence of the anodic current density with $t^{-1/2}$ was found as predicted by the Cottrell equation:

$$i = F\left(\frac{D}{\pi}\right)^{1/2} C^{*}.\, t^{-1/2},$$

with F the Faraday constant, D the diffusion coefficient and C^{*} the analytical concentration of ferrous ions in the electrolyte. From the slope of linear Cottrell plots resulted $cD^{1/2} \approx 1.3 \times 10^{-8}$ M cm^2 s^{-1} in good agreement

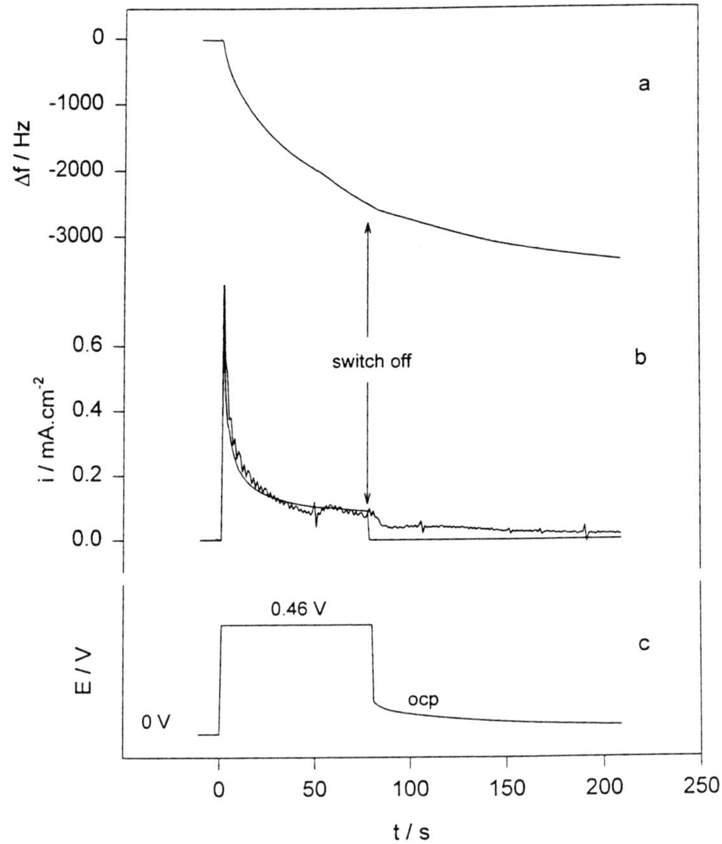

Fig. 12.1. Anodic deposition of FeOOH on Au from 5 mM $Fe(NH_4)_2SO_4$ in $(NH_4)_2SO_4$ and borax buffer of pH 8.5 at 0.7 V (NHE). (a) Time course of Δf; (b) Current density and df/dt and (c) Electrode potential. Arrows indicate switch off and open circuit potential transient.

with the result of Stimming and $D = 6.4 \times 10^{-6}\ \mathrm{cm^2\,s^{-1}}$ was obtained for $c^* = 5 \times 10^{-6}\ \mathrm{mol\,cm^{-3}}$.

The simultaneous negative frequency shift of the EQCM (Fig. 12.1(a)) indicates mass increase in agreement with Takahashi's results [44]. The apparent molar mass of deposited film, M was evaluated from the average initial slope of a linear plot—Δf vs. the anodic charge density, q in Fig. 12.2 and use of the Sauerbrey equation and Faraday's law of electrolysis.

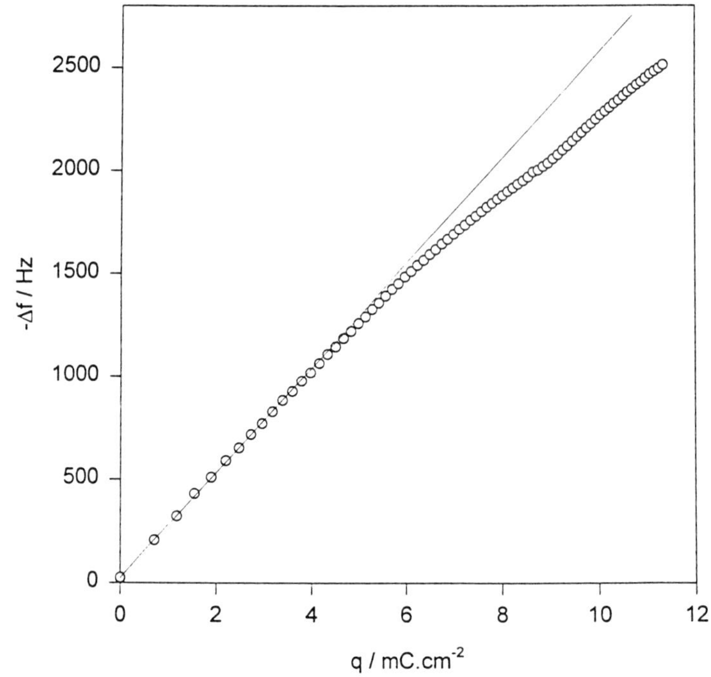

Fig. 12.2. Plot of $-\Delta f$ vs anodic charge density during deposition of γ-FeOOH on Au under the conditions of Fig. 12.1.

The average slope resulted 262 Hz cm^2 C^{-1} and yielded $M = 111$ g mol^{-1} which compares well to the molar mass of pure γ-FeOOH, 88 g mol^{-1} and the values obtained by Takahashi and co-workers, 97 and 132 g mol^{-1} in perchlorate and borate solutions respectively.

Differentiation of the Sauerbrey equation (12.1) yields the mass flux:

$$df/dt(\text{Hz s}^{-1}) = -2.26 \times 10^8\, M\, \text{g}^{-1}\,\text{cm}^3\,\text{Hz/F}.\, A \qquad (12.6)$$

where M is the molar mass of the deposit and A is the piezoelectric and electro-chemically active area. Supersimposing the resulting df/dt transient curve to the anodic current transient in Fig. 12.1(b) for the oxidation of ferrous ions on gold shows excellent agreement. The agreement of the electrochemistry and the EQCM, indicates that the mass flux of ferrous ions diffusing towards the gold electrode and the diffusion controlled anodic current due to their oxidation at the gold surface match through the entire transient as expected for the reaction:

$$Fe^{2+} + (n+2)H_2O \rightarrow FeOOH.\, nH_2O + 3H^+ + e. \qquad (12.7)$$

After about 80 s of film growth the electrode was disconnected while still recording the frequency shift. Figure 12.1 shows that after current interruption Δf continued to decrease simultaneously to a drop in the open circuit potential (ocp). Figure 12.1(b) depicts a net flux of ferrous ions towards the electrode which decays slowly after switch off. It is suggested that formation of magnetite [45] by chemical reaction of γ-FeOOH with incoming ferrous ions would consume Fe^{2+} ions from the solution even after current interruption according to [45]:

$$2\gamma - FeOOH + Fe^{2+} + 2HO^- \rightarrow Fe_3O_4 + 2H_2O. \qquad (12.8)$$

This reaction requires a stoichiometry of FeOOH to Fe^{2+} of two-to-one. Analysis of the Δf in Fig. 12.1, with the Sauerbrey equation (12.1) and the corresponding molar masses of FeOOH and Fe^{2+} confirms this hypothesis. It is to be noted that this last result could not have been possible should the EQCM had not been used.

12.6 VISCOELASTIC EFFECTS

Mason [46] first observed that the viscoelastic properties of a fluid in contact with quartz crystals can affect the resonant properties. However, Mason's work had been forgotten and for a long time there have not been studies of piezoelectric acoustic wave devices in contact with liquids until Nomura and Okuhara [15] found an empirical expression that described the changes in the quartz resonant frequency as a function of the liquid density, its viscosity and the conductivity in which the crystal was immersed. Shortly after the empirical observations of Nomura were described in terms of physical models by Kanazawa [1] and Bruckenstein [2] who derived the equation that describes the changes in resonant frequency of a loss-less quartz crystal in contact with an infinite, non conductive and perfectly Newtonian fluid:

$$\Delta f = -f_o^{2/3}\left(\frac{\rho_L \eta_L}{\pi \rho_Q \mu_Q}\right)^{1/2}, \tag{12.9}$$

and the quartz electroacoustic impedance,

$$Z_{\text{Kan}} = \frac{\sqrt{2}\omega^{3/2} L_Q}{\pi\sqrt{\mu_Q \rho_Q}}[\sqrt{\rho\eta}(1+j)], \tag{12.10}$$

with $L_Q = (h_Q^3 \rho_Q / 8Ae_{26}^2)$; ρ the density and η the viscosity of the fluid. It predicts that the frequency shift is proportional to the square root of the density-viscosity product in agreement with the empirical approach of Nomura and Okuhara [15].

More recently the treatment was extended to piezoelectric devices in contact with viscoelastic media (i.e., liquids and polymers). It was then realised that if the deposited mass was not rigidly coupled to the oscillating quartz crystal, separation of inertial mass and energy losses was not possible with the measurement of the resonant frequency alone. Quartz crystal impedance in the acoustic frequencies was introduced in order to study mass and viscoelastic changes and a full electrical characterization of the crystal behaviour near resonance was employed.

Since the unloaded QCM is an electromechanical transducer, it can be described by the Butterworth–Van Dyke (BVD) equivalent electrical circuit represented in Fig. 12.3 (box) which is formed by a series RLC circuit in parallel with a static capacitance C_o. The electrical equivalence to the mechanical model (mass, elastic response and friction losses of the quartz crystal) are represented by the inductance L, the capacitance C and the resistance, R connected in series. The static capacitance in parallel with the series motional RLC arm represents the electrical capacitance of the parallel plate capacitor formed by both metal electrodes that sandwich the thin quartz crystal plus the stray capacitance due to the connectors. However, it is not related with the piezoelectric effect but it influences the QCM resonant frequency.

In the series branch of the equivalent electrical circuit, L_q is proportional to the quartz inertial mass displaced by the shear oscillation, C_q is proportional to the energy stored in the quartz crystal during oscillation and R_q describes the energy frictional loses of the quartz crystal.

For a Newtonian liquid in contact with the quartz crystal, additional

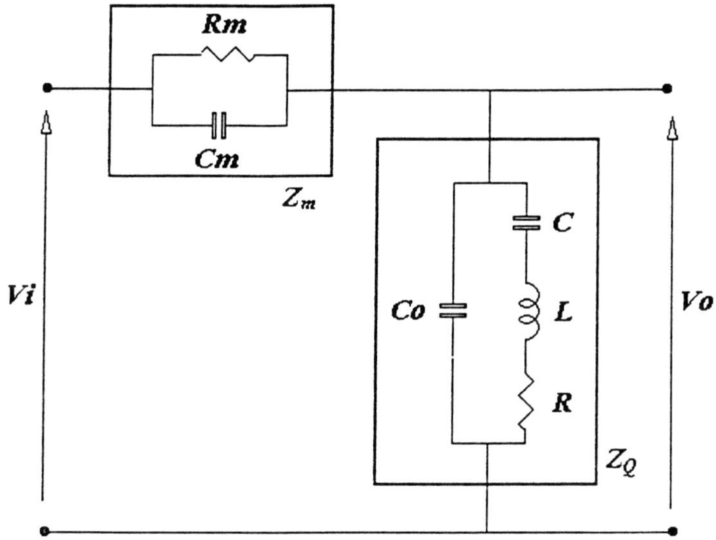

Fig. 12.3. Voltage divider measuring circuit and BVD equivalent electrical circuit (box), $\mathbf{Z}_Q$ and measuring impedance $\mathbf{Z}_m$.

inductance L_l and resistance R_l are needed in the BVD circuit in series with the motional arm in Fig. 12.3. Typical values for the electrical equivalent elements are: $C_q = 10\,\text{fF}$, $R_q = 100\,\Omega$, $L_q = 7.5\,\text{mH}$, $C_o = 1$ to $10\,\text{pF}$ and $L_l = 3.5\,\mu\text{H}$ and $R_l = 214\,\Omega$ for water.

Figure 12.4 depicts a typical admittance parametric plot for the quartz crystal resonator. Note that the effect of the static capacitance C_o in the parallel branch is to shift the admittance circle upward by ωC_o. The shift gives rise to a new resonance frequency ω_p which now depends on C_o, in addition to the series resonance frequency $\omega_s = 2\pi f_o$. Changes in the resonance frequency are related to changes in the equivalent inductance L and broadening of the admittance curve near resonance (decrease in the circle diameter $1/R$ in Fig. 12.4) are related to equivalent resistance R.

Buttry [47] measured the admittance around resonance of a quartz crystal coated with polynitrostyrene and related those measurements to the rheological changes due to film swelling. Muramatsu *et al.* used the resonant resistance in addition to the resonant frequency of the electrochemical quartz crystal microbalance (EQCM) as a criterion to evaluate the film non-rigidity for several electroactive polymer systems [6], including

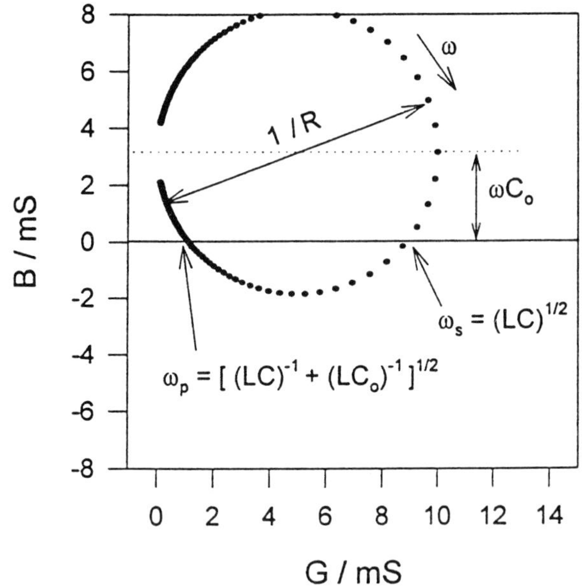

Fig. 12.4. Admittance polar plot.

polypyrrole, Nafion, Langmuir–Blodgett films [48] and other deposits [49]. Muramatsu *et al.* [6] monitored the reological changes of polypyrrole during electrodeposition and cyclic voltammetry transformations.

In the case of viscoelastic loaded QCM two approaches have been followed: one methodology is to treat the device as an acoustic transmission line with one driven piezo-electric quartz layer and one or more surface mechanical load (TLM) [50, 51]. A simpler approach is to use a lumped-element model (LEM) that represents mechanical inter-actions by their equivalent electrical BVD circuit components [52, 53].

Since the transverse shear wave may penetrate the damping surface layer and the viscous liquid, additivity of the equivalent electrical elements in the BVD circuit is only valid under certain particular conditions. Martin and Frye [53] studied the impedance near resonance of polymer film coated resonators in air with a lumped-element BVD model, modified to account for the viscoelastic properties of the film. In addition to the elements shown in Fig. 12.3 to describe the quartz crystal and the liquid, L_f and R_f were added to describe the viscoelastic film overlayer. For a small electroacoustic impedance change the total serial inductance and resistance were expressed as $L = L_q + L_m + L_l$ and $R = R_q + R_l$ respectively. In further communications Martin and Granstaff [50] developed a model in which the equivalent parameters R and L can be described in terms of the density, thickness, viscosity and elasticity of the material in contact with the quartz surface.

In most electrochemical applications the lumped element model (LEM) is a good approximation within 1% of the transmission line model (TLM) provided the quartz and film impedance condition $Z_f/Z_Q \ll 1$ is fulfilled [54].

Noel and Topart [55, 56] reported "in situ" monitoring viscoelastic/volume and mass changes during electrochemical growth and redox switch of polymer modified electrodes (polypyrrole), by simultaneously recording the damping resistance, charge and resonant frequency of an EQCM. Noel and Topart applied Martin's formalism and determined the elastic and loss moduli of polypyrrol/tetrafluoroborate conducting film during growth under slow linear sweep and potential step perturbations.

Hillman *et al.* measured the quartz crystal impedance to determine changes in rigidity, swelling and ionic exchange in conducting polymer films [57, 58] and have also used dynamic quartz crystal impedance of modified electrodes during film growth and redox conversion [57] by qualitative analysis of the acoustic admittance-frequency peak width.

Oyama and Tatsuma measured the resonant frequency and resonant resistance of quartz crystal resonators coated with several redox active polymers [58–60], DNA [61] and tungsten oxide [62].

A detailed analysis of electroacoustic quartz crystal impedance was presented by Doblhoffer and Soares [63] to study the viscoelastic properties of overlayers [64] who showed that the resonance frequency of a crystal electronic oscillator is very close to a quartz crystal natural resonant frequency.

Johannsmann *et al.* have reported the use of quartz crystal electrical impedance to study the viscoelastic properties of poly(γ-methyl-L-glutamate-co-γ-*n*-octadecyl-L-glutamate) thin films [65, 66] and Ward studied the viscoelastic characteristics of poly-styrene films with 2-chloro-toluene solvent by measuring the admittance near resonance at 5 MHz of the unloaded quartz resonator and the composite resonator [67].

Several studies of high frequency shear viscoelastic liquids have also been reported by Tozaki [68, 69] and Kanazawa [14] with the QCM.

Viscoelastic properties of polymer solutions, gels and grafted polymer layer were investigated by QCM at 9 MHz by Ivanchenko *et al.* [70].

Schumacher [71] introduced the imperfections of the crystal surface and related the change in frequency with the roughness while Urbakh [72, 73] analized the complete impedance around resonance for several rough surfaces.

Fast detection of the electro-acoustic impedance is a condition for successful kinetic studies. Soares [64] introduced a circuit to measure both resonant frequency and damping resistance R, though not as fast as simple active oscillator methods mainly used for resonant frequency measurement. Most active circuits operate in the "series frequency ω_s" although some oscillators are designed to operate in the "parallel frequency ω_p", which is slightly higher and very susceptible to the value of C_o.

In passive methods, the quartz crystal is perturbed with a sinusoidal high frequency ac voltage in a few kHz frequency range around resonance. The most commonly used instrument is the Frequency Response Analyser (FRA) at 1–10 MHz. Several commercial instruments can measure automatically a range of frequencies and evaluate the BVD circuit parameters.

The advantage of network analysers is the possibility of impedance measurement near resonance with evaluation of the parameters R, L, C and C_o and test of the equivalent electrical circuit. However frequency response and network analysers are relatively slow with 1–10 s per measurement in typical experiments. A new generation of faster instruments has come to the market like the HP E5100 Network Analyzer with 40 μs per point in the impedance spectrum which allows us to obtain the impedance of the system in less than 10 ms.

Another passive method is the transference function method (TFM) introduced by Muramatsu [6]. The method consists of an oscillator that drives a crystal through a known measuring impedance and a radiofrequency voltmeter which measures the transference modulus of the system. Muramatsu [6] neglected the effect of the parasitic capacitance and his expression for the quartz impedance resulted in a nonlinear relationship between the measured resistance R with the ac voltage divider and the value of R measured by an impedance analyser. Calvo and Etchenique [74] improved the method and introduced an analytical expression to fit the entire transfer function around resonance in order to obtain the same values of R, L and C_o as measured by a frequency response analyser.

The complex voltage divider used to measure the resonant frequency and both components of the quartz crystal Butterworth–Van Dyke equivalent circuit is shown in Figure 12.3. The ratio of input ($\mathbf{V}_i$) to output ($\mathbf{V}_o$) ac-voltage at 10 MHz is:

$$\frac{\mathbf{V}_i}{\mathbf{V}_o} = 1 + \frac{\mathbf{Z}_m}{\mathbf{Z}_Q}, \tag{12.11}$$

where $\mathbf{Z}_m$ is the measuring impedance and $\mathbf{Z}_Q$ is the electroacoustical impedance of the quartz crystal loaded with a viscoelastic fluid or film.

Note that the quantities in bold are vectors represented in the complex plane. In Fig. 12.5 a transfer function spectrum obtained in 0.25 s is shown for 100 frequencies around 10 MHz. It can be used for real time evaluation of the quartz electro-acoustical impedance when the viscoelastic properties change.

A flow sheet schematics of the real time transfer function method is depicted in Fig. 12.5. The BVD equivalent electrical circuit elements R, L and C_o are obtained by nonlinear fit of each transfer function spectra, $|V_o/V_i|(\omega)$, to the analytical expression with recovery of the complete quartz impedance near resonance (admittance, conductance and impedance). Although the voltage divider method does not measure the transfer function phase and hence it is not possible to demonstrate the validity of BVD circuit, it has the advantage of speed. Also passive methods like TFM can be applied under high viscous damping so that the shear wave phase never crosses zero and the EQCM no longer resonates.

The TFM method was further validated by comparison of the measured R and X_L with steady state data with an HP4192A Impedance Analyzer for the same crystal in contact with aqueous sucrose solutions (Fig. 12.6).

Etchenique and Calvo [75, 76] developed a fast real time measurement of R and X_L element of the electroacoustic impedance applied the LEM model of Martin to measure the rheological properties of poly(allylamine)-GOx hydrogels under fast electrochemical perturbations and derived the viscoelastic properties of the film from the quartz equivalent parameters. For thin viscoelastic films, this method was able to separate the mass due to ions and solvent exchange and the viscosity changes, when the density and the elasticity of the overlayer was known.

We can describe the contribution of the viscoelastic redox polymer layer to the total quartz impedance at 10 MHz when the acoustic wave is totally

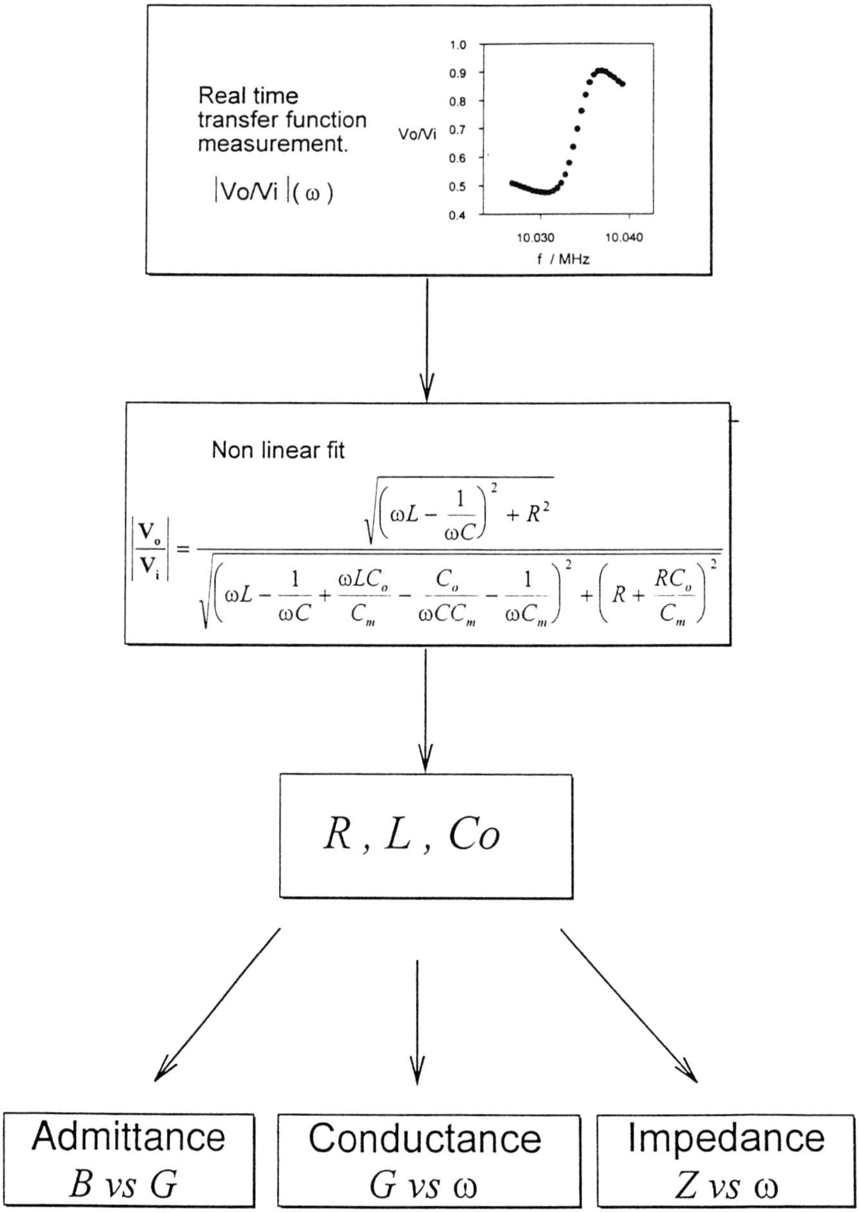

Fig. 12.5. Flowchart describing the voltage divider transfer function method (TFM): Real time measurement and nonlinear fit to the BVD equivalent electrical circuit.

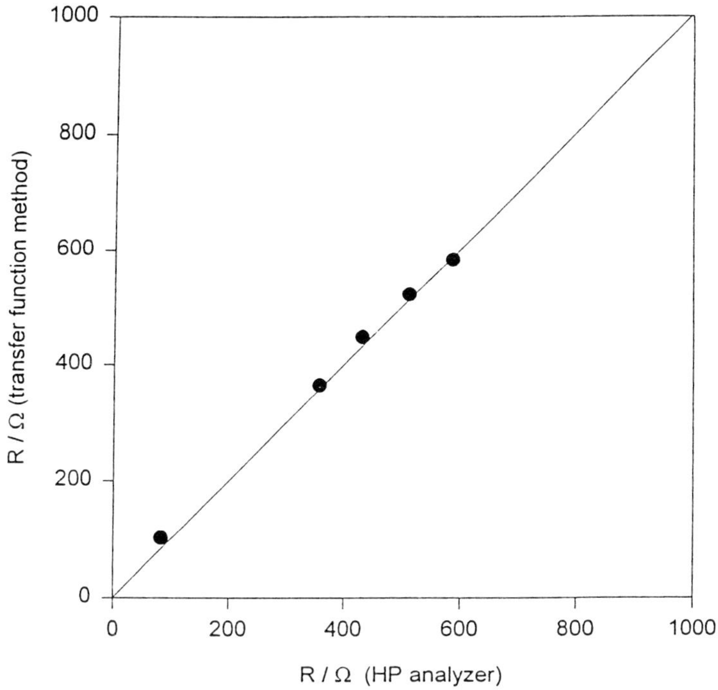

Fig. 12.6. Validation of the voltage divider TFM with a HP4194A impedance analyzer.

damped by the polymer gel before it reaches the electrolyte solution as [76]:

$$\mathbf{Z}_f = R_f + j\omega L_f = \frac{2\omega L_Q}{\pi\sqrt{\mu_Q \rho_Q}}[\sqrt{\rho_f \mathbf{G}_f}\tanh(j\omega d_f\sqrt{\rho_f \mathbf{G}_f})], \qquad (12.12)$$

where d_f is the polymer film thickness, ρ_f its density and the viscoelastic shear modulus $G_f = (G_f' + jG_f'')$ with G_f' the storage modulus and G_f'' the loss modulus of the viscoelastic film.

As a consequence of redox switch in the polymer layer several phenomena, such as ion and solvent exchange with the electrolyte may change the parameters that determine Z_f: ρ_f, $d_f = m_f/A\rho_f$ and the viscoelastic moduli G_f' and G_f'' at 10 MHz.

Impedance plane plots have been introduced to analyse the effect of

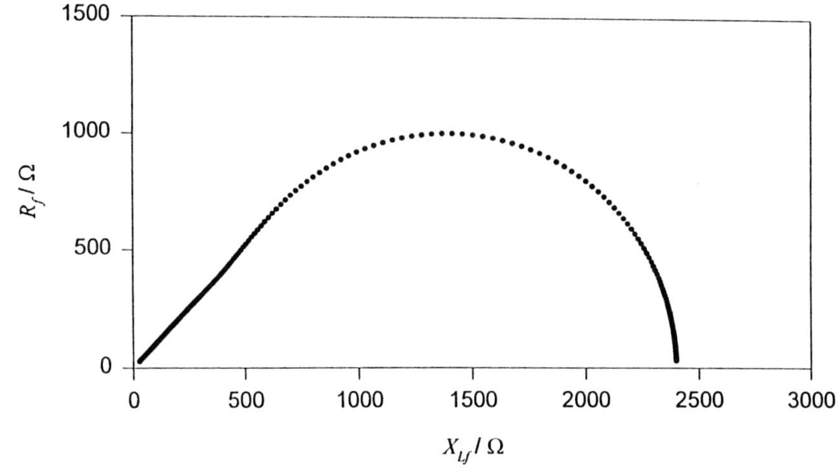

Fig. 12.7. Parametric R_f vs. X_{Lf} plot calculated with equation (12.12).

mass/volume and viscoelastic changes in the film/liquid in contact with the quartz crystal that undergoes dynamic chemical or electrochemical processes [75, 76]. Figure 12.7 shows a plot of R_f as a function of X_{Lf} calculated with equation (12.12) for a thin Newtonian fluid film of variable viscosity $\eta_f(=G''/\omega$, varying clockwise from 10^5 to $10^{10}\,\mathrm{N\,m^{-2}})$ assuming constant density ($\rho = 1\,\mathrm{g\,cm^{-3}}$), thickness ($d_f = 1\,\mu\mathrm{m}$) and $G' = 10^3\,\mathrm{N\,m^{-2}}$. At the lowest G'' values, semi-infinite Newtonian liquid is characterised by a straight line of unity slope and corresponds to the Kanazawa equation. At the largest $G'' \to \infty$, on the other hand, the Sauerbrey limit corresponds to a point on the x-axis in the Sauerbrey limit. It should be noted that the slope of these parametric film electroacoustic impedance plots can be either positive or negative, depending on the particular region. An important feature of the film impedance plots is that each point in the complex plane represents only one system and movements on the plane can be due to either changes in the density, mass or shear modulus (elastic and viscous) of the film or liquid in contact with the quartz.

An example of this analysis is a modified quartz crystal with a (poly(allylamine)-ferrocene hydrogel crosslinked with Glucose Oxidase and immersed in liquid electrolyte which has been studied by electroacoustic impedance [75]. The film impedance parameters, R_f and X_{Lf} were obtained by fitting the resonator transfer function $|V_o/V_i|$ vs. ω to a BVD equivalent circuit and analysed with the multiple non-piezoelectric layer lumped element model of Granstaff and Martin [50]. Two limiting cases were analysed for hydrogel layers of the same composition but different thickness studied by oxidizing and reducing the ferrocene/ferricenium moieties attached to the swollen polymer backbone. For the thin films the Sauerbrey approximation was valid and changes in the mass/thickness and film viscosity that resulted from the anion and water exchange were evaluated during redox conversion on the assumption of negligible storage modulus G' and a density of one. For thick gel layers, on the other hand, the penetration depth of the acoustic wave resulted far less than the film thickness and a liquid-like behaviour was apparent. In this case the film storage modulus (G') and film loss modulus (G'') were simultaneously evaluated with the cyclic voltammetry.

The spatially averaged shear modulus in the acoustic frequencies obtained simultaneously to polymer redox switching in cyclic voltammetry arises from the exchange of ions and solvent during oxidation-reduction of the ferrocene sites attached to the polymer backbone.

Another illustration of a kinetic study with the EQCM using quartz impedance is the redox switch of an osmium polymer used to mediate glucose oxidation by glucose oxidase (GOx). In this case $Os(bpy)_2$pyridine covalently attached to poly(allylamine) and crosslinked with GOx via amino groups with PEDGE has been used. The redox potential of the Os(III)/Os(II) couple in the polymer hydrogel is 0.29 V. Figure 12.8 depicts typical current transients for a potential step from the totally reduced state to the fully oxidised state. The anodic oxidation shows diffusional behaviour and from a linear Cottrell plot the quantity $cD_e^{1/2} = 5 \times 10^{-9}\,\mathrm{mol\,cm^{-2}\,s^{-1/2}}$ was derived for electron hopping between adjacent Os redox centres in the polymer. Independent evidence by ultramicroelectrode chronoamperometry yielded typical values for concentration and electron hopping diffusion constant: $c \approx 0.2\,\mathrm{M}$ and $D_e \approx 2 \times 10^{-9}\,\mathrm{cm^2\,s^{-1}}$.

In Fig. 12.8 the evolution of both quartz crystal impedance parameters during the transient electrochemical oxidation of Os(II) polymer is shown. Both components of the Os polymer film contribution to the quartz crystal impedance at 10 MHz X_{Lf} and R_f change during the transient as can be seen in Fig. 12.8(a) and 12.8(b) respectively. It may be possible that the

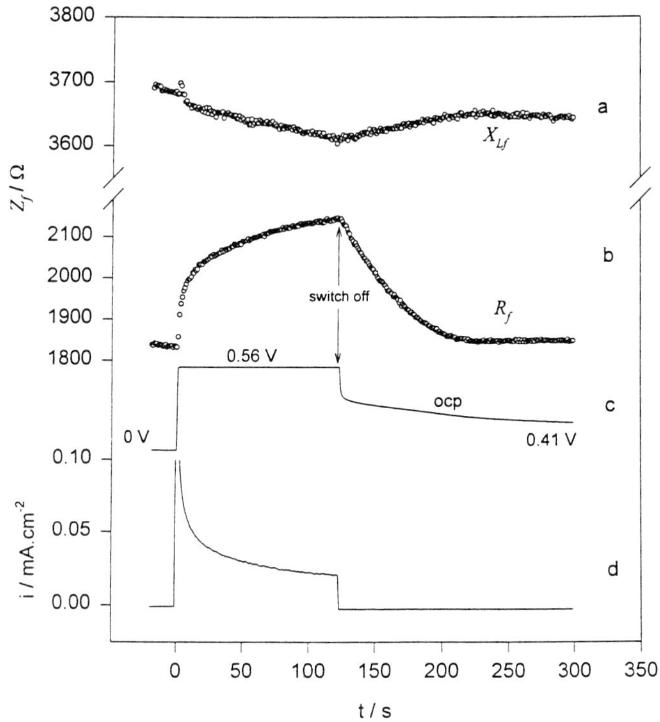

Fig. 12.8. Potentiostatic current transient (d) for the reduction of Au/quartz modified electrode with $Os^{III}(bpy)_2py$-PAA hydrogel in 0.1 M KNO_3; 50 mM TRIS buffer of pH 7.2 and 0.1 M glucose, for a potential step from 0 V to 0.56 V V (SCE) (c) and simultaneous time evolution of X_{Lf} (a) and R_f (b) components of the quartz impedance at 10 MHz. Arrows indicate switch off and open circuit potential decay.

change of redox state in the polymer affects the parameters d_f, ρ_f, G' and G'' due to ingress or egress of ions and solvent from the bathing solution with swelling or shrinking of the swollen polymer gel.

When glucose is added to the solution a catalytic anodic current due to its oxidation catalysed by GOx and mediated by the Os polymer gel is observed. After 120 s of anodic polarisation, a switch off of the electrode is produced by current interruption while the quartz impedance parameters are followed as a function of time. After switch off the open circuit potential evolved to 410 mV which corresponds to the Os polymer in its reduced form. In the absence of glucose in the solution in contact with the gel, the redox gel has memory remaining in its oxidised state seen by the ocp since there are no reducing nor oxidising species in the system.

In glucose solution, glucose reduces those Os centres in the hydrogel that "wire" or connect the enzyme GOx to the underlying gold electrode. Therefore from Fig. 12.8(a) and 12.8(b) it can be seen that enzymatic reduction of the hydrogel by glucose also results in transient changes of X_{Lf} and R_f respectively. Further studies have established that R_f and X_{Lf} depend only on the redox state of the osmium polymer and are not time dependent.

ACKNOWLEDGEMENTS

The authors wish to thank The British Council and Fundación Antorchas (Argentina) for financial support.

References

[1] C.E. Reed, K.K. Kanazawa and J.G. Gordon III, Anal. Chem. 57 (1985) 1770.
[2] S. Bruckenstein and M. Shay, Electrochim. Acta 30 (1985) 1295.
[3] D.A. Buttry and M.D. Ward, Chem. Reviews 92 (6) (1992) 1355.
[4] D. Buttry, in: H. Abruña (ed.), Electrochemical Interfaces, Chap. 10, VCH, 1991, 531.
[5] M.D. Ward, in: J. Rubinstein (ed.), Physical Electrochemistry, Chap. 7 (Marcel Dekker, 1995, 293).
[6] H. Muramatsu, X. Ye, M. Suda, T. Sakuhara and T. Ataka, J. Electroanal. Chem. 322 (1992) 311.
[7] G. Sauerbrey, Z. Phys. 155 (1959) 206.
[8] H.D. Hillier and M.D. Ward, Anal. Chem. 64 (1992) 2539.
[9] K.K. Kanazawa and J. Gordon II; Anal. Chim. Acta 175 (1985) 99.
[10] F. Josse, Z.A. Shana, D.E. Radke and D.T. Haworth, IEEE Trans. Ultrasound, Ferroelectrics, Freq. Control, 37 (1990) 359.
[11] K. E. Heusler, A. Grzegorzewski, L. Jackel and J. Pietrucha, Ber. Bunsen. Ges. Phys. Chem. 92 (1988) 1218.
[12] S.J. Martin and H.E. Hager, J. Appl. Phys. 65 (1989) 2627.
[13] Z. Lin and M.D. Ward, Anal. Chem. 67 (1995) 685.
[14] C.E. Reed, K.K Kanazawa and J.H. Kaufman, J. Appl. Phys. 68 (1990) 1883.
[15] R. Nomura and M. Okuhara, Anal. Chim. Acta 147 (1982) 365.
[16] S. Bruckenstein and S. Swathirajan, Electrochim. Acta 30 (1985) 851.

[17] M. Deakin, T. Li and O. Melroy, J. Electroanal. Chem. 243 (1988) 343.
[18] M.Benje, M.Eiermann, U.Pittermann and K.G.Weil, Ber.Bunsenges.Phys.Chem. 90 (1986) 435.
[19] Sheng-li Chen, Bing-liang Wu and Chuan-sin Cha, J. Electroanal.Chem. 420 (1997) 111.
[20] S.Bourkane, C.Gabrielli and M.Keddam, Electrochim. Acta l.34(8) (1989) 1081.
[21] A. Grzegorzewski and K. Heusler, J. Electroanal. Chem. 228 (1987) 455; K.E. Heusler and A. Grzegorzewski, Electrochim. Acta 35 (1990) 539.
[22] J. Zheng, I.M. Ritchie, S.R. Labrooy and P. Singh, Hydrometalurgy 39 (1995) (1–3) 277.
[23] C. Gabrielli and H. Perrot, J. Electrochem. Soc. 141 (9) (1994) L103.
[24] S. Gottesfeld; J. Electroanal. Chem. 294 (1990) 73.
[25] K. Shimazu, M. Yanagida and K. Uosaki, J. Electroanal. Chem. 350 (1993) 321.
[26] Q. Xu., Q. Xie, M. Hu, L. Nie and S. Yao, J. Electroanal. Chem. 389 (1995) 85; Electrochim. Acta 39 (1994) 727.
[27] A. Survila, P. Kalinauskas, E. Ivaskevic and W. Kutner, Electrochim. Acta 42 (1997) 2935.
[28] A. J. Bard, D. E. Cliffel, C. Demaille, F.-R. F. Fan and M. Tsionsky, Advances in Scanning Electrochemical Microscopy, Annali di Chimica 87 (1997) 15.
[29] C. Barbero, M.C. Miras, O. Haas and R. Kotz, J. Electroanal. Chem. 310 (1991) 437.
[30] M.C. Miras, C. Barbero,, R. Kotz and O. Haas, J. Electroanal. Chem. 338 (1992) 279.
[31] W. Kautek, S. Dieluweit and M. Sahre, J. Phys. Chem. B 101 (1997) 2709.
[32] S. Bourkane, C. Gabrielli and M. Keddam, J. Electroanal. Chem. 256 (1988) 471.
[33] S. Cordoba-Torresi, C. Gabrielli, M. Keddam, H. Takenuoti and R. Torresi, J. Electroanalyt. Chem. 290 (1990) 269.
[34] S. Bourkane, C. Gabrielli and M. Keddam, Electrochim. Acta 34 (1989) 1081.
[35] C. Gabrielli, M. Keddam, F. Minouflet and H. Perrot, Electrochim. Acta 41 (1996) 1217.
[36] C. Gabrielli and M. Keddam, Electrochim. Acta 41 (1996) 957.
[37] N. Oyama and T. Ohsaka, Prog. Polym. Sci. 20 (1995) 761.
[38] A.R. Hillman, D.C. Loveday and S. Bruckenstein, J. Electroanal. Chem. 300 (1991) 67.
[39] A.R. Hillman, Solid State Ionics 94 (1997) 151.
[40] M.D. Ward, J. Phys. Chem. 92 (1988) 2049.
[41] M.D. Ward, J. Electrochem. Soc. 135 (1988) 2747.
[42] G. Inzelt, J. Electroanal. Chem. 348 (1993) 465.
[43] U. Stimming, J. Electroanal. Chem. 130 (1982) 345.
[44] M. Seo, K. Yoshida, H. Takahashi and L. Sawamura, J. Electrochem. Soc. 139 (1992) 3108.
[45] M. Stratmann, K. Bohnenkamp and H.J. Engell, Corrosion Science 23 (1983) 969.
[46] W.P. Mason, W.O. Baker, H.J. McSkimin and J.H. Heiss, Phys. Rev. 75 (6) (1949) 936.
[47] R. Borjas and D.A. Buttry, J. Electroanal. Chem. 280 (1990) 73.

[48] H. Muramatsu and K. Kimura, Anal. Chem. (1992) 64 2502
[49] H. Muramatsu, A. Egawa and T. Ataka, J. Electroanal. Chem. (1995) 388 89.
[50] V. Grastaff and S. Martin, J. Appl. Phys. 75 (1994) 1319.
[51] R. Lucklum, C. Behling, R.W. Cernosek and S.J. Martin, J. Appl. Phys. 30 (1997) 346.
[52] S. Martin, V. Grastaff and G. Frye, Anal. Chem. 63 (1991) 2272.
[53] S.J. Martin and G.C. Frye, Proc. IEEE Ultrasonics Symp. (IEEE, N. York, 1991, pp. 393).
[54] R. W. Cernosek, S.J. Martin, A.R. Hillman and H.L. Bandey, 1997 IEEE International Frequency Control Symposium (Orlando, Fl, May 28–30, 1997).
[55] M.A. Noel and P.A. Topart, Anal. Chem. 66 (1994) 484.
[56] P.A. Topart and A.M. Noel, Anal. Chem. 66 (1994) 2926.
[57] Kelly, D., Vos J. and Hillman, R., J. Chem. Soc. Faraday Trans. 92 (1996) 4121; 89 (1993) 1695.
[58] N. Oyama, T. Tatsuma and K. Takahashi, J. Phys. Chem. 97 (1993) 10504.
[59] S. Ikeda and N. Oyama, Anal. Chem. 65 (1993) 1910.
[60] N. Oyama, K. Takada, T. Tatsuma, K. Naoi, T. Okajmia and T. Oksaka, Sensors and Actuators B, 13–14 (1993) 372.
[61] S. Yamaguchi, T. Shimomura, T. Tatsuma and N. Oyama, Anal. Chem. 65 (1993) 1925.
[62] H. Inaba, M. Iwaku, T. Tatsuma and N. Oyama; J. Electroanal. Chem. 71 (1995) 387.
[63] C. Frubose, K. Doblhofer and D.M. Soares, Ber. Bunsenges. Phys. Chem. 97 (1993) 475.
[64] D.M. Soares, W. Kautek, C. Frubose and K. Doblhofer, Ber. Bunsenges. Phys. Chem. 98 (1994) 219.
[65] D. Johannsmann, K. Mathauer, G. Werner and W. Knoll, Phys. Review B, 46 (1992) 7808.
[66] D. Johannsmann, J. Grüner, J. Wesser, K. Matlhauer, G. Wegner and W. Knoll, Thin Solid Films 210/211 (1992) 662.
[67] A. Katz and M.D. Ward, J. Appl. Phys. 80(7) (1996) 4153.
[68] K. Tozaki, M. Kimura, N. Komatsu and S. Itou, Jpn. J. Appl. Phys. 31 (1992) 2592.
[69] K. Tozaki, M. Kimura and S. Itou, Jpn. J. Appl. Phys. 33 (1994) 6633.
[70] M.I. Ivanchenko, H. Kobayashi, E.A. Kulik and N.B. Dobrova, Anal. Chim. Acta 314 (1995) 23.
[71] R. Schumacher, G. Borges and K.K. Kanazawa, Surf. Science 163 (1985) L621.
[72] M. Urbakh and Daikhin, L., Phys. Rev. B, 49 (1994) 49 4866.
[73] M. Urbakh and L. Daikhin, Langmuir 10 (1994) 2836.
[74] E.J. Calvo, C. Danilowicz and R. Etchenique, J. Chem. Soc. Faraday Trans. 91 (1005) 4083.
[75] R.A. Etchenique and E.J. Calvo, Anal. Chem. 69 (1997) 4833.
[76] E.J. Calvo, R. Etchenique, P.N. Bartlett, K. Singhal and C. Santamaria, Faraday Discussion 107 (1997) 141–157.

Chapter 13

Visualizing Ion and Solvent Transfer Processes in Electroactive Polymer Films

STANLEY BRUCKENSTEIN, A. ROBERT HILLMAN and HELEN L. BANDEY

We describe the use of the electrochemical quartz crystal microbalance (EQCM) to study electroactive polymer film dynamics. Crystal impedance measurements provide a diagnostic of film (non)rigidity. Under deposition conditions in CH_2Cl_2, oxidized poly(vinylferrocene) films are viscoelastic. Upon transfer to aqueous media, they show rigid film characteristics. Under the latter conditions, the EQCM can be used to follow redox-driven ion and solvent population changes within the film. Coulometric and gravimetric responses imply that film redox switching involves coupled electron/anion transfer, solvation changes and polymer configuration changes. The latter two processes have both reversible and irreversible components. This mechanistic complexity can be rationalized using a scheme of cubes model. Redox cycling of poly(thionine) films in aqueous acetic acid buffers involves coupled electron/proton transfer, film solvation changes and acetic acid coordination state changes. Again, the mechanistic complexities of this ECC′ system can be readily visualized using a scheme of cubes approach. In general, the scheme of cubes is well suited to explaining and visualizing a range of electroactive polymer film characteristics, notably those associated with "break-in", overpotential, electrode history and experimental time scale phenomena. This approach should be of particular value when using non-electrochemical population probes in conjunction with electrochemical control functions.

13.1 INTRODUCTION

13.1.1 Background

The modification of electrode surfaces with electroactive polymer films provides a means to control interfacial characteristics. With such a capability, one can envisage numerous possible applications, in areas as diverse as electronic devices, sensors, electrocatalysis, energy conversion and storage, electronic displays, and reference electrode systems [1, 2]. With these applications in view, a wide variety of electroactive polymeric materials have been investigated. These include both redox polymers (by which we imply polymers with discrete redox entities distributed along the polymer spine) and conducting polymers (by which we imply polymers with delocalised charge centres on the polymer spine).

The key to success is the ability to control and manipulate interfacial structure and properties in a predictable manner. As a first order strategy, the choice of polymer for a particular application is based upon the properties of a monomeric analog in solution. In general terms, it is recognized that (i) attachment of the redox entity to a polymer chain and (ii) immobilization of that chain at a solid/liquid interface under external potential control are both likely to modify the properties of the redox entity to some extent. However, there is little detailed understanding of the underlying reasons for these effects and little insight into how these phenomena might be controlled.

At an operational level, the number of successful applications of polymer modified electrodes is disappointingly small. This is particularly surprising, given the wide range of materials available and the many ways in which one may synthetically derivatise parent compounds. It is a common anecdotal observation that *qualitative* characteristics of a given electrode/polymer system are reproducible. However, *quantitative* electrode performance is variable between measurements on the same electrode, between nominally "identical" electrodes, and in the hands of different operators. This is generally fatal to applications other than those in which a highly skilled operator is allowed to make calibration measurements under highly controlled conditions. In this work we explore some fundamental mechanistic issues that help explain why this is so. These considerations may help to delineate conditions under which electrode performance may be optimized or, at worst, made reproducible.

Our approach to this problem involves a detailed mechanistic study of model systems, in order to identify the (electro)chemical parameters and the physicochemical processes of importance. This approach takes advantage of one of the major developments in electrochemical science over the last two decades, namely the simultaneous application of *non*-electrochemical techniques to study interfaces maintained under electrochemical control [3–5]. In general terms, spectroscopic methods have provided insight into the detailed structure at a variety of levels, from atomic to morphological, of surface-bound films. Other *in situ* methods, such as ellipsometry [6], neutron reflectivity [7] and the electrochemical quartz crystal microbalance (EQCM) [8–10], have provided insight into the overall penetration of mobile species (ions, solvent and other small molecules) into polymer films, along with spatial distributions of these mobile species and of the polymer itself. Of these techniques, the one upon which we rely directly here is the EQCM, whose operation and capability we now briefly review.

13.1.2 The electrochemical quartz crystal microbalance (EQCM)

The EQCM is a sensitive *in situ* technique able to monitor changes in the mass of an electrode, including any surface film. These mass changes may be associated, for example, with the deposition or dissolution of a film, or with the exchange of ions and/or solvent between the film and the bathing electrolyte. In the present context, we will primarily use the technique as a gravimetric probe of the populations of ions and solvent within the film, as a consequence of exchange processes with the bathing solution. In particular, we seek to determine these ion and solvent populations as functions of potential and/or time in both long time scale ("equilibrium") and short time scale (dynamic) experiments.

The EQCM comprises a quartz crystal oscillator, in which one of the Au exciting electrodes is also exposed to the solution and acts as the working electrode in a conventional (here, three electrode) cell. Provided any surface film is rigidly coupled to the underlying electrode changes in inertial mass (Δm) of the electrode result in crystal resonant frequency changes (Δf) that are described by the Sauerbrey equation [11]:

$$\Delta f = \left(\frac{-2f_0^2}{A(\rho_q \nu_q)}\right)\Delta m, \tag{13.1}$$

where f_0 is the unperturbed frequency (we used 10 MHz AT-cut crystals), ρ_q is the density of quartz, and ν_q is the wave velocity within quartz. Thus, the EQCM can be used to monitor mass changes due to film deposition or due to redox driven ion and solvent exchange between the film and solution *provided that the film remains rigid* [8–10].

In the event that the film is not rigid, the EQCM response is a function of both the film mass and its rheological characteristics. Application of the Sauerbrey equation under these circumstances is inappropriate; it underestimates the mass change, to an extent that is dependent on the viscoelastic properties of the film. Under these circumstances, there are two questions to be addressed: first, how does one diagnose film (non-)rigidity and, second, how does one interpret responses from a non-rigid film? The answers to both questions can be found from crystal impedance measurements. This is a technique in which one determines the admittance (or impedance) of the loaded crystal as a function of frequency in the vicinity of resonance. Effectively, one determines the shape (width and height) and position (on the frequency axis) of the resonance, rather than just its position (as in the simple EQCM technique).

The principle of the crystal impedance technique is illustrated in Fig. 13.1. Purely gravimetric changes result in a shift in resonant frequency, with no change in shape (see Fig. 13.1(a)). This would be the case for deposition of a rigid film, or exchange of ions and solvent between a rigid film and its bathing solution. Purely viscoelastic changes result in a change in peak shape (see Fig. 13.1(b)). In the general case of both gravimetric *and* viscoelastic changes, peak position *and* shape change (see Fig. 13.1(c)). This would be the case for deposition of a non-rigid film or for ion/solvent exchange that resulted in a change in film viscoelastic properties, e.g., solvent plasticisation. In this work, we use the crystal impedance method as a diagnostic of film rigidity. Where the film is rigid, we can use the Sauerbrey equation to interpret frequency changes. Where it is not, we use a more complex analysis to parameterize film rheological behaviour in terms of shear moduli.

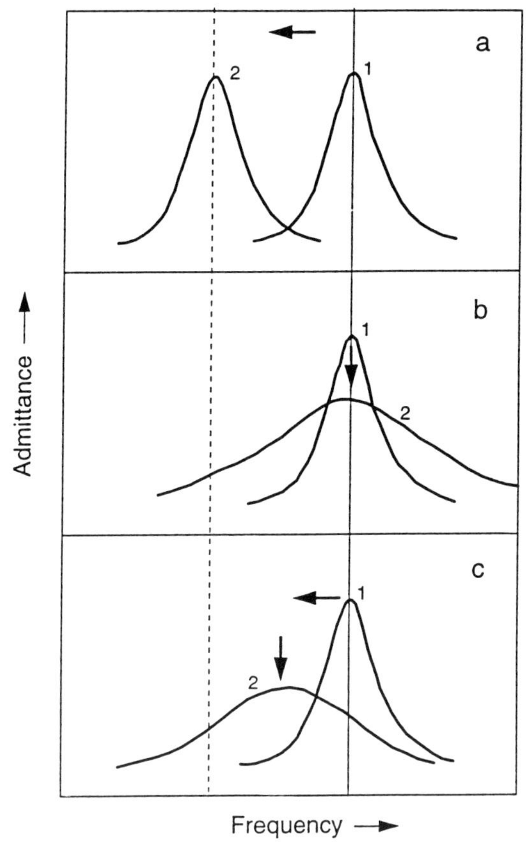

Fig. 13.1. Schematic crystal impedance responses, plotted as admittance vs. frequency, for (a) purely gravimetric, (b) purely viscoelastic and (c) simultaneous gravimetric and viscoelastic changes in resonator loading. Curves 1 represent the bare crystal and curves 2 the coated crystal responses.

13.2 VISUALIZING REDOX SWITCHING PROCESSES

13.2.1 Objective and strategy

Our goal is to develop a method for visualizing extremely complicated electroactive polymer film redox switching mechanisms. To exemplify the problem and our proposed solution, we discuss the mechanisms of two electrochemically driven processes: the redox cycling of poly(thionine) films in acetic acid buffer, and the "break-in" process and subsequent redox chemistry of poly(vinylferrocene) films.

First we review processes that ordinarily accompany the redox switching of an electroactive polymer. The key step is coupled electron/ion transfer, which converts one or more reduced forms of the polymer to one or more oxidized forms of the polymer. Solvent and neutral species transfers and polymer structural changes (reconfigurations or coordination state changes) accompany the switching process under permselective conditions. Under nonpermselective conditions, salt transfer also occurs [12].

13.2.2 Scheme of cubes

As a first step in visualizing the steps in redox switching, we assume that both totally oxidized and totally reduced states may exist in different solvated and structural (configurational or coordination) forms. Although we consider a small number of discrete states, we recognize that reality may correspond to a continuum of all these kinds of polymer forms. As a first step, we assume that only two kinds of each state exist, i.e., both the totally oxidized and totally reduced forms of the polymer exist in only two solvated and two configurational (or coordination) forms, totalling eight chemically distinct species. Consequently, chemical transformations before and after electron transfer may involve paths in which a solvation step (chemical step C) precedes a configurational step (chemical step C′), or *vice versa*. In other words, we will deal only with ECC′ mechanisms, and consider all the permutations of E, C and C′ steps. This leads to a 3D cube model of such systems. Extensions of this approach to higher dimensions are given elsewhere [13].

Figure 13.2 shows the basic $1 \times 1 \times 1$ cube; the insert is the 3D axis system which generates the cube. O and R, respectively, represent the oxidized and reduced forms of the polymer, the superscript "S" the more solvated forms of the polymer, and the subscripts "a" and "b" the two structural forms (configurational states) of the polymer. This cube is useful in describing the redox cycling of many electroactive polymers. The Frank–Condon principle makes electron/ion transfers along horizontal cube edges the most energetically favoured. Diagonal transfers across cube faces or through the cube are energetically far less favourable. Rather, the solvation and reconfiguration steps either precede or follow electron/ion transfer on a different time scale.

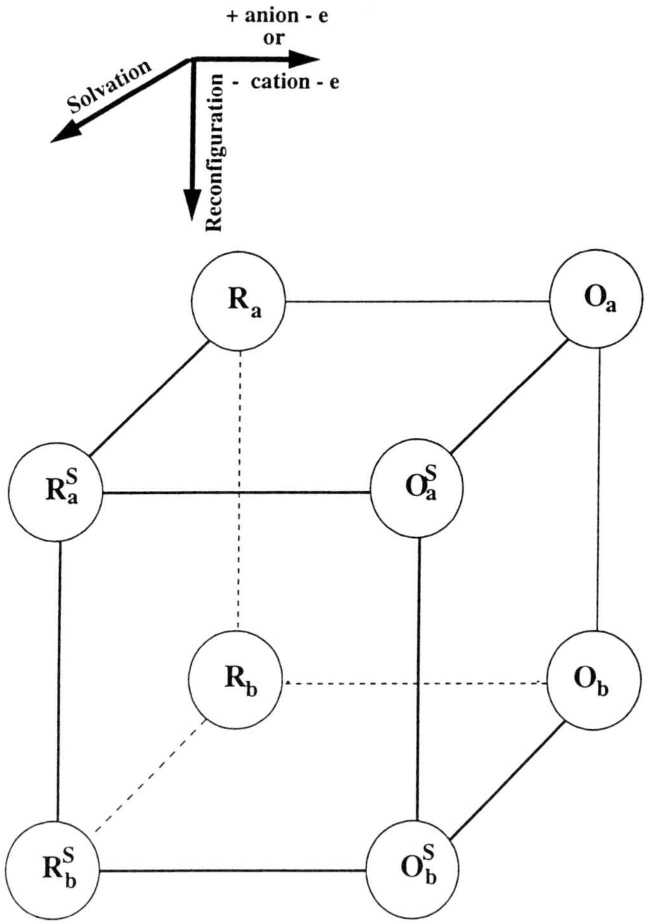

Fig. 13.2. Cube representing an electroactive polymer with two oxidation states (*R* and *O*), each having two solvation and two configuration states. Cube corners represent the eight possible species. Left/right, front/back and top/bottom planes, respectively, differ with respect to oxidation state, solvation state and configuration state. Insert shows 3D axes for the three elementary steps.

13.2.3 Overpotential effects

A key issue in the redox switching of electroactive polymers relates to the value of the potential at which coupled electron/ion transfer occurs. The importance of this issue is illustrated in Fig. 13.3. The six cubes in

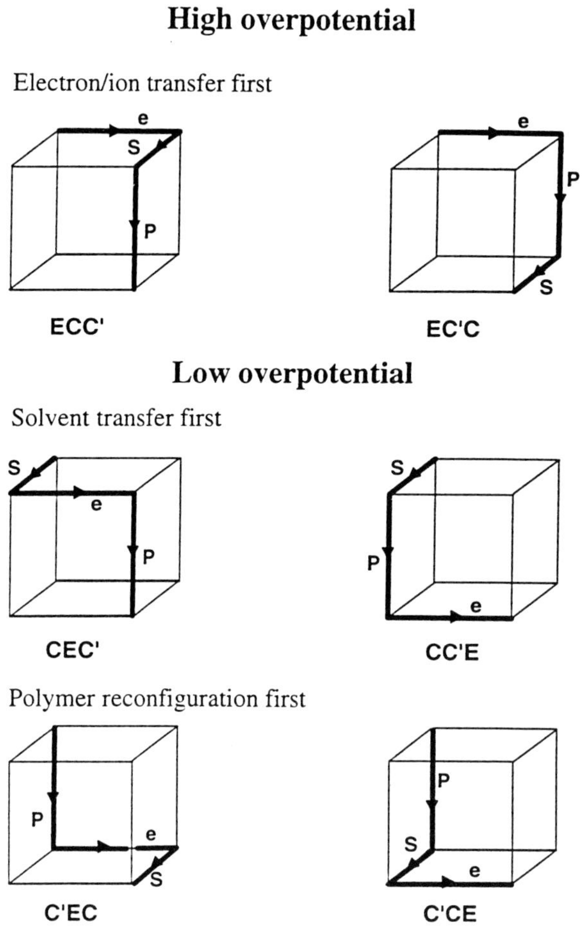

Fig. 13.3. Dependence of redox switching mechanism on overpotential and the rates of chemical steps.

this figure have species at their corners that represent the ones on the corners of the cube in Fig. 13.2. In the six cubes shown in Fig. 13.3, we assume that initially only the reduced species (R_a) exists at equilibrium, and when it is oxidized only a single product (O_b^S) will eventually exist at equilibrium.

First we consider application of a large instantaneous high overpotential step. As seen in Fig. 13.3, there are two possible high overpotential paths, one involving solvation (step C) as the first step after electron/ion transfer, followed by reconfiguration (step C′) as the second step after instantaneous electron transfer, and *vice versa*. Accordingly, the two allowed mechanisms for the conversion of R_a to O_b^S are ECC′ and EC′C.

Next we consider the case where a low overpotential exists at some point in the electrochemical switching process, as is necessarily the case in cyclic voltammetric or chronopotentiometric experiments. Now, four additional mechanisms become possible on the time scale of the electron/ion transfer process. This is so because the electron/ion transfer step, E, is no longer constrained to be the first. Consequently, either one or both of the chemical steps may precede the coupled electron/ion transfer. During the oxidation of R_a to O_b^S, if solvation of R_a is more rapid than its reconfiguration, the two mechanisms, CEC′ and CC′E may occur. If reconfiguration of R_a is more rapid than its solvation, two additional mechanisms, C′EC and C′CE, become possible.

Analogous to the oxidation process discussed above, the conversion of O_b^S back to R_a may follow six different reduction paths, depending on the overpotential. Consequently, a redox cycle may involve any of 36 possible mechanisms for the hypothetical case of a single start state and a single end state.

This simple example, involving one start state, illustrates the importance of the chosen electrochemical technique when studying a redox polymer film. It illustrates why different workers, using the same polymer and only slightly different electrochemical control functions (which necessarily have different time scales), may reach different mechanistic conclusions concerning a redox switching process. The practical reality is that there may not be a single start state for two reasons. First, thermodynamics may dictate significant equilibrium populations of more than one reactant solvation or configurational state. Second, kinetic effects may mean that attaining these equilibrium populations is not a rapid process, so that a considerable effort is required to reproduce a start state. The common practice of potential cycling to obtain reproducible current-potential curves does not accomplish this. It merely creates a steady state distribution of populations that is a function of the conditioning protocol. An example of a more complex situation involving multiple starting states is presented at the end of this work, and a fuller exposition is given elsewhere [14].

Quite generally, large overpotential techniques, followed by instantaneously open circuiting the filmed electrode, will greatly simplify solving the kinetic problem. This approach corresponds to a coulostatic experiment in which the charge injection process instantaneously converts every species in one oxidation state to its corresponding solvation and configuration form in the other oxidation state. The simplification results from assuring that the electron/ion transfers (E) are first. Under these conditions (see Fig. 13.3), starting with a single reactant (R_a), solvation and reconfiguration processes can only occur from O_a. This decreases the number of possible mechanisms for a redox half-cycle from six to two. If similar considerations apply to the reduction of O_b^S (i.e., it must proceed via R_b^S), the number of possible mechanisms for a full redox cycle is decreased from thirty six to four. We show elsewhere how to disentangle overlapping solvation and configurational processes, and how to measure their rates [14].

13.3 POLYTHIONINE (PTh)

13.3.1 Background

The monomer (in the oxidised form) can be electropolymerized as illustrated in Scheme 13.1 [15–17]. The subsequent redox chemistry of poly(-thionine) is illustrated in Scheme 13.2. The polymer has voltammetric and spectroscopic signatures that broadly mirror those of the monomer,

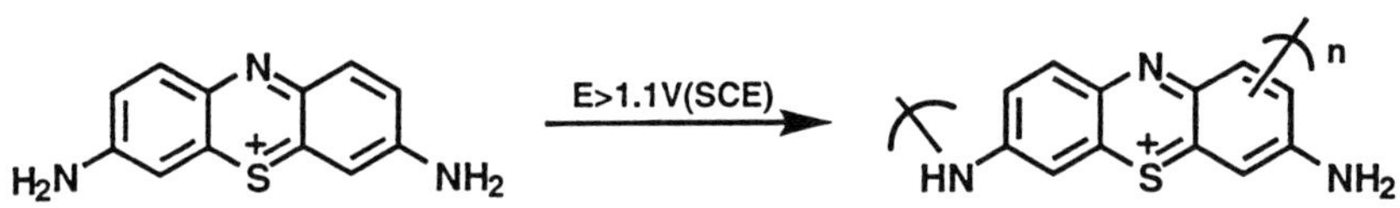

Scheme 13.1. Electrochemical polymerization of thionine.

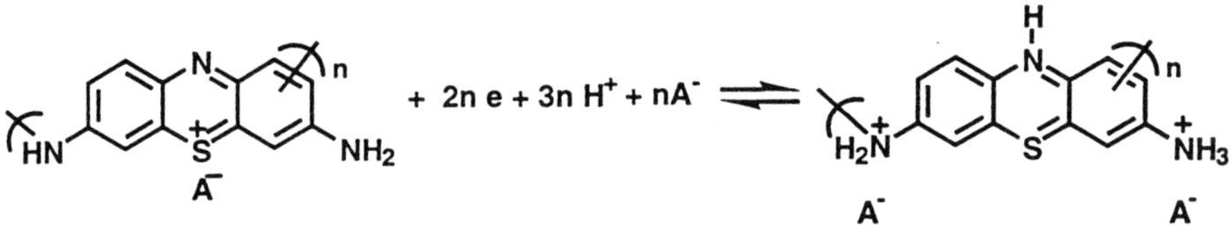

Scheme 13.2. Poly(thionine) film redox chemistry.

consistent with the view that the polymer behaves as a collection of independent thionine-like redox moieties [16, 17]. In the light of their mediated charge transfer characteristics towards solution redox species, polythionine modified electrodes have attracted considerable interest as possible selective electrodes in photogalvanic energy conversion cells [15]. Here we use this system as an archetypal organic polymer, in that it involves coupled electron and proton transfers, for mechanistic studies.

13.3.2 Experimental observations and mechanistic interpretation

Polythionine films behave as rigidly coupled systems, allowing interpretation of EQCM responses in purely gravimetric terms. Figure 13.4 shows EQCM current and mass responses accompanying redox switching of a polythionine film exposed to an aqueous solution of a weak acid, acetic acid. Slow scan voltammetric experiments established the overall normalised mass change, $\Delta MF/Q = 16\ \mathrm{g\ mol^{-1}}$ [18, 19].

Our published studies [18, 19] of poly(thionine) redox switching showed that, in acetic acid buffer, the leuco (reduced) form (L) is oxidized, losing two electrons and two protons, to generate the oxidised form (T). Furthermore, acetic acid is coordinated only to the T-form so that reduction causes T to lose an acetic acid molecule. However, reduction of T to L follows a different path than the reverse of the oxidation path. Water transfer is also involved in the redox cycle. We can (in simplified form) write the overall redox switching process as:

$$[\mathrm{TH}^{+}A^{-}.\mathrm{HA.H_2O}]_{\mathrm{P}} + 2H_S^{+} + 2e \rightleftarrows [\mathrm{LH}_4^{2+}(A^{-})_2]_P + \mathrm{H_2O}_S \quad (13.2)$$

where A represents the acetate ion, and subscripts P and S, respectively, denote polymer and solution phases. This overall process (gain of one water and loss of two protons on oxidation) is consistent with the measured overall normalized mass change (see above).

Our problem is to assign a mechanism to the implied combination of electron/proton, solvation and acetic acid coordination steps. The mechanistic details of this redox cycle were given in Figs. 6 and 7 of Ref. [19]. As we now show, the scheme of cubes is a vastly superior and more concise way to present the details of the complete redox cycle for this process.

Figure 13.5 shows the 3D-rectangular axis system that applies to the cube in Fig. 13.6, representing the poly(thionine) redox system. The species in

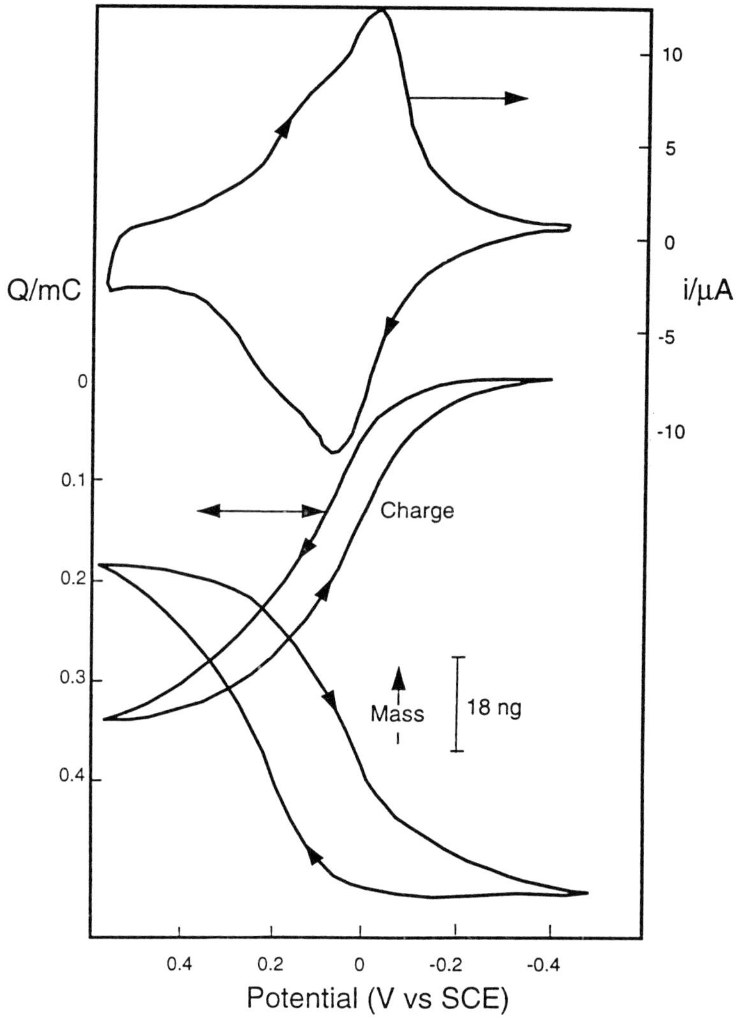

Fig. 13.4. Current vs. potential, charge vs potential and mass change vs potential curves for redox switching of a poly(thionine) film ($\Gamma = 7.7$ nmol cm^{-2}) in aqueous 0.1 mol dm-3 acetic acid (pH = 2.9). The electrode was subject to a cyclic potential scan at $5\,mV\,s^{-1}$. (Adapted from Ref. [18] with permission.)

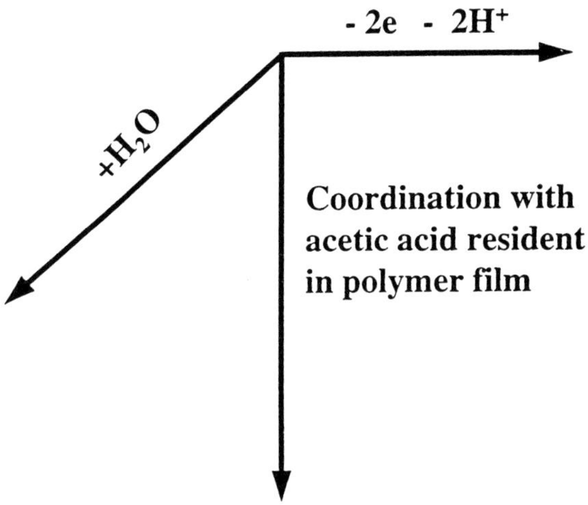

Fig. 13.5. 3D axes showing elementary steps associated with poly(thionine) redox switching in aqueous acetic acid buffer: coupled electron/ion (proton) transfer, solvation and acetic acid coordination.

the top, rear left corner corresponds to the reduced leuco form L. We show its oxidation (along the top, rear horizontal edge) to T, which then coordinates acetic acid (from the H^+ and A^- released from L upon its oxidation; see equation (13.2)) to form T_{HA} (the right, rear vertical edge). The latter then reacts with water (S) in the rate determining step and forms T_{HA}^S (bottom, right horizontal edge), to complete the oxidation half cycle. Reduction immediately converts T_{HA}^S to L_{HA}^S (bottom, front horizontal edge). Next, L_{HA}^S de-coordinates HA to produce L^S (front, left vertical edge). Finally, the reduction half cycle is completed by a de-solvation step that regenerates the starting species, L (top, left horizontal edge). At this point, the reader should accept that looking at the cube is far superior to a reading a description of the redox cycle.

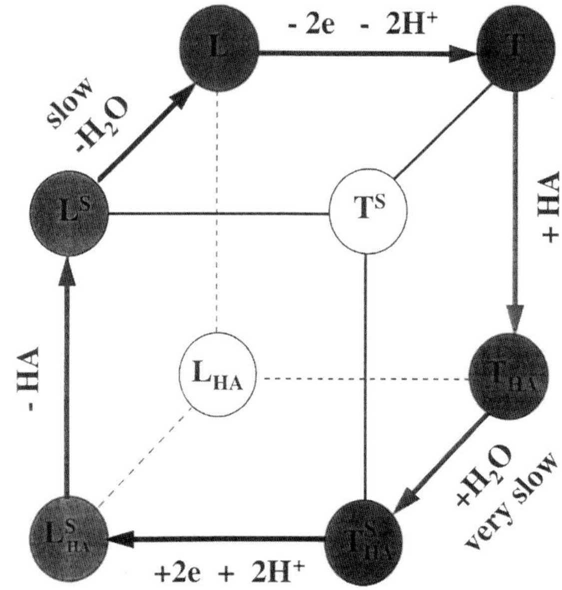

Fig. 13.6. Cube representing the poly(thionine) redox mechanism. Shaded (white) circles at the corners of the cube indicate species that are (are not) accessed in the redox cycle.

13.4 POLYVINYLFERROCENE (PVF)

13.4.1 Background

The chemical stability and electrochemical reversibility of PVF films makes them potentially useful in a variety of applications. These include electrocatalysis of organic reductions [20] and oxidations [21], sensors [22], secondary batteries [23], electrochemical diodes [24] and non-aqueous reference electrodes [25]. These same characteristics also make PVF attractive as a model system for mechanistic studies. Classical electrochemical methods, such as voltammetry [26–28] chronoamperometry [26], chronopotentiometry [27], and electrochemical impedance [29], and *in situ* methods, such as spectroelectrochemistry [30], the SECM [26] and the EQCM [31–38] have been employed to this end. Of particular relevance here are the insights they have provided on anion exchange [31, 32], permselectivity [32, 33] and the kinetics of ion and solvent transfer [34–

36]. One general observation from these studies, upon which our present studies shed light, is that mobile species movement can be markedly dependent upon film history [37, 38].

Thin films of PVF exposed to non-swelling solvents are anticipated to behave as rigidly coupled masses, so that the Sauerbrey equation (equation (13.1)) accurately describes the EQCM responses. However, experimental conditions may not always conform to this situation. For example, an early EQCM study of PVF electroprecipitation yielded film "mass" data that were not consistent with coulometric assay of the film [39]; non-rigidity of PVF films in the CH_2Cl_2 deposition solution was suggested. Also, Oyama *et al.* have studied thermoresponsive *N*-isopropylacrylamide-vinylferrocene copolymers [40] and found them to exhibit potential dependent viscoelastic characteristics. In the light of these observations, we use the crystal impedance technique (illustrated schematically in Fig. 13.1) to guide our study of PVF films.

13.4.2 PVF deposition

The deposition procedure is illustrated schematically in Scheme 13.3. It relies upon the difference in solubility between the reduced (uncharged) and oxidized (positively charged) forms of the polymer in dichloromethane solutions containing a tetraalkylammonium salt, such as tetraethylammonium tetrafluoroborate (TEAT). Upon the application of a potential step to 0.7 V (SCE), PVF^0 is oxidized to PVF^+, which deposits upon the

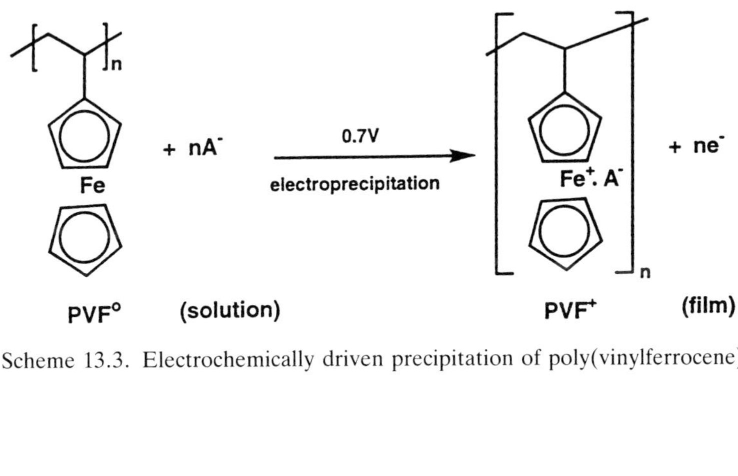

Scheme 13.3. Electrochemically driven precipitation of poly(vinylferrocene).

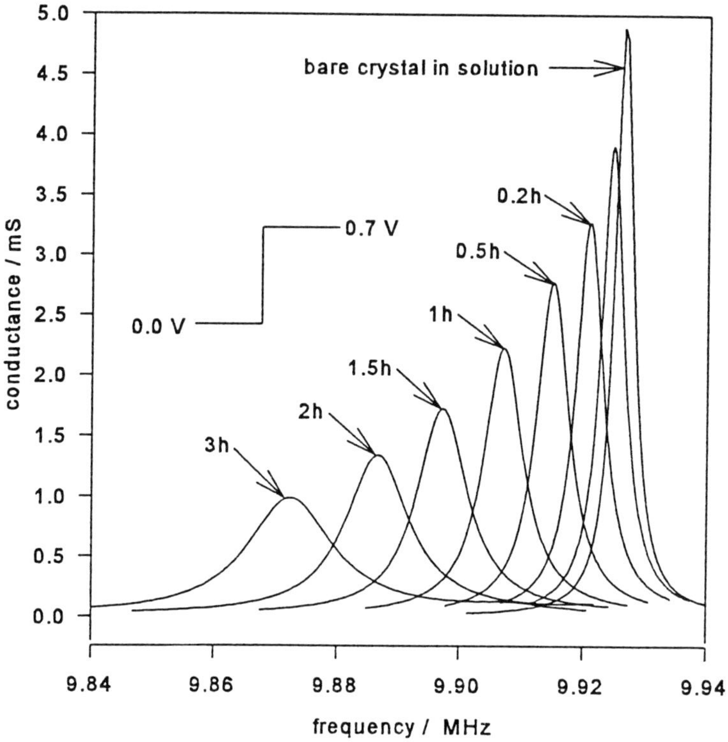

Fig. 13.7. Crystal impedance spectra acquired during the deposition of a PVF film from CH_2Cl_2/0.1 mol dm^{-3} TEAT. Inset shows the potential program. Numbers indicate deposition time. (Reproduced from Ref. [42] with permission.)

electrode surface. Polymer coverage can be controlled via the deposition time (a few minutes to 5 hours).

Figure 13.7 shows a set of crystal impedance spectra as a function of time (coverage) during electroprecipitation of a PVF film. These measurements were made using a network analyzer in reflectance mode, as described previously [41]. For the long deposition times employed here, the film is relatively thick. Furthermore, there is appreciable solvent incorporation. These two factors mean that there will be considerable shear deformation of the film as a result of crystal oscillation.

Quantitatively, we proceed via the use of equivalent circuit models. The most general model is the distributed transmission line model of Fig.

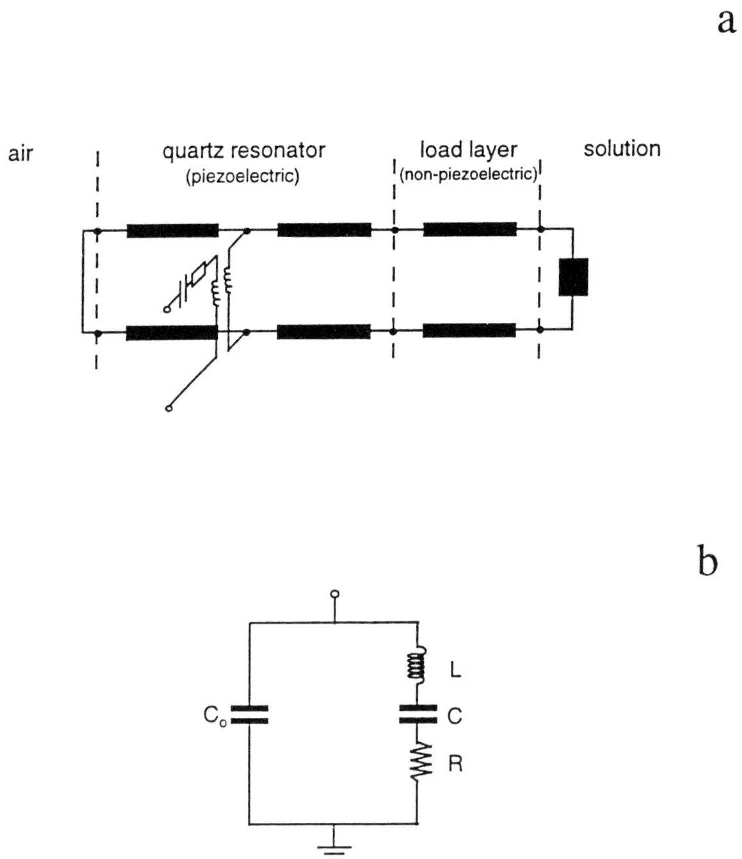

Fig. 13.8. Equivalent circuit models for crystal impedance responses: (a) transmission line model; (b) lumped element (modified Butterworth van Dyke) model.

13.8(a). Since one face of the crystal is exposed to air, the loading at that face of the crystal is negligible; the other face of the crystal has a surface mechanical impedance characteristic of the polymer and electrolyte.

One can show [42] that, when the surface mechanical impedance is not large, the distributed model in the vicinity of resonance (where we make measurements) can be reduced to the simpler lumped-element model of Fig. 13.8(b). This modified Butterworth–van Dyke (BVD) electrical equivalent circuit comprises parallel "static" and "motional" arms. The static arm consists of a capacitance (C_o) associated with the two Au electrodes on either side of the quartz and stray capacitance associated with the measurement fixture. The motional arm is associated with the electromechanical coupling of the piezoelectric quartz and the loading (polymer film and electrolyte solution). Capacitive (C), inductive (L) and resistive (R) elements, respectively, represent mechanical elasticity, inertial mass and energy dissipation within the various components of the loaded resonator. The electrical impedance (Z_e) of the lumped element model is related to the surface mechanical impedance (Z_s) and the quartz shear wave impedance (Z_q) by

$$Z_e = \frac{N\pi}{4K^2\omega_s C_0}\left(\frac{Z_s}{Z_q}\right) = R + j\omega L \tag{13.3}$$

where N is the harmonic number (here $N = 1$), K^2 is the piezoelectric coupling constant for quartz, and $\omega_s = 2\pi f_s$ (where f_s is the series resonant frequency). The resistive and inductive contributions to R and L from the crystal, film and solution are assumed to be additive (a reasonable description when the viscoelastic losses are small).

Thus, rigid films contribute zero resistance to the system, and changes in inductance associated with film mass changes cause shifts in resonant frequency with no change in peak admittance. This corresponds to the situation illustrated in Fig. 13.1(a), described by the Sauerbrey equation. Simultaneously increasing resistance corresponds to increasing energy loss, as would occur in a viscoelastic polymer; this corresponds to the situation in Fig. 13.1(c). Inspection of the raw data in Fig. 13.7 immediately provides the qualitative deduction that depositing PVF films in dichloromethane are described by this latter case. Fitting to the BVD equivalent circuit (Fig. 13.8(b)) yields substantial and systematically increasing resistance values that confirm this result. Hence we conclude that the Sauerbrey equation cannot be employed to interpret the data.

The above provides a qualitative diagnostic showing the growing films to be non-rigid, but provides no direct physical insight. We therefore now extend the method to determine the shear modulus for the viscoelastic polymer film by relating the measured surface mechanical impedance (Z_s) to the appropriate physical parameters. We need to consider three types of loading: ideal (rigid) mass, viscoelastic material and Newtonian fluid.

These correspond, respectively, to polymer or electrolyte entrapped within surface features, the polymer film, and the solution. The first of these is a minor effect when using polished crystals; the surface mechanical impedance of this contribution is $Z_s = j\omega\rho_s$, where $j = \sqrt{-1}$, $\omega = 2\pi f_0$, and ρ_s is the areal mass density of the entrapped material. For finite and semi-infinite viscoelastic layers, the surface mechanical impedance is given by $Z_s = (\mathbf{G}\rho_f)^{1/2}$ and $Z_s = (\mathbf{G}\rho_f)^{1/2}\tanh(\gamma h_f)$, respectively, where ρ_rf and h_f are the film density and thickness and $\gamma = j\omega(\rho_f/\mathbf{G})^{1/2}$. For the solution, $Z_s = (\omega\rho_s\eta_s/2)^{1/2}(1 + j)$, where ρ_s and η_s are the density and viscosity of the solution. When rigid mass, finite viscoelastic film and semi-infinite liquid loadings are all present, as in the experiment of Fig. 13.7, one can show that [42]:

$$Z_s = a\left(\frac{b\cosh(\gamma h_f) + a\sinh(\gamma h_f)}{a\cosh(\gamma h_f) + b\sinh(\gamma h_f)}\right) + j\omega\rho_s, \qquad (13.4)$$

where $a = (G\rho_f)^{1/2}$ and $b = (\omega\rho_S\eta_S/2)^{1/2}(1 + j)$.

For the films and conditions we have used, the transmission line and lumped element models give indistinguishable results. Fitting of the data of Fig. 13.7 yields **G** as a function of time. These values increase at short times (due to nucleation phenomena) to long time limiting values of $G' = 1.9 \times 10^6$ dyne cm^{-2} and $G'' = 3.0 \times 10^8$ dyne cm^{-2}. These values of the shear modulus components show that, in dichloromethane, the PVF film is a very rubbery polymer in which there is considerable viscoelastic loss when the film thickness exceeds 1 μm.

13.4.3 PVF redox cycling

After the desired coverage has been achieved (via deposition time), the PVF^+-coated electrode can be rinsed with water and transferred to aqueous 0.1 mol dm^{-3} sodium perchlorate for characterization. Since both oxidation states of the film are insoluble in aqueous media, we are able to investigate the redox switching process of surface-confined material, as schematically illustrated in Scheme 13.4. The polymer coverage (Γ/mol cm^{-2}, expressed in terms of immobilised ferrocene sites) is obtained coulometrically by integration of the current response to a slow voltammetric scan. We now discuss the behaviour of thin (<0.1 μm) PVF films after

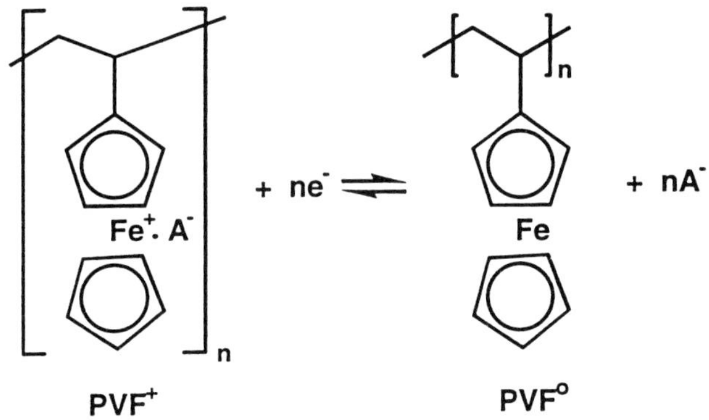

Scheme 13.4. Poly(vinylferrocene) film redox chemistry.

transfer to aqueous solution. The film viscoelastic characteristics in the two media differ markedly: in aqueous solutions the PVF films we will discuss behave as rigid solids with a shear modulus on the order of 10^{10} dyne cm^{-2}. Hence, we are able to use the Sauerbrey equation to interpret EQCM responses.

13.4.3.1 Poly(vinylferrocene) break-in: the double cube

When a freshly prepared poly(vinylferrocene) film is first oxidized in an aqueous electrolyte using cyclic voltammetry, the first cycle is remarkably different from subsequent ones [37]. Figure 13.9 shows a typical "break-in" set of current-potential and mass-potential curves obtained during an EQCM experiment. Several points are obvious. First, the potentials of the first and subsequent cycle oxidation current peaks are different. Second, the mass change during the first oxidation cycle is much larger than in subsequent cycles. Comparison of charge data with mass change data shows that the difference in mass just noted corresponds to an increased flux of water that enters the polymer when it is first oxidized. This water is retained within the film upon subsequent redox cycling.

Figures 13.10 to 13.16 present a visual explanation for the pattern of mass- and current-potential curves that is obtained in this system. They are based on a cube-system in which the three elementary steps are coupled electron/ion transfer, solvation and polymer reconfiguration. Explanation of the phenomena observed in Fig. 13.9 requires the use of two cubes, joined at the bottom front edge of one and the top back edge of the other,

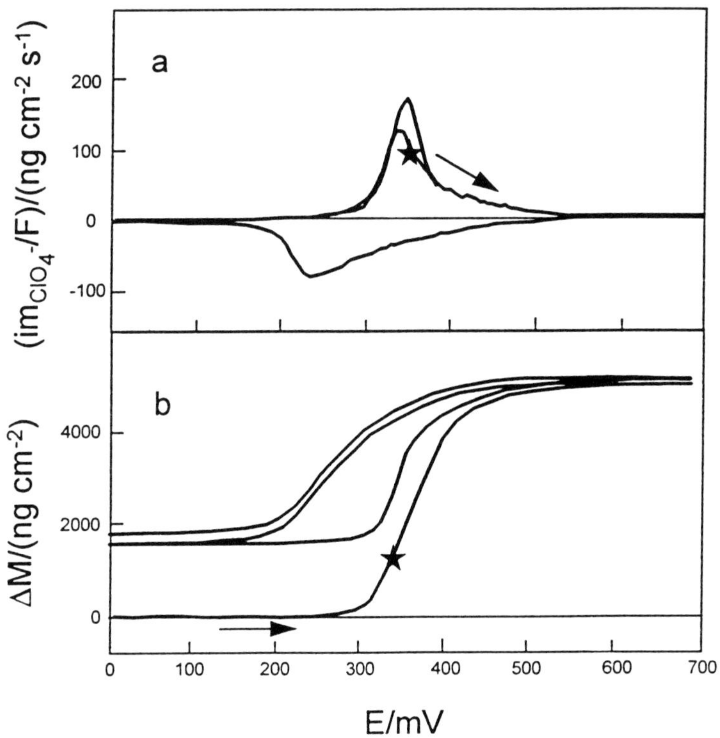

Fig. 13.9. EQCM data obtained during the "break-in" of a poly(vinylferrocene) film (Γ = 23.7 nmol cm^{-2}). Frames a and b, respectively, correspond to the current- and mass change-potential curves for the first two redox cycles in aqueous 0.1 mol dm^{-3} $NaClO_4$ immediately following deposition from CH_2Cl_2. In frame a, the current (electron flux) was converted to the equivalent mass flux of counter ions, for subsequent correlation with the total (observed) gravimetric response of frame b. Potential scan rate: 5 mV s^{-1}. The star symbol denotes the first anodic scan responses. (Adapted from Ref. [37] with permission.)

as shown in Fig. 13.10. This double cube configuration, which we present for the first time here, allows the possibility of three solvation and three configurational states for both the totally oxidized and reduced forms of poly(vinylferrocene). The superscripts S and S', respectively, represent states more and much more solvated than the starting state. The subscripts a, b, c represent the three configurational states for this system.

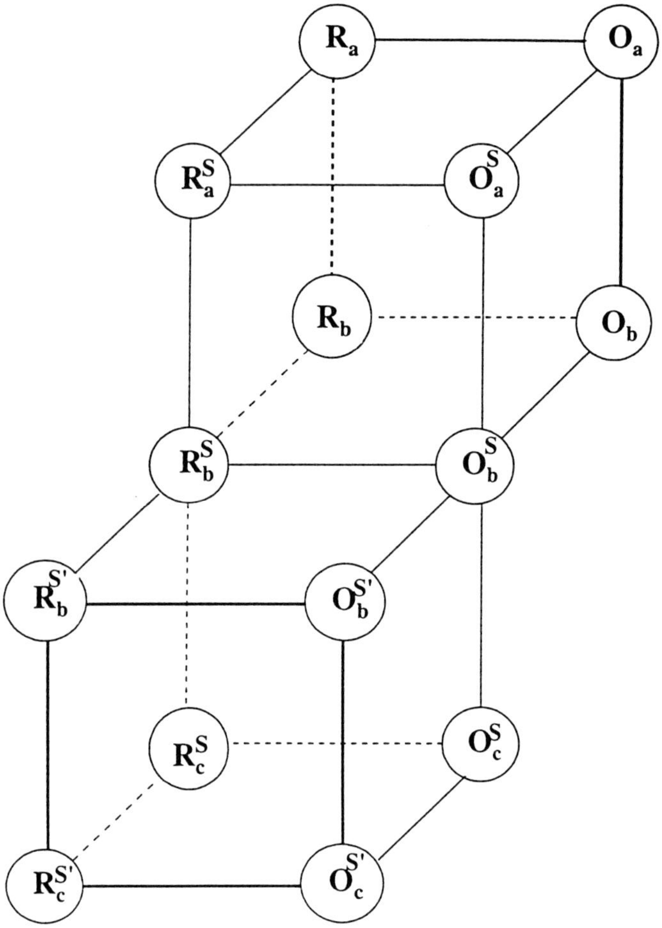

Fig. 13.10. Double cube used to explain "break-in" phenomena observed for poly(vinylferrocene) films in aqueous solution.

13.4.3.2 First oxidation half cycle

Figure 13.11 shows the creation of O_a from R_a by the initial electron/ion transfer step. Figure 13.12 illustrates the second step, a reconfiguration to form O_b. The third step is a solvation step in the upper cube to form O_b^S, as seen in Fig. 13.13. The next step forms a product in the lower cube, the reconfigured, solvated species, O_c^S (Fig. 13.14). This step is followed

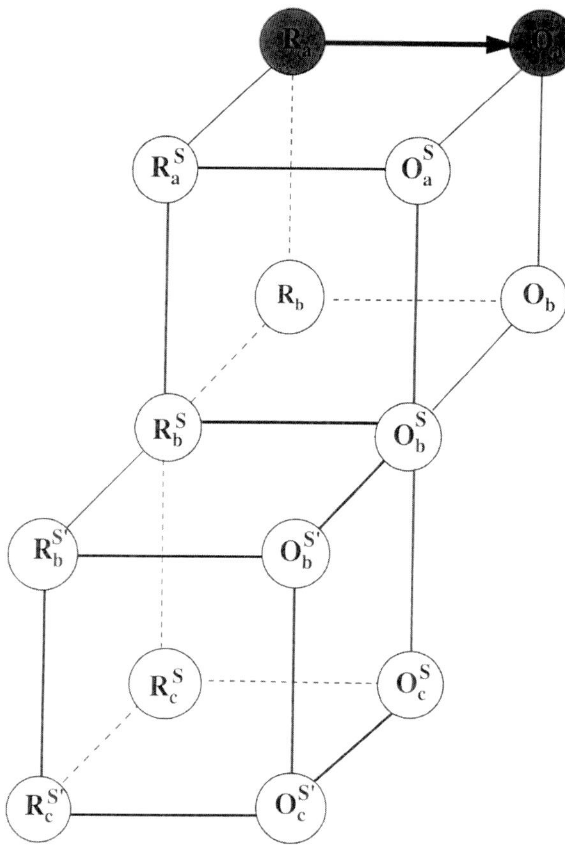

Fig. 13.11. Poly(vinylferrocene) break-in: first oxidation half cycle, first step (coupled electron/ion transfer). In this and subsequent figures, the darker arrow(s) represent(s) the step(s) under discussion and the lighter arrow(s) represent(s) the preceding step(s). Also, shaded circles represent species that either have been or are presently being accessed.

by a final solvation step to form the stable oxidized species $O_c^{S'}$, also seen in Fig. 13.14.

13.4.3.3 First reduction half cycle

The shape of the current-voltage curve for the first reduction half cycle depends on the time scale imposed on the system by the scan rate used in the cyclic voltammogram. We have not experimentally explored the scan

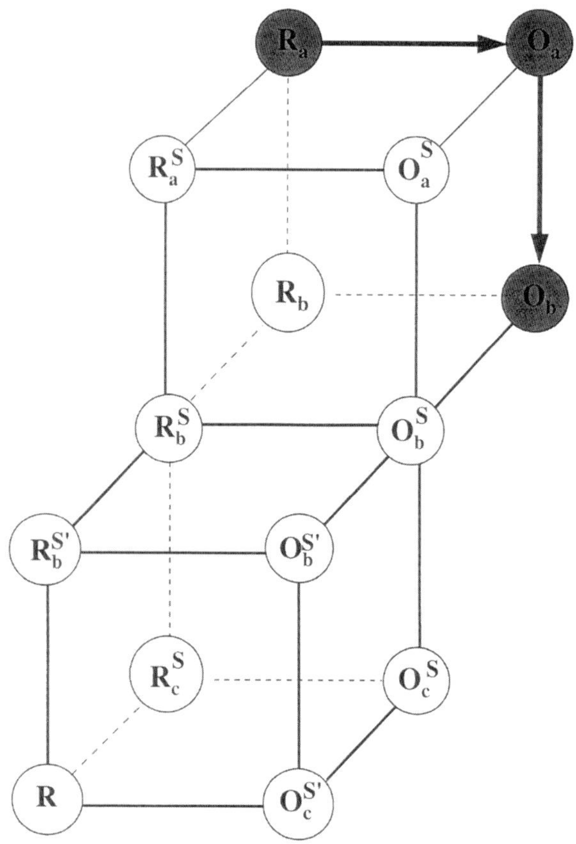

Fig. 13.12. Poly(vinylferrocene) break-in: first oxidation half cycle, second step (reconfiguration). Reconfiguration is assumed faster than solvation in the upper cube.

rate dependence of the break-in process, but suggest the following analysis of what one might expect.

Figure 13.15 shows the cyclic reduction and oxidation processes that would occur when using a very high scan rate cyclic voltammogram. $O_b^{S'}$ converts to $R_c^{S'}$ which reconverts to $O_c^{S'}$ before $R_c^{S'}$ has time to undergo a de-solvation or reconfiguration reaction. Potential cycling is confined to the two species on the lower front edge of the lower cube.

Figure 13.16 illustrates the situation for a slightly longer time scale cyclic

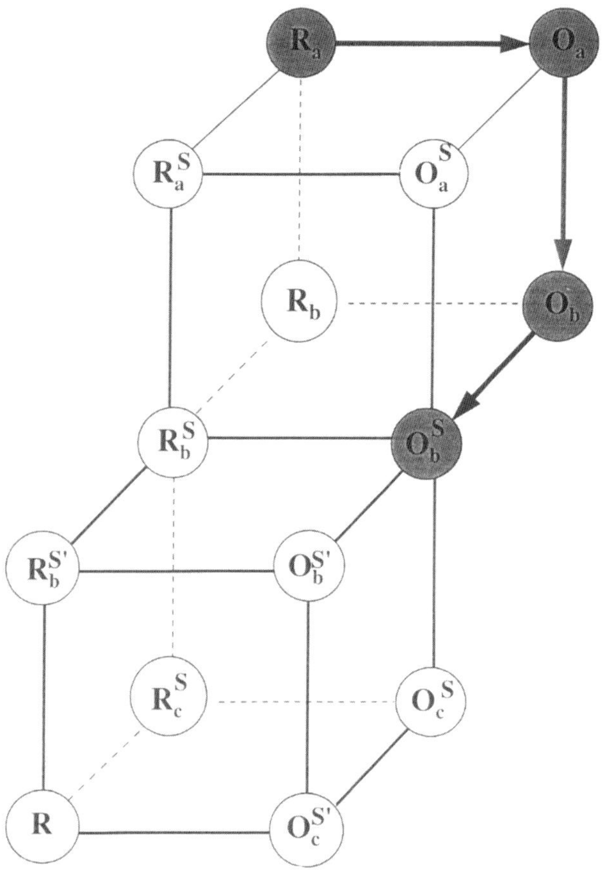

Fig. 13.13. Poly(vinylferrocene) break-in: first oxidation half cycle, third step (solvation). Reconfiguration is assumed faster than solvation in the upper cube.

voltammogram in which the reduction product $R_c^{S'}$ has sufficient time to reconfigure to $R_b^{S'}$, but not to undergo further reconfiguration or solvation before oxidation to $O_b^{S'}$. Thus subsequent cyclic voltammograms would be confined to cycling through the species $O_b^{S'}$ to $O_c^{S'}$ to $R_c^{S'}$ to $R_b^{S'}$ to $O_b^{S'}$, i.e., to the front face of the lower cube.

Figure 13.17 represents the expected path for intermediate time scales. During the first reduction half cycle, the electron/ion transfer process would produce $R_c^{S'}$, which would reconfigure to $R_b^{S'}$ and de-solvate to R_b^S.

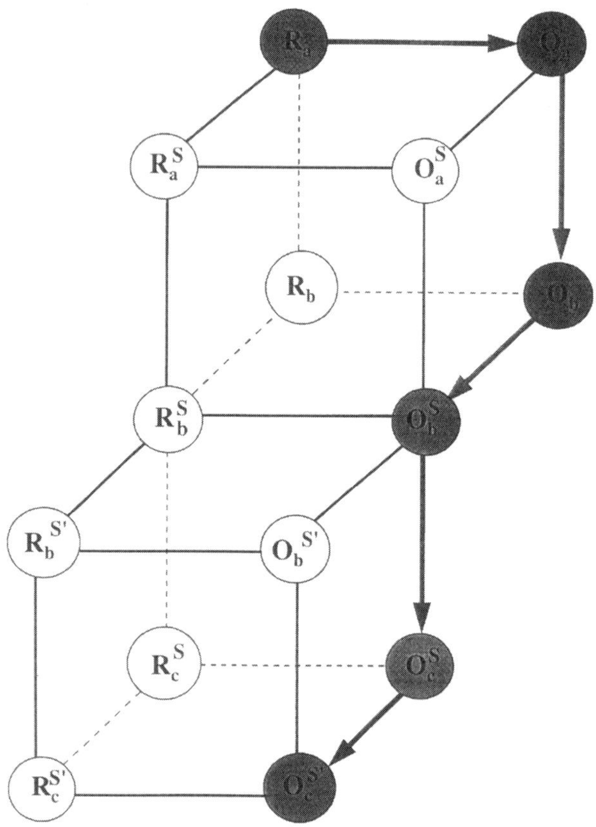

Fig. 13.14. Poly(vinylferrocene) break-in: completion of first oxidation half cycle, by a second reconfiguration followed by a second solvation step, to form the most stable reduction product (doubly solvated, doubly reconfigured $O_c^{S'}$).

This in turn would undergo oxidation to O_b^S when the potential scan was reversed. Subsequent cyclic voltammograms would then involve a path confined to the lower cube.

If progressively slower sweep rates are used, the first reduction cycle will follow the path illustrated in Fig. 13.18. Now, the system has time to explore parts of the upper (as well as lower) cubes, although the original unsolvated states (back face of the upper cube) are never accessed, consistent with the data of Fig. 13.9 and numerous similar experiments.

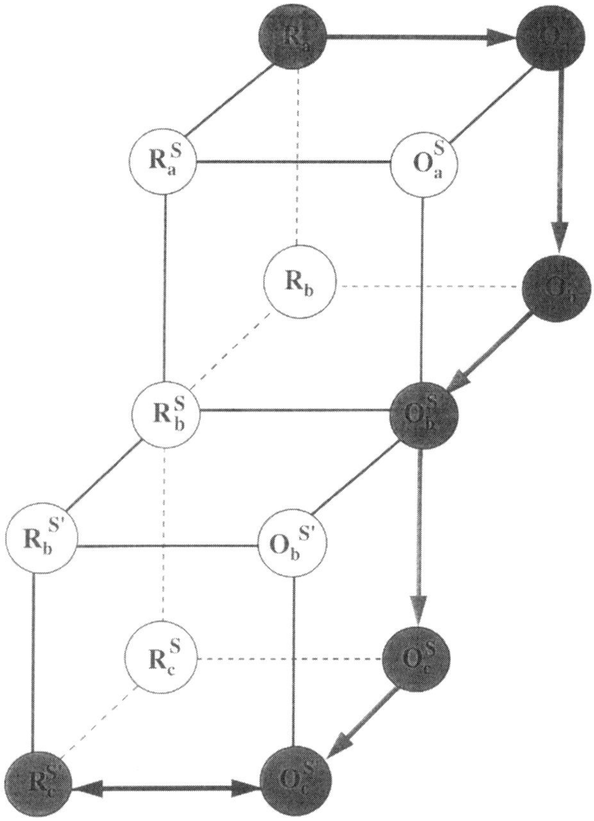

Fig. 13.15. Redox cycling of a broken-in poly(vinylferrocene) film on very short time scales.

The formal potential for the first oxidation process, R_a to O_a, differs from that of the second, $R_c^{S'}$ to $O_c^{S'}$, and subsequent ones, $R_b^{S'}$ to $O_b^{S'}$, R_b^S to O_b^S, R_a^S to O_a^S [43], which explains why the peak potential seen in Fig. 1 of reference [37] changes between the first and subsequent cycles.

Continued cycling to steady state would populate many of the species, leading to the result shown in Fig. 13.19. Here there are four redox couples accessed (R_a^S/O_a^S, R_b^S/O_b^S, $R_b^{S'}/O_b^{S'}$, and $R_c^{S'}/O_c^{S'}$, whose distinct standard potentials are related by the equilibrium constants around the cube faces [43]. This will result in a set of overlapping waves on the i–E curves, and corresponding mass changes on the ΔM–E curves, which will be practically

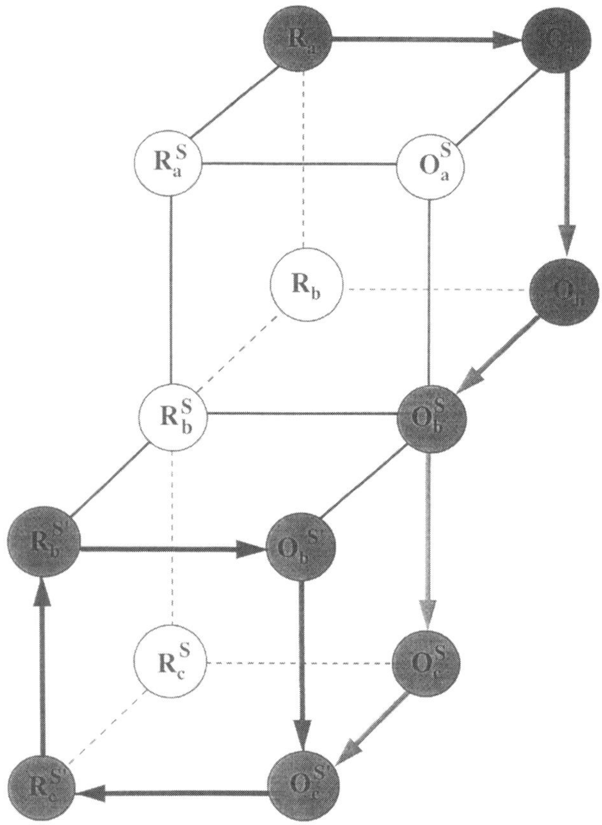

Fig. 13.16. Redox cycling of a broken-in poly(vinylferrocene) film on short time scales. Reconfiguration is assumed faster than (de-)solvation in the lower cube.

impossible to separate. We suggest that arbitrary mixtures of these processes, governed by different experimental procedures, may be the cause of differing responses for PVF and analogous redox polymer films. Therefore, we emphasize the importance of using carefully designed film history and experimental time scale protocols in order to generate defined start and end states, for which individual responses may be measured.

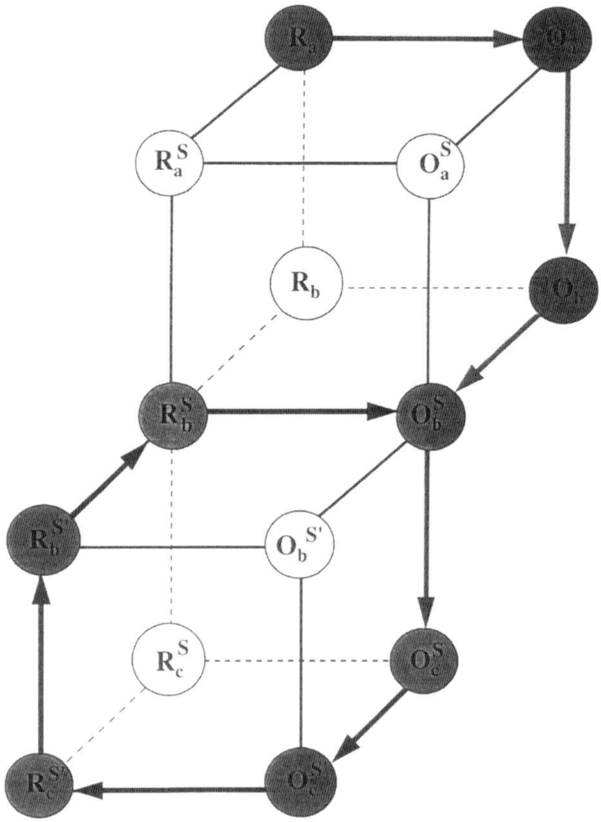

Fig. 13.17. Redox cycling of a broken-in poly(vinylferrocene) film on intermediate time scales. Reconfiguration is assumed faster than (de-)solvation in the lower cube.

13.5 CONCLUSIONS

The electrochemical quartz crystal microbalance is a versatile technique for studying several aspects of electroactive polymer film dynamics. For rigid films, it is a sensitive probe of mobile species (ion and solvent) population changes within the film in response to redox switching. For non-rigid films, it can be used to determine film shear moduli. In the former case, one simply follows changes in crystal resonant frequency. In the latter case, the frequency dependence of resonator admittance in the

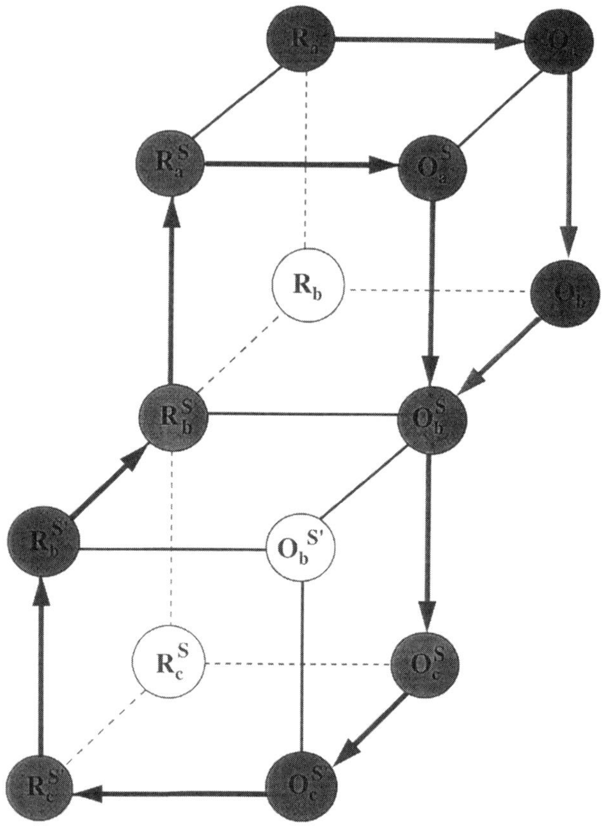

Fig. 13.18. Redox cycling of a broken-in poly(vinylferrocene) film on long time scales. Reconfiguration is assumed faster than (de-)solvation in both cubes.

vicinity of resonance is required; this also gives a diagnostic for film (non-)rigidity.

During deposition, poly(vinylferrocene) films are non-rigid. Analysis of crystal impedance data yields shear moduli that are typical of a rubbery material; this contrasts with the rigid film characteristics observed in aqueous media.

Redox cycling of poly(thionine) films in aqueous acetic acid solution involves not only electron transfer (coupled with proton transfer to maintain electroneutrality), but also film solvation and acetic acid coordination

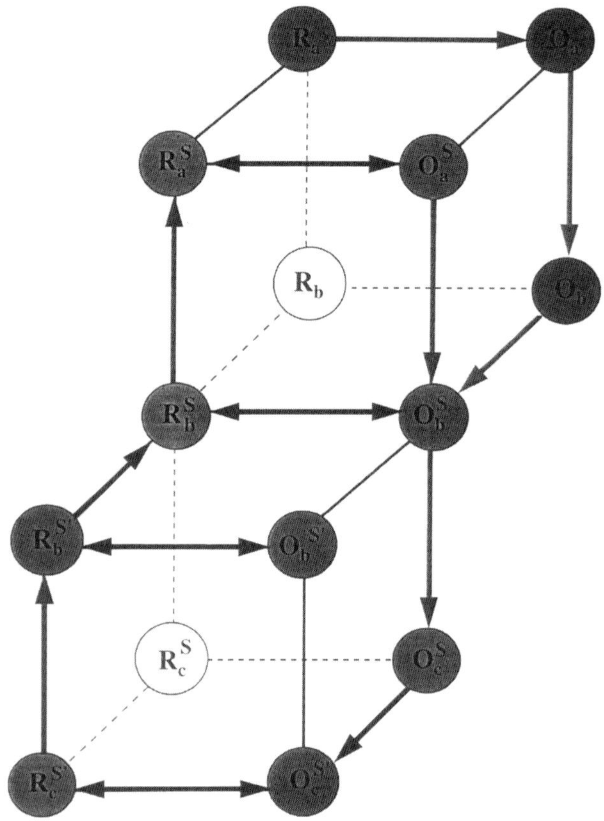

Fig. 13.19. Redox cycling of a broken-in poly(vinylferrocene) film after multiple cycles, generating a mixture of "start" states.

state changes. Mechanistic complexities associated with this ECC′ system can be rationalized and readily visualized using a scheme of cubes approach.

Redox cycling of poly(vinylferrocene) films involves not only coupled electron/anion transfer, but also solvation and polymer configuration changes. The situation is further complicated by both reversible and irreversible elements of solvation and configuration changes, commonly referred to as "break-in", immediately following deposition. Here, a double cube model, allowing for three solvation and three configuration states, is able to rationalize break-in and reversible behaviour of the film.

We conclude that the scheme of cubes is well suited to explaining and visualizing a range of electroactive polymer film characteristics, notably those associated with "break-in", overpotential, electrode history and experimental time scale phenomena. This approach should be of particular value when using non-electrochemical population probes in conjunction with electrochemical control functions.

ACKNOWLEDGEMENTS

We thank the EPSRC (GR/K29982) and the NSF (CHE-9616641) for financial support, and HLB thanks the University of Leicester for a studentship.

References

[1] R.W. Murray, ed., Molecular Design of Electrode Surfaces (John Wiley, New York, 1992).
[2] A.R. Hillman in: R.Linford, ed., Electrochemical Technology of Polymers (Elsevier Applied Science Publishers, London 1987, pp. 103 & 241).
[3] R.J. Gale, Spectroelectrochemistry (Plenum Press, New York 1988).
[4] A.T. Hubbard, ed., Handbook of Surface Imaging and Visualization (CRC Press, Boca Raton, 1995).
[5] P.A. Christensen and A. Hamnett, Techniques and Mechanisms in Electrochemistry (Blackie, Glasgow 1994).
[6] S. Gottesfeld in: A.J. Bard, ed., Electroanalytical Chemistry (Marcel Dekker, New York, 1989, vol. 15, p. 143).
[7] J. Penfold, R.M. Richardson, A. Zarbakhsh, J.R.P. Webster, D.G. Bucknall, A.R. Rennie, R.A.L. Jones, T. Cosgrove, R.K. Thomas, J.S. Higgins, P.D.I. Fletcher, E. Dickinson, S.J. Roser, I.A. McLure, A.R. Hillman, R.W.Richards, E.J. Staples, A.N. Burgess, E.A. Simister and J.W. White, Far. Trans. 93 (1997) 3899.
[8] D.A. Buttry in: A.J. Bard, ed., Electroanalytical Chemistry, vol. 17 (Marcel Dekker, New York, 1991, p. 1).
[9] S. Bruckenstein and A.R. Hillman, in: A.T. Hubbard, ed., Handbook of Surface Imaging and Visualization (CRC Press, Boca Raton, 1995, p. 101).
[10] Interaction of Acoustic Waves with Thin Films and Interfaces, Far. Disc. 107 (1997).
[11] G. Sauerbrey, Z. Physik 155 (1959) 206.
[12] S. Bruckenstein and A.R. Hillman, J. Phys. Chem. 95 (1991) 10748.

[13] A.R. Hillman and S. Bruckenstein, Far. Trans. 89 (1993) 3779.
[14] S. Bruckenstein and A.R. Hillman, J. Phys. Chem. B 102 (1998) 10826.
[15] W.J. Albery, A.W. Foulds, K.J. Hall, A.R.Hillman, R. G. Egdell and A.F. Orchard, Nature 282 (1979) 793.
[16] W.J.Albery, A.W.Foulds, K.J.Hall and A.R.Hillman, J.Electrochem.Soc. 127 (1980) 654.
[17] W.J.Albery, M.G.Boutelle and A.R.Hillman, J. Electroanal.Chem. 182 (1985) 99.
[18] S. Bruckenstein, C.P. Wilde and A.R. Hillman, J. Phys. Chem. 94 (1990) 6458.
[19] S. Bruckenstein, C.P. Wilde and A.R. Hillman, J. Phys. Chem. 97 (1993) 6853.
[20] H. Gülce, H. fzyörük and A. Yıldız, Ber. Bunsenges. Phys. Chem. 98 (1994) 228.
[21] S. Dong and G. Che, J. Electroanal. Chem. 309 (1991) 103.
[22] A.-L. Nguyen and J.H.T. Luong, Appl. Biol. and Biotechnol. 43 (1993) 117.
[23] T. Kawai, C. Iwakura and H. Yoneyama, Electrochim. Acta 34 (1989) 1357.
[24] X.-B. Wang, J.M. Bonnett, R. Pethig, P.K. Baker, O.L. Parri and A.E. Underhill, J. Molec. Electronics 7 (1991) 167.
[25] R.M. Kannuck, J.M. Bellama, E.A. Blubaugh and R.A. Durst, Anal. Chem. 59 (1987) 1473.
[26] F.-R.F. Fan, M.V. Mirkin and A.J. Bard, J. Phys. Chem. 98 (1994) 1475.
[27] P. Daum and R.W. Murray, J. Phys. Chem. 85 (1985) 389.
[28] A. Merz and A.J. Bard, J. Am. Chem. Soc. 100 (1978) 3222.
[29] W.J. Albery and A.R. Mount, Far. Trans. 89 (1993) 327.
[30] M.G. Sullivan and R.W. Murray, J. Phys. Chem. 98 (1994) 4343.
[31] P.T. Varineau and D.A. Buttry, J. Phys. Chem. 91 (1987) 1292.
[32] A.R. Hillman, D.C. Loveday and S. Bruckenstein, J. Electroanal. Chem. 274 (1989) 157.
[33] A.R. Hillman, D.C. Loveday, S. Bruckenstein and C.P. Wilde, Far. Trans. 86 (1990) 437.
[34] A.R. Hillman, D.C. Loveday, M.J. Swann, S. Bruckenstein and C.P. Wilde, Far. Trans. 87 (1991) 2047.
[35] A.R. Hillman, D.C. Loveday and S. Bruckenstein, J. Electroanal. Chem. 300 (1991) 67.
[36] A.R. Hillman, N.A. Hughes and S. Bruckenstein, Analyst 119 (1994) 167.
[37] A.R. Hillman, N.A. Hughes and S. Bruckenstein, J. Electrochem. Soc. 139 (1992) 74.
[38] A.R. Hillman and S. Bruckenstein, Far. Trans. 89 (1993) 3779.
[39] A.R. Hillman, D.C. Loveday and S. Bruckenstein, Langmuir 7 (1991) 191.
[40] N. Oyama, T. Tatsuma and T. Takahashi, J. Phys. Chem. 97 (1993) 10504.
[41] H.L. Bandey, M. Gonsalves, A.R. Hillman, A. Glidle and S. Bruckenstein, J. Electroanal. Chem. 410 (1996) 219.
[42] H. L. Bandey, A.R. Hillman and S. J. Martin, in: A.J. Ricco, M. A. Butler, P. Vanysek, G. Horvai and A.F. Silva, eds., Chemical and Biological Sensors and Analytical Electrochemical Methods, The Electrochemical Society (Pennington, NJ, 1997, p.2 18).
[43] P. Krtil, S. Bruckenstein and A.R. Hillman, J. Phys. Chem. 102 (1998) 4994.

Chapter 14

Ab Initio Prediction of the Performance of Membrane Separation Processes

W. RICHARD BOWEN, NIDAL HILAL, MOHIT JAIN, ROBERT W. LOVITT, A. WAHAB MOHAMMAD, ADEL O. SHARIF, PAUL M. WILLIAMS and CHRIS J. WRIGHT

14.1 INTRODUCTION

The development and application of membrane separation processes is one of the most significant advances in chemical and biological process engineering in recent years. Membrane processes are advanced filtration processes which utilise the separation properties of finely porous polymeric or inorganic films [1, 2]. Membrane separations are used in a wide range of industrial processes to separate biological macromolecules, colloids, ions, solvents and gases. They also have important medical uses, especially in renal dialysis. The world-wide annual sales of membranes and membrane equipment are worth in excess of $1 billion.

The work described in this chapter is especially concerned with three of the most widely used pressure driven membrane processes: microfiltration, ultrafiltration and nanofiltration. These are usually classified in terms of the size of materials which they separate, with ranges typically given as 10.0–0.1 μm for microfiltration, 0.1 μm–5 nm for ultrafiltration, and ~1 nm for nanofiltration. The membranes used have pore sizes in these ranges. Such pores are best visualised by means of atomic force microscopy (AFM) [3]. Figure 14.1 shows an example of a single pore in each of these three types of membrane. An industrial membrane process may use several hundred square meters of membrane area containing billions of such pores.

For more traditional separation processes, such as distillation, there are available reliable and general design methods. These methods allow, for example, a prediction of the design and operation of a distillation column from a knowledge of the physicochemical properties of the substances to

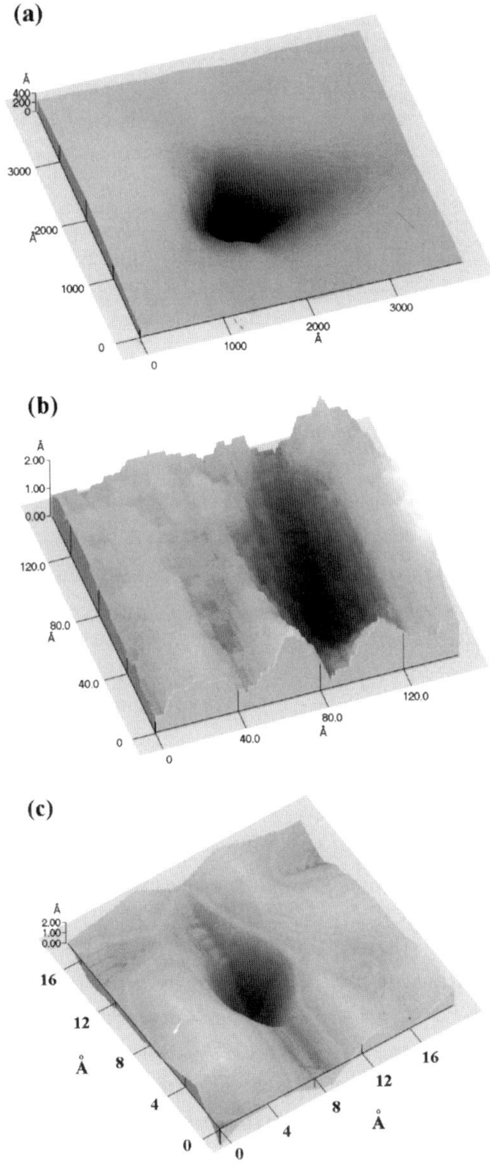

Fig. 14.1. Non-contact atomic force microscope images of membranes with pore diameters of (a) 0.2 μm, (b) 5 nm, and (c) 0.5 nm, covering the ranges of microfiltration, ultrafiltration and nanofiltration, respectively.

be separated. The design and operation of membrane separations in the process industries also requires quantitative methods for predicting separation performance, especially filtration rate and rejection. Although there are a number of such methods available, they have until recently had a relatively high empirical content. As a result, the commonly used methods can be limited in scope in the sense that the cannot be applied to the separation of a new process stream with a high degree of confidence. Hence, in assessing a new application for membrane processes it is often necessary to carry out extensive testing, which can be both expensive and time-consuming. There is, therefore, a need for the development of more general predictive methods. By analogy with distillation, the starting point for such calculations should be the physicochemical properties of the materials to be separated.

A practically useful predictive method must provide quantitative process prediction from accessible physical property data. Such a method should be physically realistic and require a minimum number of assumptions. A method which is firmly based on the physics of the separation is likely to have the widest applicability. It is also an advantage if such a method does not involve mathematics which is tedious, complicated or difficult to follow. For the pressure driven processes of microfiltration, ultrafiltration and nanofiltration, such methods must be based on the microhydrodynamics and interfacial events occurring at the membrane surface and inside the membrane. This immediately points to the requirement for understanding the colloid science of such processes. Any such method must account properly for the electrostatic, dispersion, hydration and entropic interactions occurring between the solutes being separated and between such solutes and the membrane.

The aim of the present chapter is to provide an overview of recent work on the development of quantitative predictive methods for membrane processes at the Centre for Complex Fluids Processing, University of Wales Swansea. The main aim of such work is the development of predictive methods that require no adjustable parameters—that is methods that are truely *ab initio*. Three theoretical aspects will be illustrated: the prediction of rates of ultrafiltration, the prediction of rejection in microfiltration and ultrafiltration, and the prediction of the rejection in nanofiltration. These examples show the importance of bridging the gap between physical chemistry and process engineering in the development of new process technologies. Finally, the use of AFM in the quantification of the adhesion of solutes to membranes will be described. Minimisation of such adhesion is an essential feature of optimised membrane processes.

14.2 PREDICTION OF ULTRAFILTRATION RATES

There have been many models, both simple and sophisticated, that describe the operating patterns of ultrafiltration processes [4]. Most of these models describe how the rate of ultrafiltration is controlled by the properties of a region of very high solute concentration, a filter cake or concentration polarised layer, close to the membrane surface. Relatively few of these models have a genuinely predictive capability. Remarkably, only a very few [5–7] of these models consider the most important feature of the solutes being separated by ultrafiltration—that they fall in the colloidal size range. For colloidal materials, the properties of the filter cake or concentration polarised layer will be controlled by the interparticle interactions in such a region. The important interactions which need to be taken into account are [8]:

Electrostatic. Virtually all colloids in solution acquire a surface charge and hence an electrical double layer. When particles interact in a concentrated region their double layers overlap resulting in a repulsive force which opposes further approach. Any theory of filtration of colloids needs to take into account the multi-particle nature of such interactions. This is best achieved by using a Wigner–Seitz cell approach combined with a numerical solution of the non-linear Poisson–Boltzmann equation, which allows calculation of a configurational force that implicitly includes the multi-body effects of a concentrated dispersion or filter cake.

Dispersion. Dispersion or London-van der Waals forces are ubiquitous. The most rigorous calculations of such forces are based on an analysis of the macroscopic electrodynamic properties of the interacting media. However, such a full description is exceptionally demanding both computationally and in terms of the physical property data required. For engineering applications there is a need to adopt a procedure for calculation which accurately represents the results of modern theory yet has more modest computational and data needs. An efficient approach is to use an effective Lifshitz–Hamaker constant for flat plates with a Hamaker geometric factor [9]. Effective Lifshitz–Hamaker constants may be calculated from readily available optical and dielectric data [10].

Entropic. The packing of particles at a high concentration leads to an entropic pressure tending to disperse them. Such entropic pressures may be precisely calculated using molecular dynamic approaches, and such calculations may be well-represented for both low and high volume fractions by analytical expressions of which the most accurate is a Padé approximation [11].

In some cases, it may be necessary to take into account more specific interactions, such as hydration forces in the example of silica colloids.

Having provided accurate descriptions of solute-solute interactions, there are two approaches to the formulation of a predictive method for ultrafiltration:

Cake filtration approach. Descriptions of interparticle interactions are used to calculate the hydraulic resistance of filter cakes through a recognition that the local compressive drag pressure at any position in such a cake is equal to the net local repulsive pressure (osmotic pressure) arising from interparticle interactions [4, 12]. Such cakes have specific resistances which vary with the depth of the cake and with time. They are compressible due to the "softness" of the interactions. As an example, Fig. 14.2 shows a comparison of such a calculation with experimental data for the dead-end filtration of the protein bovine serum albumin (BSA). (In dead-end filtration the entire fluid flow is normal to the membrane surface.) The figure shows a wide variation in rate of filtration with variation of pH and ionic strength, these being the two major factors leading to the variation of the magnitude of protein-protein interactions. There is excellent agreement between prediction and experiment whereas the classical theory assuming a hard sphere interaction between particles is seriously in error.

Boundary layer formulation. Many membrane processes are operated in cross-flow mode, in which the pressurised process feed is circulated at high velocity parallel to the surface of the membrane, thus limiting the accumulation of solutes (or particles) on the membrane surface to a layer which is thin compared to the height of the filtration module [2]. The decline in permeate flux due to the hydraulic resistance of this concentrated layer can thus be limited. A boundary layer formulation of the convective diffusion equation can give predictions for concentration polarisation in cross-flow filtration and, therefore, predict the flux for different operating conditions. Interparticle force calculations are used in two ways in this approach. Firstly, they allow the direct calculation of the osmotic pressure at the membrane. This removes the need for difficult and extensive experi-

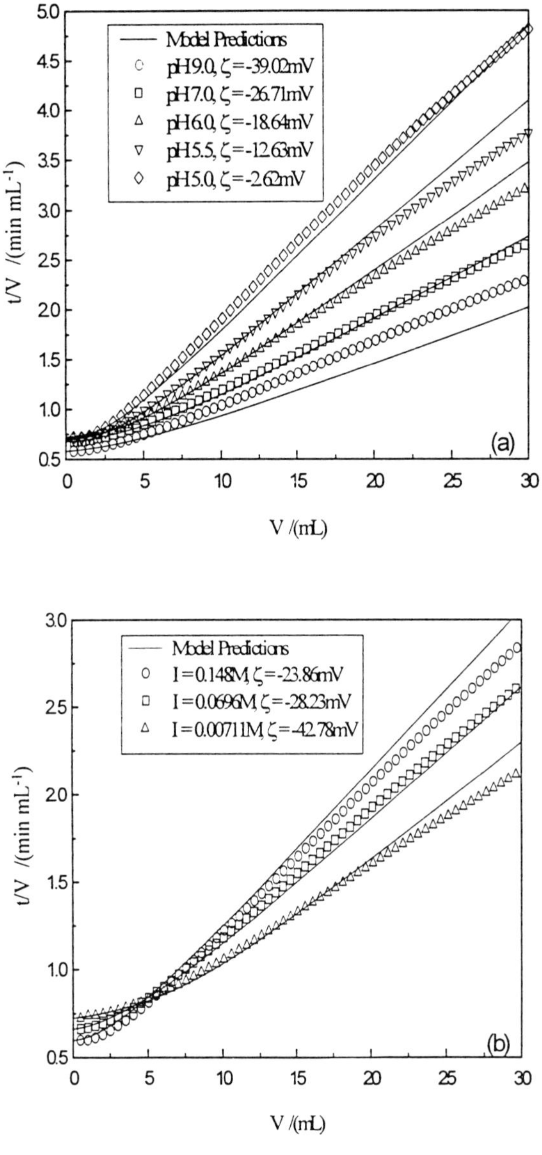

Fig. 14.2. Comparison of theory and experiment for the dead-end filtration of the protein bovine serum albumin (BSA). (a) Time (t) over permeate volume (V) vs. permeate volume for varying pH at an ionic strength of 0.03 M. (b) Time over permeate volume vs. permeate volume for varying ionic strength at a pH of 8.0. The model predictions contain no adjustable parameters. ζ signifies zeta-potential.

mental determination of osmotic pressure. Secondly, the concentration dependent diffusion coefficient can be calculated by means of a generalised form of the Stokes–Einstein equation that modifies the Brownian value of the dilute limit by accounting for the presence of neighbouring particles. There are two contributions to such a modification: a thermodynamic contribution involving the first derivative of the osmotic pressure with respect to concentration and a hydrodynamic contribution dependent on the sedimentation coefficient [13]. Figure 14.3 shows a comparison of such a calculation with experimental steady-state flux data for the cross-flow filtration of BSA [14]. There is again good agreement between theory and experiment as the pH and ionic strength are varied. Work in progress to account for the concentration dependence of the rheological properties of the protein solutions is expected to improve the agreement between theory and experiment even further.

14.3 PREDICTION OF REJECTION IN MICROFILTRATION AND ULTRAFILTRATION

The ability to quantitatively predict rejection during membrane processing is an important requirement. Such an ability allows the selection of membrane properties and operating conditions for the recovery of any solute. It also allows selection of the best membranes and operating conditions for the tantalising goal of large scale membrane fractionation of products such a proteins. Progress toward this latter goal has been hindered by the widespread assumption that membranes separate solutes in terms of their size (as shown by the usual classifications of the various membrane processes). On the contrary, the charge of solutes is equally, and sometimes of greater, importance, as has been elegantly shown experimentally [15]. In most cases the approach and entry of a solute at a pore in an ultrafiltration membrane is controlled by the balance of three forces:

Electrostatic. In many practical situations, both membrane and solute have net negative charges. Hence, as the solute approaches a pore in the membrane it experiences an electrostatic repulsion. A quantitative theoretical description of this interaction requires solution of the non-linear Poisson–Boltzmann equation for the interacting solute and membrane followed by calculation of the resulting force by integrating the electric stress tensor on the solute surface. Due to the complexity of the geometry

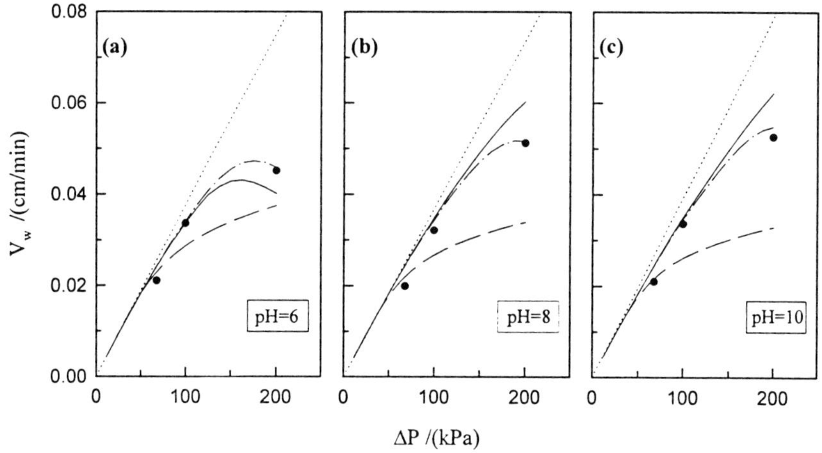

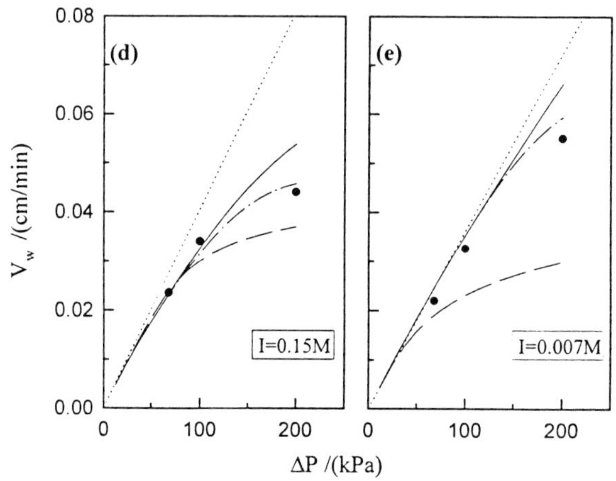

Fig. 14.3. (a)–(c) Steady-state permeate flux (V_w) vs applied pressure difference (ΔP) for BSA solutions at a constant ionic strength of 0.03 M and a variable pH: (●) experimental points; (........) experimental line for pure electrolyte; (———) calculation with variable diffusion coefficient and zeta-potential values calculated from mobility measurements; (—·—·) calculation with variable diffusion coefficient and zeta-potential values giving the best agreement with experimental filtration; (— — —) calculation with constant diffusion coefficient. (b)–(e) Steady-state permeate flux vs applied pressure difference for BSA solutions at a constant pH of 8.0 and variable ionic strength: same symbols as (a)–(c).

for such calculations they can only be carried out numerically by using, for example, a finite element approach.

Dispersion. This will usually result in an attraction between the solute and the membrane.

Hydrodynamic. For a pressure driven process such as ultrafiltration the flow of solvent towards the membrane results in a drag which carries the solute in the same direction. This drag is a function of the distance of the solute from the pore entrance. At large distances it is equal to the isolated solute value (Stokes limit), but as the solute approaches and begins to enter the pore, the drag, for a constant filtration velocity, increases due to the restriction of solvent flow. This increase depends on the ratio of solute diameter to pore diameter.

The result of the existence of these opposing forces is that there exists a "critical velocity", defined as the in-pore velocity at which the electrostatic, dispersion and hydrodynamic forces acting on the solute are balanced [16–18]. The critical velocity defines a velocity limit below which the electrostatic force is greater than the sum of the hydrodynamic and dispersion forces and the particle will not be transported through the pore. In a practical sense for a membrane process this velocity is related to the filtration flux which may be controlled by the applied pressure (giving rise to an analogous "critical pressure"). Figure 14.4 gives an example of the calculation of such a critical velocity for a solute of constant surface potential approaching membranes of differing pore sizes. Two features of these curves are of especial practical importance:

(a) There is a maximum in the curves such that a solute will not be able to come into intimate contact with, and be transported across, the membrane if the operating velocity is less than this maximum velocity. Hence, the curves define potential (charge) controlled rejection.
(b) The curves also define non-fouling operating behaviour as they show that solutes may be rejected due to repulsive interactions experienced at some distance from the membrane surface.

It is found that many current commercial membranes have potentials toward the lower end of the available range. Ultrafiltration processes could be operated with higher flux, lower fouling and greater selectivity if more highly charged membranes were developed. This poses a challenge for membrane manufacturers. Nevertheless, existing membranes allow good

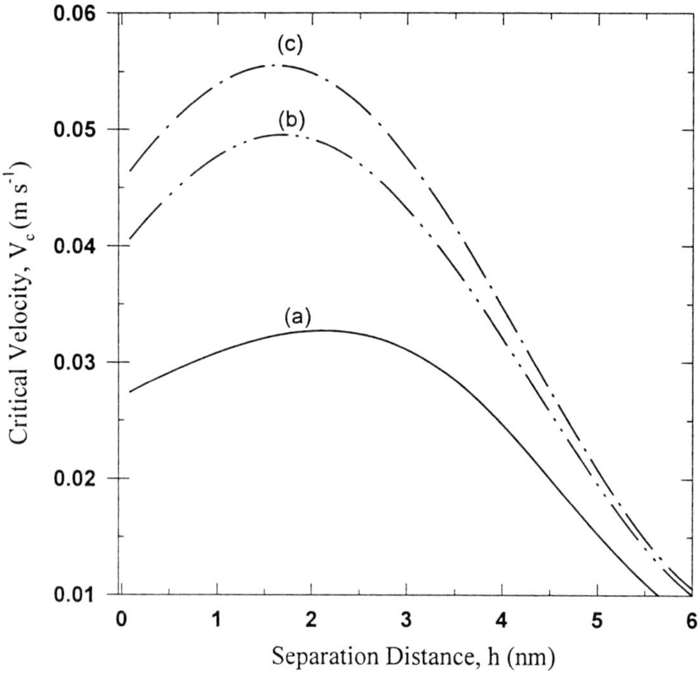

Fig. 14.4. Theoretical critical velocity as a function of separation distance for different values of solute radius/pore radius ratio (λ). Solute radius (a) = 5 nm; κa = 1.0 (κ is the Debye parameter); membrane and solute surface potential = 50 mV. Values of λ: (a) 0.8; (b) 0.6; (c) 0.4.

use to be made of these effects. Figure 14.5 shows how the rejection of the protein BSA may be varied by control of the applied transmembrane pressure difference. Operation of the process below the theoretically predicted critical pressure gives high rejection of the protein. However, operation at pressures in excess of the critical pressure gives low rejection as the hydrodynamic force drives the protein molecules through the membrane.

This balance of forces on the approach of particles to membrane pores also has substantial consequences for rates of filtration. Of special interest is the case where the particle to be removed is comparable to or only slightly greater in dimensions than the pore diameter. In the absence of colloidal interactions, this can lead in a catastrophic loss in filtration flux as such particles can plug pores highly effectively. Such a fouling mechan-

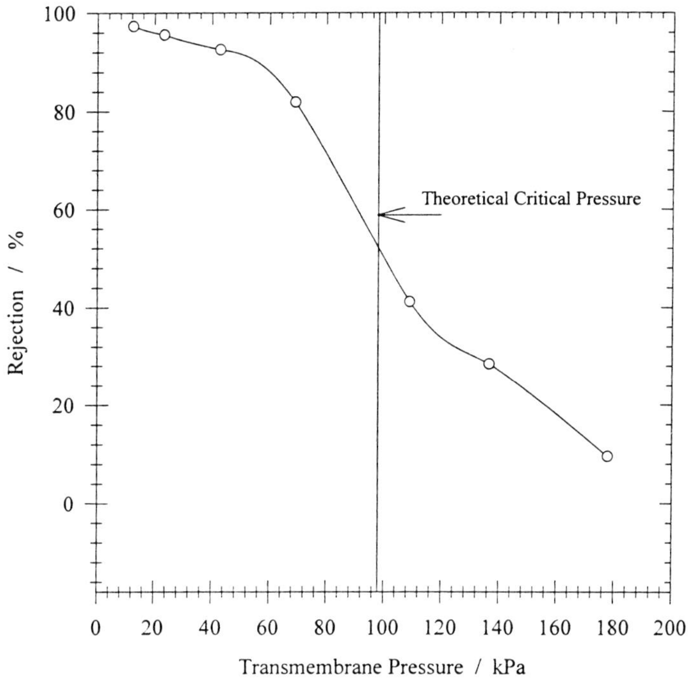

Fig. 14.5. Experimental rejection (o) and theoretical prediction of the critical pressure for filtration of BSA in 0.001 M NaCl solution at pH 9 at a membrane of mean pore diameter 84 nm. Rejection is high below the critical pressure as electrical double layer repulsion prevents the protein (effective spherical diameter ~6 nm) from entering the membrane pores. As the critical pressure is approached, hydrodynamic forces increase and drive the protein through the membrane.

ism has been termed "complete" pore blocking. However, the presence of long-range electrostatic interactions can, when coupled with appropriate operating conditions, lead to maintenance of a high rate of filtration even in this very challenging case. Figure 14.6 shows data for the filtration of silica colloids at two values of the transmembrane pressure. For the conditions shown, the critical pressure is 130 kPa. Operation above this pressure gives an initially high rate of filtration, but this rapidly declines. Operation below this pressure gives a lower rate of filtration, but only a very slight decline. In the latter case, the hydrodynamic force is lower and electrostatic forces prevent the blocking of the membrane pores.

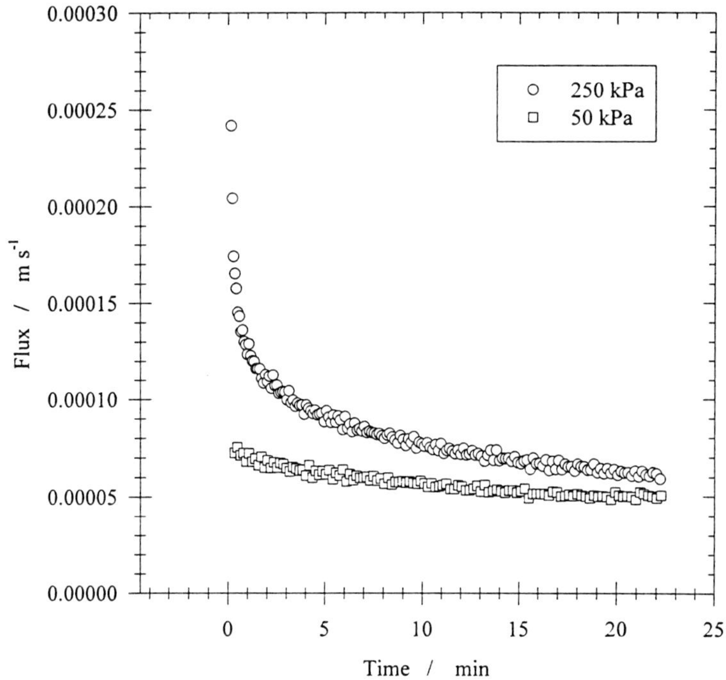

Fig. 14.6. Filtration flux as a function of time of filtration for the filtration of 0.01 g/L silica particles in 0.001 M NaCl solution at pH 6 at a membrane of mean pore diameter 84 nm. The particle size was very close to the pore size. The critical transmembrane pressure for these conditions was calculated as 130 kPa. Operation below this pressure gives only a gradual decline in filtration flux with time. Operation above this pressure gives an initially higher filtration flux which declines rapidly with time. In the latter case the intial hydrodynamic force exceeds the electrical double layer repulsion between the membrane and the particles, causing the particles to block the membrane pores.

14.4 PREDICTION OF REJECTION AT NANOFILTRATION MEMBRANES

It would not be untrue to suggest that attempting the prediction of the rejection properties of nanofiltration membranes is a suitable undertaking only for the brave or foolhardy. This is largely because the events leading to rejection at such membranes are taking place on a length scale of the order of a nanometre. This is a scale at which macroscopic descriptions of hydrodynamics and interactions are begining to break down—it is a scale not much greater than atomic dimensions. It may be asked, for example, as to whether it is sensible to describe the transport channels in such membranes as being pores—though recent atomic force microscope studies indicate that, at least for some such membranes, pores do exist with diameters slightly less than a nanometre (see Fig. 14.1, [19]).

Despite these difficulties, good progress has been made in modelling nanofiltration. Some of the most notable work [20] has made use of the extended Nernst-Planck equation. This equation decribes the transport of ions as a result of diffusion, electric field gradient and convective terms. When coupled with a suitable description of the partitioning of ions between solution and membrane, for example a Donnan expression, the Nernst–Planck equation provides a complete description of the process. However, use of the equation as a predictive method requires knowledge of two parameters, the effective membrane charge and effective membrane thickness. The best way to obtain these parameters is by measurement of the rejection of single salt solutions. When such membrane characterisation has been carried out it is possible to predict the separation of mixtures of electrolytes at the membrane.

We have developed this approach in three ways [19, 21] to formulate a method with better predictive power:

(a) A porous flow model has been formulated in which the extended Nernst–Planck equation is solved for a porous network with hindrance factors for diffusion and convection being included to allow for the transport of ions in the membrane taking place within a confined space. The values of the hindrance factors, which depend on the ionic radius/pore radius ratio, were calculated using a finite element technique [22]. The inclusion of hindrance factors not only improves the accuracy of predictions but also allows a calculation of the effective pore diameter of the membrane.
(b) The entry and exit of solutes at the membrane is decribed in terms of a partitioning depending on the membrane charge, pore size, solute charge and solute size.
(c) One of the most remarkable features of nanofiltration membranes is the way in which their charge depends on the solution with which they are in contact. The effective membrane charge density may be described in terms of a suitable isotherm. Expressing the solution

properties in terms of charge (equivalents/m^3) can result in a single isotherm for multivalent electrolytes.

An example of the excellent agreement that can be found between theoretical prediction and process performance is shown in Figure 14.7. This is for the "diafiltration" of a charged red dye. Such dyes are manufactured in solutions of high salt content but sold as products of low salt content. Removal of the salt (NaCl) can be achieved by extensive washing through a nanofiltration membrane—the dye is retained by the membrane but the salt can pass through. However, as the salt concentration decreases,

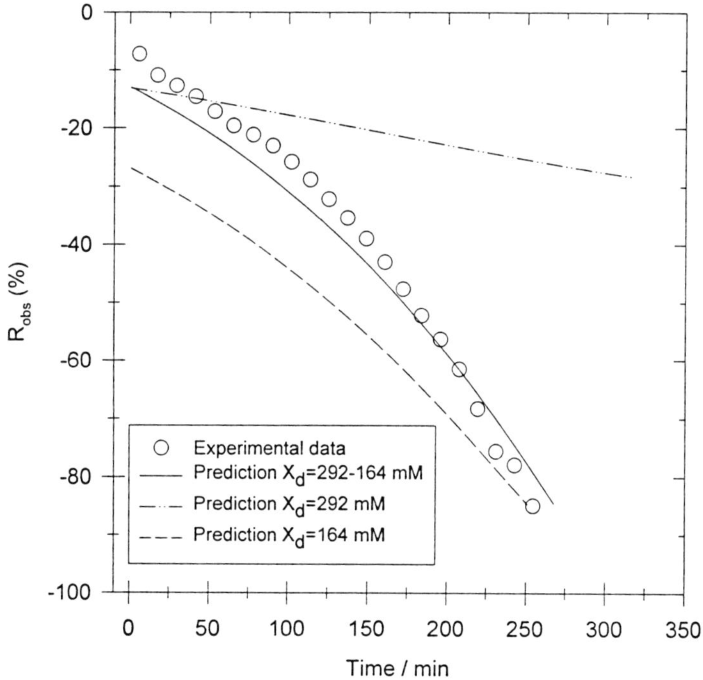

Fig. 14.7. Chloride ion rejection (R_{obs}) during the diafiltration of NaCl from a solution of red dye. During this process the charge properties of the membrane change continuously. The bold line shows the prediction allowing for this change in properties. The upper line is the prediction if it is assumed that the membrane charge remains at its initial value and the lower line is the prediction if it is assumed that the membrane charge remains at its final value. (X_d is the membrane charge.)

so the membrane charge decreases and the membrane rejection properties change (see points (b) and (c) above). Hence, predicting the performance of the industrial process must take into account the changing properties of the membrane. The figure shows that an excellent prediction is obtained when these properties are properly accounted for.

14.5 SOLUTE ADHESION—MEMBRANE FOULING

Membranes are very finely porous structures and like all such porous structures used in an industrial context are susceptible to fouling caused by adhesion of components of the materials being processed. This fouling can be minimised or avoided if suitable polymers are used in membrane manufacture. However, the selection of membrane polymers suited to particular separations has until now been a matter of experience (and failure) rather than science. However, an AFM used with the colloid probe technique [23] can provide a rapid means of assessing the adhesion of solutes to membrane materials and is hence a powerful tool for the membrane technologist.

Figure 14.8 shows a colloid probe consisting of a single polystyrene sphere (diameter 11 μm) attached to a tipless AFM cantilever [24]. In quantifying adhesion, such a probe is first brought into (momentary) contact with the membrane surface. The probe is then retracted. Resulting force curves as a function of retraction distance (piezo displacement) are shown in Fig. 14.9. The curves are shown for two types of membrane of very similar pore size and surface roughness. Curve (a) is for a conventional polyethersulphone membrane and curve (b) for a membrane made from a mixture of polymers chosen with the aim of achieving low membrane fouling. The nature of these plots can be understood by considering curve (a). From A to A′ the colloid probe and the membrane move together with no displacement relative to the piezo. From A′ to B there is a stretching of the probe and/or membrane which gives them a relative movement with respect to the piezo. From B to C the stretching continues and the contact between the colloid probe and the membrane is finally broken at C. From C to D the probe and membrane move further apart.

The difference in force between B and C is a direct measurement of the adhesive interaction, in this case 1.98 mN/m. For the modified membrane, curve (b) the adhesive interaction is just 0.38 mN/m, so the use of a mixture

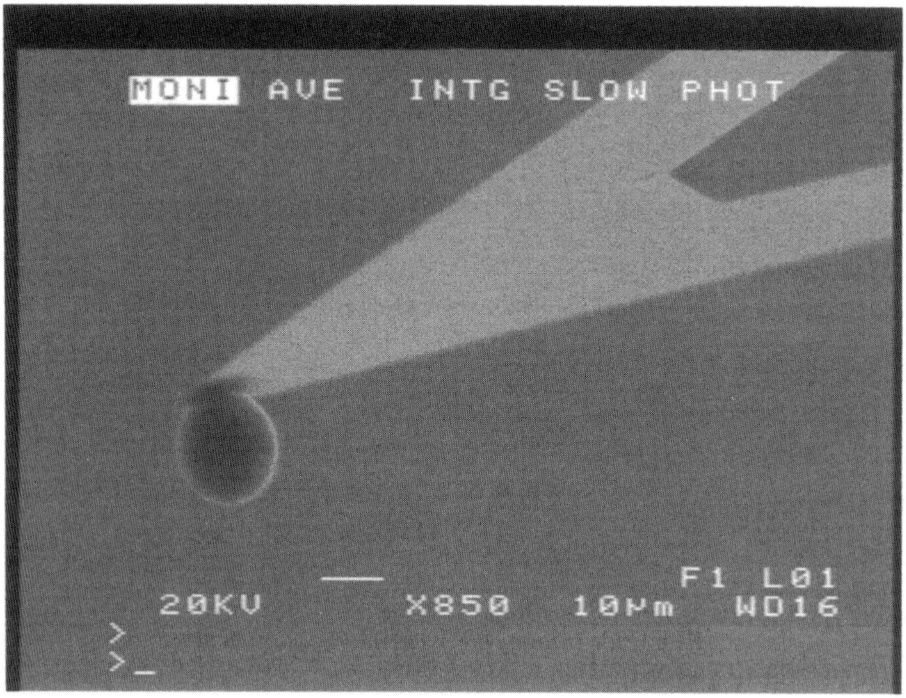

Fig. 14.8. SEM image of a 11 μm polystyrene sphere attached to an AFM cantilever—a colloid probe.

of polymers in manufacturing the membrane has achieved the required goal. It is also possible to study the adhesion of single biological cells by immobilising such a cell at the end of an AFM cantilever (a cell probe) and the adhesion of macromolecules by coating inorganic (silica) probes with the test molecules. Such tests have also shown that the modified membrane has lower adhesion properties. This technique has great potential for the technologist developing new types of membranes. A data base can be constructed as the number of membranes characterised using probe techniques increases. This database can be consulted each time a membrane separation process is designed, the measured force of adhesion for each type of probe being an important part of the membrane specification. Non-fouling membranes carry out separations faster and at a constant rate giving important technical and economic benefits.

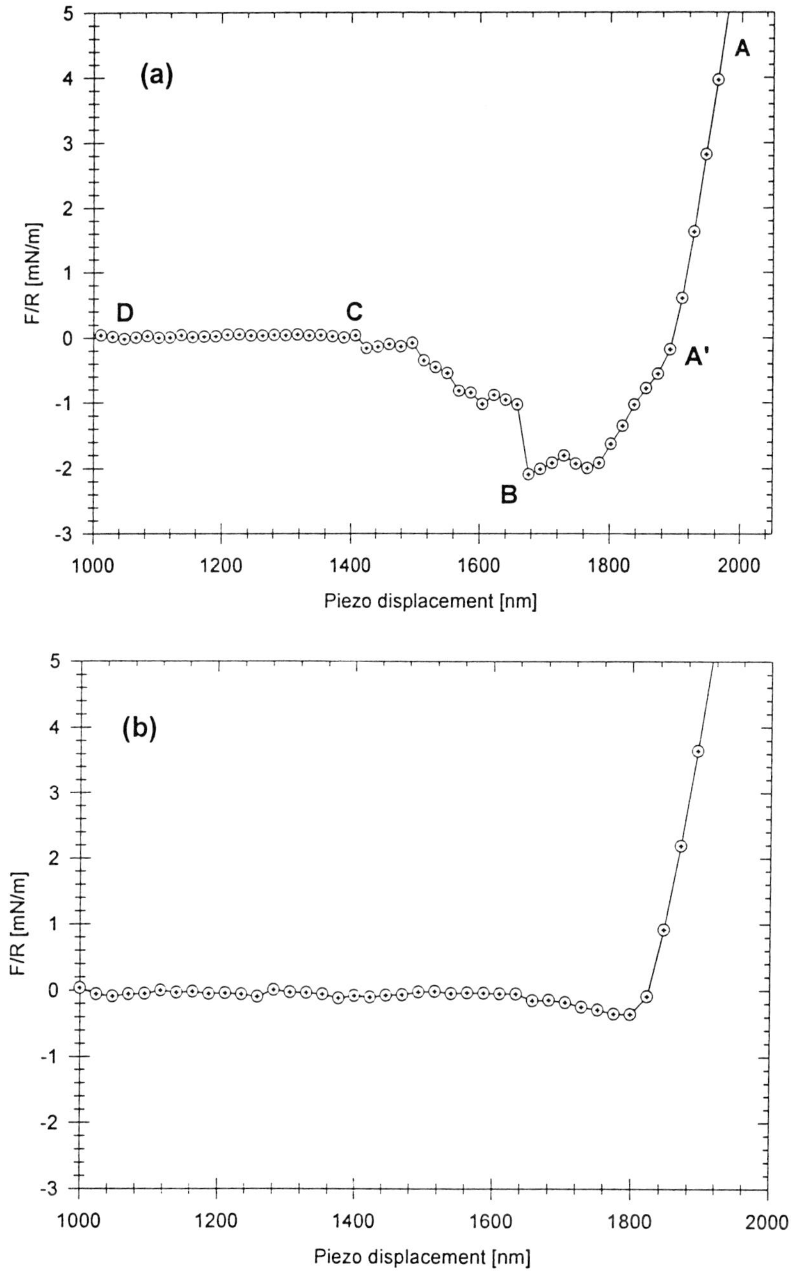

Fig. 14.9. Normalised force (F/R; R is the particle radius) vs piezo displacement plot (retraction) for a polystyrene colloid probe in 0.01 M NaCl at pH 8.0. (a) Conventional polyethersulphone membrane. (b) Modified mixed polymer membrane.

14.6 CONCLUSIONS

The development of *ab initio* methods for the prediction of the performance of membrane separation processes has made substantial developments. Sophisticated methods now exist for such prediction, and these have been experimentally verified in the laboratory. The present challenges are two-fold. Firstly, to continue the fundamental development to more complex separations. Secondly, to apply the verified methods in the design of full-scale industrial processes. The existence of good predictive methods is expected to further expand the application of membrane processes.

ACKNOWLEDGEMENTS

All research requires a source of inspiration. The original inspiration for WRB's research lies in the intellectual stimulation provided by WJA as tutor and doctoral research supervisor. This work described in this chapter was funded by the UK BBSRC, the UK EPSRC and the EC.

References

[1] M. Mulder, Basic principles of membrane technology, (Kluwer Academic Publishers, Dordrecht, 1991).

[2] W.R. Bowen, Membrane Separation Processes, Chapter 20 in Coulson and Richardson's Chemical Engineering, 4th edition (Pergamon, 1991).

[3] W.R. Bowen, N. Hilal, R.W. Lovitt and C.J. Wright, Atomic force microscope studies of membrane surfaces, in Surface chemistry and electrochemistry of membrane surfaces, ed. T.S. Sørensen (Marcel Dekker, in press).

[4] W.R. Bowen and F. Jenner, Theoretical descriptions of membrane filtration of colloids and fine particles: an assessment and review, Adv. Colloid Interface Sci. 56 (1995) 141–200.

[5] R.M. McDonogh, C.J.D. Fell and A.G. Fane, Surface charge and permeability in the ultrafiltration of non-flocculating colloids, J. Membrane Sci. 21 (1984) 285–294 (and subsequent papers).

[6] P. Bacchin, P. Aimar and V. Sanchez, Model for colloidal fouling of membranes, AIChEJ 41 (1995) 368–376.

[7] A.-S Jönsson and B. Jönsson, Ultrafiltration of colloidal dispersions—a theoretical model of the concentration polarisation, J. Colloid Interface Sci. 180 (1996) 504–518.

[8] W.R. Bowen and F. Jenner, Dynamic ultrafiltration model for charged colloidal dispersions: a Wigner-Seitz cell approach, Chem. Eng. Sci. 50 (1995) 1707–1736.
[9] J. Mahanty and B.W. Ninham, Dispersion forces (Academic Press, London, 1976).
[10] W.R. Bowen and F. Jenner, The calculation of dispersion forces for engineering applications, Adv. Colloid Interface Sci. 56 (1995) 201–243.
[11] K.R. Hall, Another hard sphere equation of state, J. Chem. Phys. 57 (1972) 2252–2254.
[12] W.R. Bowen and P.M. Williams, Dynamic ultrafiltration model for proteins: a colloidal interaction approach, Biotechnol. Bioeng. 50 (1996) 125–135.
[13] G.K. Batchelor, Brownian diffusion of particles with hydrodynamic interaction, J. Fluid Mech. 74 (1976) 1–29.
[14] W.R. Bowen, A. Mongruel and P.M. Williams, Prediction of the rate of cross-flow ultrafiltration: a colloidal interaction approach, Chem. Eng. Sci. 51 (1996) 4321–4333.
[15] S. Saksena and A.L. Zydney, Effect of solution pH and ionic strength on the separation of albumin from immunoglobulins (IgG) by selective filtration, Biotech. Bioeng. 43 (1994) 960–968.
[16] W.R. Bowen and A.O. Sharif, The hydrodynamic and electrostatic interactions on the approach and entry of a charged spherical particle to a charged cylindrical pore in a charged planar surface—with implications for membrane separation processes, Proc. R. Soc. Lond. A 452 (1996) 2121–2140.
[17] W.R. Bowen and A.O. Sharif, Adaptive finite element solution of the non-linear Poisson-Boltzmann equation—a charged spherical particle at various distances from a charged cylindrical pore in a charged planar surface, J. Colloid Interface Sci. 187 (1997) 363–374.
[18] W.R. Bowen and A.O. Sharif, Hydrodynamic and colloidal interaction effects on the rejection of a particle larger than a pore in microfiltration and ultrafiltration membranes, Chem. Eng. Sci. 53 (1998) 879–890.
[19] W.R. Bowen, A.W. Mohammad and N. Hilal, Characterisation of nanofiltration membranes for predictive purposes—use of salts, uncharged solutes and atomic force microscopy, J. Membrane Sci. 126 (1997) 91–105.
[20] T. Tsuru, S. Nakao and S. Kimura, Calculation of ion rejection by extended Nernst-Planck equation with charged reverse osmosis membranes for single and mixed electrolytes, J. Chem. Eng. Japan 24 (1991) 511–517.
[21] W.R. Bowen and H. Mukhtar, Characterisation and prediction of separation performance of nanofiltration membranes, J. Membrane Sci. 112 (1996) 263–274.
[22] W.R. Bowen and A.O. Sharif, Transport through microfiltration membranes—particle hydrodynamics and flux reduction, J. Colloid Interface Sci. 168 (1994) 414–421.
[23] W.A. Ducker, T.J. Senden and P.M. Pashley, Measurement of forces in liquids using a force microscope, Langmuir 8 (1992) 1831–1836.
[24] W.R. Bowen, N. Hilal, R.W. Lovitt and C.J. Wright, A new technique for membrane characterisation: direct measurement of the force of adhesion of a single particle using an atomic force microscope, J. Membrane Sci. 139 (1998) 269–274.

Chapter 15

A New Approach to the Prediction of Diffusion Coefficients

P.R. FISK and P. JONATHAN

New models for the prediction of molecular diffusion coefficients are described, and compared to previously established ones. These are based on solute molecular size, solvent viscosity, solvent molecular size, and temperature. The data set of diffusion coefficients used was primarily the one developed by Wilke and Chang and upon which their commonly used diffusion model is based (A.I.Ch.E. Journal, 1 (1955), 264).

The new diffusion models are successful in the description they provide, and are particularly useful because they require only the chemical structure as an input, no experimental data being necessary. Molecular size is described by the use of the calculated molar refractivity. A new model for the temperature dependence of the viscosity of water was required and developed.

Some compound class-specific outliers are described, and an explanation of why they are predicted incorrectly is given. The interpretation of the results shows that there are serious inadequacies in the development of prior models such as the Wilke-Chang equation.

15.1 INTRODUCTION

Diffusion is defined as "the net movement of molecules from a region of high concentration to one of low concentration under fixed conditions and without the application of an external force" [1]. It is a fundamentally important process in many chemical and biological contexts.

The objective was to develop a mathematical model allowing the prediction of diffusion coefficients for different solutes in various solvents, based in part on the Wilke–Chang models in the literature [2, 3], but introducing several important differences described in detail below. These are, principally, the use of a calculated volume-related parameter and a thorough statistical analysis. It was particularly desirable to be able to predict values for any substance, based on theory only.

15.2 THEORY

In the following text subscript A refers to the solute and subscript B to the solvent.

15.2.1 Introduction

The diffusion coefficient is the proportionality constant in the rate equation for diffusion, ordinarily defined as

$$N_A = -D\frac{\mathrm{d}C_A}{\mathrm{d}L}. \tag{15.1}$$

This is Fick's first law of diffusion, where N_A = the rate of diffusion of component A through an area perpendicular to the direction of diffusion; C_A = the concentration of A; L = the distance in the direction of diffusion; D = the diffusion coefficient.

Eyring [4] suggests the following relationship

$$(D\eta_B)/T = \text{a constant for a given solute and solvent pairing,} \tag{15.2}$$

where η_B = the viscosity of the solvent; T = absolute temperature; k = Boltzmann constant; r_A = the radius of the (spherical) solute molecule.

The Stokes–Einstein equation [5] states that for large solute molecules in a solvent comprised of small molecules

$$D\eta_B/T = k/6\pi r_A, \tag{15.3}$$

where k = the Boltzmann constant; r_A = the radius of the (spherical) solute molecule.

15.2.2 Previous empirical models

In addition to the relationship with η_B, T and r_A, two empirical relationships are found experimentally [3]

$$D \propto M_B^{0.5} \tag{15.4}$$

and

$$D \propto V_A^{-0.6} \tag{15.5}$$

where M_B = the molecular weight of the solvent; V_A is the molar volume of the solute.

V_A may be estimated by the calculation method of Le Bas [2] estimates of D based on this method are reviewed in Reference [6].

Molar refractivity is related to the size of a molecule. The advantage of using molar refractivity in a model (rather than V_A) is that calculation is readily available via for example, the program CMR within the Pomona Medicinal Chemistry Project software [8]. Errors in results from CMR are considered by the present authors to be small compared to the errors in measured D values.

Calculations of volume based on molecular orbital calculations are possible; they are, however, not easily accessible to the non-specialist, so CMR is preferred.

15.2.3 Molar refractivity

The electromagnetic wave theory of light enables the definition of the molar refractivity

$$R = \frac{\mu^2 - 1)M}{(\mu^2 + 2)\rho},$$

where μ is the refractive index, ρ the density of a liquid, M its molar mass.

It can be shown that R is the volume of the molecules, as distinct from M/ρ which is the apparent volume [7].

R (or CMR) is an additive and constitutive property and has been used widely in determination of molecular structure, before the advent of spectroscopic methods.

The Pomona MedChem program CMR [8] is used in this paper to generate molar refractivity values; it is based on literature values of R [9]. Atoms are ascribed fragment values and small corrections for the bonds present are made. The statistical approach and inclusion of data for solids are described in Reference [8].

15.2.4 Approach

The above theory suggests that the following relationships exist and it is with these in mind we can begin to build a model:

$$D \propto T; \qquad D \propto 1/\eta_B; \qquad D \propto 1/r_A.$$

Because $r_A \propto (\text{Volume}_A)^{1/3}$ which implies that $r_A \approx \propto(\text{CMR}_A)^{1/3}$, hence $D \propto (1/\text{CMR}_A)^{1/3}$, where CMR_A = the calculated molar refractivity of the solute. Trials [3] suggest that $D \propto (M_B)^{\beta}$, where β is approximately 0.5 and D is proportional to $V_A^{-0.6}$.

15.3 METHOD

15.3.1 Solute properties

Solute properties in Table 15.1 include CMR_A, M_A, D and T (the temperature of the measurement) for the following solvents (with the number of data points in parentheses): water (116), methanol (61), carbon tetrachloride (13), ethanol (5), acetone (13), benzene (58) and toluene (15).

The D values are taken from Refs. [2] and [3], with the inclusion of further data [10]. Thus, this work does not cover data significantly different in scope to the original studies.

TABLE 15.1
Solute properties

Solute	Solvent	M_A	CMR_A	Temp/K	Diffusion coefficient $\times 10^9\ m^2\ s^{-1}$
Acetamide	Water	59.1	1.509	293	1.05
Acetamide	Water	59.1	1.509	298	1.32
Acetanilide	Methanol	135.2	4.021	288	1.50
Acetic Acid	Acetone	60.1	1.294	298	3.31
Acetic acid	Acetone	60.1	1.294	313	4.04
Acetic acid	Acetone	60.1	1.294	288	2.92
Acetic acid	Benzene	60.1	1.294	298	2.08
Acetic acid	Benzene	60.1	1.294	279	1.59
Acetic acid	Benzene	60.1	1.294	288	1.92
Acetic acid	Carbon tetrachloride	60.1	1.294	298	1.49
Acetic acid	Carbon tetrachloride	60.1	1.294	280	1.15
Acetic acid	Carbon tetrachloride	60.1	1.294	313	1.78
Acetic acid	Carbon tetrachloride	60.1	1.294	288	1.27
Acetic acid	Water	60.1	1.294	286	0.91
Acetic acid	Water	60.1	1.294	298	1.19
Acetic acid	Toluene	60.1	1.294	280	1.66
Acetic acid	Toluene	60.1	1.294	298	2.27
Acetic acid	Toluene	60.1	1.294	288	1.91
Acetone	Water	58.1	1.605	298	1.28
Acetonitrile	Methanol	41.1	1.119	288	2.64
Acetonitrile	Water	41.1	1.119	288	1.26
Acetyldiphenylamine	Benzene	211.1	6.532	288	0.90
Acetyldiphenylamine	Methanol	211.1	6.532	288	0.98
Adipic acid	Water	146.1	3.338	298	0.74
Alanine	Water	89.1	2.126	298	0.91
Alloxan	Water	142.1	2.735	293	0.67
Allylalchohol	Methanol	58.1	1.696	288	1.80
Allylalcohol	Water	58.1	1.696	288	0.90
Aminobenzoic acid	Water	137.1	3.710	298	0.84
Aminobutyric acid	Water	103.1	2.590	298	0.83
Ammonia	Water	17.0	0.546	298	2.28
Aniline	Ethanol	93.1	3.057	292	2.70
Aniline	Methanol	93.1	3.057	288	1.49
Arabinose	Water	150.1	3.085	293	0.70
Benzaldehyde	Benzene	106.1	3.188	288	1.73
Bennzaldehyde	Methanol	106.1	3.188	288	1.66
Benzene	Water	78.1	2.689	298	1.09
Benzoic acid	Acetone	122.1	3.341	298	2.62
Benzoic acid	Acetone	122.1	3.341	286	2.37

TABLE 15.1 (Contd.)

Solute	Solvent	M_A	CMR_A	Temp/K	Diffusion coefficient $\times 10^9$ $m^2 s^{-1}$
Benzoic acid	Acetone	122.1	3.341	313	3.05
Benzoic acid	Benzene	122.1	3.341	298	1.38
Benzoic acid	Benzene	122.1	3.341	288	1.17
Benzoic acid	Benzene	122.1	3.341	313	1.76
Benzoic acid	Benzene	122.1	3.341	288	1.36
Benzoic acid	Carbon tetrachloride	122.1	3.341	288	0.78
Benzoic acid	Carbon tetrachloride	122.1	3.341	298	0.91
Benzoic acid	Carbon tetrachloride	122.1	3.341	314	1.17
Benzoic acid	Toluene	122.1	3.341	298	1.49
Benzoic acid	Toluene	122.1	3.341	289	1.29
Benzoic acid	Toluene	122.1	3.341	313	1.85
Trichlorobenzene	Toluene	195.5	4.627	281	1.32
Benzyl alcohol	Water	108.1	2.842	298	0.93
Bromine	Benzene	159.8	1.731	285	2.00
Bromine	Water	159.8	1.731	285	0.90
Bromoaniline	Benzene	172.0	3.834	288	1.41
Bromoaniline	Methanol	172.0	3.834	288	1.41
Bromobenzene	Benzene	157.0	3.466	288	1.86
Bromobenzene	Benzene	157.0	3.466	280	1.41
Bromobenzene	Methanol	157.0	3.466	288	1.75
Bromoform	Acetone	252.8	2.972	293	2.74
Bromoform	Benzene	252.8	2.972	291	1.62
Bromoform	Benzene	252.8	2.972	293	1.77
Bromoform	Ethanol	252.8	2.972	293	0.97
Bromonaphthalene	Benzene	207.1	5.154	288	1.30
Bromonaphthalene	Ethanol	207.1	5.154	293	0.76
α-Bromonaphthalene	Benzene	207.1	5.154	280	1.04
α-Bromonaphthalene	Toluene	207.1	5.154	281	1.24
Bromonitrobenzene	Benzene	202.0	4.860	288	1.33
Bromonitrobenzene	Methanol	202.0	4.860	288	1.43
Bromophenol	Benzene	173.0	4.212	288	1.34
Bromophenol	Methanol	173.0	4.212	288	1.34
Bromopropionic acid	Methanol	153.0	2.535	288	1.35
Bromopropylene (3)	Methanol	120.9	2.320	288	2.22
Bromotoluene (3)	Toluene	170.9	3.929	281	1.52
Butane	Water	58.1	2.033	298	0.97
Butanol	Water	74.1	2.186	288	0.77
Butanol	Water	74.1	2.186	298	0.97
Butanol	Water	74.1	2.186	298	0.98
Butryamide	Water	86.1	2.437	298	1.07
Butyric acid	Water	88.1	2.221	298	0.92

TABLE 15.1 (Contd.)

Solute	Solvent	M_A	CMR_A	Temp/K	Diffusion coefficient $\times 10^9\ m^2\ s^{-1}$
Caffeine	Water	194.2	4.990	283	0.42
Caproic acid	Water	116.2	3.149	298	0.78
Carbon dioxide	Water	44.0	0.713	298	1.92
Carbon dioxide	Water	44.0	0.713	291	1.71
Carbon monoxide	Water	28.0	0.440	298	2.30
Carbon tetrabromide	Benzene	331.7	3.749	280	1.12
Carbon tetrachloride	Benzene	153.8	2.607	298	2.00
Carbon tetrachloride	Carbon tetrachloride	153.8	2.607	298	1.41
Carbon tetrachloride	Methanol	153.8	2.607	288	1.70
Carbon tetrachloride	Toluene	153.8	2.607	298	2.19
Chloral hydrate	Methanol	165.5	2.858	288	1.16
Chlorine	Water	70.9	1.160	298	1.48
Chloroacetic acid	Benzene	94.5	1.785	288	1.48
Chloroacetic acid	Methanol	94.5	1.785	288	1.52
Chloroacetic acid	Water	94.5	1.785	298	1.04
Chloroaniline	Benzene	127.6	4.142	288	1.56
Chloroaniline	Methanol	127.6	4.142	288	1.37
Chlorobenzene	Benzene	112.6	3.180	288	1.42
Chlorobenzoic acid	Methanol	156.6	3.833	288	1.29
Chlorobromomethane	Methanol	129.4	1.998	288	2.50
Chloroethene	Water	62.5	1.571	298	1.34
Chloroform	Benzene	85.0	2.115	288	2.11
Chloroform	Methanol	85.0	2.115	288	2.07
Chlorohydrin(1)	Methanol	110.5	2.366	288	1.30
Chloromethane	Water	50.5	1.133	298	1.49
Chloronaphthalene	Benzene	162.6	4.868	288	1.20
Chloronitrobenzene	Benzene	157.6	3.906	288	1.70
Chloronitrobenzene	Methanol	157.6	3.906	288	1.68
Chlorophenol	Benzene	128.6	3.333	288	1.42
Chlorophenol	Methanol	128.6	3.333	288	1.32
Cinnamic acid	Acetone	148.2	4.547	298	2.41
Cinnamic acid	Benzene	148.2	4.547	298	1.21
Cinnamic acid	Carbon tetrachloride	148.2	4.547	298	0.76
Cinnamic acid	Toluene	148.2	4.547	298	1.18
Cyclohexane	Water	84.2	2.783	298	0.90
Cyclopentane	Water	70.1	2.319	298	1.04
Dextrose	Water	180.2	3.701	298	0.68
Dibenzylamine	Methanol	197.3	6.496	288	0.86
Dibromobenzene	Benzene	235.9	4.243	288	1.37
Dibromobenzene	Methanol	235.9	4.243	288	1.55
Dibromonaphthalene	Benzene	286.0	5.931	288	1.25

TABLE 15.1 (Contd.)

Solute	Solvent	M_A	CMR_A	Temp/K	Diffusion coefficient $\times 10^9\ m^2\ s^{-1}$
Dibromonaphthalene	Methanol	286.0	5.931	288	1.33
Dichloroacetic acid	Methanol	128.9	2.277	288	1.36
Dichlorobenzene(4)	Benzene	147.0	3.671	288	1.90
Dichlorohydrin(1,1)	Methanol	145.0	2.858	288	1.36
Dichloronaphthalene	Benzene	97.1	5.359	288	1.40
Dichloronaphthalene	Methanol	197.1	5.359	288	1.52
Dichloropropanol(1,1)	Water	129.0	2.705	288	0.75
Diethylamine	Water	73.1	2.401	298	1.11
Diglycine	Water	132.0	2.994	298	0.79
Dihydroxypropane(1,2)	Water	75.0	1.875	288	0.76
Diiodomethane	Methanol	267.8	3.254	288	1.68
Dinitrobenzene(3)	Benzene	168.1	4.228	288	1.54
Dinitrobenzene(3)	Methanol	168.1	4.228	288	1.56
Dinitronaphthalene	Benzene	219.0	5.828	288	1.23
Dinitronaphthalene	Methanol	219.0	5.828	288	1.32
Dinitrophenol(2,4)	Methanol	184.1	4.293	288	1.40
Ethane	Water	30.1	1.105	298	1.38
Ethanol	Water	46.1	1.258	288	1.00
Ethanol	Water	46.1	1.258	298	1.24
Ethene	Water	28.1	1.080	298	1.55
Ethyl acetate	Water	88.1	2.221	298	1.12
Ethyl benzene	Water	106.0	3.616	298	0.90
Ethyl benzoate	Acetone	150.2	4.269	293	2.47
Ethyl bromide	Methanol	109.0	1.882	288	1.95
Ethyl ether	Benzene	74.1	2.186	288	2.21
Ethyl iodide	Methanol	156.0	2.411	288	1.56
Ethyl nitrate	Methanol	91.0	1.984	288	2.20
Ethylene bromide	Benzene	106.9	1.857	288	2.40
Ethylene bromide	Benzene	106.9	1.857	288	1.97
Ethylene glycol	Water	62.1	1.411	298	1.16
Ethylene iodide	Methanol	153.9	2.386	288	2.16
Ethylene iodide	Methanol	153.9	2.386	288	1.40
Formamide	Water	45.0	1.046	298	1.67
Formic acid	Acetone	46.0	0.830	298	3.77
Formic acid	Acetone	46.0	0.830	292	3.27
Formic acid	Acetone	46.0	0.830	280	3.13
Formic acid	Water	46.0	0.830	298	1.52
Formic acid	Benzene	46.0	0.830	279	1.99
Formic acid	Benzene	46.0	0.830	298	2.58
Formic acid	Benzene	46.0	0.830	292	2.31
Formic acid	Carbon tetrachloride	46.0	0.830	282	1.61

TABLE 15.1 (Contd.)

Solute	Solvent	M_A	CMR_A	Temp/K	Diffusion coefficient $\times 10^9$ m^2 s^{-1}
Formic acid	Carbon tetrachloride	46.0	0.830	298	1.89
Formic acid	Carbon tetrachloride	46.0	0.830	288	1.67
Formic acid	Toluene	46.0	0.830	298	2.65
Formic acid	Toluene	46.0	0.830	287	2.46
Formic acid	Toluene	46.0	0.830	279	2.29
Furfural	Methanol	96.1	2.402	288	1.70
Furfural	Water	96.1	2.402	298	1.12
Glucose	Water	180.2	3.701	288	0.52
Glutaric acid	Water	132.1	2.874	298	0.79
Glycerol	Water	92.1	2.028	283	0.63
Glycerol	Water	92.1	2.028	298	0.93
Glycine	Water	75.1	1.663	298	1.06
Glycyl-leucine	Water	188.2	4.850	298	0.62
Histidine	Water	155.2	3.856	298	0.73
Hydrogen	Water	2.0	0.177	298	4.40
Hydroquinol	Methanol	110.0	2.995	288	1.25
Hydroxybenzylalcohol	Water	124.1	3.459	293	0.78
Hydroxy-proline	Water	131.1	3.030	298	0.83
Iodine	Benzene	253.8	2.790	293	1.95
Iodine	Carbon tetrachloride	253.8	2.790	298	1.50
Iodobenzene	Benzene	204.0	3.995	280	1.35
Iodobenzene	Methanol	204.0	3.995	288	1.65
Iodoform	Benzene	393.7	4.560	288	1.38
Iodoform	Methanol	393.7	4.560	288	1.33
Iodopropionic acid	Methanol	200.0	3.064	288	1.36
Iodopropylene(3)	Methanol	167.9	2.850	288	1.72
Isoamylalcohol	Benzene	88.2	2.649	288	1.48
Isoamylalcohol	Ethanol	88.2	2.649	293	0.78
Isoamylalcohol	Methanol	88.2	2.649	288	1.34
Isoamylalcohol	Water	88.2	2.649	288	0.69
Isobutanol	Water	74.1	2.186	298	0.93
Isobutyramide	Water	87.1	2.437	298	1.02
Isobutyric acid	Water	88.1	2.221	298	0.95
Isopropanol	Water	60.1	1.722	298	1.08
Lactic acid	Methanol	90.1	1.911	288	1.36
Lactose	Water	342.3	7.072	293	0.46
Leucine	Water	131.2	3.158	298	0.73
Leucyl-glycine	Water	188.0	4.850	298	0.61
Leucyl-glycyl-clycine	Water	245.0	6.182	298	0.55
Maltose	Water	342.3	7.072	293	0.43
Mannitol	Water	182.2	3.879	293	0.61

TABLE 15.1 (Contd.)

Solute	Solvent	M_A	CMR_A	Temp/K	Diffusion coefficient $\times 10^9\ m^2\ s^{-1}$
Methane	Water	16.0	0.641	298	1.67
Methanol	Water	32.0	0.794	288	1.28
Methanol	Water	32.0	0.794	298	1.66
Methyl cyclopentane	Water	84.1	2.783	298	0.93
Methyl iodide	Toluene	141.9	1.948	281	2.23
Methyl salicylate	Benzene	152.2	3.958	288	1.56
Neon	Water	20.2	0.280	298	3.01
Nicotine	Water	162.2	4.988	283	0.53
Nitrobenzaldehyde(3)	Methanol	151.1	3.397	288	1.24
Nitrobenzene	Acetone	123.1	3.414	293	2.94
Nitrobenzene	Benzene	123.1	3.414	288	1.84
Nitrogen	Water	28.0	0.560	298	1.99
Nitrogen	Water	28.0	0.560	295	2.02
Nitrogen monoxide	Water	30.0	0.550	298	2.34
Nitronaphthalene	Benzene	173.2	5.102	288	1.39
Nitrophenol	Methanol	139.1	3.567	288	1.38
Nitrous oxide	Water	44.0	0.827	289	1.54
Oxygen	Water	32.0	0.306	298	2.29
Oxygen	Water	32.0	0.306	298	2.60
Pentaerythritol	Water	136.2	3.109	293	0.69
Pentane	Water	72.2	2.496	298	0.97
Phenol	Benzene	94.1	2.842	278	1.27
Phenol	Methanol	94.1	2.842	288	1.40
Phenol phthalein	Methanol	318.3	8.956	288	0.78
Phenyl acetate	Methanol	136.2	3.805	288	1.62
Phenyl alanine	Water	165.2	4.638	298	0.71
Phthalic acid	Benzene	166.1	3.994	288	1.37
Phthalic acid	Methanol	166.1	3.994	288	1.30
Pimelic acid	Water	160.2	3.802	298	0.71
Proline	Water	115.1	2.877	298	0.88
Propane	Water	44.1	1.569	298	1.16
Propanol	Benzene	60.1	1.722	288	1.60
Propanol	Water	60.1	1.722	288	0.87
Propanol	Water	60.1	1.722	298	1.12
Propene	Water	42.1	1.543	298	1.44
Propionamide	Water	73.1	1.973	298	1.20
Propionic acid	Methanol	74.1	1.758	288	1.62
Propionic acid	Water	74.1	1.758	298	1.01
Propylene glycol	Water	76.1	1.875	298	1.00
Pyridine	Ethanol	79.1	2.478	293	1.12
Pyridine	Methanol	79.1	2.478	288	1.58

TABLE 15.1 (Contd.)

Solute	Solvent	M_A	CMR_A	Temp/K	Diffusion coefficient $\times 10^9\ m^2\ s^{-1}$
Pyrallol	Water	126.1	3.148	288	0.56
Quinone	Benzene	108.1	2.499	288	1.68
Raffinose	Water	504.5	10.443	293	0.36
Raffinose	Water	504.5	10.443	298	0.43
Resorcinol	Water	110.1	2.995	293	0.77
Salicylaldehyde	Benzene	122.1	3.341	288	1.78
Salol	Methanol	214.2	6.006	288	1.29
Serine	Water	105.1	2.279	298	0.88
Succinic acid	Water	118.1	2.410	298	0.86
Sucrose	Water	342.3	7.072	285	0.36
Sucrose	Water	342.3	7.072	298	0.52
Sulphur dioxide	Water	64.1	0.873	298	1.59
Tartaric acid	Water	150.1	2.716	288	0.61
Tetrachlorobenzene(1,2,4)	Benzene	215.9	4.654	281	1.24
Toluene	Water	92.1	3.152	298	0.95
Tribromoacetic acid	Methanol	296.8	3.625	288	1.23
Tribromophenol(2,4,6)	Methanol	330.8	5.173	288	1.12
Trichloroacetic acid	Methanol	163.4	2.768	288	1.45
Trichlorophenol(2,4,6)	Methanol	197.5	4.316	288	1.21
Trichloropropane(1,2,3)	Benzene	195.5	4.627	281	1.34
Triethylene glycol	Water	150.2	3.573	298	0.76
Triglycine	Water	189.0	4.326	298	0.67
Trimethylacetic acid	Water	102.1	2.685	298	0.82
Trinitrophenol(2,4,6)	Benzene	229.1	5.018	288	1.39
Trinitrophenol(2,4,6)	Methanol	229.1	5.018	288	1.32
Trinitroluene	Benzene	227.1	5.329	288	1.39
Trytophan	Water	204.2	5.666	298	0.66
Urea	Water	60.1	1.414	298	1.38
Urea	Water	60.1	1.414	285	0.99
Urethane	Methanol	89.1	2.126	288	1.41
Urethane	Water	89.1	2.126	288	0.87
Valeric acid	Water	103.1	2.685	298	0.82
Valine	Water	117.2	3.054	298	0.77
Vanillin	Methanol	152.2	3.958	288	1.00
Water	Water	18.0	0.331	298	2.45

15.3.2 Solvent properties

Table 15.2 consists of measures of viscosity (η_B, centipoise) and temperature (K) for each of the following solvents: water (101 values), methanol (6), carbon tetrachloride (5), ethanol (4), acetone (3), benzene (4) and toluene (5). CMR_B and M_B for the solvents are also given.

Again, the numbers in brackets represent the number of observations available for each solvent.

15.3.3 Models

15.3.3.1 Modelling diffusion coefficients

The Wilke–Chang model is

$$D = 7.4 \times 10^{-8}(\chi M)^{1/2}T/\eta V^{0.6}, \qquad (15.5)$$

where D is the diffusion coefficient (cm^2 s^{-1}); M is the molecular weight of solvent (g mol^{-1}); T is the absolute temperature (K); V is the molar volume (cm^3 mol^{-1}); χ is the solvent association parameter characteristic of each solvent; η is the solvent viscosity (cP).

The following model, suggested by analogy, is the initial starting point in our new analysis:

$$D = (A(M_{\mathrm{B}})^{\beta}T^{c})/(\eta_B^{b}(\mathrm{CMR}_A)^{n}), \qquad (15.6)$$

where A, β, b, c and n are constants to be estimated.

Taking logs of this expression gives the following linear model:

$$\log D = \log A + \beta \log M_B + c \log T - b \log \eta_B - n \log(\mathrm{CMR}_A). \qquad (15.7)$$

This model can now be fitted and the relevant coefficients estimated using least squares regression in SAS [11].

A comparison with the Wilke–Chang equation will be made in the Discussion.

The theory indicates that $D \propto T$ and $D \propto 1/\eta_B$, suggesting that the following models may also be appropriate:

$$D/T = (A(M_B)^{\beta})/(\eta_B^b(\text{CMR}_A)^n)$$
$$\Rightarrow \log(D/T) = \log A + \beta \log M_B - b \log \eta_B - n \log(\text{CMR}_A). \quad (15.8)$$

In this model we are simply constraining the coefficient of log(T) to be 1. Similarly,

$$D\eta_B/T = A(M_B)^{\beta}/(\text{CMR}_A)^n$$
$$\Rightarrow \log(D\eta_B/T) = \log A + \beta \log M_B - n \log(\text{CMR}_A). \quad (15.9)$$

In this model we are constraining the coefficient of log(T) to be 1 and log (η_B) to be -1. Again these are both linear models where the coefficients can be estimated using least squares.

Models including CMR_B and M_A have also been examined.

15.3.3.2 Modelling solvent viscosity

For much of the data for water as solvent, measured values of η_B at the stated temperature are available (see Table 15.2). However, a mathematical relationship between η_B and T would also be useful in the modelling, particularly in the case of solvents other than water, for which viscosity-temperature data is sparse. Alternatively, η_B could be ignored completely and the η_B term 'absorbed' into the term in T. An exponential model of the relationship between η_B and T will be reported below.

The models developed can be fitted for each solvent independently and for all the solvents simultaneously.

15.4 ANALAYSIS AND RESULTS

In all the tables of estimated coefficients in this section the values in brackets represent the standard errors of the coefficient estimates and * indicates the coefficient estimates were at least significant at the 95% level using a student t test.

15.4.1 Modelling viscosity

Many exponential models were fitted to the water viscosity data but the most successful of these was the following:

TABLE 15.2
Solvent Properties

Table 15.2.1 Water
$MW_B = 18.02$ $CMR_B = 0.331$

Temp/K	η/cP	Temp	η	Temp	η
273.0	1.7870	307.0	0.7340	341.0	0.4155
274.0	1.7280	308.0	0.7194	342.0	0.4098
275.0	1.6710	309.0	0.7052	343.0	0.4042
276.0	1.6180	310.0	0.6915	344.0	0.3987
277.0	1.5670	311.0	0.6783	345.0	0.3934
278.0	1.5190	312.0	0.6654	346.0	0.3882
279.0	1.4720	313.0	0.6529	347.0	0.3831
280.0	1.4280	314.0	0.6408	348.0	0.3781
281.0	1.3860	315.0	0.6291	349.0	0.3732
282.0	1.2460	316.0	0.6178	350.0	0.3684
283.0	1.3070	317.0	0.6067	351.0	0.3638
284.0	1.2710	318.0	0.5960	352.0	0.3592
285.0	1.2350	319.0	0.5856	353.0	0.3547
286.0	1.2020	320.0	0.5755	354.0	0.3503
287.0	1.1690	321.0	0.5656	355.0	0.3460
288.0	1.1390	322.0	0.5561	356.0	0.3418
289.0	1.1090	323.0	0.5468	357.0	0.3377
290.0	1.0810	324.0	0.5378	358.0	0.3337
291.0	1.0530	325.0	0.5290	359.0	0.3297
292.0	1.0270	326.0	0.5204	360.0	0.3259
293.0	1.0020	327.0	0.5121	361.0	0.3221
294.0	0.9779	328.0	0.5040	362.0	0.3184
295.0	0.9548	329.0	0.4961	363.0	0.3147
296.0	0.9325	330.0	0.4884	364.0	0.3111
297.0	0.9111	331.0	0.4809	365.0	0.3076
298.0	0.8904	332.0	0.4736	366.0	0.3042
299.0	0.8705	333.0	0.4665	367.0	0.3008
300.0	0.8513	334.0	0.4596	368.0	0.2975
301.0	0.8327	335.0	0.4528	369.0	0.2942
302.0	0.8148	336.0	0.4462	370.0	0.2911
303.0	0.7975	337.0	0.4398	371.0	0.2879
304.0	0.7808	338.0	0.4335	372.0	0.2848
305.0	0.7647	339.0	0.4273	373.0	0.2818
306.0	0.7491	340.0	0.4213		

TABLE 15.2 (Contd.)

Table 15.2.2 Methanol
$MW_B = 32.04$ $CMR_B = 0.794$

Temp	η
273	0.82
288	0.623
293	0.597
298	0.547
303	0.510
313	0.456

Table 15.2.3 Carbon tetrachloride
$MW_B = 153.82$ $CMR_B = 2.607$

Temp	η
273	1.329
288	1.038
293	0.969t
303	0.843
313	0.739

Table 15.2.4 Ethanol
$MW_B = 46.07$ $CMR_B = 1.258$

Temp	η
273	1.773
283	1.466
293	1.200
303	1.003

Table 15.2.5 Acetone
$MW_B = 58.08$ $CMR_B = 1.605$

Temp	η
273	0.399
288	0.337
298	0.316

Table 15.2.6 Benzene
$MW_B = 78.11$ $CMR_B = 2.689$

Temp	η
273	0.912
283	0.758
293	0.652
303	0.564

Table 15.2.7 Toluene
$MW_B = 92.14$ $CMR_B = 3.152$

Temp	η
273	0.772
290	0.610
293	0.590
303	0.526
313	0.471

$$\eta_B = a\exp(-(b/T)^3)$$

$$\text{therefore } \log \eta_B = \log a - (b/T)^3 = A - (b/T)^3, \qquad (15.10)$$

where A and b are coefficients to be estimated using least squares, and η_B is the viscosity in cpoise. The estimated coefficients were:

$$A = -2.43$$

$$b = 394.41 \text{ K}$$

and both estimates were significant using a 'studentised' t test.

The adequacy of this fit can be seen by examining Figure 15.1, showing the fitted line superimposed over the observed data. The R^2 value for this

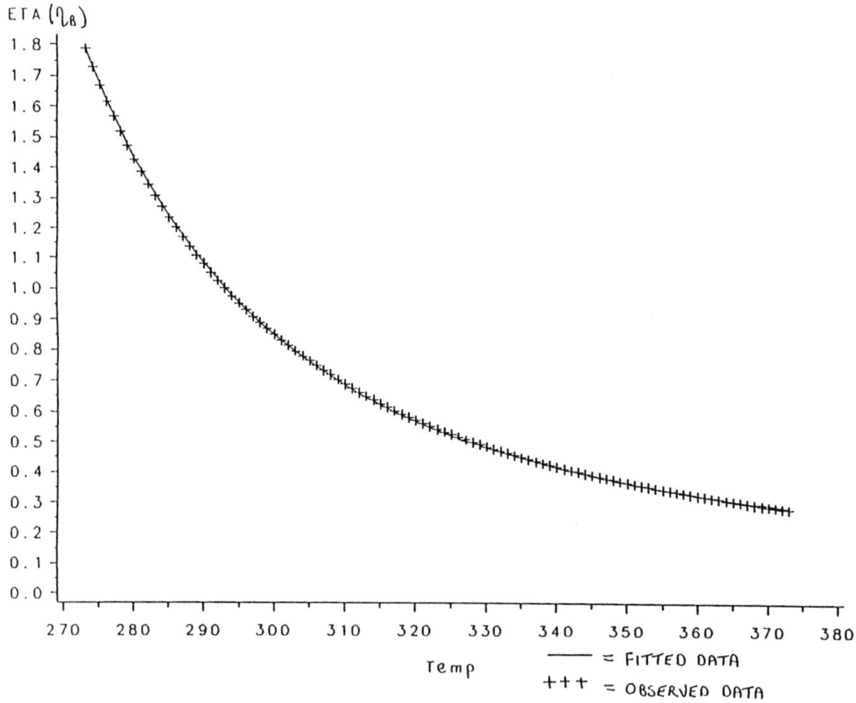

Fig. 15.1. Cubic fit of viscosity data for water.

model was 0.9998 indicating an excellent fit. The residual plot for this model shows a cyclic pattern. This is due to the serial correlation between the error terms in the model (this was checked using the Durbin Watson statistic, details of which are not given here). However given the small size of these residuals the fit is good enough for our purposes and the model is adequate, and to the authors' knowledge, is a new representation of viscosity data.

The model was assumed for the other solvents, considering the limited data available for these solvents. Also, using the same model provides consistency. The resulting coefficient estimates and R^2 values for these solvents are given below.

Solvent	A	b/K	R^2
Methanol	−1.94*	328.45*	0.9973
Carbon Tetrachloride	−1.44*	328.01*	0.9987
Ethanol	−1.54*	351.12*	0.9972
Acetone	−1.95*	275.19	0.9917
Benzene	−1.87*	330.82*	0.9997
Toluene	−1.71*	309.65*	0.9991

The estimated value for b for acetone, although not significant at the 5% level, was significant at the 10% level.

15.4.2 Modelling the diffusion coefficient D for water as solvent

15.4.2.1 Modelling D as a function of M_B, T, η_B, $(CMR)_A$

The model defined in Section 15.3.3 was fitted for water as solvent, under three sets of conditions. The first set was developed by ignoring viscosity η_B. Under this approach the model becomes:

$$\begin{aligned}\log(D) &= \log A + \beta \log M_B + c \log T - n \log(\text{CMR}_A) \\ &= \text{constant} + c \log T - n \log(\text{CMR}_A) \qquad (15.11)\end{aligned}$$

since $M_B = 18.02$ for water and is fixed.

The model was fitted using least squares regression which gave the following coefficient estimates and R^2 value:

constant	−41.12*	(3.21)
c	7.29*	(0.56)
n	0.54*	(0.01)
R^2	0.9582	

Although this model gives a high R^2, ignoring η_B is contrary to the previous work described in the Introduction.

The second model used observed viscosity η_B as an input. Since there was a reasonable amount of data for the water solvent we can include observed values of η_B in the model. Thus the model (with M_B fixed) becomes:

$$\log D = \text{constant} + c \log T - b \log \eta_B - n \log(\text{CMR}_A). \quad (15.12)$$

The least squares coefficient estimates and R^2 value for this model were:

constant	−140.61	(161.88)
c	24.80	(28.49)
n	0.54*	(0.01)
b	−2.42	(3.91)
R^2	0.9595	

Initial observation of a plot of the data and the R^2 value suggest that this is a good model. However, looking at the values of the constant, c and b and the size of their standard errors, the model is clearly inadequate and unable to separate terms.

The third approach is to use the exponential modelling of η_B. The estimated parameters A and b from the model defined above were used to generate estimated values for η_B. These were then used to model the diffusion coefficient. The coefficient estimates and R^2 values obtained were:

constant	−155.19	(186.39)
c	27.36	(32.80)
n	0.54*	(0.01)
b	−2.72	(4.45)
R^2	0.9593	

Again, the parameter estimates obtained are not stable, except for n. These studies suggest that examination of D as the dependent variable will not provide reliable models.

15.4.2.2 Modelling D/T as a function of η_B, CMR_A

In the next approach studied, we constrain the dependence of D to be proportional to temperature. The model was fitted in three ways, again for water as solvent. Firstly, ignoring η_B for

$$\log(D/T) = \text{constant} - n\log(\text{CMR}_A) \qquad (15.13)$$

$$R^2 = 0.91.$$

The model is inferior to other models (described below), at least in respect of correlation, but is potentially useful because of its simplicity. Next we use the observed viscosity η_B and the model now becomes:

$$\log(D/T) = \log A - b\log\eta_B - n\log(\text{CMR}_A). \qquad (15.14)$$

The coefficient estimates and R^2 value were:

constant	−5.38*	(0.01)
b	0.85*	(0.08)
n	0.54*	(0.01)
R^2	0.9583	

Examination of a graph, a high R^2 and the stability of the coefficient estimates suggests that this is a good model (the standard errors are small compared to the size of the coefficients). Finally, we consider exponential modelling. Estimated values for η_B were generated using the estimated values for A and b given above. These were then used to model D/T; the coefficient estimates and R^2 value were:

constant	−5.38*	(0.01)
b	0.85*	(0.08)
n	0.54*	(0.01)
R^2	0.9582	

We may now summarise the findings for the modelling of diffusion with water as solvent. Ignoring the viscosity term η_B in modelling both D and (D/T) gives reasonable fits, but the scientific validity of excluding this term is doubtful. Also, because η is a function of T, it is difficult to separate the effects of η and T on D. Modelling D/T is superior, providing stable models in agreement with existing theory. The estimate of n, the coefficient of CMR_A, is consistent (giving a value of 0.54) in all the models investigated.

15.4.3 Modelling the diffusion coefficient D for solvents other than water

Estimates of η_B were from the exponential model given above; there were insufficient viscosity values at the necessary temperatures to allow the use of measured values. Modelling D did not provide reliable models, just as described in the case of water as solvent, and is not reported further.

15.4.3.1 Modelling D/T as a function of M_B, η_B, CMR_A

Under this approach the model becomes, for a particular solvent:

$$\begin{aligned}\log(D/T) &= \log A + \beta \log M_B - b \log \eta_B - n \log(\mathrm{CMR}_A) \\ &= \text{constant} - b \log \eta_B - n \log(\mathrm{CMR}_A),\end{aligned} \tag{15.15}$$

since M_B is fixed for each solvent.

Least squares regression gives the results shown in Table 15.3. For methanol and ethanol the model gives imprecise estimates of b and n respectively.

15.4.3.2 Modelling $D\eta_B/T$ as a function of M_B, CMR_A, M_A, CMR_A

The introduction of a term in M_A brings about an element of instability in the model even though slight improvements in R^2 values can be seen. The exception to this is toluene where a significant increase in the R^2 value was found. The resulting coefficient estimates and R^2 values are shown in Table 15.4.

15.4.3.3 Modelling $D\eta_B/T$ as a function of M_B, CMR_A

Modelling for all solvents separately was done using generated η_B values, and gives the results shown in Table 15.5, for

TABLE 15.3
Modelling D/T for solvents: regression coefficients
For the model

$$\log(D/T) = \text{constant} - b \log \eta_B - n \log(\text{CMR}_A)$$

Solvent	Constant	n	b	R^2
Methanol	−4.84* (B)	0.38*	0	0.46
	(0.07)	(0.05)	(−)	
Carbon tetrachloride	−5.21*	0.47	0.76	0.81
	(0.06)	(0.07)	(0.31)	
Ethanol	−12.58*	0.35	−41.13*	0.94
	(1.40)	(0.31)	(7.43)	
Acetone	−5.35*	0.22*	0.79*	0.89
Benzene	−5.03*	0.32*	0.46*	0.62
	(0.09)	(0.04)	(0.21)	
Toluene	−4.92*	0.35*	0.19	0.76
	(0.16)	(0.06)	(0.31)	

For the above table (and those following) (B) indicates the coefficient estimate is a biased estimate.
* Means that the term is significant at 95% confidence.

TABLE 15.4
Modelling $D\eta_B/T$ as a function of M_B, CMR_A, M_A, CMR_A

$$\log(D\eta_B) = \text{constant} - \beta_2 \log M_A + n_2 \log(\text{CMR}_A)$$

The resulting coefficient estimates and R^2 values were:

Solvent	Constant	β_2	n_2	R^2
Water	−5.01*	0.10*	0.44*	0.96
	(0.11)	(0.03)	(0.03)	
Methanol	−5.46*	−0.04	0.41*	0.46
	(0.34)	(0.08)	(0.09)	
Carbon tetrachloride	−7.19*	−0.49*	0.91*	0.97
	(0.26)	(0.06)	(0.06)	
Ethanol	−5.51*	−0.01	0.43	0.43
	(1.23)	(0.31)	(0.54)	
Acetone	−5.77*	−0.05	0.26*	0.89
Benzene	−5.48*	−0.06	0.36*	0.56
	(0.26)	(0.06)	(0.06)	
Toluene	−8.79*	−0.87*	1.05*	0.93
	(0.49)	(0.12)	(0.10)	

TABLE 15.5
Modelling $D\eta_B/T$ for all solvents: regression coefficients
$\log(D\eta_B/T) = \text{constant} - n\log(\text{CMR}_A)$

Solvent	Constant	n	R^2
Water	−5.39*	0.54*	0.94
	(0.01)	(0.01)	
Methanol	−5.29*	0.38*	0.45
	(0.07)	(0.05)	
Carbon tetrachloride	−5.21*	0.49*	0.84
	(0.05)	(0.07)	
Ethanol	−4.88*	0.44	0.05
	(1.21)	(1.03)	
Acetone	−5.88*	0.22*	0.88
	(0.02)	(0.02)	
Benzene	−5.23*	0.31*	0.55
Toluene	−5.31*	0.36*	0.68
	(0.07)	(0.07)	

$$\log(D\eta_B/T) - \text{constant} - n\log(\text{CMR}_A).$$

The model is acceptable for all solvents except ethanol, which may be due to only five data points being available.

15.4.4 Modelling the diffusion coefficient D for all solvents simultaneously

As an alternative to modelling each solvent individually, an obvious approach is to model all the solvents simultaneously. In order to do this, estimated values for the viscosity η_B were generated using the coefficients given above. The models for D as a function of M_B, T, η_B, CMR_A, and D/T as a function of M_B, η_B, CMR_A are similar to those found for water alone, but the use of D/T provides smaller standard errors in the coefficients.

Several approaches are possible, and therefore two examples are given first in order to illustrate some of the possibilities, and then a third, simpler, model is shown. Thus, for the model

$$\log(D/T) = \log A + \beta \log M_B - b \log \eta_B - n \log(\mathrm{CMR}_A). \quad (15.16)$$

The coefficient estimates and R^2 value were:

$\log A$	-5.95^* (0.06)
β	0.18^* (0.02)
n	0.42^* (0.02)
b	0.93^* (0.04)
R^2	0.83

Figure 15.2 shows the observed and fitted values for this model. Statistical evidence points to this being a good model. We note that the term in η_B is close to -1; if we set it at -1 and consider the inclusion of CMR_B, we find an element of instability in the model if M_A is included. Therefore,

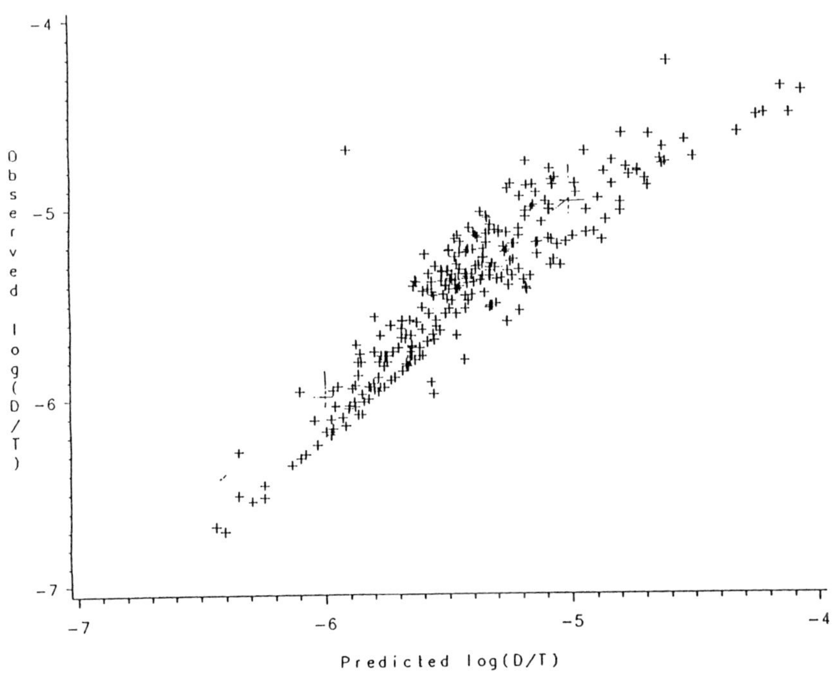

Fig. 15.2. Model fit of log(D/T), all solvents, based on M_B, B, CMR_A.

in order to examine the significance of the CMR_B term, if M_A is excluded we obtain

$$\log(D\eta_B/T) = \log A + \beta \log(M_B) + n_1 \log(CMR_B) - n_2 \log(CMR_A). \qquad (15.17)$$

The resulting coefficient estimates and R^2 value were:

$\log A$	−4.35*	(0.26)
β	0.25*	(0.07)
n_1	0.33*	(0.05)
n_2	0.44*	(0.01)
R^2	0.77	

Figure 15.3 shows the observed vs. fitted values for this model. There

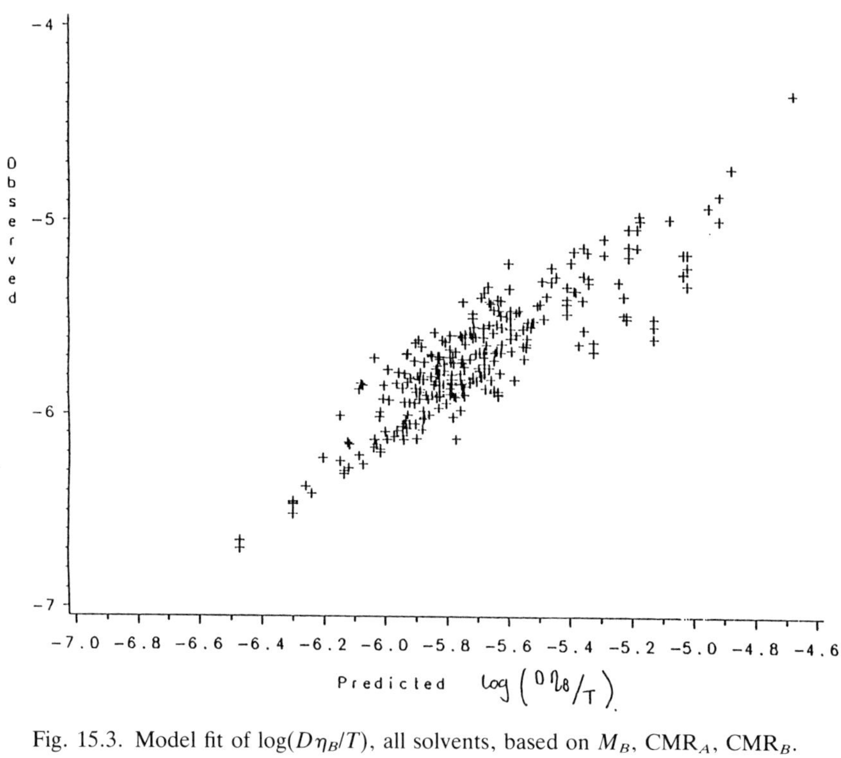

Fig. 15.3. Model fit of $\log(D\eta_B/T)$, all solvents, based on M_B, CMR_A, CMR_B.

seems to us no obvious reason for the terms in CMR_B and M_B to be significant, but they may express some physical factor which η_B alone does not.

However, three solvent properties seems unnecessary, when CMR_B and M_B are correlated, and therefore considering all solvents simultaneously via a simpler model

$$\log(D\eta_B/T) = \log A + \beta \log M_B - n \log(\mathrm{CMR}_A), \qquad (15.18)$$

the coefficient of $\log(T)$ is constrained to be 1 and the coefficient of $\log(\eta_B)$ to be -1, as suggested by the theory. The resulting coefficient estimates and R^2 value were

$\log A$	-5.93^*	(0.06)
β	0.17^*	(0.02)
n	0.42^*	(0.02)
R^2	0.70	

Figure 15.4 shows the observed and fitted values. Examination of Figure 15.4, the small size of the confidence intervals of the coefficients, and the relatively high R^2 value indicate that this is a good model; however, a problem area has been found on detailed examination of the results, which is indicated in the next section.

15.4.5 Correlation between residuals and chemical class

During inspection of the residuals from linear regression, it was noticed that there was some correlation between residual value and the chemical class of solute. Each observation can be categorised as pertaining to a *sugar*, *acid*, *small* molecule or *other* molecule. These classifications reflect the composition of the source data.

For the simplest model,

$$\log(D\eta_B/T) = \text{constant} - n \log(\mathrm{CMR}_A)$$

residuals are found to be clustered, and none of them centred at the origin.

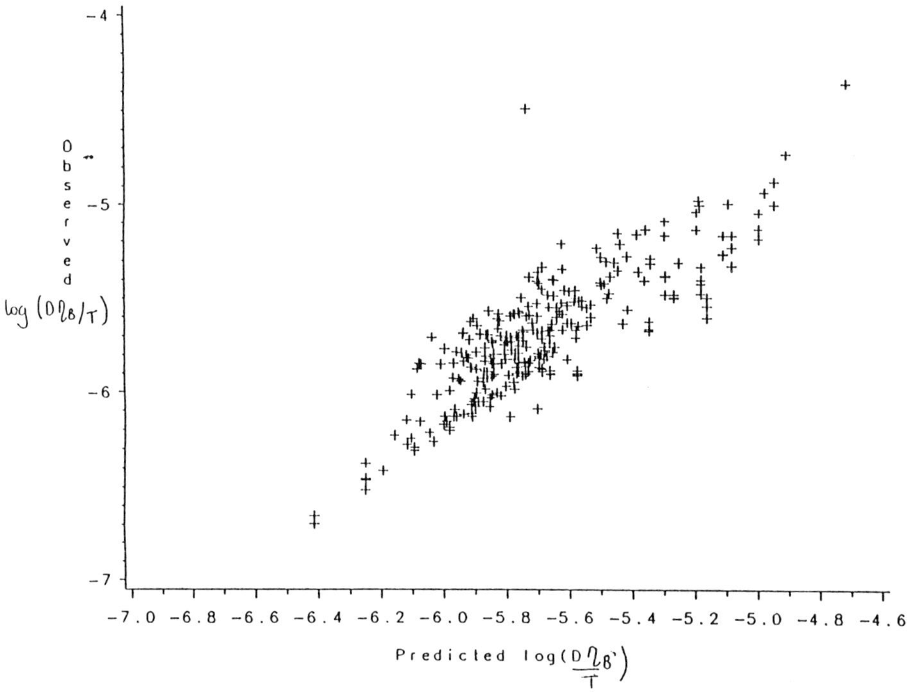

Fig. 15.4. Model fit of $\log(D\eta_B/T)$, all solvents, based on CMR_A and M_B.

Class	Mean residual
Sugar	−0.35
Acid	−0.12
Small	0.02
Other	0.09

Therefore, the difference between chemical classes appears to be important. The results show that for sugars, observed values are smaller than predicted; this applies to all models given herein. Before considering this further, we should consider whether this problem (or a similar one) exists within the Wilke–Chang model [3].

The Wilke–Chang model in its final form is

$$D = 7.4 \times 10^{-8}(\chi M)^{1/2}T/\eta V^{0.6} \quad (15.5)$$

There are several similarities to the results of this work. In log form, the Wilke–Chang equation is

$$\log(D\eta_B/T) = \text{constant} + 0.5 \log M_B - 0.6 \log V_A + 0.5 \log \chi. \quad (15.20)$$

This relationship is clearly related to equation (15.18) in particular. The term in $\log \chi$ is of note. In the original reference [3] very limited justification for its inclusion was made. Furthermore, the values of χ quoted in Reference [3] were obtained from then unpublished work, and effectively have the nature of an empirical correction, which in itself is not a major problem in the utility of the model. However, when the work was published the values of the parameter were quite different [12], yet this error in the Wilke–Chang equation has not been corrected in the literature, to the authors' knowledge.

The importance of the factor χ is of interest; this is an adjustable parameter for each solvent, considered by Wilke and Chang to account for the aggregation of the solvent molecules. In fact, it is likely that the effect of aggregation (mutual association of solvent molecules) has, in fact, already been accounted for in the viscosity term η_B. So what is the significance of the χ value? In their data set, for the classes 'sugar', 'acid' and 'small', water is usually the solvent, which has χ greater than 1. Therefore the χ value artificially increases the predicted D, which would otherwise be predicted too small (and appears to make physical sense only if it is considered as a correction to the viscosity term).

It is concluded from this work that the trend in residuals we observe in our analysis is not present in the analysis of Wilke and Chang. However, the investigation of the literature has shown that the empirical correction they use has not been properly justified, and that users of the equation of Wilke and Chang need to be aware that it is less firm in its basis than has been supposed.

A possible explanation of the fact that our models predict D values that are larger than observed is now presented. Sugars and carboxylic acids form strong hydrogen bonds. If, for example, six water molecules are hydrogen-bonded to the diffusing sugar, the CMR value would be effectively too low:

CMR of glucose 3.70
CMR of glucose plus waters 5.69
$\Rightarrow \log_e$ of the ratio of CMR values is 0.43.

This compares to the absolute residual of 0.35, and thus provides a plausible explanation of the origin of the residuals.

The same argument applies to acids. The problem now is to address how to quantify the proposed effect of the hydrogen bonded water molecules.

15.5 DISCUSSION AND CONCLUSIONS

Based on the theory section and consideration of the statistics, the models of most interest are

(i) Fitting all solvents simultaneously:

$$\log(D/T) = -5.95 + 0.18 \log M_B - 0.93 \log \eta_B - 0.42 \log(\mathrm{CMR}_A), \quad (15.14)$$

which allows the coefficient of η_B to vary.

(ii) A very simple model, fitting each solvent independently, is:

$$\log(D\eta_B/T) - \text{constant} - n \log(\mathrm{CMR}_A).$$

The coefficient estimates for each solvent are given in Table 15.6.

(iii) Fitting all the solvents simultaneously provides a robust model:

$$\log(D\eta_B/T) = -5.93 + 0.17 \log M_B - 0.42 \log(\mathrm{CMR}_A). \quad (15.18)$$

It is possible that this model can be extrapolated to solvents outside the test set. An important model which is an extension of (ii) above and which has been shown to give a good fit is:

(iv)
$$\log(D\eta_B/T) = -4.35 - 0.25 \log(M_B) + 0.33\, log(CMR_B) - 0.44 \log(\mathrm{CMR}_A). \quad (15.17)$$

Inclusion of the CMR_B term results in a negative coefficient for M_B;

although this disagrees with previous statistical modelling, it is nonetheless statistically meaningful and not in contradiction to the theory of diffusion.

The established Wilke–Chang model in log form is

$$\log(D\eta_B/T) = \text{constant} + 0.5 \log M_B - 0.6 \log V_A + 0.5 \log \chi. \quad (15.20)$$

The term in $0.5 \log \chi$ is scientifically suspect; in other respects the equation is similar to equation (15.18). The present study provides an understanding of diffusion that is statistically more rigorous than previous ones. The lack of data prevents a thorough analysis for solvents other than water. However, the similarity of the coefficients found when separate regression for each solvent are carried out suggests that a combined solvent model is satisfactory.

In future it is necessary that collection of D values outside the original set needs to be performed in order to test the new models more fully. Outliers have been identified which diffuse more slowly than predicted, and a possible cause has been identified. The hydrogen-bonded water molecules directly attached to a diffusant molecule may be firmly bound and effectively 'moving with' it, or at least be increasing its effective size.

References

[1] J. Crank, N.R. McFarlane, J.C. Newby, G.D. Paterson and J.B. Pedley, Diffusion Processes in Environmental Systems (MacMillan, London, 1981), 1.
[2] C.R. Wilke, Chemical Engineering Progress 218 (1949).
[3] C.R. Wilke and P. Chang, A.I.Ch.E. Journal 1 (1955) 264.
[4] H. Eyring, J. Chem. Phys. 4 (1936) 283.
[5] S. Glasstone, Textbook of Physical Chemistry (MacMillan, London, 1940) page 255)).
[6] W.J. Lyman, W.F. Reehl and D.M. Rosenblatt, Handbook of Chemical Property Estimation Methods, (McGraw Hill, New York, 1982).
[7] S. Glasstone, Textbook of Physical Chemistry (MacMillan, London, 1940) page 518.
[8] A. Leo and D. Weininger, MedChem Software Manual Release 3.5, Medicinal Chemistry Project (Claremont, California).
[9] G.H. Jeffery, R. Parker and A.I. Vogel, J. Chem. Soc. (1961) 570.
[10] W. Hayduk and H. Laudie, A.I.Ch.E.J. 20 (1974) 611.
[11] SAS Institute Inc, Box 8000 Cary, North Carolina 27511–8000.
[12] B. Jacobson, Acta Chem. Scand. 9 (1955), 999.

Chapter 16

Electrode Reactions in Microvolumes

CHRISTOPHER M.A. BRETT

16.1 INTRODUCTION

Miniaturization is currently an important theme in many areas of science and technology. Apart from the benefits from the lesser space required, new and more detailed types of information may become available. This is the case of electrode reactions in microvolumes, as will be seen in this chapter.

Recent advances in the fabrication of electrodes, in electrochemical instrumentation and in associated equipment have made the local probing of electrode reactions possible with high temporal and spatial resolution. Such miniaturization makes many, hitherto unavailable, experimental systems amenable to detailed study. The experimental strategy for such probing can be divided into two types of approach, as shown in Fig. 16.1. The first uses microelectrodes for making localised measurements of species within a microvolume portion close to the electrode surface of a large volume of solution. The second involves measuring electroactive species in solutions whose total volume is in the microlitre range. These two approaches have a number of similarities and features in common and can be usefully combined in many situations. Both will be reviewed and some of the exciting new possibilities indicated.

This chapter deals with electrode reactions in microvolumes, which we define as volumes of solution of 100 μL or less, either total volume or microvolume contained within a larger quantity of solution. Particular attention will be paid to the strategies used to obtain experimental results under such conditions. The theoretical analysis of the results leading to the relevant thermodynamic, kinetic and mechanistic parameters can often follow the methodologies applicable for electrode reactions in larger solution volumes; such treatments are available elsewhere, e.g., [1]. Emphasis

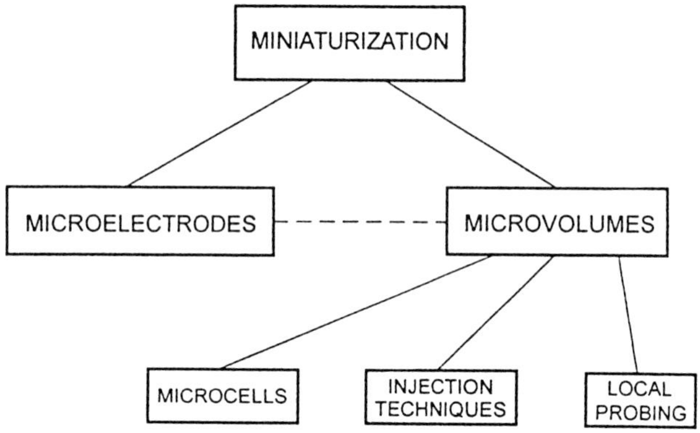

Fig. 16.1. Miniaturization strategies for the study of electrode reactions.

will be placed on recent developments in microlitre injection techniques, which, as will be shown, have many similarities with hydrodynamic electrodes in flow systems.

16.2 ELECTROCHEMISTRY IN ELECTROCHEMICAL CELLS OF SUB-MICROLITRE VOLUME

Advances in technology have permitted the miniaturization of many electrochemical systems [2]. In particular, application of modern fabrication techniques permits the manufacture of electrochemical cells of extremely small volume, Fig. 16.2. The first step in this direction was probably the development of thin-layer electrochemical cells. These contain "large" (2–3 mm) electrodes in a rectangular cell of dimensions typically 2 cm × 1 cm but of small height produced by a spacer of thickness 2–100 μm, which corresponds to a cell volume of 4–200 μL [3]. Cell volumes have been progressively reduced, so that now measurements often have to be made with microelectrodes.

One approach is to use free droplets on surfaces which contain one or all three electrodes [4–10], Fig. 16.2(a). A droplet of solution (5–10 μL) can be placed on the substrate which contains the electrode and the other electrodes positioned within the solution from above. In some cases the droplet can be pulled across the surface, by judicious movement of the

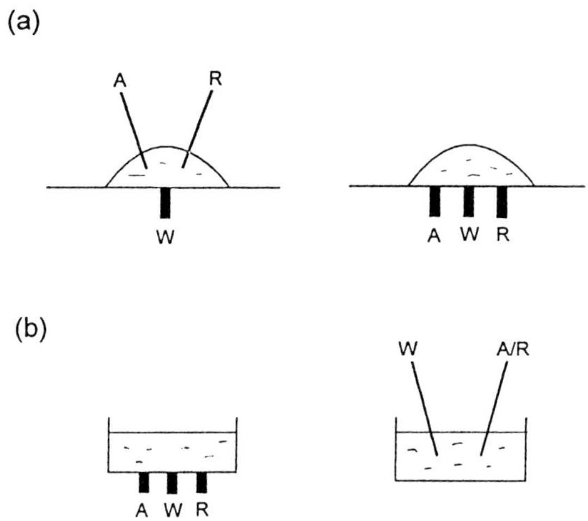

Fig. 16.2. Schematic microcell designs: W = working electrode, A = auxiliary electrode, R = reference electrode some designs use a 2-electrode cell. (a) Droplet cells with solution on a surface containing (i) one or (ii) three microelectrodes, (b) Microchamber cells with electrodes (i) on the walls or (ii) in solution.

electrodes in solution the droplet serving as electrolyte to probe the reactions occurring over a larger area of the electrode surface itself, such as demetallization reactions [4]. On the other hand, the droplet can be placed so as to completely cover a disk microelectrode; these have been used, for example, for the study and fast anodic stripping of heavy metals in 5 μL droplets [5] and for the fast cathodic stripping of anions in 10 μL samples [6]. A variant on this theme, but for larger volumes of the order of 20 μL, involves placing the droplet over a ring electrode and rotating the sample by a tangentially-applied stream of inert gas in order to produce convective transport and either increase the sensitivity through reducing the diffusion layer thickness, as in the rotating disc electrode system, or permit electrolysis to be carried out more quickly [7]. This was applied to preconcentrate mercury in the stripping analysis of mercury ions [7]. Alternatively, a droplet, of volume as little as 0.02 μL, can be applied over three microband electrodes on an inert surface [8–10]. In a lithographically-produced microcell surface, requiring 2 μL of solution in the droplet, the system was tested successfully with $Ru(NH_3)_6^{3+}$ reduction at a gold electrode [8]. A simpler constructional approach used the polished cross-section of a multilayer sandwich of metal foils (platinum, gold or silver) for the electrodes, with heat sealing film spacers and glass slides [9, 10]; the droplet cells, of volume as low as 0.02 μL were tested with various electroactive species including the oxidation of ferrocene in dimethylformamide.

The use of micromachining techniques to manufacture microchamber cells, Fig 16.2(b), rather than using droplets, can be accomplished by means such as photolithography, vacuum deposition, anisotropic etching and fusion bonding. These techniques were already used several years ago to fabricate cells of 2 μL volume in stationary mode [11] and also as a flow cell on a silicon substrate for glucose monitoring [12]. More recently a 3-electrode cell of subnanolitre dimensions, a microchamber of 200 μm diameter and 20 μm depth, has been made by lithographic patterning of metals and photopolymerizable polyimide in order to investigate the response of single living cells to hormones [13]; this required the inclusion of working and pseudo-reference electrodes at the bottom of the chamber to serve as complete electrochemical cell. In other work [14], picolitre microvials have been fabricated in polystyrene by using a lithographically generated silicon-based structure as mould, and then transferring the pattern to the polystyrene by hot pressing; the electrochemical response was characterised using ferrocenecarboxylic acid. These methods open up fantastic new possibilities since in this way one living cell can be placed in the chamber and the electrochemical response of this single cell monitored. Nevertheless, the problem of solvent evaporation must be conveniently addressed, which is a general difficulty with electrochemical cells of such tiny volume.

For the probing of these microenvironments, appropriately-sized electrochemical sensors and electrodes are necessary, often of nanometre linear dimension, unless they can be incorporated in the walls of the electrochemical cell. By etching, disc microelectrodes with radii of as small as 2 nm have been fabricated [15]; nevertheless, problems of interelectrode reproducibility can be large at such miniature electrodes.

16.3 MICROVOLUMES AND SEPARATION TECHNIQUES

Real solution samples often consist of mixtures of electroactive species in complex matrices, which usually requires the application of separation techniques before an appropriate detector can be employed. To be efficient, the detector cell must have a volume similar to that of a single theoretical plate of the system, which can reach below 1 nL in capillary HPLC and capillary electrophoresis. It can be seen from the developments described in the previous section that such micro- and nanovolume detection can be done by using appropriate electrochemical detector cells containing microelectrodes, as has been amply demonstrated [16]. The injected sample volumes can be in the microlitre or submicrolitre range.

The information that can be obtained with electrochemical detectors is not restricted to quantification. Instead of the conventional use of electrochemical detectors in amperometric mode at fixed potential, electrode arrays with each electrode held at different values of fixed potential can be used, in order to build up chronovoltammograms, three-dimensional current-voltage-time profiles. A 32-microband electrode array has been described for this purpose and applied to phenolic compounds [17] and which permits studying the electrode reaction mechanism at the same time as identification and quantification are carried out. Alternatively, fast voltammetric techniques such as fast-scan cyclic voltammetry or square wave voltammetry can be used to create chronovoltammograms of the eluted components.

In normal high pressure liquid chromatography, typical sample volumes are 20–200 μL; this can become as little as 1 nL in capillary HPLC. Pretreatment of the sample may be necessary in order to protect the stationary phase in the column from deactivation. By employing supercritical fluids such as carbon dioxide, pretreatment can be bypassed in many instances so that whole samples from industrial and environmental matrices can be introduced directly into the column. This is due to the fact that the fluid acts as both extraction solvent and mobile phase. Post-column electrochemistry has been demonstrated. For example, fast-scan cyclic voltammograms have been recorded as a function of time after injection of microgram samples of ferrocene and other compounds in dichloromethane solvent and which are eluted with carbon dioxide at pressures of the order of 100 atm and temperatures of ~50°C; the chromatogram is constructed as a plot of peak current vs. time [18].

An alternative approach, possibly better suited to tiny volumes, is that of capillary electrophoresis [19], which is highly efficient and fast, and has low operational cost. In capillary electrophoresis, an open-ended capillary, filled with the sample solution, links two buffer solutions. Voltages of the order of 10–40 kV are applied across the ends of the capillary which leads to electrophoretic separation of the charged species due to the electric field, the rate of movement depending on their charge-to-size ratios. This is superimposed on the electroosmotic flow of solvent, which can sometimes cause all solutes to elute at the cathodic end of the capillary; thus, sample introduction is done at the anodic end. Typically, capillaries, of glass or fused silica, have diameters in the region of 2–200 μm and length 10–100 cm, corresponding to a volume in the picolitre to nanolitre range, although there are reports of injections into the capillary in the femtolitre (10^{-15} L) and attolitre (10^{-18} L) ranges [20]. A detector, essentially a modified HPLC detector, is placed directly at one end of the capillary; the main difficulty is the exact positioning of the detector relative to the capillary. With respect to electrochemical amperometric and voltammetric detectors [16], this can be solved in several ways [21, 22]: by using an electrode much larger than the capillary diameter placed directly against the end of the capillary, by inserting a carbon fibre microelectrode into the end of the capillary, or by depositing a metal film, which acts as electrode, directly onto the capillary tip.

16.4 INJECTION OF MICROVOLUMES

Many injection techniques involve injection, typically of microlitre volumes, into a carrier stream. Measurements are carried out whilst the sample plug containing the electroactive species of interest is flowing over the electrode. This means that much of the theory for continuous flow systems can be applied, with appropriate modifications, to the theoretical response in these cases. In this section, appropriate ways of carrying this out will be described.

The fundamentals of the electrochemical response at electrodes operating in a regime of forced convection, hydrodynamic electrodes, and the information that can be obtained have been reviewed [23, 24]. Some of these electrodes are good candidates for direct introduction into flow systems, in particular tube/channel electrodes and impinging jet (wall-jet and wall-tube) electrodes. Particular practical advantages of these flow-past hydrodynamic electrodes are that there is no reagent depletion while the sample plug passes the electrodes, and there is no build-up of unwanted intermediates or products. Recent advances in instrumentation also mean

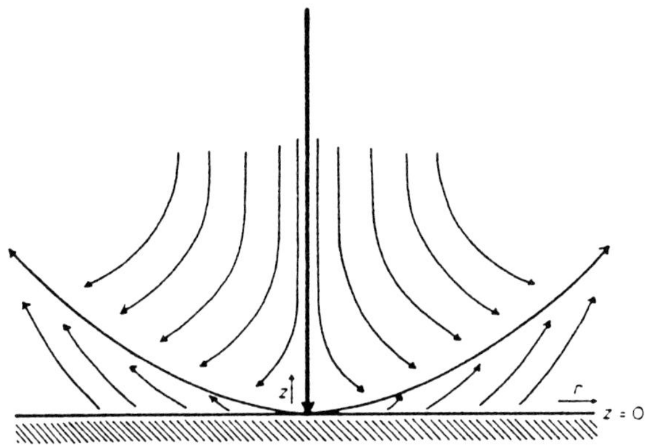

Fig. 16.3. Schematic streamlines at a wall-jet electrode illustrating the highly non-uniform accessibility of the electrode surface and that the electrode surface does not receive any solution from fluid recirculation (from Ref. [23]).

that kinetic and mechanistic, besides analytical, information can be obtained simultaneously in many experiments.

Of these hydrodynamic electrodes, the wall-jet electrode has received a lot of attention, e.g., [25, 26], partly because of its portable nature and its particular characteristics. In a wall-jet electrochemical cell, a fine jet of solution impinges perpendicularly on the centre of a disc electrode and then spreads out radially. Schematic streamlines corresponding to this situation are shown in Fig. 16.3. An important advantage that can be deduced from Fig. 16.3 is that only fresh solution from the incoming jet reaches the electrode surface—nothing from recirculation. This means that the internal volume of the wall-jet cell is irrelevant, the only important geometrical parameter being the volume (or volume flow rate) of the jet of incoming solution. Typical jet diameters are 0.3–0.5 mm and electrode diameters 3.0–5.0 mm. Note that a wall-tube cell corresponds to the situation where the jet is the same size or larger than the electrode thus guaranteeing uniform accessibility.

The wall-jet electrode is highly non-uniformly accessible as shown by the streamlines and by the equation for the limiting current, I_L

$$I_L = 1.38nFD^{2/3}\nu^{-5/12}a^{-1/2}V_f^{3/4}R^{3/4}c_\infty$$

in which D is the diffusion coefficient of electroactive species, of concentration c_∞, a is the diameter of the jet which impinges on the disc electrode of radius R at volume flow rate V_f, and ν is the kinematic viscosity of the solution. The non-uniform accessibility—the diffusion layer thickness falls off with the (5/4) power of the radial distance—and high flow rate dependence can be extremely useful in applications involving the study of the kinetics of simple and complex electrode reactions. Analytical expressions to deduce the kinetics from the voltammetric profile for a simple irreversible reaction have been deduced [27, 28], including an approximate Koutecky–Levich-type relation [28], which works best in the zone of the half-wave potential. Given the impossibility of obtaining accurate analytical expressions, a computational procedure has also been employed [29]; the exploitation of the highly non-uniform accessibility for better discriminating between electrode reaction mechanisms than at uniformly accessible electrodes has been demonstrated [30].

The wall-jet continuous flow system is readily modified to discrete injection mode as shown in Fig. 16.4. This is the electrochemical version of a general discrete injection technique known as batch injection analysis (BIA) [31] and is a hybrid between flow injection analysis and wall-jet in continuous flow.

In flow injection analysis [32] with electrochemical detection a sample is injected into an electrolyte carrier stream; dispersion of the sample plug into the carrier stream occurs so that electrolyte is effectively "added" to the sample—with consequent sample dilution—before reaching the electrode. Even so, by using a capillary flow injection system nanolitre sample volumes can be investigated [33]. In continuous flow systems, electrolyte often has to be added to the sample beforehand, also leading to sample dilution.

One of the benefits of electrochemical batch injection analysis is that dilution of the sample with electrolyte is not necessary, see below. A sample of volume $\leq 100\ \mu L$ is injected directly from a micropipette, tip diameter ~ 0.5 mm, over the centre of a macroelectrode exactly as in a wall-jet system. This is equivalent to a flow injection system with zero dispersion. During the injection, and after a short initial period to reach steady-state, the hydrodynamics is wall-jet type and a time-independent current is registered. BIA was first devised using amperometric, e.g., [31], and potentiometric, e.g., [34], detection. A typical amperometric trace is shown in Fig. 16.5. By using a programmable, motorised electronic

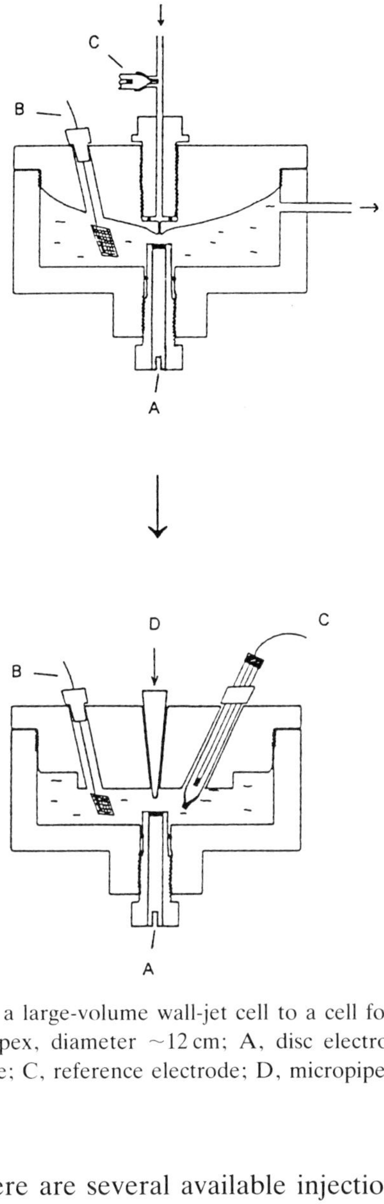

Fig. 16.4. Modification of a large-volume wall-jet cell to a cell for batch injection analysis (BIA). Cell body of perspex, diameter ~12 cm; A, disc electrode contact; B, auxiliary electrode; C, reference electrode; D, micropipette tip.

micropipette [35], there are several available injection flow rates and the possibility of programming series of consecutive injections, of equal or different volumes, up to the maximum total micropipette volume. Volume accuracy and reproducibility are excellent.

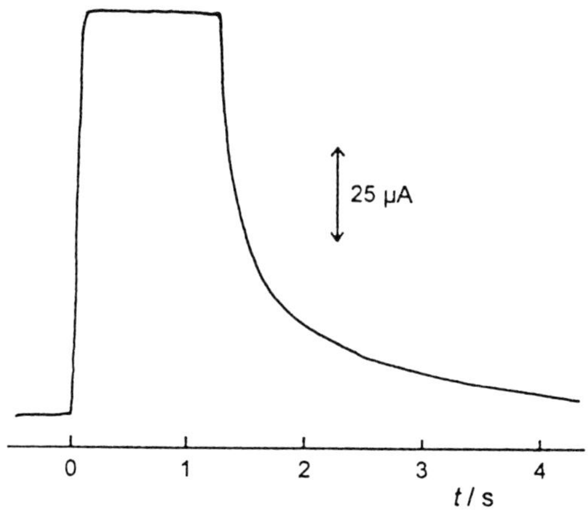

Fig. 16.5. BIA chronoamperometric transient for the oxidation of 2 mM $K_4Fe(CN)_6$ in 0.4 M K_2SO_4 electrolyte: injection of 100 μL onto Pt disc electrode (d = 3.28 mm) at +0.6 V vs. SCE. Injection flow rate 75.3 μL s^{-1}, micropipette tip internal diameter 0.47 mm. $t = 0$ corresponds to the start of the injection period.

During the injection, when the steady state has been reached, wall-jet theory can be applied to deduce information concerning the magnitude of the current response. Wall-jet hydrodynamics has been verified [36]. This wall-jet flow characteristic is extremely important, because it means that the response to the injection of a sample does not depend on the solution composition close to the electrode before the injection, i.e., there are no carry-over effects in solution, see Fig. 16.3. Additionally, the electrical resistance of a sample without added electrolyte is not a significant contribution to the total cell resistance because it is only a thin layer of solution close to the working electrode. This has also been verified [37] and means that sample dilution is not necessary, as mentioned above; besides the simplification this brings, it also avoids any change in the solution speciation due to adding electrolyte.

With this experimental arrangement, a number of voltammetric techniques become accessible through BIA, offering the possibility of obtaining information easily and without sample preparation or use of complex manifolds or sample dilution. Additionally, since the contact time of the sample with the electrode is only a maximum of several seconds, problems from poisoning of the electrode surface by components in the sample matrix are also less.

Examples of experiments which can be carried out are shown in Figs. 16.5 and 16.6 can be summarised as:

- simple amperometric measurement, for example applying a potential equal to that corresponding to the limiting current (Fig. 16.5) [36];
- recording of a point-by-point pseudo-steady-state voltammetric curve by consecutive injections during a slow linear potential scan (Fig. 16.6(a)), followed by Tafel analysis or curve fitting [37] (see also [27, 28];
- cyclic voltammetry during or after sample injection to determine the kinetics of the electrode reaction and/or concentration determination (Fig. 16.6(b)) [37];
- square wave voltammetric scan during the injection to give kinetic and analytical information (Fig. 16.6(c)) [37].

The detection limit with these simple types of experiment is 2×10^{-5} M for amperometric or pseudo-voltammetric detection down to 5×10^{-6} M for square wave voltammetry. Fitting of the curves obtained leads to the deduction of the relevant rate constants.

Use can also be made of measurement of the charge under the curve. Figure 16.7 shows examples of chronoamperometric and chronocoulometric transients for the three different possible flow rates. The charge under the chronamperometric curve is highest in the case of the lowest flow rate, a very significant contribution coming from the relaxation part of the transient when the injection has finished; this means that low flow rates are best for coulometric experiments since the percentage electrolysis is higher. The reason for this can be traced to the fact that the higher the flow rate the more the injected sample containing electroactive species will be pushed away from close to the electrode, and will be unable to diffuse back after the injection. Similar reasoning can explain why, in coulometric experiments, the signal hardly increases for injected volumes greater than 50 μL. Thus, a signal enhancement by a factor of 2.5 can be obtained by injecting 25 μL four times rather than 100 μL once. Examples in which the optimisation of the coulometric approach has been applied with success is in stripping voltammetry, where the preconcentration step is either coulometric or the equivalent in terms of adsorptive accumulation [38–40].

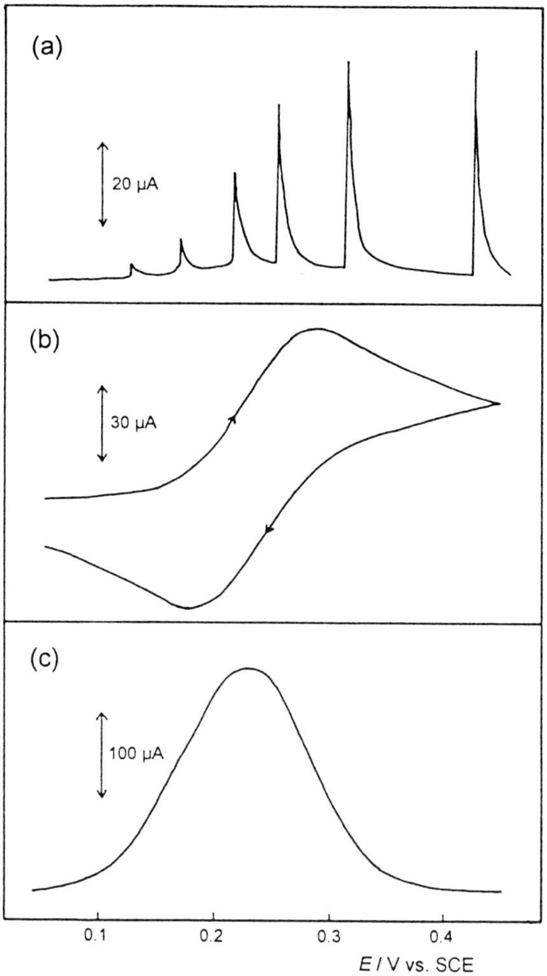

Fig. 16.6. Examples of BIA voltammetry, illustrated for the oxidation of 2 mM $K_4Fe(CN)_6$ in 0.4 M K_2SO_4 electrolyte at a Pt electrode, dispension flow rate 24.5 $\mu L\ s^{-1}$, cell parameters as in Fig. 16.5. (a) Consecutive injections of 16 μL during a linear potential sweep, scan rate 10 $mV\ s^{-1}$; (b) Background-subtracted cyclic voltammogram recorded during injection, scan rate 2 $V\ s^{-1}$; (c) Background-subtracted square wave (SW) voltammogram recorded during injection: SW amplitude 50 mV, SW increment 2 mV, frequency 100 Hz.

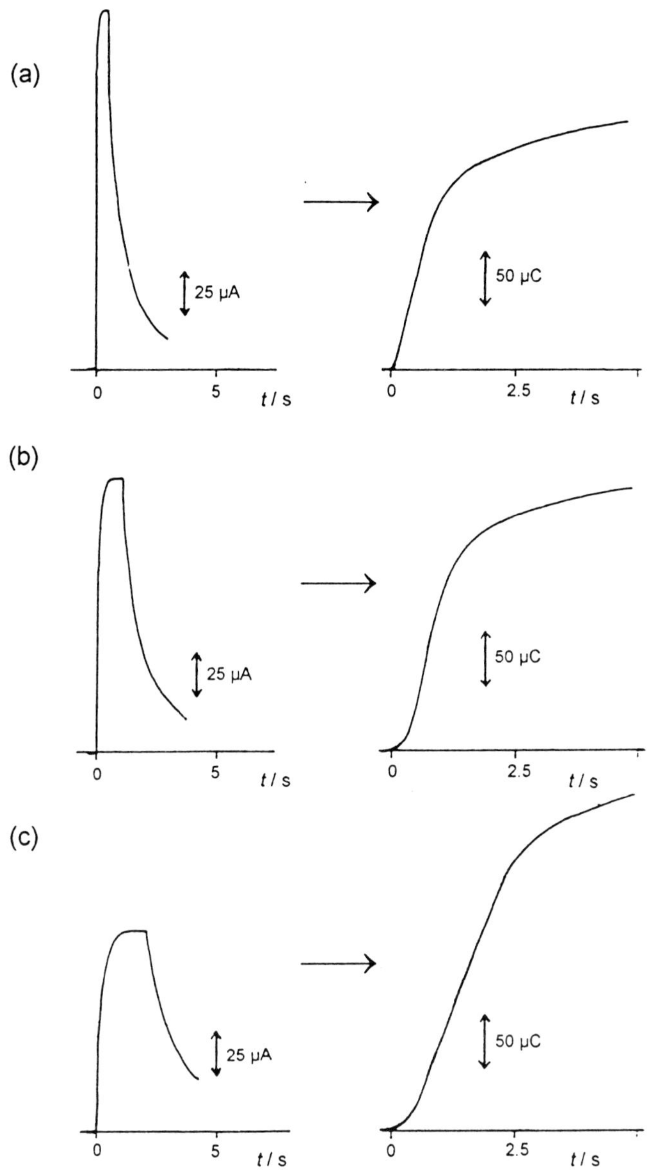

Fig. 16.7. Chronoamperometric and chronocoulometric transients for the oxidation of 2.0 mM ascorbic acid in 0.2 M pH 3.5 acetate buffer electrolyte: injection of 50 µL at applied potential +0.8 V vs. SCE at Pt disc electrode, cell parameters as in Fig. 16.5. Micropipette dispension volume flow rates: (a) 75.3, (b) 47.6 (c) 24.5 $\mu L\,s^{-1}$. $t = 0$ corresponds to the start of the injection period.

16.5 LOCAL PROBES—SCANNING ELECTROCHEMICAL MICROSCOPY

A particularly interesting approach for probing electrode reactions at the micro-level has arisen from the combination of microelectrodes and the accurate control instrumentation associated with scanning probe microscopies.

The usual objective of scanning probe microscopy techniques [41] is to provide images of a solid surface—normally topographic information—with up to atomic resolution. However, they can also be used to probe local solution composition and electrode reactions, as will be described.

Scanning probe microscopy involves scanning a tip of submicrometer dimensions over a surface, the position of the tip usually being controlled by accurate, linear piezoelectric drives; the resolution is limited by the tip size and can approach atomic resolution if the protruding end of the tip is also of atomic dimensions. Since the tip is held within several nanometres of the surface, the medium becomes unimportant and measurements can be performed in fluids, a vacuum thus becoming unnecessary—unless it is desired to exclude components which may react with the surface. The most well-known forms are scanning tunnelling microscopy (STM), in which a tunnelling current is passed between tip and conducting substrate at a distance of 2 nm or less, and atomic force microscopy (AFM), when the force between a tip, attached to a flexible cantilever, and the substrate is measured. The current or force, respectively, is held constant which leads to a topographic image of the surface; there is no chemical information.

An important extension of these techniques is when the substrate is made into an electrode in a small electrochemical cell, so that the change in surface morphology of the substrate can be monitored with time. In this way, for example, direct visualisation of electrodeposition at the submicroscopic level can be seen in real time, both nucleation and growth phases, e.g., the electrodeposition of copper in media with and without organic additives [42].

For the investigation of electrode reaction parameters and chemistry at these dimensions, another approach is necessary, in order to make the system species-selective through monitoring the electrochemistry. This involves making tip and substrate into independent electrodes; the tip is thus a microelectrode. The microelectrode tip is scanned over the surface: this is known as scanning electrochemical microscopy (SECM) [43] and due to its local probe nature can be used to probe microvolumes.

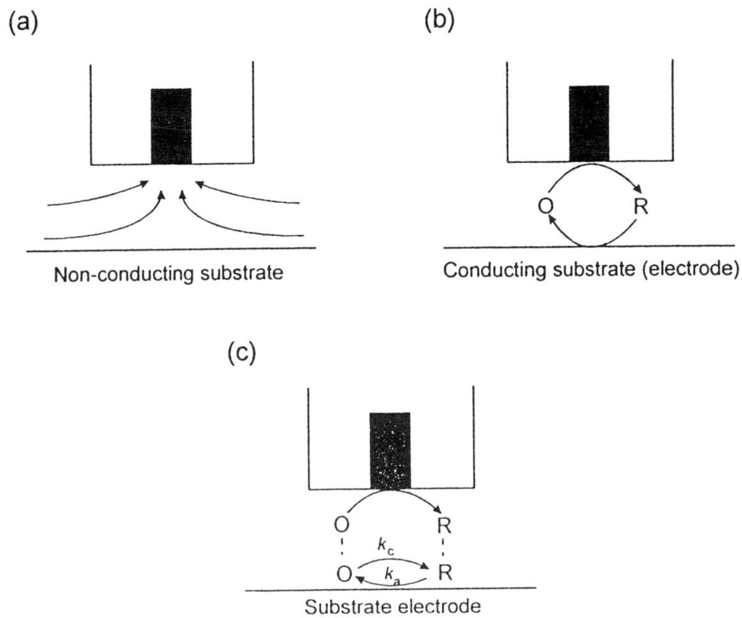

Fig. 16.8. Schematic effect of substrate on microelectrode tip response in SECM: (a) non-conducting substrate: hindered diffusion and reduction in current; (b) conducting substrate: recycling (positive feedback) and current enhancement; (c) conducting substrate: recycling with current enhancement attenuated by slow electrode kinetics at substrate electrode.

The response of SECM is distance-dependent as shown in Fig. 16.8. At large distances from the substrate the current measured is that of the microelectrode tip. The tip is a disc microelectrode, so that far from a substrate in the steady state the diffusion-limited tip current, $I_{T,\infty}$, will be measured. For a simple n-electron electrode reaction this is

$$I_{T,\infty} = 4nFDc_{\infty}a,$$

where a is the microelectrode radius and D the diffusion coefficient of electroactive species of bulk concentration c_{∞}. As the microelectrode tip is brought closer to the substrate, the current observed at the microelectrode due to electrode reaction of species present in solution may decrease owing to hindered diffusion (insulating substrate—Fig. 16.8(a)) or it may increase (conducting substrate—Fig. 16.8(b)) due to positive feedback between tip and substrate, the electroactive species being cycled between the two electrodes. With a redox species in solution, the variation of current with vertical distance can be employed to make a topographic image by maintaining the current constant as the surface is scanned.

Various types of application to probe the electrochemistry occurring on the substrate surface or in solution are also possible; some of the latest advances have recently been summarised [44]. It is possible to monitor substances produced at the substrate electrode and at electrode surfaces, from thin film to biological and polymer, at tiny tip-substrate distances. The theory for much of this has been developed and verified. Electrode reactions can be probe, since slow electrode kinetics influences the magnitude of the feedback current (Fig. 16.8(c)), representing the transition between extremes of conducting substrate (fast kinetics) and insulating substrate (slow kinetics). The Fe(III)–Fe(II) system and oxidation of $Ru(NH_3)_6^{2+}$ have been investigated at glassy carbon electrodes with platinum microelectrode tips and there is good agreement with theory; rate constants up to $20\,cm\,s^{-1}$ are measurable [45, 46]. The kinetics of adsorption/desorption phenomena with desorption of H^+ induced by the SECM, particularly on hydrous metal oxides, was studied [47]. The cases of first and second order chemical reactions following electron transfer have also been analysed theoretically and have been demonstrated experimentally [48, 49]. Other interesting systems occur where the substrate is non-conductive but incorporates a species on its surface which reacts with the product of the electrochemical reaction at the tip. Such species can be redox-active enzymes, for example glucose oxidase [50], or antigens, for example carbinoembryonic antigen [51].

Accurate electrochemical etching can be carried out, and this may become a viable alternative to other processes. Deposition of metallic structures at high resolution can be carried out on non-conductive substrates—recent work has described the formation of silver deposits in this way. An example of the latter is the deposition of silver microstructures [52].

Whereas SECM usually uses tips that pass a current in amperometric or voltammetric mode, an important related application uses passive ion-selective sensor tips which can be used for mapping two-dimensional and even three-dimensional ion distributions and concentration profiles over the surface, of species such as protons and zinc ions [53].

Recent developments have enabled the shrinking of the size of the tip down to 10–20 nm which gives excellent spatial resolution. In particular, a volume of solution can be enclosed below the tip within protruding walls made of wax which touch the substrate [54, 55]. These tips are made by covering tips of sharpened Pt/Ir wire with insulating wax and then causing a hole in the wax to be formed over the electrically conducting centre of the tip by applying a potential between tip and substrate. This leads to a recessed disc electrode enclosed within a wax sheath [56]. The tip is then brought towards the substrate until it contacts the substrate. Electrode reactions can be monitored within the enclosed volume, which is of the order of $10^{-18}\,cm^3$. The number of electroactive species present is calculated to be approximately one for millimolar concentrations, which is confirmed by experiment taking into account the amplification factor from positive feedback of approximately 10^7 [55]. This is thus single molecule detection and can provide individual rate parameters.

It can be appreciated that, now that the technical aspects are sufficiently developed, many further applications of SECM to study microvolumes (and less) of solution and microscopic characteristics of surfaces can be expected. Combination with non-electrochemical time-resolved techniques such as the quartz crystal microbalance will also be fruitful.

16.6 CONCLUSION

The strategies for probing electrode reactions described in this chapter have only become possible with recent and continuing advances in miniaturization technology and associated accurate instrumentation. It is to be expected that this progress will lead to more widespread study of electrode reactions in microvolumes or smaller in the near future. This opens up exciting possibilities of more detailed and localised information to be obtained from, for example, the detailed study of ever smaller numbers of electroactive species, the effect of tiny surface heterogeneities on electrode reactions, or of electrochemical processes occurring in single living cells.

References

[1] C.M.A. Brett and A.M. Oliveira Brett, Electrochemistry. Principles, Methods and Applications (Oxford University Press, Oxford, 1993).

[2] J.W. Schultze ed., Electrochemical Microsystem Technologies, Electrochim. Acta Special Issue 42 (1997) Nos. 20–22.
[3] F.E. Woodward and C.N. Reilley, in: eds. E. Yeager, J.O'M. Bockris, B.E. Conway and S. Sarangapani, Comprehensive Treatise of Electrochemistry (Plenum, New York, Vol. 9, 1984, pp. 353–392).
[4] M.M. Lohrengel, Electrochim. Acta 42 (1997) 3265.
[5] A.S. Baranski, Anal. Chem. 59 (1987) 662.
[6] A.R. Harman and A.S. Baranski, Anal. Chim. Acta 239 (1990) 35.
[7] A. Cserey and M. Gratzl, Anal. Chem. 69 (1997) 3687.
[8] M. Morita, M.L. Longmire and R.W. Murray, Anal. Chem. 60 (1986) 2770.
[9] W.J. Bowyer, M.E. Clark and J.L. Ingram, Anal. Chem. 64 (1992) 459.
[10] M.E. Clark, J.L. Ingram, E.E. Blakely and W.J. Bowyer, J. Electroanal. Chem. 385 (1995) 157.
[11] M. Lambrechts and W. Sansen, Biosensors: Microelectrochemical Devices (Institute of Physics Pub., New York, 1992, Chapter 5).
[12] Y. Murakami, T. Uchida, T. Takeuchi, E. Tamiya, I. Karube and M. Suda, Electroanalysis 6 (1994) 735.
[13] C.D.T. Bratten, P.H. Cobbold and J.M. Cooper, Anal. Chem. 69 (1997) 253.
[14] R.A. Clark, P.B. Hietpas and A.G. Ewing, Anal. Chem. 69 (1997) 259.
[15] Y. Shao, M.V. Mirkin, G. Fish, S. Kokotov, D. Palanker and A. Lewis, Anal. Chem. 69 (1997) 1627.
[16] A.G. Ewing, J.M. Mesaros and P.F. Gavin, Anal. Chem. 66 (1994) 527A.
[17] M.H. Chao and H.J. Huang, Anal. Chem. 69 (1997) 463.
[18] S.F. Dressman and A.C. Michael, Anal. Chem. 67 (1995) 1339.
[19] A.G. Ewing, R.A. Wallingford and T.M. Olefirowicz, Anal. Chem. 61 (1989) 292A.
[20] D.T. Chiu, A. Hsiao, A. Gaggar, R.A. Garza-López, O. Orwar and R.N. Zare, Anal. Chem. 69 (1997) 1801.
[21] S. Park, M.J. McGrath, M.R. Smyth, D. Diamond and C.E. Lunte, Anal. Chem. 69 (1997) 2994, and references therein.
[22] P.D. Voegel, W. Zhou and R.P. Baldwin, Anal. Chem. 69 (1997) 951, and references therein.
[23] C.M.A. Brett and A.M. Oliveira Brett, in: eds. R.G. Compton and C.H. Bamford, Comprehensive Chemical Kinetics (Elsevier, Amsterdam, Vol. 26, 1986, Chapter 5).
[24] C.M.A. Brett and A.M. Oliveira Brett, Electrochemistry. Principles, Methods and Applications (Oxford University Press, Oxford, 1993, Chapter 8).
[25] W.J. Albery and C.M.A. Brett, J. Electroanal. Chem. 148 (1983) 201.
[26] W.J. Albery and C.M.A. Brett, J. Electroanal. Chem. 148 (1983) 211.
[27] W.J. Albery, J. Electroanal. Chem. 191 (1985) 1.
[28] K. Aoki, K. Tokuda and H. Matsuda, J. Electroanal. Chem. 206 (1986) 37.
[29] R.G. Compton, C.R. Greaves and A.M. Waller, J. Appl. Electrochem. 20 (1990) 575.
[30] R.G. Compton, A.C. Fisher and G.P. Tyley, J. Appl. Electrochem. 21 (1991) 295.
[31] J. Wang and Z. Taha, Anal. Chem. 63 (1991) 1053.
[32] J. Ruzicka and E. Hansen, Flow Injection Analysis, 2nd ed. (John Wiley, New York, 1988).

[33] J. Wang and L. Chen, Talanta 42 (1995) 385.
[34] J. Wang and Z. Taha, Anal. Chim. Acta 252 (1991) 215.
[35] J. Wang, L. Chen, L. Angnes and B. Tian, Anal. Chim. Acta 267 (1992) 171.
[36] C.M.A. Brett, A.M Oliveira Brett and L.C. Mitoseriu, Electroanalysis 7 (1995) 225.
[37] C.M.A. Brett, A.M. Oliveira Brett and L.C. Mitoseriu, Anal. Chem. 66 (1994) 3145.
[38] C.M.A. Brett, A.M. Oliveira Brett and L. Tugulea, Anal. Chim. Acta 322 (1996) 151.
[39] C.M.A. Brett, A.M. Oliveira Brett and L. Tugulea, Electroanalysis 8 (1996) 639.
[40] C.M.A. Brett, A.M. Oliveira Brett F.-M. Matysik, S. Matysik and S. Kumbhat, Talanta 43 (1996) 2015.
[41] R. Wiesendanger, Scanning probe microscopy and spectroscopy. Methods and applications (Cambridge University Press, Cambridge, 1994).
[42] R.J. Nichols, in: ed. A.A. Gewirth and H. Siegenthaler, Nanoscale Probes of the Solid/Liquid Interface, NATO ASI Series E288 (Kluwer, Dordrecht, 1995, 163–182).
[43] A..J. Bard, F.-R.F. Fan and M.V. Mirkin, in: ed. A.J. Bard, Electroanalytical Chemistry (Dekker, New York, Vol. 18 (1993) pp. 243–273).
[44] M.V. Mirkin, Anal. Chem. 68 (1996) 177A.
[45] D. Wipf and A.J. Bard, J. Electrochem. Soc. 138 (1991) 469.
[46] A.J. Bard, M.V. Mirkin, P.R. Unwin and D.O. Wipf, J. Phys. Chem. 96 (1992) 1861.
[47] P.R. Unwin and A.J. Bard, J. Phys. Chem. 96 (1992) 5035.
[48] P.R. Unwin and A.J. Bard, J. Phys. Chem. 95 (1991) 7814.
[49] F. Zhou, P.R. Unwin and A.J. Bard, J. Phys. Chem. 96 (1992) 4917.
[50] D.F. Pierce, P.R. Unwin and A.J. Bard, Anal. Chem. 64 (1992) 1795.
[51] H. Shiku, T. Matsue and I. Uchida, Anal. Chem. 68 (1996) 1276.
[52] C. Hess, K. Borgwarth, C. Ricken, D.G. Ebling and J. Heinze, Electrochim. Acta 42 (1997) 3065.
[53] K. Toth, G. Nagy, C. Wei and A.J. Bard, Electroanalysis 7 (1995) 801.
[54] F.-R.F. Fan and A.J. Bard, Science 267 (1995) 871.
[55] A.J. Bard and F.-R.F. Fan, Acc. Chem. Res. 29 (1996) 572.
[56] F.-R.F. Fan, J. Kwak and A.J. Bard, J. Am. Chem. Soc. 118 (1996) 9669.

Chapter 17

Some Kinetic Considerations of Chemical Vapour Deposition (CVD) Processes

MICHAEL L HITCHMAN

17.1 INTRODUCTION

Materials are crucial to the quality of life, and to the economic security and competitiveness of a nation. Societies have recognised the importance of man-made materials to the quality of life for many centuries, and in many cases the security and defence of tribes and nations has substantially depended on the availability of materials. It is not surprising, therefore, that historical periods have been named after materials—the Bronze Age, the Iron Age, etc. The major criteria for materials have always been based on properties and performance, and today, in this age of advanced materials, this is no less the case. So, for example, modern society requires advanced materials in high technology products, such as communications and electronic equipment, computers and robotics, aircraft and spacecraft, military armament and rocketry, fibre optics, professional and scientific instruments, and engines of many sorts in order for them to function successfully. These materials are usually state-of-the-art developments and are generally high performance, comparatively high-cost materials. In many applications combinations or composites of different classes of materials are required to provide enhanced properties. In particular, the surfaces of bulk materials often need to be modified to meet the demands of the environment. A bulk material is usually chosen from the standpoint of structural and economic criteria, but it is the surface of the material which has to co-exist with the external world. Thus, bulk materials are often coated to improve their mechanical properties such as wear resistance and tribological characteristics, or to enhance the chemical properties, such as providing corrosion protection or giving enhanced catalytic activity.

Also surface optical properties of a material sometimes need to be changed, for example in making anti-reflection coating for lenses or reflective surfaces for CDs, the magnetic properties may need to be influenced as in the case of giving a ferroelectric surface to a plastic for magnetic recording, and, perhaps most extensively of all, the surface electrical properties need to be controlled in microelectronic devices used in computers and all modern electronic equipment.

There is a wide range of techniques for putting down films onto surfaces [1]. Typical commercial technologies available for surface treatment in solution include electroplating, anodisation, electrophoresis electroless deposition, spray pyrolysis, controlled precipitation, and dip growth. Many modern technologies, however, use surface treatment procedures involving the deposition from the gaseous phase. Examples of these techniques include evaporation, sputtering, ion plating, ion implantation, molecular beam epitaxy, recoil mixing, and chemical vapour deposition (CVD).

Whichever technique is used there are a number of criteria which need to be met to varying degrees, depending upon the application, for the film or layer which is being deposited. Generally, it is desirable to be able to control the thickness of the layer which may vary from anything from less than 10 nm up to greater than 100 μm. In addition to controlling the absolute thickness of a layer it is also essential that for most applications that the uniformity of the deposit is controlled both within a given sample and from one sample to another. Here there may be a requirement of achieving uniform growth with variations of less than ±1% in the most demanding application to perhaps ±10% in less stringent cases. Whatever the thickness requirements, though, it is clearly necessary for satisfactory surface modification for the film to adhere or to bond to the substrate. The need for good adherence also highlights the importance of compatibility between the film and the substrate, and the need not to disturb or distort the substrate unduly. Furthermore, the composition of any film needs to be exactly controlled in order to influence the surface properties in a reproducible manner. It is, however, not only the composition of the layer, but also its surface structure (i.e., the morphology) which can be important, with some applications requiring amorphous material, others polycrystalline material, and in some specialised cases, particularly for microelectronic and optoelectronic applications, single crystal material is required. Finally, as already mentioned, the purpose of depositing a thin film on a substrate is to influence one or more of a range of properties such as the electrical, optical, mechanical, or chemical characteristics. Therefore any functional improvement has to be related to the additional costs involved and thus economic factors, as well as technical and scientific ones, are important in thin film processing. One technique for thin film deposition which has been extensively used for a wide range of applications is that of chemical vapour deposition (CVD) [2]. In this article this technique will be introduced and some examples of its use and underlying principles briefly discussed. Then some considerations of reaction kinetics of CVD processes will be considered. Finally, an example of the application of CVD to the preparation of a relatively new class of materials, high temperature superconductors, and some results of probing the chemistry of the deposition process will be given.

17.2 CHEMICAL VAPOUR DEPOSITION

17.2.1 Introduction

CVD is a process whereby a thin solid film is synthesised from the gaseous phase by chemical reaction. It is this reactive process which distinguishes CVD from physical deposition processes. The development of CVD has been very closely related to development of the materials needs of society. Perhaps the oldest example of a material deposited by CVD is that of pyrolytic carbon since some prehistoric art was done on cave walls with incomplete oxidation of firewood. A similar procedure in fact formed the basis of what must be one of the earliest patents in commercial exploitation of the CVD process. This patent was issued to John Howarth and was for the preparation of carbon black as a pigment. Unfortunately just as the plant, which was mainly constructed of wood, was becoming profitable, it burned down. Carbon CVD provided the next major application with patents issued to Sawyer and Mann [3] for improvements in filaments for electric lamps. However, because even these improved filaments were far from robust the future for the pyrolytic carbon CVD industry was limited, and a few years later processes for the deposition of metals to improve the quality of lamp filaments were described. From the turn of the century through to the late 1930s a variety of CVD techniques appeared for the preparation of a number of metals, particularly refractory metals for a range of applications [4]. It was also during this period that CVD bridged the gap between science and the arts with the appearance of a detailed description of the decomposition of arsine to form an arsenic mirror in Dorothy L. Sayers "Strong Poison" [5]. This very sensitive test for arsenic was used by Lord Peter Wimsey, with Bunter's expert assistance, to identify a mysterious white powder in the presence of Miss Murchison, who discovered it. The early part of the 20th century also saw silicon first deposited by hydrogen reduction of silicon tetrachloride [6] and the use of that material for electronic applications was presaged by the preparation of silicon based photocells and rectifiers [7] just after World War II. The preparation of high purity metals, various coatings, and electronic materials have all developed significantly in the last 50 years or so. It is undoubtedly the demands and requirements of the semiconductor and microelectronics industries which have been the main driving force in the development of CVD techniques and the greater efforts to understand the fundamentals of CVD processes.

With the growth of a material by CVD, as indeed with any other technology, it is clearly desirable to be able to control the properties of the layers grown. These will depend on the growth conditions in general, and, in particular, on the chemistry of the deposition process. Hence there have been increasing efforts by many groups of workers in recent years to seek a better understanding of CVD processes. In this article, therefore, the emphasis will be on principles of understanding rather than on details of specific processes, although examples will be given to illustrate the principles. In order to set the scene for the discussion about principles of CVD it is necessary to first of all give an overview of CVD processes.

17.2.2 CVD processes

The exact arrangement of the equipment for a CVD system can vary widely depending on the particular application. The basic requirements of a CVD apparatus can, though, be illustrated by reference to a generalised system. Figure 17.1 shows a schematic arrangement for deposition of tungsten carbide coatings. The overall reaction may be written as

$$\text{(R1)} \qquad 2WF_6(g) + \frac{1}{6}C_6H_6(g) + \frac{11}{2}H_2(g) \rightarrow W_2C(s) + 12HF(g).$$

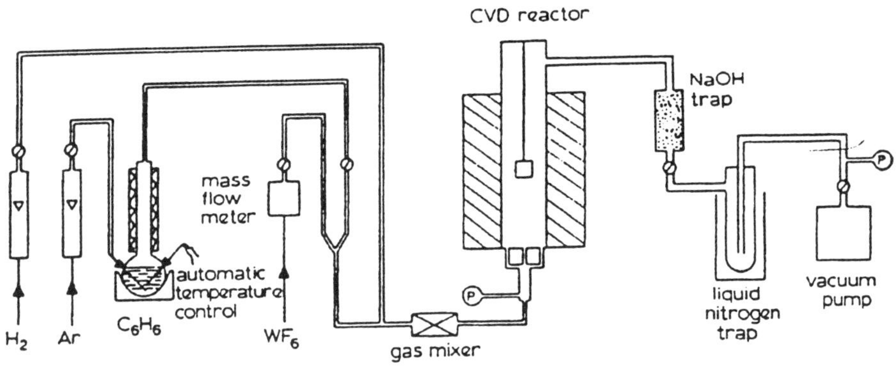

Fig. 17.1. Schematic arrangement for the CVD of tungsten carbide coatings.

The apparatus can be thought of as being made up of three main constituent parts—precursor handling, the reactor, and the exhaust of by-products.

For many years precursors for materials deposited by CVD have been restricted to those which are part of the inventory of compounds already available, and this has usually meant simple inorganic sources. However, with the increasing demand for more sophisticated deposits and the need for tighter control of deposition rates, uniformity and layer properties and quality, there has been a major interest in recent years in organometallic precursors where the metal has been made volatile by bonding it to organic ligands. Some examples of this will be given towards the end of the article. Whatever the composition of the precursor, though, the simplest form for handling purposes is one which is a gas at normal temperatures and pressures. Then all that is required is a high pressure gas cylinder and piping and valving leading to a flow meter or controller; hydrogen is such an example shown in Fig. 17.1. When it is necessary to use precursors which are liquid and even solid under ambient conditions then higher temperatures and/or lower pressures have to be employed in order to transport the vapour to the reactor with the aid of a carrier gas passing through a temperature controlled bubbler; this is illustrated by the benzene in Fig. 17.1. The tungsten hexafluoride in Fig. 17.1 is an example of a precursor where there is the possibility of condensation of vapour in pipework and valves leading to the reactor and these parts of the apparatus must be heated to an appropriate temperature.

The reactor is the heart of any CVD system and, in common with other

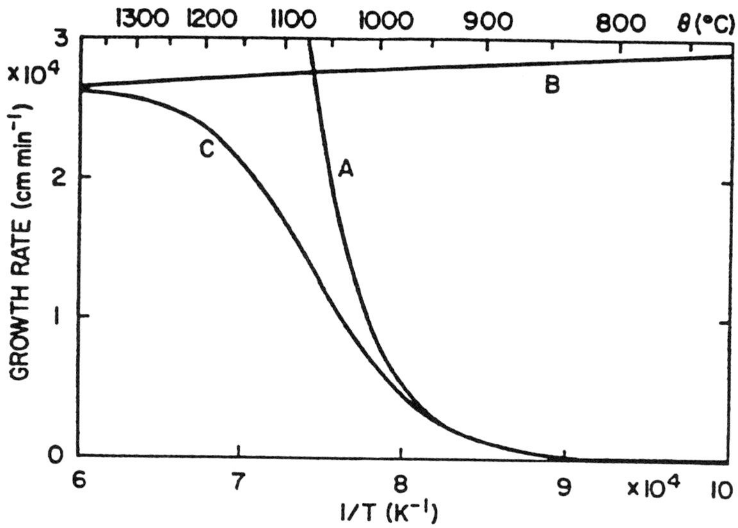

Fig. 17.2. Dependence of the growth rate of a CVD layer on temperature. A—kinetically controlled growth; B—transport controlled growth; C—overall growth [8].

hearts, a malfunction or failure of this vital organ can have disastrous consequences. Therefore a significant fraction of this article is devoted to considering what goes on in a CVD reactor. Affairs of the heart are often unpredictable, stormy and troublesome and so it is essential to try and understand what it is that makes the reactor 'heart' tick. Here we first of all illustrate some broad, general features.

The driving force for many CVD processes is the thermal energy input. However, because CVD is, by its very nature, a chemical process the deposition rate can be strongly dependent on deposition temperature through an Arrhenius relationship (curve A, Fig. 17.2). Therefore for many CVD processes, particularly those occurring in cold wall reactors, the deposition is carried out under conditions where the controlling factor is transport of the reactant to the substrate and hence the growth rate is less sensitive to temperature (curve B, Fig. 17.2) [8]. This, though, brings with it the attendant problem of layer uniformity resulting from the variation of mass transport characteristics with the geometric arrangement in the reactor. As a result, in this type of reactor it is very difficult to obtain controlled uniform coatings on a large number of samples. Furthermore, because of the large thermal gradients in the reactor, the flow patterns and temperature distributions are extremely complex and one is well away from the origin in the CVD space depicted in Fig. 17.3 [2, Ch. 1].

The practical difficulties arising from operating under transport control can often be overcome by moving into the kinetically controlled domain through a reduction in reaction pressure so that the diffusivity (D) of gas molecules increases significantly and the mass transfer coefficient (k_T), which can be represented by D/d where d is the distance which a molecule must diffuse to a surface, becomes significantly greater than the global heterogeneous rate constant (k_S) for the surface process. This can be represented by the equation

$$\frac{1}{j_G} = \frac{1}{j_K} + \frac{1}{j_T} = \frac{1}{k_S C_{AB}^{\infty}} + \frac{1}{k_T C_{AB}^{\infty}}, \tag{17.1}$$

where j_G is the overall growth rate, j_K the kinetic contribution (represented by curve A in Fig. 17.2), j_T is the contribution from transport (curve B in Fig. 17.2), and C_{AB}^{∞} is the precursor concentration some distance from the surface. Equation (17.1) is the sum of the reciprocal fluxes and the smaller the contributing flux the greater the effect. Reciprocal fluxes may be thought of as resistive elements in series with the first term in the equation being the resistance to reaction at the interface and the second term the resistance to transport to the interface.

Of course, over a wide temperature range there will be contributions to growth from both transport and kinetics (curve C, Fig. 17.2) in which case one moves along the z axis of CVD space and the system becomes even more complex. However, by reducing the reactor pressure enough to make $k_T \gg k_S$ one can operate in the truly kinetic region of the growth rate curve. Furthermore, if the cold wall reactor is replaced by a hot wall system, sufficient control of temperature can be maintained to ensure controllable and reproducible growth. Now one is close to the origin of CVD space and, provided one has some understanding of the chemistry of the process, the modeling can be somewhat easier than in a cold wall, atmospheric system. From a practical point of view, however, the significant advantage is the high packing density of substrates that can be achieved in the reactor as a result of the removal of transport limitations.

The gaseous by-products from a CVD reactor can often have unpleasant

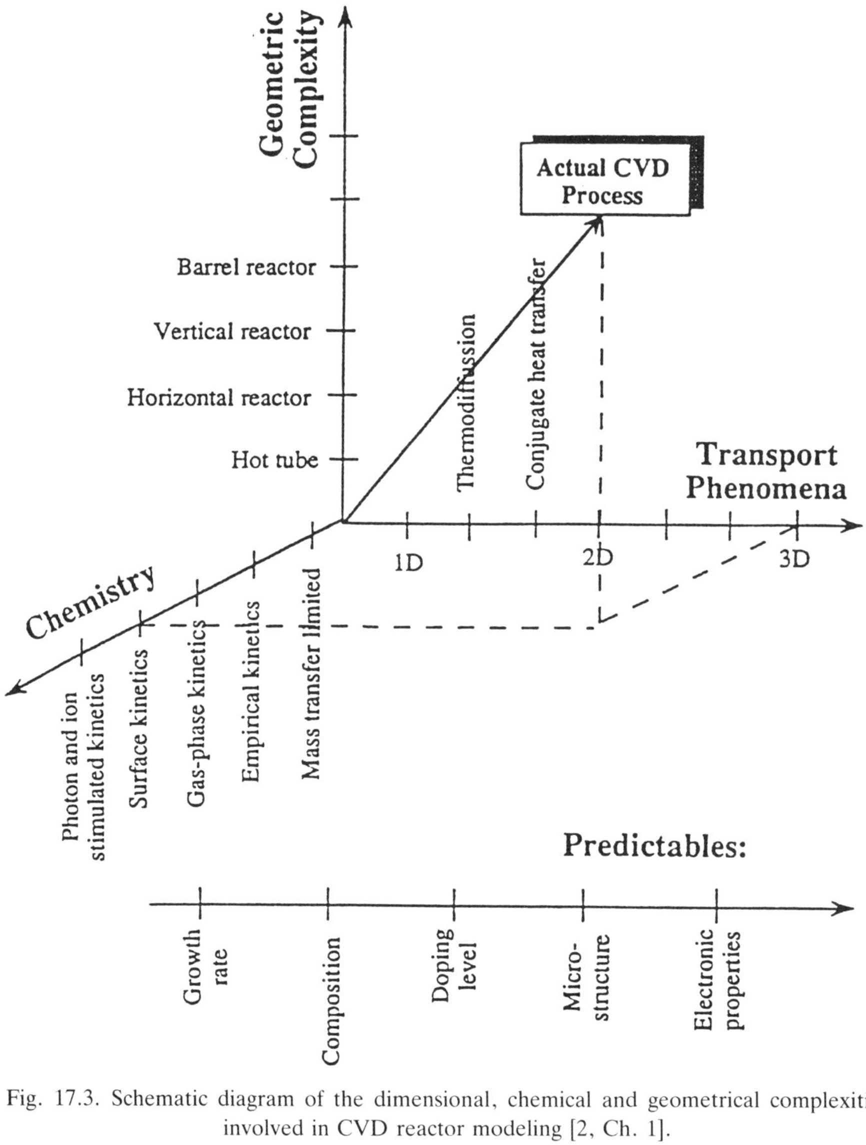

Fig. 17.3. Schematic diagram of the dimensional, chemical and geometrical complexities involved in CVD reactor modeling [2, Ch. 1].

properties. They cannot, therefore, be directly vented to the atmosphere but must be trapped in some way and disposed of in a safe and environmentally sound manner. Chemical and physical methods of removal of the product gas and any excess reactor gases may be employed, as illustrated in Fig. 17.1. The safe handling and disposal of CVD chemicals often require special precautions [9] with the monitoring of CVD reagents becoming of increasing importance [10].

Before moving on to look at one or two CVD systems in more detail, the wide range of materials that can be prepared by CVD are illustrated in Table 17.1, which gives a succint overview of layers currently prepared commercially and in development.

17.3 CVD AT LOW PRESSURE

17.3.1 Introduction

Overall stoichiometric equations for CVD processes, such as given in reaction (R1), mask considerable mechanistic complexities. This is illustrated schematically in Fig. 17.4 which shows that for a CVD process there are contributions from a number of individual steps. First, reactant materials have to be transported into the growth region of the reactor, and then from the bulk of the gas to the surface. During the latter mass transfer step there may be homogeneous reactions occurring in the gas phase; products of these reactions are, though, unlikely to be lattice species formed, for example, as small particles or droplets which would then fall onto the surface, since high quality layers would not be obtained by such a mechanism. Therefore the next step is for the reactant species to be adsorbed onto the surface followed by various surface pathways, such as heterogeneous reactions, surface diffusion and site incorporation. Any volatile products of reaction then have to be desorbed, transported away from the surface into the main gas stream and transferred out of the growth region.

The reaction sequence shown in Fig. 17.4 is characteristic of many heterogeneous processes, including electrochemical and catalytic systems and to describe in detail such a complete sequence is a formidable task. Insights and understanding can be achieved, though, by using the fundamental kinetic principle that if steps occur consecutively then any

TABLE 17.1
Examples of layers prepared by CVD

General type of material	Specific type of layer	Particular examples
Conductors	Metals	Al, Cu, W
	Silicides	$MoSi_2$, $TaSi_2$, $TiSi_2$, WSi_x
	Doped semiconductors	P/Si, B/Si
	Transparent	In_2O_3, SnO_2
	Organic	Polyacetylene, polypyrrole
	Superconductors	$YBa_2Cu_3O_{7-x}$, Bi–Sr–Ca–Cu–O, Tl–Ba–Ca–Cu–O
Semiconductors	Elemental	Si (epitaxial, polycrystalline, amorphous), Ge, Ge/Si, C
	Compound	III–V (e.g., GaAs, InP, AlGaAs, GaInP, GaInAsP)
		II–VI (e.g., ZnS, CdTe, CdHgTe
		IV–IV (e.g., SiC)
Insulators and dielectrics	Single oxides	SiO_2, Al_2O_3, TiO_2, Ta_2O_5
	Mixed oxides	Arsenosilicate glasses, borosilicate glasses, phosphosilicate glasses, borophosphosilicate glasses, aluminosilicate glasses
	Nitrides	Si_3N_4
	Oxynitrides	Si_xO_yN
Optoelectronic layers	III–V compounds	GaAs, InP, AlGaAs, GaInP, GaInAsP
	II–VI compounds	ZnS, CdTe, CdHgTe, PbS
	Others	SiO_x, C (diamond), Se, SnO_2
Coatings	Wear resistant	Carbides (e.g., TiC, W_2C, SiC, B_4C), borides (e.g., TiB_2, NiB, WB), nitrides (e.g., TiN, Si_3N_4, BN), carbonitrides (e.g., TiC_xN_{1-x}), oxides (e.g., Al_2O_3, ZrO_2)
	Tribological	TiC, TiN, Cr–C, Fe_2B
	Corrosion resistant	Oxides (e.g., Al_2O_3, Cr_2O_3, SiO_2), metals (e.g., Cr, Zn, Al, Ni), silicides (e.g., $MoSi_2$, WSi_2, VSi_2), carbides (e.g., Mo_2C, NbC)
	Decorative	Metals (e.g., Al, Cr) oxides (e.g., Ta_2O_5, TiO_2)
Optical layers	Reflective	Al, Rh, Ag, Au
	Antireflective	MgF_2, cryolite
	Absorptive	Cr, Ni, Rh, nichrome, inconel
	Spectrally selective interference filters	$Ag/MgF_2/Ag$, SiO
	Analytical	Au, SnO_2

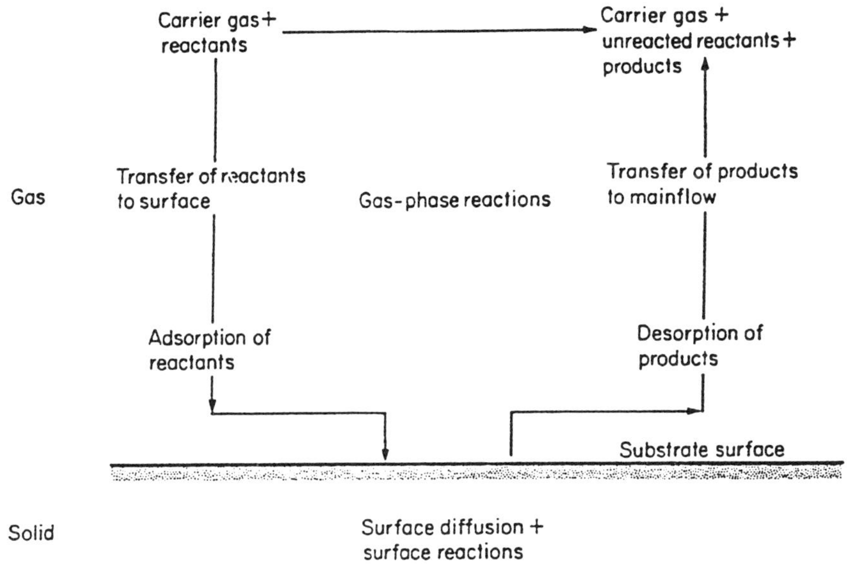

Fig. 17.4. Processes contributing to CVD growth.

one step that is inherently slower than the rest will determine the overall growth rate. This means that if one can carry out a reaction under conditions where such a rate determining step is predominant then the analysis of the overall growth under these conditions becomes simpler. As already mentioned in the previous section, one particularly useful simplification for CVD processes can be achieved by reducing the total pressure in the reactor which increases the diffusivity of reactant species, since the magnitude of the diffusion coefficient (D) of a gas is inversely proportional to gaseous pressure. Thus the mass transfer coefficient (k_T) in equation (17.1) also increases, and if it becomes significantly greater than the heterogeneous rate constant the growth process can be described solely in terms of chemical kinetics.

For many years the growth by CVD of layers of thin films was carried out in atmospheric pressure reactors. We have just seen that from the point of view of fundamental understanding there is a significant advantage to operating CVD reactors at pressures below atmospheric. The pressures used can be very approximately, although conveniently, divided into three regimes. Depositions carried out at pressures in the range ca. 100 Torr to

1 Torr (i.e., ca. 10^4 Pa to 10^2 Pa) are commonly referred to as reduced pressure CVD (RPCVD). For the range 1 Torr to 10 m Torr (ca. 10^2 Pa to 1 Pa) the term low pressure CVD (LPCVD) is used. At pressures of less than ca. 10 m Torr (less than ca. 1 Pa), we move into the realm of ultrahigh vacuum CVD (UHVCVD). Of these three regimes that corresponding to LPCVD is the most widely used, particularly in conjunction with hot wall reactors for the deposition of polycrystalline and amorphous materials. As we shall see, LPCVD can allow growth under kinetic control and so can provide the conditions for study of the growth kinetics without interference from mass transport.

The move away from atmospheric pressure CVD (APCVD) to low pressures has also occurred commercially for a number of practical reasons. For example, a detailed analysis [11] of the cost benefits of LPCVD for semiconductor applications has estimated that the per-wafer savings in power, materials, labour and capital depreciation is about 90% compared with conventional CVD. The main reason for such cost reductions is the high packing density of wafers that can be achieved in a stand-up configuration in an LPCVD reactor. In an APCVD reactor, one is limited to a low packing density, lie-down configuration in order to allow uniform accessibility of the surfaces to the reaction gases. This requirement is necessary for uniform deposition because of the effect of mass transfer control on deposition at normal pressures [8]; cf. equation (17.1). As discussed already, at low pressures there is usually no significant mass transfer control on growth, only kinetic control, so that uniform accessibility of the wafer surfaces to the reaction gases can still be maintained even with very close wafer spacing, e.g., 0.3–0.6 cm [12, 13]. This lack of mass transfer control, coupled with the good temperature uniformity achievable in a hot wall reactor, results in LPCVD-grown layers showing improved radial uniformity across a wafer compared with APCVD grown layers. For polysilicon and silicon nitride, for example, the uniformity is usually about ±1–3% for LPCVD layers compared with ±4–8% for APCVD layers [14]. Step coverage of LPCVD layers is also good, again because of the lack of mass transport limitations. Taking all these features of LPCVD as a whole, it is not at all surprising that there has been such widespread acceptance of the technique in the electronics and other industries.

Considerations of cost, layer uniformity, step coverage and cleanliness are, of course, very important for the production of materials in general.

For microelectronic applications the advent of very and ultra large-scale integration (VLSI and ULSI) has meant that the requirements imposed by high product yield and device reliability are particularly stringent and even higher quality layers are needed for new generations of multiple, vertical layer structures. In order to satisfy these requirements it is vital to be able to control properties of the layers grown, such as thickness, uniformity, structure and morphology, and electrical characteristics. All of these properties will depend upon the growth conditions in general and, in particular, on the chemistry of the deposition process and there is a need for a basic understanding which becomes particularly important when problems arise with deposition processes. For example, although silicon dioxide, semi-insulating polysilicon (SIPOS), and doped polysilicon are routinely deposited in microelectronics fabrication lines, the layer thickness on a wafer can be as much as 100% greater at the edge than at the centre [13, 15, 16]. This clearly can lead to problems with the yield of devices on a wafer. With the trend to finer and finer line structures on devices, it is important not only to control layer thickness, but also grain size. Polysilicon is an example of a material whose degree of crystallinity and grain size can vary considerably depending on deposition conditions [17]. The inability to adequately control the grain size of CVD layers of aluminium was largely responsible for impeding acceptance of that process within the industry [18]. With some processes, layer quality may not only not conform to required specifications, but may actually vary from run to run; deposition of thick layers of tungsten selectively on silicon substrates and not on surrounding oxides is one example of lack of reproducibility [19].

In some instance recipes can be found that allow layer specifications to be met. It has been reported, [17] for example, that much improved layer quality of annealed polysilicon can be achieved if deposition is carried out at lower temperatures than normal (e.g., 580°C compared with the typical 600–620°C). And the radial non-uniformity mentioned above can be dramatically improved by the use of a cage placed around the wafers [20–22]. But while recipes for controlling processes may be useful, it is still not very satisfactory if there is no real scientific understanding underlying such recipes. In the next section, modelling of a LPCVD process is reviewed, with particular reference to film growth rates and uniformity process.

17.3.2 LPCVD of polysilicon

17.3.2.1 Introduction

One of the most widely used materials for the fabrication of modern VLSI circuits is polycrystalline silicon, commonly referred to as polysilicon. It is used for the gate electrode in metal oxide semiconductor (MOS) devices, for the fabrication of high value resistors, for diffusion sources to form shallow junctions, for conduction lines, and for ensuring ohmic contact between crystalline silicon substrates and overlying metallization structures.

Typical conditions for depositing polysilicon are temperatures in the range 580–650°C and pressures in the range 25–130 Pa (0.2–1.0 Torr). The overall chemical reaction is

(R2) $$SiH_4(g) \rightarrow Si(s) + 2H_2(g)$$

The process can be used to deposit polysilicon on 100–200 wafers in one run with within-wafer and wafer-to-wafer thickness uniformities of better than a few per cent. Deposition rates are typically 10–20 nm min^{-1}.

17.3.2.2 Deposition kinetics

The actual chemistry of the CVD of polysilicon is somewhat more complex than the simple stoichiometric reaction (R2) would imply. The mechanism of deposition has been the subject of a number of studies [12, 23–27]. A general analysis of the factors affecting deposition of polysilicon from SiH_4 has shown [12] that there is no transport control of growth and that the kinetic behaviour can be explained by a Langmuir adsorption model in which the growth rate is linearly proportional to the SiH_4 partial pressure and where the adsorption of hydrogen is greater than that of SiH_4. This kinetic scheme can be represented by

(R3) $$SiH_4(g) \Leftrightarrow SiH_4(a)$$

(R4) $$SiH_4(a) \rightarrow Si(s) + 2H_2(a)$$

(R5) $$H_2(a) \Leftrightarrow H_2(g)$$

Others have reached similar conclusions, with Duchemin *et al.* [23] showing that at the low pressures used for polysilicon deposition the hydrogen is dissociated on the surface, and Cadoret and Hottier [25] suggesting that the reaction of SiH_4 molecules on a surface inhibited by the adsorbed hydrogen atoms supplies the silicon atoms necessary for cluster growth. The role of adsorbed hydrogen in silicon growth has been explored further in UHV studies by Gates and coworkers [27–29]. At higher pressures (>130 Pa) the polysilicon LPCVD process can be complicated further by mass transfer phenomena, and at higher temperatures (>700°C) gas phase nucleation and the formation of silane polymers both occur leading to radially non-uniform growth on wafers [24]. For polysilicon this is not usually a problem under the standard deposition conditions mentioned above, but polymer formation does become a problem when in-situ phosphorous doping is introduced (see Section 17.3.3(iii)).

Further evidence for the involvement of a heterogeneous kinetic process at low pressures and temperatures is supplied by the effect of temperature on growth rate. Figure 17.5 shows, in the form of an Arrhenius plot, a typical variation of growth rate of polysilicon for the temperature range 575–650°C. In plotting CVD growth rates in this way it must be remembered that the growth rate is determined not only by the global rate constant but also by the partial pressure or mole fraction of the reactant. So it is important to allow for any changes in reactant concentration with change of temperature, as can occur, for example, by depletion effects in the direction of flow in a horizontal hot wall reactor. In Fig. 17.5 the growth rates have been corrected for such depletion effects and have been normalised to a standard SiH_4 concentration. The activation energy obtained from the least-mean-squares slope is 184 kJ mol^{-1} or 1.91 eV. This value is comparable to those reported by others for deposition in atmospheric pressure reactors; e.g., 1.85 eV found by Chiang [30] and 1.9 eV found by Ban *et al.* [31]. With normal pressure CVD there is always a danger that any measured activation energy will contain a transport contribution and will, therefore, be lower than the true kinetic activation energy. The values reported by Chiang and Ban are based on measurements made at low temperature (<700°C) and probably do not contain a large transport contribution. Thus it would appear that the activation energy in low pressure and normal pressure reactors are the same, suggesting a similar rate-determining step.

For a rate-controlling step involving the gas-phase decomposition of SiH_4 one would, on the basis of Rice, Ramsperger, Kassel and Marcus (RRKM) theory, expect a difference in activation energies obtained for

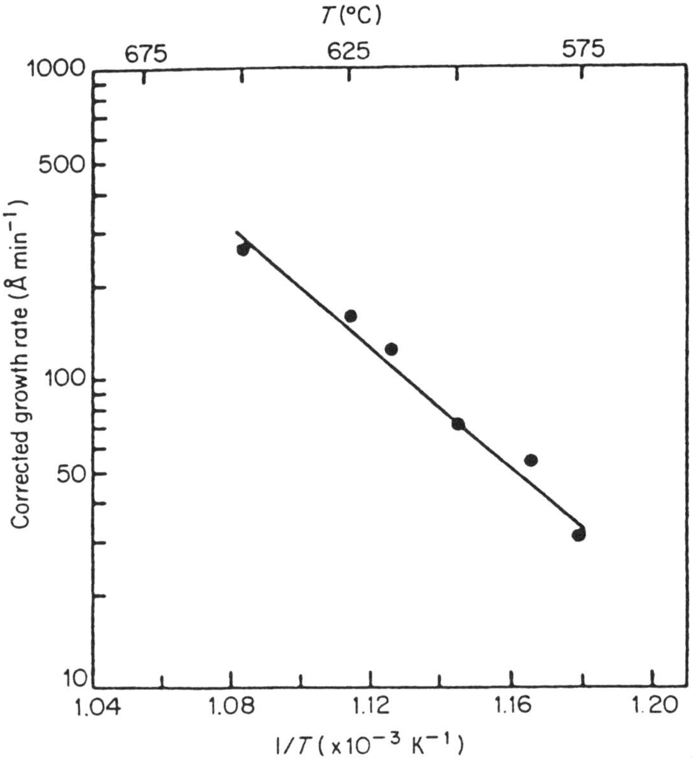

Fig. 17.5. Dependence of growth rate of polysilicon on temperature [2, Ch. 4].

low pressure and atmospheric pressure systems. However, the activation energy difference for homogeneous SiH_4 decomposition at 130 Pa and 1.33×10^5 Pa is $<10\ \mathrm{kJ\ mol^{-1}}$ (or 0.1 eV) [32] and this is comparable to the differences between the activation energy values given above. On this basis, therefore, one cannot say whether the rate-determining step for polysilicon deposition involves a gas phase or a surface process. However, when one looks at the actual value for the activation energy for homogeneous SiH_4 decomposition by the energetically most favorable route of H_2 elimination

$$\text{(R6)} \qquad SiH_4(g) \rightarrow SiH_2(g) + H_2(g)$$

it is found that $E_a = 238$ to $255\ \text{kJ mol}^{-1}$ (2.5–2.7 eV) which is significantly greater than any experimental value given above, thus suggesting a heterogeneous rate-determining step. Evidence, in fact, for a heterogeneous rate determining step is obtained from the Arrhenius plot of Fig. 17.5 if one calculates the entropy of activation, $\Delta S^{\ddagger}$, for the overall reaction. This is found to be about -80 e.u., which is approximately the same as the value one can estimate from the results of Chiang and Ban ($\Delta S^{\ddagger} \cong -85$ e.u.), again indicating a similar rate determining step for the two pressure regimes. Using statistical treatment of reaction rates the value of $\Delta S^{\ddagger}$ expected from the change in the number of degrees of freedom in going from the initial state to an activated state can be calculated for various models. Any gas phase reaction gives a value of $\Delta S^{\ddagger}$ of only a few entropy units and so again it is seen that a homogeneous rate controlling step is unlikely. For heterogeneous processes one can consider the adsorption of a molecule from the gas with loss of one or more degrees of translational and rotational freedom and calculate the entropy changes using the appropriate partition functions. When this is done the only model which gives an entropy of activation comparable to that found experimentally is that corresponding to the adsorption and complete immobilisation of a SiH_4 or SiH_2 molecule: $\Delta S^{\ddagger} \cong -75$ e.u. There is thus good evidence that a heterogeneous adsorption is the rate controlling step for polysilicon deposition from silane. Further support for this concept is given in studies by Scott *et al.* [33] delineating the pressure regions dominated by either homogeneous or heterogeneous reactions.

There is clearly now some understanding of the underlying chemistry of the LPCVD of polysilicon, but, as already indicated, it is important to relate that basic understanding to practical issues of layer characteristics.

17.3.2.3 Layer thickness

Controlling film thickness can rely simply on an empirical knowledge of the dependence of growth rate on total reactor pressure, reactant partial pressures, flow characteristics, and temperature. This information has been available from the earliest days of successful polysilicon deposition by LPCVD, but the more fundamental studies discussed above have put such results on a surer footing. So, for example, assuming the kinetic model of Langmuirian adsorption of SiH_4 with dissociative adsorption of H_2 and combining this with a description of the coupled convective and diffusive fluxes in the annular flow region and between each pair of wafers, one can

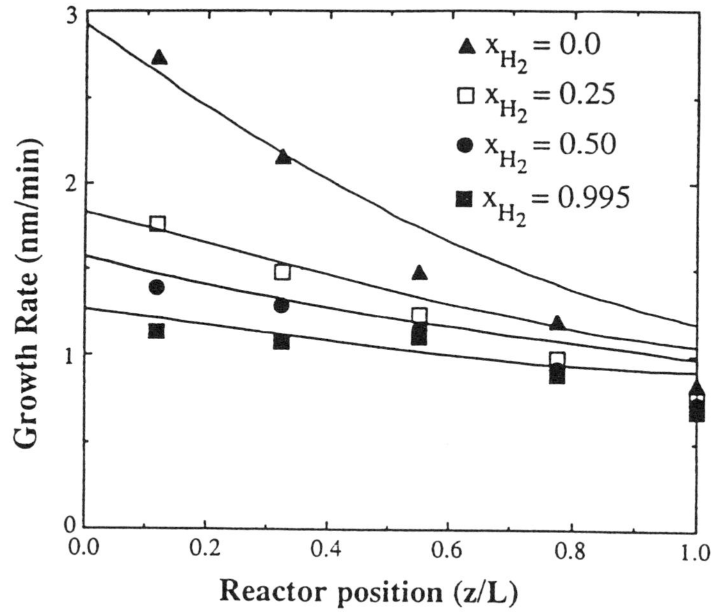

Fig. 17.6. Model predictions (curves) and experimental data (points) for polysilicon growth rates along a tubular LPCVD reactor at various H_2 feed compositions [34].

predict quantitatively both the variation of deposition rate which occurs, because of depletion of reactants, along the length of a tubular LPCVD reactor and the inhibition of H_2 on the polysilicon deposition rate (Fig. 17.6) [34]. Furthermore, this combination of kinetics and mass balances allows one to calculate the temperature profile required along a reactor to counterbalance this effect of wafer-to-wafer thickness non-uniformity. It also leads to alternative postulates for improving wafer-to-wafer uniformity of recycling of reactor effluent or the use of moving bed reactors.

17.3.2.4 Radial thickness uniformity

The very good within-wafer uniformity of film thickness is, as already mentioned, one of the notable features of LPCVD polysilicon layers. This good uniformity arises, also as already mentioned, because of the lack of any mass transfer limitations on growth. This can be shown to be the case very simply if we consider the rate of heterogeneous kinetics (j_s), for polysilicon deposition in terms of an overall heterogeneous rate constant (k_s') and the gas phase concentration (n_s) of silane; i.e., $j_s = k_s' n_s$. With a typical growth rate for polysilicon at 600°C of 10 nm min^{-1}, or $j_s \cong 10\ cm^{-2}\ s^{-1}$ under a silane partial pressure of ca. 20 Pa or $n_s \cong 10^{15}\ cm^{-3}$, the global rate constant k_s', will have a value of ca. 1 cm s^{-1}. Under the same deposition conditions of temperature and pressure D_s has a value of ca. $4 \times 10^3\ cm^2\ s^{-1}$. With a typical diffusion distance of several centimetres from the edge to the centre of the wafer the mass transfer coefficient (k_T) will have a value ca. $10\ cm^3\ s^{-1}$ and so $k_T \gg k_s'$ and deposition will be under kinetic control (cf. equation (17.1)).

17.3.3 LPCVD of doped polysilicon

17.3.3.1 Introduction

In Section 17.3.2(i) we have indicated a range of uses for polysilicon. For all of these applications there is a need to control layer structure. In addition, careful control of doping levels is also necessary. This can be done with post-deposition doping but this requires a high temperature (typically >900°C) in order to move the dopant atoms from the surface and to drive them into the layer. This high temperature diffusion step can cause thermal damage. One solution is to use *in situ* doping at the normal temperature of polysilicon deposition; i.e., ca. 600°C.

The procedure is the same as for depositing polysilicon except that there is an additional flow of the dopant gas. For example, for the incorporation of phosphorous, which of various possible impurity dopants has been the most extensively studied to date, phosphine diluted in an inert gas, such as nitrogen, is introduced into the reactor. The overall process can be represented by

(R7) $$SiH_4 + xPH_3/N_2 \rightarrow Si/P_x + xN_2 + (1.5x + 2)H_2$$

Deposition temperatures of typically 550–650°C and reactor pressures in the range 15–130 Pa (0.1–1 Torr) are comparable to those used for undoped polysilicon deposition. Flow rate ratios (γ) of PH_3/SiH_4 from 10^{-5} to 10^{-1} have been reported, but the most commonly used range is 10^{-3}–10^{-2}.

17.3.3.2 Deposition kinetics

As in the case of undoped polysilicon the actual chemistry of deposition is much more complex than the simple overall reaction (R7) would suggest. A consideration of the known gaseous and surface chemistry of silicon in the presence of phosphine implies an overall reaction scheme as shown in Table 17.2 and pictorially in Fig. 17.7 [16]. Reactions (R10), (R15) and (R14) are exactly the same as (R3), (R4) and (R5) discussed earlier for the case of undoped polysilicon. Phosphine with a lone pair of electrons might be expected to interact strongly with dangling bonds at a silicon surface. Evidence for the strong dissociative adsorption of PH_3 has been obtained [35, 36] by studies of the uptake of PH_3 on clean silicon surfaces. This strong adsorption of PH_3 effectively blocks much of the weakly adsorbing SiH_4 with its low sticking coefficient [36]. However, deposition can still occur via (R8), (R11) and (R17) because even though a large percentage of the surface area is occupied by PH_3, silylene could be expected to compete effectively with PH_3 for surface sites because of its unpaired electrons. Certainly in the gas phase it is known that SiH_2 inserts itself into Si–H and P–H bonds with near unity reaction probability [36]. For doped polysilicon deposition, therefore, one can expect a significant contribution to growth from silylene. Silylene is chosen as a likely candidate for an alternative precursor for silicon deposition because, while the activation energy for the forward reaction of (R8) is relatively high (e.g., ca. 250 kJ mol^{-1} under typical deposition conditions) [26, 37, 38] it is much less than that required for the formation of other silicon containing species such as SiH_3 [26, 27, 38–40].

Another gas-phase species which is likely to play an important role in doped polysilicon deposition is disilane. Early results of John and Purnell [41] indicated that the formation of Si_2H_6 from SiH_2 and SiH_4 was extremely rapid and more recent results by two independent groups [42, 43] have led to even higher rate constants for disilane formation. Furthermore, deposition of polysilicon from disilane is much more rapid than from monosilane and growth does not appear to be inhibited by the presence of PH_3 [44]. In addition, other work has shown [28, 28] that Si_2H_6 is much more strongly adsorbed on silicon than is SiH_4. Also the activation energy for the sticking coefficient of SiH_4 is ca. 0 kJ mol^{-1}, indicating direct chemisorption with no molecular precursor state, while that for Si_2H_6 the activation energy is ca. -6 kJ mol^{-1}, suggesting a molecular precursor prior to chemisorption lying below the initial state. These effects are

TABLE 17.2
Reaction scheme for polysilicon deposition

Gas-phase reactions	
$SiH_4(g) \underset{\overleftarrow{k}_R}{\overset{\overrightarrow{k}_R}{\Longleftrightarrow}} SiH_2(g) + H_2(g)$	(R8)
$SiH_4(g) + SiH_2(g) \underset{\overleftarrow{k}_D}{\overset{\overrightarrow{k}_D}{\Longleftrightarrow}} SiH_2(g) + H_2(g)$	(R9)
Adsorption equilibria	
$SiH_4(g) \underset{k_{dS}}{\overset{k_{aS}}{\Longleftrightarrow}} SiH_4(a)$	(R10)
$SiH_2(g) \underset{k_{dR}}{\overset{k_{aR}}{\Longleftrightarrow}} SiH_2(a)$	(R11)
$Si_2H_6(g) \underset{k_{dD}}{\overset{k_{aD}}{\Longleftrightarrow}} Si_2H_6(a)$	(R12)
$PH_3(g) \underset{k_{dP}}{\overset{k_{aP}}{\Longleftrightarrow}} PH_3(a)$	(R13)
$H_2(g) \underset{k_{dH}}{\overset{k_{aH}}{\Longleftrightarrow}} H_2(a)$	(R14)
Growth reactions	
$SiH_4(a) \xrightarrow{k'_4} 2H_2(a)$	(R15)
$Si_2H_6(a) \xrightarrow{k'_3} 2Si(s) + 3H_2(a)$	(R16)
$SiH_2(a) \xrightarrow{k'_2} Si(s) + 2H_2(a)$	(R17)

attributed to the Si–Si bonding orbitals of Si_2H_6 being closer in energy to silicon surface states than Si–H molecular orbitals. All these results would indicate that disilane is a very likely precursor for doped polysilicon deposition.

Overall, then, a reaction scheme of the form shown in the equations in

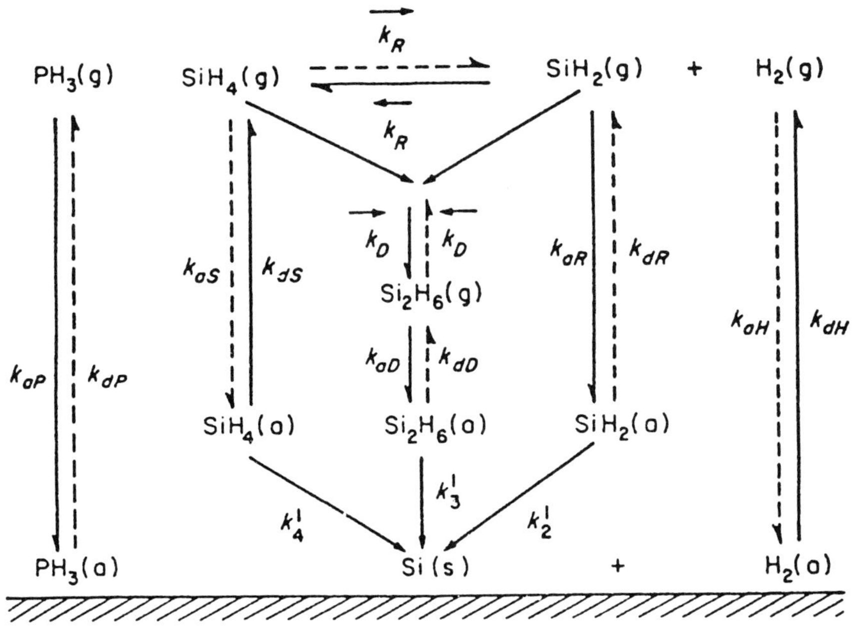

Fig. 17.7. Reaction scheme for the deposition of doped polysilicon [16].

Table 17.2 and in Fig. 17.7 is suggested. The ways in which the chemistry influences the properties of doped polysilicon layers is discussed in the next section.

17.3.3.3 Layer thickness

Figure 17.8 shows growth rates measured at the centre of wafers plotted against the PH_3/SiH_4 flow ratio. A detailed analysis has shown [13] that such a dramatic decrease in growth rate cannot simply be a dilution effect because keeping deposition conditions the same and substituting a N_2 flow for the PH_3/N_2 flow does not produce any marked fall in deposition rate. One is therefore led to the postulate already discussed that the PH_3 is blocking surface sites and inhibiting growth. An analytical consideration [16] of general gas-phase equilibria such as (R8) and (R9) together with general adsorption equilibria of the form (R10) and (R11) and heterogeneous decomposition of silane and a general silane species SiH_i allows the

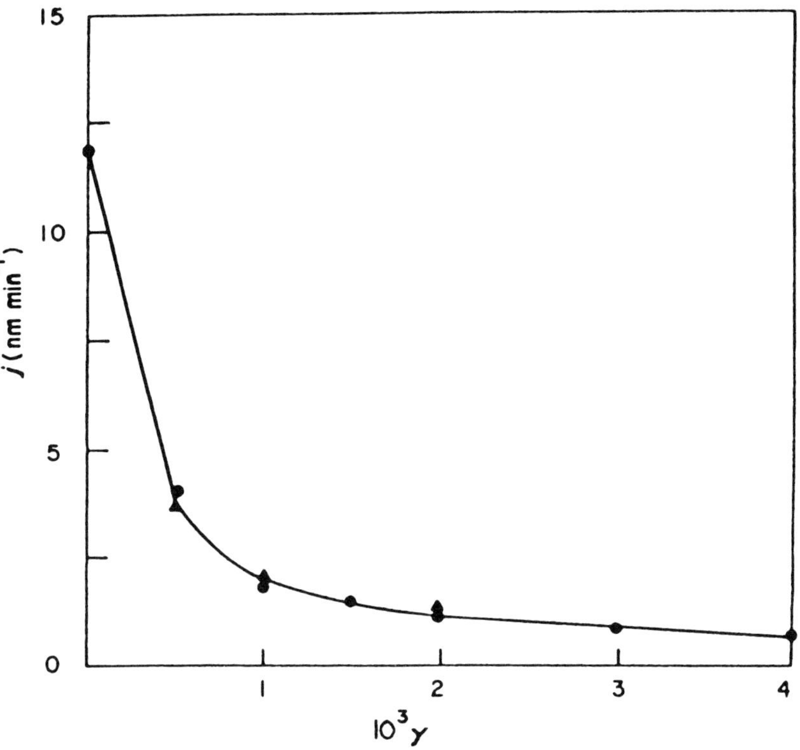

Fig. 17.8. Variation of growth rate of doped polysilicon with PH_3/SiH_4 flow ratio: deposition temperature 650°C; wafer spacing 5 mm [16].

growth rate, j, to be expressed as a function of the various thermodynamic and kinetic parameters and the gas flow rate ratio, γ:

$$\frac{1}{j} \cong \frac{1}{k_4' + b}\left(a\gamma + \frac{c}{n_s}\right), \tag{17.2}$$

where a, b and c contain thermodynamic constants for the various processes, k_4' is the heterogeneous rate constant for the decomposition of SiH_4 (R15), and n_s is the gas-phase concentration of SiH_4.

On the basis of equation (17.2), a plot of the inverse growth rate against γ should be linear with an intercept corresponding to the deposition of

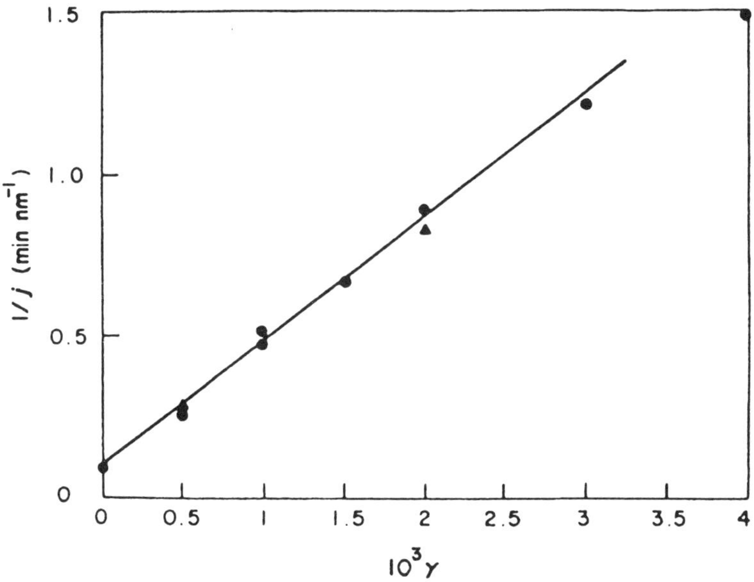

Fig. 17.9. Test of equation (17.2) for data taken from Fig. 17.8 [16].

undoped polysilicon. Figure 17.9 shows such a plot for data taken from Fig. 17.8. Up to $\gamma = 3 \times 10^{-3}$ the plot is linear with a correlation coefficient of 0.995. For $\gamma > 3 \times 10^{-3}$ there is a deviation from linearity, but this is not unexpected since there are indications that the PH_3 fractional coverage of the surface reaches a limiting value less than unity at a given temperature; hence the poisoning effect will fall away and the growth rate will tend to level off.

A numerical approach to the same overall reaction scheme as given above combined with the effect of mass transfer processes in the annular reactor flow region and the space between the wafers allows not just an analysis of results of growth rate, but also enables a prediction of rates to be made [20]. This is illustrated in Fig. 17.10. In this figure growth rates are plotted as a function of radial position on the wafer, and this is dealt with in the next section, but the reasonable agreement between theory and experiment of growth in the central region is apparent. The contribution to the overall growth rate from the SiH_4 surface reaction is shown shaded in the figure. Since this growth process is kinetically controlled the contribution is independent of wafer spacing.

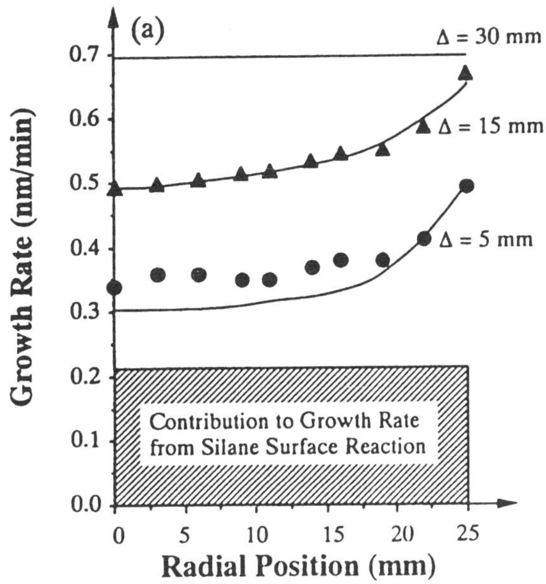

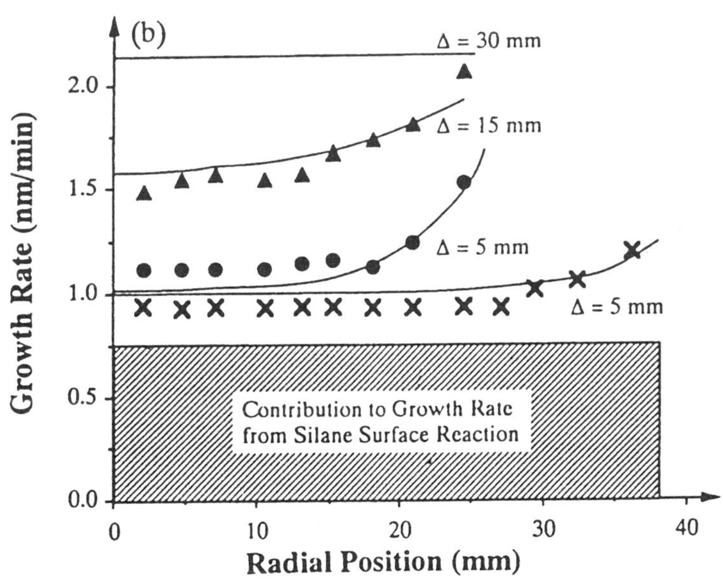

Fig. 17.10. Predicted and experimental growth rates for doped polysilicon deposition for various reaction conditions: (a) deposition temperature 625°C, $\gamma = 4 \times 10^{-3}$; (b) deposition temperature 650°C, $\gamma = 2 \times 10^{-3}$ [20].

17.3.3.4 Radial thickness uniformity

Figure 17.10 also shows a typical variation of growth rate with radial position on a wafer for *in situ* phosphorous doped polysilicon. The simple consideration of transport and kinetics given above shows that such a variation cannot be due to depletion of SiH_4 with radial distance in the interwafer spacing. In the presence of phosphorous doping, however, when there is strong adsorption of PH_3 and blocking of surface sites the adsorption of SiH_4 will, as already noted, be very effectively blocked because of the very low sticking coefficient for this species. As we have seen, this reduces the growth rate dramatically. In addition, however, it will change the route by which growth occurs. This postulate involves the very reactive intermediate SiH_2 generated by the gas-phase reaction (R8). The mass transfer coefficient for this species can be expected to be very similar to that for SiH_4 since the diffusion coefficient will not be too different and the diffusion length will be just the same. However, by comparing reaction probabilities it is estimated that [45] the rate constant k_s'' for growth from SiH_2 will be about 10^4 times that for growth from SiH_4 (k_s'): $k_s'' \cong 10^4\ \mathrm{cm\ s^{-1}}$. Therefore, growth from SiH_2 will be dominated by mass transport rather than being controlled by the kinetics of the deposition reaction.

In the absence of phosphorous doping this will not be too important since under the usual conditions of temperature and pressure for deposition the equilibrium (R8) lies well over to the left [38] and the main route for deposition will be via the adsorption and decomposition of SiH_4, which we have seen is not transport limited. Thus, layers will be of uniform thickness radially.

In conclusion, it can be said that the model depicted in Fig. 17.7 gives a good description of the observed results. A numerical analysis based on the same model is seen from Fig. 17.10 to also allow the radial growth rate dependence to be predicted and the agreement with experimental observations is reasonable.

17.4 *IN SITU* MONITORING OF CVD PROCESSES

17.4.1 Introduction

The chemical complexity of CVD processes in general has been illustrated in Figs. 17.2–17.4, while Fig. 17.7 has shown, in particular, the rich homogeneous chemistry which precedes the heterogeneous formation of polysilicon from the relatively simple molecule SiH_4. This phenomenon of pre-nucleation reactions is likely to be common to most, if not all, CVD processes since before a precursor molecule reaches the reaction surface it will experience a thermal energetic input. In a hot wall reactor, such as used in the LPCVD of polysilicon, the gas phase and the substrate surfaces are all at the same temperature and so molecular reorganisation and/or bond breaking could be expected, as indeed we have seen in the case for the silane chemistry. In a cold wall reactor, as is often used for the CVD of high temperature superconductors discussed below, precursor molecules will have to traverse a thermal boundary layer close to the substrate surface. The least this will do is to bring about a change in the physical properties of the gas (e.g., density, viscosity, diffusion coefficient), but in addition thermal molecular modifications are likely. If one is to begin to understand CVD processes in more detail then clearly it is necessary to be able to have some knowledge about this gas phase thermochemistry. For simple molecules *ab initio* calculations can provide both qualitative and quantitative insights into decomposition mechanisms, but for more complex species this is more difficult to achieve theoretically. Therefore there is a need for *in situ* studies of homogeneous CVD precursor chemistry and in recent years there has been an increasing amount of work done in this area [2, Ch.3]. It is likely that in the not too distant future all self respecting CVD practitioners will have one or more *in situ* chemical diagnostic monitors on their CVD equipment, be it for obtaining fundamental information about reaction mechanisms or for carrying out process analysis and control of commercial processes. One can illustrate the usefulness of in situ monitoring by reference to the decomposition of organometallic compounds used in the CVD of high temperature superconductors.

High temperature superconductors (HTS) were discovered by Bednorz and Mueller in 1986) [46] and this was followed by the preparation a year later [47] of $YBa_2Cu_3O_{7-x}$ (YBCO) with a value of T_c above the boiling point of liquid nitrogen. The first paper describing the preparation of thin films of YBCO by CVD appeared in 1988 [48]. Subsequently, the discoveries of an HTS containing Bi, Sr, Ca and Cu (BSCCO) and one containing Ti, Ba, Ca, and Cu (TBCCO) in 1988 were rapidly followed by methods describing the CVD of these materials [49, 50]. Since that time there have been several hundred papers on the CVD of YBCO and tens of papers on BSCCO and TBCCO. At the same time as the research of the new HTS materials has developed there has been intense commercial interest in them. In 1998 it was predicted [51] that the first commercial applications of HTS would be in electronic devices, and this was fulfilled in 1990 with a description of the use of a superconductor transistor in microwave amplifiers, oscillators, phase shifters and signal mixers [52], and with the first semiconductor-CMOS semiconductor circuits being described in 1993 [53]. By the year 2005 it is estimated that the world market for electronic devices based on HTS will be in excess of \$300 M [54].

In order to build electronic HTS devices it will be necessary to produce thin films of HTS by techniques which are compatible with semiconductor fabrication lines, which allow close control of layer properties, which produce conformal thin films over large areas and have a high level of intra- and inter-sample uniformity, and which give a high sample throughput. As has been indicated earlier, CVD is just such a technique, and in addition it has the features of allowing a range of oxidants to be used (e.g., O_2, N_2O, O_3) of permitting low temperature deposition (e.g., plasma enhanced CVD, photoCVD), and of maintaining a high partial pressure of oxidant, which is often desirable [55]. However, as we have seen, the C in CVD reminds us that the technique is very dependent on precursor chemistry, deposition chemistry, and reaction parameters, and the CVD of YBCO and other HTS is also a multiparameter problem. Much of what has been reported in the literature has been of an empirical nature and there is a need for understanding both precursor and deposition chemistry. Some preliminary steps in this direction have been taken in the last few years.

For the CVD of YBCO, β-diketonates with the general molecular formula shown below have been found to be the most suitable precursors [56, 57].

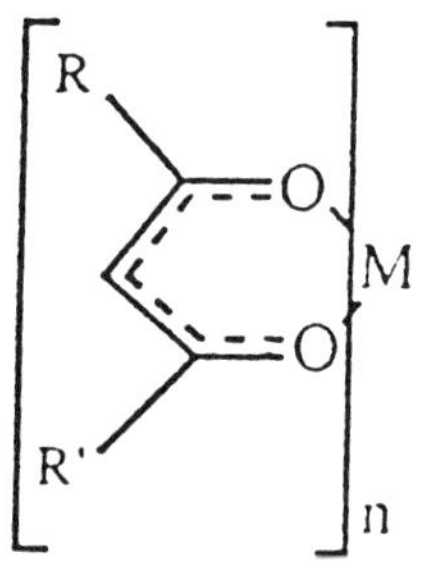

M = Y	n = 3
M = Ba	n = 2
M = Cu	n = 2

For Y and Cu metal complexes of HTMHD, where TMHD = 2,2,6,6-tetramethylheptane-3,5-dione (i.e., R = R′ = ^{t}Bu) are frequently used because of their good volatility and stability. For barium, however, no such stable volatile precursor was available [56] until recently. Now several have been reported [58–61], one of which is [Ba(TDFND)$_2$.tetraglyme] [59, 60], where TDFND = 1,1,1,2,2,3,3,7,7,8,8,9,9,9-tetradecafluorononane-4,6-dione (i.e., R = R′ = C_3F_7). Little has been reported, though, on the thermal decomposition of any of these compounds, especially in the presence of oxygen which is necessary for the growth of HTS layers. The only studies which have been carried out have been by mass spectrometry for the thermal decomposition of [Cu(TMHD)$_2$] and [Y(TMHD)$_3$] [62–63]. In situ FTIR spectroscopy would be a useful complementary technique since it would probe the decomposition directly in the reactor and at the same time give information about bonding changes. Therefore in order to try and obtain a better understanding of the chemical decomposition of [Y(TMHD)$_3$], [Cu(TMHD)$_2$] and [Ba(TDFND)$_2$.tetraglyme], which we have previously used as precursors for the MOCVD of HTS thin films [59, 64], we have studied their thermal decomposition in the gas phase in the absence and presence of oxygen under typical deposition conditions [65–67], using both FTIR and mass spectrometry, although here we only describe results for the decomposition of the yttrium precursor.

The experimental arrangement used for the FTIR in this study has already been described elsewhere [65]. Briefly, it consisted of an FTIR spectrometer, a gas-handling facility, a reactor made up of an evaporator, a reactor tube and a thermostatted jacket, and a vacuum system. The yttrium precursor was heated in a stainless steel evaporator and a typical temperature [64] of 145°C was used. The precursor vapour was carried to the reactor by a flow of argon. To prevent condensation of the vapour before it reached the reactor the wall of the jacket was heated to 200°C. The reactor was made of quartz with a diameter of 2.5 cm and a total length of 27 cm. The temperature range investigated in the reaction zone (15 cm) was 200 to 700°C and the total pressure was 1.33×10^3 Pa (10 Torr). Several spectra of the vapour in the decomposition zone were obtained at each temperature for the spectral region extending from 4500 cm^{-1} to 700 cm^{-1}.

Initial attempts to obtain mass spectra of precursors used for the MOCVD of HTS thin films were unsuccessful because a number of serious problems were encountered in introducing the heavy molecular weight complexes into the mass spectrometer. These problems were partly associated with contamination of the quadrupole, but, more significantly, with condensation of the precursors at the inlet of the spectrometer. To overcome the condensation problem a heated box at the entrance zone of the spectrometer for the purpose of maintaining the sampling and analyser heads above 100°C at all times proved to be successful in preventing the orifice becoming blocked. An extension, which was also heated, was built onto this hot box to include a small reactor and a precursor pot, but valve units were eliminated in this extension in order to prevent any long term memory effects. The quartz reactor had a diameter of 0.8 cm and was ca. 45 cm long. The precursor vapour was carried to the reactor by a flow of Ar. For experiments with oxygen there was a mixed flow of this gas with Ar. The temperature range investigated in the reaction zone was 160 to 600°C and the total pressure was 3.33×10^2 Pa (2.5 Torr). Several spectra were obtained at each temperature for the m/e^- range 40 to 640.

A typical IR spectrum for $[Y(TMHD)_3]$ at 200°C is shown in Fig. 17.11. It is interesting to note the sensitivity of the FTIR instrument of about 0.01%. Characteristic spectral bands appear at 2967 cm^{-1}, which can be suggested to be due to the $C—H_x$ stretch of the tertiary butyl groups, at

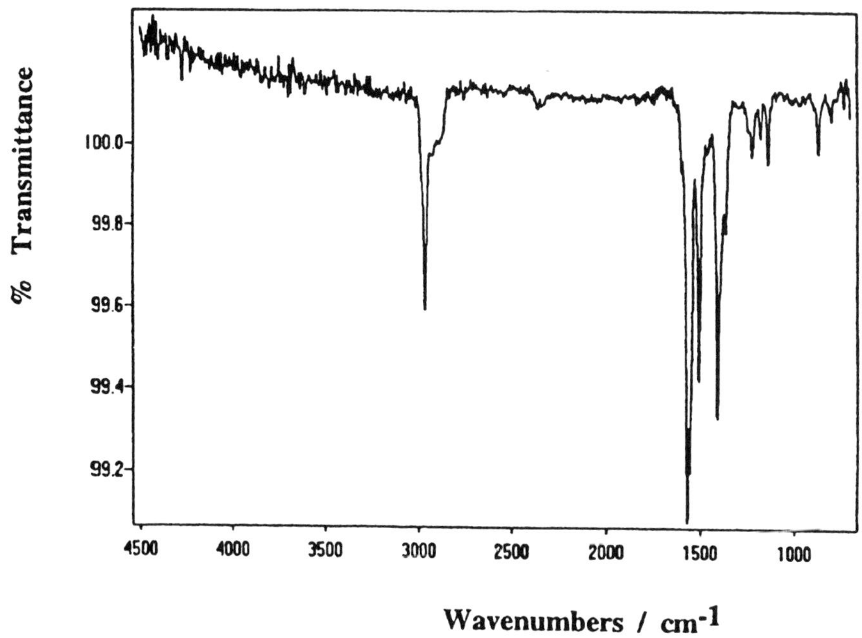

Fig. 17.11. Typical IR spectrum of $[Y(TMHD)_3]$ [65].

1571–1556 cm^{-1} arising from C=C or C=O stretching in the rings with the higher frequency corresponding to a C=C stretching [68], and at 1508 cm^{-1} and 1408 cm^{-1} also due to ring C=C stretching. The absorption coefficient (α) was calculated from the relation $\alpha = [\log(I_0/I)]/d$ where I_0 is the background intensity, I the signal intensity and d the length of the tube between the two windows. Values of the absorption coefficient normalised to the corresponding mean value obtained at 200°C are plotted as a function of temperature in Fig. 17.12.

Up to about 600°C and in the absence of oxygen (Fig. 17.12(a)) no significant change is observed of the relative absorption coefficient with temperature for the absorptions due to vibrations from both the ring and the ring substituent, t-Bu. Above about 600°C the absorptions arising from the C=C stretches begin to decrease slightly more rapidly than the absorptions attributed to the C–H stretch. This probably indicates that the structure of the rings is less stable than the hydrocarbon moiety. The stability of the C–H bonds is not surprising since it is well known [69] that thermal cracking of such bonds requires considerably higher temperatures than those used here.

In the presence of oxygen, up to about 400°C again little variation of absorption coefficient with temperature is observed for all the four main bands (Fig. 17.12(b)). Above ca. 400°C there is a much more dramatic decrease in the relative absorption coefficient with temperature, and above ca. 650°C none of the four initial vibrational modes could be detected. The appearance of a peak in the spectrum at 2356 cm^{-1} due to CO_2 for temperatures >450°C shows that oxidation of the entire organic structure is occurring, although again the hydrocarbon moiety seems to be slightly more resistant to decomposition than the rings.

By varying the temperature of the precursor pot, different partial pressures of [Y(TMHD)$_3$] could be generated, as was reflected by different values of absorbances at 200°C. Comparison of absorbance values at a temperature where decomposition was occurring, such as at 460°C (cf. Fig. 17.12(b)), with the corresponding values at 200°C then allowed determination of the order of reaction with respect to [Y(TMHD)$_3$]. From the data plotted in Fig. 17.13 it can be seen that the reaction is first order. The data is plotted for the C=C stretch at 1571 cm^{-1}, but a similar result was obtained for the other absorbance frequencies. For a fixed precursor partial pressure (i.e., at a constant precursor pot temperature) variation of the oxygen flow rate allowed determination of the reaction order with

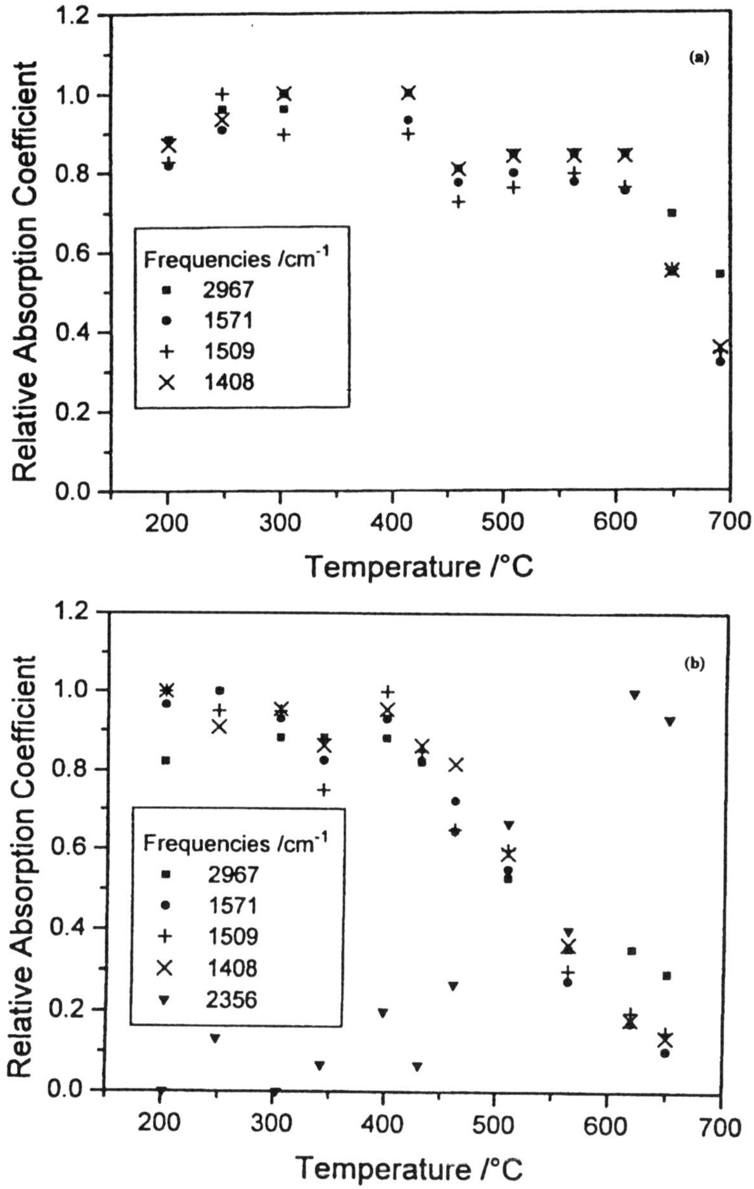

Fig. 17.12. Plots of relative absorption coefficients as a function of temperature for $[Y(TMHD)_3]$ vapour under a total pressure of 1.33×10^3 Pa (10 Torr) (a) without oxygen and (b) with oxygen [65].

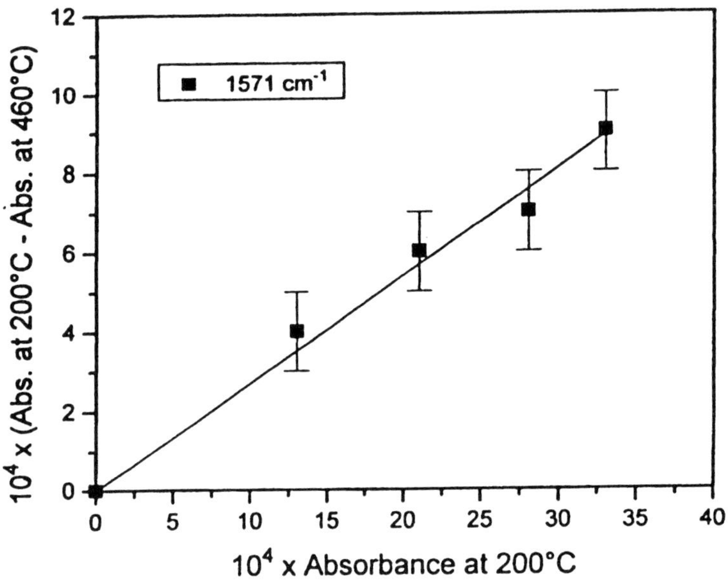

Fig. 17.13. Plot to determine reaction order with respect to precursor partial pressure [66].

respect to the oxidant, and again this was found to be first order—Fig. 17.14. Further experiments are underway to obtain similar plots to Figs. 17.13 and 17.14 at different temperatures in order to obtain the activation energies and to allow further consideration of the decomposition mechanism. This will be assisted by the results of mass spectrometric analysis.

Figure 17.15 shows a typical mass spectrum for $[Y(TMHD)_3]$ at 200°C. The four major peaks of highest mass are those at m/e^- of 638, 581, 455 and 183. The first peak corresponds to the single positively charged parent compound $[Y(TMHD)_3]$, the second to the parent with the loss of a single t-Bu, the third to the parent with the loss of one ligand (i.e., $[Y(TMHD)_2]^+$), and the fourth to a single ligand moiety—$TMHD^+$.

Figure 17.16 shows the temperature variation in the absence of oxygen of the four signals mentioned above. Above about 450°C there is clear evidence of decomposition of the parent precursor and of the formation of free ligand, TMHD, with an indication that at 600°C this also starts to break up. All this is consistent with the FTIR data which is summarised at the top of the figure. As noted earlier, these data show no significant

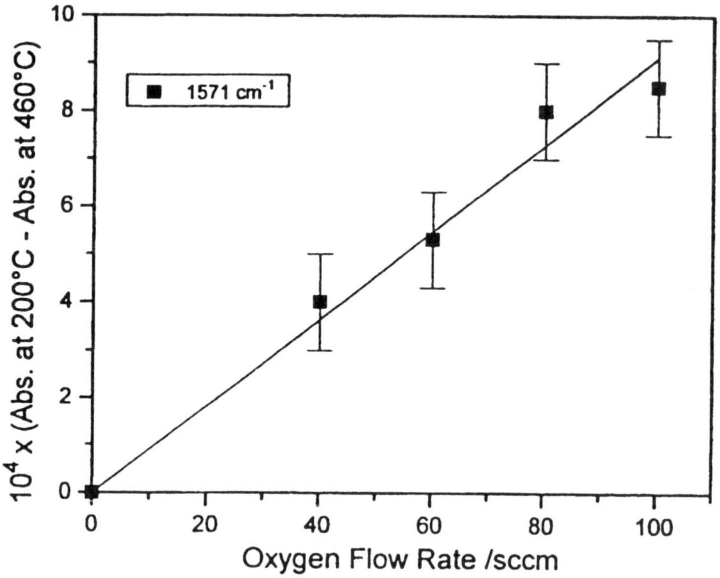

Fig. 17.14. Plot to determine reaction order with respect to oxygen partial pressure [66].

variation up to 600°C which indicates that the C=C bonding in the ligand is being retained.

Results for the temperature variation of the mass spectrometric signals in the presence of oxygen are shown in Fig. 17.17. It can be seen that the precursor starts to decompose above about 400°C, but now at the same temperature the TMHD ligand also decomposes. The FTIR data, again summarised at the top of the figure, show the break-up of the TMHD ligand as well. However, whereas the FTIR results indicate a steady decline with temperature of the amount of TMHD in the gas phase, the mass spectrometric data suggest that above 500°C TMHD is being formed faster than it is being oxidised. This effect may be a result of different conditions in the two reactor systems. For example, the space time for the mass spectrometric reactor system was about twice that of the FTIR system, and so there is a correspondingly longer time required to treat a reactor volume of feed in the former case. The FTIR results presented above have shown that the oxidation of the precursor is first order in the precursor itself and in oxygen. If we consider the reaction to occur in consecutive

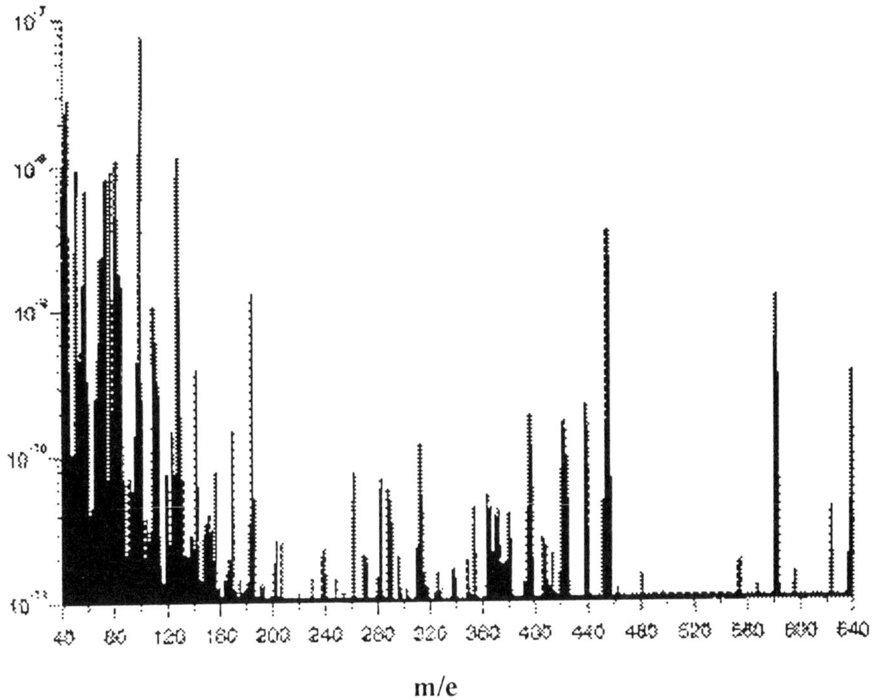

Fig. 17.15. Typical mass spectrum of $[Y(TMHD)_3]$ taken at 200°C [66].

steps of ligand release followed by oxidation, and if the oxidation is slow compared with free ligand formation, then for plug flow systems, such as we have here, the amount of intermediate (i.e., the TMHD ligand) can be higher in the system with the larger space time [70]. This point is being investigated further, but it illustrates the importance of considering reactor design in addition to decomposition kinetics.

17.5 CONCLUSIONS

This review has been primarily concerned with the kinetics and mechanisms of reactions occurring at or near to a heated surface during the formation of thin films by CVD. My interest in these processes was engen-

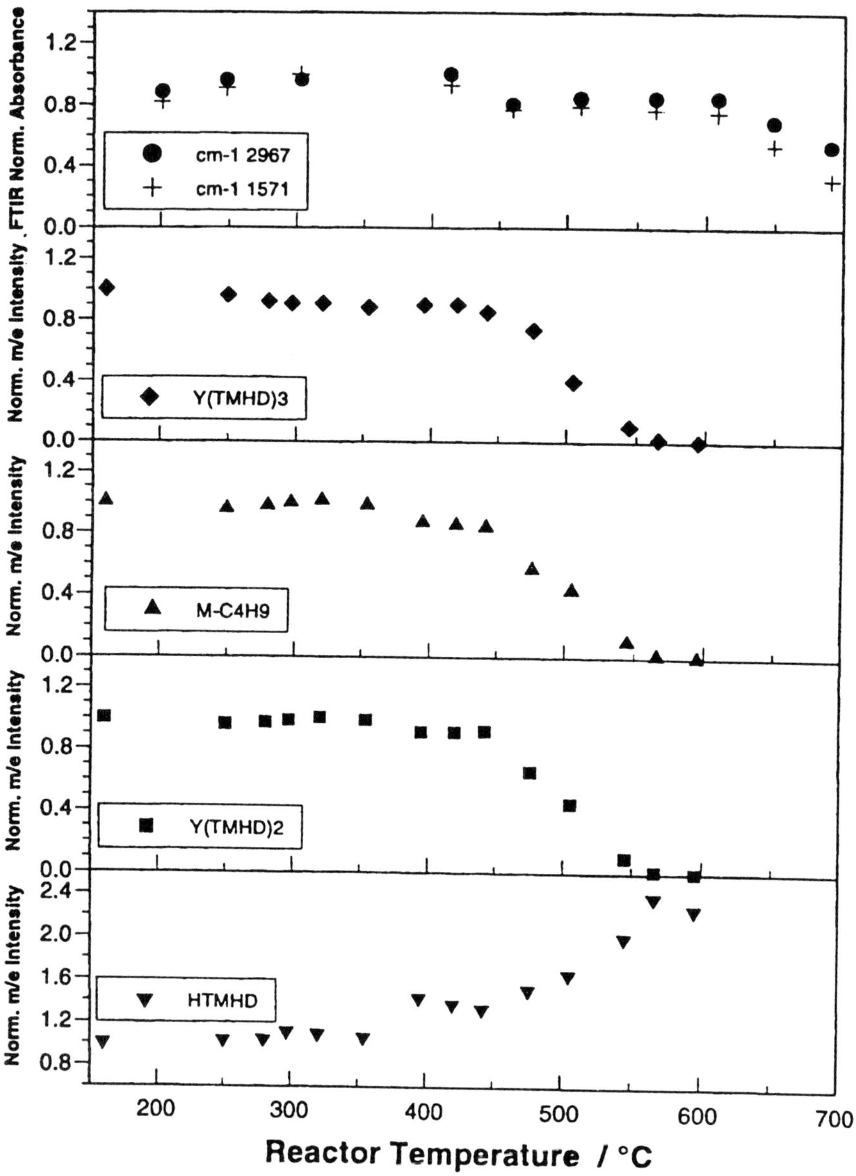

Fig. 17.16. Plots of mass spectral intensities and relative absorption co-efficients as a function of temperature for [$Y(TMHD)_3$] vapour without oxygen [66].

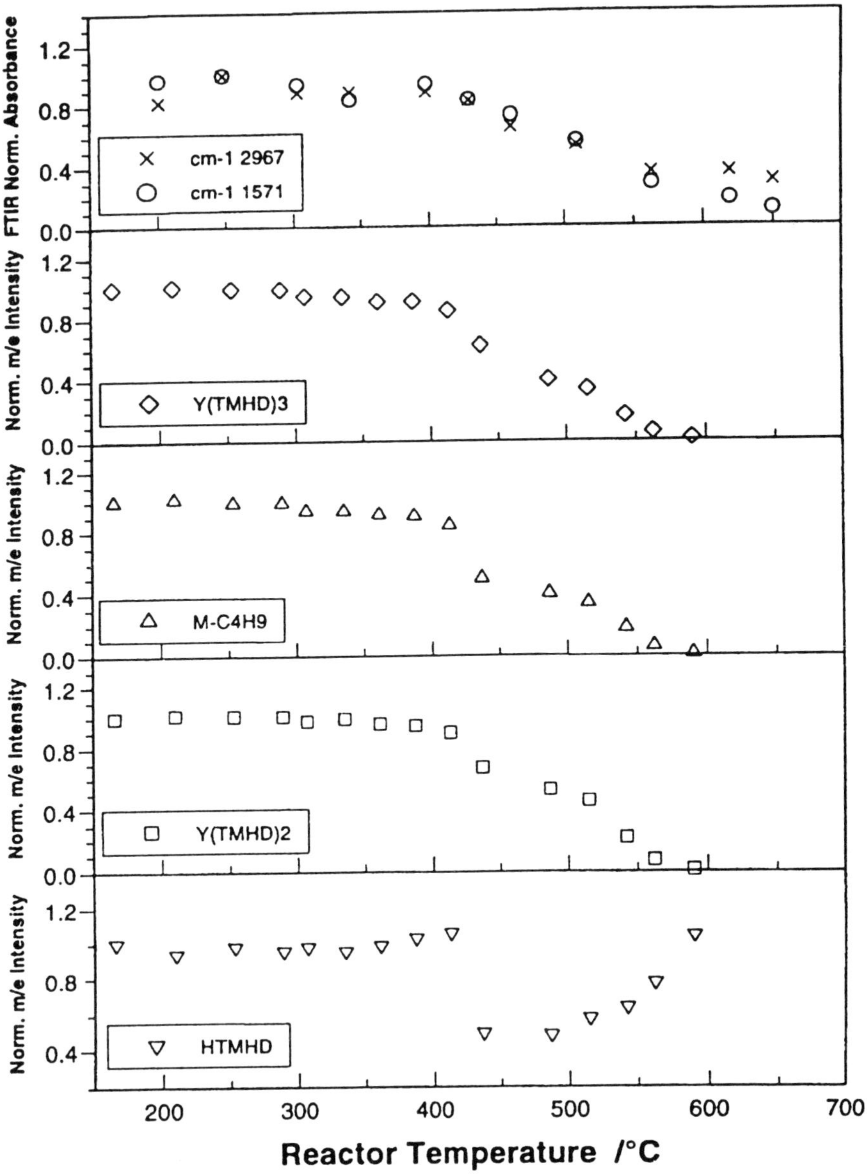

Fig. 17.17. Plots of mass spectral intensities and relative absorption co-efficients as a function of temperature for [$Y(TMHD)_3$] vapour with oxygen [66].

dered by earlier electrochemical studies using rotating electrodes under the stimulating supervision of John Albery. In fact, the first CVD system I built was a rotating disk operating at ca. 1300°C for investigating the growth of epitaxial silicon by hydrogen reduction of $SiCl_4$ [71–73]. Clearly, there are many similarities between electrochemical and CVD systems. Both involve homogeneous phases, both require transport of reactants and products through the homogeneous phase to and from the heterogeneous phase, both can operate under kinetic, transport or mixed control, and both can be expected to show kinetic complexities. In the case of CVD there are likely to be rather more such complexities occurring in the homogeneous phase because of the extension of the energetic field at the reaction surface into the homogeneous phase. This article has attempted to give an idea of the extent of the kinetics occurring in CVD processes, as a result of the thermal field near the reaction surface. There is, I believe, a strong overlap with the interests of John Albery which have always been with homogeneous and heterogeneous kinetics, and it is a pleasure, therefore, to dedicate this review to him.

ACKNOWLEDGEMENTS

Many of the results described in this review have been obtained over the last 25 years by a large number of co-workers and colleagues with the financial support and help from a wide range of government, industrial and academic funding. I should like to acknowledge all the work and aid which have made the studies possible and so enjoyable.

References

[1] J.L. Vossen and W. Kern (eds.), Thin Film Processes II (Academic, London 1991).
[2] M.L. Hitchman and K.F. Jensen (eds.), Chemical Vapor Deposition (Academic, London, 1993).
[3] W.E. Sawyer and A. Mann, (1880), U.S. Patent 229, 335.
[4] E.M. Sherwood and J.M. Blocher, J. Metals 17 (1965) 594.
[5] D.L. Sawyers, Strong Poison (Gollancz, London, 1930).
[6] R. Hoelbling, Z. Agnew. Chem. 40 (1927) 655.
[7] G.K. Teal, J.R. Fisher and A.W. Treptow, J. Appl. Phys. 17 (1946) 879.
[8] M.L. Hitchman, Progr. Crystal Growth Characterisation 4 (1981) 249.

[9] W. Kern and K.K. Schuegraf in: K.K. Schuegraf, ed., Handbook of Thin Film Deposition Processes and Techniques, Ch. 1 (Noyes Publications, 1988).
[10] E. Tucker, European Semiconductor, March (1990) 53.
[11] M.L. Hammond and R.J. Gieske, Tempress Microelectronics, Los Gatos, CA, Technical note 501, 1976.
[12] M.L. Hitchman, J. Kane and A.E. Widmer, Thin Solid Films 59 (1979) 231.
[13] M.L. Hitchman and J. Kane, J. Cryst. Growth 55 (1981) 485.
[14] R.S. Rosler, Solid State Technol. 20(4) (1977) 63.
[15] M.L. Hitchman, Vacuum 34 (1984) 377.
[16] M.L. Hitchman, W. Ahmed, S. Shamlian and M. Trainor, Chemtronics 2 (1987) 147.
[17] G. Harbeke, L. Krausbauer, E.F. Steigmeier, A.E. Widmer, H.F. Kappert and G. Neugebauer, RCA Rev. 44 (1983) 187.
[18] R.A. Levy and M.L. Green, J. Electrochem. Soc. 134 (1987) 37C.
[19] M.L. Hitchman, A.D. Jobson and L.F.T. Kwakman, App. Surf. Sci. 38 (1989) 312.
[20] K.F. Jensen, M.L. Hitchman and W. Ahmed in: J.-O. Carlsson and J. Lindstrom, eds., Proc. 5th Eur. Conf. CVD (University Press, Uppsala, Sweden, 1985, p.144).
[21] A.J. Learn, J. Electrochem. Soc. 132 (1985) 390.
[22] A. Yeckel, S. Middleman and A.Y. Hochberg, J. Electrochem. Soc. 136 (1989) 2038.
[23] M.J.-P. Duchemin, M.M. Bonnet and M.F. Koelsch, J. Electrochem. Soc. 125 (1978) 637.
[24] C.H.J. van den Brekel and L.J.M. Bollen, J. Cryst. Growth 54 (1981) 310.
[25] R. Cadoret and F. Hottier, J. Cryst. Growth 64 (1983) 583.
[26] J.M. Jasinski, B.S. Meyerson and B.A. Scott, Ann. Rev. Phys. Chem. 38 (1987) 109.
[27] J.M. Jasinski and S.M. Gates, Acc. Chem. Res. 24 (1991) 9.
[28] S.M. Gates, Surf. Sci. 195 (1988) 307.
[29] S.M. Gates, C.M. Greenlief, S.K. Kulkani and H.H. Sawin, J. Vac. Sci. Technol. A8 (1990) 2965.
[30] Y.S. Chiang, in: H.R. Tuft and R.R. Burgess, eds., Semiconductor Silicon (ECS, Pennington, N.J., 1973, p. 285).
[31] V. Ban, H. Tsuchikawa and K. Maeda, in: H.R. Tuft and R.R. Burgess, eds, Semiconductor Silicon (ECS, Pennington, N.J., 1973 p. 292).
[32] K.F. Roenigk, K.F. Jensen and R.W Carr, J. Phys. Chem. 91 (1987), 5732.
[33] B.A. Scott, R.D. Estes and J.M. Jasinski, J. Chem. Phys. 89 (1988) 2544.
[34] K.F. Roenigk and K.F. Jensen, J. Electrochem. Soc. 132 (1985) 448.
[35] B.S. Meyerson and W. Olbricht, J. Electrochem. Soc. 131 (1984) 2361.
[36] B.S. Meyerson and M.L. Yu, J. Electrochem. Soc. 131 (1984) 2366.
[37] J.H. Purnell and R. Walsh, Proc. Roy. Soc. Ser. A 293 (1966) 543.
[38] C.G. Newman, H.E. ONeal, M.A. Ring, F. Leska and N. Shipley, Int. J. Chem. Kinet. 21 (1979) 1167.
[39] M.E. Coltrin, R.J. Kee and J.A. Miller, J. Electrochem. Soc. 131 (1984) 425.
[40] M.E. Coltrin, R.J. Kee and G.H. Evans, J. Electrochem. Soc. 136 (1989) 819.
[41] P. John and J.H. Purnell, J. Chem. Soc. 69 (1973) 1455.
[42] G. Inoue and M. Suzuki, Chem. Phys. Lett. 122 (1985) 361.
[43] J.M. Jasinski and J.O. Chu, J. Chem. Phys. 88 (1988) 1678.

[44] S. Nakayama, H. Yonezawa and J. Murota, Jpn. J. Appl. Phys. 23 (1984) 4493.
[45] R.F.C. Farrow, J. Electrochem. Soc. 121 (1974) 899.
[46] J.G. Bednorz and K.A. Mueller, Z. Phys. B 64 (1986) 189.
[47] M.K. Wu, J.R. Ashburn, C.J. Torng, P.H. Horr, R.L. Meng, L. Gao, Z. J. Huang, Y.Q. Wang and C.W. Chu, Phys. Rev. Letts 58 (1987) 908.
[48] A.D. Berry, K.D. Gaskill, R.T. Holm, E.J. Cukauskas, R. Kaplan and R.L. Henry, Appl. Phys. Lett. 52 (1988) 1743.
[49] A.D. Berry, R.T. Holm, E.J. Cukauskas, M. Fatemi, D.K. Gaskill, R. Kaplan and W.B. Fox, J. Crystal Growth 92 (1988) 344.
[50] D.S. Richeson, L.M. Tonge, J. Zhao, J. Zhang, H.O. Marcy, T.J. Marks, B.W. Wessels and C.R. Kannewurf, Appl. Phys. Lett. 54 (1989) 2154.
[51] W. Gosling, in: A. Conway, ed., Superconductivity: The Opportunities, Implications and Applications (IBC Technical Services Ltd, London 1988, p. 25).
[52] Science and the Citizen, Scientific American 264 (1991) 15.
[53] Industry News, Semiconductor International, April 1993, p. 18.
[54] News Item, Chem. Eng. P. (Sept. 1989), p. 44.
[55] J. Zhao, C.S. Chern, Y. Q. Li, D.W. Noh, P. Norris, P. Zawadzki, B. Kear and B. Gallois, J. Cryst. Growth 107 (1991) 699.
[56] M.L. Hitchman, D.D. Gilliland, D.J. Cole-Hamilton and S.C. Thompson, in: D. Holland, ed., New Materials and their Applications (Inst. Phys. Conf. Ser. III), (Institute of Physics, Bristol, 1991 p. 305).
[57] M. Leskela, H. Molsa and L. Niinisto, Supercond. Sci. Technol. 6 (1993) 627.
[58] S.C. Thompson, D.J. Cole-Hamilton, D.D. Gilliland, M.L. Hitchman and J.C. Barnes, Adv. Mat. Opt. Electron. 1 (1992) 81.
[59] S.H. Shamlian, M.L. Hitchman, S.L. Cook and B. C. Richards, J. Mater. Chem. 4 (1994) 81.
[60] M.L. Hitchman, S.H. Shamlian, D.D. Gilliland, D.J. Cole-Hamilton, J.A.P. Nash, S.C. Thompson and S.L. Cook, J. Mater. Chem. 5 (1995) 47.
[61] K. Timmer, K.I.M.A. Spee, A. Mackor and H.A. Meinema, Inorg. Chim. Acta 190 (1992) 109.
[62] A.F. Bykov, P.P. Semyannikov and K. Igumenov, J. Thermal Anal. 38 (1992) 1463.
[63] A.F. Bykov, P.P. Semyannikov and K. Igumenov, J. Thermal Anal. 38 (1992) 1477.
[64] B.C. Richards, S.L. Cook, D.L. Pinch, G.W. Andrews, G. Lengeling, B. Schulte, H. Juergenson, Y.Q. Shen, P. Vase, T. Freltoft, C.I.M.A. Spee, J.L. Linden, M.L. Hitchman, S.H. Shamlian and A. Brown, Physica C 252 (1995) 229.
[65] A.Y. Kovalgin, F. Chabert-Rocabois, M.L. Hitchman, S.H. Shamlian and S.E. Alexandrov, J. Phys. IV, Colloque C5 (1995) 357.
[66] M.L. Hitchman, S.H. Shamlian, G.G. Condorelli and F. Chabert-Rocabois, J. Alloys Compd. 251 (1997) 297.
[67] G.G. Condorelli, M.L. Hitchman, A.Y. Kovalgin and S.H. Shamlian, in: M.D. Allendorf and C. Bernard, eds., Chemical Vapor Deposition (The Electrochemical Society, Pennington, N.J., 1997, p. 901).
[68] K. Nakamoto, Infrared Spectra of Inorganic and Co-ordination Compound (John Wiley, N.Y., 1963, p. 216).

[69] P. Wiseman, An Introduction to Industrial Organic Chemistry (Applied Science Publishers, London, 1979, p. 69).
[70] O. Levenspiel, Chemical Reaction Engineering, 2nd edn. (John Wiley, N.Y., 1962, p. 177).
[71] M.L. Hitchman, J. Crystal Growth 48 (1980) 394.
[72] M.L. Hitchman and B.J. Curtis, Progr. Crystal Growth Characterisation 4 (1981) 283.
[73] M.L. Hitchman and B.J. Curtis, J. Crystal Growth 60 (1982) 43.

Chapter 18

The Mechanism of Methanol Electro-oxidation

ANDREW HAMNETT

18.1 INTRODUCTION

The electro-oxidation of methanol is not only an important reaction in its own right, since it is one of the primary reactions in modern fuel-cell technology, but it has important generic characteristics, representing, as it does, a whole class of electrocatalytic processes. Although the literature was reviewed extensively by Parsons and Van der Noot [1] in 1988, and more recently by the author [2] there has been an explosion of knowledge since then, largely driven by modern *in situ* spectroscopic and spectrometric techniques, and this work is summarised for platinum below. The structure of this review is to consider, in order, the process of chemisorption, and the identification of adsorbed intermediates, the kinetics of the adsorption process itself, the processes following adsorption, and in particular those processes leading to CO_2 formation, and the action of morphological features in enhancing these processes.

18.2 THE ADSORPTION OF METHANOL AND THE NATURE OF THE ADSORBATE

18.2.1 Electrochemical studies

The chemisorption process of methanol on clean platinum surfaces has been found to be far faster than the steady-state oxidation rate for methanol, with the result that the initial electro-oxidation current is found rapidly to decay with time. The origin of this decrease is now known to be the accumulation of adsorbed intermediates on the surface whose further oxidation to CO_2 is slow, and the rate differences allow us to separate the process of chemisorbate formation from that of the subsequent oxidation of the chemisorbed intermediates to CO_2. The nature of the intermediates can, in principle, be studied purely electrochemically by measuring separately the charge, Q_{ads}, required to chemisorb methanol, the charge Q_H required to cover the remaining the surface with adsorbed hydrogen (which can also be measured for the intermediate-free surface, as Q_H^0, and used to estimate the total surface area assuming a charge of 0.21 mC per true cm^2) and the charge Q_{ox} required completely to oxidise the adsorbed fragments.

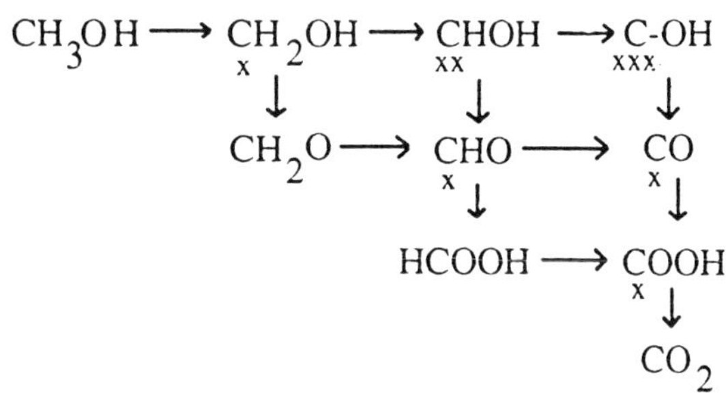

Given that there are a limited number of possible intermediates, we can draw up a scheme of squares of the form shown above [3], where the "x" symbolises a Pt—C bond, and which summarises the following reactions:

$$CH_3OH + Pt(s) \rightarrow Pt—CH_2—OH + H^+ + e^- \quad (18.1)$$

$$Pt\text{-}CH_2—OH + Pt(s) \rightarrow Pt_2CHOH + H^+ + e^-, \quad (18.2)$$

$$Pt_2CHOH + Pt(s) \rightarrow Pt_3COH + H^+ + e^-, \quad (18.3)$$

$$Pt_3COH \rightarrow Pt\text{-}CO + 2Pt(s) + H^+ + e^-, \quad (18.4)$$

$$Pt(s) + H_2O \rightarrow Pt—OH + H^+ + e^-, \quad (18.5)$$

$$Pt—OH + Pt—CO \rightarrow Pt—COOH, \quad (18.6a)$$

or

$$Pt—CO + H_2O \rightarrow Pt—COOH + H^+ + e^-, \quad (18.6b)$$

$$Pt—COOH \rightarrow Pt(s) + CO_2 + H^+ + e^-. \quad (18.7)$$

Additional reactions described in the scheme above include:

$$\text{Pt—CH}_2\text{OH} \rightarrow \text{Pt(s)} + \text{HCHO} + \text{H}^+ + \text{e}^-, \tag{18.8}$$

$$\text{Pt}_2\text{CHOH} \rightarrow \text{Pt(s)} + \text{Pt—CHO} + \text{H}^+ + \text{e}^-, \tag{18.9}$$

$$\text{Pt—CHO} + \text{Pt—OH} \rightarrow 2\text{Pt(s)} + \text{HCOOH}, \tag{18.10a}$$

or

$$\text{Pt—CHO} + \text{H}_2\text{O} \rightarrow \text{Pt(s)} + \text{HCOOH} + \text{H}^+ + \text{e}^-, \tag{18.10b}$$

$$\text{Pt}_3\text{C—OH} + \text{Pt—OH} \rightarrow 3\text{Pt(s)} + \text{Pt—COOH} + \text{H}^+ + \text{e}^-, \tag{18.11}$$

or

$$\text{Pt}_3\text{C—OH} + \text{H}_2\text{O} \rightarrow 2\text{Pt(s)} + \text{Pt—COOH} + 2\text{H}^+ + 2\text{e}^- \tag{18.12}$$

The difficulties faced in analysing a scheme of this complexity by purely electrochemical means are clearly formidable, but, at the least, potential step measurements of the type described above should give unambiguous and reproducible results provided: (a) that the potential protocol adopted to clean the surface prior to each measurement does not alter the morphology of the surface; (b) that the surface, once cleaned, will remain clean indefinitely, save for such processes as we wish to study; (c) the *only* contribution to the anodic charge observed on stepping the potential from the anodic limit usually chosen for cleaning the surface (~1.2–1.6 V) is the chemisorption of methanol to form adsorbed fragments.

All three of these assumptions have been questioned: Itaya and coworkers have shown [4] that cycling of a Pt(111) surface up to 1.5 V gives rise to substantial roughening, in which all long-range coherence is lost and relaxation of the resultant disordered surface is slow. The second assumption requires very high degrees of purity of reagents and also requires the solutions to be changed very frequently to prevent contamination by residual partially oxidised organic species building up in solution. The third assumption is usually justified on the basis that the measured steady-state current, once chemisorption is complete, is far smaller than the currents observed immediately after the potential step, but this is not unambiguous: oxidation of methanol by parallel routes, one leading to complete oxidation to CO_2 and the other to the formation of relatively unreactive intermediates may be taking place, and there is strong evidence that this is indeed the case for formic acid oxidation.

These different caveats go far to explain the difficulty experienced in the early literature in reconciling results from different groups. Measurements by Bagotzky and Vassilyev [5] at 0.4 V, using an upper potential limit of 1.2 V for 20 sec to clean the electrode, and an electrolyte of 0.1 M CH_3OH in 0.5 M H_2SO_4 suggest that the same charge is required to oxidise the chemisorbed intermediate as to form it. Since six electrons are required fully to oxidise each methanol molecule, this implies that the intermediate must require three electrons to form. In addition, their measurements showed that the number of electrons per platinum site required to oxidise the chemisorbate, n_{eps}, is one, implying that the intermediate must be co-ordinated to three Pt sites. This led Bagotzky and Vassilyev to postulate that the methanol chemisorption process must involve reactions (1)–(3) only above, with further oxidation of the Pt_3C–OH being slow.

Unfortunately, whilst the Russian group strongly favoured the Pt_3C–OH species as being the dominant intermediate, others have reported different results. The results obtained by Biegler and co-workers [6, 7] show a more complex behaviour, with the results being consistent with a substantial fraction of the coverage at low potential derived from a species such as Pt_3C–OH but at higher potentials/coverages, where $Q_{ads}/Q_{ox} \rightarrow 2$, the results being more consistent with Pt–CO as the main intermediate. This observation was supported by Vielstich and co-workers [8–10], who also showed that Q_{ads}/Q_{ox} values varied from unity at low coverages to close to 2 at high coverages.

Biegler's results were also confirmed by Lamy and co-workers: their data for polycrystalline platinum [11] showed that n_{eps} was close to 2 for all adsorption times τ_{ads} in concentrated methanol solution (0.1 M), with the coverage rising very roughly linearly with $\ln(\tau_{ads})$ (Fig. 18.1). However, in more dilute methanol solution (0.001 M), where coverages are far lower, n_{eps} was observed to rise from values close to 1 at shorter times (<1 s) to values close to *three* before decreasing slowly towards 2 at high adsorption times. Values of n_{eps} near three imply the formation of a *formyl* species, Pt–CHO on the surface at intermediate times and low methanol concentrations, a result that again strongly implies that the nature of the intermediate changes according both to coverage and methanol concentration. The lower values of n_{eps} at low coverages and methanol concentrations were also confirmed by Leiva and Giordano [12], with Pt–CO forming at higher coverages and methanol concentrations.

Bagotzky *et al.* [5] have also reported some kinetic studies for methanol chemisorption on polycrsytalline platinum. Given that the current transient

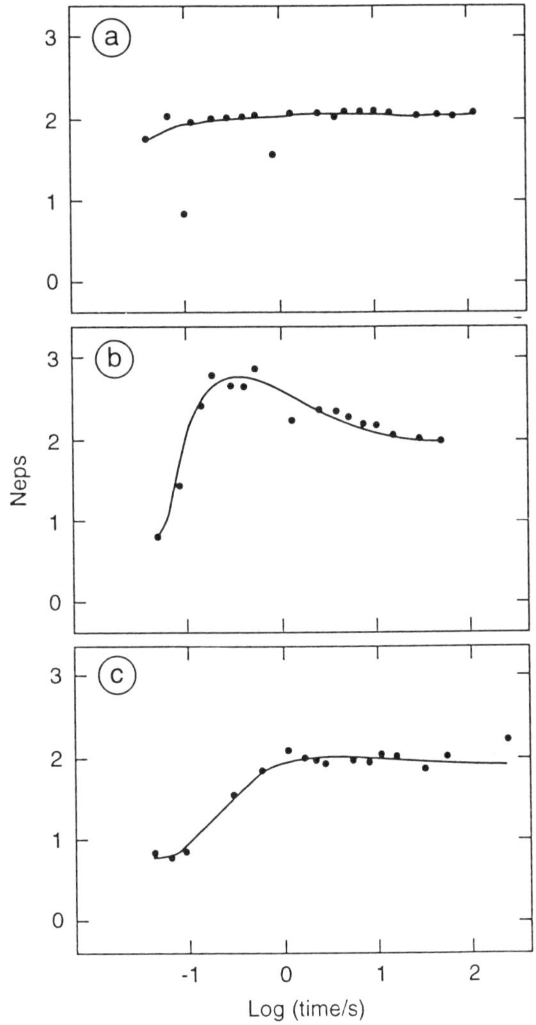

Fig. 18.1. Number of electrons per site, n_{eps} vs. the logarithm of adsorption time for (a) 0.1 M MeOH at polycrystalline Pt; (b) 0.001 M MeOH at polycrystalline Pt; (c) 0.1 M MeOH at a Pt(100) single crystal. [From C. Lamy and J.M. Leger; Bull. Electrochem. 7 (1991) 493].

must be dominated below *c.* 0.5 V by formation of the chemisorbate, the current

$$i \sim k' \frac{\mathrm{d}\theta_{\mathrm{ads}}}{\mathrm{d}t}.$$

If we assume that the adsorption isotherm for the chemisorbate is of the Temkin form, then $i \sim k'' \cdot c_{\mathrm{MeOH}} \cdot \exp(-\alpha f \theta_{\mathrm{ads}})$, where α is the transfer coefficient, with a value usually close to 0.5, and f is the inhomogeneity factor of the Pt surface, with a value of 10–11. Integrating this equation assuming that c_{MeOH} and E are constant, we obtain

$$\theta_{\mathrm{ads}} \sim \mathrm{const.} + \frac{1}{\alpha f} \ln t,$$

whence $i \sim \mathrm{const.}/t$. A plot of $\ln(i)$ vs. $\ln(t)$ should be linear, as indeed it is over the mid-range of coverage [5, 10, 11]. The justification for assuming that the adsorption isotherm was Temkin in form comes from earlier work by the Russian group [13–15] in which the equation

$$\theta_{\mathrm{ads}} \sim \mathrm{const.} + \frac{1}{f} \ln c_{\mathrm{MeOH}}$$

was established. As we shall see below, this has important implications for the overall steady-state electrochemical kinetics.

The complex picture emerging from these studies on polycrystalline platinum impelled several groups to re-examine the problem using single-crystal platinum surfaces, with the aim of working with well-defined surfaces, which would be highly sensitive to any perturbation. The earliest papers [16–18] demonstrated clearly that with single-crystals in which long coherence lengths are maintained, the different platinum surfaces have very different activities for methanol oxidation, at least as measured by potentiodynamic sweeps, and also show marked sensitivity to competitive anion adsorption. The activity order established by Clavilier and co-workers showed that Pt(110) had the highest initial activity, but this activity decayed to low levels with time and potential cycling; the Pt(100) surface had a lower initial activity, but was able to sustain this activity with time, and the Pt(111) surface had a rather low absolute activity. These authors also reported that methanol (at 0.1 M concentration) and CO both showed similar behaviour towards oxidation on the different Pt surfaces, adding plausibility to the assignment of CO_{ads} as an important intermediate (since CO_{ads} is known to form when Pt is exposed to CO dissolved in the electrolyte). In addition, it was established that there was a considerable difference in activities for the Pt(111) surface in perchloric and sulphuric acids, with the latter showing strong anion adsorption that severely inhibited methanol oxidation. Finally, this group established that for methanol oxidation, the behaviour of a poly-faceted Pt-bead electrode can, to a first approximation, be described as the weighted sum of the activities of the three low-index platinum surfaces.

The marked effect of sulphate anions on oxidation of methanol on Pt(111) and other low-index surfaces has also been studied by Kita and co-workers [19], and by Markovic and Ross [20]. There is considerable evidence that HSO_4^- is strongly adsorbed on Pt(111); the cyclic voltammetry of this surface is quite different in background sulphuric and perchloric acids, with the marked structure near 0.5 V in the former electrolyte being shifted to higher potentials (near 0.8 V) and substantially modified in the latter. The adsorption of HSO_4^- on Pt(111) has also been studied both spectroscopically and by radio-tracer techniques [21, 22]: the results show that the onset of HSO_4^- adsorption correlates well with the appearance of "anomalous" structure in the Pt(111) CV in H_2SO_4 electrolyte near 0.35 V: according to Gamboa-Aldeco and co-workers [21], the coverage by bisulphate reaches a maximum near 0.75 V but then *decreases* at higher potentials, reaching a value of essentially zero above 1 V. Bisulphate adsorption is considerably greater on the Pt(111) surface than on Pt(100), a result that can be ascribed to the near perfect symmetry match between the triangular terminals of the bisulphate oxygens and the hexagonal Pt(111) surface geometry [23, 24]. However, this geometric model does not explain the *desorption* at higher potentials, which the authors ascribe to the formation of a strongly bonded water monolayer which they believe to be partially discharged.

For Pt(111), the hydrogen region is comparatively little disturbed during methanol oxidation in sulphuric acid, strongly suggesting that methanol does not chemisorb on Pt(111) in this electrolyte, at least on the time scale of the dynamic experiments carried out, but the methanol oxidation currents are very low, reaching a maximum of only *c.* $0.15\,mA\,cm^{-2}$.

Currents are significantly higher on Pt(111) if the sulphuric acid is replaced by perchloric acid: in the presence of methanol, the hydrogen region between 0.0 and 0.4 V is now significantly suppressed (by about 40%) and the "butterfly" at 0.8 V is absent, suggesting that in perchloric acid there much more rapid and complete chemisorption of methanol.

Results for the Pt(100) face show first that the effect of sulphate adsorption is much less, with the currents being reduced at the maximum by a factor of two on going from perchloric acid electrolyte to sulphuric acid, in qualitative agreement with the adsorption isotherm for bisulphate on Pt(100) presented above; secondly, the hydrogen region is appreciably affected, with coverages of chemisorbed fragments being *c.* 0.4 in H_2SO_4 and 0.6 in $HClO_4$. Results for Pt(110) are the most extreme, with very high initial activity, which rapidly decays during cycling, with the hydrogen region being almost completely eliminated. What is clear from their paper is the fact that co-adsorption of anions must play an important role in the formation of the methanol chemisorbate, and may well account for some of the rather subtle differences between CO_{ads} formed from methanol and from CO dissolved in the electrolyte.

Kita *et al.* [19] draw attention to the puzzling decrease in activity for methanol oxidation at higher potentials: unlike polycrystalline Pt surfaces, where such a decrease in activity correlates well with the formation of surface oxides, no such simple correlation exists for the single-crystal surfaces, a result that the authors interpret in a similar manner to the interpretation of bisulphate desorption given above. Interestingly, Kita *et al.* [19] report that the rate of oxidation of hydrogen gas also reaches a maximum on Pt(111) between 0.4 and 0.6 V, and then decreases at more positive potentials; this effect is extremely marked in sulphate electrolyte, but is also seen in perchloric acid: the maximum oxidation rate of H_2 with potential is, in fact, seen on all the Pt low-index surfaces, and is can again be understood in terms of a layer of chemisorbed and partially discharged water molecules.

The most thoroughgoing purely electrochemical study of adsorption onto single-crystal surfaces has been carried out by Sun and Clavilier [25]. These authors utilised a wide range of surfaces including not only the low-index surfaces discussed above but higher-index surfaces, which can be thought of as possessing terraces of a low-index surface interspersed with steps. The results for the measurement of n_{eps}, θ and Q_{ox} show that for both perchloric and sulphuric acid electrolytes, the coverage of intermediate on these surfaces is almost independent of the size of the (100) terrace. Moreover, whilst the values of n_{eps} are significantly lower than 2, save for the (100) surface itself in sulphuric acid, in perchloric acid, values of n_{eps} close to 2 are found for all (100)-derived surfaces, suggesting that adsorbed linearly-bonded CO dominates on these surfaces in the absence of any anion specific adsorption. This sensitivity to specific adsorption of anions is also found for the (111) face, which has n_{eps} close to 2 in perchloric acid (though with a very low coverage of 0.37) but only 1.57 in sulphuric acid. The general observation of $n_{eps} \leqslant 2$ is, however, reinforced by this study, and this can be ascribed either to simultaneous adsorption of both Pt_3C–OH and Pt–CO or, as the authors point out, to a mixture of linearly and *bridge* bonded CO, the latter having the structure Pt_2CO or even Pt_3CO for CO bridging two Pt atoms on the surface or adsorbed in a 3-fold site. Studies of the variation of n_{eps} with time of adsorption have also been carried out for different crystal surfaces: for Pt(100), Papoutsis *et al.* [11, 26] showed that n_{eps} rises from a value of about 0.9 at short adsorption times to a value close to 2 at higher times, the latter figure being in agreement with Sun and Clavilier. The very low value of n_{eps} at short times was again ascribed to the formation of bridge-bonded CO.

The third assumption above is that methanol chemisorbs at lower potentials essentially solely to give the chemisorbed intermediate(s), with only a negligible quantity of charge being associated with either any parallel oxidation pathway or with further conversion of the intermediate to CO_2. This problem was first tackled by Breiter [27–30], who suggested a parallel mechanism rather than a serial one at lower potentials, with the formation of a second "active" intermediate in a scheme of the form

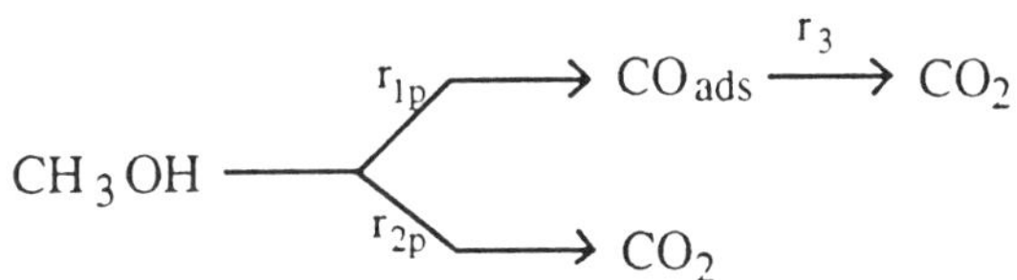

This scheme was further investigated by Matsui [31, 32] who assumed initially a series reaction scheme (for which r_{2p} was assumed small, but no assumptions were made about the constancy of rate r_3 in the scheme above). Matsui found that there was clear evidence at long times for a parallel route for potentials as low as 0.55 V, with the parallel route even more marked at higher potentials (between 0.6 and 0.8 V) [33], where it becomes dominant. The molecular mechanism dominating this parallel reaction was not elucidated by Matsui, who suggested that it might be associated with the formation of chemically stable intermediates such as HCHO and HCOOH in solution, that were then rapidly further oxidised; certainly there is cogent IR evidence for formic acid formation at higher potentials, as we shall see below, but there is also evidence that CO adsorbs to form islands, and that the CO_{ads} on the edges of these islands shows a significantly higher susceptibility to oxidation.

At lower potentials, the importance of the parallel pathway is far less easy to establish. In a careful study of the instantaneous current transients on low-index platinum surfaces, Herrero and co-workers [34] were able to establish (a) that all the surfaces showed kinetic isotope effects, these being more pronounced at lower potentials (~0.4 V), consistent with the rate-determining-step being C–H bond scission on the clean surfaces; (b) the rate of CO_{ads} formation deduced from an analysis that assumed a parallel pathway was found to be strongly correlated with the surface activity as deduced from the magnitude of $i_{t=0}$; (c) the effects of different surfaces and different anions on $i_{t=0}$ are both extremely significant, and in particular, there is a good inverse correlation between adsorption of the anion and initial current density. In agreement with the results of Kita *et al.* [19], the authors report that at high potentials, the values of $t_{t=0}$ actually decrease with E, though increased ordering of the anion layer is suggested as the cause. The effects of different anions are qualitatively very similar to those reported earlier by Sobkowski and co-workers [35], who also showed that $H_2PO_4^-$ and CO_3^{2-} also strongly inhibited methanol oxidation on polycrystalline Pt.

Although the data of Herrero *et al.* [34] were interpreted in terms of a parallel reaction scheme model, such a model is certainly not established by their treatment, and Vielstich and Xia [36] have criticised such a model on the basis of their Differential Electrochemical Mass Spectroscopy (DEMS) data [37]. At least below a potential of 420 mV, the very sensitive DEMS technique detects no CO_2 evolved from a polycrystalline particulate Pt electrode surface on chemisorption of methanol; indeed, the only product detected other than adsorbed CO, in very small yield (one or two orders of magnitude smaller), is methyl formate from the intermediate oxidation product HCOOH. This is graphically illustrated in Fig. 18.2 in which the clean electrode is maintained at 50 mV, a 0.2M methanol/0.1M $HClO_4$ electrolyte introduced, and the electrode swept at 10 mV s^{-1} anod-

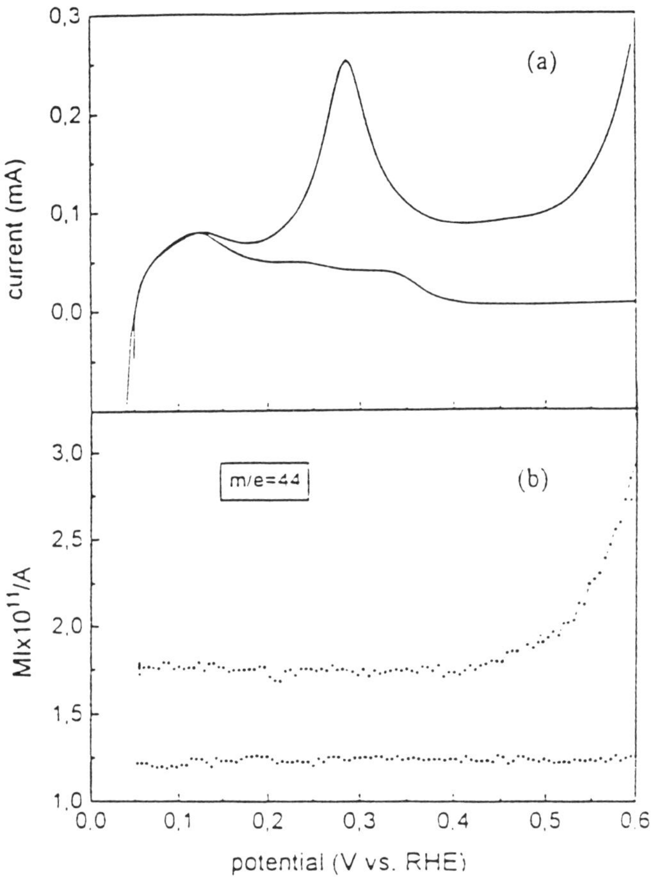

Fig. 18.2. First potential scan at a porous platinum electrode in 0.2 M MeOH/0.1 M $HClO_4$ starting at 50 mV with 10 mV s^{-1}; (a) current as a function of potential; (b) simultaneously recorded mass signal for CO_2 (*m/e* 44). The lower curves are from a blank experiment without methanol in solution. [From W. Vielstich and X.H. Xia; J. Phys. Chem. 99 (1995) 10421].

ically. The peak at 0.25 V is not accompanied by *any* indication in the DEMS data of CO_2 formation at or near this potential and must arise purely from methanol chemisorption. In response, Herrero *et al.* [38] calculated the integrated charge passed on stepping the potential to various values and measuring the complete current transient. Their results show that at least for Pt(100) the parallel mechanism appears to be significant even at as low a potential as 0.42 V, though the evidence from the other two surfaces is less clear-cut. Even so, a parallel mechanism must be operating above 0.5 V for Pt(111) and above 0.6 V for Pt(110). Finally, a recent report by Jarvi *et al.* [39] again confirms on the basis of charge transient measurements that on Pt(100) the parallel mechanism becomes significant above 0.45 V and dominant above 0.6 V. It should be pointed out that the analysis by these authors *assumes* that the only intermediate present on the surface is CO_{ads}, in agreement with the earlier study by Papoutsis *et al.* [11].

Summarising this first section, the use of current transient techniques has been able to give considerable insight into the nature of the chemisorbed species forming on platinum. Provided the surface can be prepared and cleaned in a reproducible fashion, the electrochemical data suggest that there is probably an evolution in the nature of the chemisorbate with time and coverage on all surfaces, but that linearly bonded CO would appear to be the final product on almost every surface. However, although a value of n_{eps} of 2 is difficult to reconcile with any other intermediate than linearly adsorbed CO, the same is not true of lower values of n_{eps} found in many studies: a value of 1, for example, is consistent with both triply bridged ≡C–OH and doubly bridged =C=O, and other techniques must be sought to resolve this problem. The nature of the chemisorbate also changes with the concentration of methanol in solution, with apparently a Temkin isotherm describing the coverage at any given potential.

18.2.2 Spectroscopic and non-electrochemical studies

Although the electrochemical studies discussed above do point clearly towards linearly or bridge-bonded CO_{ads} as significant and possibly dominant components of the methanol chemisorbate on platinum, particularly at high adsorption times and high methanol concentrations, the first definitive proof that adsorbed CO was present came from in situ Infra-Red experiments carried out by Beden and co-workers [40]. Their spectrum showed a bipolar band centred at 2068 cm^{-1} which can be assigned, on the basis of well-established gas-phase studies, to linearly adsorbed CO. A second peak was also seen at 1850 cm^{-1}; the monopolar nature of this peak at least at 0.5 M methanol concentrations indicates that it derives from a species that is present to a greater degree at lower potentials, and its frequency would suggest, on analogy with gas-phase spectra, that it derived from a bridged CO species that was favoured at lower potentials. Interestingly, at lower methanol concentrations, this peak is reported as bipolar in later papers. Lamy and co-workers have also extended their work to single-crystal platinum surfaces [41–43] (Fig. 18.3), and have shown that different low-index surfaces do show quite different behaviour in EMIRS. The Pt(100) surface, for example, shows, in addition to a bipolar peak at 2040 cm^{-1}, a monopolar band at 1890 cm^{-1} and a rather complex shaped band at 1700 cm^{-1}, which was particularly prominent at low adsorption times. This latter is assigned to a formyl species, Pt–CHO, and is also seen as a weak structure on polycrystalline platinum. Formyl would give n_{eps} values of three, which has been reported above for Pt(100) at low adsorption times, and the authors tentatively assign this formyl species to an "active" intermediate that would account for a parallel pathway for methanol oxidation. The authors were able to show that the behaviour of polycrystalline Pt could, to a first approximation, be understood in terms of a weighted contribution from the three low-index Pt surfaces. The very remarkable results from the French group thus established that linearly and bridge-bonded CO were present on the Pt surface but two problems remained: the technique needed to be *quantified* so that the coverage of CO could be determined, and it could be argued (and indeed was!) that the process of potential modulation could lead to the formation of species such as CO_{ads} that did not form under the normal conditions of electrode operation.

Both these concerns were addressed by the development of modified IR techniques. In the technique of Subtractively Normalised Fourier Transform IR Spectroscopy (SNIFTIRS) or Potential Difference IR (SPAIRS or PDIR) [37], the increased stability and sensitivity of Fourier Transform IR is exploited, allowing usable spectra to be obtained by simple subtraction and ratioing of spectra obtained at two potentials without the need for potential modulation or repeated stepping. A second technique which does not call for potential modulation, but actually modulates the polarisation direction of the incoming IR beam is termed Photo-elastically Modulated Infra-Red Reflectance Absorption Spectroscopy (PM-IRRAS); this was applied to the methanol chemisorption problem by Russell and co-workers [44], and Beden's assignments verified, including the potential-induced shift model for CO_{ads}.

The quantification of the detection of CO_{ads} was carried out by Kuni-

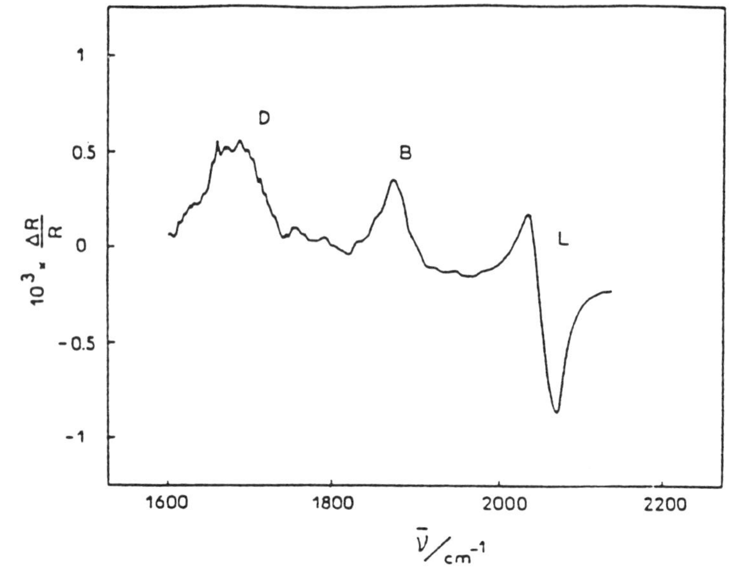

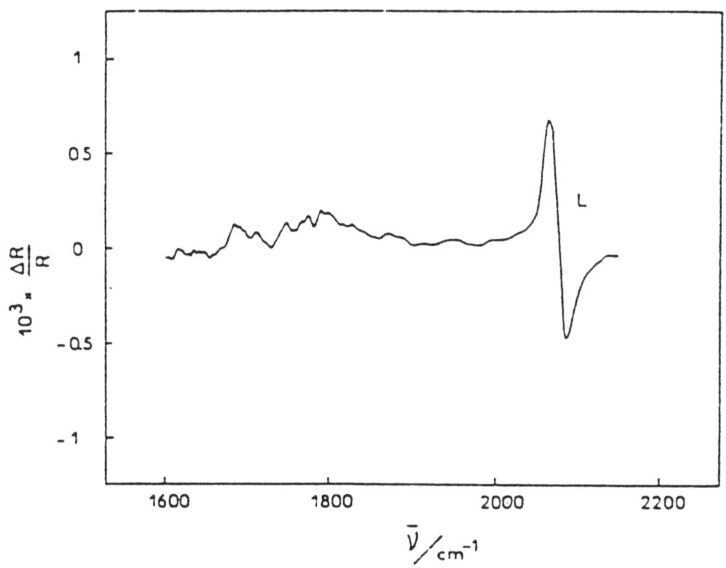

Fig. 18.3. (a) EMIRS spectrum of the adsorbed species resulting from the chemisorption of 0.1 M MeOH in 0.5 M $HClO_4$ at a Pt(100) single-crystal electrode: the modulation amplitude is 0.4 V at 13.6 Hz and the mean electrode potential 0.35 V; (b) the same for a Pt(110) single crystal electrode; (c) the same for a Pt(111) single-crystal electrode. [From B. Beden, S. Juanto, J.M. Leger and C. Lamy; J. Electroanal. Chem. 238 (1987) 323].

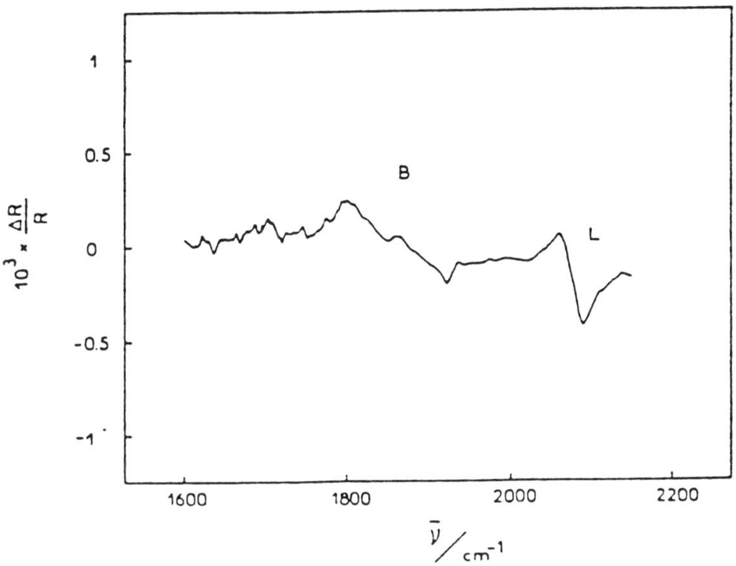

Fig. 18.3. (Contd.)

matsu and co-workers [45–47] using PM-IRRAS, and the intensity of the signal compared at equal coverages to that derived from the adsorption of CO itself from the electrolyte. The SNIFTIRS spectrum for methanol oxidation from Kunimatsu's group is shown in Fig. 18.4 and a comparison of the PM-IRRAS spectra for CO and methanol chemisorption in Fig. 18.5. Interestingly, as Fig. 18.5 shows, there are differences in the position of the IR-absorption peak depending on whether CO_{ads} is derived from CO(aq.) or methanol, the latter showing a slightly lower absorption frequency. Moreover, as the total coverage of organic chemisorbate rises, there is an increasing discrepancy between the intensity of the linearly bonded CO_{ads} peak from methanol chemisorption and CO adsorption. Some of this discrepancy is certainly due to the presence of bridge-bonded species, but some may be due to different packing arrangements on the surface and/or to the presence of additional organic adsorbates of a non-carbonyl form, a point that we shall return to below. Kunimatsu's work does, however, very clearly indicate that on well equilibrated surfaces, linearly bonded CO is the major adsorbed species. PM-IRRAS also allows

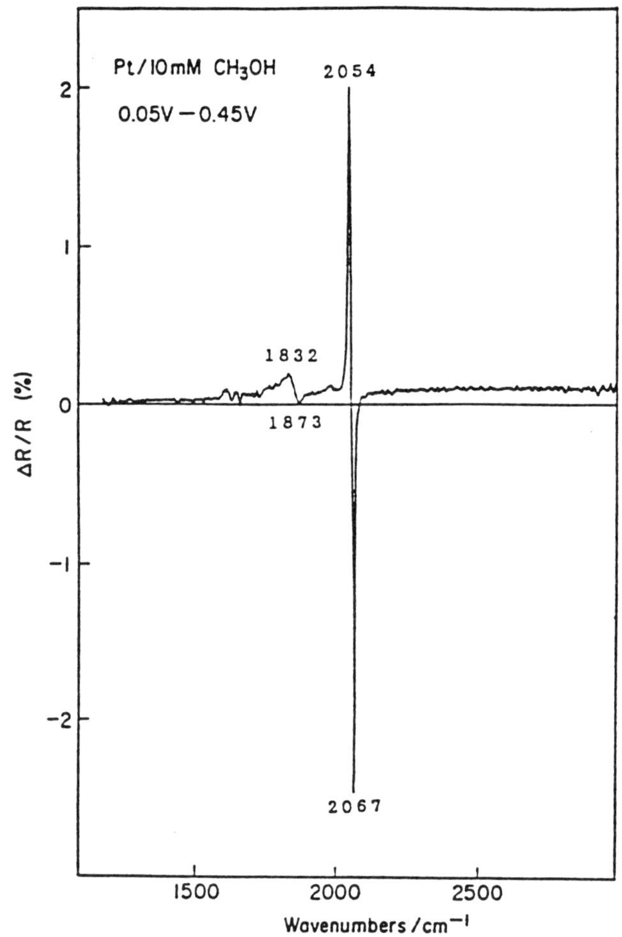

Fig. 18.4. SNIFTIRS spectrum observed between 0.05 V and 0.45 V on a polycrystalline platinum electrode in 0.01 M MeOH/0.5 M H_2SO_4. [From K. Kunimatsu; Ber. Buns. Phys. Chem. 94 (1990) 1025].

us to follow the chemisorption process in more detail; Fig. 18.6 shows the dependence of the signal from the linear CO_{ads} species as a function of potential, starting with a near clean surface at 0.05 V. There is evidently a substantial shift in peak position as the potential and coverage are both increased, with 100 cm^{-1} V^{-1}. There is clear evidence that the frequency

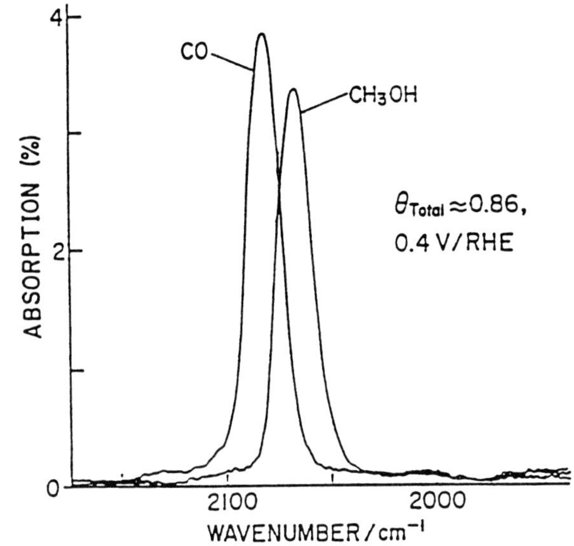

Fig. 18.5. Comparison of the polarisation modulated (PM-IRRAS) spectra of linear CO_{ads} derived from adsorption of CO and of methanol from solution onto a Pt surface from, respectively, CO(sat)/0.5 M H_2SO_4 and 1 M MeOH/0.5 M H_2SO_4 at 0.4 V for 15 min. The polarisation modulation was conducted at 80 Hz by using a rotating polariser, and spectra are referred to 0.8 V. The total adsorbate coverage of 0.86 is common to the two spectra. [From K. Kunimatsu; Ber. Buns. Phys. Chem. 94 (1990) 1025].

of the CO_{ads} shows marked sensitivity to coverage, and this is has attracted some considerable attention recently (*v.i.*). Thus, intensities must be treated with some care, but nevertheless, the increase in intensity of the CO_{ads} band with time for an electrode maintained at 0.4 V in 0.01 M methanol/0.5 M H_2SO_4 was found accurately to obey the expression derived by Bagotzky [5]:

$$\theta_{ads} - \text{const.} + \frac{1}{\alpha f}\ln t.$$

Normal SNIFTIRS or FTIR Potential Step experiments also show a steady increase in coverage of CO_{ads} with increasing potential and time (Fig. 18.7) [48–51]. Simple potential step experiments on polycrystalline Pt surfaces show clear peaks at 2050 cm^{-1} and 1836 cm^{-1}, which can be

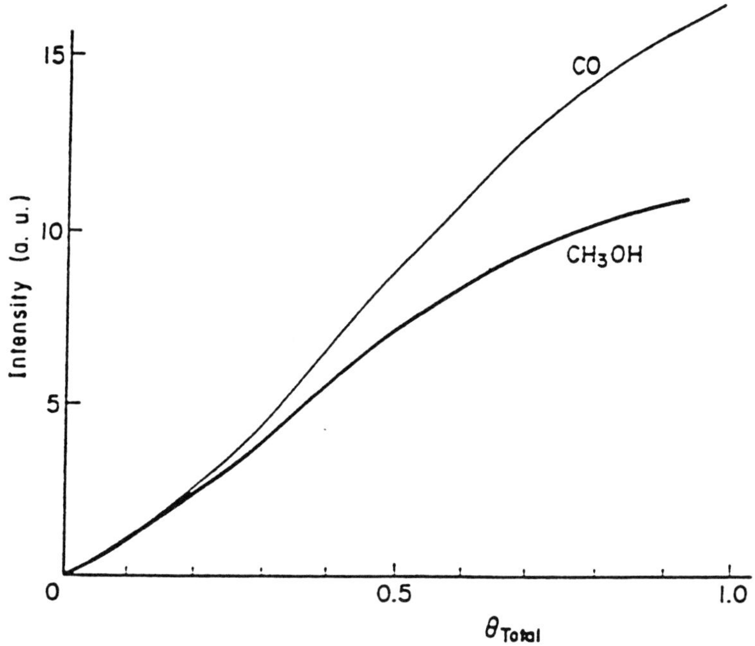

Fig. 18.6. Depedence of the PM-IRRAS spectrum of linear CO_{ads} derived from methanol adsorbed onto Pt in 0.01 M MeOH/0.5 M H_2SO_4 on the initial adsorption potential. [From K. Kunimatsu; Ber. Buns. Phys. Chem. 94 (1990) 1025].

assigned to linear and bridged CO, together with a small peak at 1976 cm^{-1} [52]. These experiments also provide some evidence that bridged species are more prevalent at lower potentials [53], a result in general agreement with the studies on CO adsorption on Pt surfaces reported below. A significant advantage of SNIFTIRS is that the formation of products can be clearly monitored; Fig. 18.7 also shows, for example, the development not only of CO_2 and adsorbed CO at lower potentials, but the formation of significant quantities of formic acid at higher current densities (see above). There are also methanol loss features at 2840 and 2950 cm^{-1} in the ν(C–H) region, and at 1476, 1450 and 1405 cm^{-1} in the ν(C–O) and δ(COH) regions, and a gain due to formic acid at 1435 cm^{-1} (*v.i.*).

The Potential Step FTIR method has now become the preferred approach for exploring the nature of chemisorbates at the surface, and an intensive attack has been carried out using this and related techniques on

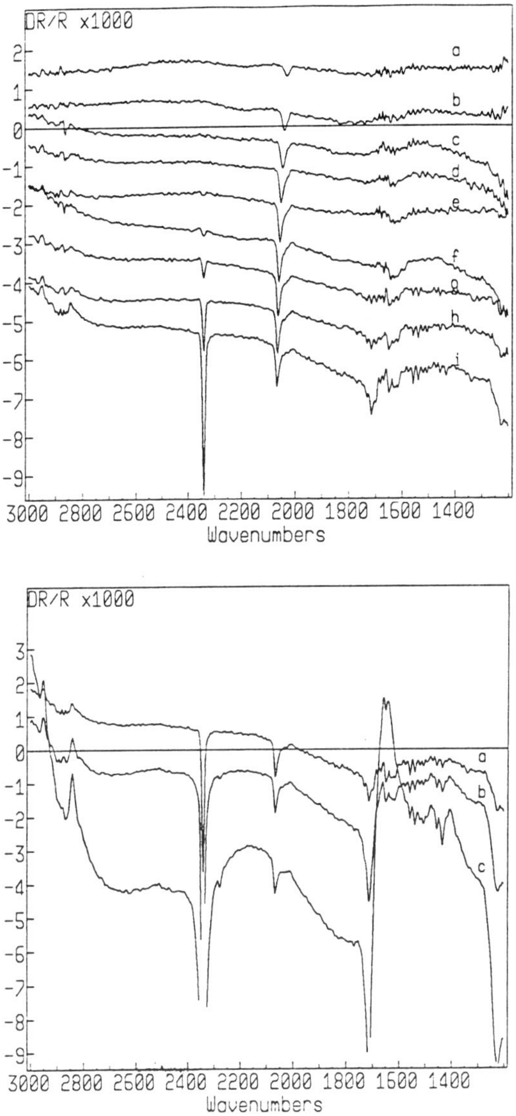

Fig. 18.7. (a) SPAIRS spectra derived from methanol oxidation on Pt starting from a reference clean Pt surface at 0.1 V and recording spectra at 0.05 V steps from *a*. 0.1 V to *i*. 0.55 V; (b) the same for steps from *a*. 0.55 V to *c*. 0.65 V in 0.05 V steps. [From P.A. Christensen, A. Hamnett and G.L. Troughton; J. Electroanal. Chem. 362 (1993) 207].

single-crystal Pt surfaces in order better to understand the processes taking place. These studies have frequently included comparative investigations of CO adsorption and oxidation, since there are strong similarities and intriguing differences between these processes, and we, therefore, turn to a discussion of adsorbed CO on platinum first. The adsorption of CO on platinum has both structural and electronic aspects: Wiechowski and co-workers [54, 55] have reported changes in the relative ratios of bridge to terminal CO from structural studies, and thee have been confirmed in extensive studies by Weaver and co-workers [56, 57]. For Pt(111), for example, in 0.1 M $HClO_4$/CO, the results at a constant coverage of 0.65 show that below ~0.4 V, there is a very low frequency vibration (~1770–1790 cm^{-1}) which is assigned to *triply* bonded CO, and above 0.4 V, there is a jump to higher frequencies (1840–1850 cm^{-1}) assigned to doubly bonded CO, and in general for this surface, the proportion of terminal to bridged CO increases monotonically with θ. Evidence from a variety of spectroscopic and structural studies, which is reviewed in more detail below, strongly suggests that at lower coverages, the CO does *not* adsorb randomly on the Pt(111) surface, but forms close-packed islands, a process that is particularly evident if a monolayer of CO_{ads} is formed and then partially oxidised. Under these circumstances, island formation is very marked.

Qualitatively similar results are found for Pt(100) [58], and there is an apparently general preference for the bridged form of CO_{ads} at lower potentials, which may reflect the effect of co-adsorbed hydrogen or water: neither of the surfaces prepared electrochemically is completely covered with CO, and particularly at lower coverages, there is substantial co-adsorption of H at the lower potentials. It may also reflect a real electronic effect: the enhanced d_π–p_π bonding expected at lower potentials should preferentially stabilise the bridged form [58], a result that can be mimicked under UHV conditions by co-adsorption of ions and dipolar species [59, 60], and can also be explored over wider potential ranges in non-aqueous solvents [61]. There is evidence, moreover, that island formation is less marked on Pt(100), where the islands appear kinetically more easy to dissociate. On Pt(110), island formation is complicated by the lifting of the normal reconstruction on CO adsorption.

More recent combined STM/IRRAS studies [62] of CO adsorption on Pt(111) from $HClO_4$ solution have shown that for saturation coverage, and for $E < 0.25$ V, a (2×2)-3CO adlayer forms possessing both terminal and triply bonded CO (in the ratio 1:2), and with an IRRAS spectrum showing peaks at 2066 and 1850 cm^{-1} (with intensity 2:1). However, above 0.25 V, a surface phase transition takes place to yield a $(\sqrt{19} \times \sqrt{19})$-R23.4° structure with 13 CO molecules per unit cell. It is intersting to note that whilst the relative intensities of bridged and terminal adsorbed CO from IRRAS are in qualitative agreement with these changes, there is no quantitative agreement, since strong dipole-dipole coupling effects lead to intensity borrowing by the higher frequency peak, and a general upshift in all band frequencies.

A good illustration of this can be seen for CO adsorption on the high-index Pt(533) surface [63–65]. As explained below, such surfaces can serve as interesting models for platinum particles, the (533) surface itself being [4(111) × (100)] in character. It is well established that the step edge sites are preferentially occupied by adsorbed CO at low coverage, with terrace bonding only taking place at higher coverages. At low coverages in 0.1 M $HClO_4$ ($\theta_{CO} < 0.3$), an IRRAS band is seen at 2006–2025 cm^{-1}, being shifted up at constant coverage by ~15 cm^{-1} in the range 0.05 to 0.35 V. This is assigned to terminally bonded CO at step edge sites. Surprisingly, there is very little change in frequency for this band as the coverage increases from 0.0 to 0.3, which the authors ascribe to co-adsorbed hydrogen disrupting the expected dipole–dipole coupling effects. A bridge-bonded CO is also seen, with a much more marked shift to higher frequency with coverage (from 1799 to 1821 cm^{-1} at 0.35 V as θ/θ_{max} increases from 0.11 to 0.35, and from 1788 to 1805 cm^{-1} for the same increase in θ/θ_{max} at 0.05 V). The position of this CO_B is unclear, but models derived from EELS and TPD data under UHV conditions place this CO on the edge sites with the terminal CO in a structure in which bridge and terminal sites alternate. For $\theta/\theta_{max} > 0.35$, marked changes are seen in the IRRAS spectra: there is a marked upshift in the (CO) for terminally bonded CO, with the appearance of a shoulder to lower frequency that finally disappears for $\theta/\theta_{max} > 0.6$. There is no doubt that CO molecules populate the terraces under these conditions, with the higher terrace-bonded CO frequency now borrowing intensity from the lower edge frequency, though the suddenness of the upshift may also reflect the very enhanced dynamic dipole–dipole interactions on the terraces as measured by isotopic dilution experiments [65]. A very broad band is seen in the bridging CO region, partially resolved into two peaks at −1790 and 1860 cm^{-1}; at the highest coverages this low-frequency component is lost, presumably by intensity transfer.

A review article from Chang and co-workers in 1991 [66] summarises earlier work from Weaver's group on the kinetics of CO oxidation on low-index surfaces of Pt. It is well established that owing to dipole–dipole coupling effects, increasing CO coverage through dosing gives rise to a pronounced shift in ν(CO) to higher frequencies, but re-oxidation of the CO_{ads} layer, once formed, does *not* give rise to any immediate bathochromic shift, at least if the CO is adsorbed on Pt(111) or Pt(110) at relatively positive potentials [67], since it appears that the CO_{ads}, once formed, prefers to become aggregated into islands. This is marked contrast to time-equilibrated data on polycrystalline Pt [68] where oxidation of the adsorbed CO led to a steady decrease in ν(CO) as expected (*v.i.*). Support for this suggestion of island formation on single-crystal surfaces comes from isotope measurements: if a ^{13}CO monolayer is formed and then partially oxidised, and then ^{12}CO adsorbed, the ^{12}CO is preferentially oxidised to $^{12}CO_2$, suggesting that the CO-islands, once formed, are not involved in subsequent nucleation and growth. In addition. CO_{ads} *sub*-monolayers oxidise at lower potentials confirming the view that co-adsorbed water facilitates the oxidation of CO (*v.i.*).

The authors also report that rapid re-oxidation of CO_{ads}, especially on Pt(100), does, however, give rise to a *time-dependent* CO frequency, a result also found for polycrystalline Pt, with a timescale of 10–15 min, suggesting that islands on this surface can relax to a more disordered state, presumably driven by entropy effects [69]. Time dependent effects may also be seen if CO is adsorbed at low potentials [67]. The mobility of CO_{ads} on Pt surfaces has been little considered until recently by the electrochemical community, though it is a well-established concept in gas-phase work [70]; this mobility is certainly a function of coverage, since it is known that strongly repulsive interactions exist between adsorbed CO on adjacent sites [71], with the desorption energy, E_d, decreasing by a factor of two as the coverage increases; the decrease is especially marked at very high coverage, leading to adsorbed CO molecules that are substantially less stable and more mobile. Finally, the multiple oxidation peaks seen for CO_{ads} on platinum are not, in Weaver's view, due to bridged vs. terminal CO, (in contrast to earlier studies [72] which had suggested such an effect for Pt(111) but not Pt(100)), but arise from the presence of CO domains of differing structure: for example, on Pt(110), the two oxidation peaks appear to be associated with terminal CO, $(CO_{ads})_t$, in unreconstructed (1×1) and reconstructed (1×2) domains.

The nature of methanol oxidation on various low-index surfaces as studied by FTIR has been reported by Weaver and co-workers [73, 74]; in contrast to formic acid, or CO itself, methanol oxidation (0.05 M in 0.1 M $HClO_4$) was found to lead to only moderate ($\leqslant 0.2$) CO coverages, with the CO predominantly in the terminal configuration, though a weak band at 1820–1860 cm^{-1} is found for methanol oxidation on Pt(111). By measuring the hydride coverage, H, electrochemically, the authors were able to show that ($\theta_H + \theta_{CO}$) was comparable to the saturation hydride or CO coverage on Pt(111) (~0.7), but this total on Pt(100) or Pt(110) (~0.6) was significantly smaller than saturation chemisorption values on these latter surfaces. These results are consistent with the presence of appreciable additional adsorbed species not detected in the carbonyl region of the IR spectrum. In agreement with earlier work, the authors found substantial CO_{ads} coverage only at potentials positive of the classical hydride region (~0.3 V) on Pt(111), with a steady increase in both current and θ_{CO} at higher potentials. In comparison to the ν_{CO} data for adsorption from CO solutions, the frequency at comparable coverage and potential for terminal CO was intermediate, suggesting that small islands of CO_{ads} were forming, but not on the scale found for high coverage of CO_{ads} described above. Interestingly, no bridged CO was found for methanol oxidation on Pt(111) at the concentrations used (0.05 M to 5 M). For Pt(100), the situation is rather different: both bridged and terminal CO are found, the former being present even at quite low potentials (0 V), with terminal CO being found above 0.2 V. It is also noteworthy that on Pt(100), only *bridged* species are found at these low coverages for adsorption of CO from solution; clearly the co-adsorption of other species is stabilising the terminal form of the CO on Pt(100) in methanol.

A thoroughgoing LEED/IRRAS comparative study of the oxidation of CO and methanol on Pt(111) in sulphuric acid (as distinct from the perchloric acid used in the studies described above) [75] showed that for CO oxidation on Pt(111) in sulphuric acid, adsorption of CO at 0.05 V gave peaks at 2128 and 1338 cm^{-1} corresponding to terminally bonded CO and adsorbed HSO_4^-; at 0.4 V, four peaks were seen, at 2131 (terminal CO_{ads}), 1865 (bridged CO_{ads}), 1339 and 1250 cm^{-1}. The last of these corresponds to an ordered HSO_4^- overlayer $(\sqrt{3} \times \sqrt{3})R30°$, though the LEED pattern is rather distorted, and the former assigned to re-orientation of HSO_4^- caused by interaction with CO_{ads}, the evidence for this being that at higher potentials, when CO_{ads} is much reduced in coverage, this peak disappears.

If the electrode potential is now swept sufficiently positive for the CO_{ads} to disappear completely, and then reversed, a clear ($\sqrt{3} \times \sqrt{3}$)R30° LEED pattern is observed. By contrast, in $HClO_4$, there is little evidence of adsorbed anion structure. For methanol oxidation in sulphuric acid on Pt(111), no structure is seen until $E > 0.3$ V, when LEED showed a disordered ($\sqrt{7} \times \sqrt{7}$)R19.1° pattern, with bands at 1249 and 1343 cm^{-1} associated with HSO_4^-. At more positive potentials (>0.5 V), the 1250 cm^{-1} band disappears and the 1343 cm^{-1} band becomes very sharp, as does the $\sqrt{7}$ structure. Above 0.9 V, however, the current begins to decrease (*v.s.*) and the $\sqrt{7}$ structure also begins to deteriorate. There does appear to be some correlation between this structure and activity to methanol oxidation on Pt(111); in addition, the ratio $\theta_{HSO_4}/\theta_{CO}$ is *always* higher for methanol oxidation than CO oxidation.

We may summarise this sub-section by re-iterating the fact that the CO adlayers formed on oxidation of methanol clearly closely resemble those formed by adsorption of CO from aqueous solution, but they also show differences. Not only are coverages usually lower, but there is at least some indications that in addition to the terminal and bridge-bonded CO_{ads} species expected, there are probably other carbon-containing species on the surface that affect the way in which methanol is oxidised. In addition, whilst there is some evidence for the formation of islands of adsorbed CO on chemisorption of methanol, these islands appear to be smaller than those that form on adsorption of CO itself, and particularly in the case of sulphuric acid, appear to be strongly associated with adsorbed anions.

The realisation that not all spectroscopic measurements converged on linearly bonded CO has come not only from FTIR studies, but from measurements using additional techniques, particularly those based on mass spectrometry. The most compelling evidence has come from DEMS data [76–78]: by calibrating electrodes by adsorption of CO from solution, these authors were able to establish that in the absence of any adventitious oxygen the main adsorbate on their porous Pt electrodes required three electrons per molecule to generate one molecule of CO_2. This showed that the adsorbate must be ≡C–OH or –CHO, at least under the conditions of the experiment. Interestingly, in the presence of oxygen, the number of electrons required to generate one molecule of CO_2 fell to between 2 and 2.5. The DEMS data also established, as indicated above, that no CO_2 formed below 0.42 V, but there was a small signal due to methyl formate, whose magnitude was at most 1% of the CO_2 signal. The authors also investigated the effects of chemisorbing deutero-methanol from D_2O and oxidising the resultant species in H_2O and doing the reverse (i.e., chemisorbing methanol and then oxidising it in D_2O); the results were initially inconclusive [76] but by using dilute acid electrolyte [79] it was established that the intermediate at low coverage was apparently ≡C–OH and not –CHO, in direct contradiction to the suggestions by the Poitier group. Further strong evidence for the fact that hydrogen is associated with the has been obtained using Electrochemical Thermal Desorption Mass Spectroscopy (ECTDMS) [10, 80]: in this experiment, the electrode is transferred from electrochemical cell to UHV chamber without exposure to air, and the chemisorbed layer thermally desorbed. After calibration, the ratio of hydrogen detected to CO gives an indication of the relative amounts of ≡C–OH and CO adsorbed on the surface; the results are shown in Fig. 18.8, and clearly confirm the contention from electrochemical data that at higher coverages, adsorbed CO dominates, but at lower coverages, ≡C–OH plays a very significant role. The ECTDMS results are, of course, also consistent with –CHO as a low coverage intermediate, but the isotope data do not support this latter species. Clearly, if ≡C–OH is the intermediate formed at low coverage, it must also be very stable on the surface, since it differs little in energy from CO, a fact which would tend to rule it out as the "active" intermediate in any low-potential parallel pathway.

The existence of ≡C–OH as an adsorbate is clearly signalled by these studies, but unambiguous identification by spectroscopic methods has proved much more difficult. Calculations [81] suggested that ν(C–O) might be found at 1380 cm^{-1} for ≡C–OH and 1900 cm^{-1} for –CHO at the Pt-gas interface, but the observed frequencies in aqueous electrolyte are likely to be lower by at least 100 cm^{-1}. The first report to claim to have observed this species with FTIR was Vielstich and co-workers [82], who reported a small signal near 1230 cm^{-1} which developed at higher potentials. However, this signal could also be assigned to methyl formate [83]. Signals at 1320 cm^{-1} and 1425 cm^{-1} in addition to a signal at 1215 cm^{-1} were also reported by Nichols and Bewick [84], but again these were obtained only at higher potentials (~0.7 V); moreover, these signals, which were assigned by Nichols and Bewick to ν(C–O) and δ(C–OH) of ≡C–OH and ν(C–O) of (–CH_xOH) respectively, might also be assigned to formic acid. However, by carefully monitoring the adsorption of 1 M methanol in 0.1 M $HClO_4$ as a function of time on a clean reference surface at 0.05 V, Iwasita and Nart [85] were able to observe at 0.35 V two small peaks at 1200 cm^{-1}

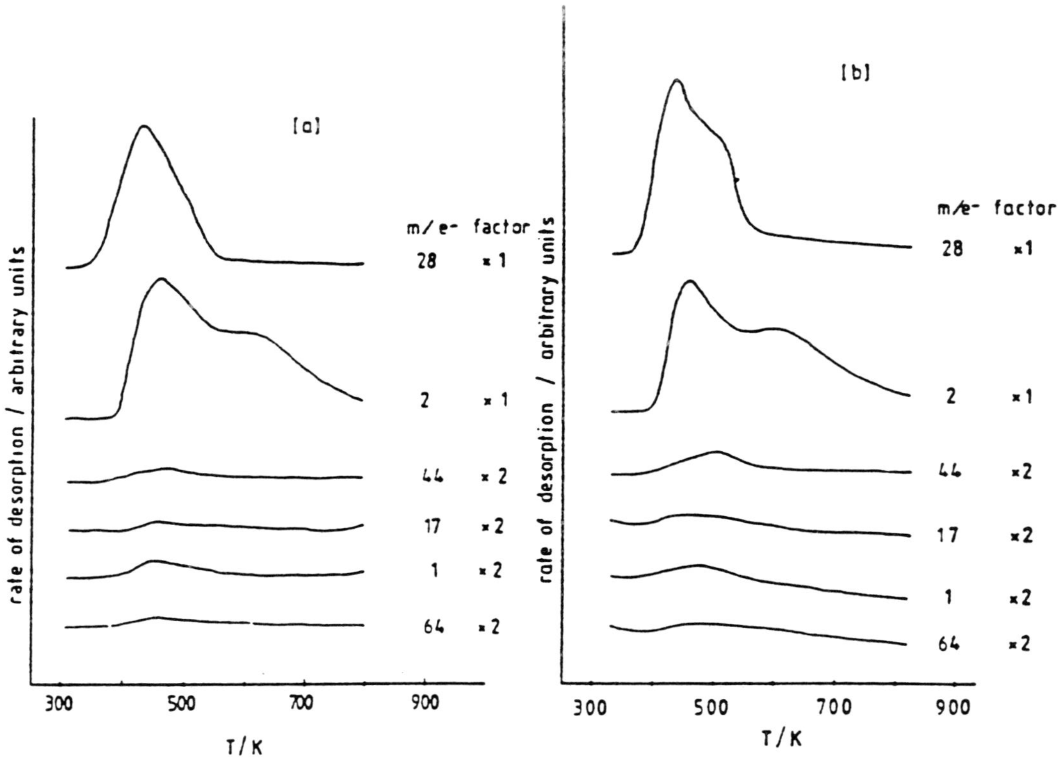

Fig. 18.8. (a) ECTDMS of methanol adsorbate obtained by exposure of the electrode to [a] 0.005 M MeOH/0.5 M H_2SO_4; and [b] 0.5 M MeOH/0.5 M H_2SO_4. Desorption traces given for *m/e* = 28 (CO); 44 (CO_2); 17 and 1 (H_2O) and 64 (SO_2); (b) Mole fraction of two methanol adsorbates, CO_{ads} and ≡C—OH obtained from the type of curves shown in (a): (●) adsorption from 0.005 M MeOH; (+) adsorption from 0.5 M MeOH. [From S. Wilhelm, T. Iwasita and W. Vielstich; J. Electroanal. Chem. 238 (1987) 383].

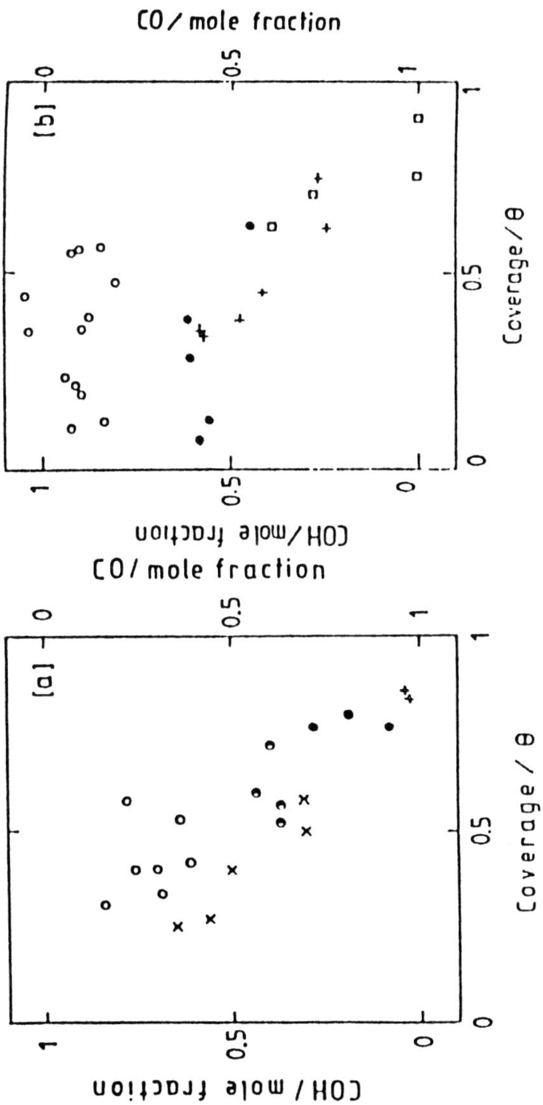

Fig. 18.8. (Contd.)

and 1273 cm^{-1}, in addition to rather larger signals in the linear and bridged carbonyl region. On adsorption of CD_3OH, the signal at 1200 cm^{-1} disappears, presumably being down-shifted out of the range of the window/detector system used, and the authors assign the peak at 1273 cm^{-1} to ν(C–O) of ≡C–OH and the peak at 1200 cm^{-1} tentatively to δ(C–OH) of adsorbed =CHOH.

Finally, in addition to studies on polycrystalline and low-index single-crystal electrodes, IR studies on methanol oxidation have also been carried out on Pt particles and on higher-index Pt surfaces. It is well known that high-index surfaces can serve as good models for platinum particles [86], with the (533) surface [4(111) × (100)] serving as a model for edge atoms in a truncated octahedron and the stepped-kinked (432) surface as a model for corner atoms in the same type of particle. Comparative studies of methanol oxidation on Pt(111) and Pt(533) surfaces [87] by stepped potential FTIR show that on the former, methanol chemisorption to give CO_{ads} only takes place above *c.* 0.3 V, once the normal adsorbed hydrogen atoms have been stripped off, whereas for the latter, CO_{ads} is seen even in the hydride region, albeit rather weakly. Adsorption from 0.1 M methanol/0.1 M $HClO_4$ at 0.15 V gives rise to clearly distinguishable peaks due to CO_{ads} on terrace and edge sites, the former being seen at 2059 cm^{-1} and the latter at 2009 cm^{-1}; in addition, a peak can be seen at 1860 cm^{-1}. This peak is assigned to bridging CO at edge sites, since it is near in frequency to bridged CO observed on Pt(100) sites, which exist only at the edges of the Pt(533) surface. Increasing the potential in the same electrolyte caused the frequency of the signal from the edge-site linearly bonded CO to increase rapidly and to lose relative intensity, eventually being lost (Fig. 18.9), a process associated with increasingly strong coupling between terrace and edge CO_{ads} species as coverage increases. Decreasing the methanol concentration to 0.05 M gives rise to much weaker structure at 0.15 V, with only a CO_{ads} peak at 2049 cm^{-1} being seen. At higher potentials, a doublet is again seen in the CO_{ads} region and a weaker peak at 1846 cm^{-1} also appears.

The first observations, on platinum roughened by repeated cycling, were made by Beden and co-workers [88] who observed that whilst linearly bonded CO could clearly be seen as a bipolar band in EMIRS in 1 M methanol/0.5 M $HClO_4$, the signal due to this species disappeared on dilution to 0.1 M methanol in the same potential range ($\bar{E} = 0.2$ V, $\Delta E =$ 0.4 V), suggesting that adsorption of CO was significantly reduced on

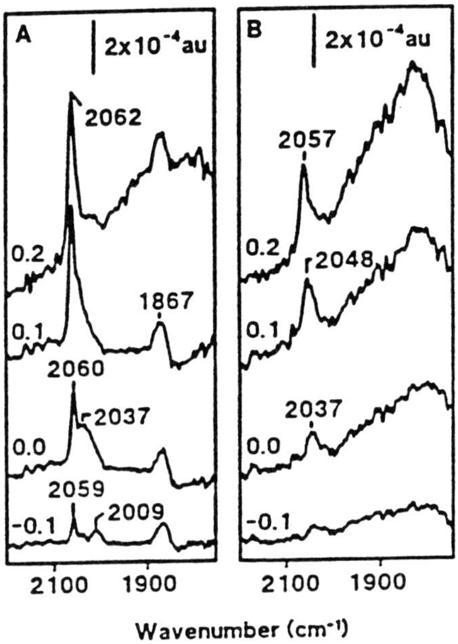

Fig. 18.9. IR spectra of CO_{ads} formed during oxidation of methanol at (A) Pt(335) and (B) Pt(111) electrodes. The electrodes were in aqueous 0.1 M $HClO_4$ electrolyte solution with 0.1 M MeOH. spectra were obtained whilst the electrodes were stepped from an initial potential of −0.25 V vs. SCE, and the reference spectra were obtained at that potential. [From J. Shin and C. Korzeniewski; J. Phys. Chem. 99 (1995) 3419].

these higher surface area samples, or, more probably in the light of the above discussion, that transformation to such species as ≡C–H has taken place. These studies were extended in a second paper [89], in which roughened platinum electrodes again prepared by potential cycling followed by electro-reduction showed both little CO_{ads} coverage, especially at low methanol concentrations, and a band between 1640 and 1700 cm^{-1} that was particularly marked at lower adsorption times and was assigned to –CHO or –COOH.

Rather different data were reported by Christensen and co-workers [90, 91]: the spectrum shown in Fig. 18.10, which was obtained under similar conditions to those of Fig. 18.7 but using Pt particles dispersed on a carbon electrode, is much richer at lower potentials, showing not only the CO_2 peak, but a *bipolar* peak in the adsorbed CO region, with a loss peak near

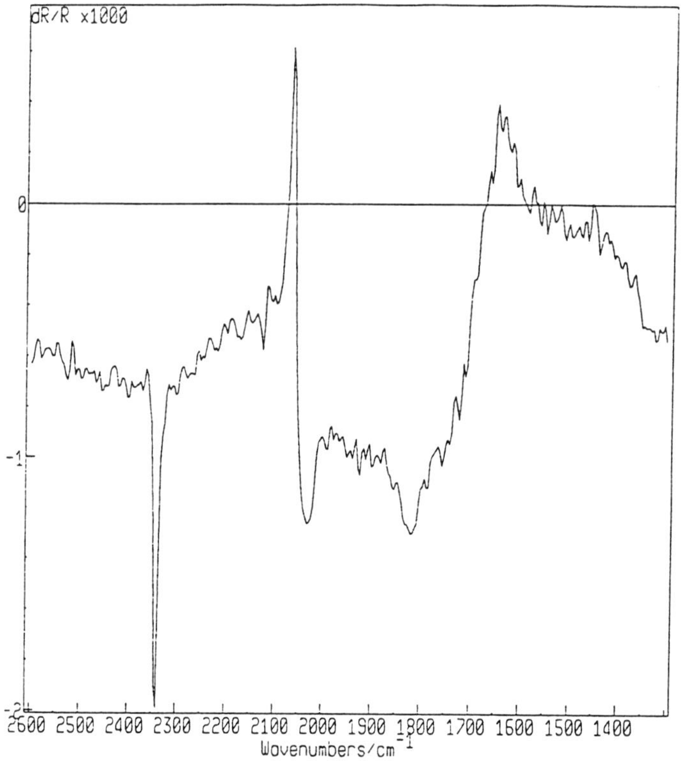

Fig. 18.10. *In situ* FTIR difference spectrum collected at +400 mV vs. RHE during an experiment in which the potential of the platinum particulate electrode (dispersed on carbon), which is initially immersed in 1.0 M H_2SO_4, is first fixed at +50 mV vs. RHE. 1.0 M CH_3OH is then added to the electrolyte, and the potential was increased from +50 mV vs. RHE in successive 50 mV steps to 400 mV. All the spectra (8 cm^{-1} resolution, 100 co-added averaged scans requiring ca. 60 scans per second) were normalised to the reference spectrum taken at the base potential of 50 mV.

2060 cm^{-1} and a gain at 2030 cm^{-1}. There is also a clear peak at 1820 cm^{-1}, which can be ascribed to multiply bonded CO, and there is a loss peak at 1640 cm^{-1} as well as a broad gain feature from 1680 to ca. 2000 cm^{-1}.

Turning to this latter first, an experiment was carried out in which the platinum electrode was held at 50 mV vs. RHE immersed in 0.1 M H_2SO_4 and a reference spectrum was taken. Keeping the potential constant, 1.0 M H_2SO_4 was then admitted to the cell and diffused into the thin layer. The

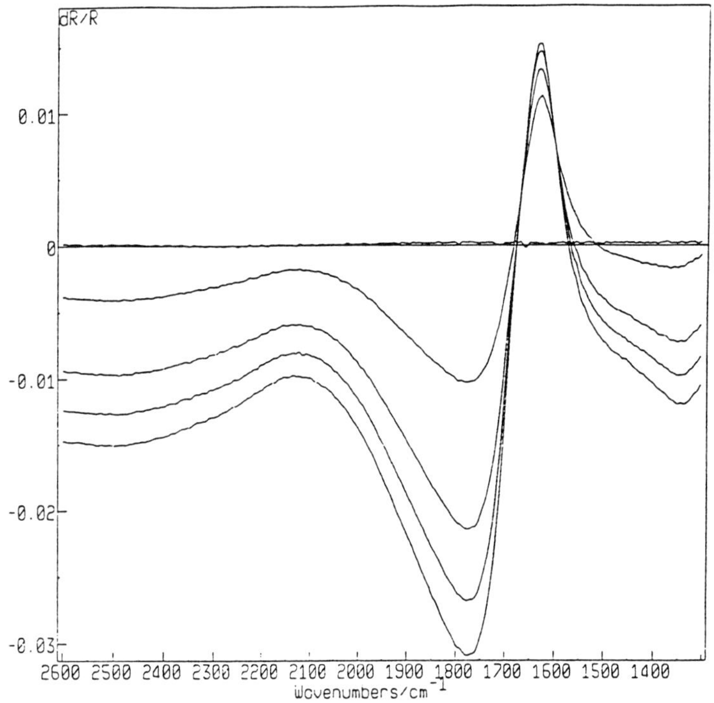

Fig. 18.11. Spectra collected as a function of time immediately before and after the admission of 1.0 M H_2SO_4 to the cell. The reference spectrum was taken prior to the admission of the 1.0 M H_2SO_4, when the cell contain 0.1 M H_2SO_4. The spectrum close to the baseline is obtained before the exchange of electrolyte, and the four other spectra every minute after the exchange of the electrolyte, giving rise to stronger and stronger absorptions. The loss band at 1640 cm^{-1} is due to loss of free water, whereas the broad band between 1680 cm^{-1} and ca. 2000 cm^{-1} is due to water of solvation of the additional protons.

results of this experiment are shown in Fig. 18.11. It is clear that the loss feature at 1640 cm^{-1} and the broad gain at higher frequencies are mimicked in this experiment, and we can, with some confidence, ascribe the loss at 1640 cm^{-1} to the loss of water as it becomes complexed to protons, and the broad gain then corresponds to absorptions of H_3O^+ species. This pattern is also evident in Fig. 18.10; it would seem likely, therefore, that the structure near 1700 cm^{-1} can be associated with the formation of protons. This in turn suggests that the platinum particles are significantly

TABLE 18.1

CO assignment	Pt(533)	Pt particles
CO_T (111 terrace)	2043 cm^{-1} (0.05 V) 2054 cm^{-1} (0.35 V)	2065 cm^{-1}
CO_T (step) 2013 cm^{-1} (0.05V) 2037 cm^{-1} (0.35 V)	2030 cm^{-1}	
CO_B (111 terrace)	1834 cm^{-1} (0.05 V) 1853 cm^{-1} (0.35 V)	1820 cm^{-1}
CO_B (step)	1793 cm^{-1}	not seen

CO_T is terminally bonded CO and CO_B is bridge bonded CO.

more active as catalysts than polycrystalline planar platinum, a result confirmed by the significantly larger CO_2 gain in the FTIR.

The bipolar band near 2045 cm^{-1} is much more puzzling at first sight. It is well known that CO adsorbed on platinum does show a bipolar band as a function of potential, but for a positive potential step, the gain invariably lies to *higher* frequency, not lower frequency as seen here. This shift to higher frequencies as the potential is increased is well understood as indicated above. The shift in Fig. 18.10 is very much more difficult to understand, and assignment may best be made by considering the spectra of CO adsorbed on the model Pt(533) surface as described above. These data are given in Table 18.1, and compared to the data on the Pt particles at 0.4 V. Remarkable agreement can be seen between the data for the terminally bonded CO on both particle and Pt(533) surface, confirming not only that this surface is a good model for Pt particles, but strongly suggesting that it is migration from terrace to step sites on the Pt particles that is responsible for the spectrum seen. Whilst the fit for the bridge-bonded CO is less good, it is difficult to see what else the peak at 1820 cm^{-1} can be but a mutiply bonded CO species.

A final and important point about Fig. 18.10 in comparison with Fig. 18.7 is that in the latter case, adsorbed CO is seen as a gain only; at the potential at which methanol is added to the electrolyte (+0.05 V vs. RHE), adsorption of methanol onto planar polycrystalline platinum is extremely slow, but the rate increases rapidly with potential, giving rise to a CO_{ads} gain in the IR. By contrast, adsorption of methanol on platinum particles has clearly already taken place even at 0.05 V; this very rapid chemisorption suggests that the step sites on the Pt particles are key to methanol adsorption, a result in agreement with data on the Pt(533) surface [87].

We are then faced with the fact that the CO adsorbed on the Pt particles appears to migrate from the terraces to the steps, and that these latter are important in the initial chemisorption process. A possible resolution of this conundrum arises from some recent electrochemical mass spectrometric data [91]. This technique allows us to probe the formation of volatile products at the platinum particle electrode as a function of sweep potential; Fig. 18.12 shows the cyclic voltammogram of the Pt electrode, with the electrode re-activation clearly seen during the negative potential sweep. Figure 18.13 shows the mass spectrometric signal due to mass 44 (CO_2) species, which clearly follows the main current signal, with both showing an onset near 0.5 V vs. RHE during the positive sweep. However, Fig. 18.14 shows a similar plot of the signal for mass 60 (methyl formate), and whilst there is a peak in much the same position as that seen for the CO_2 signal on the positive sweep, there is also a marked shoulder with an onset near 0.3 V, and this clearly corresponds to a small rise in current near this same potential. This signal suggests that oxidation of some species on the surface to HCOOH has taken place, this latter then forming a volatile ester with the solution-phase methanol.

The most natural explanation of this is that doubly bonded >CHOH on the surface of the platinum particles is oxidised to formic acid; given that subsequent to this process, CO bonded at step sites is observed to form, this suggests that the >CHOH species are actually formed at the steps at low potentials; indeed, it would appear likely that the step sites may well facilitate the oxidative chemisorption of methanol, and the doubly bonded >CHOH species are then intermediates in the overall chemisorption to adsorbed CO. The removal of these species from the step sites by oxidation leads to migration of CO from the terraces to the steps, and the reduction in CO coverage on the terraces then leads to conversion of some of the linearly bonded CO to multiply bonded forms.

We can conclude this second section by asserting the following: chemisorption of methanol gives rise to linear and bridge bonding CO, the relative proportions depending on the crystallographic face exposed. The

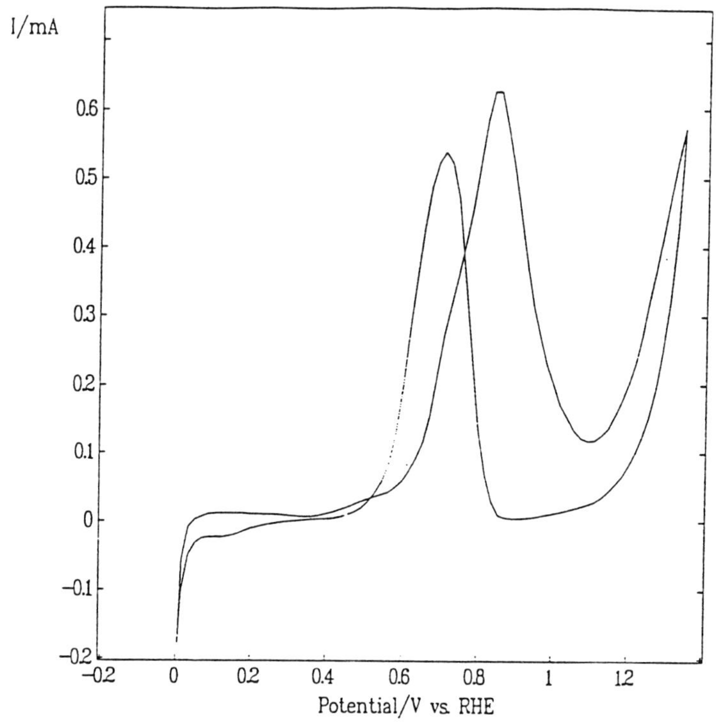

Fig. 18.12. Cyclic voltammogram of a platinum particulate electrode in 0.5 M H_2SO_4/1.0 M MeOH in a DEMS cell.

surface is only covered, however, to a relatively low level with these adsorbates unless high methanol concentrations are used and long equilibration times are allowed. In general, under normal operating conditions, at least one additional adsorbate is present, with the bulk of current evidence favouring ≡C–OH, particularly at lower methanol concentrations. There is some evidence that more than one pathway exists for methanol oxidation, in addition to oxidation of adsorbed CO, particularly on Pt(100) above 0.5 V. Such a "parallel" pathway may go through solution or very weakly adsorbed intermediates, such as HCHO and/or formic acid, or more probably may go through an adsorbate such as CO^*_{ads} adsorbed on metastable site; evidence below suggests that one possibility may be a site

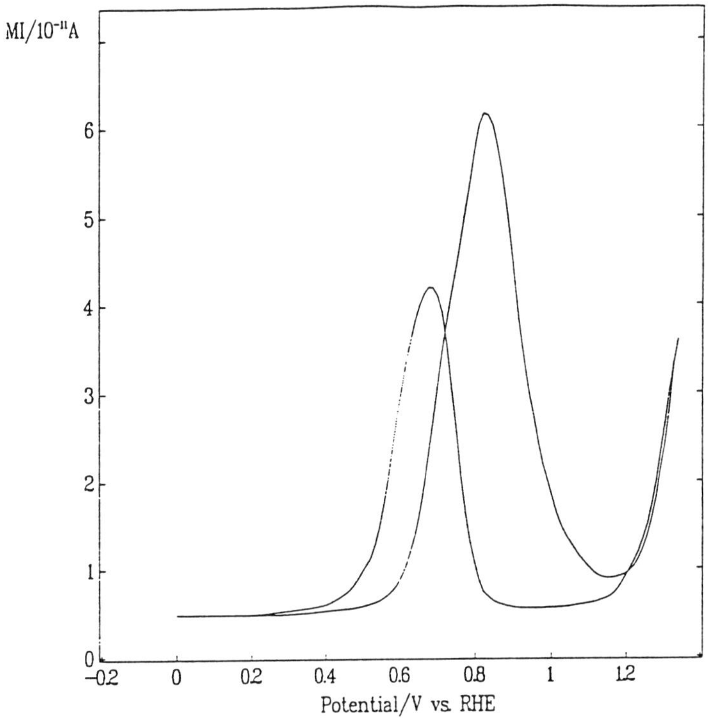

Fig. 18.13. Mass spectrometric signal as a function of the potential sweep on the electrode as shown in Fig. 18.12, with the mass spectrometer tuned to mass number 44 (CO_2).

at the edge of a CO_{ads} island, since the evidence for formation of such islands is very strong. It is also clear that CO_{ads}, at least at the edges of islands, is quite mobile on the Pt surface, and a great deal of experimental data can be understood by assuming that over periods of a few minutes, migration of CO_{ads} can take place on Pt surfaces to randomise the islands formed on saturation chemisorption, once partial oxidation of these islands takes place. On platinum particles, this migration of CO appears to play a very important role in methanol oxidation, particularly at lower potentials. We finally turn to a consideration of the kinetics of methanol chemisorbate oxidation, in order to understnad more fully the nature of the oxidation process.

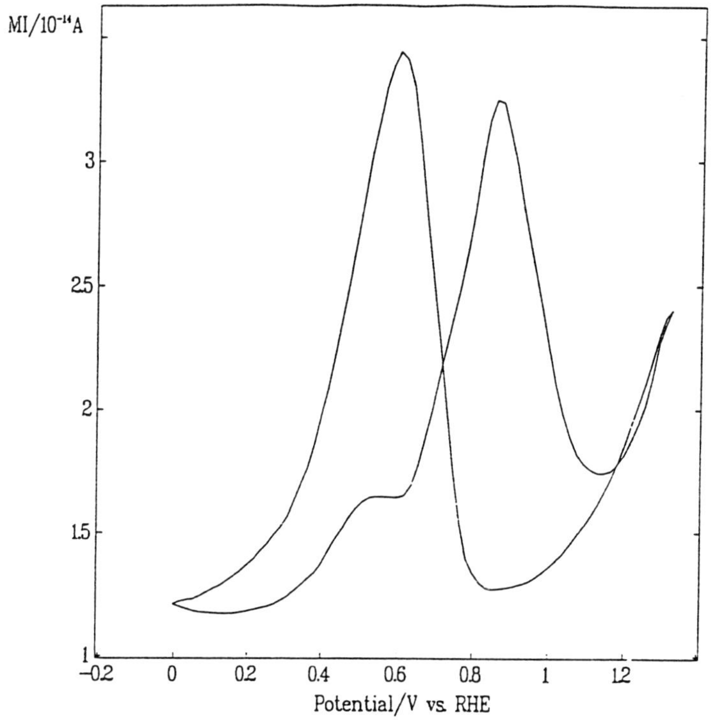

Fig. 18.14. Mass spectrometric signal as a function of the potential sweep on the electrode as shown in Fig. 18.12, with the mass spectrometer tuned to mass number 60 (methyl formate).

18.3 THE KINETICS OF OXIDATION OF METHANOL TO CO_2 ON PLATINUM

We have seen that the process of oxidation of methanol involves the formation of chemisorbed fragments, predominantly CO_{ads} and (probably) ≡C–OH. At lower potentials ($E < 0.5$ V), chemisorption of methanol on a clean platinum surface is faster than subsequent oxidation of the chemisorbed fragments to CO_2, but all investigators have reported that a steady-state can be established, in which a small residual current flows. It is less clear what the rate-limiting step is for this residual current, and intensive studies were first carried out by Bagotzky and Vassiliev [5] to attempt to distinguish the mechanism. For 1 M methanol/0.05 M H_2SO_4 on smooth polycrystalline platinum, these authors found a Tafel slope of 55–60 mV for $0.42 < E < 0.55$ V and a slope of 110 mV for $0.55 < E < 0.72$ V, however, lower concentrations of methanol only showed the higher Tafel slope. At higher potentials, or in more dilute methanol solutions, there are pronounced deviations from Tafel behaviour. Given that the rate of adsorption of methanol on clean platinum surfaces shows only a relatively weak dependence on potential in the double-layer region, whereas the rate of oxidation of methanol rises strongly, it is found that for $E > \sim 0.65$ V, the former will certainly become rate limiting. Indeed, above 0.8 V methanol oxidation rates decrease strongly: in fact, if θ_O is the coverage of oxide on platinum, if is possible to show that: $i \sim k\ \exp(-\alpha' f \theta_O)$, where $\alpha' f \sim 7.7$–8.4. This is consistent with oxygen chemisorption affecting the rate of chemisorption of methanol, rather than any subsequent electrochemical step. If oxidation of the methanol chemisorbate is rate-limiting, then assuming a Temkin isotherm, we would expect to see $i \sim k \exp(\beta f \theta_R)$, where θ_R is the coverage of the chemisorbed fragment. We have already seen that as a consequence of Temkin's isotherm, we expect $\theta_R \sim \text{const.} + (1/f) \ln c$ and $i \sim k' c^{\beta}$, where c is the concentration of methanol. Experimentally, Bagotzky finds $\beta \sim 0.5$. Experimentally, it is also found that for θ_R constant, $i \sim k \exp(\beta' FE/RT)$, giving a general rate equation $i \sim k \exp(\beta f \theta_R) \exp(\beta' FE/RT)$, with β' having the value 1 at lower potentials and 0.5 at higher potentials. Now, oxidation of the chemisorbed fragment can either take place through

$$\text{Pt–CO}_{ads} + \text{OH}_{ads} \rightarrow \text{Pt–COOH} \rightarrow \text{products} \tag{18.13}$$

or

$$\text{Pt–CO}_{ads} + \text{H}_2\text{O} \rightarrow \text{Pt–COOH} + \text{H}^+ + \text{e}^- \rightarrow \text{products} \tag{18.14}$$

In the first case, the coverage of OH_{ads} will be given by a Nernstian expression of the form

$$E_{OH} = E^0_{OH} + \frac{RT}{F} \ln\left(\frac{a_{OH} \cdot a_{H^+}}{a_{H_2O}}\right),$$

giving rise to a value of $\beta' \sim 1$. The second possible rate-limiting step will give rise to a value of β' closer to 0.5. In principle, this could account for the observed behaviour, particularly if there is a limit to θ_{OH} imposed by the structure/morphology of the surface (i.e., if OH_{ads} can only form at a limited number of active sites at the surface). This basic mechanistic analysis has been extended by a number of authors: Kauranen and co-workers used it to model the behaviour of particulate electrodes successfully [92]. Similar kinetic results were also reported by Inada and co-workers [93], with these authors reporting Tafel slopes of 100 mV in the range $0.55 < E < 0.7$ V. Below 0.55 V, these authors reported substantial time dependency in their results; nonetheless, they report that the oxidation current actually *decreased* in the steady state with coverage below 0.55 V, though it increases, as expected, above this potential. Inada and co-workers therefore accept that reaction between OH_{ads} and Pt–CO_{ads} is rate-limiting for $E > 0.55$ V, but suggest that chemisorption of methanol may be rate-limiting below this potential.

Kinetic analyses were also reported by Christensen and co-workers [51] who used the rate of generation of CO_2 rather than the current as a measure of the steady-state reaction kinetics. In agreement with Inada and co-workers, they report Tafel slopes on smooth platinum of 95 mV at higher concentrations of methanol, with rather poorly defined Tafel regions for concentrations below 0.25 M. These authors also explored the consequences of deliberately roughening the platinum, finding that for higher concentrations of methanol, no well-defined Tafel region was generated: instead, behaviour of the type shown in Fig. 18.15 was obtained, in which the current becomes almost potential independent between 0.4 and 0.55 V. For lower concentrations of methanol, well-defined Tafel slopes were obtained, even for these roughened surfaces, and the authors also reported that if the platinum was roughened in the absence of methanol, and methanol then introduced and current/potential curves obtained, behaviour characteristic of the roughened or smooth surfaces was found depending on the rapidity with which the experiments were carried out as shown in Fig. 18.16. Evidently cycling the potential in the presence of methanol stabilised the rough surface, which otherwise would relax over a period of tens of minutes, and this rough surface possesses a finite number of active sites. The authors suggested that diffusion of CO_{ads} or ≡C–OH_{ads} to these sites was rate-limiting, and suggested that such sites could be low-co-ordination Pt atoms, possibly with either OH_{ads} or activated adsorbed H_2O_{ads}, a suggestion also made by Burke and co-workers [94]. Similar conclusions as to the importance of CO_{ads} diffusion to sites

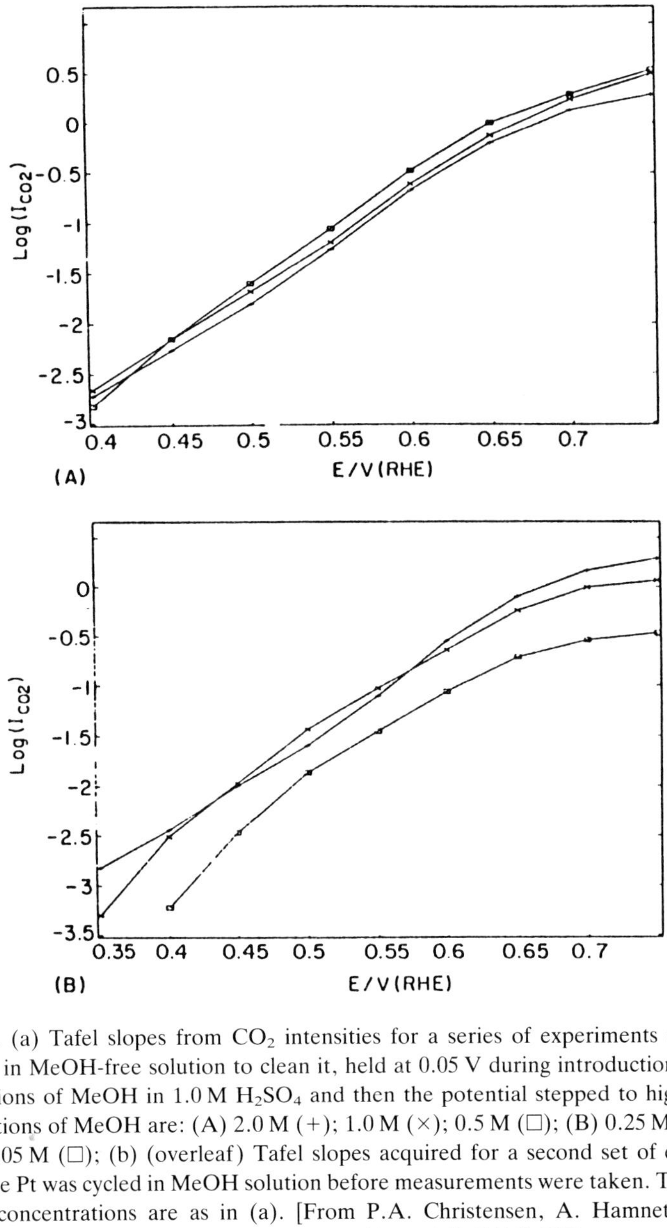

Fig. 18.15. (a) Tafel slopes from CO_2 intensities for a series of experiments in which Pt was cycled in MeOH-free solution to clean it, held at 0.05 V during introduction of varying concentrations of MeOH in 1.0 M H_2SO_4 and then the potential stepped to higher values. Concentrations of MeOH are: (A) 2.0 M (+); 1.0 M (×); 0.5 M (□); (B) 0.25 M (+); 0.1 M (×) and 0.05 M (□); (b) (overleaf) Tafel slopes acquired for a second set of experiments in which the Pt was cycled in MeOH solution before measurements were taken. The different methanol concentrations are as in (a). [From P.A. Christensen, A. Hamnett and G.L. Troughton; J. Electroanal. Chem. 362 (1993) 207].

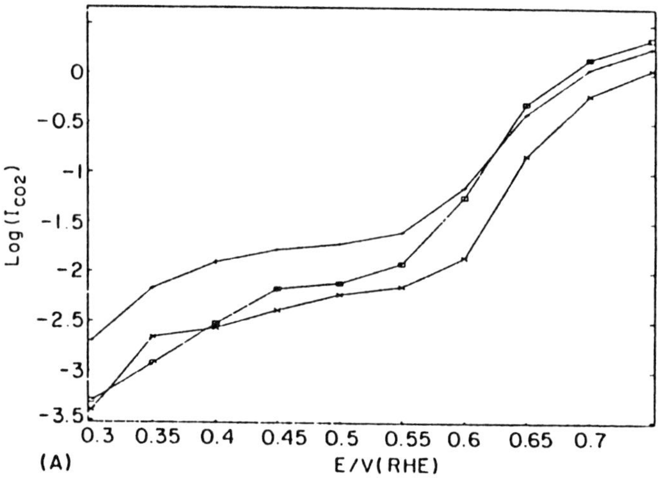

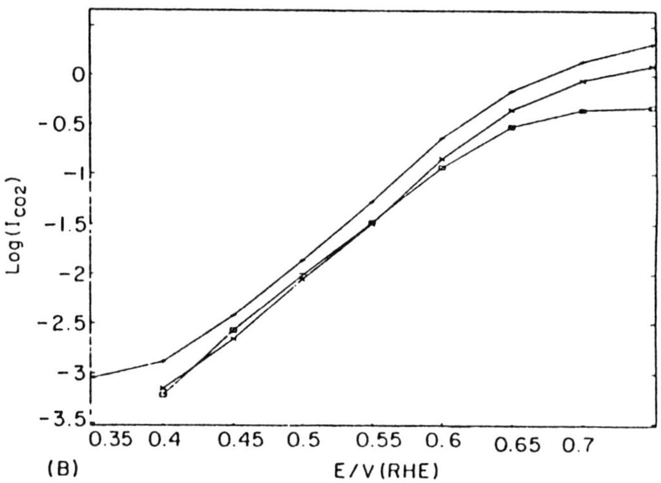

Fig. 18.15. (Contd.)

of OH_{ads} adsorption were also reached by Gasteiger and co-workers [95] in their consideration of the catalytic activity of adsorbed Ru on Pt. The diffusion of adsorbed CO on both Pt and Ru *in vacuo* is actually quite facile: diffusion coefficients of order 10^{-9} cm^2 s^{-1} have been reported for both metals [96], and whilst these will certainly be rather smaller in electro-

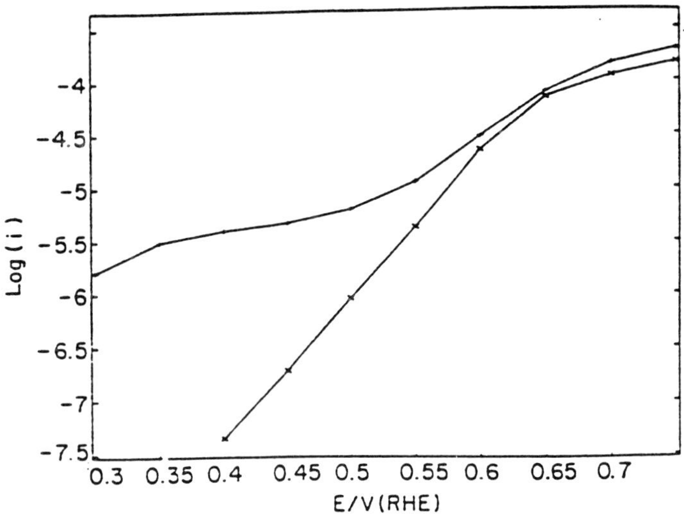

Fig. 18.16. Comparison of Tafel slopes obtained from current measurements in which the Pt was initially cycled in methanol free solutions and held at 0.05 V during introduction of 2.0 M MeOH as above with the current measurements being taken (+) after no delay and (×) after a delay of 45 minutes between cycling the electrode and the polarisation measurements. [From P.A. Christensen, A. Hamnett and G.L. Troughton; J. Electroanal. Chem. 362 (1993) 207].

lyte solution, there is a high probability of the surface mobility playing a substantial role in methanol oxidation. The importance of platinum surface morphology has also been emphasised by Pletcher and Solis [97], who showed that much higher transient currents were observed on chemisorption of methanol in the double-layer region if the Pt surface is first rapidly cycled between 0.55 and 1.6 V.

It remains to consider whether oxidation of the adsorbed CO_{ads} or $\equiv C{-}OH_{ads}$ is taking place through H_2O or OH_{ads}. This has proved a thorny issue, partly because of the sensitivity of the process to surface morphology and to the formation conditions of the adsorbed monolayer itself, but also because kinetic analysis has given ambiguous answers. In principle, both Tafel slope and pH dependence should differ for reactions (18.13) and (18.14) above, but whilst Bagotzky and Vassiliev's data [5] are most naturally interpreted in terms of reaction (14) above, the current equation derived from the studies of Inada *et al.* [93], $i \sim kc^{0.32}_{\mathrm{MeOH}}c^{-0.6}_{\mathrm{H^+}} \exp(0.6FE/RT)$

is more consistent with a pre-equilibrium process of the type Pt + $H_2O \leftrightarrow$ Pt-OH + H^+ + e^-, for which a Tafel slope of 60 mV and an inverse dependence on c_{H^+} is expected. However, Inada's results are also not conclusive.

If reactions such as (18.14) are important, then some consideration must be given to the details of the reaction. The original suggestion by Gilman [98] that attack was by *adsorbed* water has been the subject of semi-empirical calculation by Anderson and co-workers [99], who found that the Tafel slope for this reaction would actually have the wrong sign, since as the potential increases, H_2O_{ads} becomes much more strongly bound to the surface; given that in any reasonable transition state, the water molecule must be prised off the surface by ~1 Å to attack the carbon of the CO_{ads}, the increasingly unfavourable enthalpy associated with this process essentially offsets the expected increase in reaction rate at higher potentials. An alternative is the direct attack on CO_{ads} by H_2O in the outer Helmholtz layer [100]; the evidence for this is two-fold: there is a weak isotope effect that would not be expected for attack by OH_{ads}, and there is only a small entropy change for the reaction on platinised platinum, the latter being inconsistent with the expected large negative entropy associated with the formation of solvated proton and Pt–OH_{ads}. Again, however, calculation gives rather ambiguous results [99], though more convincing than attack by adsorbed water.

Attempts to calculate the potential dependence of the reaction $CO_{ads} + OH_{ads} \rightarrow$ products [101] led initially to similar results as for H_2O_{ads}, namely that the increasing adsorption strength of OH_{ads} at increasing potential essentially negated the expected increase in rate derived from increasing coverage. However, the authors were also able to show that at higher CO_{ads} coverages, attack by OH_{ads} became much more facile, a result also presumably true for reaction (18.14), though the authors do not comment on this. At sufficiently high coverage of CO_{ads} the calculated activation energy does, in fact, become smaller than that found experimentally, leading the authors to suppose that at the highest coverages, the rate limiting step is probably the actual formation of OH_{ads}, in agreement with the ideas above. Further calculations of Pt–OH_{ads} have been reported recently [102], suggesting that this species will form preferentially on adsorption at Pt(111) one-fold sites, rather than at Pt(100) whcrc the two-fold sites are less stable. The authors suggest that this might account for some of the differences between oxidation of CO_{ads} on the two surfaces, but some caution is needed here, since there are other kinetic effects operating.

Some help in the resolution of these difficulties has been provided by isotope measurements. Studies of methanol oxidation below 0.5 V show that neither ^{13}CO nor $\equiv^{13}C–OH$ formed on the surface from electrosorption of $^{13}CH_3OH$ exchange on immersion in $^{12}CH_3OH$, even during oxidation of the $^{12}CH_3OH$ to $^{12}CO_2$ [103, 104]. Not only does this strongly support the suggestion above that methanol adsorption leads to islands of adsorbed CO [105], but it also suggests, in agreement with Christensen *et al*. [51], that oxidation to CO_2 takes place only at the *edges* of these islands. This may well be the resolution of the enigma of the parallel reaction, and brings the original suggestion of Leiva and Giordano [12] back into focus: the adsorption of CO undoubtedly leads to the formation of islands that are stable to potentials above 0.6 V, potentials well above the point at which rapid oxidation of solution methanol to CO_2 is taking place. This oxidation probably involves the rapid chemisorption of methanol at higher potentials on the partially free surface of Pt to give CO_{ads} that migrates to the edges of islands to maintain their extension, but which is subsequently oxidised to form CO_2. The CO_{ads} at the centre of the islands thus remains unaffected, at least over short periods, by the oxidation of methanol, but the so-called "active intermediate" is a CO_{ads} molecule at the edge of the island. This argument would not be significantly different even if the chemisorbed intermediate did not become completely dehydrogenated, provided it was sufficiently mobile to migrate to the edges of adsorbed CO islands rapidly compared to further oxidation to form CO_2.

The situation with regard to the actual oxidant does, however, remain unresolved: on smooth platinum it would appear that either direct attack on CO_{ads} by water molecules, presumably at the edges of CO_{ads} islands, or attack by OH_{ads} is possible; in the latter case, given that the coverage of OH_{ads} on Pt is very low at lower potentials, we must consider two possibilities: either there is a fast pre-equilibrium, or there are specific surface sites at which OH_{ads} forms with higher probability than that expected from the Nernst equation. In the latter case, we must, as pointed out by Christensen and co-workers and Gasteiger and co-workers, consider diffusion as a possible rate-limiting process. For either type of reaction, however, there will be a severe kinetic inhibition at very high CO_{ads} coverages, since nucleation of free or Pt–OH surface must take place before either reaction can commence.

18.4 CONCLUSIONS

There does now seem to be consensus about the following aspects of the methanol oxidation reaction on Pt and Pt:Ru electrodes:

(1) Methanol strongly dissociatively chemisorbs on platinum surfaces, but the extent of adsorption depends critically on the surface morphology, the crystal face exposed, and the presence of other anions in solution that themselves can competitively co-adsorb.
(2) Adsorption on a clean surface shows a strong kinetic isotope effect, implying that the initial chemisorption process involves C–H bond scission. Adsorption appears to be facilitated by the presence of step-edge sites, since adsorbed intermediates are seen at low potentials on both polycrystalline Pt and on high-index Pt(335) surfaces.
(3) The final products of chemisorption of methanol are ≡C–OH and singly and multiply bonded CO_{ads}, the relative proportions being dependent on the methanol concentration, adsorption time and the nature of the anion. Long adsorption times, concentrated methanol solutions and co-adsorbed anions favour linearly bonded CO_{ads}, whereas short adsorption times and low methanol concentrations favour multiply bonded species. The spatial distribution of the CO_{ads} can also change with time, with stable islands forming over long times. Only the CO molecules adsorbed at the edges of these islands appear to be electroactive at lower potentials. On surfaces possessing step-edge sites, such as polycrystalline Pt and Pt(335), CO_{ads} species adsorbed both on the terraces and on the step sites can clearly be distinguished.
(4) Below ~0.45 V the chemisorbed intermediates formed on methanol adsorption are stable on any smooth platinum surface, with the steady-state current for methanol oxidation being extremely small. Above this potential, oxidation of methanol takes place at a rate that increases exponentially with potential, with the product being primarily CO_2. In addition, above potentials of approximately 0.6 V, the surface is steadily stripped of adsorbed carbon-containing species, with the loss of such species being complete near 0.8 V. It would seem likely on most surfaces that it is oxidation of CO_{ads} or ≡C–OH in a sequential reaction pathway that leads to CO_2, but more active intermediates, such as CO adsorbed at less stable sites, such as those at the edges

of the CO_{ads} islands, are implicated as intermediates at lower and intermediate potentials.

(5) The CO_{ads} islands are electro-inactive save at the periphery, where attack by water or OH_{ads} is possible. The size of these islands remains uncertain, and clearly where they are small, as on platinum particles, or less tightly packed, as on Pt(100), movement of CO_{ads} and relaxation of the surface structure can take place even at high coverage.

(6) Platinum itself can be made more active in a number of ways, including surface roughening, but the most successful electrodes for methanol oxidation are those in which a second metal is added as a promoter. On roughened platinum, migration of CO_{ads} species to active sites, such as low-co-ordination Pt* atoms may become rate limiting.

ACKNOWLEDGEMENTS

It is a pleasure to acknowledge the help of Dr. Paul Christensen and Dr. Jens Munk, without whose experimental skills this work could not have been carried out. I would also like to acknowledge innumerable discussions with Prof. Dr. Wolf Vielstich, whose work in this area has been seminal. I would also like to acknowledge the financial assistance of EPSRC and the Unviersity of Newcastle.

References

[1] R. Parsons and T. Van der Noot, J. Electroanal. Chem. 257 (1988) 1.
[2] A. Hamnett, Catalysis Today 38 (1997) 445.
[3] V.S. Bagotzky, Yu. B. Vassilyev and O.A. Khazova, J. Electroanal. Chem. 81 (1977) 229.
[4] K. Itaya, S. Sugawara, K. Sashikata and N. Furuya, J. Vac. Sci. Tech. A8 (1990) 515.
[5] V.S. Bagotzky and Yu.B. Vassilyev, Electrochimica Acta 12 (1967) 1323.
[6] T. Biegler and D.F.A. Koch, J. Electrochem. Soc. 114 (1967) 904.
[7] T. Biegler, J. Phys. Chem. 72 (1968) 1571.
[8] W. Vielstich, ACS publication, 1987, 394.
[9] T. Iwasita, W. Vielstich and E. Santos, J. Electroanal. Chem. 229 (1987) 367.
[10] S. Wilhelm, T. Iwasita and W. Vielstich, J. Electroanal. Chem. 238 (1987) 383.
[11] A. Papoutsis, J.-M. Leger and C. Lamy, J. Electroanal. Chem. 234 (1987)315.
[12] E.P.M. Leiva and M.C. Giordano, J. Electroanal. Chem. 158 (1983) 115.
[13] O.A. Khasova, Yu.B. Vassiliev and V.S. Bagotzky, Elektrokhimia 1 (1965) 82.

[14] S.S. Beskorovaynaya, Yu. B. Vassiliev and V.S. Bagotzky, Elektrokhimia 1 (1965) 1029.
[15] V.S. BAgotzky and Yu. B. Vassiliev, Electrochim. Acta 11 (1966) 1439.
[16] J. Clavilier, C. Lamy and J.M. Leger, J. Electroanal. Chem. 125 (1981) 249.
[17] R.R. Adzic, A.B. Tripkovic and W. O'Grady, Nature (London) 296 (1982) 137.
[18] C. Lamy, J.M. Leger, J. Clavilier and R. Parsons, J. Electroanal. Chem. 150 (1983) 71.
[19] H. Kita, Y. Gao, T. Nakato and H. Hattori, J. Electroanal. Chem. 373 (1994) 177.
[20] N. Markovic and P.N. Ross, J. Electroanal. Chem. 330 (1992) 499.
[21] M.E. Gamboa-Aldeco, E. Herrero, P.A. Zelaney and A. Wiechowski, J. Electroanal. Chem. 348 (1993) 451.
[22] P. Faguy, N. Markovic, R. Adzic, C. Fieiro and E. Yeager, J. Electroanal. Chem. 289 (1990) 1247.
[23] R. Michaelis, M.S. Zhai and D.M. Kolb, J. Electroanal. Chem. 339 (1992) 299.
[24] O. Magnussen and R.J. Behm, Farad. Disc. 94 (1992) 330.
[25] S.G. Sun and J. Clavilier, J. Electroanal. Chem. 236 (1987) 95.
[26] C. Lamy and J.M. Leger, Bull. Electrochem. 7 (1991) 493.
[27] M.W. Breiter, J. Electroanal. Chem. 14 (1967) 407.
[28] M.W. Breiter, J. Electroanal. Chem. 15 (1967) 221.
[29] M.W. Breiter, in: eds. J.O'M. Bockris and B.E. Conway, Modern Aspects of Electrochemistry, vol. 10, p. 205 (Plenum Press, New York, 1975).
[30] M.W. Breiter, Disc. Farad. Soc. 45 (1968) 79.
[31] H. Matsui and T. Hisano, Bull Chem. Soc. Jap. 60 (1987) 863.
[32] H. Matsui, T. Hisano and T. Terazawa, Bull Chem. Soc. Jap. 58 (1985) 911.
[33] H. Matsui, Bull. Chem. Soc. Jap. 61 (1988) 3295.
[34] E. Herrero, K. Franeszczuk and A. Wiechowski, J. Phys. Chem. 98 (1994) 5074.
[35] J. Sobkowski, K. Franaszczuk and K. Dobrowolska, J. Electroanal. Chem. 330 (1992) 529.
[36] W. Vielstich and X.H. Xia, J. Phys. Chem. 99 (1995) 10421.
[37] P.A. Christensen and A. Hamnett, Techniques and Mechanisms in Electrochemistry (Blackie, London, 1994).
[38] E. Herrero, W. Chrzanowski and A. Wiechowski, J. Phys. Chem. 99 (1995) 10423.
[39] T.D. Jarvi, S. Sriramulu and E.M. Struve, J. Phys. Chem. B 101 (1997) 3649.
[40] B. Beden, A. Bewick, K. Kunimatsu and C. Lamy, J. Electroanal. Chem. 121 (1981) 343.
[41] S. Juanto, B. Beden, F. Hahn, J.M. Leger and C. Lamy, J. Electroanal. Chem. 237 (1987).
[42] B. Beden, S. Juanto, J.M. Leger and C. Lamy, J. Electroanal. Chem. 238 (1987) 323.
[43] B. Beden, F. Hahn, C. Lamy, J.M. Leger, N.R. de Taconi, R.O. Lezna and A.J. Arvia, J. Electroanal. Chem. 261 (1989) 401.
[44] J.W. Russell, J. Overend, K. Scanlon, M. Severson and A. Bewick, J. Phys. Chem. 86 (1982) 3066.
[45] K. Kunimatsu, J. Electroanal. Chem. 213 (1986) 148.
[46] K. Kunimatsu and H. Kita, J. Electroanal. Chem. 218 (1987) 155.

[47] K. Kunimatsu, Ber. Buns. Phys. Chem. 94 (1990) 1025.
[48] P.A. Christensen, A. Hamnett and S.A. Weeks, J. Electroanal. Chem. 250 (1988) 127.
[49] T. Iwasita and W. Vielstich, J. Electroanal. Chem. 250 (1988) 451.
[50] D.S. Corrigan and M.J. Weaver, J. Electroanal. Chem. 241 (1988) 143.
[51] P.A. Christensen, A. Hamnett and G.L. Troughton, J. Electroanal. Chem. 362 (1993) 207.
[52] T. Iwasita and F. Nart, J. Electroanal. Chem. 317 (1991) 291.
[53] F. Kitamura, M. Takahashi and M. Itoh, J. Phys. Chem. 92.
[54] M. Wasberg, L. Palaikis, S. Wallen, M. Kamrath and A. Wiechowski, J. Electroanal. Chem. 256 (1988) 51.
[55] D. Zurawski, M. Wasberg and A. Wiechowski, J. Phys. Chem. 94 (1990) 2076.
[56] S.-C. Chang and M.J. Weaver, Surf. Sci. 238 (1990) 142.
[57] S.-C. Chang and M.J. Weaver, J. Chem. Phys. 92 (1990) 4582.
[58] S.-C. Chang and M.J. Weaver, J. Phys. Chem. 94 (1990) 5095.
[59] S.-C. Chang and M.J. Weaver, J. Phys. Chem. 95 (1991) 5391.
[60] I. Villegas and M.J. Weaver, J. Phys. Chem. 101 (1997) 5842.
[61] S.-C. Chang, X. Jiang, J.D. Roth and M.J. Weaver, J. Phys. Chem. 95 (1991) 5378.
[62] J. Villegas and M.J. Weaver, J. Chem. Phys. 101 (1994) 1648.
[63] C.S. Kim, W.J. Tornquist and C. Korzeniewski, J. Phys. Chem. 97 (1993) 6484.
[64] C.S. Kim, W.J. Tornquist and C. Korzeniewski, J. Chem. Phys. 100 (1994) 628.
[65] C.S. Kim, W.J. Tornquist and C. Korzeniewski, J. Chem. Phys. 101 (1994) 9113.
[66] S.C. Chang, A. Hamelin and M.J. Weaver, J. Chim. Phys. 88 (1991) 1615.
[67] N. Furuya, S. Motoo and K. Kunimatsu, J. Electroanal. Chem. 239 (1988) 347.
[68] K. Kunimatsu, H. Seki, W.G. Golden, M.R. Philpott and J.G. Gordon II, Langmuir 2 (1986) 464.
[69] S.C. Chang, J.D. Roth and M.J. Weaver, Surf. Sci. 244 (1991) 113.
[70] C.T. Campbell, G. Ertl, H. Kuipers and J. Segner, Surf. Sci. 107 (1981) 207.
[71] R.W. McCabe and L.D. Schmidt, Surf. Sci. 86 (1977) 101.
[72] L. Palaikis, D. Zurawski, M. Hourani and A. Wiechowski, Surf. Sci. 199 (1988) 183.
[73] S.-C. Chang, L.-W.H. Leung and M.J. Weaver, J. Phys. Chem. 94 (1990) 6013.
[74] S.-C. Chang, Y. Ho and M.J. Weaver, Surf. Sci. 265 (1992) 81.
[75] H. Ogasawaru and M. Ito, Chem. Phys. Lett. 245 (1995) 304.
[76] J. Willsau, O. Wolter and J. Heitbaum, J. Electroanal. Chem. 185 (1985) 163.
[77] T. Iwasita and W. Vielstich, J. Electroanal. Chem. 201 (1986) 403.
[78] J. Willsau and J. Heitbaum, Electrochimica Acta 31 (1986) 943.
[79] T. Iwasita, W. Vielstich and E. Santos, J. Electroanal. Chem. 229 (1987) 367.
[80] S. Wilhelm, W. Vielstich, H.W. Buschmann and T. Iwasita, J. Electroanal. Chem. 229 (1987) 377.
[81] J.M. Bowman, J.S. Bittman and L.B. Harding, J. Chem. Phys. 85 (1986) 911.
[82] W. Vielstich, P.A. Christensen, S.A. Weeks and A. Hamnett, J. Electroanal. Chem. 242 (1988) 327.
[83] T. Iwasita and W. Vielstich, J. Electroanal. Chem. 250 (1988) 451.
[84] R.J. Nichols and A. Bewick, Electrochim. Acta 33 (1988) 1691.

[85] T. Iwasita and F. Nart, J. Electroanal. Chem. 317 (1991) 291.
[86] R.G. Greenler, K.D. Burch, K. Kretzschmar, R. Klauser, A.M. Bradshaw and B.E. Hayden, Surf. Sci. 152/3 (1985) 338.
[87] J. Shin and C. Korzeniewski, J. Phys. Chem. 99 (1995) 3419.
[88] B. Beden, F. Hahn, C. Lamy, J.M. Leger, N.R. de Taconi, R.O. Lezna and A.J. Arvia, J. Electroanal. Chem. 261 (1989) 401.
[89] B. Beden, F. Hahn, J.M. Leger, C. Lamy, C.L. Perdriel, N.R. de Tacconi, R.O. Lezna and A.J. Arvia, J. Electroanal. Chem. 301 (1991) 129.
[90] P.A. Christensen, A. Hamnett, J. Munk and G.L. Troughton, J. Electroanal. Chem. 370 (1994) 251.
[91] J. Munk, P.A. Christensen, A. Hamnett and E. Skou, J. Electroanal. Chem. 401 (1996) 215.
[92] P.S. Kauranen, E. Skou and J. Munk, J. Electroanal. Chem. 404 (1996) 1.
[93] R. Inada, K. Shimazu and H. Kita, J. Electroanal. Chem. 277 (1990) 315.
[94] L.D. Burke and J.K. Casey, Electrochim. Acta 37 (1992) 1817, and refs. therein.
[95] H.A. Gasteiger, N. Markovic, P.N. Ross and E.J. Cairns, J. Phys. Chem. 98 (1994) 617.
[96] V.J. Kwasniewski and L.D. Schmidt, Surf. Sci. 274 (1992) 329.
[97] D. Pletcher and V. Solis, Electrochim. Acta 27 (1982) 775.
[98] S. Gilman, J. Phys. Chem. 68 (1964) 70.
[99] P. Shiller and A. B. Anderson, J. Electroanal. Chem. 339 (1992) 201.
[100] A. Wiechowski, J. Electroanal. Chem. 78 (1977) 229.
[101] A.B. Anderson and E. Grantscharova, J. Phys. Chem. 99 (1995) 9143.
[102] S. Seong and A.B. Anderson, J. Phys. Chem. 100 (1996) 11744.
[103] L.-W.H. Leung and M.J. Weaver, Langmuir 6 (1990) 223.
[104] J.D. Roth and M.W. Weaver, J. Electroanal. Chem. 307 (1991) 119.
[105] J. Lu and A. Bewick, J. Electroanal. Chem. 270 (1989) 225.

Chapter 19

Thermodynamics and Kinetics of Spontaneous Formation and Breakdown of Vesicles in Aqueous Media*

MISNI MISRAN, SEYDA BUCAK and BRIAN H. ROBINSON

19.1 INTRODUCTION

Kinetic studies have been important in many areas of chemistry, but, to date, classic kinetic approaches have not played a major rôle in the area of colloidal science related to self-organisation of surfactants. Although micellar kinetics were investigated in some detail in the 1970s, related work involving other self-assembly systems, e.g., vesicles, lyotropic liquid-crystalline phases, have been limited.

In this article, some of the fundamental kinetic ideas relating to spontaneous vesicle formation and breakdown by surfactants in aqueous media are described. The work is related to liposome formation and breakdown using phospholipids, and the induced rupture of membranes.

Surfactant molecules in water can spontaneously self-assemble to form a wide range of aggregate structures depending on the molecular architecture of the surfactant, and the experimental conditions, particularly the surfactant concentration: e.g., micelles (spherical and non-spherical), lamellar or hexagonal lyotropic liquid-crystalline structures and vesicles or liposomes (single or multi-walled) are readily formed. In water, single-chain surfactants readily form micelles at a concentration exceeding the critical micelle concentration (cmc). Normally these micelles are small and essentially monodisperse (aggregation number ~100), but at higher concentrations of surfactant the micelles grow and elongated structures-rods and discs- can be present, which are precursors to the formation of

---

We are delighted to dedicate this article to our mentor, John Albery.

various lyotropic liquid crystal structures (hexagonal and lamellar) at even higher concentrations. These spherical to rod -shape transformations can also be induced for single-chain surfactants by addition of electrolyte (e.g., NaCl) to the solution. For synthetic surfactants, as opposed to phospholipids, sonication is not required.

Considerations of the packing parameter concept of Israelachvili *et al.* [1] suggest that double-chain surfactants, which form the basis of measurements described in this article, cannot readily form spherical micelles. With double-chain surfactants, a more likely aggregate structure is the formation of bilayer vesicles, which can be also thought of as a dispersed lamellar phase ($L\alpha$); as such the vesicular dispersed form is likely to be preferentially formed at low concentrations ($\sim$1 mmol dm^{-3}) of surfactant. Furthermore, it is necessary to consider the possibility, unlike in the case of micelles, that such vesicles, formed by self-assembly of surfactant monomers, will not be thermodynamically stable. The instability is then likely to be in the direction of growth to a thermodynamically-stable lamellar phase from the vesicles. This process will be driven, at least initially, by fusion of two vesicles.

In our model study reported in this contribution, we have chosen two double-chained C-13 alkylbenzenesulphonate surfactants (SLABS) of closely-related structure, which form micelles in aqueous solution in the absence of salt. However, when small amounts of electrolyte are added (e.g., ~20 mM NaCl), vesicles are spontaneously formed over a time period of seconds/minutes. These vesicle structures are then reasonably stable over a period of hours/days. The onset of vesicle formation can be readily characterised by the determination of the critical salt concentration (csc), needed to induce the formation of vesicles, from smaller aggregates or monomers. This parameter is easily determined experimentally from the increase in light scattering associated with self-assembly. It has now been determined for a number of electrolyte systems.

The effect of temperature on vesicle formation has also been studied, and for our systems at an appropriate salt concentration, we can effect a transition in the direction of vesicle formation on lowering the temperature. We can also affect the stability of the system by changing the solvent composition, e.g., by adding small volume fractions of ethanol to water. Addition of ethanol has the effect of changing both the charge interactions at the surface of the vesicle and the solvophobicity of the tail groups.

It is not perhaps surprising that relatively few kinetic studies had been made on vesicle systems until quite recently. Indeed for many years it was supposed, but for no obvious reason, that synthetic surfactants could not form vesicles even though phospholipids can readily form structures of this type, more usually referred to as liposomes. However, some twenty years ago, Kunitake observed vesicle formation with the cationic surfactant didodecyldimethylammonium bromide [2]. However, in water, more stable systems are formed by mixing in approximately equal proportions of a cationic and anionic surfactant system, and it is possible that as a result these systems may indeed be thermodynamically stable [3], such that meaningful phase diagrams can be constructed. Vesicles can also be easily formed from micelles by reducing the pH into a narrow pH range; e.g., using oleate/oleic acid, where vesicle formation is evident in the pH range 7–9. In this situation, the head groups are partially charged rather than fully ionised. In both of the above situations, the stability will be due in part to the option of having a different composition of the two surfactants on the inside and the outside of the vesicle bilayer, to produce a film which has the correct unstrained curvature. However, the use of a combination of surfactants does not appear to be a necessary requirement for spontaneous self-assembly to vesicles. In this model study, a single surfactant is used and there are other known cases. For example, Corti has reported spontaneous vesicle formation from a single ganglioside, which is a glucosidic amphiphile of biological origin[4]. Vesicles form at concentrations as low as $\sim 0.1\ \text{mmol dm}^{-3}$. A common method to induce the breakdown of vesicles is to add a single-chain surfactant, e.g., sodium dodecylsulphate (SDS) to a vesicular assembly. The result will be the formation of micelles containing a mixture of double- and single-chain surfactants. The onset of the vesicle$\Leftrightarrow$ micelle transition for a SLABS/SDS system, as the concentration of SDS is increased, is rather abrupt, and indicates co-operative phase-transition behaviour. A detailed analysis of systems of this type has recently been carried out [5]. The kinetic analysis gives a Hill coefficient of ~ 8, which means that at least eight single-chain surfactants have co-operated in the bilayer to disrupt the vesicles. The associated thermodynamic analysis reveals that one in ten of the surfactant molecules in the bilayer are single-chain at the critical concentration of added SDS required to break down the vesicles. Related studies have also been carried out for phospholipids and natural membranes mixed with single-chain surfactants [6–9], and for double-chain cationic surfactants mixed with single-chain surfactants (e.g., didodecyldimethylammonium bromide with dodecyltrimethylammonium bromide).

Vesicle structures are more 'permanent' structures than micelles. Although micellar structures are thermodynamically stable, there are two dynamic processes which can be identified in micellar solutions, characterised by the so-called fast and slow relaxation times. Micelles are often portrayed as static spherical aggregates of ordered rigid rods, but the now generally accepted view is that the head-group/water interface is rough and the surfactant chains are significantly kinked. The micellar surfactant is in dynamic equilibrium with monomers in bulk solution such that surfactant in solution is being continuously exchanged with surfactant in the micelle, with the entry rate constant being essentially diffusion-controlled, and the exit rate constant varying proportionately with the cmc. These processes determine the fast relaxation time, which is typically measured in the microsecond time range. As a result of these exchanges, the aggregation number of an individual micelle will fluctuate in real time. When, as a result of these fluctuations, the micelle aggregation number drops below a critical size for stability, the micelle will rapidly disintegrate and another micelle must spontaneously form in another part of the solution to keep the overall concentration of micelles unchanged. This relaxation time occurs in the ms-s region, and is conventionally followed by small-perturbation temperature- and pressure-jump relaxation methods [10]. Both relaxation times are related to the overall stability of the micelle, which is conventionally measured through the cmc. The lower the cmc, the more stable is the micelle. An understanding of these kinetic processses is essential for the interpretation of dynamic surface tension data, and the kinetic information aids our understanding of such processes as wetting and solubilisation (detergency).

In the case of vesicles, there is an obvious attraction in the utilisation of these systems for delivery and targeting of molecules and there has been much interest in using vesicle systems for drug delivery. Attempts have also been made to polymerise vesicles to give them added structural rigidity and stability, but these efforts have not been particularly successful so far.

We have studied in this contribution the kinetics of formation and breakdown of vesicles on changing the salt and surfactant concentration with a view to elucidating the mechanism of the reaction., and the intermediate structures which are likely to be involved.

In the model system, the reaction has been kept as simple as possible by investigating self-assembly in a solution containing a single surfactant. The surfactants sodium *p*-6-tridecylbenzenesulphonate (6-SLABS) and sodium *p*-7-tridecylbenzenesulphonate (7-SLABS) (Fig. 19.1(a) and 19.1(b)) were pure samples obtained as a gift from Dr. Peter Garrett of Unilever, Port Sunlight Laboratory, UK. All other chemicals were standard reagents and were used without further purification.

Characterisation of vesicles was achieved using a combination of methods, including photon-correlation spectroscopy, video-enhanced and cryo-electron microscopy. Measurements of the cmc of the surfactants (in the absence of salt) were made using uv-visible spectrophotometry and electrical conductivity (κ). For cmc measurements, there is a convenient change in the extinction coefficient of the benzene chromophore at 262 nm. The onset of vesicle formation, and hence the determination of the csc, can be measured by 90° scattering and 180° optical turbidity measurements at 300 nm.

Kinetic measurements were carried out using a Hewlett-Packard diode-array spectrophotometer and/or a Hi-Tech small-volume single-mixing stopped-flow instrument (fastest mixing-time ~5 ms) in absorbance (turbidity) mode.

19.2 RESULTS

19.2.1 Thermodynamic aspects

The molecular structures of the two **S**odium **L**ong-chain **A**lkyl**B**enzene**S**ulphonate surfactants (SLABS) are given in Fig. 19.1.

These surfactants can form micelles provided that the salt concentration is low. The cmc of 6-SLABS was readily determined by spectrophotometry at 262 nm, and the discontinuity in the plot is clearly shown in Fig. 19.2(a) from which the cmc can be obtained as 0.0014 mol dm^{-3}. It should be noted that the cmc is found to be essentially independent of temperature over the experimentally-measured range of 15–30°C. This is generally found for micellisation involving ionic surfactants in water, so that the enthalpy change on transferring monomer from aqueous solution to the micelle is approximately zero. There is a 20% decrease in extinction coefficient of the benzene ring chromophore on transfer from an aqueous

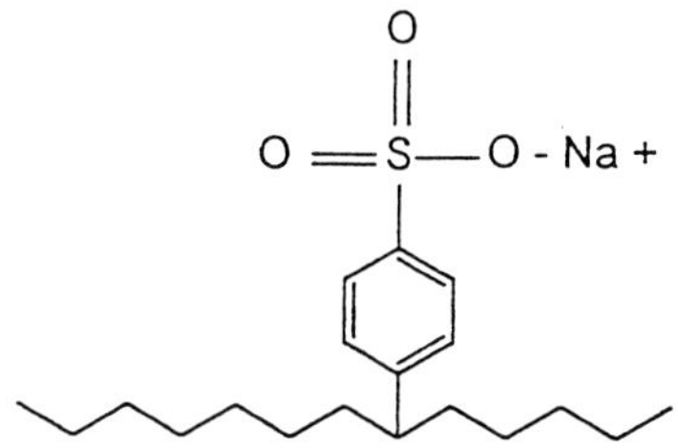

6-SLABS (a)

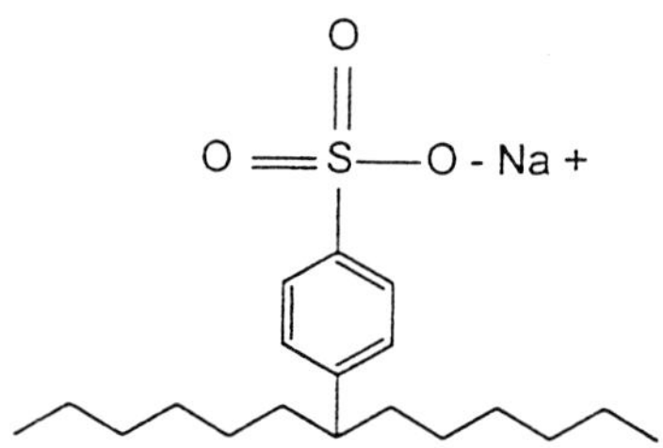

7-SLABS (b)

Fig. 19.1.

environment to the micelle, which can therefore be used as an intrinsic indicator of micellisation.

The same value of the cmc for 6-SLABS was determined from measurements of the electrical conductivity (the decrease in the conductivity above the cmc being 24%). Data obtained for 7-SLABS by conductivity is given in Fig. 19.2(b). The cmc is found to be significantly increased—to 0.0020 mol dm^{-3}, which is consistent with the more rectangular structure of 7-SLABS, which therefore disfavours micelle formation as compared with 6-SLABS. Previous measurements of the cmc of sodium *p*-(*x*-decyl)-benzenesulphonate surfactants, where x was varied from 1 to 5, showed a steady increase in the cmc as the benzenesulphonate group was moved to the middle of the chain, the cmc of the surfactant with $x = 5$ being 0.008 mol dm^{-3} at 25°C, whereas the value for the surfactant with $x = 1$ was 0.004 mol dm^{-3} at 50°C [11]. Confirmation that small micelles form

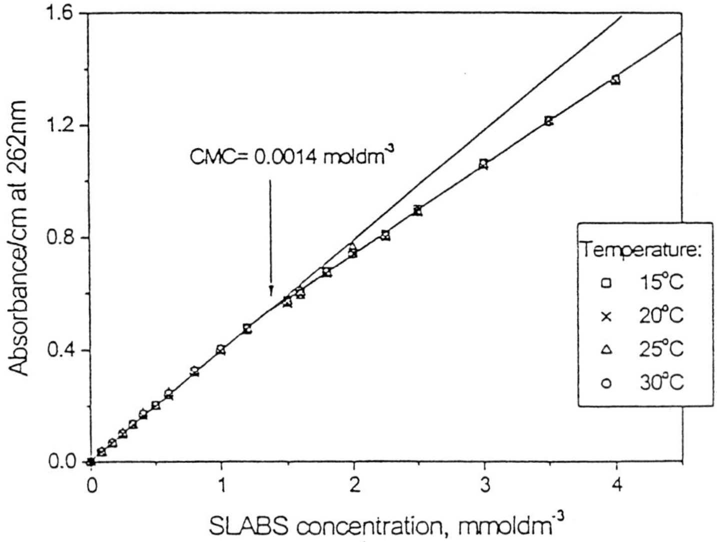
Absorbance/cm at 262nm
CMC= 0.0014 moldm⁻³
Temperature:
15°C
20°C
25°C
30°C
SLABS concentration, mmoldm⁻³

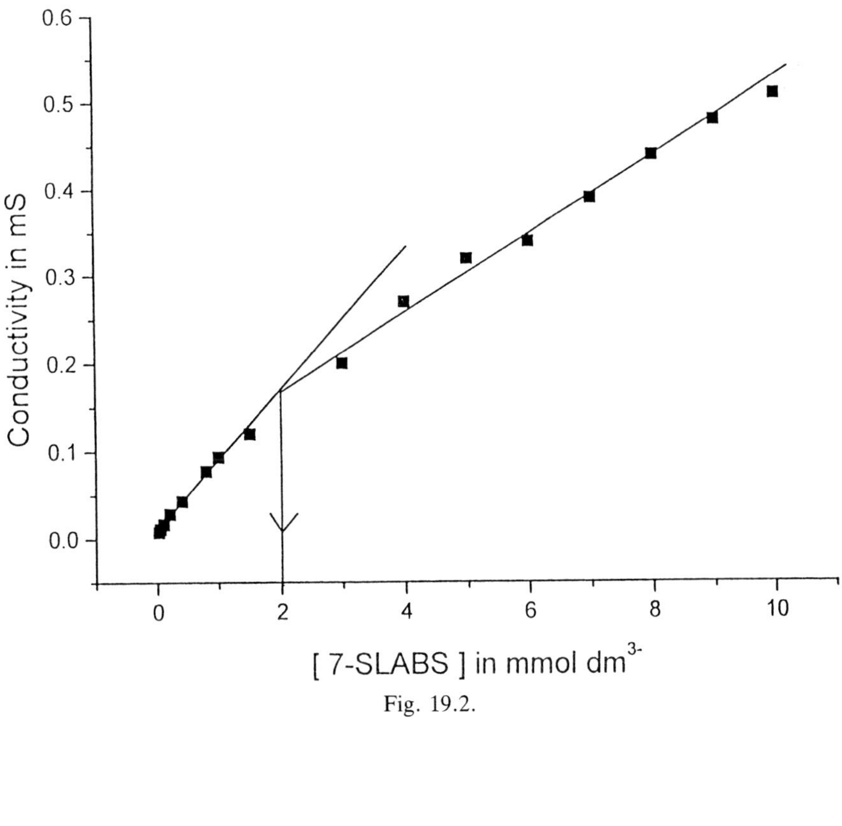
cmc of 7- SLABS
Conductivity in mS
[7-SLABS] in mmol dm³⁻

Fig. 19.2.

in water in the case of the SLABS surfactants was provided by translational diffusion coefficient measurements using fluorescence correlation spectroscopy (where the aggregate (micelle) contained a solubilised molecule of Rhodamine 6G). It was found that the size of the diffusing entity in water was comparable to that of a sodium dodecylsulphate micelle [12].

On addition of electrolyte, e.g., 30 mmol dm^{-3} NaCl in the case of 6-SLABS, there is a change in the uv-visible spectrum as shown in Fig. 19.3(a). For a concentration of surfactant < cmc in water, there is a broad additional component to the absorbance spectrum which becomes manifest on addition of salt, extending through the visible region, which is characteristic of conventional light scattering from the sample. There is no significant light absorption by the SLABS surfactant at wavelengths >290 nm, so 300 nm was chosen as a convenient wavelength for monitoring changes in turbidity. The dramatic increase in scattering which is observed indicates that very large structures, relative to micelles, have formed in the solution. Cryoelectron-microscopy measurements clearly demonstrated that vesicles were present. Vesicles are considerably larger than micelles, and have an extensive aqueous core region separated from the external water medium by a bilayer skin of surfactant, as shown in Fig. 19.3(b). We found the diameter of the SLABS vesicles to be typically 250 nm using photon correlation spectroscopy, dependent to some extent on the SLABS concentration.

Vesicles can be readily prepared by spontaneous self-assembly on mixing a salt solution with a SLABS surfactant solution at a concentration below the cmc in water. The vesicles form spontaneously over a few minutes, no sonication being required. Previously it has been reported that sodium *p*-heptylnonylbenzenesulphonate could form vesicles following sonication, and these were stable for many weeks before reverting back to a liquid-crystal form [13]. Kaler *et al.* reported on the ageing and stability of vesicular dispersions formed by the same surfactant, and concluded again that the systems were unstable over long times (24 h), but more stable in the presence of salt (NaCl) due to a lowering of the surfactant solubility in water [14]. It was therefore of interest to establish whether the SLABS systems, which form vesicles spontaneously and without the input of energy, e.g.,by sonication, are thermodynamically stable. The stability of our system was therefore checked by measuring the turbidity over a period of days, gently shaking the solution before absorbance/turbidity measurements were taken. As can be seen from Fig. 19.4(a), the turbidity is

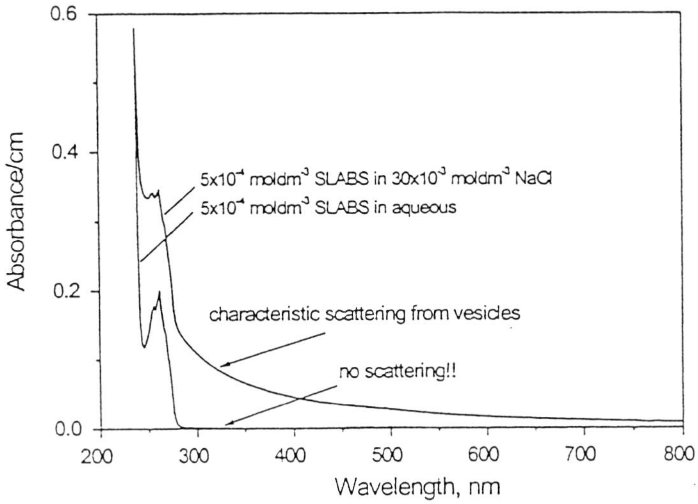
0.6
0.4
0.2
0.0
Absorbance/cm
5x10⁻⁴ moldm⁻³ SLABS in 30x10⁻³ moldm⁻³ NaCl
5x10⁻⁴ moldm⁻³ SLABS in aqueous
characteristic scattering from vesicles
no scattering!!
200
300
400
500
600
700
800
Wavelength, nm

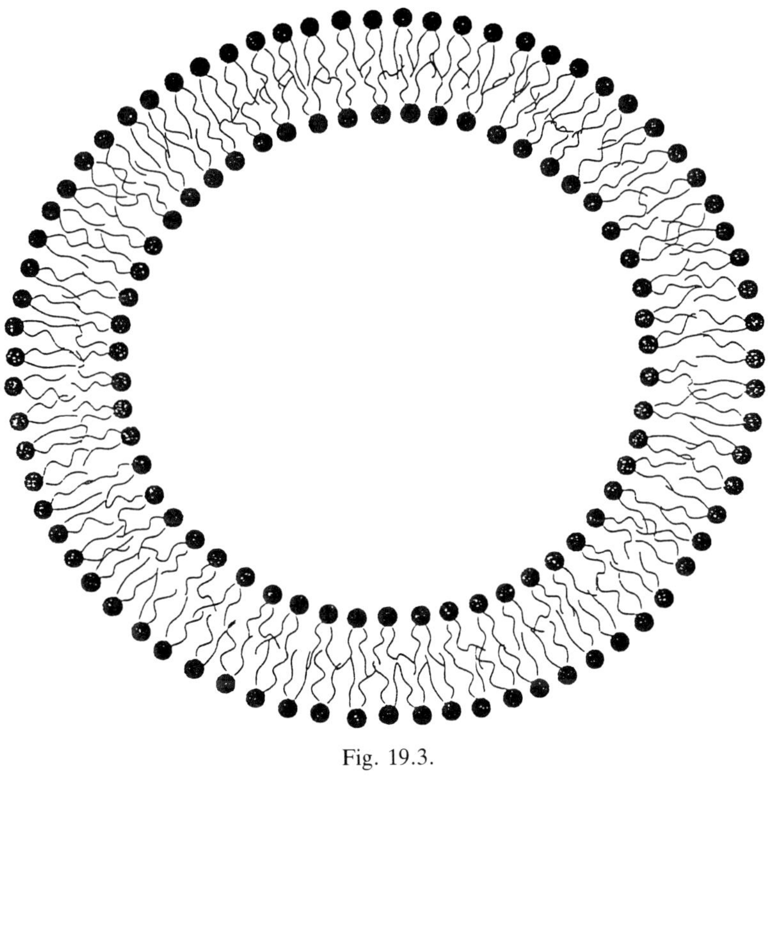

Fig. 19.3.

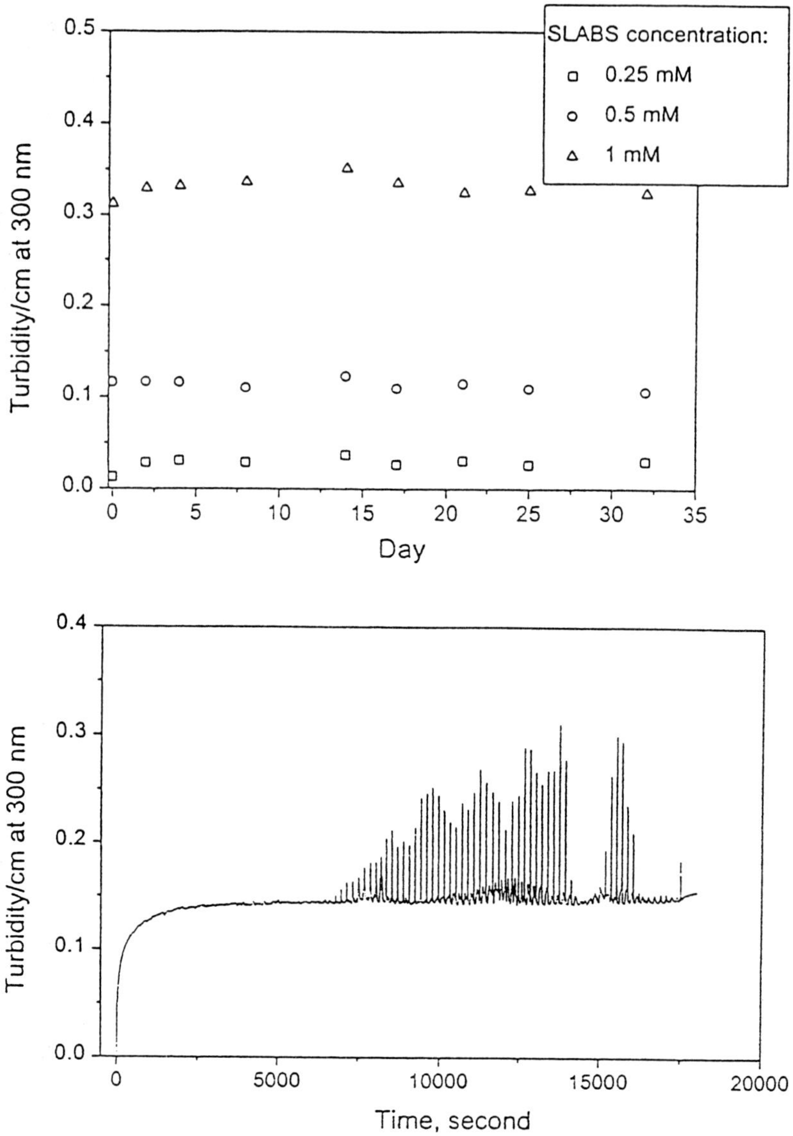
SLABS concentration:
0.25 mM
0.5 mM
1 mM
Turbidity/cm at 300 nm
Day
Time, second

Fig. 19.4.

essentially unchanged. However, if a sample is left undisturbed in the spectrophotometer cell over the full period, changes in the solution are observed and the formation of a small number of considerably larger structures is indicated by the appearance of sharp lines on the turbidity-time plots (Fig. 19.4(b)). This is associated with the onset of formation of dispersed lamellar phase. Indeed, careful visual inspection of the cuvette indicated the presence of phase-separated material. Examination under a microscope showed that this material was in the form of flocculated vesicles. This material was readily redispersed by the application of hand-shaking, and this explains why the spectrum in Fig. 19.4(a) did not show any evidence for phase separation, since the vesicle flocs are unstable on shaking. These results tend to confirm our belief that even when SLABS vesicles form spontaneously, they will not generally be stable over the long term, in that instability initially takes the form of formation of flocculated vesicles which will be followed by the fusion of vesicles to give a liquid-crystalline phase which would not be expected to be so readily redispersed.

The onset of vesicle formation as the electrolyte concentration is increased occurs at a particular concentration of salt, which we have assigned as the critical salt concentration (or csc value). Typical plots for 6-SLABS and 7-SLABS are shown in Fig. 19.5. The csc for 7-SLABS is at a significantly lower concentration than for 6-SLABS, which is consistent with vesicle formation being easier with the more symmetrical 7-SLABS. The csc value can be conveniently taken as the mid-point of the transition region, and the csc values are ~20 mmol dm^{-3} for 6-SLABS and ~7 mmol dm^{-3} for 7-SLABS. The sharpness of the transition indicates that at the csc there will be equal amounts of monomer and vesicular surfactant present. It should be noted that the turbidity for electrolyte concentrations > csc was constant indicating that the size of the vesicles is unchanged with addition of salt. This was confirmed by photon-correlation spectroscopy measurements of the translational diffusion coefficient. The csc is also independent of the total concentration of SLABS, providing that the concentrations are all in the millimolar region, as shown in Fig. 19.6.

However, it is found that temperature has a rather dramatic effect on the csc, the value decreasing from 28 mmol dm^{-3} at 35°C to 10 mmol dm^{-3} at 10°C, as shown in Fig. 19.7 where the data are shown for 6-SLABS. This means that if a salt concentration of 20 mmol dm^{-3} is employed, at 10°C the system will contain 100% vesicles, whereas at 35°C the system will be 100% monomers or small weakly-scattering aggregates. This beha-

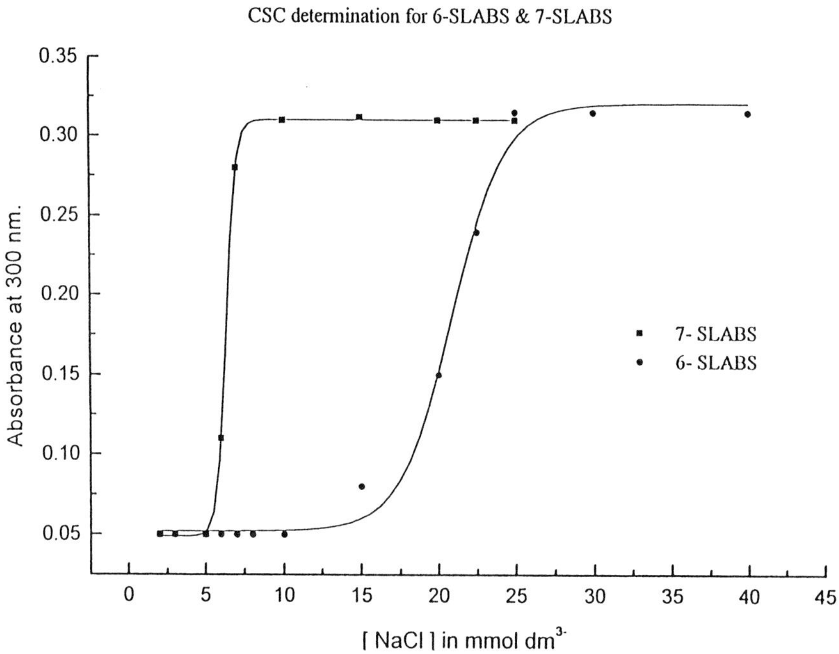

Fig. 19.5.

viour is shown in Fig. 19.7. It is interesting to speculate on the state of the system just before the onset of turbidity. There is some evidence, from other measurements, that small micellar-type aggregates might be present, in the form of small rods or discs. These structures would be precursors of vesicles and so may be expected to be relevant to mechanistic considerations involved in the self-assembly/disassembly of vesicles.

The effect of other electrolytes is also of interest. Addition of divalent cations, e.g., in the form of calcium and magnesium chloride salts, results in precipitation of the anionic vesicles, which is also observed for micelles in water. There is, however, a significant trend within the alkali metal cations, e.g., from Na^+ to Rb^+. The larger rubidium ion has a promotional effect on the formation of vesicles, and indeed with Cs^+, vesicles are already formed on the addition of trace amounts of salt. The relevant data are shown in Fig. 19.8.

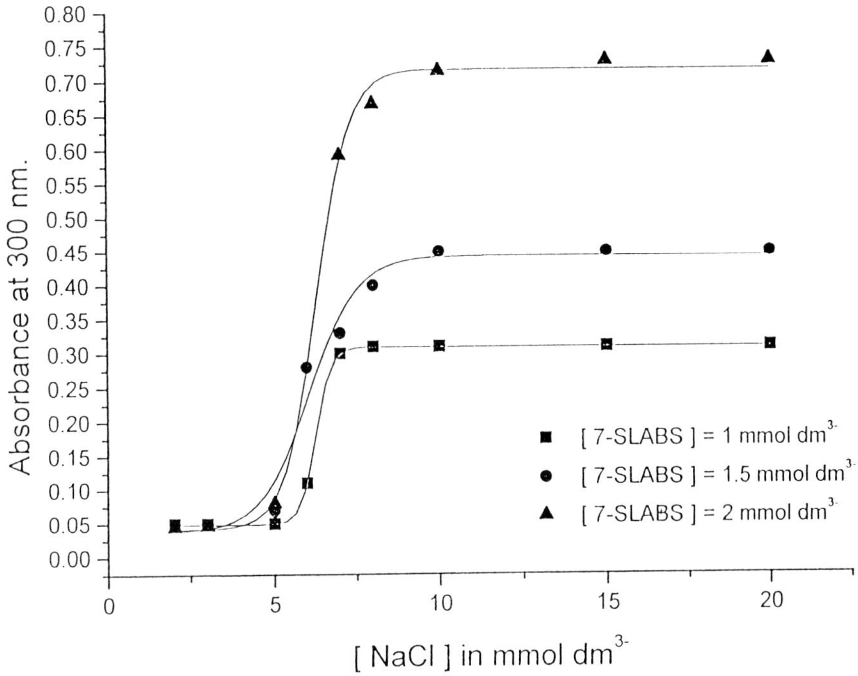

Fig. 19.6.

As well as addition of salt, the addition of a polar co-solvent e.g., ethanol, can also affect the thermodynamic properties of the system. On addition of 5% v/v ethanol to water, the cmc (in the absence of salt) is reduced from 1.4 mmol dm^{-3} to 1.0 mmol dm^{-3}. However, addition of higher concentrations of ethanol will lead to an increase in the cmc. Data for the temperature dependence of the csc, in the presence of 5% ethanol, are shown in Fig. 19.9. There is a dramatic shift of the csc to higher salt concentrations which means that it is harder to form vesicles in the ethanol-containing medium. This is consistent with the lowering of the cmc in the presence of ethanol which means that the small aggregate (micellar-type) structures are more stable. The addition of ethanol is expected to affect, in particular, the charge interactions in the head-group region, the effect of hydrophobicity being essentially the same for micelles and vesicles.

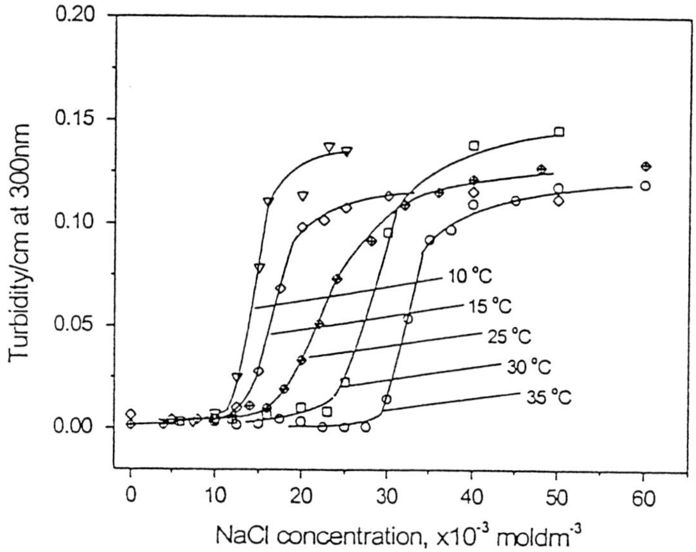

Fig. 19.7.

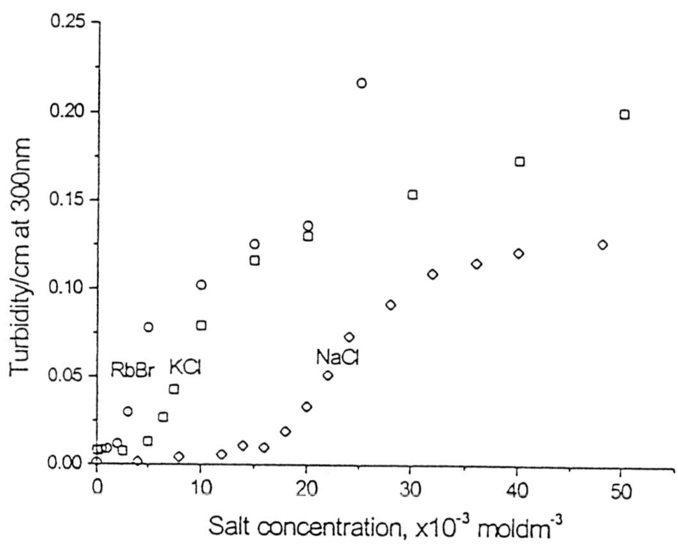

Fig. 19.8.

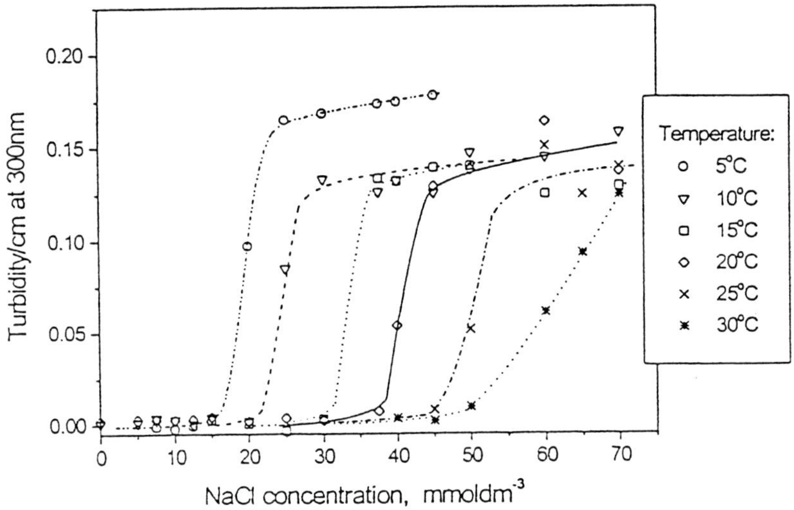

Fig. 19.9.

19.2.2 Kinetic aspects

The kinetics of formation of vesicles can be readily followed by the increase in turbidity of the sample solution with time. Slow changes can be followed using a conventional spectrophotometer, whereas faster changes (<10 s) can be monitored using the same detection system, but using the stopped-flow method. A simple experiment is to start with monomers/micelles in water and 'jump' the ionic strength (electrolyte concentration) into a region where the vesicles are stable. A reaction scheme for this process for 6-SLABS (Scheme 1) is shown below. Note that I is the ionic strength of the medium (applied through the added salt concentration). The contribution of the surfactant to the ionic strength has been ignored.

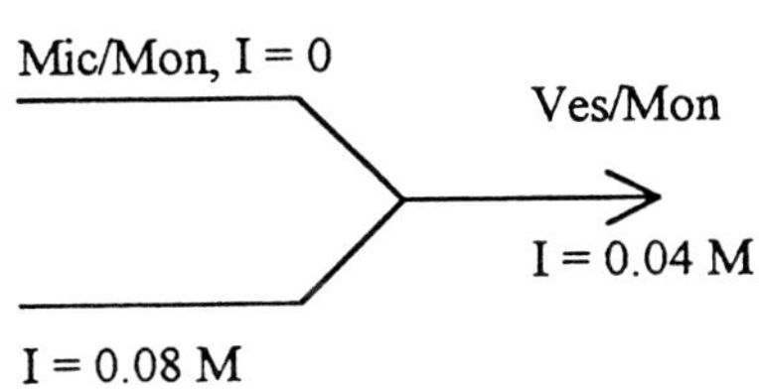

Scheme 19.1.

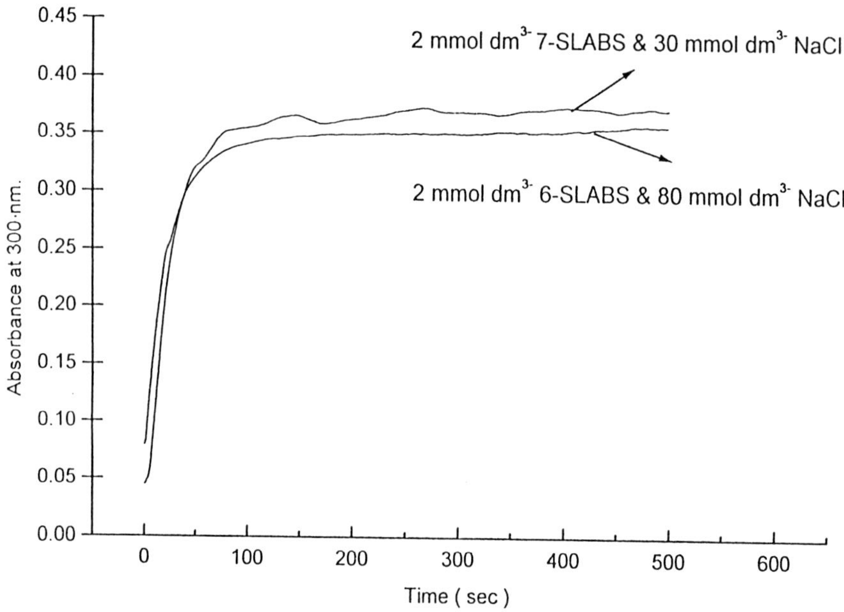

Fig. 19.10.

A typical kinetic profile for the reaction leading to spontaneous formation of vesicles is shown in Fig. 19.10, both for 6-SLABS and 7-SLABS. It can be seen that the kinetics are relatively simple in this configuration and the formation of vesicles occurs on the time scale of about a minute.

The simplest kinetic treatment would be based on the determination of a half-life for the reaction, where $t_{1/2}$ is the time taken for the turbidity to reach half the maximum value. Since the curves approximate well to a single exponential, it is then possible to define an apparent first-order rate constant such that:

$$k_{\mathrm{app}} = \ln 2/t_{1/2}$$

It is then of interest to investigate how the final salt concentration influences the kinetics of vesicle formation. The increase in salt concentration provides the *driving force* for the reaction, and it is apparent that the larger the salt jump, the greater the tendency will be to form vesicles, which we would expect to be reflected in the observed rates. It should be

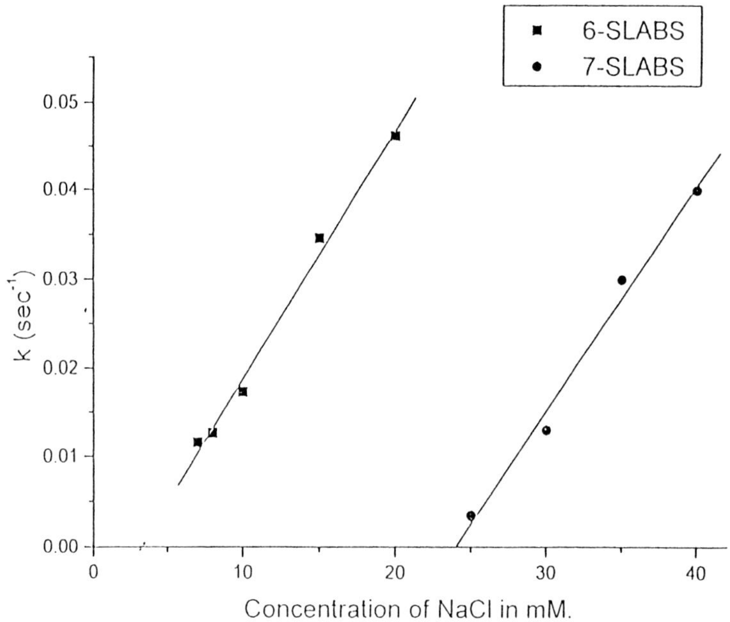

Fig. 19.11.

recalled that the size of the SLABS vesicles is unchanged over the salt range from 25 mmol dm^{-3} to 50 mmol dm^{-3} when NaCl is used. A plot of k_{app} vs. the final salt concentration shows that the rate of formation of vesicles does indeed depend on the NaCl concentration, and the form of the plot is essentially linear for both forms of SLABS, as shown in Fig. 19.11. If very high concentrations of salt are employed (>100 mmol dm^{-3}), this linear relationship breaks down, and this is likely to be associated with the formation, in the reaction medium,of lamellar structures rather than vesicles, at these much higher salt concentrations.

The salt is essentially operating through a medium effect and in this experiment we have a particularly dramatic salt effect in which the free energy barrier for the assembly reaction is systematically decreased as the salt concentration is increased. As the temperature is lowered, it is found that the rate of the self-assembly reaction increases. This can be understood from considerations of Fig. 19.7. As the temperature is lowered, the csc decreases, so if the salt concentration is kept constant, there will be a larger driving force at lower temperatures. This effect apparently overrides the usual effect of the rate of a reaction increasing with increase in temperature. The transition from monomers to vesicles has the appearance of a co-operative transition and, as for the thermodynamics, the kinetics of such phase changes have recently been considered for the change from bilayer vesicles to mixed micelles induced by addition of a single-chain surfactant such as SDS to SLABS [5].

A kinetic scheme for the kinetics of vesicle breakdown is shown in the scheme below:

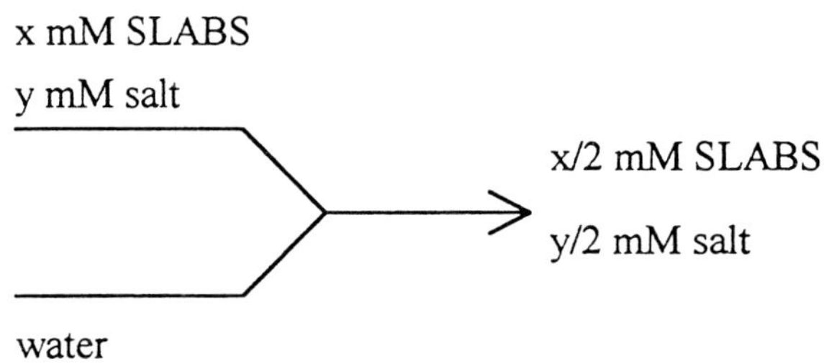

Scheme 19.2.

The reaction can again be monitored on two different time scales: stopped-flow for the period 10 ms–10 s and conventional spectrophotometry for longer time scales. However, we observe in practice two time regimes for disruption of vesicles—a first step (Fig. 19.12(a)) which has $t_{1/2} < 100$ ms and which represents only 1% of the total amplitude, and a slow step which represents the bulk of the change and is shown in Fig. 19.12(b). The value of $t_{1/2}$ for this slow step increases dramatically as the salt concentration y is increased in the region 27–40 mmol dm^{-3} for 6-SLABS, i.e., as the final salt concentration ($y/2$) approaches the csc. As for the vesicle formation step, as the salt driving force for the reaction is decreased, the reaction rate slows down. However, the rate does not depend significantly on the SLABS concentration x in the region 0.5–2 mmol dm^{-3}. The rapid process ($t_{1/2} < 100$ ms) can be tentatively identified with a size redistribution in the direction of smaller vesicles. In conclusion, it is clear that both the kinetics of vesicle formation and breakdown are strongly dependent on the precise salt conditions.

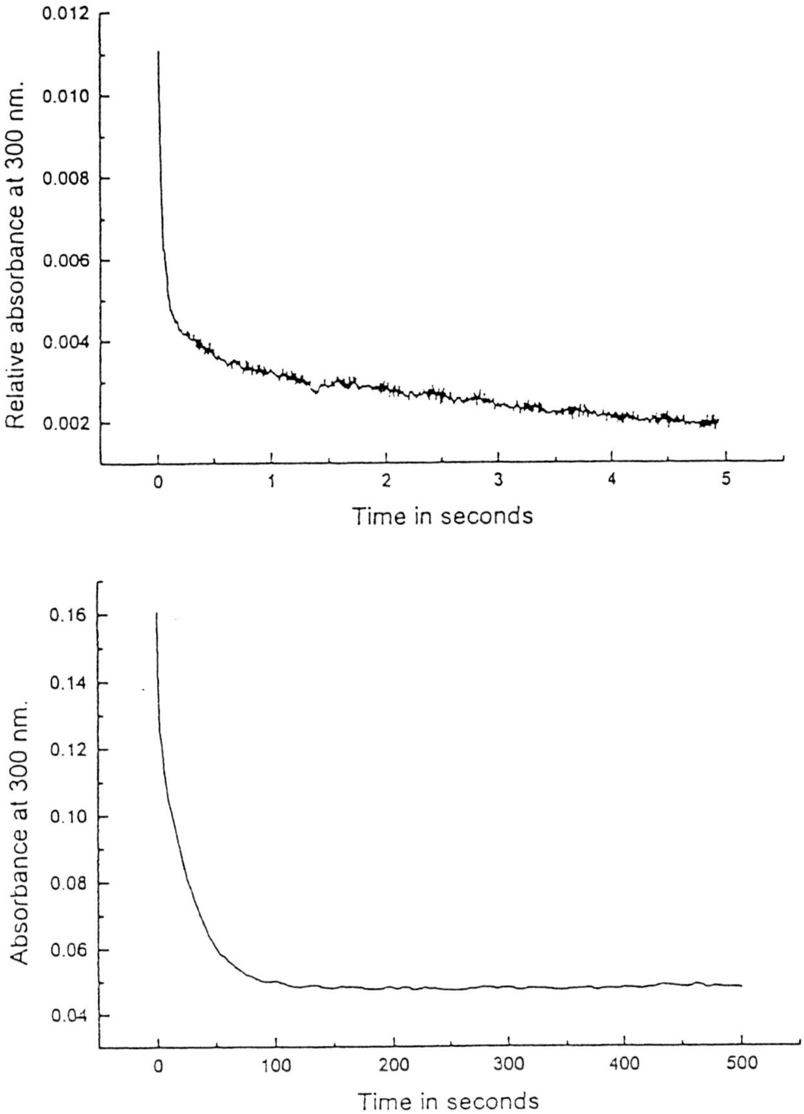

0.012
0.010
0.008
0.006
0.004
0.002
Relative absorbance at 300 nm.
0
1
2
3
4
5
Time in seconds
0.16
0.14
0.12
0.10
0.08
0.06
0.04
Absorbance at 300 nm.
0
100
200
300
400
500
Time in seconds

Fig. 19.12.

19.3 DISCUSSION

It is interesting to speculate on the intermediates which might be involved in these complex kinetic pathways which involve the cooperative rearrangement of large numbers of surfactant molecules. It is easiest to visualise the breakdown process and there are two possibilities. A likely first step on decreasing the salt concentration to below the csc is that the surfactant will be more stable in the water and so the free surfactant concentration will increase, such that it will be progressively lost from the outside wall of the vesicle. After a certain time, this imbalance of surfactant on the two sides of the bilayer will result in the development of a critically unstable vesicle which is then likely to rupture into fragments, probably disc-like. One could expect that there will also be interesting internal pressure effects in the bilayer during this process. Depending on the final value of the ionic strength we would expect these fragments, which might be slightly distorted micelles, to be relatively stable, and indeed we are presently probing the structure of these entities using the fluorescence correlation technique referred to previously.

Recent studies with 'giant' vesicles, which can be directly visualised by video-microscopy techniques, have shown that when placed in an unstable environment, vesicles can either progressively shrink or experience bilayer undulations followed by rupture [15]. In our system, we feel the latter mechanism is more likely.

The process of spontaneous assembly of vesicles is not so easy to visualise. Again the first step is likely to involve the assembly of surfactant into rods or, more likely, discs. These could then join together and eventually one might hope to form closed bilayer spheres. There is a driving force for this as the edges of the sheets will have a high free energy and so there will be a strong tendency for the edges to seal. An alternative scenario might involve the disjoining of large disc-like bi-layer structures followed by re-assembly of the bilayer by addition of surfactant to the inner exposed layer formed as a result of disjoining. In any event, the system will be driven by the need to minimise the contact between the hydrocarbon chains and water and considerations of the bending energy/elasticity of the bilayer. One could also speculate that the assembly and disassembly processes might go via different pathways. Since these changes to the system are induced by large perturbations (through changes in the salt concentration), it is indeed possible for the reactions to proceed by routes involving different energy surfaces, and so the intermediates need not necessarily be the same in the assembly and disassembly processes.

The other interesting aspect of these reactions is that if the vesicles are not thermodynamically stable, curious behaviour such as hysteresis effects might be observed when the vesicle-containing media are heated and then cooled, or when the synthesis protocol is varied. These possibilities are currently under investigation in our laboratory.

ACKNOWLEDGEMENTS

We would like to acknowledge lively discussions with WJA on the mechanism of these reactions.

References

[1] J. Israelachvili, D.J. Mitchell and B.W. Ninham, J. Chem. Soc. Faraday Trans II, 72 (1976) 1525].
[2] T. Kunitake and Y. Okahata, J. Amer. Chem. Soc. 99 (1977) 3860.
[3] E.W. Kaler, A.K. Murthy, B.E. Rodriguez and J.A.N. Zasadzinski, Science 245 (1989) 1371.
[4] L. Cantu, M. Corti, M. Musolino and P. Salina, Progress in Colloid and Polymer Science 84 (1991) 21.
[5] U. Brinkmann, E. Neumann and B.H. Robinson, J. Chem. Soc Faraday Soc. 94 (1998) 1281.
[6] D. Lichtenberg, Biochim. Biophys. Acta 821 (1985) 470.
[7] A. De La Maza and J.L. Parra, Biochem J. 303 (1994) 907.
[8] T. Schurholz, Biophys. Chem. 58 (1996) 87.
[9] A. Helenius and K. Simons, Biochim. Biophys. Acta. 415 (1975) 29.
[10] E.A. Aniansson, S.N. Wall, M. Almgren, H. Hoffmann, L. Kielmann, W. Ulbricht, R. Zana and C. Tondre, J. Phys. Chem. 80 (1976) 905.
[11] N.M. Van Os, M.G.J. Daane and T.A.B.M. Bolsman, J. Colloid Interface Sci. 115 (1987) 402.
[12] B.H. Robinson and G.R. Burnett, unpublished results (University of East Anglia).
[13] E.I. Franses, Y. Talmon, L.E. Scriven, H.T. Davis and W.G. Miller, J. Colloid Interface Sci. 86 (1982) 449.
[14] H. Madani and E.W. Kaler, Langmuir 6 (1990) 125.
[15] P.L. Luisi, ETH Zurich, personal communication.

Subject Index